Lecture Notes in Computer Science 4392

Commenced Publication in 1973
Founding and Former Series Editors:
Gerhard Goos, Juris Hartmanis, and Jan van Leeuwen

T0189662

Salil P. Vadhan (Ed.)

Theory of Cryptography

4th Theory of Cryptography Conference, TCC 2007
Amsterdam, The Netherlands, February 21-24, 2007
Proceedings

 Springer

Volume Editor

Salil P. Vadhan
Harvard University
Division of Engineering & Applied Sciences (DEAS)
33 Oxford Street, Cambridge, MA 02138, USA
E-mail: salil@eecs.harvard.edu

Library of Congress Control Number: 2007920469

CR Subject Classification (1998): E.3, F.2.1-2, C.2.0, G, D.4.6, K.4.1, K.4.3, K.6.5

LNCS Sublibrary: SL 4 – Security and Cryptology

ISSN   0302-9743
ISBN-10   3-540-70935-5 Springer Berlin Heidelberg New York
ISBN-13   978-3-540-70935-0 Springer Berlin Heidelberg New York

Springer is a part of Springer Science+Business Media

springer.com

© International Association for Cryptologic Research 2007
Printed in Germany

Typesetting: Camera-ready by author, data conversion by Scientific Publishing Services, Chennai, India
Printed on acid-free paper   SPIN: 12021475   06/3142   5 4 3 2 1 0

Preface

TCC 2007, the Fourth Theory of Cryptography Conference, was held in Amsterdam, The Netherlands, from February 21 to 24, 2007, at Trippenhuis, the headquarters of the Royal Dutch Academy of Arts and Sciences (KNAW). TCC 2007 was sponsored by the International Association for Cryptologic Research (IACR) and was organized in cooperation with the Cryptology and Information Security Group at CWI, Amsterdam; the Mathematical Institute, Leiden University; and DIAMANT, the Dutch national mathematics cluster for discrete interactive and algorithmic algebra and number theory. The General Chair of the conference was Ronald Cramer.

The conference received 118 submissions, of which the Program Committee selected 31 for presentation at the conference. These proceedings consist of revised versions of those 31 papers. The revisions were not reviewed, and the authors bear full responsibility for the contents of their papers. The Best Student Paper Award was given to Saurabh Panjwani for his paper "Tackling Adaptive Corruptions in Multicast Encryption Protocols."

The conference program also included a tutorial on "Quantum Cryptography", given by Renato Renner, and a special event on "The Assumptions for Cryptography", consisting of a few short talks and a panel discussion. In addition, the Program Committee decided to augment the traditional rump session to include short informal presentations of not only new results, but also open problems and future research directions.

One of the things that has made my job as Program Chair a pleasure is the wonderful dedication our community has to the success of TCC. I am grateful to the many people who have contributed to the organization and content of the conference. First and foremost, this includes the authors of all submitted papers, whose research efforts are the raison d'etre for TCC. I am also indebted to my extremely dedicated Program Committee. They were faced with a larger than expected workload due to a jump in the number of submissions, yet they carried out the review process with extraordinary thoroughness and care for the high standards and integrity of TCC. I also thank the many external reviewers who assisted the Program Committee in its work.

I thank the Steering Committee of TCC for entrusting me with this responsibility, and its Chair, Oded Goldreich, for being available as a source of wisdom throughout the process. I also benefited from the experience and advice of the past TCC chairs, Moni Naor, Joe Kilian, Tal Rabin, and Shai Halevi. I am especially indebted to Shai, who wrote a wonderful software package that I used for handling the submissions, the PC discussions, and these proceedings and provided rapid-response customization and technical support throughout.

I am very grateful to Ronald Cramer, TCC 2007 General Chair; his Co-chairs, Serge Fehr, Dennis Hofheinz, and Eike Kiltz; and Wilmy van Ojik, the CWI

Conference Organizer, for all the work they have put into hosting the conference and managing its logistics. Thanks also to Microsoft for a generous donation that supported the conference in various ways, including stipends to help students attend.

My work as Program Chair was supported in part by grants from the National Science Foundation (CNS-0430336) and office of Naval Research (N00014-04-1-0478).

I appreciate the assistance provided by the Springer LNCS editorial staff, including Alfred Hofmann, Frank Holzwarth, and Anna Kramer, in assembling these proceedings. Finally, I thank Carol Harlow for the administrative help she provided here at Harvard.

TCC 2007 would not have been possible without the efforts of all the people I have mentioned here, as well as the many that I have surely forgotten (to whom I apologize).

December 2006   Salil Vadhan

TCC 2007
The 4th Theory of Cryptography Conference

KNAW Trippenhuis, Amsterdam, The Netherlands
February 21–24, 2007

Sponsored by *The International Association for Cryptologic Research*
Organized in cooperation with *Centrum voor Wiskunde en Informatica (CWI)*
and *Mathematisch Instituut, Universiteit Leiden*
With financial support from *Microsoft Corporation*

General Chair

Ronald Cramer, CWI Amsterdam and Leiden University

Program Committee

Mihir Bellare	University of California, San Diego
Ran Canetti	IBM T.J. Watson Research Center
Ivan Damgård	University of Aarhus
Cynthia Dwork	Microsoft Research
Serge Fehr	CWI Amsterdam
Yuval Ishai	The Technion
Jonathan Katz	University of Maryland
Rafael Pass	MIT and Cornell University
Oded Regev	Tel Aviv University
Omer Reingold	Weizmann Institute of Science
Ronen Shaltiel	University of Haifa
Victor Shoup	New York University
Yael Tauman Kalai	MIT and Weizmann Institute of Science
Salil Vadhan (Chair)	Harvard University
Bogdan Warinschi	INRIA-Lorraine

TCC Steering Committee

Mihir Bellare	University of California, San Diego
Ivan Damgård	University of Aarhus
Oded Goldreich (Chair)	Weizmann Institute of Science
Shafi Goldwasser	MIT and Weizmann Institute of Science
Johan Håstad	Royal Institute of Technology
Russell Impagliazzo	University of California, San Diego
Ueli Maurer	ETH Zürich
Silvio Micali	Massachusetts Institute of Technology
Moni Naor	Weizmann Institute of Science
Tatsuaki Okamoto	NTT Laboratories

External Reviewers

Michel Abdalla
Benny Applebaum
Michael Backes
Boaz Barak
Adam Barth
Amos Beimel
Avraham Ben-Aroya
Michael Ben-Or
Eli Ben-Sasson
Bruno Blanchet
Alexandra Boldyreva
Jan Camenisch
Claude Carlet
Dario Catalano
Rafi Chen
Martin Cochran
Anupam Datta
Giovanni Di Crescenzo
Yan Zong Ding
Yevgeniy Dodis
Orr Dunkelman
Stefan Dziembowsky
Nelly Fazio
Marc Fischlin
Matthias Fitzi
Jun Furikawa
Ariel Gabizon
Rosario Gennaro
Craig Gentry
Jens Groth
Dan Gutfreund
Joshua Guttman
Robbert de Haan

Stuart Haber
Iftach Haitner
Shai Halevi
Goichiro Hanaoka
Danny Harnik
Prahladh Harsha
Avinatan Hassidim
Johan Håstad
Ishay Haviv
Alex Healy
Jonathan Herzog
Dennis Hofheinz
Susan Hohenberger
Nicholas J. Hopper
Nick Howgrave-Graham
Antoine Joux
Nathan Keller
Eike Kiltz
Chiu-Yuen Koo
Eyal Kushilevitz
Yehuda Lindell
Anna Lysyanskaya
Frank McSherry
Ilya Mironov
David Molnar
Tal Moran
Gregory Neven
Jesper Buus Nielsen
Adam O'Neill
Shien Jin Ong
Ivan Osipkov
Rafail Ostrovsky
Saurabh Panjwani

Kenny Paterson
Chris Peikert
Benny Pinkas
Manoj Prabhakaran
Tal Rabin
Zulfikar Ramzan
Leo Reyzin
Andrei Romashchenko
Alon Rosen
Guy Rothblum
Amit Sahai
Kazue Sako
Louis Salvail
Christian Schaffner
Gil Segev
Hovav Shacham
abhi shelat
Vladimir Shpilrain
Tomas Toft
Marten Trolin
Eran Tromer
Boaz Tsaban
Vinod Vaikuntanathan
Ivan Visconti
Shabsi Walfish
Brent Waters
John Watrous
Hoeteck Wee
Enav Weinreb
Douglas Wikström
Andreas Winter
David P. Woodruff

Table of Contents

Encryption I

Does Privacy Require True Randomness? 1
    Carl Bosley and Yevgeniy Dodis

Tackling Adaptive Corruptions in Multicast Encryption Protocols 21
    Saurabh Panjwani

Universally Composable Security

Long-Term Security and Universal Composability 41
    Jörn Müller-Quade and Dominique Unruh

Universally Composable Security with Global Setup 61
    Ran Canetti, Yevgeniy Dodis, Rafael Pass, and Shabsi Walfish

Arguments and Zero Knowledge

Parallel Repetition of Computationally Sound Protocols Revisited 86
    Krzysztof Pietrzak and Douglas Wikström

Lower Bounds for Non-interactive Zero-Knowledge 103
    Hoeteck Wee

Perfect NIZK with Adaptive Soundness 118
    Masayuki Abe and Serge Fehr

Notions of Security

Security Against Covert Adversaries: Efficient Protocols for Realistic
Adversaries 137
    Yonatan Aumann and Yehuda Lindell

On the Necessity of Rewinding in Secure Multiparty Computation 157
    Michael Backes, Jörn Müller-Quade, and Dominique Unruh

On Expected Probabilistic Polynomial-Time Adversaries: A Suggestion
for Restricted Definitions and Their Benefits 174
    Oded Goldreich

Obfuscation

On Best-Possible Obfuscation 194
    Shafi Goldwasser and Guy N. Rothblum

Obfuscation for Cryptographic Purposes 214
  Dennis Hofheinz, John Malone-Lee, and Martijn Stam

Securely Obfuscating Re-encryption 233
  *Susan Hohenberger, Guy N. Rothblum, abhi shelat, and
  Vinod Vaikuntanathan*

Secret Sharing and Multiparty Computation

Weakly-Private Secret Sharing Schemes 253
  Amos Beimel and Matthew Franklin

On Secret Sharing Schemes, Matroids and Polymatroids 273
  Jaume Martí-Farré and Carles Padró

Secure Linear Algebra Using Linearly Recurrent Sequences 291
  Eike Kiltz, Payman Mohassel, Enav Weinreb, and Matthew Franklin

Towards Optimal and Efficient Perfectly Secure Message
Transmission 311
  *Matthias Fitzi, Matthew Franklin, Juan Garay, and
  S. Harsha Vardhan*

Signatures and Watermarking

Concurrently-Secure Blind Signatures Without Random Oracles or
Setup Assumptions 323
  Carmit Hazay, Jonathan Katz, Chiu-Yuen Koo, and Yehuda Lindell

Designated Confirmer Signatures Revisited 342
  Douglas Wikström

From Weak to Strong Watermarking 362
  Nicholas Hopper, David Molnar, and David Wagner

Private Approximation and Black-Box Reductions

Private Approximation of Clustering and Vertex Cover 383
  Amos Beimel, Renen Hallak, and Kobbi Nissim

Robuster Combiners for Oblivious Transfer 404
  Remo Meier, Bartosz Przydatek, and Jürg Wullschleger

One-Way Permutations, Interactive Hashing and Statistically Hiding
Commitments 419
  Hoeteck Wee

Towards a Separation of Semantic and CCA Security for Public Key
Encryption   434
     Yael Gertner, Tal Malkin, and Steven Myers

Key Establishment

Unifying Classical and Quantum Key Distillation   456
     *Matthias Christandl, Artur Ekert, Michał Horodecki,
     Paweł Horodecki, Jonathan Oppenheim, and Renato Renner*

Intrusion-Resilient Key Exchange in the Bounded Retrieval Model   479
     *David Cash, Yan Zong Ding, Yevgeniy Dodis, Wenke Lee,
     Richard Lipton, and Shabsi Walfish*

(Password) Authenticated Key Establishment: From 2-Party to
Group   499
     *Michel Abdalla, Jens-Matthias Bohli,
     María Isabel González Vasco, and Rainer Steinwandt*

Encryption II

Multi-authority Attribute Based Encryption   515
     Melissa Chase

Conjunctive, Subset, and Range Queries on Encrypted Data   535
     Dan Boneh and Brent Waters

How to Shuffle in Public   555
     Ben Adida and Douglas Wikström

Evaluating Branching Programs on Encrypted Data   575
     Yuval Ishai and Anat Paskin

Author Index   595

Does Privacy Require True Randomness?

Carl Bosley and Yevgeniy Dodis*

New York University
{bosley,dodis}@cs.nyu.edu

Abstract. Most cryptographic primitives require randomness (for example, to generate their secret keys). Usually, one assumes that perfect randomness is available, but, conceivably, such primitives might be built under weaker, more realistic assumptions. This is known to be true for many authentication applications, when entropy alone is typically sufficient. In contrast, all known techniques for achieving privacy seem to fundamentally require (nearly) perfect randomness. We ask the question whether this is just a coincidence, or, perhaps, privacy inherently requires true randomness?

We completely resolve this question for the case of (information-theoretic) private-key encryption, where parties wish to encrypt a b-bit value using a shared secret key sampled from some imperfect source of randomness $\mathscr{S}$. Our main result shows that if such n-bit source $\mathscr{S}$ allows for a secure encryption of b bits, where $b > \log n$, then one can deterministically extract nearly b almost perfect random bits from $\mathscr{S}$. Further, the restriction that $b > \log n$ is nearly tight: there exist sources $\mathscr{S}$ allowing one to perfectly encrypt $(\log n - \log\log n)$ bits, but not to deterministically extract even a single slightly unbiased bit.

Hence, to a large extent, *true randomness is inherent for encryption*: either the key length must be exponential in the message length b, or one can deterministically extract nearly b almost unbiased random bits from the key. In particular, *the one-time pad scheme is essentially "universal"*.

Our technique also extends to related *computational* primitives which are *perfectly-binding*, such as perfectly-binding commitment and computationally secure private- or public-key encryption, showing the necessity to *efficiently* extract almost b *pseudorandom* bits.

1 Introduction

Randomness is important in many areas of computer science. It is especially indispensable in cryptography: secret keys must be random, and many cryptographic tasks, such as public-key encryption, secret sharing or commitment, require randomness for every use. Typically, one assumes that all parties have access to a perfect random source, but this assumption is at least debatable, and the question of what kind of *imperfect random sources* can be used for various applications has attracted a lot of attention.

EXTRACTION. The easiest such class of sources consists of *extractable* sources for which one can deterministically extract nearly perfect randomness, and then use

---

* Supported by NSF Grants #0515121, #0133806, #0311095.

S.P. Vadhan (Ed.): TCC 2007, LNCS 4392, pp. 1–20, 2007.

it in any application. Although various examples of such non-trivial sources are known (see [TV00, KRVZ06] and the references therein), most natural sources, such as the so called entropy sources[1] [SV86, CG88, Zuc96], are easily seen to be non-extractable. One can then ask the natural question of whether perfect randomness is indeed inherent for the considered application, or perhaps one can do with weaker, more realistic assumptions. Clearly, the answer depends on the application.

POSITIVE RESULTS. For one such application domain, a series of celebrated results [VV85, SV86, CG88, Zuc96, ACRT99] showed that entropy sources are sufficient for simulating probabilistic polynomial-time algorithms — namely, problems which do not *inherently* need randomness, but which could potentially be sped up using randomization. Thus, extremely weak imperfect sources can still be tolerated for this application domain. This result was later extended to interactive protocols by Dodis et al. [DOPS04].

Moving to cryptographic applications, entropy sources are typically sufficient for authentication applications, since entropy is enough to ensure unpredictability. For example, in the non-interactive (i.e., one-message) setting Maurer and Wolf [MW97] show that, for a sufficiently high entropy rate (specifically, more than $1/2$), entropy sources are indeed sufficient for unconditional one-time authentication (while Dodis and Spencer [DS02] showed that smaller rate entropy sources are not sufficient to authenticate even a single bit). Moreover, in the interactive setting, Renner and Wolf [RW03] show information-theoretic authentication protocols capable of tolerating any constant-fraction entropy rate. Finally, Dodis et al. [DOPS04] consider the existence of computationally secure digital signature (and thus also message authentication) schemes, and, under (necessarily) strong, but plausible computational assumptions, once again showed that entropy sources are enough to build such signature schemes. From a different angle, [DS02] also show that for all entropy levels (in particular, below $1/2$) there exist "severely non-extractable" imperfect sources which are nevertheless sufficient for non-trivial non-interactive authentication. Thus, good sources for authentication certainly do not require perfect randomness.

RANDOMNESS FOR PRIVACY? The situation is much less clear for privacy applications, whose security definitions include some kind of indistinguishability. Of those, the most basic and fundamental is the question of (private-key) encryption, whose definition requires that the encryptions of any two messages are indistinguishable. (Indeed, this will be the subject of this work.)

With one exception (discussed shortly), all known results indicate that true randomness might be inherent for privacy applications, such as encryption. First, starting with Shannon's one-time scheme [Sha49], all existing methods for building secure encryptions schemes, as well as other privacy primitives, crucially

---

[1] Informally, entropy sources guarantee that every distribution in the family has a non-trivial amount of entropy (and possibly more restrictions), but do not assume independence between different symbols of the source. Thus, they are the most general sources one would wish to tolerate, since cryptography clearly requires entropy.

depend on perfect randomness somewhere in their design. And this is true even in the computational setting. For example, the Goldreich-Levin [GL89] reduction from unpredictability to indistinguishability, as well the the entire theory of pseudorandomness, crucially use a random seed to obtain the desired constructions. Second, attempts to build secure encryption schemes (and other privacy primitives) based on known "non-extractable" sources, such as various entropy sources, *provably failed*, indicating that such sources are indeed insufficient for privacy. For example, McInnes and Pinkas [MP90] showed that unconditionally secure symmetric encryption cannot be based on entropy sources, even if one is restricted to encrypting a single bit. This result was subsequently strengthened by Dodis et al. [DOPS04], who showed that entropy sources are not sufficient even for *computationally* secure encryption (as well as essentially any other task involving "privacy", such as commitment, zero-knowledge and others).

The only reassuring result in the other direction is the work of Dodis and Spencer [DS02], who considered the setting of symmetric encryption, where the shared secret key comes from an imperfect random source, instead of being truly random. In this setting, they constructed a particular non-extractable imperfect source, nevertheless allowing one to perfectly encrypt *a single bit*. By itself, this result is not surprising. For example, a uniform distribution on $\{0, 1, 2\}$ allows one to encrypt a bit (by addition modulo 3), but not to extract a bit, which is obvious. Indeed, the actual contribution of [DS02] was not to show that the separation between one bit encryption and extraction *exists* — as we just saw, this is trivial — but to show that a very strong separation still holds even if one additionally requires all the distributions in the imperfect source to have high entropy (in fact, very close to n). In practice, however, we typically care about encrypting considerably more than a single bit. In such cases, it is certainly unreasonable to expect that, say, encryption of b bits will necessarily imply extraction of *exactly* b bits (which was indeed disproved by [DS02] for $b = 1$). One would actually *expect* that an implication, if true, would lose at least a few bits (perhaps depending on the statistical distance ε from the uniform distribution that we want our extraction to achieve).

In particular, the results of [DS02] leave open the following extreme possibilities: (a) perhaps any source encrypting already two bits must be extractable; or (b) perhaps there exists an n-bit source allowing one to perfectly encrypt almost n bits, and yet not to extract even a single bit. Clearly, possibility (a) would strongly indicate that true randomness *is* inherent for encryption, while possibility (b) that it is *not*. As we will see shortly, both (a) and (b) happen to be false, but our point is that the results of [DS02] regarding *one-bit* encryption and extraction do not answer what we feel is the more appropriate question:

Assume an imperfect source allows for a secure private-key encryption of b bits. Does this necessarily imply one can deterministically extract at least one (and, hopefully, close to b) nearly perfect bits from this source?

OUR RESULT. We resolve the above question. Our main result shows that if an n-bit source $\mathscr{S}$ allows for a secure (and even slightly biased) encryption of

b bits, where $b > \log n$, then one can deterministically extract almost b nearly perfect random bits from $\mathscr{S}$; see Theorem 1(a) for the precise bound. Moreover, the restriction that $b > \log n$ is essentially tight: there exist imperfect sources allowing one to perfectly encrypt $b \approx \log n - \log\log n$ bits, from which one cannot deterministically extract even a single slightly unbiased (let alone random!) bit; see Theorem 1(b).[2] Hence, to a large extent, *true randomness is inherent for (information-theoretic) private-key encryption*:

Either the key length n must be exponential in the message length b, or One can deterministically extract almost b nearly random bits from the key.

In particular, in the case when b is large enough, so that it is infeasible to sample more than 2^b (imperfect) bits for one's secret key, our result implies the following. In order to build a secure b-bit encryption scheme, one must come up with a source of randomness from which one can already deterministically extract almost b nearly random bits! Notice, since such extracted bits can then be used as a one-time pad, we get that any b-bit encryption scheme can in principle be converted to a "one-time-pad-like" scheme capable of encrypting nearly b bits! In this sense, our results show that, *for the purpose of information-theoretically encrypting a "non-trivial" number of bits, the one-time pad scheme is essentially "universal"*.

EXTENSIONS. Our result can be extended in several ways.

First, the basic extractor we construct is inefficient, even if the encryption scheme is efficient (i.e., runs in time polynomial in n). However, using the technique of Trevisan and Vadhan [TV00] (see also [DSS01, Dod00]), we can obtain the following marginally weaker result which maintains efficiency: if a source $\mathscr{S}$ enables an *efficient* encryption of $b > \log n$ bits, then there exists an *efficient* deterministic extractor allowing one to extract roughly $(b - \log n)$ nearly perfect bits from $\mathscr{S}$. Despite the small loss of $\log n$ bits, we still get the same pessimistic conclusion: unless the key is exponential in the message length, efficient encryption implies efficient extraction of nearly the same number of bits.

Second, our technique extends to computationally secure privacy primitives which are *perfectly (or statistically) binding*, which includes perfectly-binding commitment (which, therefore, must be computationally hiding) and computationally secure private- or public-key encryption. Specifically, let λ be the security parameter, $n = \mathsf{poly}(\lambda)$ be the number of random bits coming from the imperfect source $\mathscr{S}$, and assume that $\mathscr{S}$ is good enough to *efficiently* (i.e., in time polynomial in λ) implement the required *computationally secure* (but perfectly-binding) primitive on $b = \omega(\log \lambda)$ bits. Then we show that there exists an *efficient* extractor capable of extracting $b(1 - o(1))$ *pseudorandom* bits from $\mathscr{S}$. Of course, at this point one can also apply a pseudorandom generator, whose existence is typically implied by the existence of the corresponding

---

[2] This result is a non-trivial extension of the separation of [DS02] from 1-bit to (roughly) $(\log n)$-bit encryption. Indeed, without the entropy constraints, our proof is considerably more involved than that of [DS02]. See also Section 4.5.

computational primitive, to stretch the extracted (pseudo)randomness further by any polynomial amount. Also, since every *individual* pseudorandom bit must actually be *statistically* random (otherwise, the distinguisher succeeds by simply outputting this bit), we still get that any of the above computationally secure primitives on $b = \omega(\log \lambda)$ bits requires *at least some* nearly perfect randomness.

To summarize, non-trivial computationally secure primitives which are perfectly binding require some *efficiently extractable true randomness*.

ORGANIZATION. We define the needed notation in Section 2, which also allows us to formally state our main result (Theorem 1). In Section 3 we prove that encryption of $b > \log n$ bits using an n-bit key implies extraction of roughly b random bits, and mention the "computational" extensions of this result. In Section 4, which is the main technical section, we show that encryption of up to $(\log n - \log\log n)$ bits does not necessarily imply extraction of even a single bit. Finally, in Section 5 we conclude and state some open problems.

2 Notation and Definitions

We use calligraphic letters, like $\mathcal{X}$, to denote finite sets. The corresponding large letter X is then used to denote a random variable over $\mathcal{X}$, while the lowercase letter x denotes a particular element from $\mathcal{X}$. $U_{\mathcal{X}}$ denotes the uniform distribution over $\mathcal{X}$. A source $\mathscr{S}$ over $\mathcal{X}$ is a set of distributions over $\mathcal{X}$. We write $X \in \mathscr{S}$ to state that $\mathscr{S}$ contains a distribution X.

The statistical distance $\mathsf{SD}(X_1, X_2)$ between two random variables X_1, X_2 is

$$\mathsf{SD}(X_1, X_2) = \frac{1}{2} \sum_{x \in \mathcal{X}} \left| \Pr[X_1 = x] - \Pr[X_2 = x] \right| \tag{1}$$

$$= \max_{\mathcal{T} \subseteq \mathcal{X}} \left(\Pr[X_1 \in \mathcal{T}] - \Pr[X_2 \in \mathcal{T}] \right) \tag{2}$$

If $\mathsf{SD}(X_1, X_2) \leq \varepsilon$, this means that no (even computationally unbounded) distinguisher D can tell apart a sample from X_1 from a sample from X_2 with an advantage greater than ε.

Definition 1. *A random variable R over $\mathcal{R}$ is ε-fair if $\mathsf{SD}(R, U_{\mathcal{R}}) \leq \varepsilon$. Given a source $\mathscr{S}$ over some set $\mathcal{K}$, a function $\mathsf{Ext} : \mathcal{K} \to \mathcal{R}$ is an $(\mathscr{S}, \varepsilon)$-extractor if for all $K \in \mathscr{S}$, $\mathsf{Ext}(K)$ is ε-fair:*

$$\mathsf{SD}(\mathsf{Ext}(K), U_{\mathcal{R}}) \leq \varepsilon \tag{3}$$

If such Ext exists for $\mathscr{S}$, we say that $\mathscr{S}$ is $(\mathcal{R}, \varepsilon)$-extractable. ◇

Definition 2. *An encryption scheme $\mathcal{E}$ over message space $\mathcal{M}$, key space $\mathcal{K}$ and ciphertext space $\mathcal{C}$ is a pair of algorithms $\mathsf{Enc} : \mathcal{K} \times \mathcal{M} \to \mathcal{C}$ and $\mathsf{Dec} : \mathcal{K} \times \mathcal{C} \to \mathcal{M}$, which for all keys $k \in \mathcal{K}$ and messages $m \in \mathcal{M}$ satisfies $\mathsf{Dec}(k, \mathsf{Enc}(k, m)) = m$.*

Given a source $\mathscr{S}$ over $\mathcal{K}$, we say that the encryption scheme $\mathcal{E}$ is $(\mathscr{S}, \delta)$-secure if for all messages $m_1, m_2 \in \mathcal{M}$ and all distributions $K \in \mathscr{S}$ we have

$$\mathsf{SD}(\mathsf{Enc}(K, m_1), \mathsf{Enc}(K, m_2)) \leq \delta \tag{4}$$

If $\mathscr{S}$ admits some $(\mathscr{S},\delta)$-secure encryption $\mathcal{E}$ over $\mathcal{M}$, we say that $\mathscr{S}$ is $(\mathcal{M},\delta)$-encryptable. When $\delta = 0$, we say that $\mathcal{E}$ is perfect on $\mathscr{S}$, and $\mathscr{S}$ is perfectly encryptable (on $\mathcal{M}$). ◇

Throughout we will use the following capital letters to denote the cardinalities of various sets: key set cardinality $|\mathcal{K}| = N$, message set cardinality $|\mathcal{M}| = B$, ciphertext set cardinality $|\mathcal{C}| = S$, and extraction space cardinality $|\mathcal{R}| = L$. Although our results are general, for historical reasons it is customary to translate the results into "bit-notation". To accommodate these conventions, we let $b = \log B$, $\ell = \log L$, $n = \log N$ (here and elsewhere, all the logarithms are base 2), and will use the terms "b-bit encryption", "ℓ-bit extraction" or "n-bit key" with the obvious meanings attached. Moreover, we will slightly abuse the terminology and say that a source $\mathscr{S}$ is (1) n-*bit* if it is over a set $\mathcal{K}$ and $|\mathcal{K}| = N$; (2) (ℓ,ε)-*extractable* if it is $(\mathcal{R},\varepsilon)$-extractable and $|\mathcal{R}| = L$, and (2) (b,δ)-*encryptable* if it is $(\mathcal{M},\delta)$-encryptable and $|\mathcal{M}| = B$. Clearly, when b, ℓ or n are integers, this terminology is consistent with our intuitive understanding.

With this in mind, our main result can be restated as follows:

Theorem 1. *Secure encryption of b bits with an n-bit key requires nearly perfect randomness (in fact, almost b random bits!) if and only if b is greater than $\log n$. More precisely,*

(a) $\forall \varepsilon > 0$*, if $\mathscr{S}$ is (b,δ)-encryptable, and $b > \log n + 2\log\left(\frac{1}{\varepsilon}\right)$, then $\mathscr{S}$ is $\left(b - 2\log\left(\frac{1}{\varepsilon}\right), \varepsilon+\delta\right)$-extractable. Further, if the encryption scheme is* efficient *(i.e., polynomial in n), then there exists an efficient extractor outputting $\left(b - \log n - 2\log\left(\frac{1}{\varepsilon}\right) - 2\right)$ bits within statistical distance $(\varepsilon+\delta)$ from uniform. Thus, encryption of $b > \log n$ bits implies extraction of almost b nearly perfect bits.*

(b) *For any $b \le \log n - \log\log n - 2$,[3] there exists a source $\mathscr{S}$ which is $(b,0)$-encryptable, but not $(1,\varepsilon)$-extractable, where $\varepsilon = \frac{1}{2} - 2^{(2b-\frac{n}{2b})} \ge \frac{1}{2} - \frac{1}{16n^2}$. Thus, even perfect encryption of nearly $\log n$ bits does not imply extraction of even a single slightly unbiased bit.*

3   Encryption $\Rightarrow$ Extraction if $b > \log n$

In this section we prove the implication given in Theorem 1(a), which shows that encryption of b bits implies extraction of nearly b bits. Assume $\mathcal{E} = (\mathsf{Enc}, \mathsf{Dec})$ is $(\mathscr{S},\delta)$-secure over message space $\mathcal{M}$, ciphertext space $\mathcal{C}$ and key space $\mathcal{K}$. For convenience, let us identify the message space $\mathcal{M}$ with $\{1,\ldots,B\}$. Also, let ℓ (to be specified later) denote the number of bits we wish to extract, $L = 2^{\ell}$, and $\mathcal{R}$ be an arbitrary set of cardinality L.

We start constructing the needed extractor $\mathsf{Ext} : \mathcal{K} \to \mathcal{R}$ by showing that it is sufficient to construct a good extractor $\mathsf{Ext'} : \mathcal{C} \to \mathcal{R}$ for an auxiliary source $\mathscr{S'}$, defined by

$$\mathscr{S'} = \{\mathsf{Enc}(k, U_{\mathcal{M}}) \mid k \in \mathcal{K}\}$$

---

[3] The formula also holds for $b = \log n - \log\log n - 1$, but yields a slightly smaller $\varepsilon = \frac{1}{2} - \frac{1}{4\log n}$.

Lemma 1. *If $\mathscr{S}'$ is (ℓ, ε)-extractable and $\mathcal{E}$ is $(\mathscr{S}, \delta)$-secure, then $\mathscr{S}$ is $(\ell, \varepsilon+\delta)$-extractable. In fact, if Ext' is the assumed extractor for $\mathscr{S}'$, then the following extractor Ext is the claimed extractor for $\mathscr{S}$:*

$$\mathsf{Ext}(k) = \mathsf{Ext}'(\mathsf{Enc}(k, 1)) \tag{5}$$

Proof. Take any distribution $K \in \mathscr{S}$, and let $p_k = \Pr[K = k]$. Also, let Ext' be the assumed $(\mathscr{S}', \varepsilon)$-extractor. Thus, $\mathsf{SD}(\mathsf{Ext}'(\mathsf{Enc}(k, U_\mathcal{M})),\ U_\mathcal{R}) \leq \varepsilon$ for all $k \in \mathcal{K}$. Then, using definition of Ext in Equation (5), we have

$$\begin{aligned}
\mathsf{SD}(\mathsf{Ext}(K),\ U_\mathcal{R}) &= \mathsf{SD}(\mathsf{Ext}'(\mathsf{Enc}(K, 1)),\ U_\mathcal{R}) \\
&\leq \mathsf{SD}(\mathsf{Enc}(K, 1),\ \mathsf{Enc}(K, U_\mathcal{M})) + \mathsf{SD}(\mathsf{Ext}'(\mathsf{Enc}(K, U_\mathcal{M})),\ U_\mathcal{R}) \\
&\leq \delta + \sum_k p_k \cdot \mathsf{SD}(\mathsf{Ext}'(\mathsf{Enc}(k, U_\mathcal{M})),\ U_\mathcal{R}) \\
&\leq \delta + \sum_k p_k \cdot \varepsilon\ =\ \delta + \varepsilon
\end{aligned}$$

The first inequality follows from the triangle inequality on statistical distance. The second — from the δ-security of the encryption (stating that encryption of 1 is δ-close to the encryption of a random message $U_\mathcal{M}$) and the convexity of statistical distance (when expanding K as the convex combination of "point" distributions). Finally, the last inequality follows from the fact that Ext' is an ε-fair extractor for $\mathscr{S}'$. □

The point of this reduction (which is the only place in our argument using the δ-security of $\mathcal{E}$) is to reduce the task of constructing an extractor for our (potentially infinite) source $\mathscr{S}$ to an extractor for a source $\mathscr{S}'$ containing "only" N distributions. Moreover, every distribution $D_k \stackrel{\text{def}}{=} \mathsf{Enc}(k, U_\mathcal{M})$ in $\mathscr{S}'$ contains b bits of entropy. Indeed, for any $k \in \mathcal{K}$ and $m_1 \neq m_2$, we have $\mathsf{Enc}(k, m_1) \neq \mathsf{Enc}(k, m_2)$, since otherwise one would not be able to recover the message from the ciphertext.[4] Thus, each D_k is a uniform distribution on some B-element subset of the ciphertext space $\mathcal{C}$: we call such distributions *b-flat*. It turns out that this is the only thing we need to know to ensure the existence of a good extractor for $\mathscr{S}'$!

Lemma 2. *Assume $\mathscr{S}' = \{D_k \mid k \in \mathcal{K}\}$ is any collection of b-flat distributions of cardinality N over some space $\mathcal{C}$, where $b > \log\log N + 2\log\left(\frac{1}{\varepsilon}\right)$. Then $\mathscr{S}'$ is $\left(b - 2\log\left(\frac{1}{\varepsilon}\right), \varepsilon\right)$-extractable.*

Proof. Let $\ell = b - 2\log\left(\frac{1}{\varepsilon}\right)$, so that $L = \varepsilon^2 B$. We show that a completely *random* function $f : \mathcal{C} \to \mathcal{R}$ gives a required *deterministic* extractor Ext' with non-zero (in fact, overwhelming!) probability, implying that the claimed Ext' exists. Take

---

[4] This is the only place where we use the existence of the decryption algorithm. This is why our result will later extend to any perfectly (or statistically) binding primitive.

any fixed $k \in \mathcal{K}$ and any fixed subset $\mathcal{T} \subseteq \mathcal{R}$. Let $p \stackrel{\text{def}}{=} |\mathcal{T}|/|\mathcal{R}|$ be the density of $\mathcal{T}$. For any fixed f, define the quantity

$$\Delta_f(k, \mathcal{T}) \stackrel{\text{def}}{=} \Pr[f(D_k) \in \mathcal{T}] - \Pr[U_{\mathcal{R}} \in \mathcal{T}] \tag{6}$$

and let us estimate $\Pr_f[\Delta_f(k, \mathcal{T}) > \varepsilon]$ as follows. First, it is clear that $\Pr[U_{\mathcal{R}} \in \mathcal{T}] = p$. Second, assume D_k is a uniform distribution over some set $\{c_1, \ldots, c_B\} \subseteq \mathcal{C}$, and let X_m denote an indicator random variable which is 1 if and only if $f(c_m) \in \mathcal{T}$. Clearly, if f is random, we have $\Pr_f[X_m = 1] = p$. Also, letting $\hat{X} = \frac{1}{B} \cdot \sum_m X_m$ be the average of B independent indicator variables X_m, for any fixed f we get $\Pr[f(D_k) \in \mathcal{T}] = \frac{1}{B} \cdot \sum_m X_m = \hat{X}$. Thus, recalling the definition of $\Delta_f(k, \mathcal{T})$ from Equation (6), using $\mathbb{E}[\hat{X}] = p = \Pr[U_{\mathcal{R}} \in \mathcal{T}]$, and applying the standard additive Chernoff bound to $\hat{X}$, we get

$$\Pr_f[\, \Delta_f(k, \mathcal{T}) > \varepsilon \,] = \Pr_f[\, \hat{X} - p > \varepsilon \,] \leq e^{-2\varepsilon^2 B}$$

We now take a union bound over all $\mathcal{T} \subseteq \mathcal{R}$ and all $k \in \mathcal{K}$. Recalling definition of $\Delta_f(k, \mathcal{T})$ (Equation (6)), using $b > \log\log N + 2\log\left(\frac{1}{\varepsilon}\right)$ (so $N < 2^{\varepsilon^2 B}$) and $\ell = b - 2\log\left(\frac{1}{\varepsilon}\right)$ (so $2^L = 2^{\varepsilon^2 B}$), we conclude that

$$\Pr_f[\exists\, k, \mathcal{T} \text{ s.t. } \Pr[f(D_k) \in \mathcal{T}] - \Pr[U_{\mathcal{R}} \in \mathcal{T}] > \varepsilon] \leq N \cdot 2^L \cdot e^{-2\varepsilon^2 B} = 2^{-\Omega(\varepsilon^2 B)} \ll 1$$

Thus, there exists a specific f such that $\Pr[f(D_k) \in \mathcal{T}] - \Pr[U_{\mathcal{R}} \in \mathcal{T}] \leq \varepsilon$, for *all* subsets $\mathcal{T}$ and keys k. Using the definition of statistical distance (Equation (2)), this means that $\mathsf{SD}(f(D_k), U_{\mathcal{R}}) \leq \varepsilon$ for all $k \in \mathcal{K}$, completing the proof. $\square$

The first assertion of Theorem 1(a) follows immediately by combining Lemma 1 and Lemma 2. In the following subsections we mention the extensions to efficient extraction and other computational primitives which are perfectly-binding.

3.1   Efficient Encryption Implies Efficient Extraction

Using Lemma 1 (and, in particular, Equation (5)), we see that when the encryption algorithm Enc is efficient (i.e., runs in time polynomial in n), to construct an efficient extractor Ext for $\mathscr{S}$ it suffices to construct an efficient extractor Ext$'$ for the source $\mathscr{S}'$ consisting of N efficiently samplable b-flat distributions $D_k = \mathsf{Enc}(k, U_{\mathcal{M}})$, where $k \in \mathcal{K}$. Unfortunately, the extractor Ext$'$ that we built for $\mathscr{S}'$ via Lemma 2 was generally inefficient. Luckily, we can build an efficient extractor for $\mathscr{S}'$ using the technique of Trevisan and Vadhan [TV00], which was later explored in more detail by [Dod00].

The idea is to sample the function f (which will define Ext$'$) at random from any family $\mathcal{F}_t$ of t-*wise independent functions* from $\mathcal{C}$ to $\mathcal{R}$. Recall, such families have the property that for any distinct $c_1 \ldots c_t \in \mathcal{C}$, the values $f(c_1) \ldots f(c_t)$ are random and independent from each other, if f is chosen at random from $\mathcal{F}_t$. Also, one can construct t-wise independent function families where each f can be evaluated in time polynomial in t and s, where s is the length of an element of $\mathcal{C}$. Since the encryption scheme is efficient, s is polynomial in n. Thus, as long as t is polynomial in n, every member $f \in \mathcal{F}_t$ will be efficiently computable. As was shown by [TV00, Dod00], setting $t = O(n)$ is already enough: the following Lemma (essentially from [Dod00]) is proven for self-containment and because it uses a slightly different parameter setting.

Lemma 3 ([Dod00]). *Assume $\ell \leq b - \log n - 2\log\left(\frac{1}{\varepsilon}\right) - 2$, and f is chosen at random from a family of $2n$-wise independent functions from $\mathcal{C}$ to $\mathcal{R}$, where $|\mathcal{R}| = L = 2^{\ell}$. Then for any collection $\mathscr{S}' = \{D_k \mid k \in \mathcal{K}\}$ of b-flat distributions of cardinality 2^n over $\mathcal{C}$, $\Pr_f[\ f$ is not an $(\mathscr{S}', \varepsilon)$-extractor $] < 2^{-n}$.*

Proof. The first attempt to prove this result would be to use the same proof template as in Lemma 2. Namely, to prove that for any subset $\mathcal{T} \subseteq \mathcal{R}$ and any b-flat distribution $D_k \in \mathscr{S}'$, $\Pr_f[f(D_k) \in \mathcal{T}]$ is unlikely to be different from its expectation $\Pr[U_{\mathcal{R}} \in \mathcal{T}]$ by more then ε. Unfortunately, with "only" a t-wise independent function f, the tail bound we would get for this undesirable event is not strong enough to take the union bound over all subsets $\mathcal{T}$ (unless t is exponential in b, which was the case when a truly random f was chosen in Lemma 2). Instead, we will only consider "singleton" sets $\mathcal{T} = \{r\}$, for $r \in \mathcal{R}$, but will prove a stronger bound on $\Delta_f(k, \{r\}) \overset{\text{def}}{=} (\Pr_f[f(D_k) = r] - \frac{1}{L})$ when $\ell \leq b - 2\log\left(\frac{1}{\varepsilon}\right) - \log n - 2$. This stronger bound will enable us to use Equation (1) (rather than Equation (2)) when bounding the statistical distance, and then take a union bound over "only" L singleton sets $\{r\}$ instead of 2^L subsets $\mathcal{T}$. Details follow.

We fix any $k \in \mathcal{K}$, $r \in \mathcal{R}$, and estimate $\Pr_f[\ |\Delta_f(k, \{r\})| > \frac{2\varepsilon}{L}\ ]$. We do it similarly to Lemma 2. Assume D_k is a uniform distribution over some set $\{c_1, \ldots, c_B\} \subseteq \mathcal{C}$, and let X_m denote an indicator random variable which is 1 if and only if $f(c_m) = r$. Since f is $2n$-wise independent, so are the variables $\{X_m\}$: any $2n$ of them are random and independent from each other. Let $X = \sum_m X_m$. Then $\Pr_f[X_m = 1] = \Pr_f[f(c_m) = r)] = \frac{1}{L}$, and $\mathbb{E}[X] = \frac{B}{L}$. Also,

$$\Delta_f(k, \{r\}) = \frac{1}{B} \cdot \sum_m \Pr[f(c_m) = r] - \frac{1}{L} = \frac{1}{B} \cdot (X - \mathbb{E}[X]) \qquad (7)$$

Next, we use the tail bound for the sum X of t-wise independent random variables from [Dod00] (Theorem 5, page 48). It says that if $t \geq 8$ is an even integer and $\varepsilon < \frac{1}{2}$, then $\Pr(|X - \mathbb{E}[X]| \geq 2\varepsilon \cdot \mathbb{E}[X]) \leq \left(\frac{t}{4\varepsilon^2 \mathbb{E}[X]}\right)^{t/2}$. In our case, $t = 2n$, $\mathbb{E}[X] = \frac{B}{L}$, and we get by Equation (7)

$$\Pr_f\left[\ |\Delta_f(k, \{r\})| > \frac{2\varepsilon}{L}\ \right] = \Pr_f[\ |X - \mathbb{E}[X]| > 2\varepsilon \cdot \mathbb{E}[X]\ ] \leq \left(\frac{2nL}{4\varepsilon^2 B}\right)^n \leq 2^{-3n}$$

where the last inequality used $\ell \leq b - 2\log\left(\frac{1}{\varepsilon}\right) - \log n - 2$. Taking now the union bound over all $k \in \mathcal{K}$ and $r \in \mathcal{R}$, we get that with probability at least $(1 - 2^{-n})$ over the choice of f, we have $|\Delta_f(k, \{r\})| \leq \frac{2\varepsilon}{L}$ for all $k \in \mathcal{K}$ and $r \in \mathcal{R}$. In other words, for any $k \in \mathcal{K}$, $f(D_k)$ hits *every* element $r \in \mathcal{R}$ with probability between $(1\pm 2\varepsilon)/L$. Using the definition of statistical distance in Equation (1), this implies that with probability at least $(1 - 2^{-n})$ over the choice of f, $\mathsf{SD}(f(D_k), U_{\mathcal{R}}) \leq \varepsilon$ for all $k \in \mathcal{K}$, which completes the proof. ∎

The above lemma immediately gives a *constructive probabilistic method* for showing the existence of an efficient *deterministic* extractor claimed by the second part of Theorem 1(a). Namely, combining Lemma 1 and Lemma 3 we get a concrete family of efficient functions most of which are guaranteed to be good deterministic extractors for $\mathscr{S}$. However, to actually fix a concrete extractor, one must either directly look at the source $\mathscr{S}$ in question, or choose the extractor *obliviously* by sampling it (using good randomness) from our family *once and for all*, or rely on non-uniformity. Alternatively, in case the length s of the ciphertext c is only slightly larger than the length b of the plaintext m, we can use an explicit deterministic extractor of Trevisan and Vadhan [TV00] for the efficiently samplable source $\mathscr{S}'$. Assuming some strong complexity assumptions (see [TV00]), this would give us an explicit way to deterministically extract $\Omega(b)$ bits, provided $s < (1 + \gamma)b$ for a small enough constant γ.

3.2 Other Perfectly-Binding Computational Primitives

We now extend our results above to handle *computationally* secure privacy primitives which are *perfectly binding*, which includes perfectly-binding commitment (which, therefore, must be computationally hiding) and computationally secure private- or public-key encryption.

Let λ be the security parameter, $n = \mathsf{poly}(\lambda)$ be the number of random bits coming from the imperfect source $\mathscr{S}$, and assume that $\mathscr{S}$ is good enough to *efficiently* (i.e., in time polynomial in λ) implement the required *computationally secure* (but perfectly-binding) primitive P on $b = \omega(\log \lambda)$ bits. Trying to unify all the above examples into one template, this means that there exists a polynomial-time algorithm Enc, which takes input $m \in \mathcal{M}$ and "randomness" $k \in \mathcal{K}$, and outputs a perfectly-binding "commitment" c to m. Here k denotes *all* the randomness needed to evaluate Enc once. For example, for secret- or public-key encryption, k includes the randomness used to sample the secret and/or public key, and, if required, the local randomness used to encrypt the message. On the other hand, for commitment, k includes the randomness used to set-up the global commitment parameters, as well as the randomness used to commit to the messages.

We assume that c is *perfectly-binding* in the following sense: for any randomness k and any $m_1 \neq m_2$, we have $\mathsf{Enc}(k, m_1) \neq \mathsf{Enc}(k, m_2)$. Notice, we do not require any efficient "decryption" algorithm recovering m from c and k (which we have in the case of encryption, but not commitment). Clearly, this includes the perfectly-binding encryption and commitment applications above. In fact, it even includes some primitives which are traditionally *not* considered perfectly-binding. For example, Pedersen's commitment [Ped91] computes $\mathsf{Enc}((r, g, h, p), m) = g^r h^m \bmod p$, where $k = (r, g, h, p)$ includes a prime p, two generators g and h of some large-enough subgroup G of $\mathbb{Z}_p^*$ of prime order q, and local randomness $r \in \mathbb{Z}_q$ used to mask the message $m \in \mathbb{Z}_q$. Traditionally, this commitment scheme is considered *perfectly-hiding* (in the setting of ideal randomness), since for any m, the value $\mathsf{Enc}((r, \ldots), m)$ is uniformly distributed for a *random* r. However, it is *perfectly-binding* according to our definition, since for any *fixed* value of r, the value of m is (inefficiently but) uniquely determined given c (and g, h, p). Thus, our notion of perfect binding is a weaker restriction than what might originally appear.

Also, in terms of computational security of P w.r.t. a source of randomness $\mathscr{S}$, we require that for any distribution $K \in \mathscr{S}$ and any $m \in \mathcal{M}$, no efficient attacker A can distinguish $\mathsf{Enc}(K, m)$ from $\mathsf{Enc}(K, U_\mathcal{M})$ with non-negligible probability (in λ). Finally, we say that an efficient algorithm Ext extracts ℓ *pseudorandom* bits from some source $\mathscr{S}$, if for any $K \in \mathscr{S}$ and any efficient attacker A, A has at most a negligible in λ chance of telling apart a sample of $\mathsf{Ext}(K)$ from a sample of U_ℓ. Needless to say, any ε-fair "statistical" extractor satisfies this definition *as long as ε is negligible in λ*.

With these clarifications in mind, we can generalize Lemma 1 and Lemma 3 as follows. Lemma 1 trivially extends to show that if some *efficient* Ext' extracts b' *pseudorandom* bits from the source $\mathscr{S}' \overset{\text{def}}{=} \{\mathsf{Enc}(k, U_\mathcal{M})\}$, then $\mathsf{Ext}(k) \overset{\text{def}}{=} \mathsf{Ext}'(\mathsf{Enc}(k, 1))$ also extracts b' *pseudorandom* bits from $\mathscr{S}$. This is the only place using the computational security of P, the rest of the proofs stays information-theoretic. As for Lemma 3, it stays the same, but we use it with any value ε which is negligible in λ, but still such that $\log\left(\frac{1}{\varepsilon}\right) = o(b)$. This is possible since we assumed that $b = \omega(\log \lambda)$. Then Lemma 3 implies the existence of an efficient extractor Ext' for $\mathscr{S}'$ (since $n = \mathsf{poly}(\lambda)$, so that one can efficiently evaluate a $2n$-wise independent function) which extracts $b - 2\log\left(\frac{1}{\varepsilon}\right) - \log n - O(1) = b - o(b) - O(\log \lambda) = b(1 - o(1))$ bits of negligible statistical distance ε from the uniform distribution, implying that these $b(1 - o(1))$ bits are also pseudorandom.

To summarize, for any perfectly-binding primitive P on $b = \omega(\log \lambda)$ bits, we get the possibility of efficiently extracting $b(1 - o(1))$ pseudorandom bits.

4 Encryption $\not\Rightarrow$ Extraction if $b < \log n - \log\log n$

In this section we prove the non-implication given in Theorem 1(b), which shows that even perfect encryption of up to $(\log n - \log\log n)$ bits does not necessarily imply extraction of even a single bit. For that we need to define a specific b-bit encryption scheme $\mathcal{E} = (\mathsf{Enc}, \mathsf{Dec})$ and a source $\mathscr{S}$, such that $\mathscr{S}$ is perfect on $\mathcal{E}$, but "non-extractable". The proof will proceed in several stages.

4.1 Defining Good Encryption $\mathcal{E}$

As the first observation, we claim that we only need to define the encryption scheme $\mathcal{E}$, and then let the source $\mathscr{S} = \mathscr{S}(\mathcal{E})$ be the set of all key distributions K making $\mathcal{E}$ perfect:

$$\mathscr{S}(\mathcal{E}) = \{K \mid \forall\, m_1, m_2 \in \mathcal{M}, c \in \mathcal{C} \Rightarrow \Pr[\mathsf{Enc}(K, m_1) = c] = \Pr[\mathsf{Enc}(K, m_2) = c]\}$$

Indeed, $\mathscr{S}(\mathcal{E})$ is the largest source which is $(b,0)$-encryptable by means of $\mathcal{E}$, so it is the hardest one to extract even a single bit from. We call distributions in $\mathscr{S}(\mathcal{E})$ *perfect* (for $\mathcal{E}$).

Although we are not required to do so, let us intuitively motivate our choice of $\mathcal{E}$ before actually defining it. For that it is very helpful to view our key space $\mathcal{K}$ in terms of the encryption scheme $\mathcal{E}$ as follows. Given any $\mathcal{E} = (\mathsf{Enc}, \mathsf{Dec})$, we identify each key $k \in \mathcal{K}$ with an ordered B-tuple of ciphertexts $(c_1, \dots, c_B)$, where $\mathsf{Enc}(k, m) = c_m$. Notice, some B-tuples might not correspond to valid keys. For example, this is the case when $c_i = c_j$ for some $i \neq j$, since then encryptions of i and j are the same under this key. Intuitively, however, the larger is the set of valid B-tuples of ciphertexts, the more variety we have in the set of perfect distributions $\mathscr{S}(\mathcal{E})$, and the harder it would be to extract from $\mathscr{S}(\mathcal{E})$. This suggests that every B-tuple $(c_1, \dots, c_B)$ of ciphertexts should correspond to a potential key, except for the necessary constraint that all the c_m's must be distinct to enable unique decryption.

A bit more formally, we assume that N can be written as $N = S(S-1) \dots (S-B+1)$ for some integer S.[5] Then we define the set $\mathcal{C} = \{1, \dots S\}$ to be the set of ciphertexts, $\mathcal{M} = \{1, \dots, B\}$ be the set of plaintexts, and view the key set $\mathcal{K}$ as the set of distinct B-tuples over $\mathcal{C}$:

$$\mathcal{K} = \{k = (c_1, \dots c_B) \mid \forall\, i \neq j \Rightarrow c_i \neq c_j\}$$

We then define $\mathsf{Enc}((c_1 \dots c_B), m) = c_m$, while $\mathsf{Dec}((c_1, \dots, c_B), c)$ to be the (necessarily unique) m such that $c_m = c$, and arbitrarily if no such m exists. Notice, $N < S^B$, so that $S > N^{1/B}$, which is strictly greater than B when $b < \log n - \log\log n$. Thus, S contains enough ciphertexts to allow for B distinct encryptions.

4.2   Excluding 0-Monochromatic Distributions

Let us now take an arbitrary bit extractor $\mathsf{Ext} : \mathcal{K} \to \{0,1\}$ and argue that it is not very good on the set of perfect distributions $\mathscr{S}(\mathcal{E})$. We say that a distribution K is 0-*monochromatic* if $\Pr[\mathsf{Ext}(K) = 0] = 1$. Clearly, if the set of perfect distributions $\mathscr{S}(\mathcal{E})$ contains a 0-monochromatic distribution K, then $\mathsf{SD}(\mathsf{Ext}(K), U_1) = \frac{1}{2}$ (here and below, U_1 is the uniform distribution of $\{0,1\}$), and we would be done. Thus, for the remainder of the proof we assume that $\mathscr{S}(\mathcal{E})$ *does not contain a 0-monochromatic distribution*. The heart of the proof then will consist of designing a perfect encryption distribution K such that

$$\Pr[\mathsf{Ext}(K) = 0] \leq \frac{B^2}{S} \tag{8}$$

Once this is done, recalling that $S > N^{1/B} = 2^{n/2^b}$ we immediately get

$$\mathsf{SD}(\mathsf{Ext}(K), U_1) = \left| \frac{1}{2} - \Pr[\mathsf{Ext}(K) = 0] \right| \geq \frac{1}{2} - 2^{(2b - \frac{n}{2^b})}$$

---

[5] If not, take largest S such that $N \geq S(S-1) \dots (S-B+1)$, and work on the subset of $N' = S(S-1) \dots (S-B+1)$ keys, but this will not change our bounds.

as claimed by Theorem 1(b). Thus, we concentrate on building a perfect distribution K satisfying Equation (8). For that, in the following subsections we will (1) characterize perfect distributions using linear algebra; (2) use this characterization to understand the implication of the lack of 0-monochromatic perfect distributions; and, finally, (3) use this implication to construct the required perfect distribution K.

4.3   Characterizing Perfect Distributions

Let K be any distribution on $\mathcal{K}$. Given a key $k = (c_1 \ldots c_B)$, let $p_k = p_{(c_1 \ldots c_B)} = \Pr[K = (c_1 \ldots c_B)]$ and p be the N-dimensional column vector whose k-th component is equal to p_k. Notice, being a probability vector, we know that $\sum p_k = 1$ and $p \geq 0$ (which is a shorthand for $p_k \geq 0$ for all k). Conversely, any such p defines a unique distribution K.

Assume now that K is a perfect encryption distribution for $\mathcal{E}$. This adds several more constraints on p. Specifically, a necessary and sufficient condition for a perfect encryption distribution is to require that for all $c \in \mathcal{C}$ and all $m > 1$, we have

$$\Pr[c_1 = c \mid (c_1 \ldots c_B) \leftarrow K] = \Pr[c_m = c \mid (c_1 \ldots c_B) \leftarrow K] \tag{9}$$

We can translate this into a linear equation by noticing that the left probability is equal to $\sum_{\{(c_1 \ldots c_B):c_1=c\}} p_{(c_1 \ldots c_B)}$, while the second — to $\sum_{\{(c_1 \ldots c_B):c_m=c\}} p_{(c_1 \ldots c_B)}$. Thus, Equation (9) can be rewritten as

$$\sum_{\{(c_1 \ldots c_B):c_1=c\}} p_{(c_1 \ldots c_B)} - \sum_{\{(c_1 \ldots c_B):c_m=c\}} p_{(c_1 \ldots c_B)} = 0 \tag{10}$$

We can then rewrite all these constraints on p into a more compact notation by defining a *constraint matrix* $V = \{v_{i,j}\}$, which has $(1 + (B-1)S)$ rows (corresponding to the constraints) and N columns (corresponding to keys). The first row of V will consist of all 1's: $v_{1,k} = 1$ for all $k \in \mathcal{K}$. This will later correspond to the fact that $\sum p_k = 1$. To define the rest of V, which would correspond to $(B-1)S$ constraints from Equation (10), we first make our notation more suggestive. We index the N columns of V by tuples $(c_1, \ldots c_B)$, and the remaining $(B-1)S$ rows of V by tuples (m, c), where $m \in \{2, \ldots B\}$ and $c \in \{1 \ldots S\}$. Then, we define

$$v_{(m,c),(c_1,\ldots,c_B)} = \begin{cases} 1, & c = c_1, \\ -1, & c = c_m, \\ 0, & \text{otherwise.} \end{cases} \tag{11}$$

Now, Equation (10) simply becomes $\sum_k v_{(m,c),k} \cdot p_k = 0$. Finally, we define a $(1 + (B-1)S)$-column vector e by $e_1 = 1$ and $e_i = 0$ for $i > 1$. Combining all this notation, we finally get

Lemma 4. *An N-dimensional real vector p defines a perfect distribution K for $\mathcal{E}$ if and only if $Vp = e$ and $p \geq 0$.*

4.4   Using the Lack of 0-Monochromatic Distributions

Next, we use Lemma 4 to understand our assumption that no perfect distribution K is 0-monochromatic with respect to Ext. Before that, we remind the reader of a well known Farkas Lemma (e.g., see [Str80]):

Farkas Lemma. *For any matrix A and column vector e, the linear system $Ax = e$ has no solution $x \geq 0$ if and only if there exists a row vector y s.t. $yA \geq 0$ and $ye < 0$.*

Now, let $Z = \{k \mid \textsf{Ext}(k) = 0\}$ be the set of "0-keys" under Ext, and let A denote $(1 + (B-1)S) \times |Z|$-matrix equal to the constraint matrix V restricted its $|Z|$ columns in Z. Take any real vector p such that $p_k = 0$ for all $k \notin Z$. By Lemma 4, p corresponds to a (necessarily 0-monochromatic) perfect distribution K if and only if $Vp = e$ and $p \geq 0$. But since $p_k = 0$ for all $k \notin Z$, the above conditions are equivalent to saying that the $|Z|$-dimensional restriction $x = p|_Z$ of p to its coordinates in Z satisfies $Ax = e$ and $x \geq 0$. Conversely, any x satisfying the above constraints defines a 0-monochromatic perfect distribution p by letting $p|_Z = x$ and $p_k = 0$ for $k \notin Z$.

Thus, Ext defines no 0-monochromatic perfect distributions if and only if the constraints $Ax = e$ and $x \geq 0$ are unsatisfiable. But this is exactly the precondition to the Farkas' Lemma above! Using the Farkas Lemma on our A and e, we get the existence of the $(1 + (B-1)S)$-dimensional row vector y such that $yA \geq 0$ and $ye < 0$. Just like we did for the rows of V, we denote the first element of y by y_1, and use the notation $y_{(m,c)}$ to denote the remaining elements of y. We now translate the constraints $yA \geq 0$ and $ye < 0$ using our specific choices of A and e.

Notice, since $e_1 = 1$ and $e_i = 0$ for $i > 1$, it means that $ye = y_1$, so the constraint that $ye < 0$ is equivalent to $y_1 < 0$. Next, recalling that A is just the restriction of V to its columns in Z, and that the first row of V is the all-1 vector, we get that $yA \geq 0$ is equivalent to saying that for all $(c_1, \ldots, c_B) \in Z$ we have

$$y_1 + \sum_{m>1} \sum_c y_{(m,c)} \cdot v_{(m,c),(c_1,\ldots,c_B)} \geq 0 \tag{12}$$

Notice, since $y_1 < 0$, this equation implies that the double sum above is *strictly* greater than 0. Thus, recalling the definition of $v_{(m,c),(c_1,\ldots,c_B)}$ given in Equation (11), we conclude that for all $k = (c_1, \ldots, c_B)$, such that $\textsf{Ext}(k) = 0$, we have

$$\sum_{m>1} \left(y_{(m,c_1)} - y_{(m,c_m)} \right) > 0 \tag{13}$$

The last equation finally allows us to derive the implication we need:

Theorem 2. *Assume Ext defines no 0-monochromatic perfect distributions. Then there exist real numbers $\{y_{(m,c)} \mid m \in \{2 \ldots B\}, c \in \{1 \ldots S\}\}$ such that the following holds. If a key $k = (c_1, \ldots, c_B)$ is such that*

$$y_{(m,c_1)} - y_{(m,c_m)} \leq 0 \qquad \text{for all } m > 1, \tag{14}$$

then $\textsf{Ext}(k) = 1$.

Proof. Summing Equation (14) for all $m > 1$ we get a contradiction to Equation (13), which means that $\mathsf{Ext}(k) \neq 0$; i.e., $\mathsf{Ext}(k) = 1$.   □

4.5   Developing Intuition: Special Case $b = 1$

To get some intuition, we take a momentary detour and consider the special case $b = 1$, therefore reproving the result of [DS02]. Theorem 2 tells us that if Ext cannot be fixed to 0, there exists real numbers $y_1 \ldots y_S$ such that $y_i \leq y_j$ implies that the key $k = (i, j)$ gets mapped to 1 by Ext. Thus, by rearranging the y's in the non-decreasing order $y_1 \leq y_2 \leq \ldots \leq y_S$, we get that $\mathsf{Ext}((i,j)) = 1$ for any $i < j$. In particular, the uniform distribution on S keys $\{(1,2),(2,3),\ldots,(S-1,S),(S,1)\}$ is easily seen to define a perfect encryption distribution K (as both $\mathsf{Enc}(K,1)$ and $\mathsf{Enc}(K,2)$ sample a uniformly random ciphertext) at most one of whose components — the key $(S,1)$ — could conceivably get mapped to 0 by Ext. Thus, $\Pr[\mathsf{Ext}(K) = 0] \leq 1/S$, showing (even stronger) Equation (8) and thus completing this special case.

Interestingly, Dodis and Spencer [DS02] used a simpler "graph-theoretic" method to show the existence of exactly the same perfect distribution K as above. They viewed ciphertexts as vertices of the complete directed graph G on S vertices, and keys $k = (c_1, c_2)$ (where $c_1 \neq c_2$) — as directed edges connecting $c_1 = \mathsf{Enc}(k,1)$ to $c_2 = \mathsf{Enc}(k,2)$. With this notation, it is easy to see that a uniform distribution on any cycle in this graph defines a perfect encryption distribution. Now, considering first 2-cycles $\{(c_1,c_2),(c_2,c_1)\}$, the fact that none of them is 0-monochromatic implies that at least one of $\mathsf{Ext}((c_1,c_2)) = 1$ or $\mathsf{Ext}((c_2,c_1)) = 1$ is true, for any $c_1 \neq c_2$. Taking one such edge from every 2-cycle yields what is called a *tournament* graph, every one of whose edges extracts to 1. Now, a well known (and simple to prove) result in graph theory states that every tournament graph has a Hamiltonian path. In other words, there exists an ordering of ciphertexts $c_1 \ldots c_S$ such that every edge (c_i, c_j) belongs to the 1-monochromatic tournament subgraph whenever $i < j$; i.e., $\mathsf{Ext}((c_i,c_j)) = 1$ if $i < j$. Completing this Hamiltonian path to a Hamiltonian cycle (by adding the edge (c_S, c_1)) yields the same kind of perfect distribution K we built earlier using Theorem 2.

Unfortunately, it seems hard to extend this graph-theoretic argument to "hypergraphs" corresponding to $b > 1$. Instead, we chose to rely on linear algebra (i.e., Theorem 2) to get a better handle on the problem. Still, our proof below for general $b > 1$ is quite more involved than the proof above for $b = 1$.

4.6   Building Non-extractable Yet Perfect K

Returning to the general case, we build a special perfect distribution K which contains many keys satisfying Equation (14), meaning that $\mathsf{Ext}(K)$ is very biased towards 1. We will construct such K having a very special form below.

Definition 3. *Assume $\pi_1, \ldots, \pi_d : \mathcal{C} \to \mathcal{C}$ are d permutations over the ciphertext space $\mathcal{C} = \{1 \ldots S\}$. We say that $\pi_1, \ldots, \pi_d$ are d-valid if for every $c \in \mathcal{C}$, and distinct $i, j \in \{1 \ldots d\}$, we have $\pi_i(c) \neq \pi_j(c)$.*   ◇

The reason for this terminology is the following. Given any B-valid $\pi_1, \ldots, \pi_B$, where recall that $B = |\mathcal{M}|$, we can define S valid keys $k_1, \ldots, k_S \in \mathcal{K}$ by $k_c = (\pi_1(c), \ldots, \pi_B(c))$, where the B-validity constraint precisely ensures that all the B ciphertexts inside k_c are distinct, so that k_c is a legal key in $\mathcal{K}$. Now, we denote by $K_{(\pi_1, \ldots, \pi_B)}$ the uniform distribution over these S keys $k_1, \ldots, k_S$.

Lemma 5. *If $\pi_1, \ldots, \pi_B$ are B-valid permutations, then $K_{(\pi_1, \ldots, \pi_B)}$ is a perfect encryption distribution.*

Proof. For any message m, $\mathsf{Enc}(K_{(\pi_1, \ldots, \pi_B)}, m)$ is equivalent to outputting $\pi_m(U_{\mathcal{C}})$, where $U_{\mathcal{C}}$ is the uniform distribution over $\mathcal{C}$. Since each π_m is a permutation over $\mathcal{C}$, this is equivalent to $U_{\mathcal{C}}$. Thus, encryption of every message m yields a truly random ciphertext $c \in \mathcal{C}$, which means that $K_{(\pi_1, \ldots, \pi_B)}$ is perfect. $\square$

CHOOSING GOOD PERMUTATIONS. We will construct our perfect distribution $K = K_{(\pi_1, \ldots, \pi_B)}$ by carefully choosing a B-valid family $(\pi_1, \ldots, \pi_B)$ such that $\mathsf{Ext}(K)$ is very biased towards 1. We start by choosing π_1 to be the identity permutation $\pi_1(c) = c$ (for all c), and proceed by defining $\pi_2 \ldots \pi_B$ iteratively. After defining each π_d, we will maintain the following invariants which clearly hold for the base case $d = 1$:

(i) $\pi_1, \ldots, \pi_d$ are d-valid.
(ii) There exists a large set T_d of "good" ciphertexts (where, initially, $T_1 = \mathcal{C}$) of size $q_d > S - d^2$, which satisfies the following equation for all $c \in T_d$ and $1 < m \le d$:[6]

$$y_{(m,c)} - y_{(m, \pi_m(c))} \le 0 \tag{15}$$

Now, assuming inductively that we have defined $\pi_1 = id, \pi_2, \ldots, \pi_d$ which satisfy properties (i) and (ii) above, we will construct π_{d+1} still satisfying (i) and (ii).

This inductive step is somewhat technical, and we will come back to it in the next subsections. But first, assuming it is true, we show that we can easily finish our proof. Indeed, we apply the induction for $B - 1$ iterations and get B permutations $\pi_1, \ldots, \pi_B$ satisfying properties (i) and (ii) above. Then, property (i) and Lemma 5 imply that $K_{(\pi_1, \ldots, \pi_B)}$ is a perfect encryption distribution. On the other hand, property (ii) and the definition of $k_c = \{c, \pi_2(c), \ldots, \pi_B(c)\}$ imply that any key $k_c \in T_B$ satisfies Equation (14). Thus, by Theorem 2 we get that $\mathsf{Ext}(k_c) = 1$ for every $c \in T_B$. Since, $|T_B| > S - B^2$, we get that at most B^2 out of S keys k_c extract to 0. Thus, since $K_{(\pi_1, \ldots, \pi_B)}$ is uniform over its S keys, we get

$$\Pr[\mathsf{Ext}(K_{(\pi_1, \ldots, \pi_B)}) = 0] \le \frac{B^2}{S}$$

which shows Equation (8) and completes our proof (modulo the inductive step).

---

[6] To get some intuition, we will see shortly that "good" ciphertexts c will lead to keys k_c satisfying Equation (14), so that $\mathsf{Ext}(k_c) = 1$ by Theorem 2.

4.7   Preparing for Induction: Detour to Matchings

Before doing the inductive step, we recall some basic facts about bipartite graphs, which we will need soon. A (balanced) bipartite graph G is given by two vertex sets L and R of cardinality S and an edge set $E = E(G) \subseteq L \times R$. A *matching* P in G is a subset of node-disjoint edges of E. P is *perfect* if $|P| = S$. In this case every $i \in L$ is matched to a unique $j \in R$ and vice versa.

We say that a subset $L' \subseteq L$ is *matchable* (in G) if there exists a matching P containing L' as the set of its endpoints in L. In this case we also say that L' is *matchable with R'*, where $R' \subseteq R$ is the set of P's endpoints in R. (Put differently, L' is matchable with R' precisely when the subgraph induced by L' and R' contains a perfect matching.) The famous Hall's marriage theorem gives a necessary and sufficient condition for L' to be matchable.

Hall's Marriage Theorem. *L' is matchable if and only if every subset A of L' contains at least $|A|$ neighbors in R. Notationally, if $\mathcal{N}(A)$ denotes the set of elements in R containing an edge to A, then L' is matchable iff $|\mathcal{N}(A)| \geq |A|$, for all $A \subseteq L'$.*

We will only use the following two special cases of Hall's theorem.

Corollary 1. *Assume every vertex $v \in L \cup R$ has degree at least $S-d$: $\deg_G(v) \geq S - d$. Then, for any $L' \subset L$ and $R' \subset R$ of cardinality $2d$, we have that L' is matchable with R'.*

Proof. Let us consider the $2d \times 2d$ bipartite subgraph G' of G induced by L' and R'. Clearly, that every vertex $v \in L' \cup R'$ has degree at least d in G', since each such v is not connected to at most d opposite vertices in the entire G, let alone G'. We claim that L' meets the conditions of the Hall's theorem in G'. Consider any non-empty $A \subseteq L'$. If $|A| \leq d$, then any vertex v in A had $\deg_{G'}(v) \geq d \geq |A|$ neighbors, so $|\mathcal{N}(A)| \geq |A|$. If $d < |A| \leq 2d$, let us assume for the sake of contradiction that $|\mathcal{N}(A)| < |A|$. Consider now any vertex $v \in R \backslash \mathcal{N}(A)$. Such v exists as $|\mathcal{N}(A)| < |A| \leq 2d = |R'|$. Then no element in A can be connected to v, since $v \notin \mathcal{N}(A)$. Thus, the degree of v can be at most $2d - |A| < d$, which is a contradiction. □

Corollary 2. *Assume L contains a subset $L' = \{c_1, \ldots, c_\ell\}$ such that $\deg_G(c_i) \geq i$, for $1 \leq i \leq \ell$. Then L' is matchable in G. In particular, G contains a matching of size at least ℓ.*

Proof. We show that L' satisfies the conditions of Hall's theorem. Assume $A = \{c_{i_1}, \ldots, c_{i_a}\}$, where $1 \leq i_1 < i_2 < \ldots < i_a \leq \ell$. Notice, this means $i_j \geq j$ for all j. Then the neighbors of A at least include the neighbors of i_a, so that $|\mathcal{N}(A)| \geq \deg_G(c_{i_a}) \geq i_a \geq a = |A|$. □

4.8   Mapping Induction into a Matching Problem

We return to our induction. Recall, we are given permutations $\pi_1 = id, \pi_2, \ldots, \pi_d$ satisfying properties (i) and (ii), and need to construct π_{d+1} also satisfying properties (i) and (ii). We translate this task into some graph matching problem, starting with the property (i) first.

For every $c \in C$, we define the "forbidden" set $F_c = \{c, \pi_2(c), \ldots, \pi_d(c)\}$. Then, the $(d+1)$-validity constraint (i) is equivalent to requiring $\pi_{d+1}(c) \notin F_c$ for all $c \in C$. Next we define a bipartite "constraint graph" G on two copies L and R of C containing all the non-forbidden edges: $(c, c') \in E(G)$ if and only if $c' \notin F_c$. We observe two facts about G. First,

Lemma 6. *Every vertex* $v \in L \cup R$ *has degree at least* $S - d$: $\deg_G(v) \geq S - d$. *In particular, by Corollary 1 every two 2d-element subsets of L and R are matchable with each other in G.*

Proof. The claim is obvious for $v \in L$ as $|F_v| = c$. It is also true for $v \in R$, since any value $v \in R$ is forbidden by exactly d (necessarily distinct) elements $v, \pi_2^{-1}(v), \ldots, \pi_d^{-1}(v)$. □

Second, any perfect matching P of G uniquely defines a permutation π on S elements such that $P = \{(c, \pi(c))\}_{c \in L}$. Since, by definition, $\pi(c) \notin F_c$, it is clear that this π will always satisfy constraint (i). Thus, we only need to find a perfect matching P for G which will define a permutation π_{d+1} satisfying condition (ii).

Notice, our inductive assumption implies the existence of a subset T_d of L (recall, L is just a copy of C) of size $q_d > S - d^2$ such that Equation (15) is satisfied for all $c \in T_d$ and $1 < m \leq d$. Irrespective of the permutation π_{d+1} we will construct later, we will restrict T_{d+1} to be a *subset* of T_d. This means that Equation (15) will already hold for all $c \in T_{d+1}$ and $1 < m \leq d$. Thus, we will only need to ensure this equation for $m = d + 1$; i.e., that for all $c \in T_{d+1}$

$$y_{(d+1,c)} - y_{(d+1,\pi_{d+1}(c))} \leq 0 \tag{16}$$

This constraint motivates us to define a subgraph G' of our constraint graph G as follows. As edge $(c, c') \in E(G')$ if and only if $(c, c') \in E(G)$ (i.e., $c' \notin F_c$) and $y_{(d+1,c)} - y_{(d+1,c')} \leq 0$. In other words, we only leave edges (c, c') which will satisfy Equation (16) if we were to define $\pi_{d+1}(c) = c'$. The key property of G' turns out to be

Lemma 7. *G' contains a matching P' of size at least $S - d$.*

Proof. We will use Corollary 2. Let us sort the vertices $v_1 \ldots v_S$ of L and R in the order of non-decreasing $y_{(d+1,\cdot)}$ values; i.e.

$$y_{(d+1,v_1)} \leq y_{(d+1,v_2)} \leq \cdots \leq y_{(d+1,v_S)}$$

Then, the edge (v_i, v_j) satisfies $y_{(d+1,v_i)} - y_{(d+1,v_j)} \leq 0$ whenever $i \leq j$. Thus, such (v_i, v_j) belongs to G' if and only if it also belongs to the larger constraint graph G; i.e., $v_j \notin F_{v_i}$. But since each v_i has at most d forbidden edges in G, and $|\{j \mid j \geq i\}| = S - i + 1$, we have that $\deg_{G'}(v_i) \geq (S - i + 1) - d$. In particular, $\deg_{G'}(v_{S-d}) \geq 1, \ldots, \deg_{G'}(v_1) \geq S - d$. By Corollary 2, $\{v_{S-d}, \ldots, v_1\}$ is matchable in G', completing the proof. □

4.9 Finishing the Proof

Finally, we can collect all the pieces together and define a good matching P in G (corresponding to π_{d+1}). With an eye on satisfying property (ii), we start with a large (but not yet perfect) matching P' of G' of size at least $S - d$, guaranteed by Lemma 7. Ideally, we would like to extend P' to some perfect matching in the full graph G, by somehow matching the vertices currently unmatched by P'. Unfortunately, we do not know how to argue that such extension is possible, since there are at most d vertices unmatched, and we can only match arbitrary sets of size at least $2d$ by Lemma 6. So we simply take an arbitrary sub-matching P'' of P' of size $S - 2d$, just throwing away any $|P'| - (S - 2d)$ edges of P'.

Notice, P'' is also a matching of G which has exactly $2d$ unmatched vertices on both sides. By Lemma 6, we know that we can always match these missing vertices, and get a perfect matching P of the entire G. We finally claim that this perfect matching P defines a permutation π_{d+1} on $\mathcal{C}$ satisfying properties (i) and (ii).

Property (i) is immediate since P is a perfect matching of G. As for property (ii), let L' denote the $S - 2d$ endpoints of P'' in L. Now, every $c \in L'$ satisfies Equation (16), since this is how the graph G' was defined and $(c, \pi_{d+1}(c)) \in P'' \subseteq E(G')$. Thus, we can inductively define $T_{d+1} = T_d \cap L'$ and have T_{d+1} satisfy property (ii). We only need to argue that T_{d+1} is large enough, but this is easy. Since L' misses only $2d$ ciphertexts, we get by induction that

$$|T_{d+1}| \geq |T_d| - 2d > S - d^2 - 2d > S - (d + 1)^2$$

completing the induction and the whole proof.

5 Conclusions and Open Problems

We study the question of whether true randomness is inherent for achieving privacy, and show a largely positive answer for the case of information-theoretic private-key encryption, as well as computationally secure perfectly-binding primitives. The most interesting question is to study other privacy primitives (either information-theoretic or computational) not immediately covered by our technique. For example, what about 2-out-2 secret sharing (which is strictly implied by private-key encryption [DPP06]) or computationally binding commitment schemes? Do they still require true randomness?

More generally, we hope that our result and techniques will stimulate further interest in understanding the extent to which cryptographic primitives can be based on imperfect randomness.

Acknowledgments. We would like to thank Amit Sahai, Salil Vadhan and the anonymous referees for suggesting most of the "computational" extensions of our result. We would also like to thank Shien Jin Ong and Salil Vadhan for suggesting to use a better Chernoff bound in the proof of Theorem 1(a).

References

[ACRT99]   Alexander Andreev, Andrea Clementi, Jose Rolim, and Luca Trevisan. Dispersers, deterministic amplification, and weak random sources. *SIAM J. on Computing*, 28(6):2103–2116, 1999.

[CG88]   Benny Chor and Oded Goldreich. Unbiased bits from sources of weak randomness and probabilistic communication complexity. *SIAM J. on Computing*, 17(2):230–261, 1988.

[Dod00]   Yevgeniy Dodis. Exposure-Resilient Cryptography (PhD Thesis). *MIT PhD Thesis*, 2000.

[DOPS04]   Yevgeniy Dodis, Shien Jin Ong, Manoj Prabhakaran, and Amit Sahai. On the (im)possibility of cryptography with imperfect randomness. In *Proc. 45th IEEE FOCS*, pages 196–205, 2004.

[DPP06]   Yevgeniy Dodis, Krzysztof Pietrzak and Bartosz Przydatek. Separating Sources for Encryption and Secret-Sharing. In *Proc. Theory of Cryptography Conference (TCC)*, pages 601–616, 2006.

[DS02]   Yevgeniy Dodis and Joel Spencer. On the (non-)universality of the one-time pad. In *Proc. 43rd IEEE FOCS*, pages 376–388, 2002.

[DSS01]   Yevgeniy Dodis, Amit Sahai, and Adam Smith. On perfect and adaptive security in exposure-resilient cryptography. In *Proc. EUROCRYPT'01*, pages 301–324, 2001.

[GL89]   Oded Goldreich and Leonid Levin. A Hard-Core Predicate for all One-Way Functions. In *Prof. STOC*, pp. 25–32, 1989.

[KRVZ06]   Jesse Kamp, Anup Rao, Salil Vadhan and David Zuckerman. Deterministic extractors for small-space sources. In *Proc of STOC*, pp. 691–700, 2006.

[MP90]   James L. McInnes and Benny Pinkas. On the impossibility of private key cryptography with weakly random keys. In *Proc. CRYPTO'90*, pages 421–436, 1990.

[MW97]   Ueli Maurer and Stefan Wolf. Privacy amplification secure against active adversaries. In *Proc. CRYPTO'97*, pages 307–321, 1997.

[Ped91]   Torben P. Pedersen Non-Interactive and Information-Theoretic Secure Verifiable Secret Sharing. In *Proc. of CRYPTO*, pp. 129–140, 1991.

[RW03]   Renato Renner and Stefan Wolf. Unconditional authenticity and privacy from an arbitrary weak secret. In *Proc. CRYPTO'03*, pages 78–95, 2003.

[SV86]   Miklos Santha and Umesh V. Vazirani. Generating quasi-random sequences from semi-random sources. *JCSS*, 33(1):75–87, 1986.

[Sha49]   Claude Shannon. Communication Theory of Secrecy systems. In *Bell Systems Technical J.*, 28:656–715, 1949.

[Str80]   Gilbert Strang. Linear Algebra and Its Applications. *Academic Press*, London, 1980.

[TV00]   Luca Trevisan and Salil Vadhan. Extracting randomness from samplable distributions. In *Proc. 41st IEEE FOCS*, pages 32–42, 2000.

[vN51]   John von Neumann. Various techniques used in connection with random digits. *National Bureau of Standards, Applied Mathematics Series*, 12:36–38, 1951.

[VV85]   Umesh V. Vazirani and Vijay V. Vazirani. Random polynomial time is equal to slightly-random polynomial time. In *Proc. 26th IEEE FOCS*, pages 417–428, 1985.

[Zuc96]   David Zuckerman. Simulating BPP using a general weak random source. *Algorithmica*, 16(4/5):367–391, 1996.

Tackling Adaptive Corruptions in Multicast Encryption Protocols[*]

Saurabh Panjwani

University of California, San Diego
http://www-cse.ucsd.edu/users/spanjwan

Abstract. We prove a computational soundness theorem for symmetric-key encryption protocols that can be used to analyze security against adaptively corrupting adversaries (that is, adversaries who corrupt protocol participants *during* protocol execution). Our soundness theorem shows that if the encryption scheme used in the protocol is semantically secure, and encryption cycles are absent, then security against adaptive corruptions is achievable via a reduction factor of $O(n \cdot (2n)^l)$, with n and l being (respectively) the *size* and *depth* of the key graph generated during any protocol execution. Since, in most protocols of practical interest, the depth of key graphs (measured as the longest chain of ciphertexts of the form $\mathcal{E}_{k_1}(k_2), \mathcal{E}_{k_2}(k_3), \mathcal{E}_{k_3}(k_4), \cdots$) is much smaller than their size (the total number of keys), this gives us a powerful tool to argue about the adaptive security of such protocols, without resorting to non-standard techniques (like non-committing encryption).

We apply our soundness theorem to the security analysis of multicast encryption protocols and show that a variant of the Logical Key Hierarchy (LKH) protocol is adaptively secure (its security being quasi-polynomially related to the security of the underlying encryption scheme).

Keywords: Adaptive Corruptions, Encryption, Multicast, Selective Decryption.

1 Introduction

Imagine a large group of users engaged in a private virtual conversation over the Internet. The group is monitored by a group manager who ensures that at all points in time, users share a common secret key which is used for secure communication within the group (e.g., for encrypting all data that is exchanged between group members). Over time, the composition of the group changes—users can leave and/or join it at various (a priori unknown) instants—and, accordingly, the manager sends "update" messages to the group which enable all and only current participants to acquire the common secret. At some calamitous hour, a

[*] This material is based upon work supported by the National Science Foundation under the Grant CNS-0430595. A full version of the paper is available from the author's webpage.

S.P. Vadhan (Ed.): TCC 2007, LNCS 4392, pp. 21–40, 2007.

large number of user terminals get hijacked (e.g., an Internet worm infects half the Windows users in the group) and all information possessed by these users gets compromised. Clearly, this results in the compromise of group data that was exchanged while these ill-fated participants were part of the group. The question is—can one be sure that the data for other instants (that is, instants when affected participants were all outside the group) is still secure?

Answering such a question in the affirmative, even for simple security protocols (based on conventional, symmetric-key encryption alone) is often beset with tough challenges. The possibility of user corruptions occurring during protocol execution, and in a manner that is *adaptively* controlled by the attacker, increases the threat to a protocol's security and makes the task of *proving* protocols secure an unnerving task. It is known that, in general, protocols proven secure against non-adaptive attacks may actually turn insecure once an adversary is allowed to corrupt participants adaptively. (See [5] for a simple separation result for protocols based on secret sharing.) The situation is especially annoying for protocols that make use of encryption—adversaries can spy on ciphertexts exchanged between two honest parties, and later, at will, corrupt one of the parties, acquire its internal state, and use such information to "open" all ciphertexts which were previously sent or received by that party. While trying to prove security of such a protocol, one must argue that all "unopened" ciphertexts (those that cannot be decrypted trivially using the compromised keys) leak essentially no information to the adversary (that is, appear as good as encryptions of random bitstrings). The heart of the problem lies in the fact that one does not a priori know *which* ciphertexts are going to be opened by the adversary since these decisions are made only as the protocol proceeds. Besides, every ciphertext is a binding commitment to the plaintext it hides—one cannot hope to "fool" the adversary by sending encryptions of random bitstrings every time and then, when he corrupts a party, somehow convince him that the ciphertexts he saw earlier on (and which he can now open) were, in fact, encryptions of real data.

PREVIOUS APPROACHES. In the past, security analysis of encryption-based multiparty protocols against adaptive adversaries has largely been conducted using three approaches. The first (and the simplest) involves bypassing adaptive security altogether—if you cannot prove a protocol adaptively secure, then so be it. (That is, rest your minds with non-adaptive security.) The second approach attempts to solve the problem, but by studying it in the "erasure" model [3], in which all honest parties are assumed to delete their past state the moment they enter a new state configuration (wherein keys are generated afresh). Proving adaptive security of protocols in such a model is easy because adversaries are trivially disallowed from opening previously-sent ciphertexts—the corresponding decryption keys are assumed to have been erased from the system! However, the model itself is quite unrealistic: an honest party could simply forget to erase its previous states, or else, internally deviate from the rules of the game (that is, store all keys and behave in an "honest-but-curious" manner). Besides, some cryptographic protocols, for the sake of efficiency, *require* users to store keys received in the past and such protocols (an example will be discussed in this paper) would need to be re-designed in order to comply with the model.

The third approach, and perhaps the most compelling one, to adaptive security has been to develop non-standard notions of security of an encryption scheme. This corresponds to a line of research initiated by Canetti *et al.* [5], who introduced a cryptographic primitive, called *non-committing encryption*, specifically to address the problem of adaptive corruptions in multiparty protocols. Non-committing encryption schemes have the unusual property that ciphertexts created using them need not behave as binding commitments on the corresponding plaintexts (hence the name "non-committing"). That is, it is possible that an encryption of '0' collide with an encryption of '1' (or, more generally, encryption of real data be the same as encryption of a random bitstring). However, such collisions occur with only negligible probability—the chances of encrypting '0' and obtaining a ciphertext which can later be opened as '1' are very small. At the same time, these schemes allow to sample "ambiguous" ciphertexts (those that can be opened as either '0' or '1') efficiently and to *convince* an adversary of such a ciphertext being an encryption of '0' or of '1', as the situation demands. Encryption protocols implemented with non-committing encryption can be easily proven to achieve adaptive security—in the security proof, one just simulates the real protocol by transmitting ambiguous ciphertexts and upon corruption of a party, convinces the adversary that the ciphertexts he saw earlier were indeed the encryptions of the revealed data. Non-committing encryption schemes, though interesting in their own right, have their share of limitations—they are typically too inefficient for practical applications, and require bounding (a priori) the number of message bits that can be encrypted using any single key (usually, the number of bits that can be encrypted with a key cannot be more than the size of the key itself, which is highly prohibitive for real applications)[1].

OUR CONTRIBUTION. In this paper, we show that it is possible to argue about the adaptive security of a large class of encryption protocols, without requiring erasures and without resorting to primitives like non-committing encryption, while simultaneously achieving efficiency that meets practical requirements. We focus on protocols built generically from symmetric-key encryption (no other primitives are involved) and where every ciphertext is created by encrypting a key or a data element, with a single other key (no nesting of the encryption operation). We show that for a large variety of such protocols if keys are generated independently of each other, then protocols can be proven adaptively secure, *even under the assumption that the encryption scheme is semantically secure*, with very reasonable assurances on the strength of the protocol against adaptive corruptions.

Our main contribution is a general computational soundness theorem for encryption protocols which works as follows. Consider an abstract game played

---

[1] As shown by Nielsen [16], any non-committing encryption scheme that has a non-interactive encryption procedure must use a decryption key that is at least as long as the total number of bits to be decrypted.

between an adversary and a challenger, both being given access to a semantically secure symmetric-key encryption algorithm $\mathcal{E}$. Initially, the challenger generates n independent keys $k_1, k_2, \cdots, k_n$ and keeps them secret from the adversary. During the game, the adversary gradually and adaptively builds a directed graph G over n nodes labeled 1 through n. He arbitrarily introduces edges into the graph and for each such edge $i \rightarrow j$ he asks the challenger to provide an encryption of the key k_j under the key k_i, that is, $\mathcal{E}_{k_i}(k_j)$. (Thus, creation of the edge $i \rightarrow j$ in G depicts the fact that given k_i, the adversary can recover k_j, via the decryption operation corresponding to $\mathcal{E}$.) The adversary can also (again adaptively) decide to "corrupt" some nodes in the graph—from time to time, he instructs the challenger to reveal the key associated with the ith node in G (for any arbitrary i) and the challenger must answer with k_i in such a situation. We refer to G as the *key graph* generated by the adversary and the nodes in G that correspond to the revealed keys are called *corrupt nodes*. Note that any node i' in G that is reachable from a corrupt node i is also effectively corrupt; the adversary can recover the corresponding key using successive decryptions along the path from i to i'. The question is—can we prove that, at the end of the game, keys corresponding to nodes that are *not* reachable from any of the corrupt nodes, are still pseudorandom?

This simple game (formalized further in Section 2) provides an effective abstraction for many of the challenges a security analyst can expect to face when proving protocols secure against adaptive corruptions. The power to corrupt nodes in an adaptive fashion models the ability of attackers to compromise keys of users during the execution of the protocol. The power to decide the structure of all ciphertexts abstracts the fact that the execution flow of the protocol is indeterminable at design time and can potentially be influenced by the adversary during run-time. (A slight variant of the game would be one in which the adversary can also acquire ciphertexts formed by encrypting arbitrary messages of his choice. We will discuss this variant further in Section 2.) Note that we allow the creation of ciphertexts even after nodes have been corrupted (that is, the compromise of a key at some point in the protocol should not hamper security of ciphertexts created using future uncompromised keys). Likewise, the security of keys transmitted in the past must be preserved even if other keys are compromised in the future.

A naive first step to proving security in the game we just described would be to *guess*, a priori, the set of nodes that the adversary is going to corrupt during the execution and for every edge issuing from such a node, reply with a real ciphertext while for the other edges reply with encryptions of random bitstrings. Any security reduction seeded with such an idea would give us a reduction factor that is exponential in n (that is, we would end up proving a statement like "*if the encryption scheme is ϵ-secure then security in the game is guaranteed with probability $2^n \epsilon$*"). Such a reduction would be completely impractical; in most applications, n would be of the order of the number of protocol participants, which can be extremely large.

In this paper, we prove security in this game using a significantly different approach, and one that is of much better practical value. We show that if the key graph G generated by the adversary is acyclic[2] and if its *depth* (defined as the length of the longest path in G) is upper bounded by a parameter l, then security in our game can be proven via a reduction factor of $O(n \cdot (2n)^l)$. Here, by "security in our game" we mean that keys that (a) cannot be trivially recovered by the adversary (that is, are not reachable from corrupt nodes in G) and (b) are not used to encrypt other keys[3], remain pseudorandom at the end of it. That is, we prove that the security of a semantically secure encryption scheme can degrade in the face of adaptive attacks (as those captured by our game) by a factor of at most $O(n \cdot (2n)^l)$ but not by worse.

AN APPLICATION. So what is this reduction good for? At first glance, it would appear that it is much worse than the naive solution—l could potentially be of the order of n and $n \cdot (2n)^n$ is obviously no more consoling than 2^n. Well, for arbitrary key graphs, this is indeed the case. However, in practice, key graphs are much smaller (in fact, orders of magnitude smaller) in depth than in total size. For example, the key graphs generated in the execution of most broadcast encryption protocols (those falling under the subset-cover framework introduced by Naor *et al.* [15]) have depth 1 and their depth remains fixed for arbitrarily long runs of the protocol. All encryption-based group key distribution protocols (designed for secure multicast over the Internet, and also called *multicast encryption* protocols) generate key graphs that have depth at most logarithmic in the total number of users in the system (again, the depth remains fixed for arbitrarily long runs of the protocol, once the total space of users has been ascertained). In general, in all encryption protocols, the depth of key graphs created in any execution is likely to be related to the number of decryptions performed by users in order to be able to recover certain keys while their total size to the number of users themselves; it is reasonable to expect that protocol designers, for the purpose of efficiency, would strive to keep the former smaller than the latter.

We exemplify the power of our soundness result by applying it to the security analysis of the Logical Key Hierarchy (LKH) protocol [17]. LKH is a protocol originally developed for secure communication in multicast groups on the Internet (applications of the form we discussed in the first paragraph) and has since then attracted a lot of interest from both cryptographers and researchers in the networking community. Surprisingly, even though the protocol gets mentioned in a lot of papers on cryptography, there has been little effort from within our community towards analyzing its security (adaptive or otherwise) rigorously or to make any claims to the contrary.

The original LKH protocol has a security flaw in it [12]. Although this flaw is quite easy to spot, we are not aware of any work (prior to ours) that

---

[2] Acyclicity of key graphs is an almost-inescapable criterion required in security proofs of protocols based on encryption. We will discuss this issue further in Section 2.

[3] This is a necessary criterion if our goal is to guarantee pseudorandomness of these keys.

rectifies this flaw in a provably secure manner. (In [12], a fix is suggested but not proven secure.) In Section 3 of this paper, we present a variant of LKH which is not only as efficient as the original protocol, but also enjoys strong guarantees of security *against adaptive adversaries*. In particular, we use our soundness theorem to show that the security of the improved protocol is related to the semantic security of the underlying encryption scheme via a reduction factor that is quasi-polynomial in the number of protocol participants. Concretely, our reduction factor is of the order of $\tilde{n}^{\log(n)+2}$, where n is the number of users in the protocol and $\tilde{n} = O(n)$.

This reduction factor, though not strictly polynomial in n, is still quite reasonable from a practical perspective. For example, in a system with 128 users, one is guaranteed that an execution of our protocol provides at least 65 bits of adaptive security when implemented with 128-bit AES in counter mode (for a run with upto 64 key updates)[4]. Our result practically eliminates the need for using expensive techniques like non-committing encryption to build adaptively secure multicast encryption protocols, and it does this while matching the efficiency of existing schemes.

RELATION WITH SELECTIVE DECRYPTION. The abstract game used in our soundness theorem is reminiscent of the well-studied (though largely unresolved) problem of *selective decryption*. In this problem (like in ours), an adversary interacts with a challenger who initially generates a set of plaintexts $m_1, \cdots, m_n$ and a corresponding set of keys $k_1, \cdots, k_n$. (We stress here that the plaintexts are not chosen by the adversary, but generated by the challenger using some fixed distribution.) The adversary first wants to see the encryptions of all the plaintexts, $\{\mathcal{E}_{k_i}(m_i)\}_{i=1}^n$, and later "open" some of them adaptively; that is, he queries an arbitrary set $I \subseteq [n]$ and the challenger replies with $\{k_i\}_{i \in I}$. The question now is to show that plaintexts corresponding to all unopened ciphertexts are still "safe", in the sense that the adversary cannot learn any more information about them than what he could learn from the revealed plaintexts. In our soundness theorem, we are essentially generalizing this game to a setting in which the adversary can ask for not only *single* ciphertexts but *chains* of ciphertexts of the form $\mathcal{E}_{k_1}(k_2), \mathcal{E}_{k_2}(k_3), \mathcal{E}_{k_3}(k_4), \cdots$ and he is also allowed to open such chains adaptively (as above). Plus, we allow the adversary to interleave his "encrypt" and "open" queries arbitrarily. (Indeed, the fact that ciphertexts can be asked for in an adaptive manner, possibly depending upon past corruptions, is responsible for much of the complication in our proof.) It is for this reason that we refer to our game (detailed in Section 2) as the *generalized selective decryption (GSD)* game.

Does this paper solve the selective decryption problem? Not really. A crucial ingredient of that problem is the distribution from which the plaintexts $m_1, \cdots, m_n$ are drawn by the challenger. It has been shown [8] that if this distribution is such that each plaintext can be generated independently of the others

---

[4] These numbers are computed assuming the protocol is implemented using a binary hierarchy of keys; for non-binary hierarchies, the security guarantee is actually better.

then the unopened ciphertexts indeed remain secure and the adversary learns essentially no partial information about the plaintexts they hide from his interaction with the challenger. In the GSD game, too, we require all keys, even those which are not used for further encryption, to be generated independently of each other, and this "independence property" is crucial in our proof[5]. Our soundness theorem essentially builds up on this positive result for selective decryption and extends it to the more general scenario of arbitrarily (and adaptively) generated key graphs. The question of solving selective decryption without the independence assumption on plaintexts still remains open.

We remark that independence of all keys is not just a simplifying assumption in our theorem; it is almost a requirement for the security of the protocols we are interested in analyzing. A multicast encryption protocol that uses related group keys across key updates may not guarantee good security at all.

RELATED WORK. The notion of computational soundness theorems was introduced by Abadi and Rogaway [1], and has since then found applications in the security analysis of various cryptographic tasks, including key exchange [7,6], mutual authentication [13,6], XML security [2] and multicast key distribution [11,12]. Although most of the literature on computational soundness theorems deals with protocols that make use of encryption as the fundamental primitive, to the best of our knowledge, none of these works prove soundness in the presence of adaptively corrupting adversaries. Recently, Gupta and Shmatikov [10] developed a symbolic logic that allows reasoning about a weak variant of adaptive security for the case of key exchange protocols; however, the protocols they analyze, do not make use of encryption (and instead use Diffie-Hellman exponentiation coupled with signatures).

The soundness result of this paper is of a very different flavor than those in previous works in the area. The protocol model we use is relatively simpler— in the protocols we consider, every message generated during an execution is either a key or an encryption of a key under a key or else, a sequence of values with one of these types[6]. Symbolic analysis of such protocols can be effectively conducted using graph-theoretic terminology: keys can be interpreted as nodes, ciphertexts as edges, and Dolev-Yao attacks on protocols can be expressed in terms of reachability from adversarial nodes (corresponding to corrupted keys). As such, all discussions on symbolic analysis in this paper take place within a graph-theoretic framework (as illustrated by the GSD game). This simplifies our presentation considerably and brings us quickly to the crux of the matter.

---

[5] Jumping ahead, we remark that even in the variant in which the adversary can acquire encryptions of arbitrary messages of his choice, we need only keys to be independent of each other, and *not* the messages.

[6] We remark that extending our result to protocols that use nested encryption is also possible, but the soundness theorem and the corresponding proof become much more complex. We avoid nested encryption largely for the sake of simplicity (and partly because most existing multicast encryption protocols don't use nesting).

Lastly, a few words comparing the result of this paper with our previous work, joint with Micciancio [11,12], on the computationally sound analysis of encryption protocols are in order. Although both our works address adaptive attacks on encryption protocols, the adversarial model used in the current work is stronger: we not only allow the adversary to adaptively modify the execution flow of the protocol (as in our past work) but also to corrupt participants in an adaptive manner. Tackling the latter type of attacks is significantly more non-trivial, and forms the central theme of this paper. Another difference is that our previous soundness results applied only to protocols that satisfied certain syntactic conditions *besides* acyclicity of key graphs. Informally, these conditions require protocols to use every key in two phases—a *distribution* phase in which keys are used as plaintexts, followed by a *deployment* phase in which the distributed keys are used for encrypting other keys or messages. Key distribution is not allowed to succeed key deployment. Our new result, while incorporating adaptive corruptions, also does away with this restriction. The downside, however, is that this result provides security guarantees in a manner that is dependent on the depth of protocol key graphs, and it is not meaningful for protocols that could potentially generate key graphs with arbitrary depth. We believe that improving the result of this paper to overcome this limitation is non-trivial, but a worthy direction for future research; in particular, obtaining an analogous result with a reduction factor smaller than $\Theta(n^l)$ would be quite remarkable, and could lead to even newer techniques to address adaptive corruptions in security protocols.

2   The Main Result

Fix a symmetric-key encryption scheme $\Pi = (\mathcal{E}, \mathcal{D})$[7]. We use the standard notion of indistinguishability against chosen plaintext attacks (Ind-CPA) for encryption schemes as defined by Bellare *et al.* [4]. Specifically, let $\mathcal{O}_{k,b}^{\Pi}$ denote a left-or-right oracle for Π which first generates a key k uniformly at random from $\{0,1\}^\eta$ (η being the security parameter) and subsequently, responds to every query of the form $(m_0, m_1) \in \{0,1\}^* \times \{0,1\}^*$ (such that $|m_0| = |m_1|$) with $\mathcal{E}_k(m_b)$—the encryption of m_b under key k. For any adversary (that is, any arbitrary probabilistic Turing machine) A, let $\mathsf{A}^{\mathcal{O}_{k,b}^{\Pi}}$ denote the random variable corresponding to the output of A when interacting with such an oracle.

Definition 1. Let $t \in I\!N^+$ and $0 < \epsilon < 1$. An encryption scheme Π is called (t, ϵ)-Ind-CPA secure if for every adversary A running in time t: $|\mathbf{P}[\mathsf{A}^{\mathcal{O}_{k,b}^{\Pi}} = 1 | b = 0] - \mathbf{P}[\mathsf{A}^{\mathcal{O}_{k,b}^{\Pi}} = 1 | b = 1]| \leq \epsilon$

---

[7] In this paper, we consider encryption schemes where key generation is defined by picking a uniformly random bitstring from the set $\{0,1\}^\eta$ with η being the security parameter. Thus, the key generation algorithm is implicit in the definition of encryption schemes. We also assume that the encryption scheme allows to encrypt arbitrary bitstrings; so, keys themselves can always be used as plaintexts.

THE GSD GAME. Consider the following game, which we call the *generalized selective decryption (GSD)* game, played between an adversary A and a challenger B. Both parties are given blackbox access to the algorithms $\mathcal{E}$ and $\mathcal{D}$. In the beginning, A specifies an integer n, and the challenger generates a set of keys, $k_1, k_2, \cdots, k_n$, each key being sampled independently and uniformly at random from the set $\{0, 1\}^{\eta}$ (where η is the security parameter). B also generates a *challenge* bit b (uniformly at random from $\{0, 1\}$), which A is required to guess in the end. It stores the generated values for the rest of the game, and uses them to answer all of A's queries.

A can make three types of queries to B:

1. encrypt: At any point, A can make a query of the form $\texttt{encrypt}(i, j)$, in response to which B creates a ciphertext $c \leftarrow \mathcal{E}_{k_i}(k_j)$ (using fresh coins for the encryption operation each time) and returns c to A.
2. corrupt: A can also ask for the value of any key initially generated by B; it does this by issuing a query of the form $\texttt{corrupt}(i)$, in response to which it receives k_i.
3. challenge: Finally, A can issue a query of the form $\texttt{challenge}(i)$. The response for such a query is decided based on the bit b: if $b = 0$, B returns the key k_i to A, whereas if $b = 1$, it generates a value r_i uniformly at random from $\{0, 1\}^{\eta}$, and sends r_i to A[8].

Multiple queries of each type can be made, interleavingly and adaptively. We stress here that A can make more than one challenge queries in the game and it can choose to interleave its challenge queries with the other two types of queries. (This is a slight generalization of the setting described in the introduction.) Giving the adversary the power to make multiple challenge queries models the requirement that keys linked with challenge nodes be "jointly" pseudorandom (as opposed to individual keys being pseudorandom by themselves). Allowing it to interleave challenge's with other queries means that such keys are required to retain their pseudorandomness even after more corruptions or ciphertext transmissions have occurred.

We think of the queries of A as creating a directed graph over n nodes (labeled $1, 2, \cdots, n$), edge by edge, and in an adaptive fashion. Each query $\texttt{encrypt}(i, j)$ corresponds to creating an edge from i to j, denoted $i \rightarrow j$, in this graph. For any adversary A, the graph created by its queries in this manner is called the *key graph* generated by A and is denoted $G(\mathsf{A})$. A node i in $G(\mathsf{A})$ for which A issues a query $\texttt{corrupt}(i)$ is called a *corrupt node* while one for which A issues a query $\texttt{challenge}(i)$ is referred to as a *challenge node*. The set of all corrupt nodes is denoted $V^{\mathsf{corr}}(\mathsf{A})$ and that of all challenge nodes is denoted $V^{\mathsf{chal}}(\mathsf{A})$. Note that $G(\mathsf{A}), V^{\mathsf{corr}}(\mathsf{A})$ and $V^{\mathsf{chal}}(\mathsf{A})$ are all random variables depending on the coins used by both A and B.

---

[8] If A issues multiple challenge queries with argument i and if b equals 1, B must return the same value r_i everytime.

LEGITIMATE ADVERSARIES. There is a trivial way in which any adversary can win in the GSD game—by corrupting a node i in $G(A)$ and making a query challenge(j) for any j that is reachable from i, A can easily guess the challenge bit b. The interesting case to consider is, thus, one in which A is constrained *not* to issue queries of this form, that is, where A is restricted to make queries in a manner such that no challenge node is reachable from a corrupt node in $G(A)$.

Our intuition suggests that if the encryption scheme is secure (in the Ind-CPA sense), then the chances of such an adversary being able to decipher b correctly are no better than half. However, translating this intuition into a proof is far from easy. For one, it is not even possible to do this without further restrictions on the adversary's queries: if a key k_j is used to encrypt other keys (that is, there exists an edge issuing from j in $G(A)$), then k_j cannot be guaranteed to remain pseudorandom, even if j is not reachable from the corrupt nodes. In other words, we can hope to prove pseudorandomness of keys associated with challenge nodes only as long as these nodes have no outgoing edge in $G(A)$. Secondly, arguing about the security of encryption schemes in the presence of key cycles is a gruelingly hard problem; in particular, it is currently not known whether an arbitrary Ind-CPA-secure encryption scheme can be proved to retain its security in a situation where ciphertexts of the form $\mathcal{E}_{k_1}(k_2), \mathcal{E}_{k_2}(k_3), \cdots, \mathcal{E}_{k_{t-1}}(k_t), \mathcal{E}_{k_t}(k_1)$, for some $t > 1$, are created using it. Standard techniques do not allow to prove such statements and counterexamples are not known either. Given this state of affairs, our only hope to prove security in the GSD game is to forbid the creation of key cycles altogether. The following definition formalizes all our requirements from the adversary:

Definition 2. An adversary A is called *legitimate* if in any execution of A in the GSD game, the values of $G(A), V^{\mathsf{corr}}(A)$ and $V^{\mathsf{chal}}(A)$ are such that:

1. For any $i \in V^{\mathsf{corr}}(A)$ and any $j \in V^{\mathsf{chal}}(A)$, j is unreachable from i in $G(A)$.
2. $G(A)$ is a DAG and every node in $V^{\mathsf{chal}}(A)$ is a sink in this DAG.

THE RESULT. Let A be any legitimate adversary playing the GSD game. We say that A is an (n, e, l)-adversary if in any execution in the game, the number of nodes and edges in the key graph generated by A are bounded from above by n and e respectively and the *depth* of the graph (the length of the longest path in it) is at most l. We denote the random variable corresponding to the output of A in the game by $A^{B_b^{\Pi}}$.

Definition 3. Let $t, n, e, l \in \mathbb{N}^+$ and $0 < \epsilon < 1$. An encryption scheme Π is called (t, ϵ, n, e, l)-GSD secure if for every legitimate (n, e, l)-adversary A running in time t: $|\mathbf{P}[A^{B_b^{\Pi}} = 1 \mid b = 0] - \mathbf{P}[A^{B_b^{\Pi}} = 1 \mid b = 1]| \leq \epsilon$

Here, probabilities are taken over the random choices made by both A and B (including the randomness used by B in creating ciphertexts). The following is the main result of this paper:

Theorem 4. Let $t, n, e, l \in \mathbb{N}^+$ and $0 < \epsilon < 1$. If an encryption scheme Π is (t, ϵ)-Ind-CPA secure, then it is (t', ϵ', n, e, l)-GSD secure for quantities t' and ϵ' defined as:

$$\epsilon' = \epsilon \cdot \frac{3n}{2} \cdot (n+1) \cdot (2n+1)^{l-1}$$

$$t' = t - (O(n) \cdot t_{\mathsf{GenKey}} + e \cdot t_{\mathsf{Encrypt}})$$

where t_{GenKey} (resp. t_{Encrypt}) denotes the time taken to perform key generation (resp. encryption) in Π.

OVERVIEW OF THE PROOF. The starting point of the proof of our theorem is the positive result on the selective decryption problem (more precisely, the selective *decommitment* problem) due to Dwork *et al.* [8]. Consider first the GSD game for the case $l = 1$. The graph $G(\mathsf{A})$ in this case is a directed bipartite graph mapping a set of sources to a set of sinks. (In the problem studied in [8], the map from sources to sinks is one-to-one. In our case, it could be many-to-many; plus, it could be adaptively generated based on previous corruptions.) How can we argue about security in this case? Intuitively, an attacker's ability to differentiate between real and random values for *all* nodes in $V^{\mathsf{chal}}(\mathsf{A})$ translates into its ability to differentiate between the two values for *some* node (say the jth one) in $V^{\mathsf{chal}}(\mathsf{A})$; that is, such an adversary can effectively differentiate between two worlds, one in which the reply to each of the first $j - 1$ queries of the form challenge(i) is r_i (and for the rest, it is k_i), and the other in which the reply to each of the first j queries of this form is r_i (and that for the rest is k_i).

Let us call these worlds $\mathsf{World}_j(0)$ and $\mathsf{World}_j(1)$ respectively. Let us assume that the argument specified in A's jth challenge query is known a priori (it can be guessed with success probability $1/n$) and equals i_j. Let $I(i_j)$ denote the set of nodes i_s for which there exists an edge $i_s \rightarrow i_j$ in $G(\mathsf{A})$. Now consider this modified version of the game: While generating keys in the beginning of the game, B also generates a random key $\tilde{k}_{i_j}$, independently of all other keys. It replies to the adversary's queries in one of two worlds again, but now the worlds are defined as follows. Each query of the form encrypt(i_s, i_j) is replied to with the real ciphertext $\mathcal{E}_{k_{i_s}}(k_{i_j})$ in the first world, $\mathsf{World}'_j(0)$, but with a *fake* one, namely $\mathcal{E}_{k_{i_s}}(\tilde{k}_{i_j})$, in the other one, $\mathsf{World}'_j(1)$. All other encrypt queries are replied to with real ciphertexts in both worlds. For the challenge queries the replies always have the same distribution—r_i for the first $j - 1$ challenge queries and k_i for the rest. (In particular, the reply for challenge(i_j) is always k_{i_j}.) It is easy to see that the distribution on the challenger's replies in $\mathsf{World}'_j(0)$ is exactly the same as in $\mathsf{World}_j(0)$. The key observation to make here is that the distribution on the replies in $\mathsf{World}'_j(1)$ is also the same as that in $\mathsf{World}_j(1)$! This is true because the keys k_{i_j}, $\tilde{k}_{i_j}$ and r_{i_j} are generated by the challenger independently of each other, and so, replying to encrypt(i_s, i_j) with $\mathcal{E}_{k_{i_s}}(k_{i_j})$ and challenge(i_j) with r_{i_j} (as done in $\mathsf{World}_j(1)$) produces the same distribution as replying to the former with $\mathcal{E}_{k_{i_s}}(\tilde{k}_{i_j})$ and the latter with k_{i_j} (as done in $\mathsf{World}'_j(1)$). Thus, our adversary can differentiate between $\mathsf{World}_j(0)$ and $\mathsf{World}_j(1)$ with the same probability as it can differentiate between $\mathsf{World}'_j(0)$ and $\mathsf{World}'_j(1)$.

Why are the two worlds $\mathsf{World}'_j(0)$ and $\mathsf{World}'_j(1)$ indistinguishable? Because the encryption scheme is Ind-CPA-secure. If the adversary can distinguish between two sets of ciphertexts $\{\mathcal{E}_{k_{i_s}}(k_{i_j})\}_{i_s \in I(i_j)}$ (the real ones) and $\{\mathcal{E}_{k_{i_s}}(\tilde{k}_{i_j})\}_{i_s \in I(i_j)}$ (the fake ones) then it must be able to tell the difference between $\mathcal{E}_{k_{i_s}}(k_{i_j})$ and $\mathcal{E}_{k_{i_s}}(\tilde{k}_{i_j})$ for *some* node $i_s \in I(i_j)$. (A standard hybrid argument applies here.) This goes against the Ind-CPA-security of Π.

Going beyond $l = 1$. In the general setting, a node i_s, pointing at any node $i_j \in V^{\mathsf{chal}}(\mathsf{A})$ need not be a source—there could be other edges incident upon each such i_s and extending the above argument to this general setting requires more work. In order to be able to make a statement like *"the ciphertext $\mathcal{E}_{k_{i_s}}(k_{i_j})$ is indistinguishable from $\mathcal{E}_{k_{i_s}}(\tilde{k}_{i_j})$"*, one must first argue that every ciphertext of the form $\mathcal{E}_{k_{i'_s}}(k_{i_s})$ (where $i'_s \rightarrow i_s$ is an edge in $G(\mathsf{A})$) looks the same as one of the form $\mathcal{E}_{k_{i'_s}}(\tilde{k}_{i_s})$ (a fake ciphertext). But every such $k_{i'_s}$ could, in turn, be encrypted under other keys (that is, the node i'_s could have other edges incident on it). There could be a lot of nodes ($O(n)$, in general) from which i_j is reachable in $G(\mathsf{A})$ and at some point or the other, we would need to argue that replying with real ciphertexts created under each of these nodes is the same as replying with fake ones. Worse still, we do not a priori know the set of nodes from which i_j can be reached in $G(\mathsf{A})$ since the graph is created adaptively; so we must make guesses in the process.

It is easy to come up with an argument where the amount of guesswork involved is exponential in n (simply guess the entire set of nodes from which there is a path to i_j). In our proof, however, we take a radically different approach. We first define a sequence of hybrid distributions on the replies given to A such that in each of the distributions, the replies corresponding to some of the edges in the key graph are fake, and these "faked" edges are such that their end-points lie on a single path ending in i_j. (Henceforth, we will refer to every edge for which the corresponding reply is fake, as a *faked* edge.) The extreme hybrid distributions are defined as in the two worlds $\mathsf{World}'_j(0)$ and $\mathsf{World}'_j(1)$ for $l = 1$: in one extreme, the replies corresponding to all edges are real, and in the other extreme, the replies corresponding to all edges incident on i_j are fake (while the rest of the replies are still real). Intermediate to these extremes, however, are several distributions in which edges other than those incident on i_j are faked. For any two adjacent distributions in the sequence of distributions, the following properties are always satisfied:

(a) The distributions differ in the reply corresponding to a *single* edge $i_s \rightarrow i_t$; the reply is real in one distribution while fake in the other.
(b) In both distributions, for every $i_r \in I(i_s)$, the edge $i_r \rightarrow i_s$ is faked.
(c) There exists a path from i_t to i_j in the key graph and in both distributions, "some" of the edges incident upon this path are faked, the faked edges being the same in both distributions.
(d) No other edge in the key graph is faked in either of the distributions.

Properties (a) and (b) are meant to ensure that any two adjacent hybrids can be simulated using a single left-or-right encryption oracle (and so, A's capability to distinguish between them would imply that the encryption scheme is not Ind-CPA-secure); properties (c) and (d) enable the simulation to be carried out by guessing a path (that goes from i_s to i_t to i_j) as opposed to guessing all the nodes from which i_j is reachable. (This partly explains why our reduction factor is exponential in the depth, rather than the size, of the key graph.) In order to simultaneously achieve all these properties, we order the hybrid distributions such that *(i)* when the reply for any edge $i_s \to i_t$ is changed (from real to fake or vice versa) in moving from one hybrid to another, all edges of the form $i_r \to i_s$ have already been faked in previous hybrids; and *(ii)* after changing the reply for $i_s \to i_t$, there is a sequence of hybrids in which the replies for all edges $i_r \to i_s$ are, step by step, *changed back from fake to real.* This is done in order to satisfy property (d) above (particularly, to make sure that it is satisfied when the replies for edges issuing from i_t are changed in a subsequent hybrid).

Thus, if we scan the sequence of hybrid distributions from one extreme to the other, we observe both "real-to-fake" and "fake-to-real" transitions in the replies given to A, taking place in an oscillating manner. The oscillations have a recursive structure—for every oscillation in replies (transition from real to fake and back to real) for an edge $i_s \to i_t$, there are two oscillations (transition from real to fake to real to fake to real) for every edge $i_r \to i_s$ incident upon i_s. Simulating these hybrid distributions (using a left-or-right oracle) and subsequently, *proving* that the simulation works correctly is the most challenging part of the proof. After developing an appropriate simulation strategy, we prove its correctness using an inductive argument—assuming that, for some $l' \leq l$, the simulation behaves correctly whenever i_s is at depth smaller than l' in the key graph, we show that the simulation is correct also when i_s is at depth smaller than $l' + 1$; this simplifies our analysis considerably. Details of the entire proof are given in the full version of the paper.

OTHER VARIANTS. A natural variant of the GSD game would be one in which the adversary is allowed to acquire encryptions of messages of its choice (besides receiving encryptions of keys, as in the original game). Consider the following modified version of the game: A issues encrypt and corrupt queries, as before, but instead of making challenge queries, it makes queries of the form encrypt_msg(i, m_0, m_1) (such that $m_0, m_1 \in \{0, 1\}^*$ and $|m_0| = |m_1|$). In return for each such query, the challenger sends it the ciphertext $\mathcal{E}_{k_i}(m_b)$. A legitimate adversary in this modified game would be one whose key graph is always a DAG and for whom every query encrypt_msg(i, m_0, m_1) is such that i is unreachable from the corrupt nodes in the DAG. We remark that a result analogous to Theorem 4 can also be proven for this modified game, and with only a slight modification to the proof of that theorem. Specifically, we can show that if Π is (t, ϵ)-Ind-CPA secure, then for any t'-time (n, e, l) adversary A (t' as defined in Theorem 4),

$$|\mathbf{P}[\mathsf{A}^{\mathsf{B}_b^\Pi} = 1 \mid b = 0] - \mathbf{P}[\mathsf{A}^{\mathsf{B}_b^\Pi} = 1 \mid b = 1]| \leq \epsilon \cdot \frac{3n}{2} \cdot (2n + 1)^l$$

A different variant of our game would be one in which A is provided encryptions of messages, but these messages are sampled by the *challenger* using some fixed distribution known to A. In this variant, the messages themselves can be thought of as nodes (more specifically, sinks) in the key graph, whose values are hidden from A but whose probability distribution is defined differently from that of the keys. The goal now would be to argue that from A's perspective, all "unopened" messages (that is, messages that are not reachable from corrupt nodes in the key graph) appear as good as fresh samples from the same message space. If we assume that messages are sampled independently of each other, then security in this variant can also be proven, and with almost the same reduction factor as in Theorem 4. (Specifically, the reduction factor would be $(3/2) \cdot M(n+1)(2n+1)^l$, where M is an upper bound on the total number of messages that are encrypted.) However, in the absence of this assumption, it becomes considerably more challenging to prove the same claim. The techniques developed in this paper do not allow us to argue about security in such a setting, not even in the case where the key graph has depth 0 (only messages, and not keys, are used as plaintexts)[9].

3   The Application

In this section, we illustrate how our result from Section 2 applies to the security analysis of multicast encryption protocols.

MULTICAST ENCRYPTION. A group of n users, labeled $U_1, \cdots, U_n$, share a broadcast channel and wish to use it for secure communication with each other. At any point in time t, only a subset of users, labeled S_t, are "logged in" to the network, that is, are authorized to receive information sent on the channel. We would like to ensure that for all t, only the users in S_t (called *group members*) be able to decipher the broadcasts. We assume the existence of a central group manager C who shares a unique long-lived key k_{U_i} with each user U_i[10] and runs a key distribution program, KD, in order to accomplish the said task. The manager (or, equivalently, the program KD) receives user login and logout requests and for the request at time t, sends out a set of *rekey messages*, $\mathcal{M}_t$, on the channel. These rekey messages carry information about a key $k[t]$ (the *group key* for t), and are such that only the group members can decipher them (and,

---

[9] Here, by "argue about security" we mean the following: Consider an adversary A who makes only encrypt and corrupt queries in the above variant of the GSD game. At the end of the game, provide A with one out of two sets of values: in one world, reveal the real values of all unopened messages; in the other, provide an equal number of messages, sampled from the same probability space *conditioned* on the values of the opened messages. Now show that A cannot tell the two worlds apart. This problem is essentially the same as the selective decryption problem where plaintexts are allowed to be mutually dependent. We don't know of a solution to this problem yet.

[10] In practice, such long-lived keys could be established during the first login request made by users using, say, public-key based approaches.

subsequently, recover $k[t]$). The key $k[t]$ can then be used to carry out all group-specific security tasks until the next login/logout request arrives, which, we assume, happens at time $t + 1$. For example, it can be used for ensuring privacy of all data sent between time t and $t + 1$ and/or guaranteeing "group authenticity" of data (that is, enabling members to verify that the sender of the data is a group member at time t, and not an outsider). To ensure security of any such task, it is important to guarantee that $k[t]$ appears pseudorandom to users *not in* S_t (the non-members) for all instants t, even when such users can collude with each other and share all their information. The problem is to design the program KD in a manner such that this guarantee is achieved.

Fiat and Naor [9] were the first to define this problem formally and they introduced it under the title of *broadcast* encryption—a formulation in which all users are assumed to be stateless and group members are required to be able to recover $k[t]$, given only $\mathcal{M}_t$ and their long-lived keys. Subsequent work (for example, [14,17]) lifted the problem to the more general setting of stateful users, and studied it in the context of ensuring privacy in multicast groups on the Internet (hence the name *multicast* encryption). LKH is a protocol that relies on the statefulness assumption.

THE PROTOCOL. A trivial approach to multicast key distribution would be to have the center generate a new, purely random key $k[t]$ for every group membership change, and to let $\mathcal{M}_t$ (the rekey messages for time t) be the set of ciphertexts obtained by encrypting $k[t]$ individually under the long-lived keys of *every* user in S_t, that is, the set $\{\mathcal{E}_{k_{U_i}}(k[t])\}_{U_i \in S_t}$. This, however, is an unscalable solution since it involves a linear communication overhead per membership change, which is prohibitive for most applications that use multicast.

The LKH protocol betters the above trivial approach by distributing to users, in addition to the group key, a set of *auxiliary* keys, with each auxiliary key being given to some subset of the current group members. All keys in the system are organized in the form of a *hierarchy*—the group key is associated with the root node in the hierarchy, the long-lived keys of users with the leaves, and the auxiliary keys with internal nodes. At each point in time t, a user $U_i \in S_t$ knows all keys on the path from the leaf node corresponding to k_{U_i} to the root node (which corresponds to $k[t]$). The protocol maintains this property as an invariant across membership changes.

Rekey Messages. For simplicity, we illustrate the protocol using an example where $n = 8$ and the hierarchy is binary. We assume that all parties (including the center) have blackbox access to a symmetric-key encryption scheme $\Pi = (\mathcal{E}, \mathcal{D})$ with key space $\{0,1\}^\eta$ for some fixed security parameter η. In our description, we use the terms "keys" and "nodes" interchangeably (the relation between them is obvious in the current context) and depict transmission of a ciphertext $\mathcal{E}_{k_1}(k_2)$ with an edge $k_1 \to k_2$ in the figures.

Suppose that initially ($t = 0$), the set of group members $S_0 = \{U_1, U_2, U_3, U_6, U_7\}$ as shown in Figure 1(a). The center's key distribution program KD generates the initial group key $k[0] = k_\epsilon$ (the root node) and all auxiliary keys

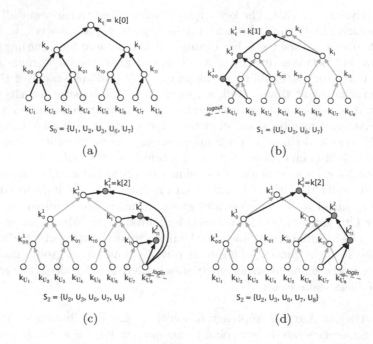

Fig. 1. LKH and rLKH: Figure 1(a) shows how key distribution to the initial set of users S_0 is performed while figure 1(b) demonstrates the rekeying process for user `logout` (both these procedures are the same in LKH and rLKH). Figure 1(c) shows how rekeying for user `login` works in LKH and fig. 1(d) illustrates the same for rLKH.

(internal nodes) which are supposed to be given to users in S_0. For example, since k_{00} and k_0 lie on the path from k_{U_1} to $k[0]$, these keys must be generated afresh and sent securely to U_1. KD transmits the keys to the designated users by sending the ciphertexts shown by dark edges in the figure. So, for example, user U_1 can obtain all the keys it is supposed to know (k_{00}, k_0, k_ϵ) by decrypting, in order, the ciphertexts $\mathcal{E}_{k_{U_1}}(k_{00}), \mathcal{E}_{k_{00}}(k_0)$ and $\mathcal{E}_{k_0}(k_\epsilon)$.

Now suppose that at time $t = 1$, user U_1 logs out of the group. That is, $S_1 = \{U_2, U_3, U_6, U_7\}$. The program KD should re-generate the group key k_ϵ, and the auxiliary keys which were known to U_1 at $t = 0$ (k_{00} and k_0) and distribute the new values in a manner such that U_1 cannot recover them but other users who are required to do so (according to the protocol invariant) still can. Specifically, it generates new keys k_{00}^1, k_0^1 and $k_\epsilon^1 =: k[1]$ (independently and uniformly at random) and sends out the ciphertexts shown in figure 1(b). Thus, every rekey operation for a user `logout` requires sending logarithmically many (specifically, $2\log_2(n) - 1$) ciphertexts; in our example, this number is 5.

THE FLAW AND THE FIX. The flaw in the original LKH protocol lies in the way it implements rekeying for user `login` operations. Suppose U_8 sends a `login` request at time $t = 2$. The center must now re-generate keys k_{11}, k_1, k_ϵ and send them securely to all the designated users (including U_8). The protocol does this by transmitting the ciphertexts shown in figure 1(c). ($k_{11}^2, k_1^2, k_\epsilon^2$ denote the newly generated keys.)

Note that the group key at $t = 1$, $k[1] = k_\epsilon^1$, is used to encrypt the group key at $t = 2$, k_ϵ^2. This is a problem since our initial goal was to guarantee pseudo-randomness of all group keys but deploying $k[1]$ in this manner clearly fails that purpose. In principle, if $k[1]$ is used in keying other applications (for example, in a message authentication scheme) at $t = 1$, and is also used for rekeying in the manner shown, then the protocol could be completely subverted (both $k[1]$ and $k[2]$ fully recovered) *even by a passive eavesdropper on the channel.* Of course, this does not mean that the protocol is broken for *any* secure implementation of the encryption scheme; but for *some*, it is.

We propose to fix the LKH protocol by changing the rekeying procedure for user logins as shown in Figure 1(d). (We remark that this fix is different from the one suggested in [12].) Notice that the communication cost incurred is the same as in the original protocol ($2 \log_2(n)$ ciphertexts for a user space of size n). Notice also that the structure of the rekey messages is now similar to that of the messages sent upon a user logout request (figure 1(b)). We refer to this modified version of LKH as "rLKH" (the r stands for "repaired"). The protocol can be easily generalized to work with arbitrary hierarchies; in particular, when the key hierarchy is a d-ary tree (so its height equals $\lceil \log_d(n) \rceil$), the communication complexity (number of ciphertexts transmitted) of rekeying would be $d\lceil \log_d(n) \rceil$ for user logins and $d\lceil \log_d(n) \rceil - 1$ for user logouts. An implementation of rLKH with n users and a d-ary hierarchy is referred to as the (n, d)-*instance* of the protocol.

One could conceive other ways of fixing the user login process of LKH (possibly as secure and as efficient as the one we propose). We prefer this fix for two reasons: (a) the key hierarchy in rLKH has the nice property that at all instants, every auxiliary key (and even the group key) is transmitted to the legitimate recipients by encrypting it under its two children only (and no other keys). This property could potentially simplify implementation of the protocol in practice; (b) more importantly, our fix ensures that the depth of the key graph generated in any execution of the protocol is independent of the number of protocol rounds; this property is useful in arguing about the protocol's adaptive security.

ADAPTIVE SECURITY. Let KD be an n-user multicast key distribution program. We define adaptive security of KD using the following game (which we call the MKD game) played between an adversary A and a challenger B. Initially, B generates the long-lived keys of all users $k_{U_1}, \cdots, k_{U_n}$ (randomly, independently from the underlying key space) and also generates a random challenge bit b. A specifies the initial set of group members, S_0, in response to which KD is invoked and the initial key distribution messages, $\mathcal{M}_0$, returned to A. Subsequently, A issues multiple queries to B, each query being either:

1. a *rekey* query—at any instant t, A can issue a query of the form rekey(command, U_i) where command is either login or logout. In response, B runs KD based on the membership change command specified and returns the set of rekey messages $\mathcal{M}_t$ to A; OR

2. a *corrupt* query—A can also issue queries of the form $\texttt{corrupt}(U_j)$, in return for which B sends it the key k_{U_j}; OR
3. a *challenge* query—finally, A can issue a $\texttt{challenge}$ query at any instant t; in response, it is given the key $k[t]$ if $b = 0$, or a fresh key $r[t]$ (sampled independently and uniformly at random from $\{0,1\}^\eta$) if $b = 1$.

All queries can be issued interleavingly and adaptively. Let $U^{\text{corr}}(A)$ be the set of all users corrupted by A during the game. Let $T^{\text{chal}}(A)$ be the set of instants t at which A issues a $\texttt{challenge}$ query. We say that A is *legitimate* if in every execution of A in the MKD game, for all $t \in T^{\text{chal}}(A)$, $S_t \cap U^{\text{corr}}(A) = \emptyset$. Let $A^{B_b^{\text{KD}}}$ denote the random variable corresponding to the output of A in the game, conditioned on the event that B selects b as the challenge bit.

Definition 5. Let $t, r \in \mathbb{N}^+$ and $0 < \epsilon < 1$. A multicast key distribution program KD is (t, r, ϵ)-secure against adaptive adversaries if for every legitimate adversary A that runs in time t, and makes r $\texttt{rekey}$ queries: $|\mathbf{P}[A^{B_b^{\text{KD}}} = 1 \mid b = 0] - \mathbf{P}[A^{B_b^{\text{KD}}} = 1 \mid b = 1]| \leq \epsilon$.

On the lines of the above definition, one can also define the problem of multicast encryption (or, for that matter, any security task based on multicast key distribution). For example, consider a multicast encryption protocol ME constructed using a key distribution program KD and an encryption scheme $\overline{\Pi} = (\overline{\mathcal{E}}, \overline{\mathcal{D}})$ as follows: the protocol distributes rekey messages for every group membership change just as KD but besides this, it also encrypts arbitrary messages—upon receiving a message m to encrypt at time t, the protocol outputs $\overline{\mathcal{E}}_{k[t]}(m)$. Security of such a scheme can be defined using a game similar to the MKD game, but with one change—every time the adversary issues a $\texttt{challenge}$ query, it also specifies two messages (m_0, m_1) $(m_0, m_1 \in \{0,1\}^*, |m_0| = |m_1|)$ and the challenger replies with $\overline{\mathcal{E}}_{k[t]}(m_b)$ ($k[t]$ being the current group key). It is possible to show that if KD is (t, r, ϵ)-secure against adaptive adversaries, and $\overline{\Pi}$ is (t, ϵ')-Ind-CPA secure, then ME is $(O(t), r, 2\epsilon + \epsilon')$-secure against adaptive adversaries.

In general, the problems of multicast key distribution and multicast encryption are equivalent to each other but studying the key distribution problem is more natural since it allows to generically build protocols for any security task (not necessarily multicast encryption) that can be accomplished using shared group keys. For this reason, we have focussed our attention on the key distribution problem alone, and discuss the security of rLKH in the same context.

Theorem 6. Let $n, d, t, r' \in \mathbb{N}^+$ such that $1 < d \leq n$. Let $0 < \epsilon < 1$. The (n, d)-instance of rLKH, when implemented using a (t, ϵ)-Ind-CPA secure encryption scheme Π, is (t', r', ϵ')-secure against adaptive adversaries for

$$\epsilon' = \epsilon \cdot \frac{3}{2} (\tilde{n} \cdot (\tilde{n} + 1) \cdot (2\tilde{n} + 1)^{\lceil \log_d(n) \rceil - 1})$$

$$t' = t - (O(\tilde{n}) \cdot t_{\text{GenKey}} + (r'd\lceil \log_d(n) \rceil) \cdot t_{\text{Encrypt}})$$

Here, $\tilde{n} = \max\{n, d^{\lceil \log_d(n) \rceil - 1} + r'\}$ and t_{GenKey} (resp. t_{Encrypt}) is the time taken to perform key generation (resp. encryption) in Π.

The proof of this theorem follows almost immediately from our soundness result of Section 2, given that (a) the key graph generated by any execution of rLKH is acyclic; (b) all group keys correspond to sinks in the protocol key graph; (c) the depth of the graph remains $\lceil \log_d(n) \rceil$ throughout; and (d) for any r'-round execution of the protocol, and for all $t \le r'$, the group key $k[t]$ can be reached from a long-lived key k_{U_i} if and only if $U_i \in S_t$. (The last part can be proven using a straightforward inductive argument, with the induction being performed on r'.) The reduction factor given in the theorem is slightly better than what one gets using a direct invocation of Theorem 4: this is achieved using the fact that in any r'-round execution of the rLKH protocol, (a) a key at depth i in the key graph (that is, at distance i from some source) is encrypted only by keys at depth $i - 1$ and (b) there are at most $d^{\lceil \log_d(n) \rceil - 1} + r'$ keys at any depth in the graph (and at most n sources in it). Note that our reduction factor is exponential in $\lceil \log_d(n) \rceil$ which is independent of the number of rounds the protocol is executed for. That is, the adaptive security of rLKH degrades polynomially (and not exponentially) with the number of rounds in the protocol execution.

Changing the hierarchy structure in rLKH involves a natural trade-off between efficiency and security: If we increase the arity d of the hierarchy (and correspondingly, reduce the height), the communication efficiency of the protocol suffers, but we get a better guarantee on its adaptive security. The extreme case is the n-ary hierarchy that has a linear rekeying communication complexity but provides adaptive security via a reduction factor of only $O(\tilde{n}^2)$. (Note that this is exactly the trivial approach to key distribution we discussed earlier on.) Whether or not one can further improve this trade-off between efficiency and security across different instances of rLKH, and, in particular, prove its adaptive security via a reduction factor smaller than the one given in Theorem 6, assuming only the semantic security of Π, is a question left open by this work.

Acknowledgements

Thanks to Daniele Micciancio, Thomas Ristenpart and Scott Yilek, for commenting on an earlier draft of the paper. Thanks also to the anonymous referees.

References

1. Martin Abadi and Philip Rogaway. Reconciling two views of cryptography (the computational soundness of formal encryption). *Journal of Cryptology*, 15(2):103–127, 2002.
2. Martin Abadi and Bogdan Warinschi. Security analysis of cryptographically controlled access to xml documents. In *Proceedings of the 24th ACM Symposium on Principles of Database Systems (PODS)*, pages 108–117, Baltimore, Maryland, June 2005. ACM.
3. Donald Beaver and Stuart Haber. Cryptographic protocols provably secure against dynamic adversaries. In Rainer A. Rueppel, editor, *Advances in Cryptology – EUROCRYPT'92*, volume 658 of *Lecture Notes in Computer Science*, pages 307–323. Springer-Verlag, May 1992.

4. Mihir Bellare, Anand Desai, Eric Jokipii, and Phillip Rogaway. A concrete security treatment of symmetric encryption. In *38th Annual Symposium on Foundations of Computer Science*, pages 394–403. IEEE Computer Society Press, October 1997.
5. Ran Canetti, Uriel Feige, Oded Goldreich, and Moni Naor. Adaptively secure multiparty computation. In *28th Annual ACM Symposium on Theory of Computing*, pages 639–648. ACM Press, May 1996.
6. Ran Canetti and Jonathan Herzog. Universally composable symbolic analysis of mutual authentication and key exchange protocols. In Shai Halevi and Tal Rabin, editors, *TCC '06: Third Theory of Cryptography Conference*, volume 3876 of *Lecture Notes in Computer Science*, pages 380–403. Springer-Verlag, 2006.
7. Anupam Datta, Ante Derek, John Mitchell, and Bogdan Warinschi. Computationally sound compositional logic for key exchange protocols. In *19th IEEE Computer Security Foundations Workshop (CSFW '06)*, pages 321–334. IEEE Computer Society, 2006.
8. Cynthia Dwork, Moni Naor, Omer Reingold, and Larry Stockmeyer. Magic functions. *Journal of the ACM*, 50(6):852–921, 2003.
9. Amos Fiat and Moni Naor. Broadcast encryption. In Douglas R. Stinson, editor, *Advances in Cryptology – CRYPTO'93*, volume 773 of *Lecture Notes in Computer Science*, pages 480–491. Springer-Verlag, August 1993.
10. Prateek Gupta and Vitaly Shmatikov. Key confirmation and adaptive corruptions in the protocol security logic. In *FCS-ARSPA 2006 (Joint Workshop on Foundations of Computer Security and Automated Reasoning for Security Protocol Analysis)*, 2006.
11. Daniele Micciancio and Saurabh Panjwani. Adaptive security of symbolic encryption. In J. Kilian, editor, *Theory of Cryptography Conference, TCC 2005*, volume 3378 of *Lecture Notes in Computer Science*, pages 169–187, Cambridge, MA, USA, February 2005. Springer-Verlag, Berlin, Germany.
12. Daniele Micciancio and Saurabh Panjwani. Corrupting one vs. corrupting many: The case of broadcast and multicast encryption. In *Automata, Languages, and Programming: 33rd International Colloquium, ICALP 2006, Proceedings, Part II*, volume 4052 of *Lecture Notes in Computer Science*. Springer-Verlag, January 2006.
13. Daniele Micciancio and Bogdan Warinschi. Soundness of formal encryption in the presence of active adversaries. In Moni Naor, editor, *TCC 2004: 1st Theory of Cryptography Conference*, volume 2951 of *Lecture Notes in Computer Science*, pages 133–151. Springer-Verlag, February 2004.
14. Suvo Mittra. Iolus: A framework for scalable secure multicasting. In *Proceedings of ACM SIGCOMM*, pages 277–288, Cannes, France, September 14–18, 1997.
15. Dalit Naor, Moni Naor, and Jeffery Lotspiech. Revocation and tracing schemes for stateless receivers. In Joe Kilian, editor, *Advances in Cryptology – CRYPTO 2001*, volume 2139 of *Lecture Notes in Computer Science*, pages 41–62. Springer-Verlag, August 2001.
16. Jesper Buus Nielsen. Separating random oracle proofs from complexity theoretic proofs: The non-committing encryption case. In Moti Yung, editor, *Advances in Cryptology – CRYPTO 2002*, volume 2442 of *Lecture Notes in Computer Science*, pages 111–126. Springer-Verlag, August 2002.
17. Chung Kei Wong, Mohamed Gouda, and Simon S. Lam. Secure group communications using key graphs. *IEEE/ACM Transactions on Networking*, 8(1):16–30, February 2000.

Long-Term Security and Universal Composability

Jörn Müller-Quade[1] and Dominique Unruh[2],[*]

[1] IAKS, Universität Karlsruhe (TH), Germany
[2] Saarland University, Saarbrücken, Germany

Abstract. Algorithmic progress and future technology threaten today's cryptographic protocols. Long-term secure protocols should not even in future reveal more information to a—then possibly unlimited—adversary.

In this work we initiate the study of protocols which are long-term secure *and* universally composable. We show that the usual set-up assumptions used for UC protocols (e.g., a common reference string) are not sufficient to achieve long-term secure *and* composable protocols for commitments or general zero knowledge arguments. Surprisingly, nontrivial zero knowledge protocols are possible based on a coin tossing functionality: We give a long-term secure composable zero knowledge protocol proving the knowledge of the factorisation of a Blum integer.

Furthermore we give practical alternatives (e.g., signature cards) to the usual setup-assumptions and show that these allow to implement the important primitives commitment and zero-knowledge argument.

Keywords: Universal Composability, long-term security, zero-knowledge, commitment.

1   Introduction

Computers and algorithms improve over time and so does the power of an adversary in cryptographic protocols. The VENONA project is an example where NSA and GCHQ stored Russian ciphertexts over years until they could eventually be cryptanalysed. Official key length recommendations, e.g. by the Federal Office for Information Security (BSI) in Germany, usually do not exceed six years and future technology like quantum computers could render even paranoid choices for the key length obsolete.

Everlasting security from assumptions which have to hold only during the protocol execution would be an ideal solution to this problem. In this work we combine the notions of universal composability and long-term security. For the first time we investigate protocols which are long-term secure *and* exhibit a composition theorem which allows a modular design of such protocols. In particular, we investigate commitment protocols and zero knowledge schemes which are composable and robust against future improvements of the adversary's computing technology.

---

[*] Most of the work was done while the second author was at the IAKS, Universität Karlsruhe (TH).

S.P. Vadhan (Ed.): TCC 2007, LNCS 4392, pp. 41–60, 2007.

To capture the threat of an adversary with increasing power we introduce the security notion of *long-term universal composability* (long-term-UC) with the intuition that the adversary becomes unlimited at some point of time after termination of the protocol. The protocols do not run after this point of time, but all information stored from past executions should not reveal any additional information to the then unlimited adversary. A surprising consequence of our work is that unconditionally hiding universally composable commitments [11] are not necessarily long-term-UC.

Long-term-UC is preserved under composition, i.e., idealised building blocks can be replaced by long-term-UC protocols while preserving the long-term security of the complete application. The security notion of long-term-UC lies strictly between information theoretical security, where the adversary is unlimited from the start, and computational security, where for a concrete security parameter the computational power of the adversary must be limited for all times to come.

The idea of everlasting security has been considered with respect to memory bounded adversaries. Key exchange protocols and protocols for oblivious transfer have been developed in the bounded storage model [5,4]. These protocols can be broken by an adversary with more memory than assumed, however they cannot be broken in retrospect even by an unlimited adversary. A scheme using distributed servers of randomness (virtual satellites) to achieve everlasting security has been implemented [22]. In this scheme the access of the adversary to the communication of the parties is limited during the key exchange. It was shown by [12] that in the bounded-storage model composability cannot be taken for granted. They gave a key-exchange protocol that is secure in the bounded-storage model even if the initial key leaks after protocol termination, and then showed that if the initial key was generated by a computationally secure key exchange protocol, the resulting protocol is insecure. However, theirs was a purely negative result in that they did not give any criteria under which composition would be possible.

Long-term security has been investigated in quantum cryptography. It is generally accepted (even though not formally proven) that an only computationally secure authentication of a quantum key exchange yields a long-term secure key. Bit commitment and oblivious transfer quantum protocols which become unconditionally secure, but rely on temporary computational assumptions have been searched, but are now known to be impossible[1] (see, e.g. [3]).

Zero knowledge proofs where the verifier cannot (ever) break the protocol and the prover can only on-line break the protocol where given in [2]. In [20] protocols achieving long-term security were stated, however, only secure function evaluation with constant input size was considered.

Another related topic is that of forward security, where it is demanded that past session keys remain computationally secure even if a long-term secret is given to the adversary. This notion is related to but less strict than long-term-UC as the session keys will not remain secure forever.

---

[1] Unless additional assumptions are made, such as bounded quantum storage or the availability of a piece of trusted hardware.

With exception of [12], previous work on long-term security did not take the problem of composability into account. When composability is required the situation changes drastically. E.g., an unconditionally hiding UC commitment is not long-term-UC and a straightforward adaption of e.g., the protocol of [2] using an unconditionally hiding UC commitment does not yield long-term-UC zero knowledge arguments.

In this work we thoroughly investigate under which assumptions long-term-UC commitments and long-term-UC zero knowledge arguments exist. We prove that a common reference string or a coin toss functionality are not sufficient for realising long-term-UC commitments. To be more general we define a functionality $\mathcal{F}$ to be only temporarily secret for a party P if, roughly speaking, every secret known to P and $\mathcal{F}$ can in principle (but not necessarily efficiently) be computed from the communication of $\mathcal{F}$ with all the other parties. Coin tossing and a common reference string are only temporarily secret for all parties and we show that long-term-UC commitments are impossible given any functionality which is only temporarily secret for the committer.

In contrast to this impossibility of commitments there exist nontrivial languages for which zero knowledge protocols are possible even with an only temporarily secret functionality. More concrete we give a zero knowledge proof of knowledge of the factorisation of a Blum integer using a helping coin toss functionality. This is astonishing as such a proof is not possible using a common reference string instead of a coin toss (unless factoring of Blum integers is easy for nonuniform machines). More generally we prove that no nonuniformly nontrivial language has a zero knowledge argument with the help of any functionality which works "offline" in the sense that it needs, like a common reference string, only be invoked before the start of the protocol and which is only temporarily secret for both parties. For example, most PKI are of this form and hence do not allow any nontrivial long-term-UC zero knowledge protocols.

Further we give two helping functionalities which are motivated from (temporarily) tamper proof hardware which allow to implement an unlimited number of long-term-UC commitments and zero knowledge arguments for all in NP. One of these functionalities resembles a trusted device which is computationally indistinguishable from a random oracle and the other a smart card which can generate digital signatures, but from which the secret key cannot be extracted. Note however that in contrast to the classical (i.e., not long-term secure) UC definition, commitments and ZK are not sufficient to implement any functionality.

1.1   Preliminaries

Notation. We call a function f *negligible*, if for any polynomial p and sufficiently large k, $f(k) \leq 1/p(k)$. We call f *overwhelming*, when $1 - f$ is negligible.

A *PPT-algorithm (probabilistic polynomial time)* is a uniform probabilistic algorithm that runs in polynomial-time in the length of its inputs.

We call a relation R on $\{0,1\}^* \times \{0,1\}^*$ *poly-balanced* if there is a polynomial p, s.t. $|w| \le p(|x|)$ for all x, w with xRw. We call R an *NP-relation* if it is poly-balanced and deciding $(x, w) \in R$ is in P. We call R an *MA-relation* if it is poly-balanced and deciding $(x, w) \in R$ is in BPP. The language L_R associated with R is $L_R := \{x \in \{0,1\}^* : \exists w : xRw\}$. We usually call x the *statement* and w with xRw the *witness* for x. We call a MA-relation R *(uniformly) trivial* if there is a PPT-algorithm that upon input $x \in L_R$ outputs a witness for x with overwhelming probability. We call R *nonuniformly deterministically trivial* there is a nonuniform deterministic polynomial-time algorithm that upon input $x \in L_R$ outputs a witness for x.

An integer $n > 0$ is called a *Blum-integer*, if $n = pq$ for two primes p, q with $p \equiv q \equiv 3 \bmod 4$.

Cryptographic tools. In [21], it is shown that assuming the existence of a one-way permutation, an unconditionally hiding commitment scheme exists. This scheme has the additional properties that the unveil-phase consists of only one message, and that given the message, the committed value v, and the transcript of the interaction in the commit phase, there is a deterministic polynomial-time algorithm that checks whether the verifier accepts the value v.

Using that commitment-scheme in the zero-knowledge proof-system for graph-3-colourability from [16], we get a statistically witness indistinguishable argument of knowledge for any NP-relation given any one-way permutation.[2] Using a statistically witness indistinguishable argument of knowledge for any NP-relation and a unconditionally hiding commitment scheme, we can easily construct a statistically witness indistinguishable argument of knowledge for any MA-relation using any one-way permutation.[3]

2   Modelling Long-Term UC

We now present our modelling of universally composable long-term security (short long-term UC). We build on the Universal Composability framework [7]. In that modelling, a computationally limited entity called the environment has to distinguish between an execution of the protocol (with some adversary) and an execution of an ideal functionality (with some simulator). To define long-term

---

[2] The resulting scheme is of course also zero-knowledge, but we do not need that property here.

[3] Let B be a PPT-algorithm s.t. $B(w, x) = 1$ with overwhelming probability for xRw and with negligible probability otherwise. Such an algorithm exists for any MA-relation R. To prove a statement $x \in L_R$, the prover first commits to the witness w, then commits to randomness r'. The verifier sends to the prover randomness r''. Then the prover proves using a statistically witness indistinguishable argument of knowledge that he knows a witness, s.t. $B(w, x) = 1$ with random-tape $r := r' \oplus r''$. Since the latter statement is in NP, this can be done given a one-way permutation.

security, we have to add the requirement that even if some entity gets unlimited computational power after the execution of the protocol, security is maintained. In the Universal Composability framework, this is quite easily done: We simply require that *after* the execution of the protocol (which is still performed against computationally limited adversaries) even an unlimited entity could not distinguish between an execution of the real protocol or of the functionality, i.e., we require that the output of the environment is *statistically* indistinguishable.[4]

Definition 1 (Long-term UC). *Let* $\mathrm{EXEC}_{\pi,\mathcal{A},\mathcal{Z}}(k,z)$ *denote the output of* $\mathcal{Z}$ *in an execution of the protocol* π *with adversary* $\mathcal{A}$ *and environment* $\mathcal{Z}$, *where* k *is the security parameter and* z *the auxiliary input of the environment* $\mathcal{Z}$. $\mathrm{EXEC}_{\mathcal{F},\mathcal{A},\mathcal{Z}}(k,z)$ *is defined analogously.*[5]

A protocol π long-term-UC *realises a functionality* $\mathcal{F}$, *if for any polynomial-time adversary* $\mathcal{A}$ *there exists a polynomial-time simulator* $\mathcal{S}$, *s.t. for any polynomial-time environment*[6] $\mathcal{Z}$ *the families of random variables* $\{\mathrm{EXEC}_{\pi,\mathcal{A},\mathcal{Z}}(k,z)\}_{k\in\mathbb{N},z\in\{0,1\}^{poly(k)}}$ *and* $\{\mathrm{EXEC}_{\mathcal{F},\mathcal{S},\mathcal{Z}}(k,z)\}_{k\in\mathbb{N},z\in\{0,1\}^{poly(k)}}$ *are statistically indistinguishable.*

Note that the Universal Composition Theorem from [7] applies with a virtually unmodified proof.

Conventions. In all our results we assume that secure channels are given for free (i.e., we are in the secure-channel network-model).[7] Further, security always denotes security with respect to static adversaries, i.e. parties are not corrupted *during* the protocol execution. However, we believe that our results can be adapted to adaptive adversaries.

We consider the case without an honest majority, since given an honest majority we could use information-theoretically secure protocols.

2.1   On the Minimality of the Security Notion

At this point one might wonder whether this definition is possibly stricter than necessary, especially in view of the various impossibility results presented below. However, if one is willing to accept stand-alone security (i.e., simulation-based security *without* an environment, see e.g. [15]), with the extra requirement that the outputs of the parties and the adversary/simulator are *statistically* indistinguishable in real and ideal model (long-term stand-alone security), as a minimal

---

[4] Note that we can w.l.o.g. assume that the output of the environment contains the whole view of that environment.

[5] See [7] for details.

[6] Not limited to environments with single bit output.

[7] This much simplifies the presentation. Since all our results concern the two-party case, it is easy to adapt our results to authenticated channels, if one adapts the definitions of the functionalities accordingly (e.g., the commitment functionality would then send the value of an unveil to the adversary as well as to the adversary). However, we cannot expect to use a key exchange protocol to make the authenticated channels secure, since such an approach would not be long-term secure.

security notion, we can argue as follows: If we want this minimal security *and* composability simultaneously, the proof from [18][8] states that the minimal security notion satisfying these two requirements is a security notion similar to Definition 1, with the only difference that the simulator is allowed to depend on the environment (specialised-simulator long-term UC). Since all our impossibility results also apply for this weaker notion (we never use the fact that the simulator does not depend on the environment), we see that we cannot find an essentially more lenient security notion than Definition 1 if we accept long-term stand-alone security as a minimal security notion.

2.2 Functionalities

In this section, we define some commonly used functionalities that we will investigate in the course of this paper.

We assume the following conventions in specifying functionalities:

We always assume that the adversary is informed of every invocation of the functionality, and the functionality only delivers its output when the adversary has triggered that delivery. So a phrase like "upon input x from P_1, $\mathcal{F}$ sends y to P_2" should be understood as "upon input y from P_1, $\mathcal{F}$ sends (*i-th input from P_1*) to the adversary, and upon a message (*deliver i*) from the adversary, $\mathcal{F}$ sends y to P_2". For better readability, we use the shorter formulation.

Most of the functionalities defined here are parametrised by a function m giving the length of their input and outputs. We will often omit explicitly stating this m if it is clear from the context.

When a functionality receives an invalid input from some party, it simply forwards that input to the adversary.

The first functionality used in this paper is the common reference string (CRS). Intuitively, the CRS denotes a random string that has been chosen by some trusted party or by some natural process, and that is known to all parties prior to the start of the protocol.

Definition 2 (Common Reference String (CRS)). *Let $\mathcal{D}_k$ ($k \in \mathbb{N}$) be an efficiently samplable distribution on $\{0,1\}^*$. At its first activation the functionality $\mathcal{F}_{\mathrm{CRS}}^{\mathcal{D}}$ chooses a value r according to the distribution $\mathcal{D}_k$ (k being the security parameter). Upon any input from P_i, send r to the adversary and to P_i (in particular, all parties P_i get the same r).*

If $\mathcal{D}_k$ is the uniform distribution on $\{0,1\}^{m(k)}$ for any k, we speak of a uniform CRS *of length m. We then write $\mathcal{F}_{\mathrm{CRS}}^m$ instead of $\mathcal{F}_{\mathrm{CRS}}^{\mathcal{D}_k}$.*

The second functionality is the coin toss. At a first glance, the coin toss looks very similar to the CRS, since also the coin toss consists of a random string that is given to both parties involved (and to the adversary). However, the coin toss guarantees that no party can learn the coin toss before *both* parties agree to toss

---

[8] With minor modifications: simply replace computational indistinguishability by statistical indistinguishability.

the coin.[9] As we will see below, a coin toss is more powerful than a CRS in the context of long-term UC.[10]

Definition 3 (Coin Toss (CT)). *When both P_1 and P_2 have given some input, the functionality $\mathcal{F}_{\mathrm{CT}}^m$ chooses a uniformly distributed $r \in \{0,1\}^{m(k)}$ and sends r to the adversary, to P_1, and to P_2.*

The next functionality models the setup assumption, that there is a trusted (pre-distributed) public key infrastructure, which provides each party with a secret key and attests the corresponding public key to any interested party.

Definition 4 (Public Key Infrastructure (PKI)). *Let G be a PPT-algorithm that upon input 1^k outputs two string sk and pk.[11] When $\mathcal{F}_{\mathrm{PKI}}^G$ runs with parties $P_1, \ldots, P_n$, upon its first activation it chooses independent key pairs $(sk_i, pk_i) \leftarrow G(1^k)$ for $i = 1, \ldots, n$ and sends $(pk_1, \ldots, pk_n)$ to the adversary. When receiving any input from P_i, send $(sk_i, pk_1, \ldots, pk_n)$ to P_i.*

The next two functionalities are well-known cryptographic building blocks that find application in the construction of many protocols.

Definition 5 (Commitment (COM)). *Let C and R be two parties. The functionality $\mathcal{F}_{\mathrm{COM}}^{C \to R, m}$ behaves as follows: Upon (the first) input $x \in \{0,1\}^{m(k)}$ from C send (committed) to R. Upon input (unveil) from C send x to R.*
We call C the sender and R the recipient.

Definition 6 (Zero-Knowledge (ZK)). *Let R be a MA-relation, and let P and V be two parties. The functionality $\mathcal{F}_{\mathrm{ZK}}^{R, P \to V, m}$ behaves as follows: Upon the first input of (x, w) from P satisfying xRw and $|x| \le m(k)$, send x to V.[12]*
We call P the prover and V the verifier.

3   Commitment

In this section we will examine the possibility of long-term-UC realising commitments. It will turn out, that commitment cannot be long-term-UC realised using CRS or coin-toss, nor with an arbitrary PKI. In particular unconditionally

---

[9] This can be illustrated by the following example: Alice and Bob want to know which of them pays the bill. So Alice and Bob agree: "We toss a coin, if the outcome is 1, Bob pays, otherwise Alice pays." Of course, if they were to use a CRS instead of a coin toss they could not use this simple protocol, because the outcome of the CRS is known before the start of the protocol.

[10] Although, in contrast, a UC secure (without long-term) coin toss can be realised using a CRS under reasonable complexity assumptions, see [9].

[11] I.e., G is a key generation algorithm.

[12] The resulting functionality $\mathcal{F}_{\mathrm{ZK}}$ is not polynomial-time if R is not an NP-relation. However, in that case $\mathcal{F}_{\mathrm{ZK}}$ can be replaced by an efficient implementation that uses a BPP-algorithm for checking xRw and errs only with negligible probability. The resulting functionality is then indistinguishable from $\mathcal{F}_{\mathrm{ZK}}$.

hiding UC commitments, which are possible with a CRS [11], are not necessarily long-term UC.[13] Note that the incompleteness of the CRS stands in stark contrast to the situation of (non-long-term) UC. In [10] it was shown that given a CRS, any functionality has a UC secure realisation. Furthermore, in [1] it was shown that the same holds for a PKI.[14] However, given a ZK functionality, commitments can be realised even with respect to long-term UC.

To state the impossibility results in a more general fashion, we first need the following definition:

Definition 7 (Only temporarily secret). *We say a functionality $\mathcal{F}$ is only temporarily secret (OTS) for party P, if the following holds in any protocol: Let trans denote the transcript of all communication between $\mathcal{F}$ and the other machines (including the adversary). Let trans $\setminus P$ denote the transcript of all communication between $\mathcal{F}$ and all machines except P. Then there is a deterministic function f (not necessarily efficiently computable) s.t. with overwhelming probability we have trans $= f(k, \text{trans} \setminus P)$.*

The intuition behind this definition is that if $\mathcal{F}$ is only temporarily secret (OTS) for P, then any secrets that P and $\mathcal{F}$ share may eventually become public. The following lemma gives some examples:

Lemma 1. *Coin toss ($\mathcal{F}_{CT}$) and CRS ($\mathcal{F}_{CRS}^{D}$ with any D) are OTS for all parties. Commitment ($\mathcal{F}_{COM}$) and ZK ($\mathcal{F}_{ZK}$) are OTS for the recipient/verifier. If G is a key generation algorithm, s.t. the secret key depends deterministically on the public key (e.g., RSA, ElGamal[15]), the PKI $\mathcal{F}_{PKI}^{G}$ is OTS for all parties.*

Proof. In the case of coin toss and CRS the adversary learns the random value r when if some party learns it, so all communication can be deduced from the communication with the adversary. In case of Commitment and ZK the communication with the recipient/verifier can be deduced from the communication with the sender. (In these cases, the function f is even efficiently computable.) All secret keys chosen by $\mathcal{F}_{PKI}^{G}$ can be calculated from the public keys $pk_1, \dots, pk_n$ sent to the adversary. □

Using this definition, we can prove that using a CRS, coin-toss or other functionalities that are OTS for the sender, one cannot long-term-UC realise a commitment:

Theorem 1 (Impossibility of commitment with OTS functionalities). *Let $\mathcal{F}$ be a functionality that is OTS for party C. Then there is no nontrivial*

---

[13] The intuitive reason being that the simulator may choose a value for the CRS which is only *computationally* indistinguishable from the uniform distribution without loosing the unconditional hiding property.

[14] Their definition $\mathcal{F}_{krk}$ of a PKI is somewhat different to ours. However, their proof directly carries over to $\mathcal{F}_{PKI}$.

[15] Under the condition, that in the secret key, group elements are always given using a *unique* representative (e.g., the secret exponent e in RSA is chosen smaller than $\varphi(n)$).

protocol that long-term-UC realises commitment with sender C ($\mathcal{F}_{COM}^{C \to R}$) *in the* $\mathcal{F}$*-hybrid model.*

If one is willing to assume NP $\not\subseteq$ P/poly, this theorem is an immediate consequence of Lemma 4 stating that $\mathcal{F}_{ZK}^{SAT,C \to R}$ (ZK for SAT with the sender C being the prover) is possible from $\mathcal{F}_{COM}^{C \to R}$, and Corollary 2 stating that $\mathcal{F}_{ZK}^{SAT,C \to R}$ cannot be realised using $\mathcal{F}$ (both shown in Section 4). However, in the full version [19] we give a direct proof (similar in spirit to that of Theorem 2) for this theorem that does not depend on NP $\not\subseteq$ P/poly.

An interesting corollary from this theorem is that long-term-UC commitments cannot be turned around, i.e. using one (or many) long-term-UC commitments from A to B, one cannot long-term-UC realise a commitment from B to A.

Corollary 1 (Commitments cannot be turned around). *There is no nontrivial protocol long-term-UC realising* $\mathcal{F}_{COM}^{A \to B}$ *using any number of instances of* $\mathcal{F}_{COM}^{B \to A}$.

Proof. Immediate from Lemma 1 and Theorem 1. □

In contrast to the impossibility results above, it is possible to get long-term-UC secure commitments using a ZK functionality:

Lemma 2 (Commitment from ZK). *Assume that a one-way permutation exists. Then there is a nontrivial protocol* π *that long-term-UC realises* $\mathcal{F}_{COM}^{C \to R}$ *(commitment with sender C) and that uses two instances of* $\mathcal{F}_{ZK}^{SAT,C \to R}$ *(ZK for SAT with the sender C being the prover).*

The protocol π looks as follows:

- *To commit to v*, the sender C first commits to v using an unconditionally hiding commitment scheme.
- Then C proves (using the first instance of $\mathcal{F}_{ZK}$) that he knows v and matching unveil information u.[16]
- *To unveil*, the sender C sends v to the recipient and proves (using the second instance of $\mathcal{F}_{ZK}$) that he knows matching unveil information u.

The long-term-UC security of this protocol stems from the following two facts. Equivocability: the simulator can unveil to any value v' since he controls the second instance of $\mathcal{F}_{ZK}$. Extractability: Since the sender cannot (efficiently) compute different unveil informations u and u', the message v given to the first instance of $\mathcal{F}_{ZK}$ must be the same as that used in the unveil phase. Since the simulator controls the first instance of $\mathcal{F}_{ZK}$, he learns that message v during the commit phase.

The actual proof is given in the full version [19].

---
[16] I.e., unveil information that would convince the verifier.

4 Zero-Knowledge

In the present section we examine to what extend long-term-UC secure zero-knowledge proofs can be implemented using various functionalities. Besides several impossibility results, we also have a quite surprising possibility result (Theorem 3).

4.1 Using OTS Functionalities

First, analogous to our investigations concerning commitments in Section 3, we are now going to examine whether long-term-UC secure ZK can be realised using functionalities that are OTS for one of the parties.

Whether long-term-UC realising ZK for some relation R is possible strongly depends on the relation R under consideration. The following definition specifies a class of relations which is going to play an important role in our results:

Definition 8 (Essentially unique witnesses). *A MA-relation R has essentially unique witnesses if there is a PPT-algorithm U_R (the witness unifier), that has the following properties:*

- *If w is a witness for x, $U_R(1^k, x, w)$ outputs a witness for x with overwhelming probability, formally: for sequences w_k, x_k with $x_k R w_k$ the probability $P(x_k R U_R(1^k, x_k, w_k))$ is overwhelming in k.*
- *If w is a witness for x, the output of $U_R(1^k, x, w)$ is almost independent of w, formally: for sequences w_k^1, w_k^2, x_k with $x_k R w_k^1$ and $x_k R w_k^2$, the families of random variables $U_R(1^k, x_k, w_k^1)$ and $U_R(1^k, x_k, w_k^2)$ are statistically indistinguishable.*

A possible way to interpret the witness unifier is as a statistically witness indistinguishable proof, that simply sends a witness in the clear.

It is most likely that relations without essentially unique witnesses exist:

Lemma 3. *If one-way-functions (secure against uniform adversaries) exist, or if $NP \nsubseteq P/poly$, then SAT does not have essentially unique witnesses.*

The proof is given in the full version [19].

We are now ready to present the first impossibility result concerning long-term-UC secure ZK:

Theorem 2 (Impossibility of ZK with OTS functionalities). *Let R be a MA-relation without essentially unique witnesses. Let $\mathcal{F}$ be a functionality that is OTS for party P. Then there is no nontrivial protocol that long-term-UC realises ZK for the relation R with prover P ($\mathcal{F}_{ZK}^{R, P \to V}$) in the $\mathcal{F}$-hybrid model.*

The rough idea of the proof is as follows: Clearly, if π was to be long-term-UC secure, the interaction between prover P and verifier V must be (almost) statistically independent from the witness V received from the environment. Further, a simulator that is able to simulate convincingly in case of a corrupted prover must be able to extract a witness $\tilde{w}$ from the communication with that prover, which is (almost) statistically independent from the witness w. So in particular, $\tilde{w}$ is (almost) statistically independent from w. Therefore, combining the prover and the simulator into one algorithm, we get an algorithm that given one witness w returns another almost independent one, in other words, a witness unifier in the sense of Definition 8. Therefore R must have essentially unique witnesses, which gives the desired contradiction.

The proof is given in the full version [19].

Note that we cannot expect an analogous result in the case that $\mathcal{F}$ is OTS for the verifier V, since commitments are OTS for the recipient and Lemma 4 show that $\mathcal{F}_{\mathrm{ZK}}^{R,P \to V}$ can be long-term-UC implemented using commitments with the verifier V as recipient.

Combining the results in this section, we get the impossibility of long-term-UC secure ZK for SAT:

Corollary 2. *Let $\mathcal{F}$ be a functionality that is OTS for party P. If one-way-functions (secure against* uniform *adversaries) exist, or if $NP \not\subseteq P/poly$, there is no nontrivial long-term-UC secure protocol for ZK with prover P for SAT in the $\mathcal{F}$-hybrid model.*

Proof. Immediate from Lemma 3 and Theorem 2. □

At this point one might ask why our impossibility result needs the restriction to relations without essentially unique witnesses. Would not the following argumentation show that given a, say, coin-toss, there is no long-term-UC ZK protocol π for any nontrivial relation: The simulator is able to extract a witness w from the interaction with the prover. Therefore w must information-theoretically already be "contained" in the interaction. On the other hand, in an interaction between simulator and verifier, the witness w cannot be "contained" in the interaction, since the simulator does not know w. However, since the interaction in both cases must be statistically indistinguishable from the interaction in the uncorrupted case, that latter both "contains" and does not "contain" w, which gives a contradiction. Surprisingly, this intuition is not sound as shows the following possibility result:

Theorem 3 (ZK for Blum-Integers using coin toss). *Assume that a one-way permutation exists. Let $nR(p,q)$ if $n = pq$, p,q prime and $p \equiv q \equiv 3 \bmod 4$. There is a nontrivial protocol using two instances of $\mathcal{F}_{\mathrm{CT}}$ that long-term-UC realises $\mathcal{F}_{\mathrm{ZK}}^{R}$ in the coin toss hybrid model.*

To construct such a protocol, we have to achieve two seemingly contradictory goals simultaneously. If the prover or verifier is corrupted, the simulator may choose the value r the coin-toss functionality returns. First, since the simulator should be able to extract a witness (p,q) (i.e., a factorisation of n in this case) in case of the corrupted prover, the simulator should be able to choose r having a trapdoor X s.t. it is possible to extract (p,q) under knowledge of that trapdoor. However, in the case of long-term-UC the value r should be statistically indistinguishable from uniform randomness. So the trapdoor should be present (but possibly unknown) even if r is chosen randomly. Further, if the verifier is corrupted, the simulator should be able to simulate the proof without knowing a witness. However, since also in this case r is almost uniformly distributed, the trapdoor X is also present. So by finding that trapdoor X we could extract a witness from the proof although the simulator never used that witness in constructing the proof. This can only be realised, if finding the witness can be reduced to finding the trapdoor.

In the case of factoring n, an example for such a trapdoor is the knowledge of random square roots modulo n. Given an oracle that finds square roots modulo n, we can factor n. So if the trapdoor X consists of the square roots of r (when we consider r as a sequence of integers modulo n) finding the trapdoor is as hard as factoring n, so there is no contradiction in the fact that by finding the trapdoor we can extract a witness (p, q) from an interaction that was produced without knowledge of (p, q).

This leads us to the following simplified version of our protocol:

- The prover sends n to the verifier.
- Prover and verifier invoke the coin-toss. The result r of that coin-toss is considered as a sequence $r_1, \ldots, r_k$ of integers modulo n.
- For each i, the prover chooses a random s_i with $s_i^2 = r_i$. It sets $s_i := \bot$ if r_i does not have a square root.[17]
- The prover sends $s_1, \ldots, s_k$ to the verifier.
- The verifier checks, whether $s_i^2 = r_i$ for all $s_i \neq \bot$, and whether at least $\frac{1}{5}$ of all $s_i \neq \bot$.

This protocol is not yet a long-term-UC realisation of $\mathcal{F}_{ZK}^R$, since it fails if n is not a Blum-integer, but it will demonstrate the main point. So why is this protocol long-term-UC secure if we guarantee that n is a Blum-integer? First, we see that if prover and verifier are both honest, the verifier will always accept. This is due to the fact that for a Blum-integer n, a random residue is a square with probability at least $\frac{1}{4}$.

Now we consider the case that the *verifier is corrupted*. In this case, the simulator has to produce coin-toss values $r_1, \ldots, r_n$ that are indistinguishable from the uniform distribution, and a proof that is statistically indistinguishable from the proof given by the prover. In other words, the simulator needs to simultaneously produce (almost) uniformly distributed $r_1, \ldots, r_n$, and for each r_i a random square root s_i modulo n if such s_i exists. Fortunately, if n is a Blum-integer, there is an efficient algorithm Q for choosing such r_i and s_i. So the simulator can successfully simulate by simply choosing the r_i and s_i using Q. Note that for this, it is vital that the simulator knows n before having to send the coin-toss result $r_1, \ldots, r_n$ to the environment. This is why we let the prover send n to the verifier *before* they invoke the coin-toss. In particular, we could not use a CRS here, because then the simulator might have to choose the r_i before the environment sends n to the prover.

---

[17] This is feasible given the factorisation of n.

Now for the case that the *prover is corrupted*. In this case, the simulator needs to interact with the environment incorporating the prover and to extract the witness (p, q) if the prover's proof would convince the honest verifier. To do this, the simulator again chooses the coin-toss $r_1, \ldots, r_n$ using the algorithm Q and therefore knows random square roots $\tilde{s}_i$ of all r_i that are quadratic residues. Now the environment sends s_i to the simulator. The uncorrupted verifier would only accept if at least $k/5$ of these s_i satisfy $s_i^2 = r_i$. Therefore after receiving the s_i from the environment, the simulator knows $k/5$ independently chosen pairs $(s_i, \tilde{s}_i)$ of square roots of r_i. For each such pair the probability of $s_i \not\equiv \tilde{s}_i \bmod n$ is $\frac{1}{2}$ (we ignore the finer detail of non-invertible r_i at this point), and in this case we get a factor of n by evaluating $\gcd(s_i \pm \tilde{s}_i, n)$. This happens with overwhelming probability, so the simulator is successful in extracting a factor and therefore the witness (p, q).

However, the protocol as described so far has a major flaw: If n is not a Blum-integer, the above security proof does not work. So we must ensure that n is in fact a Blum-integer. If the verifier is corrupted, the simulator gets n from the functionality $\mathcal{F}_{ZK}^R$ which ensures (by definition of R) that n is a Blum-integer. So in this case there is no problem. However, if the prover is corrupted, the simulator will have to choose the coin-toss $r_1, \ldots, r_n$. If n is not a Blum-integer, he might learn this later on (since he learns (p, q) in case of a successful proof), but then it might already be too late, because the simulator sends the r_i to the environment before the end of the proof (the algorithm Q does not guarantee $r_1, \ldots, r_n$ to be (almost) uniformly distributed if n is not a Blum-integer). To overcome this difficulty, we add an additional step to the beginning of the protocol. *Before the coin-toss is invoked*, the prover proves that n is indeed a Blum-integer. If the prover succeeds in this proof, the simulator can use the algorithm Q without danger, otherwise the simulator may abort (since the verifier would have done so, too). However, this introduces the additional difficulty that in case of a corrupted verifier, the simulator has to perform that proof, too, and without knowledge of the witness. To achieve this, we make use of the FLS-technique [13]: Prover and verifier first invoke another instance of the coin-toss functionality (in this case, a CRS would be sufficient, too) and then the prover proves using a statistically witness indistinguishable argument of knowledge to the verifier that either n is a Blum-integer or that he knows a the preimage of the coin-toss t under a one-way permutation f. Then the simulator can simulate this proof by simply choosing $t = f(u)$ for uniform u. Since $f(u)$ is uniformly distributed, this is indistinguishable from what an honest prover knowing the witness would produce. After having successfully performed this first step, prover and verifier proceed with the protocol as described above.

The actual proof for Theorem 3 is given in the full version [19].

Furthermore, given a commitment, long-term-UC secure ZK for any NP-relation is (unsurprisingly) possible:

Lemma 4 (ZK from commitment). *Let R be a NP-relation. Then there is a long-term-UC secure protocol π for ZK with relation R (i.e., $\mathcal{F}_{ZK}^{R, P \to V}$) using a polynomial number of commitments from prover P to verifier V (i.e., $\mathcal{F}_{COM}^{P \to V}$).*

Proof. [9] gives a UC secure protocol that realises $\mathcal{F}_{\text{ZK}}^{R,P \to V}$ using $\mathcal{F}_{\text{COM}}^{P \to V}$ where R is the relation for the Hamilton cycle problem. Their result even holds unconditionally (i.e., even when the environment is unlimited *during* the execution of the protocol) and therefore in particular with respect to long-term UC. Since the Hamilton cycle problem is NP-complete, the lemma follows.   □

Note that we cannot expect a similar result using commitments from verifier to prover, since $\mathcal{F}_{\text{COM}}$ is OTS for the recipient and thus Theorem 2 applies.

4.2   Using Offline Functionalities

In the preceding section, we saw that using a coin toss, long-term-UC secure ZK for the factorisation of Blum-integer can be realised. It is therefore a natural question to ask whether something similar is also possible using a CRS, which can be seen as the offline variant of a coin-toss. Unfortunately, the answer is no. To state this result in greater generality, let us first formalise what we mean by an offline functionality.

Definition 9 (Offline functionalities). *We call a functionality $\mathcal{F}$ offline, if it has the following form: When $\mathcal{F}$ runs with parties $P_1, \ldots, P_n$, upon its first activation, it chooses values $(c, c_{P_1}, \ldots, c_{P_n})$ according to a fixed distribution and sends c to the adversary. When receiving any input from P_i, send c_{P_i} to P_i.*

Lemma 5. *CRS and PKI are offline functionalities.*

Proof. For $\mathcal{F}_{\text{CRS}}$, set $c := c_i := r$ (cf. Definition 2), and for $\mathcal{F}_{\text{PKI}}$, set $c := (pk_1, \ldots, pk_n)$ and $c_i := (sk_i, pk_1, \ldots, pk_n)$ (cf. Definition 4).   □

The following result shows that a CRS as well as a PKI where the secret key is information-theoretically determined by the public key (cf. Lemma 1) cannot be used for long-term-UC secure ZK for any relation R unless that relation is trivial for nonuniform algorithms anyway.

Theorem 4 (Impossibility of ZK with OTS offline functionalities). *Let R be a nonuniformly deterministically nontrivial MA-relation.[18] Let $\mathcal{F}$ be an offline functionality that is OTS for party P and for party V. Then there is no nontrivial protocol that long-term-UC realises ZK for relation R with prover P and verifier V (i.e., $\mathcal{F}_{\text{ZK}}^{R,P \to V}$) in the $\mathcal{F}$-hybrid model.*

To understand the proof idea, assume that $\mathcal{F}$ is a CRS. Assume that there is a protocol π for $\mathcal{F}_{\text{ZK}}^R$. Then there is a simulator $\mathcal{S}_1$ that is able to choose the CRS r_1 and calculate a corresponding trapdoor T_1, s.t. he can simulate the prover and convince the verifier using this trapdoor (without knowledge of a witness). Furthermore, there is another simulator $\mathcal{S}_2$ that is able to choose the CRS r_2 and calculate a corresponding trapdoor T_2, s.t. he can simulate the

---

[18] I.e., there is no nonuniform deterministic polynomial-time algorithm that finds witnesses for R.

verifier and — if the verifier accepts — extract a witness w. Since both r_1 and r_2 are statistically indistinguishable from an honestly chosen CRS, it follows that an honestly chosen CRS always already "contains" such trapdoors T_1 and T_2 (however, given a CRS it can be infeasible to find these trapdoors). Therefore, if we provide S_1 and S_2 with a CRS and with trapdoors T_1 and T_2, S_1 will be able to produce a convincing proof (due to trapdoor T_1), and S_2 will be able to extract a witness from this convincing proof. Since S_1 and S_2 are polynomial-time, and CRS and trapdoors can be given as an auxiliary input, it follows that a nonuniform polynomial-time algorithm can find witnesses for R in contradiction to the nontriviality of R. Functionalities other than a CRS are handled almost identically, see the full proof.

The full proof is given in the full version [19].

A natural question arising in this context is whether this impossibility result can be made stronger. In particular, one might ask whether such an impossibility result already holds if $\mathcal{F}$ is OTS for P *or* for V. Further one might ask, whether the theorem can be strengthened to state impossibility of ZK for *uniformly non-trivial* relations. These questions are discussed in the full version.

Lemma 1 tells us that at least for some commonly used encryption schemes, $\mathcal{F}_{\mathrm{PKI}}^G$ is OTS for all parties (here and in the following G denotes the key generation algorithm) and therefore cannot be used for long-term-UC realising commitment or zero-knowledge[19]. However, in general this is not the case. As we show in the full version, there exist special public key schemes for which a PKI can be used for constructing ZK and commitment protocols.

5   Other Setup-Assumptions

As the preceding sections have shown, trying to design long-term-UC secure protocols using a CRS, coin toss or PKI is a futile endeavour. Therefore, in the following sections we will investigate alternative setup-assumptions that are more fruitful in the context of long-term-UC.

5.1   Trusted Devices Implementing a Random Oracle

A very powerful assumption in the context of universally composable security is the random oracle. It may therefore seem worthwhile to investigate whether a random oracle can be used to realise long-term-UC secure commitment and ZK. However, a closer look shows that in the context of long-term-UC security the random oracle is a very unrealistic assumption due to the following fact: Real-life implementations of the random oracle have to be done via some efficiently computable function (e.g., using trusted hardware that calculates some pseudorandom function with a secret seed). In the context of long-term-UC, this function could be "broken" by an unlimited adversary after protocol execution. In contrast, a random oracle functionality ensures, that even for an unlimited

---

[19] Except for nonuniformly trivial relations, see Theorem 4.

adversary, the function looks completely random. Therefore, we advocate that in the context of long-term-UC, instead of a random oracle one should use a functionality that evaluates a pseudorandom function with a secret seed (representing e.g. a (temporarily) trusted device).

We now give a definition of such a functionality $\mathcal{F}_{\mathrm{TPF}}$. Note however, that all possibility results given in this section also hold (with identical proofs) when using a random oracle instead of $\mathcal{F}_{\mathrm{TPF}}$.

Definition 10 (Trusted pseudorandom function (TPF)). *Let f_s be an efficiently computable family of deterministic functions $f_s : \{0,1\}^{l(|s|)} \to \{0,1\}^{l(|s|)}$ with polynomially bounded l.*

Then, the functionality trusted pseudorandom function (TPF) $\mathcal{F}_{\mathrm{TPF}}^f$ *is defined as follows: Upon its first activation, it chooses a uniformly random $s \in \{0,1\}^k$. When receiving a message $x \in \{0,1\}^{l(k)}$ from a party P or the adversary, it sends $f_s(x)$ to P or the adversary, respectively.*

At this point, one should note that the UC definition (and therefore our variant, too) implicitly assumes that when using a TPF, that TPF is accessed only by the protocol (and the adversary), but that it cannot be directly accessed by the environment. This in particular rules out that different protocols share a single TPF. A more detailed analysis of the consequences of this assumption can be found in [17,8]. However, we show that using a single TPF we can perform an *arbitrary number* of zero knowledge arguments or commitments, so that at least we do not need a large number of TPFs when constructing a larger protocol that performs many ZK arguments or commitments.

Theorem 5 (ZK from TPF). *Assume that a one-way permutation exists. Let f_s be a pseudorandom function (as in [14]), and R an NP-relation. Then there is a nontrivial protocol π using one instance of $\mathcal{F}_{\mathrm{TPF}}^f$ that long-term-UC realises unlimited number of instances of $\mathcal{F}_{\mathrm{ZK}}^R$ (i.e., ZK for the relation R).*

We give the proof idea here. First a commitment scheme is constructed which is computationally binding, unconditionally hiding and extractable (however, this commitment is not necessarily UC). The extractable commitment is constructed from a given commitment which is unconditionally hiding. To commit to a value v one first commits to $v, f_s(v)$. Then one commits to $u, f_s(u)$ where u is the unveil information for the first commitment. As the function $f_s(.)$ can only be evaluated by using the functionality $\mathcal{F}_{\mathrm{TPF}}$ a simulator can extract the committed value v from the calls which are placed to $\mathcal{F}_{\mathrm{TPF}}$.

Using this extractable commitment we modify the zero knowledge protocol for graph-3-colourability of [16]. Instead of letting the prover commit to a colouring and then let the verifier choose a random edge e for which the colours are unveiled and checked we let the verifier commit to e before the prover commits to the colouring.

In this protocol the simulator can, if the prover is corrupted, extract a witness from the commitments of the simulated real adversary or the protocol will fail and is then easily simulated. In case of a corrupted verifier the simulator can extract the edge which will later be investigated before committing to the colouring. So the simulator can easily commit to a fake colouring and still pass the test at the edge in question.

In both cases the communication between the parties, the adversary and the environment are statistically indistinguishable in the real protocol and in this simulation and we achieve a long-term-UC zero knowledge argument for graph-3-colouring and hence for all NP-statements. The complete proof can be found in the full version [19].

According to Lemma 2 one commitment can be obtained from two invocations of a zero knowledge scheme and we can hence conclude:

Corollary 3 (Commitments from TPF). *Assume that a one-way permutation exists. Let f_s be a pseudorandom function. Then there is a nontrivial protocol π using one instance of $\mathcal{F}_{\mathrm{TPF}}^{f}$ that long-term-UC realises an unlimited number of instances of $\mathcal{F}_{\mathrm{COM}}$ (i.e., commitments).*

Proof. Immediate from Lemma 2 and Theorem 5.   □

5.2   Signature Cards

One disadvantage of the TPF-assumption from the foregoing section is that trusted hardware implementing a pseudorandom function are unlikely to be available for practical use.[20] However, another kind of trusted device is already available commercially today: the signature card. A signature card is a tamperproof device with an built-in secret key. Upon request, this card signs an arbitrary document, but *never* reveals the secret key. The corresponding public key can be obtained from some certification authority. These properties are required e.g. from the German signature law [23].

These properties are captured by the following ideal functionality (based on [17]):

Definition 11 (Signature Card (SC)). *Let $\mathfrak{S} = (KeyGen, Sign, Verify)$ be a signature scheme. Let H be a party. Then the functionality $\mathcal{F}_{\mathrm{SC}}^{H,\mathfrak{S}}$ (signature card for scheme $\mathfrak{S}$ with holder H) behaves as follows: Upon the first activation, $\mathcal{F}_{\mathrm{SC}}^{H,\mathfrak{S}}$ chooses a public/secret key pair (pk, sk) using the key generation algorithm $KeyGen(1^k)$. Upon a message (pk) from a party P or the adversary, send pk to that party or the adversary, resp. Upon a message $(sign, m)$ from the holder H, produce a signature σ for m using the secret key sk and send σ to H.[21]*

As was the case with TPFs, our definition implicitly assumes that the environment has no direct access to the signature card. See the discussion after

---

[20] Not because of technical difficulties, but simply and plainly due to the forces of supply and demand.

[21] The definition from [17] additionally provides the possibility of locking the card (called *seize* and *release* there). These however are not needed in our protocols, so we omit them.

Definition 10. However, in [17] techniques where introduced that allow to share a single signature card in different protocols. It would be interesting to explore whether their approach can also be applied to our scenario.

It was shown in [17] that signature cards are powerful assumptions in the context of universal composability. Using an adaption of their technique, we can show that these signature cards are also very useful for long-term-UC security:

Theorem 6 (ZK from a signature card). *Assume that a one-way permutation exists. Let $\mathfrak{S}$ be an EF-CMA secure signature scheme. Let R be any MA-relation. Then there is a nontrivial protocol π that long-term-UC realises an unbounded number of instances of $\mathcal{F}_{ZK}^{R,P \to V}$ (i.e., ZK for the relation R with prover P) using a single instance of $\mathcal{F}_{SC}^{\mathfrak{S},P}$ (i.e., a signature card for $\mathfrak{S}$ with P as the holder).*

The idea of the proof is as follows: To prove the existence of a witness w for some statement x, the prover P signs x using his signature card (resulting in a signature σ) and then performs a statistically witness indistinguishable argument of knowledge that one of the following holds: (i) he knows a w and a σ, so that xRw and σ is a valid signature for w, or (ii) he knows a secret key sk' matching the public key pk provided by the signature card functionality.

Consider the case of a corrupted prover. Since $\mathfrak{S}$ is EF-CMA secure, it is infeasible to get a secret key sk' matching the public key pk chosen by the signature card (since the signature card allows only black-box access to the signing algorithm). So the prover has to show the knowledge of a signature σ of the witness w. The only way to obtain such a signature σ is to sign the witness w using the signature card. Since in the ideal model, the signature card $\mathcal{F}_{SC}$ is simulated by the simulator, the simulator learns that witness w. So the simulator is able to extract w while honestly simulating verifier and $\mathcal{F}_{SC}$.

In case the verifier is corrupted, the simulator knows the secret key sk matching the public key pk. So the simulator can prove (ii) instead of (i). Since the proof system we use is statistically witness indistinguishable, the resulting interaction is statistically indistinguishable.

The full proof is given in the full version [19].

Corollary 4 (Commitments from a signature card). *Assume that a one-way permutation exists. Let $\mathfrak{S}$ be an EF-CMA secure signature scheme. Then there is a nontrivial protocol π that long-term-UC realises an unbounded number of instances of $\mathcal{F}_{COM}^{C \to R}$ (i.e., commitment with sender C) using a single instance of $\mathcal{F}_{SC}^{\mathfrak{S},P}$ (i.e., a signature card for $\mathfrak{S}$ with P as the holder).*

Proof. This is an immediate consequence of Theorem 6 and Lemma 2.   □

6   Conclusions

We have examined the notion of long-term UC which allows to combine the advantages of long-term security (i.e., security that allow for unlimited adversaries after protocol end) and Universal Composability. We saw that the usual set-up assumptions used for UC protocols (e.g., CRS) are not sufficient any more in the case of long-term UC. However, we could show that there are other practical alternatives to these setup-assumptions (e.g., signature cards) that allow to implement the important primitives commitments and zero-knowledge proofs.

Further research in this directions might include the following:

- Which protocol tasks can or cannot be long-term-UC realised using commitments and zero-knowledge proofs.
- What other setup-assumptions might be useful in the context of long-term UC. In particular, under which assumptions can OT (and therefore any functionality) be realised?
- Our investigations were in the secure-channels communication-model. If only authenticated channels are present, the important issue of key exchange occurs. What setup-assumptions are necessary to implement the latter?
- The protocols presented here were not optimised for efficiency. To what extend can efficient protocols be found for the tasks discussed in this work?
- In [17] techniques were presented that allow to share a single signature card between different protocols. Can these techniques be applied to our setting, too?
- Much work on unconditional and long-term security has been done in the field of quantum cryptography. How does long-term UC behave in the presence of quantum communication. Can some of the impossibility results given in this work be avoided? In particular, quantum communication could solve the problem of key exchange mentioned above.

Acknowledgements. We thank the anonymous referees for many helpful suggestions.

References

1. Boaz Barak, Ran Canetti, Jesper Buus Nielsen, and Rafael Pass. Universally composable protocols with relaxed set-up assumptions. In *45th Symposium on Foundations of Computer Science, Proceedings of FOCS 2004, 17-19 October 2004, Rome, Italy*, pages 186–195. IEEE Computer Society, October 2004.
2. Gilles Brassard, David Chaum, and Claude Crépeau. Minimum disclosure proofs of knowledge, 1988. JCSS, 37:156-189.
3. Gilles Brassard, Claude Crépeau, Dominic Mayers, and Louis Salvail. Defeating classical bit commitments with a quantum computer. Los Alamos preprint archive quant-ph/9806031, May 1999.
4. Christian Cachin, Claude Crépeau, and Julien Marcil. Oblivious transfer with a memory-bounded receiver. In *34th Annual ACM Symposium on Theory of Computing, Proceedings of STOC 2002*, pages 493–502. ACM Press, 2002.
5. Christian Cachin and Ueli Maurer. Unconditional security against memory-bounded adversaries. In Burton S. Kaliski Jr., editor, *Advances in Cryptology, Proceedings of CRYPTO '97*, volume 1294 of *Lecture Notes in Computer Science*, pages 292–306. Springer-Verlag, 1997.

6. Ran Canetti. Universally composable security: A new paradigm for cryptographic protocols. In *42th Annual Symposium on Foundations of Computer Science, Proceedings of FOCS 2001*, pages 136–145. IEEE Computer Society, 2001.
7. Ran Canetti. Universally composable security: A new paradigm for cryptographic protocols. IACR ePrint Archive, December 2005. Full and revised version of [6].
8. Ran Canetti, Yevgeniy Dodis, Rafael Pass, and Shabsi Walfish. Universally composable security with global setup. These proceedings.
9. Ran Canetti and Marc Fischlin. Universally composable commitments. In Joe Kilian, editor, *Advances in Cryptology, Proceedings of CRYPTO '01*, volume 2139 of *Lecture Notes in Computer Science*, pages 19–40. Springer-Verlag, 2001.
10. Ran Canetti, Yehuda Lindell, Rafail Ostrovsky, and Amit Sahai. Universally composable two-party and multi-party secure computation. In *34th Annual ACM Symposium on Theory of Computing, Proceedings of STOC 2002*, pages 494–503. ACM Press, 2002. Extended abstract.
11. Ivan Damgård and Jesper Buus Nielsen. Perfect hiding and perfect binding universally composable commitment schemes with constant expansion factor. In Moti Yung, editor, *Advances in Cryptology, Proceedings of CRYPTO '02*, volume 2442 of *Lecture Notes in Computer Science*, pages 581–596. Springer-Verlag, 2002.
12. Stefan Dziembowski and Ueli Maurer. On generating the initial key in the bounded-storage model. In Christian Cachin and Jan Camenisch, editors, *Advances in Cryptology, Proceedings of EUROCRYPT '04*, volume 3027 of *Lecture Notes in Computer Science*, pages 126–137. Springer-Verlag, 2004.
13. Uriel Feige, Dror Lapidot, and Adi Shamir. Multiple non-interactive zero knowledge proofs under general assumptions. *SIAM Journal on Computing*, 29(1):1–28, 1999.
14. Oded Goldreich. *Foundations of Cryptography – Volume 1 (Basic Tools)*. Cambridge University Press, August 2001.
15. Oded Goldreich. *Foundations of Cryptography – Volume 2 (Basic Applications)*. Cambridge University Press, May 2004.
16. Oded Goldreich, Silvio Micali, and Avi Wigderson. Proofs that yield nothing but their validity or all languages in NP have zero-knowledge proof systems. *Journal of the ACM*, 38(3):690–728, 1991.
17. Dennis Hofheinz, Jörn Müller-Quade, and Dominique Unruh. Universally composable zero-knowledge arguments and commitments from signature cards. In *Proceedings of the 5th Central European Conference on Cryptology, MoraviaCrypt '05*, 2005.
18. Yehuda Lindell. General composition and universal composability in secure multi-party computation. In *44th Annual Symposium on Foundations of Computer Science, Proceedings of FOCS 2003*, pages 394–403. IEEE Computer Society, 2003.
19. Jörn Müller-Quade and Dominique Unruh. Long-term security and universal composability, 2006. Full version of this paper, IACR ePrint 2006/422.
20. Jörn Müller-Quade. Temporary assumptions—quantum and classical. In *The 2005 IEEE Information Theory Workshop On Theory and Practice in Information-Theoretic Security*, 2005. abstract.
21. Moni Naor, Rafail Ostrovsky, Ramarathnam Venkatesan, and Moti Yung. Perfect zero-knowledge arguments for NP using any one-way permutation. *Journal of Cryptology*, 11(2):87–108, March 1998.
22. Michael O. Rabin. Hyper-encryption by virtual satellite. Science Center Research Lecture Series, December 2003.
23. Gesetz über Rahmenbedingungen für elektronische Signaturen. Bundesgesetzblatt I 2001, 876, May 2001.

Universally Composable Security with Global Setup

Ran Canetti[1], Yevgeniy Dodis[2], Rafael Pass[3], and Shabsi Walfish[2]

[1] IBM Research
canetti@csail.mit.edu
[2] New York University
{dodis,walfish}@cs.nyu.edu
[3] Cornell University
rafael@cs.cornell.edu

Abstract. Cryptographic protocols are often designed and analyzed under some *trusted set-up* assumptions, namely in settings where the participants have access to global information that is trusted to have some basic security properties. However, current modeling of security in the presence of such set-up falls short of providing the expected security guarantees. A quintessential example of this phenomenon is the *deniability* concern: there exist natural protocols that meet the strongest known composable security notions, and are still vulnerable to bad interactions with rogue protocols that use the same set-up.

We extend the notion of universally composable (UC) security in a way that re-establishes its original intuitive guarantee even for protocols that use globally available set-up. The new formulation prevents bad interactions even with adaptively chosen protocols that use the same set-up. In particular, it guarantees deniability. While for protocols that use no set-up the proposed requirements are the same as in traditional UC security, for protocols that use global set-up the proposed requirements are significantly stronger. In fact, realizing Zero Knowledge or commitment becomes provably impossible, even in the Common Reference String model. Still, we propose reasonable alternative set-up assumptions and protocols that allow realizing practically any cryptographic task under standard hardness assumptions *even against adaptive corruptions*.

1 Introduction

The trusted party paradigm is a fundamental methodology for defining security of cryptographic protocols. The basic idea (which originates in [24]) is to say that a protocol securely realizes a given computational task if running the protocol amounts to "emulating" an ideal process where all parties secretly hand their inputs to an imaginary "trusted party" who locally computes the desired outputs and hands them back to the parties. One potential advantage of this paradigm is its strong "built in composability" property: The fact that a protocol π emulates a certain trusted party $\mathcal{F}$ can be naturally interpreted as implying that any system that includes calls to protocol π should, in principle, behave the same if the calls to π were replaced by ideal calls to the trusted party $\mathcal{F}$.

S.P. Vadhan (Ed.): TCC 2007, LNCS 4392, pp. 61–85, 2007.
© International Association for Cryptologic Research 2007

Several formalizations of the above intuitive idea exist, e.g. [23,27,3,9,20, 31,10,30]. These formalizations vary in their rigor, expressibility, generality and restrictiveness, as well as security and composability guarantees. However, one point which no existing formalism seems to handle in a fully satisfactory way is the security requirements in the presence of "global trusted setup assumptions", such as a public-key infrastructure (PKI) or a common reference string (CRS), where all parties are assumed to have access to some global information that is trusted to have certain properties. Indeed, as pointed out in [28], the intuitive guarantee that "running π has the same effect as having access to the trusted party" no longer holds.

As a first indication of this fact, consider the "deniability" concern, namely, allowing party A to interact with party B in a way that prevents B from later "convincing" a third party C that the interaction took place. Indeed, if A and B interact via an idealized "trusted party" that communicates only with A and B then deniability is guaranteed in a perfect, idealized way. Thus, intuitively, if A and B interact via a protocol that emulates the trusted party, then deniability should hold just the same. When the protocol in question uses no global setup, this intuition works, in the sense that emulating a trusted party (in most existing formalisms) automatically implies deniability. However, when global setup is used, this is no longer the case: There are protocols that emulate such a trusted party but do *not* guarantee deniability.

For instance, consider the case of Zero-Knowledge protocols, *i.e.* protocols that emulate the trusted party for the "Zero-Knowledge functionality": Zero-Knowledge protocols in the plain model are inherently deniable, but most Zero-Knowledge protocols in the CRS model are completely *un*deniable whenever the reference string is public knowledge (see [28]). Similarly, most authentication protocols (i.e., most protocols that emulate the trusted party that provides ideally authenticated communication) that use public key infrastructure are not deniable, in spite of the fact that ideal authenticated communication via a trusted party is deniable.

One might think that this "lack of deniability" arises only when the composability guarantees provided by the security model are weak. However, even very strong notions of composability do not automatically suffice to ensure deniability in the presence of global setup. For example, consider the Universal Composability (UC) security model of [10], which aims to achieve the following, very strong composability guarantee:

A UC-secure protocol π implementing a trusted party $\mathcal{F}$ does not affect any other protocols more than $\mathcal{F}$ does — even when protocols running concurrently with π are maliciously constructed.

When $\mathcal{F}$ is the Zero-Knowledge functionality, this property would seem to guarantee that deniability will hold even when the protocol π is used in an arbitrary manner. Yet, even UC-secure ZK protocols that use a CRS are *not* deniable whenever the reference string is globally available. This demonstrates that the UC notion, in its present formulation, does *not* protect a secure protocol π from a protocol π' that was maliciously designed to interact badly with π, in the case where π' can use the *same setup* as π.

Deniability is not the only concern that remains un-captured in the present formulation of security in the CRS model. For instance, even UC-secure Zero-Knowledge proofs in the CRS model may not be "adaptively sound" (see [22]), so perhaps a malicious prover can succeed in proving false statements after seeing the CRS, as demonstrated in [1]. As another example, the protocol in [15] for realizing the single-instance commitment functionality becomes *malleable* as soon as *two* instances use the same reference string (indeed, to avoid this weakness a more involved protocol was developed, where multiple commitments can explicitly use the same reference string in a specific way). Note that here, a UC-secure protocol can even affect the security of another UC-secure protocol if both protocols make reference to the same setup.

This situation is disturbing, especially in light of the fact that *some* form of setup is often *essential* for cryptographic solutions. For instance, most traditional two-party tasks cannot be UC-realized with no setup [15,10,16], and authenticated communication is impossible without some sort of setup [12]. Furthermore, providing a *globally available* setup that can be used throughout the system is by far the most realistic and convenient way to provide setup.

A new formalism. This work addresses the question of how to formalize the trusted-party definitional paradigm in a way that preserves its intuitive appeal even for those protocols that use globally available setup. Specifically, our first contribution is to generalize the UC framework to deal with global setup, so as to explicitly guarantee that the original meaning of "emulating a trusted party" is preserved, even when the analyzed protocol is using the *same setup* as other protocols that may be maliciously and adaptively designed to interact badly with it. In particular, the new formalism called simply generalized UC (GUC) security guarantees deniability and non-malleability even in the presence of global setup. Informally,

> *A GUC-Secure protocol π implementing a trusted party $\mathcal{F}$ using some global setup does not affect any other protocols more than $\mathcal{F}$ does — even when protocols running concurrently with π are maliciously constructed, and even when all protocols use the same global setup.*

In a nutshell, the new modeling proceeds as follows. Recall that the UC framework models setup as a "trusted subroutine" of the protocol that uses the setup. This implicitly means that the setup is local to the protocol instance using it, and cannot be safely used by any other protocol instance. That modeling, while mathematically sound, certainly does not capture the real-world phenomenon of setup that is set in advance and publicly known throughout the system. The UC with joint state theorem ("JUC Theorem") of [18] allows several instances of specifically-designed protocols to use the same setup, but it too does not capture the case of public setup that can be used by arbitrary different protocols at the same time.

To adequately capture global setup our new formalism models the setup as an additional (trusted) entity that interacts not only with the parties running the protocol, but also with other parties (or, in other words, with the external environment). This in particular means that the setup entity exists not only as part of the protocol execution, but also in the *ideal process*, where the protocol is replaced by the trusted party. For instance, while in the current UC framework the CRS model is captured as a trusted setup entity that gives the reference string only to the adversary and the parties running the actual protocol instance, here the reference string is globally available, i.e. the trusted setup entity also gives the reference string directly to other parties and the external environment. Technically, the effect of this modeling is that now the simulator (namely, the adversary in the ideal process) cannot choose the reference string or know related trapdoor information.

In a way, proofs of security in the new modeling, even with setup, are reminiscent of the proofs of security without setup, in the sense that the only freedom enjoyed by the simulator is to control the local random choices of the uncorrupted parties. For this reason we often informally say that GUC-secure protocols that use only globally available setup are "fully simulatable". We also remark that this modeling is in line with the "non-programmable CRS model" in [28].

One might thus suspect that achieving GUC-security "collapses" down to UC-security *without any setup* (and its severe limitations). Indeed, as a first result we extend the argument of [15] to show that no two-party protocol can GUC-realize the ideal commitment functionality $\mathcal{F}_{com}$ (namely, emulate the trusted party that runs the code of $\mathcal{F}_{com}$ according to the new notion), *even in the CRS model, or in fact with any global setup that simply provides public information.* On the one hand this result is reassuring, since it means that those deniable and malleable protocols that are secure in the (old) CRS model can no longer be secure according to the new notion. On the other hand, this result brings forth the question of whether there exist protocols for commitment (or other interesting primitives) that meet the new notion under *any* reasonable setup assumption. Indeed, the analyses of all existing UC-secure commitment protocols seem to use in an essential way the fact that the simulator has control over the value of the setup information.

New setup and constructions. Perhaps surprisingly, we answer the realizability question in the affirmative, in a strong sense. Recall that our impossibility result shows that a GUC protocol for the commitment functionality must rely on a setup that provides the parties with some *private* information. We consider two alternative setup models which provide such private information in a *minimal* way, and show how to GUC-realize practically any ideal functionality in any one of the two models.

The first setup model is reminiscent of the "key registration with knowledge (KRK)" setup from [5], where each party registers a public key with some trusted authority in a way that guarantees that the party can access the corresponding secret key. However, in contrast to [5] where the scope of a registered key is only a single protocol instance (or, alternatively, several instances of specifically designed protocols), here the registration is done once per party throughout the lifetime of the system, and the public key can be used in all instances of all the protocols that the party might run. In particular, it is directly accessible by the external environment.

We first observe that one of the [5] protocols for realizing $\mathcal{F}_{com}$ in the KRK model can be shown to satisfy the new notion, even with the global KRK setup, as long as the adversary is limited to *non-adaptive* party corruptions. (As demonstrated in [17], realizing $\mathcal{F}_{com}$ suffices for realizing *any* "well-formed" multi-party functionality.) However, when adaptive party corruptions are allowed, and the adversary can observe the past internal data of corrupted parties, this protocol becomes insecure. In fact, the problem seems inherent, since the adversary is now able to eventually see *all* the secret keys in the system, even those of parties that were uncorrupted when the computation took place.

Still, we devise a new protocol that realizes $\mathcal{F}_{com}$ in the KRK model even in the presence of adaptive party corruptions, and without any need for data erasures. The high level idea is to use the [15] commitment scheme with a new CRS that is chosen by the parties per commitment. The protocol for choosing the CRS will make use of the public keys held by the parties, in a way that allows the overall simulation to go through even when the same public keys are used in multiple instances of the CRS-generation protocol. Interestingly, our construction does not realize a CRS that is "strong" enough for the original analysis to go through. Instead, we provide a "weaker" CRS, and provide a significantly more elaborate analysis. The protocol is similar in spirit to the coin-tossing protocol of [19], in that it allows the generated random string to have different properties depending on which parties are corrupted. Even so, their protocol is not adaptively secure in our model.

Augmented CRS. Next we formulate a new setup assumption, called "augmented CRS (ACRS)" and demonstrate how to GUC-realize $\mathcal{F}_{com}$ in the ACRS model, in the presence of adaptive adversaries. As the name suggests, ACRS is reminiscent of the CRS setup, but is somewhat augmented so as to circumvent the impossibility result for plain CRS. That is, as in the CRS setup, all parties have access to a short reference string that is taken from a pre-determined distribution. In addition, the ACRS setup allows corrupted parties to obtain "personalized" secret keys that are derived from the reference string, their public identities, and some "global secret" that's related to the public string and remains unknown. It is stressed that *only corrupted parties* may obtain their secret keys. This means that the protocol may not include instructions that require knowledge of the secret keys and, therefore, the protocol interface tn the ACRS setup is identical to that of the CRS setup.

The main tool in our protocol for realizing $\mathcal{F}_{com}$ in the ACRS model is a new *identity-based trapdoor commitment (IBTC)* protocol. IBTC protocols are constructed in [2,32], in the Random Oracle model. In the full version of this paper [13], we provide a construction of IBTC in the standard model (assuming only one-way functions), using the Σ-protocol based commitment technique of Feige [21], where the committer runs the *simulator* of the Σ-protocol.

Realizing the setup assumptions. "Real world implementations" of the ACRS and KRK setups can involve a trusted entity (say, a "post office") that only publicizes the public value. The trusted entity will also agree to provide the secret keys to the corresponding parties upon request, with the understanding that once a party gets hold of its key then it alone is responsible to safeguard it and use it appropriately (much as in the case of standard PKI). In light of the impossibility of a completely non-interactive setup (CRS), this seems to be a minimal "interactiveness" requirement from the trusted entity.

Another unique feature of our commitment protocol is that it guarantees security even if the "global secret" is compromised, as long as this happens *after the commitment phase is completed.* In other words, in order to compromise the overall security, the trusted party has to be *actively malicious during the commitment phase.* This point further reduces the trust in the real-world entity that provides the setup.

Despite the fact that the trusted entity need not be constantly available, and need not remain trustworthy in the long term, it may still seem difficult to provide such an interactive entity in many real-world settings. Although it is impossible to achieve true GUC security with a mere CRS, we observe that the protocols analyzed here do satisfy some notion of security even if the setup entity remains non-interactive (*i.e.* when our ACRS setup functionality is instead collapsed to a standard CRS setup). In fact, although we do not formally prove a separation, protocols proven secure in the ACRS model seem intuitively *more secure* than those of [15,17] *even when used in the CRS model!* Essentially, in order to simulate information that could be obtained via a real attack on the protocols of [15,17], knowledge of a "global trapdoor" is required. This knowledge enables the simulator to break the security *of all parties* (including their privacy). On the other hand, simulating the information obtained by real attacks on protocols that are proven secure in the ACRS model merely requires some specific "identity-based trapdoors". These specific trapdoors used by the simulate allow it to break only the security *of corrupt parties who deviate from the protocol.* Of course, when using a CRS setup in "real life" none of these trapdoors are available to anyone, so one cannot actually simulate information obtained by an attacker. Nevertheless, it seems that the actual advantage gained by an attack which *could* have been simulated using the more minimal resources required by protocol simulators in the ACRS model (*i.e.* the ability to violate the security only of corrupt parties, as opposed to all parties) is intuitively smaller.

A New Composition Theorem. We present two formulations of GUC security: one formulation is more general and more "intuitively adequate", while the other is simpler and easier to work with. In particular, while the general notion directly considers a multi-instance system, the simpler formulation (called EUC) is closer to the original UC notion that considers only a single protocol instance in isolation. We then demonstrate that the two formulations are equivalent. As may be expected, the proof of equivalence incorporates much of the argumentation involved in the proof of the universal composition theorem. We also demonstrate that GUC security is preserved under universal composition.

Related work. Relaxed variants of UC security are studied in [30,8]. These variants allow reproducing the general feasibility results without setup assumptions other than authenticated communication. However, these results provide significantly weaker security properties than UC-security. In particular, they do not guarantee security in the presence of arbitrary other protocols, which is the focus of this work.

Alternatives to the CRS setup are studied in [5]. As mentioned above, the KRK setup used here is based on the one there, and the protocol for GUC-realizing $\mathcal{F}_{com}$ for non-adaptive corruptions is taken from there. Furthermore, [5] informally discuss the deniability properties of their protocol. However, that work does not address the general concern of guaranteeing security in the presence of global setup. In particular, it adopts the original UC modeling of setup as a construct that is internal to each protocol instance.

In a concurrent work, Hofheinz et. al [25] consider a notion of security reminiscent of EUC, with similar motivation to the motivation here. They also formulate a new setup assumption and show how to realize any functionality given that setup. However, their setup assumption is considerably more involved than ours, since it requires the trusted entity to interact with the protocol in an on-line, input-dependent manner. Also, they do not consider adaptive corruptions.

Future work. This work develops the foundations necessary for analyzing security and composability of protocols that use globally available setup. It also re-establishes the feasibility results for general computation in this setting. Still, there are several unexplored research questions here.

One important concern is that of guaranteeing *authenticated communication* in the presence of global PKI setup. As mentioned above, this is another example where the existing notions do not provide the expected security properties (e.g., they do not guarantee deniability, whereas the trusted party solution is expressly deniable). We conjecture that GUC authentication protocols (namely, protocols that GUC-realize ideally authentic communication channels) that use a global PKI setup can be constructed by combining the techniques of [26,15]. However, we leave full exploration of this problem out of scope for this work.

The notions of key exchange and secure sessions in the presence of global PKI setup need to be re-visited in a similar way. How can universal composability (and, in particular, deniability) be guaranteed for such protocols? Also, how can existing protocols (that are not deniable) be proven secure with globally available setup?

2   Generalized UC Security

In this section we will provide a high-level overview of our new Generalized UC (GUC) framework, as well as a useful simplification of GUC called the Externalized UC (EUC) framework. We begin with a brief review of the concepts behind the original UC framework of [10] (henceforth referred to as "Basic UC") before proceeding to outline our new security frameworks. To keep our discussion at a high level of generality, we will focus on the notion of protocol "emulation", wherein the objective of a protocol π is to emulate another protocol ϕ. Here, typically, π is an implementation (such as the actual "real world" protocol) and ϕ is a specification (where the "ideal functionality" $\mathcal{F}$ that we wish to implement is computed directly by a trusted entity). Throughout our discussion, all entities and protocols we consider are "efficient" (*i.e.* polynomial time bounded Interactive Turing Machines, in the sense detailed in [11]).

The Basic UC Framework. At a very high level, the intuition behind security in the basic UC framework is that any adversary $\mathcal{A}$ attacking a protocol π should learn no more information than could have been obtained via the use of a simulator $\mathcal{S}$ attacking protocol ϕ. Furthermore, we would like this guarantee to be maintained even if ϕ were to be used a subroutine of (*i.e.* composed with) arbitrary other protocols that may be running concurrently in the networked environment, and we plan to substitute π for ϕ in all instances. Thus, we may set forth a challenge experiment to distinguish between actual attacks on protocol π, and simulated attacks on protocol ϕ (referring to these protocols as the "challenge protocols"). As part of this challenge scenario, we will allow adversarial attacks to be orchestrated and monitored by a distinguishing environment $\mathcal{Z}$ that is also empowered to control the inputs supplied to the parties running the challenge protocol, as well as to observe the parties' outputs at all stages of the protocol execution. One may imagine that this environment represents all other activity in the system, including the actions of other protocol sessions that may influence inputs to the challenge protocol (and which may, in turn, be influenced by the behavior of the challenge protocol). Ultimately, at the conclusion of the challenge, the environment $\mathcal{Z}$ will be tasked to distinguish between adversarial attacks perpetrated by $\mathcal{A}$ on the challenge protocol π, and attack simulations conducted by $\mathcal{S}$ with protocol ϕ as the challenge protocol instead. If no environment can successfully distinguish these two possible scenarios, then protocol π is said to "UC emulate" the protocol ϕ.

Specifying the precise capabilities of the distinguishing environment $\mathcal{Z}$ is crucial to the meaning of this security notion. We must allow $\mathcal{Z}$ to choose the challenge protocol inputs and observe its outputs (which models the influence of the environment on the users of the protocol, and vice versa). We must also grant $\mathcal{Z}$ the ability to interact with the attacker (which will be either the adversary, or a simulation). As demonstrated in [10], granting precisely these capabilities to $\mathcal{Z}$ (even if we allow it to invoke only a *single session* of the challenge protocol) is sufficient to achieve the strong guarantees of *composition theorem*, which states that any arbitrary instances of the ϕ that may be running in the network can be safely substituted with a protocol π that UC emulates ϕ. Thus, even if we *constrain* the distinguisher $\mathcal{Z}$ to interactions only with the adversary and a *single session* of the challenge protocol (without allowing $\mathcal{Z}$ to invoke other protocols at all), we can already achieve the strong security guarantees we intuitively desired. Notably, although the challenge protocol may invoke subroutines of its own, it was not necessary to grant $\mathcal{Z}$ any capability to interact with such subroutines.

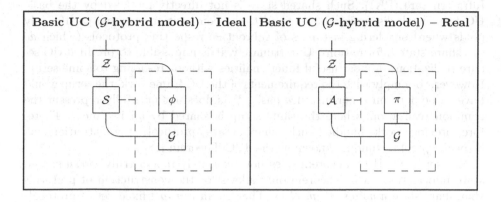

| Basic UC ($\mathcal{G}$-hybrid model) – Ideal | Basic UC ($\mathcal{G}$-hybrid model) – Real |

Fig. 1. The Basic UC Experiment in the $\mathcal{G}$-hybrid model. A simulator $\mathcal{S}$ attacks a single session of protocol ϕ running with an ideal subroutine $\mathcal{G}$, whereas an arbitrary "real" adversary $\mathcal{A}$ attacks a session of π running with an ideal subroutine $\mathcal{G}$. The dashed box encloses protocols where $\mathcal{S}$ or $\mathcal{A}$ control the network communications, whereas the solid lines represent a direct Input/Output relationship. (In a typical scenario, ϕ would be the ideal protocol for a desired functionality $\mathcal{F}$, whereas π would be a practical protocol realizing $\mathcal{F}$, with $\mathcal{G}$ modeling some "setup" functionality required by π. Observe that the environment can never interact directly with $\mathcal{G}$, and thus, in this particular scenario, $\mathcal{G}$ is never invoked at all in the ideal world since we are typically interested in the case where ideal protocol for $\mathcal{F}$ does not make use of $\mathcal{G}$.)

In order to conceptually modularize the design of protocols, the notion of "hybrid models" is often introduced into the basic UC framework. A protocol π is said to be realized "in the $\mathcal{G}$-hybrid model" if π invokes the ideal functionality $\mathcal{G}$ as a subroutine (perhaps multiple times). (As we will soon see below, the notion of hybrid models greatly simplifies the discussion of UC secure protocols that require "setup".) A high-level conceptual view of UC protocol emulation in a hybrid model is shown in Figure 1.

Limitations of Basic UC. Buried inside the intuition behind the basic UC framework is the critical notion that the environment $\mathcal{Z}$ is capable of utilizing its input/output interface to the challenge protocol to mimic the behavior of other (arbitrary) protocol sessions that may be running in a computer network. Indeed, as per the result of [10] mentioned in our discussion above, this would seem to be the case when considering challenge protocols that are essentially "self-contained". Such self-contained protocols, which do not make use of any "subroutines" (such as ideal functionalities) belonging to other protocol sessions, are called *subroutine respecting* protocols – and the basic UC framework models these protocols directly. On the other hand, special considerations would arise if the challenge protocol utilizes (or transmits) information that is also shared by other network protocol sessions. An example of such information would be the use of a global setup, such as a public "common reference string" (CRS) that is reused from one protocol session to the next, or a standard Public Key

Infrastructure (PKI). Such shared state is not directly modeled by the basic UC framework discussed above. In fact, the composition theorem of [10] only holds when considering instances of subroutine respecting protocols (which *do not* share state information). Unfortunately, it is impossible to produce UC secure realizations of most useful functionalities without resorting to some setup. However, to comply with the requirements of the UC framework, the setup would have to be done on a per-instance basis. This does not faithfully represent the common realization, where the same setup is shared by all instances. Therefore, previous works handled such "shared state" protocol design situations via a special proof technique, known as the JUC Theorem [18].

Yet, even the JUC Theorem does not accurately model truly *global* shared state information. JUC Theorem only allows for the construction of protocols that share state *amongst themselves*. That is, an *a-priori* fixed set of protocols can be proven secure if they share state information *only* with each other. No security guarantee is provided in the event that the shared state information is also used by other protocols which the original protocols were not specifically designed to interact with. Of course, malicious entities may take advantage of this by introducing new protocols that use the shared state information if the shared state is publicly available. In particular, protocols sharing global state (*i.e.* using global setups) which are modeled in this fashion may not resist adaptive chosen protocol attacks, and can suffer from a lack of deniability, as we previously mentioned regarding the protocols of [15], [17], and as is discussed in further detail in Section 3.2.

The Generalized UC Framework. To summarize the preceding discussion, the environment $\mathcal{Z}$ in the basic UC experiment is unable to invoke protocols that share state in any way with the challenge protocol. This limitation is unrealistic in the case of global setup, when protocols share state information with each other (and indeed, it was shown to be impossible to realize UC-secure protocols without resort to such tactics [15,10,16]). To overcome this limitation, we propose the Generalized UC (GUC) framework. The GUC challenge experiment is similar to the basic UC experiment, only with an *unconstrained* environment. In particular, we will allow $\mathcal{Z}$ to actually invoke and interact with arbitrary protocols, and even multiple sessions of its challenge protocol (which may be useful to $\mathcal{Z}$ in its efforts to distinguish between the two possible challenge protocols). Some of the protocol sessions invoked by $\mathcal{Z}$ may share state information with challenge protocol sessions, and indeed, they can provide $\mathcal{Z}$ with information about the challenge protocol that it could not have obtained otherwise. The only remaining limitation on $\mathcal{Z}$ is that we prevent it from directly observing or influencing the network communications of the challenge protocol sessions, but this is naturally the job of the adversary (which $\mathcal{Z}$ directs). Thus, the GUC experiment allows a very powerful distinguishing environment capable of truly capturing the behavior of arbitrary protocol interactions in the network, *even if protocols can share state information with arbitrary other protocols*. Of course, protocols that are GUC secure are also composable (this fact follows almost trivially from a greatly simplified version of the composition theorem proof of

[11], the simplifications being due to the ability of the unconstrained environment to directly invoke other protocol sessions rather than needing to "simulate" them internally).

The Externalized UC Framework. Unfortunately, since the setting of GUC is so complex, it becomes extremely difficult to prove security of protocols in our new GUC framework. Essentially, the distinguishing environment $\mathcal{Z}$ is granted a great deal of freedom in its choice of attacks, and any proof of protocol emulation in the GUC framework must hold even in the presence of other arbitrary protocols running concurrently. To simplify matters, we observe that in practice protocols which are designed to share state do so only in a very limited fashion (such as via a single common reference string, or a PKI, etc.). In particular, we will model shared state information via the use of "shared functionalities", which are simply functionalities that may interact with more than one protocol session (such as the CRS functionality). For clarity, we will distinguish the notation for shared functionalities by adding a bar (*i.e.* we use $\bar{\mathcal{G}}$ to denote a shared functionality). We call a protocol π that *only* shares state information via a single shared functionality $\bar{\mathcal{G}}$ a $\bar{\mathcal{G}}$-*subroutine respecting* protocol. Bearing in mind that it is generally possible to model "reasonable" protocols that share state information as $\bar{\mathcal{G}}$-subroutine respecting protocols, we can make the task of proving GUC security simpler by considering a compromise between the constrained environment of basic UC and the unconstrained environment of GUC. An $\bar{\mathcal{G}}$-*externally constrained* environment is subject to the same constraints as the environment in the basic UC framework, only it is additionally allowed to invoke a single "external" protocol (specifically, the protocol for the shared functionality $\bar{\mathcal{G}}$). Any state information that will be shared by the challenge protocol must be shared via calls to $\bar{\mathcal{G}}$ (*i.e.* challenge protocols are $\bar{\mathcal{G}}$-subroutine respecting), and the environment is specifically allowed to access $\bar{\mathcal{G}}$. Although $\mathcal{Z}$ is once again constrained to invoking a single instance of the challenge protocol, it is now possible for $\mathcal{Z}$ to internally mimic the behavior of multiple sessions of the challenge protocol, or other arbitrary network protocols, by making use of calls to $\bar{\mathcal{G}}$ wherever shared state information is required. Thus, we may avoid the need for JUC Theorem (and the implementation limitations it imposes), by allowing the environment direct access to shared state information (*e.g.* we would allow it to observe the Common Reference String when the shared functionality is the CRS functionality). We call this new security notion Externalized UC (EUC) security, and we say that a $\bar{\mathcal{G}}$-subroutine respecting protocol π $\bar{\mathcal{G}}$-EUC-emulates a protocol ϕ if π emulates ϕ in the basic UC sense with respect to $\bar{\mathcal{G}}$-externally constrained environments. We show that if a protocol π $\bar{\mathcal{G}}$-EUC-emulates ϕ, then it also GUC emulates ϕ (and vice versa, provided that π is $\bar{\mathcal{G}}$-subroutine respecting).

Theorem 1. *Let π be any protocol which invokes no shared functionalities other than (possibly) $\bar{\mathcal{G}}$, and is otherwise subroutine respecting (i.e. π is $\bar{\mathcal{G}}$-subroutine respecting). Then protocol π GUC-emulates a protocol ϕ, if and only if protocol π $\bar{\mathcal{G}}$-EUC-emulates ϕ.*

That is, provided that π only shares state information via a single shared functionality $\bar{\mathcal{G}}$, if it merely EUC-emulates ϕ with respect to that functionality, then π is a full GUC-emulation of ϕ! As a special case, we obtain that all basic UC emulations (which may not share *any* state information) are also GUC emulations.

Corollary 1. *Let π be any subroutine respecting protocol. Then protocol π GUC-emulates a protocol ϕ, if and only if π UC-emulates ϕ.*

The corollary follows by letting $\bar{\mathcal{G}}$ be the null functionality, and observing that the $\bar{\mathcal{G}}$-externally constrained environment of the EUC experiment collapses to become the same environment as that of the basic UC experiment when $\bar{\mathcal{G}}$ is the null functionality. Thus, it is sufficient to prove basic UC security for protocols with no shared state, or $\bar{\mathcal{G}}$-EUC security for protocols that share state only via $\bar{\mathcal{G}}$, and we will automatically obtain the full benefits of GUC security.

Figure 2 depicts the differences in the experiments of the UC models we have just described, in the presence of a single shared functionality $\bar{\mathcal{G}}$ (of course, the GUC framework is not inherently limited to special case of only one shared functionality). We further elaborate the technical details of these new models, and provide the proof of Theorem 1, in the full version of the paper [13].

We are now in a position to state a strong new composition theorem, which will directly incorporate the previous result (that proving EUC security is sufficient for GUC security). Let ρ be an arbitrary protocol (not necessarily subroutine respecting!) which invokes ϕ as a sub-protocol. We will write $\rho^{\pi/\phi}$ to denote a modified version of ρ that invokes π instead of ϕ, wherever ρ had previously invoked ϕ. We prove the following general theorem in the full version [13]:

Theorem 2 (Generalized Universal Composition). *Let ρ, π, ϕ be PPT multi-party protocols, and such that both ϕ and π are $\bar{\mathcal{G}}$-subroutine respecting, and π $\bar{\mathcal{G}}$-EUC-emulates ϕ. Then $\rho^{\pi/\phi}$ GUC-emulates protocol ρ.*

We stress that π must merely $\bar{\mathcal{G}}$-EUC-emulate ϕ, but that the resulting composed protocol $\rho^{\pi/\phi}$ fully GUC-emulates ρ, even for a protocol ρ that is not subroutine respecting.

3   Insufficiency of the Global CRS Model

In this section we demonstrate that a global CRS setup is *not* sufficient to GUC-realize even the basic two-party commitment functionality. We then further elaborate the nature of this insufficiency by considering some weaknesses in the security of previously proposed constructions in the CRS model. Finally, we suggest a new "intuitive" security goal, dubbed *full simulatability*, which we would like to achieve by utilizing the GUC-security model (and which was not previously achieved by any protocols in the CRS model).

3.1   Impossibility of GUC-Realizing $\mathcal{F}_{com}$ in the $\bar{\mathcal{G}}_{gcrs}$ Model

Recall that many interesting functionalities are unrealizable in the UC framework without any setup assumption. For instance, it is easy to see that the

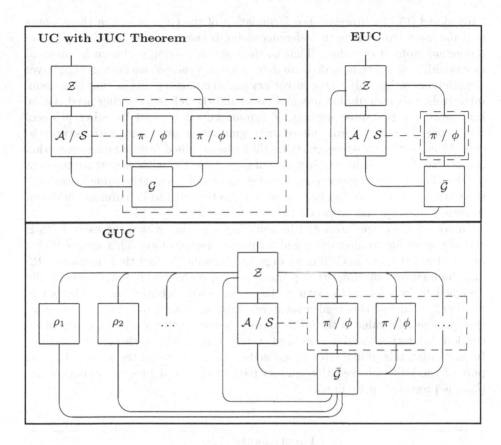

Fig. 2. Comparison of models. Using Basic UC with JUC Theorem to share state, only copies of the challenge protocol (or other protocols which may be jointly designed a priori to share $\mathcal{G}$) are allowed to access the common subroutine $\mathcal{G}$, and $\mathcal{Z}$ may only interact with the "multi-session" version of the challenge protocol. In the EUC paradigm, only a single session of the challenge protocol is running, but the shared functionality $\bar{\mathcal{G}}$ it uses is accessible by $\mathcal{Z}$. Finally, in the GUC setting, we see the full generality of arbitrary protocols $\rho_1, \rho_2, \ldots$ running in the network, alongside multiple copies of the challenge protocol. Observe that both $\mathcal{Z}$, and any other protocols invoked by $\mathcal{Z}$ (such as ρ_1), have direct access to $\bar{\mathcal{G}}$ in the GUC setting. Intuitively, the GUC modeling seems much closer to the actual structure of networked protocol environments.

ideal authentication functionality, $\mathcal{F}_{auth}$, is unrealizable in the plain model. Furthermore, many two party tasks, such as Commitment, Zero-Knowledge, Coin-Tossing, Oblivious Transfer and others cannot be realized in the UC framework by two-party protocols, even if authenticated communication is provided [15,16,10].

As a recourse, the common reference string (CRS) model was used to reassert the general feasibility results of [24] in the UC framework. That is, it was shown that any "well-formed" ideal functionality can be realized in the

CRS model [15,17]. However, the formulation of the CRS model in these works postulates a setting where the reference string is given *only to the participants in the actual protocol execution*. That is, the reference string is chosen by an ideal functionality, $\mathcal{G}_{crs}$, that is dedicated to a given protocol execution. $\mathcal{G}_{crs}$ gives the reference string only to the adversary and the participants in that execution. Intuitively, this formulation means that, while the reference string need not be kept secret to guarantee security, it cannot be safely used by other protocol executions. In other words, no security guarantees are given with respect to executions that use a reference string that was obtained from another execution rather than from a dedicated instance of $\mathcal{G}_{crs}$. (The UC with joint state theorem of [18] allows multiple executions of certain protocols to use the same instance of the CRS, but it requires all instances that use the CRS to be carefully designed to satisfy some special properties.)

In contrast, we are interested in modeling a setting where the same CRS is globally available to all parties and all protocol executions. This means that a protocol π that uses the CRS must take into account the fact that the same CRS may be used by *arbitrary* other protocols, even protocols that were specifically designed to interact badly with π. Using the GUC security model defined in Section 2, we define this weaker setup assumption as a *shared* ideal functionality that provides the value of the CRS not only to the parties of a given protocol execution, but rather to all parties, and even directly to the environment machine. In particular, this global CRS functionality, $\bar{\mathcal{G}}_{gcrs}$, exists in the system both as part of the protocol execution and as part of the ideal process. Functionality $\bar{\mathcal{G}}_{gcrs}$ is presented in Figure 3.

Functionality $\bar{\mathcal{G}}_{gcrs}$

Parameterized by a distribution D, $\bar{\mathcal{G}}_{gcrs}$ proceeds as follows, when activated by any party:

1. If no value has been previously recorded, choose a value $d \xleftarrow{R} D$, and record the value d.
2. Return the value d to the activating party.

Fig. 3. The Global Common Reference String functionality. The difference from the Common Reference String functionality $\mathcal{G}_{crs}$ of [10,15] is that $\mathcal{G}_{crs}$ provides the reference string only to the parties that take part in the actual protocol execution. In particular, the environment does not have direct access to the reference string.

We demonstrate that $\bar{\mathcal{G}}_{gcrs}$ is insufficient for reproducing the general feasibility results that are known to hold in the $\mathcal{G}_{crs}$ model. To exemplify this fact, we show that no two-party protocol that uses $\bar{\mathcal{G}}_{gcrs}$ as its only setup assumption GUC-realizes the ideal commitment functionality, $\mathcal{F}_{com}$ (presented in Figure 4). The proof, which we provide in the full version of this work [13], follows essentially the same steps as the [15] proof of impossibility of realizing $\mathcal{F}_{com}$ in the plain model.

The reason that these steps can be carried out even in the presence of $\bar{\mathcal{G}}_{gcrs}$ is, essentially, that the simulator obtains the reference string from an external entity ($\bar{\mathcal{G}}_{gcrs}$), rather than generating the reference string by itself. We conjecture that most other impossibility results for UC security in the plain model can be extended in the same way to hold for GUC security in the presence of $\bar{\mathcal{G}}_{gcrs}$.

Functionality $\mathcal{F}_{com}$

Commit Phase: Upon receiving a message ($\texttt{commit}, sid, C, V, b$) from party
C, where $b \in \{0, 1\}$, record the value b and send ($\texttt{receipt}, sid, C, V$) to
V and the adversary. Ignore any future $\texttt{commit}$ messages.
Reveal Phase: Upon receiving a message ($\texttt{reveal}, sid$) from C: If a value
b was previously recorded, then send the message ($\texttt{reveal}, sid, b$) to V
and the adversary and halt. Otherwise, ignore.

Fig. 4. The Commitment Functionality $\mathcal{F}_{com}$ (see [15])

Theorem 3. *There exists no bilateral, terminating protocol π that GUC-realizes $\mathcal{F}_{com}$ and uses only the shared functionality $\bar{\mathcal{G}}_{gcrs}$. This holds even if the communication is ideally authentic.*

In fact, it can be shown that the above impossibility result extends beyond the mere availability of $\bar{\mathcal{G}}_{gcrs}$ to any circumstance where the shared functionality will only provide information globally (or, yet more generally, the impossibility holds whenever all the shared information available to protocol participants can also be obtained by the environment). For instance, this impossibility will hold even in the (public) random oracle model, which is already so strong that it cannot truly be realized without the use of a fully interactive trusted party. Another interpretation of this result is that no completely *non-interactive* global setup can suffice for realizing $\mathcal{F}_{com}$. The next section studies the problem of realizing $\mathcal{F}_{com}$ using setup assumptions with minimal interaction requirements.

3.2   Deniability and Full Simulatability

To demonstrate that the problems with using a global CRS to realize $\mathcal{F}_{com}$, in the fashion of [17], are more than skin deep technicalities that arise only in the GUC framework we now consider the issue of deniability. Intuitively, a protocol is said to be "deniable" if it is possible for protocol participants to deny their participation in a protocol session by arguing that any "evidence" of their participation (as obtained by other, potentially corrupt protocol participants) could have been fabricated.

Recalling the intuition outlined in the introduction, we would like realized protocols to guarantee the same security as the ideal functionalities they realize, meaning that the adversary will learn nothing more from attacking the protocol than could be learned from attacking its corresponding ideal functionality. Protocols realized with such a guarantee are inherently *deniable*, since a protocol participant can accurately argue that any information sent during the protocol session could have been obtained by an adversary using only the output from the ideal functionality[1] in an attack simulation conducted entirely without his or her actual participation.

For instance, if we consider the ideal functionality for Zero Knowledge (ZK), we expect that any secure realization of that functionality should reveal no information to the adversary beyond the output of the ideal functionality (which contains only a single bit). In particular, that output can easily be generated entirely without the help of the prover, and thus the prover should be able to deny his participation in the protocol, since it reveals no information that could not have been obtained independently. However, we already know from the result of [28] that it is impossible to achieve such deniability for ZK in the CRS model. Indeed, we may see that the UC simulator for ZK functionality in [17] chooses a fresh CRS, and generates the simulated protocol transcripts with respect to that, instead of the published real-world CRS. Thus, if a protocol transcript makes use of the real-world CRS, it could not have been obtained via simulation (so a successful prover is indeed incriminated by the transcript).

When there is no deniability, the adversary is truly able to obtain some valuable information by observing protocol interactions that would not by revealed by the ideal functionality. Thus we have found a practical example of security loss that directly results from the relaxations of UC security inherent in the CRS technique of [17]. We can now clearly see that the impossibility of realizing $\mathcal{F}_{com}$ via the CRS model in the GUC setting is due to a meaningful strengthening of security guarantees (since deniability is guaranteed in the GUC setting, and that guarantee is not achieved by protocols realized in the CRS model).

On an intuitive level, it might be helpful to consider the issue of deniability in light of the "real world" resources required in order to run the GUC simulator to simulate a given protocol session. If the resources required to simulate a protocol session are readily available, then the protocol is *plausibly deniable* (since it is plausible that information obtained from the protocol was the result of a simulation). If the resources are difficult or impossible to obtain, then the protocol is not plausibly deniable. We wish to employ simulation techniques that require only minimal resources to conduct a simulation, increasing the plausibility of denials (as well as decreasing the value of any information that an adversary might obtain by attacking a secure protocol). Thus, we use the term *fully simulatable* to refer to any plausibly deniable protocol realized in the GUC framework.

From this vantage point, we observe that the resource required to conduct the protocol simulations in [17] is a "trapdoor" for the CRS. In particular, the CRS must be "rigged" with such a trapdoor *a priori*. Such rigging is certainly not plausible when there is a trusted party choosing the CRS, and this is in fact the root of the deniability problem for the CRS model. Furthermore, knowledge of

---

[1] Of course, if the output of the ideal functionality "incriminates" a user by revealing some of his secrets, the resulting protocol does not meet our intuitive understanding of the word "deniable". Still, the protocol itself may be said to be "as deniable" as the functionality it realizes.

this trapdoor implies the ability to completely violate security of any protocol constructed using the techniques of [17], and thus there would be no security against any entity capable of simulating protocols. Similarly, in the "imaginary angel" model of [30], the simulator requires access to super-polynomial time functionalities that are certainly not plausibly available in the real world (and thus, the deniability problem arises there as well). Indeed, if the "imaginary angels" of [30] were to somehow be made practical in the real world, all security would be lost.

We comment that, although we do not make any attempt to formalize a "general" notion of deniability here, the guarantee we seek to provide is that protocols are "as deniable" in the real world as they would have been in the ideal world (past works did not satisfy even this basic requirement). In fact, as we will see, our particular realization of fully simulatable security will guarantee that even "on line" (interactive) deniability is preserved, since the simulator can very practically be run in real time. Indeed, as long as an honest party P never deviates from the protocol, it is not possible for other (even corrupt) protocol participants to conclusively demonstrate P's participation in the protocol session to a third party, *even while the protocol is ongoing*!

4   Fully Simulatable General Computation

We now turn our attention to the problem of *constructing* fully simulatable GUC-secure protocols. That is, we would like it to be possible for a real-world adversary to simulate the effects of any attack on a protocol, without actually attacking the protocol (instead utilizing only the information that would be revealed by an ideally secure realization). The result of Section 3 implies that we cannot do this in the CRS model (if we correctly model a globally available CRS). Thus, we must consider alternative models if we hope to achieve our goal.

To that end, we would like to find reasonable alternative global setup assumptions that allow for realizing interesting tasks. That is, we are looking for shared functionalities $\bar{\mathcal{G}}$ (as defined in Section 2), so that on the one hand $\bar{\mathcal{G}}$ will be implementable in reality with reasonable trust assumptions, and on the other hand we will have protocols that GUC-realize interesting functionalities and still use no setup (*i.e.*, no ideal functionalities) other than $\bar{\mathcal{G}}$. We say that such GUC-secure protocols are "fully simulatable" since the GUC-simulator for attacking the ideal protocol can, in a very practical sense, be run directly by the adversary. This allows the adversary to simulate *the same information* that can be gotten by attacking any session of the real protocol, without actually performing any attack.

We first observe that if the system is equipped with a "fully interactive trusted party" that realizes, say, $\mathcal{F}_{mcom}$, the multi-session variant of $\mathcal{F}_{com}$, by interacting separately and privately with each session, then we can directly use the protocol of [17] to GUC-realize any "well-formed" functionality. However, we would like to find more reasonable global setup assumptions, and in particular assumptions that require less interaction from the trusted entity. (Indeed, this realization requires the trusted party to perform strictly more work than it would by directly computing the desired functionalities, *i.e.* the trivial realization of ideal model functionalities). Although it is clear that we can achieve fully simulatable protocols by using highly interactive trusted parties to compute functionalities, it seems to be a more difficult problem to realize GUC-secure protocols using an "offline" shared functionality. Indeed, by our earlier impossiblity results, *some* degree of interaction would seem to be essential, so we begin by considering the idea of limiting the interaction to a "registration phase".

4.1 The KRK Model

We observe that the "key registration with knowledge (KRK)" setup of [5], can be modified to serve as a shared functionality, allowing us to realize any "well-formed" ideal functionality against non-adaptive ("static") adversaries using the techniques of that work. Although the setup phase is interactive (parties must register their public keys with registration authorities), it is possible to show (with some minor modifications) that the protocol of [5] can allow the trusted party to remain "offline" for all subsequent protocol activity.

Functionality $\mathcal{G}_{krk}$

$\mathcal{G}_{krk}$ proceeds as follows, given a (deterministic) key generation function Gen (with security parameter λ), running with parties $P_1, \ldots, P_n$ and an adversary $\mathcal{S}$:

Registration: When receiving a message (register, sid, r) from party P_i that has not previously registered, compute $(PK_i, SK_i) \leftarrow \mathsf{Gen}^\lambda(r)$ and record the tuple (P_i, PK_i, SK_i).

Retrieval: When receiving a message (retrieve, sid, P_i) from party P_j (where $j \neq i$), if there is a previously recorded tuple of the form (P_i, PK_i, SK_i), then return (sid, P_i, PK_i) to P_j. Otherwise return (sid, $P_i, \bot$) to P_j. When receiving a message (retrieve, sid, P_i) from party P_i, if there is a previously recorded tuple of the form (P_i, PK_i, SK_i), then return (sid, P_i, PK_i, SK_i) to P_i. Otherwise, return (sid, $P_i, \bot$) to P_i.

Fig. 5. The Knowledge-based Key Registration Functionality (similar to that of [5]). Note that each instance of $\mathcal{G}_{krk}$ can only be invoked by the parties of a single protocol session (*i.e.* with a fixed sid). After converting this ideal functionality to a shared functionality, $\bar{\mathcal{G}}_{krk}$, and restricting retrieval of private keys to *corrupt parties only*, it is possible GUC-realize any functionality using only a single public key per-party.

Recall that the KRK setup of [5] is an ideal functionality $\mathcal{G}_{krk}$ (shown in Figure 5), that chooses a private and public key pair for each registered party and lets all parties know the value of the public key. In the natural version of the KRK setup, parties are also allowed to retrieve their own secret keys.

Since $\mathcal{G}_{krk}$ is not a shared functionality, each instance of a protocol will have its own instance of $\mathcal{G}_{krk}$, which does not lend itself to easy implementation. To fix this, we re-formulate $\mathcal{G}_{krk}$ as a shared functionality, $\bar{\mathcal{G}}_{krk}$, that chooses public keys only *once* per party – used by all instances. (This modeling makes $\bar{\mathcal{G}}_{krk}$ significantly easier to implement in a real system, since only one key is required for each party.) Furthermore, we add a simple modeling restriction: we only allow parties to learn their own secret keys if they are corrupt.[2] Using this new $\bar{\mathcal{G}}_{krk}$ setup, the protocol of [5] works (with minor modifications) even in the GUC-security model, provided that (a) party corruptions are non-adaptive, and (b) all parties with the same PID (party identity) are corrupted together – we call such corruption pattern PID-wise.

Theorem 4. *The [5] protocol GUC-realizes $\mathcal{F}_{zk}$, even when given access only to $\bar{\mathcal{G}}_{krk}$, as long as the party corruptions are non-adaptive and PID-wise.*

The proof of this theorem is by a natural extension of the proof in [5] to the EUC framework (which is, of course, equivalent to GUC), but surprisingly, we can achieve a much stronger goal than non-adaptive GUC security with interactive setup.

4.2   The Augmented CRS Model

Although it may seem that *at least* an interactive "registration phase" is required in order to avoid our earlier impossibility result, we show that something even *less interactive* will suffice. We propose a further simplification of $\bar{\mathcal{G}}_{krk}$, denoted $\bar{\mathcal{G}}_{acrs}$, and a protocol that GUC-realizes $\mathcal{F}_{com}$ (and thus any well-formed functionality) having access only to $\bar{\mathcal{G}}_{acrs}$. Unlike $\bar{\mathcal{G}}_{krk}$, the $\bar{\mathcal{G}}_{acrs}$ shared functionality does not *require* any interaction (much like $\mathcal{G}_{crs}$), but merely offers a one-time use interactive "key retrieval" service to those who choose to use it. Therefore, we refer to this new setup assumption as the *Augmented CRS* (ACRS) model. In particular, protocols realized in the ACRS model will not actually make use of the key retrieval service, since the model only allows corrupt parties to retrieve their keys. Thus, we are assured that honest parties need never communicate interactively with $\bar{\mathcal{G}}_{acrs}$.

Somewhat counter-intuitively, it is even crucial that uncorrupted parties in ACRS model never "bother" to obtain their secret keys from the trusted authority (since even an honest party may inadvertently execute a rogue protocol, which might expose the secret key). Similarly, it is crucial that corrupted parties have access to their secret keys, since otherwise they would be unable to conduct attack simulations. (On a side note, security is still guaranteed to honest parties who obtain their keys and use them to conduct attack simulations *provided that* they only use their keys for simulation purposes.) To enforce the protocol design criteria that honest parties should not require access to their secret keys, we directly define the $\bar{\mathcal{G}}_{acrs}$ functionality so that it refuses to supply secret keys to

---

[2]   This modeling restriction is discussed in further detail in Section 4.2.

honest parties. (Of course, a direct realization of $\bar{\mathcal{G}}_{acrs}$ by a trusted party cannot actually determine which parties are honest, yet intuitively this modeling should still suffice. In fact, it is not problematic even if the real-world trusted party gives keys to honest parties, as long as they are careful to protect their own security by keeping their keys secret.)

More formally, our new shared functionality $\bar{\mathcal{G}}_{acrs}$ is parameterized by two functions, Setup and Extract. It first chooses a random secret value MSK and a public value $PK \leftarrow$ Setup(MSK), and publicizes PK (as a CRS). Next, whenever a *corrupted* party P asks for its secret key, $\bar{\mathcal{G}}_{acrs}$ returns the value $SK_P \leftarrow$ Extract($PK; P; MSK$). The functionality is presented in Figure 6.

Functionality $\bar{\mathcal{G}}_{acrs}^{\text{Setup,Extract}}$

Initialization Phase: At the first activation, compute a Common Reference String $(PK) \leftarrow$ Setup(MSK) for a randomly chosen λ-bit value MSK, and record the pair (PK, MSK).

Providing the public value: Whenever activated by a party requesting the CRS, return PK to the requesting party and the adversary.

Dormant Phase: Upon receipt of a message (retrieve, sid, P) from a *corrupt* party P, return the value $SK_P \leftarrow$ Extract($PK; P; MSK$) to P. (Receipt of this message from honest parties is ignored.)

Fig. 6. The Identity-Based Augmented CRS Shared Functionality

Comparing $\bar{\mathcal{G}}_{krk}$ and $\bar{\mathcal{G}}_{acrs}$. The main difference between $\bar{\mathcal{G}}_{acrs}$ and $\bar{\mathcal{G}}_{krk}$ (the global variant of $\mathcal{G}_{krk}$) is that in $\bar{\mathcal{G}}_{acrs}$ there is a single public value, whereas in $\bar{\mathcal{G}}_{krk}$ an extra public value must be given per party identity. Using a paradigm analogous to the identity-based encryption of [6], we avoid the use of per-party public keys and replace them with a single *short* "master public key" (and indeed our constructions use short public keys that depend only on the security parameter). This property, combined with the fact that the parties who follow their protocols never obtain their secret keys, makes $\bar{\mathcal{G}}_{acrs}$ very close in spirit to a global CRS setup as in $\bar{\mathcal{G}}_{gcrs}$. In fact, in light of the far-reaching impossibility result for $\bar{\mathcal{G}}_{gcrs}$, $\bar{\mathcal{G}}_{acrs}$ can be regarded as a "minimum interaction" global setup.

We note that, as pointed out in [5], $\bar{\mathcal{G}}_{krk}$ can be naturally implemented by multiple "registration authorities", where no single authority needs to be fully trusted by all. (However, we once again stress that $\bar{\mathcal{G}}_{krk}$ requires *all* parties, *even those who honestly follow their protocols*, to interactively register with a *some* authority and obtain a public key.) Similarly, multiple instances of $\bar{\mathcal{G}}_{acrs}$ may be run by different trusted authorities. Unlike $\bar{\mathcal{G}}_{krk}$, however, parties may participate in protocols while placing their trust in an arbitrary trusted authority, without ever having registered with *any* authority. This is extremely useful for settings where PKIs are not desirable or easy to implement, and where no single "global" authority is available (see *e.g.* [4]).[3]

In the full version of this work [13], we prove the following result:

Theorem 5. *There exists a protocol that GUC-realizes $\mathcal{F}_{com}$ given access to $\bar{\mathcal{G}}_{acrs}$. Party corruptions can be adaptive (and in the non-erasure model), as long as they are PID-wise.*

Finally, we note that a GUC secure realization of $\mathcal{F}_{com}$ is indeed sufficient to GUC-realize any "well-formed" *multi-party* functionality. This may be accomplished by first using $\mathcal{F}_{com}$ to realize $\mathcal{F}_{zk}$ (as in [17]), and then using $\mathcal{F}_{zk}$ to realize the "one-to-many" Zero-Knowledge functionality, $\mathcal{F}_{zk}^{1:M}$ (via the technique of [29]). The protocol compiler from [17] can then be used to yield a UC-secure realization of any well-formed multi-party functionality in the $\mathcal{F}_{zk}^{1:M}$-hybrid model, without using any shared state (thus it is also a GUC-secure realization by Corollary 1).

5   GUC-Realizing $\mathcal{F}_{com}$ Using the $\bar{\mathcal{G}}_{acrs}$ Global Setup

We now describe the construction of a protocol satisfying the conditions of Theorem 5, above. When combined with the compiler from [17], such a *fully simulatable* realization of $\mathcal{F}_{com}$ yields a fully simulatable realization of any well-formed two-party or multi-party functionality. Furthermore, we show that, in addition to requiring only the more minimal $\bar{\mathcal{G}}_{acrs}$ setup, our protocol achieves significantly stronger properties than the fully simulatable protocol from [5] realized in the $\bar{\mathcal{G}}_{krk}$ model. (Of course, our protocol can also be trivially modified for use in the $\bar{\mathcal{G}}_{krk}$ model, where it will enjoy the same strengthened security guarantees.)

Firstly, our protocol realizing $\mathcal{F}_{com}$ remains secure even in the presence of adaptive corruptions (whereas the protocol of [5] does not). Intuitively, adaptive security seems to be difficult to attain in either the $\bar{\mathcal{G}}_{krk}$ or $\bar{\mathcal{G}}_{acrs}$ models, since an adaptive adversary is eventually able to learn nearly all secrets in the system (save only for the random coins of the trusted party), yet the simulator must make use of these secrets. Our protocol essentially skirts this difficulty by using some additional interactivity. Remarkably, the same technique also enables it to maintain the security of past executions even when the trusted party implementing $\bar{\mathcal{G}}_{acrs}$ is later corrupted (revealing the random coins used to generate the CRS, leaving the overall system with no secrets at all)! That is, our protocol guarantees that past transcripts of protocol interactions can *never* be used

---

[3] In fact, the protocol we will describe in Section 5 can also support a "graceful failure" approach similar to that outlined in [5], in the scenario where protocol participants do not mutually trust any single authority. That is, by using suitable "graceful" tools (in the case of our protocol, a "graceful" IBTC) , we can ensure full GUC security if trustworthy authorities are used by all parties, and ordinary stand-alone security for party P in the case where only party P's authority is trustworthy (even if party P's own authority is made completely unavailable after publishing its reference string, and/or is later corrupted subsequent to the completion of the protocol).

to compromise the security or deniability of honest parties *even if the trusted party is later corrupted.* Security is only lost when the trusted party acts maliciously *prior to,* or *during* protocol execution. This kind of "forward security" with respect to the trusted party further minimizes the trust assumptions required to realize $\bar{\mathcal{G}}_{acrs}$ in the real-world. For instance, an adversary cannot later coerce the trusted party into breaking the security of an honest party after the completion of the protocol. Such forward security cannot be achieved using the protocol of [5] since knowledge of the secret key allows "extraction" from past commitments, breaking privacy. Similarly, the protocol of [17] also loses privacy of past transcripts if the trusted party implementing the CRS setup later reveals a trapdoor.

5.1 High-Level Description of the Protocol

Our protocol for realizing $\mathcal{F}_{com}$ in the $\bar{\mathcal{G}}_{acrs}$ shared hybrid model, which we call Protocol UAIBC (for UC Adaptive Identity-Based Commitment), relies on two new techniques. First, we construct an *identity-based* trapdoor commitment (IBTC) which enjoys adaptive security. Then we provide a general transformation from any IBTC into a protocol that securely implements $\mathcal{F}_{com}$.

Constructing IBTC. In the setting of IBTC a single "master-key" is made public. Additionally, all parties can obtain a private-key that is associated to their party identifier. (Note that this setting corresponds exactly to the interface of $\bar{\mathcal{G}}_{acrs}$.) Intuitively, an IBTC is a commitment scheme with the additional property that a committer who *knows* the receiver's secret-key can *equivocate* commitments (*i.e.*, it can open up commitments to any value, breaking the binding property). Furthermore, an adversary that obtains the secret-keys of multiple parties still should not be able to violate the binding property of commitments sent to parties for which it has not obtained the secret-key.

Constructions of IBTCs were previously known in the Random Oracle Model [2,32]. Here we provide a conceptually simple approach to constructing an adaptively secure IBTC from any one-way function, in the standard model. Our approach relies on the use of Σ-protocols [14], in an approach based on that of [21] (and perhaps surprisingly can result in a very practical protocol). On a very high-level (and very oversimplified) the general idea is as follows: 1) let the master-key be a public-key for a signature scheme, 2) let the secret-key for a party be a signature on its party identifier, and 3) construct a commitment scheme where the reveal phase consists of a "proof" that *either* the revealed value is consistent with the value committed to, *or* the committer knows a signature on the receiver's party identifier (this "proof" must also "hide" which of these two statements actually holds). We mention that the actual instantiation of this idea is somewhat more involved, in order to guarantee adaptive security, and we provide the full details of our construction in [13].

From IBTC to GUC Commitments. Recall that a protocol for realizing $\mathcal{F}_{com}$ must intuitively satisfy two properties (in addition to the traditional binding and hiding properties of any commitment scheme): 1) it must be equivocable, and 2) it must be extractable. We show how to transform any "equivocable" commitment scheme (such as an IBTC) into a protocol for securely realizing $\mathcal{F}_{com}$ (for single bit commitments). Previously similar types of transformations have appeared in the literature (e.g., [17], [7]). Unfortunately all such transformations either require some additional *non-global* setup (and are thus not applicable in out setting), or only work in the case of static security. We now turn our focus to the protocol UAIBC, which GUC-realizes the $\mathcal{F}_{com}$ functionality via a novel transformation of an IBTC from a mere equivocable commitment (in the standard model), to an equivocable *and* extractable commitment secure against adaptive corruptions in the GUC-security model. We remark that our transformation technique can be employed by substituting *any* merely equivocable commitment scheme (such as standard public key based trapdoor commitments) in place of the IBTC in our protocol, and will yield a scheme that is both equivocable and extractable, a general approach that may prove useful in many other contexts.

On a high-level, protocol UAIBC proceeds as follows. The committer P_i and receiver P_j first perform a coin-tossing to generate a public-key K for a dense crypto-system. This coin-tossing requires the receiver to use an IBTC, and has

Step #	$P_i(b)$		P_j
Commit			
(1)		$\xrightarrow{\text{commit},sid,P_i,P_j}$	
(2)		$\xleftarrow{\quad c_k \quad}$	$k_1 \leftarrow_r \{0,1\}^\lambda$ $(c_k, d_k) = \mathsf{Com}(P_i; k_1)$
(3)	$k_2 \leftarrow_r \{0,1\}^\lambda$	$\xrightarrow{\quad k_2 \quad}$	
(4)		$\xleftarrow{\quad d_k \quad}$	$K = k_1 \oplus k_2$
(5)	$k_1' = \mathsf{Open}(P_i; c_k, d_k)$ $K = k_1' \oplus k_2$ $(c, d) = \mathsf{Com}(P_j; b)$ if $b = 0$ $e = E_K(r; d)$ if $b = 1$ $e = $ random	$\xrightarrow{\quad c,\ e \quad}$	
Reveal			
(1&2)	if $b = 0$ if $b = 1$	$\xrightarrow{b=0,\ d,\ r}$ $\xrightarrow{b=1,\ d}$	if $b = 0$ $E_K(r; d) \overset{?}{=} e$ $\mathsf{Open}(P_j; c, d) \overset{?}{=} b$

Fig. 7. Operation of Protocol UAIBC, with party P_i committing bit b to party P_j. Note that Com and Open are operations for an IBTC (the first input is the identity of the recipient), and E_K is a Dense OT-PRC secure encryption using key K (the first input is the random coins fed to the encryption operation, and the second is the plaintext). Steps 2 to 4 of the **Commit** phase are essentially a coin-tossing protocol, whereas the subsequent steps are similar to the protocol of [17].

the property that if the committer is corrupted, the outcome of the coin-tossing can be set to any value. After a completed coin-tossing, the committer commits to a single bit b using an IBTC (let c denote this commitment), and additionally sends an auxiliary string e: e is either a random string in case $b = 1$, and an encryption to the decommitment information of c if $b = 0$. (We here require that the encryption scheme used has *pseudo-random ciphertexts*.) In the reveal phase, the committer is required to provide correct decommitment information for c, and additionally reveal the value encrypted in e in case $b = 0$. We graphically illustrate the operation of this protocol in Figure 7. In the full version of this work [13], we prove that UAIBC GUC-realizes the $\mathcal{F}_{com}$ ideal functionality in a *fully simulatable* manner (even for adaptive adversaries in the non-erasure setting), and in addition features the aforementioned "forward security" property.

Acknowledgments

The authors would like to thank Philip MacKenzie, Johan Håstad, and Silvio Micali for insightful observations, as well as various anonymous referees for helpful comments and suggestions.

References

1. M. Abe, and S. Fehr. Perfect NIZK with Adaptive Soundness. In *Proc. of TCC*, 2007.
2. G. Ateniese and B. de Medeiros. Identity-based Chameleon Hash and Applications. *Proc. of Financial Cryptography*, 2004. Available at http://eprint.iacr.org/2003/167/.
3. D. Beaver. Secure Multi-party Protocols and Zero-Knowledge Proof Systems Tolerating a Faulty Minority. in *J. Cryptology*, vol 4., pp. 75–122, 1991.
4. B. Barak, R. Canetti, Y. Lindell, R. Pass and T. Rabin. Secure Computation Without Authentication. In *CRYPTO 2005*, Springer-Verlag (LNCS 3621), pages 361-377, 2005.
5. B. Barak, R. Canetti, J. Nielsen and R. Pass. Universally composable protocols with relaxed set-up assumptions. In *Proc. of FOCS*, 2004.
6. D. Boneh, and M. Franklin. Identity Based Encryption from the Weil Pairing. In *Proc. of Crypto*, 2001.
7. B. Barak and Y. Lindell. Strict Polynomial-time Simulation and Extraction. In *SIAM J. Comput.*, 33(4), pp. 783-818, 2004.
8. B. Barak and A. Sahai, How To Play Almost Any Mental Game Over the Net - Concurrent Composition via Super-Polynomial Simulation. In *Proc. of FOCS*, 2005.
9. R. Canetti. Security and composition of multi-party cryptographic protocols. *Journal of Cryptology*, Vol. 13, No. 1, winter 2000.
10. R. Canetti. Universally Composable Security: A New paradigm for Cryptographic Protocols. In *Proc. of FOCS*, pages 136–145, 2001.
11. R. Canetti. Universally Composable Security: A New paradigm for Cryptographic Protocols. In *Cryptology ePrint Archive*, Report 2000/067, **revised edition from Dec. 2005**. Available at: http://eprint.iacr.org/2000/067

12. R. Canetti. Universally Composable Signature, Certification, and Authentication. In *Proc. of CSFW*, p. 219, 2004.

13. R. Canetti, Y. Dodis, R. Pass, and S. Walfish. Universally Composable Security with Global Setup. In *Cryptology ePrint Archive*, Report 2006/432. Available at: http://eprint.iacr.org/2006/432

14. R. Cramer, I. Damgard, B. Schoenmakers. Proofs of Partial Knowledge and Simplified Design of Witness Hiding Protocols. In *Proc. of CRYPTO*, pp. 174–187, 1994.

15. R. Canetti and M. Fischlin. Universally Composable Commitments. In *Proc. of Crypto*, pages 19–40, 2001.

16. R. Canetti, E. Kushilevitz and Y. Lindell. On the Limitations of Universally Composable Two-Party Computation Without Set-Up Assumptions. In *Proc. of Eurocrypt*, Springer-Verlag (LNCS 2656), pp. 68–86, 2003.

17. R. Canetti, Y. Lindell, R. Ostrovsky and A. Sahai. Universally Composable Two-Party and Multi-Party Secure Computation. In *Proc. of STOC*, pp. 494–503, 2002.

18. R. Canetti and T. Rabin. Universal Composition with Joint State. In *Proc. of Crypto 2003*, Springer-Verlag, pp. 265-281, 2003.

19. I. Damgard and J. Nielsen. Perfect Hiding and Perfect Binding Universally Composable Commitment Schemes with Constant Expansion Factor. In *Proc. of Crypto*, Springer-Verlag, pp. 581–596, 2002.

20. Y. Dodis and S. Micali. Parallel Reducibility for Information-Theoretically Secure Computation. In *Proc. of Crypto*, Springer-Verlag (LNCS 1880), pp. 74–92, 2000.

21. U. Feige. Alternative Models for Zero Knowledge Interactive Proofs. Ph.D. thesis, Weizmann Institute of Science, Rehovot, Israel, 1990.

22. U. Feige, D. Lapidot, and A. Shamir. Multiple Non-Interactive Zero-Knowledge Proofs Based on a Single Random String. In *Proc. of FOCS*, 1990.

23. S. Goldwasser, and L. Levin. Fair Computation of General Functions in Presence of Immoral Majority. *CRYPTO '90, LNCS 537*, 1990.

24. O. Goldreich, S. Micali, and A. Wigderson. How to Solve any Protocol Problem. In *Proc.of STOC*, 1987.

25. D. Hofheinz, J. Muller-Quade, and D. Unruh. Universally Composable Zero-Knowledge Arguments and Commitments from Signature Cards. In *Proc. of the 5th Central European Conference on Cryptology MoraviaCrypt 2005*, June 2005.

26. M. Jakobsson, K. Sako, and R. Impagliazzo. Designated Verifier Proofs and their Applications. In *Proc. of Eurocrypt*, Springer-Verlag, 1996.

27. S. Micali and P. Rogaway. Secure Computation. unpublished manuscript, 1992. Preliminary version in *CRYPTO '91, LNCS 576*, 1991.

28. R. Pass. On Deniabililty in the Common Reference String and Random Oracle Model. In *Proc. of Crypto*, LNCS 2729, pp. 216–337, 2003.

29. R. Pass. Bounded-Concurrent Secure Multi-Party Computation with a Dishonest Majority. In *Proc. of STOC*, pp. 232–241, 2004.

30. M. Prabhakaran and A. Sahai. New Notions of Security: Achieving Universal Composability without Trusted Setup. In *Proc. of STOC*, 2004.

31. B. Pfitzmann and M. Waidner. Composition and Integrity Preservation of Secure Reactive Systems. In *Proc. of ACM CCS*, pages 245–254, 2000.

32. F. Zhang, R. Safavi-Naini and W. Susilo. ID-Based Chameleon Hashes from Bilinear Pairings. Available at http://eprint.iacr.org/2003/208/.

Parallel Repetition of Computationally Sound Protocols Revisited

Krzysztof Pietrzak[1] and Douglas Wikström[2]

[1] Ecole Normale Supérieure, Département d'informatique,
pietrzak@di.ens.fr
[2] ETH Zürich, Department of Computer Science,
douglas@inf.ethz.ch

Abstract. Parallel repetition is well known to reduce the error probability at an exponential rate for single- and multi-prover interactive *proofs*.

Bellare, Impagliazzo and Naor (1997) show that this is also true for protocols where the soundness only holds against computationally bounded provers (e.g. interactive arguments) if the protocol has at most three rounds.

On the other hand, for four rounds they give a protocol where this is no longer the case: the error probability does not decrease below some constant even if the protocol is repeated a polynomial number of times. Unfortunately, this protocol is not very convincing as the communication complexity of each instance of the protocol grows linearly with the number of repetitions, and for such protocols the error does not even decrease for some types of interactive *proofs*. Noticing this, Bellare et al. construct (a quite artificial) oracle relative to which a four round protocol exists whose communication complexity does not depend on the number of parallel repetitions. This shows that there is no "black-box" error reduction theorem for four round protocols.

In this paper we give the first computationally sound protocol where k-fold parallel repetition does not decrease the error probability below some constant for any polynomial k (and where the communication complexity does not depend on k). The protocol has eight rounds and uses the universal arguments of Barak and Goldreich (2001). We also give another *four* round protocol relative to an oracle, unlike the artificial oracle of Bellare et al., we just need a generic group. This group can then potentially be instantiated with some real group satisfying some well defined hardness assumptions (we do not know of any candidate for such a group at the moment).

1 Introduction

INTERACTIVE PROOFS. In a (single prover) interactive proof a prover P tries to convince a computationally bounded verifier V that their common input x is in a language L. The soundness of such a protocol is an upper bound on the error probability of V, i.e. the probability that V accepts P's claim, even though

S.P. Vadhan (Ed.): TCC 2007, LNCS 4392, pp. 86–102, 2007.

$x \notin L$. In order to lower the error probability one can repeat the interactive proof k times, where V accepts the claim if it accepts in all k runs. The protocol can be repeated either *sequentially*, here V and P start the ith run of the protocol only after finishing the $(i-1)$th, or *in parallel*. Although the computational and communication complexity of parallel and sequential repetition is the same, parallel repetition has the big advantage of not increasing the round complexity. For single prover interactive proofs, sequential and parallel repetition reduce the error at an exponential rate: if a protocol with soundness ϵ is repeated k times sequentially or in parallel, the error probability drops to ϵ^k.

In general, parallel repetition is more problematic than sequential repetition. For example: parallel repetition does not preserve the zero-knowledge property of a protocol [8], and there are two-prover proofs where running the proof twice in parallel does not decrease the error at all [6]. On the positive side, Raz [12] shows that k-fold parallel repetition of a two-prover two-round proof system with soundness ϵ does decrease the error to $\epsilon^{\alpha k}$ where $\alpha > 0$ is some constant depending only on the proof system.

COMPUTATIONAL SOUNDNESS. Interactive *arguments* are defined like interactive proofs, but where the soundness of the protocol only holds against computationally bounded provers. Damgård and Pfitzmann [4] show that sequential repetition lowers the error probability of arguments at an exponential rate.

Bellare et al. [2] show that parallel repetition reduces the error of computationally sound protocols with three rounds or less at an exponential rate. On the negative side, they give, for any k, a four round protocol where k-fold parallel repetition does not decrease the error at all. The communication complexity of this protocol is linear in k, which leaves open the possibility that parallel repetition does reduce the error if the communication complexity is not allowed to depend on the number of repetitions. This is a possibility one should consider, as the before-mentioned constant α in Raz's theorem is inverse in the communication complexity of the protocol, and this dependence is necessary [7]. So for protocols where the communication complexity grows linearly in k, parallel repetition does not imply error reduction at all for two-prover two-round proofs. Observing that the four-round protocol of Bellare et al. can be restated as a two-round two-prover protocol (without loosing the property that parallel repetition does not decrease the error), makes the possibility that unbounded communication complexity is necessary here even more likely.

Noticing this possibility, Bellare et al. propose another four-round protocol with fixed communication complexity, which has the property that *relative to an oracle* repeating the protocol any polynomial number of times in parallel, does not decrease the error. This shows that there is no "black-box" error reduction theorem for this protocol. Bellare et al. see this result as evidence that parallel repetition does not decrease the error of computationally sound protocols. Another interpretation of this result could be that parallel repetition does always reduce the error, and the reason why there's no proof of this is that such a proof would require non black-box techniques. We show that under standard assumptions the interpretation of Bellare et al. is indeed correct for protocols with eight rounds or more, and we give much stronger evidence that this is also true for protocols with four rounds.

THE VERIFIER'S SECRET. Except for Section 3, throughout we consider protocols where the verifier holds no secret and thus its strategy is efficiently computable. The reason is that otherwise there are trivial protocols where parallel repetition does not decrease the error as observed by Bellare et al. [2], we extend their observation in Section 3.

1.1 Our Contribution

For n a security parameter, we present the first computationally sound protocol where $k(n)$-fold parallel repetition does not decrease the error for any polynomial $k(\cdot)$. To achieve this we start with the protocol of Bellare et al. whose k-fold parallel repetition does not decrease the error, but we modify it such that k is chosen by the prover (in particular, if the prover has to run the protocol $k(n)$ times in parallel, he can set $k = k(n)$). As in this protocol the the length of the second message from the prover to the verifier is linear in k, we must allow a verifier V_{super} which runs in super-polynomial time, in order for the protocol to work for *any* polynomial $k(\cdot)$. We then transform this protocol into one with a fixed polynomial time verifier V_{poly} using the universal arguments due to Barak and Goldreich [1]. Loosely speaking, the long message is replaced by a hash value, which then is followed by an interactive proof to V_{poly} which shows that V_{super} would have accepted the message. We get the following theorem.

Theorem 1. *There exists an overwhelmingly complete eight round protocol with error probability $3/4$ such that $k(\cdot)$-fold parallel repetition does not reduce its error probability below $1/17$ for any polynomially bounded $k(\cdot)$, under the assumption that collision-free family of hash functions and CCA2-secure cryptosystem with respect to superpolynomial adversaries exists.*

Unfortunately, the use of an universal argument increases the round complexity of the protocol from the optimal four to eight.

In Section 5 we propose a new four round protocol relative to an oracle, where $k(n)$-fold parallel repetition does not decrease the error for any polynomial $k(\cdot)$. Unlike the artificial oracle used by Bellare et al., we only need a generic group which potentially can be instantiated with a concrete group satisfying some clearly defined hardness assumptions (basically, it must be hard to compute the inverse of a random element).

More precisely, let $p \in [2^n, 2^{n+1}]$ be a randomly chosen prime, let $\phi' : \mathbb{Z}_p \to [0, 2K - 1]$ be a randomly chosen injection and $\phi(x) \stackrel{\text{def}}{=} \phi'(x \bmod p)$ its natural extension to the whole of $\mathbb{Z}$. Then denote by $\mathcal{O}$ the oracle defined by $\mathcal{O}(x) = \phi(x)$ and $\mathcal{O}(X, Y) = \phi(\phi^{-1}(X) + \phi^{-1}(Y))$ if $X, Y \in \phi(\mathbb{Z}_p)$ and $\perp$ otherwise. We prove the following theorem.

Theorem 2. *There exists an overwhelmingly complete four round protocol relative the oracle $\mathcal{O}$ with error probability $1/2 + \mathsf{negl}(n)$ such that $k(\cdot)$-fold parallel repetition does not reduce its error probability below $1/2 + \mathsf{negl}(n)$ for any polynomially bounded $k(\cdot)$.*

2   Preliminaries

2.1   Notation

We use $\mathbb{Z}$ to denote the integers and $\mathbb{Z}_p$ to denote the integers modulo p. We use log to denote the logarithm in base two. We denote by TM the set of Turing machines. We denote by PT and PT* the set of uniform and non-uniform polynomial time Turing machines respectively. The corresponding sets of oracle machines are denoted by adding a superscript, e.g. PT$^{\mathcal{O}}$. We use n to denote the security parameter, and say that a function $\epsilon(n)$ is negligible if for every constant c there exists a constant n_0 such that $\epsilon(n) < n^{-c}$ for $n > n_0$. We use $\mathsf{negl}(n)$ to denote a fixed but unspecified non-negative negligible function. A function $f(n)$ is overwhelming if $1 - f(n)$ is negligible. If $\nu : \mathbb{N} \to \mathbb{N}$ is a function we denote by PT$^*_\nu$ the set of non-uniform Turing machines that executes in time $\nu(n)p(n)$ for some polynomial p. We say that ν is polynomial-time computable if there exists a Turing machine M_ν that on input $x \in \{0,1\}^n$ outputs $\nu(x)$ using at most $p(n)$ steps, for some polynomial p.

We say that a family of hash functions is PT$^*_\nu$-collision-free if it is collision-free with respect to adversaries in PT$^*_\nu$. Similarly, we say that a cryptosystem is PT$^*_\nu$-CCA2-secure, if it is CCA2-secure with respect to adversaries in PT$^*_\nu$.

We denote by $\langle V(x), P(y) \rangle(z)$ the output of V on private input x and common input z after interacting with P on private input y and common input z. We denote by kV the sequential repetition of k copies of V and we denote by V^k the parallel repetition of k copies of V. In both cases identical private and common inputs are given to each instance and the combined verifier accepts if and only if all instances accept.

2.2   Computationally Sound Protocols

We consider the setting introduced in [2]. Two parties, a prover P and a verifier V, are communicating. They are both given an initial context $\lambda \in \{0,1\}^*$ and the length of this string serves as the security parameter. The initial context could be the output of another protocol or some string in a set-up assumption. Since we do not mention λ explicitly below, we replace it by the security parameter in unary representation 1^n, but our results hold in the more general setting.

Both parties are also given a common input x which is generated together with some secret information w by a probabilistic polynomial time instance generator I that is given input 1^n. The secret information w is given to P at the start of the protocol.

2.3 Universal Arguments

Barak and Goldreich [1] introduce the notion of universal arguments as a special variant of Micali's computationally sound proofs [9]. They define the relation $\mathcal{R}_U$ as the set of pairs $((M, x, t), w)$ such that the Turing machine M outputs 1 on input (x, w) within t steps. Denote by $T_M(x, w)$ the number of steps made by M on input (x, w). A key property of their definition is that t is given in binary. We are mainly interested in two properties of universal arguments: (1) the complexity of the verifier depends only on the size of the common input and not on the size of the witness, and (2) the witness used by the prover can be extracted in a weak sense. The actual definition given by Barak and Goldreich [1] is duplicated below.

Definition 1 (Universal Argument). *A universal-argument system is a pair of strategies, denoted (P, V) that satisfies the following properties:*

1. **Efficient verification.** *There exists a polynomial p such that for any $y = (M, x, t)$, the total time spent by the probabilistic verifier strategy V, on common input y, is at most $p(|y|)$. (In particular, all messages exchanged in the protocol have length smaller than $p(|y|)$.)*
2. **Completeness by a relatively efficient prover.** *For every $((M, x, t), w)$ in $\mathcal{R}_U$ we have $\Pr[(P(w), V)(M, x, t) = 1] = 1$. Furthermore, there exists a polynomial p such that the total time spent by $P(w)$ on common input (M, x, t) is at most $p(T_M(x, w)) \leq p(t)$.*
3. **Computational soundness.** *For every polynomial-size circuit family $\{P_n^*\}_{n \in \mathbb{N}}$, and every $(M, x, t) \in \{0, 1\}^n \setminus \mathcal{R}_U$ $\Pr[(P_n^*, V)(M, x, t) = 1] < \mu(n)$ for some negligible function $\mu(n)$.*
4. **Weak proof of knowledge.** *For every positive polynomial p there exists a positive polynomial p' and a probabilistic polynomial-time oracle machine E such that for every polynomial-size circuit family $\{P_n^*\}_{n \in \mathbb{N}}$, and every sufficiently long $y = (M, x, t) \in \{0, 1\}^*$, if $\Pr[(P_n^*, V)(y) = 1] > \frac{1}{p(|y|)}$, then*

$$\Pr_r[\exists w \cap \{0, 1\}^t \; \forall i \in \{1, \ldots, t\} : (x, w) \in \mathcal{R}_U \wedge E_r^{P_n^*}(y, i) = w_i] > \frac{1}{p'(|y|)} \; .$$

Theorem 3 ([1]). *If there exists a family of collision-free hash functions, then there exists universal arguments with 4 rounds.*

3 When the Verifier Holds a Secret

In this section we show that parallel repetition does not decrease the error probability of computationally sound protocols when the verifier gets any private information.

Bellare et al. [2] give the following simple example of such a protocol: The common input is an RSA modulus $N = pq$ and the secret of the verifier is the factors p and q. The verifier flips a coin. If it is heads it gives the factors to the prover and otherwise not. It accepts if the prover's reply is (p, q). An even simpler example is the following one-round protocol: The verifier has a secret bit b, and accepts if the message from the prover is b.

Clearly, parallel repetition does not decrease the error probability for the two protocols above (in fact, for the first protocol it increases), but neither does sequential repetition. This leaves open the interesting possibility that parallel repetition does always decrease the error probability of computationally sound protocols where the verifier can hold a secret, for all protocols where sequential repetition does reduce the error. Below we show that this is not the case by giving a natural (four-round) protocol that when repeated sequentially lowers the error probability, but if repeated in parallel gives error probability essentially one. Here $\mathcal{CS} = (\mathsf{Kg}, \mathsf{Enc}, \mathsf{Dec})$ denotes a public key cryptosystem.

Protocol 1 (Don't Do In Parallel (Verifier Holds a Secret))
Common input: Public key pk.
Private input to both prover and verifier: Private key sk.

1. V *chooses* $b \in \{0, 1\}$ *randomly, computes* $B = \mathsf{Enc}_{pk}(b)$, *and hands* B *to* P.
2. P *chooses* $c \in \{0, 1\}$ *randomly, computes* $C = \mathsf{Enc}_{pk}(c)$, *and hands* C *to* V.
3. *If* $C \neq B$, *then* V *hands* $c = \mathsf{Dec}_{sk}(C)$ *to* P *and otherwise* $\perp$.
4. P *computes* $b' = \mathsf{Dec}_{sk}(B)$ *and hands* b' *to* V.
5. V *accepts if and only if* $b = b'$.

The next two propositions are proved in Appendix A for completeness.

Proposition 1 (Single Instance). *The protocol is overwhelmingly complete and has 4 rounds. If the cryptosystem $\mathcal{CS}$ is CCA2-secure, then for every prover $P^* \in \mathrm{PT}^*$:* $\Pr_{(pk,sk),s}[\langle V_s(sk, pk), P^*(pk)\rangle = 1] < \frac{1}{2} + \mathsf{negl}(n)$.

Proposition 2 (Sequential Repetition). *If the cryptosystem $\mathcal{CS}$ is CCA2-secure, then for every polynomially bounded $k(\cdot)$ and every prover $P^* \in \mathrm{PT}^*$:* $\Pr_{(pk,sk),s}[\langle kV_s(sk, pk), P^*(pk)\rangle = 1] < (\frac{1}{2})^k + \mathsf{negl}(n)$.

Proposition 3 (Parallel Repetition). *For every polynomially bounded $k(\cdot)$ there exists a prover $P^* \in \mathrm{PT}$ such that* $\Pr_{(pk,sk),s}[\langle V_s^k(sk, pk), P^*(pk)\rangle = 1] \geq 1 - \mathsf{negl}(n)$.

Proof. The prover P^* does the following. It waits for B_i from V_i. Then it defines $C_i = B_{i+1 \bmod k}$ and hands it to V_i. With overwhelming probability $C_i \neq B_i$, so it is given $b'_{i+1 \bmod k} = \mathsf{Dec}_{sk}(C_i)$ from V_i. Then it returns b'_i to V_i. Thus, with overwhelming probability $b_i = b'_i$, each V_i accepts, and V^k accepts with overwhelming probability as well, since k is polynomial. $\square$

4   When the Verifier Holds No Secret

From now on we consider computationally sound protocols where the verifier holds no secret. In this section we give an eight-round computationally sound protocol where parallel repetition does not decrease the error.

The Example of Bellare et al. Before we give our counter example we recall the counter example given by Bellare et al. [2] on which our example is based. The idea of the protocol is to explicitly allow the prover to make several instances of it dependent if run in parallel.

Protocol 2 (Don't Do In k-Parallel, [2])
Common input: Public key pk.
Private input to prover: Private key sk.

1. V *chooses* $b \in \{0,1\}$ *and* $r \in \{0,1\}^n$ *randomly, computes* $B = \mathsf{Enc}_{pk}(b,r)$, *and sends* B *to* P.
2. P *computes* $b = \mathsf{Dec}_{sk}(B)$. *Then it chooses* $b'_i \in \{0,1\}$ *and* $r'_i \in \{0,1\}^n$ *for* $i = 1, \ldots, k-1$ *randomly under the restriction that* $b = \bigoplus_{i=1}^{k-1} b'_i$, *computes* $C_i = \mathsf{Enc}_{pk}(b'_i, r'_i)$, *and hands* $(C_1, \ldots, C_{k-1})$ *to* V.
3. V *hands* (b,r) *to* P.
4. P *hands* $((b'_1, r'_1), \ldots, (b'_{k-1}, r'_{k-1}))$ *to* V.
5. V *accepts if* $C_i = \mathsf{Enc}_{pk}(b'_i, r'_i)$, $B \neq C_i$, *and* $\bigoplus_{i=1}^{k-1} b'_i = b$.

We have modified the protocol slightly to be more consistent with our counter example below. In the original the test is $b \neq \bigoplus_{i=1}^{k-1} b'_i$ and this is needed if given a ciphertext the cryptosystem allows construction of a new ciphertext of an identical plaintext. If we require that the cryptosystem used in the protocol is CCA2-secure this is not an issue.

Intuitively, if a single instance of the protocol is run, then a prover without access to sk can only convince the honest verifier with probability $1/2$, since it must commit itself to a guess $\bigoplus_{i=1}^{k-1} b'_i$ of b before receiving (b,r) and the cryptosystem is non-malleable (recall that CCA2-security implies non-malleability). On the other hand, if k instances of the protocol are run in parallel, then the prover can send the tuple $(C_{i,1}, \ldots, C_{i,k-1}) = (B_1, \ldots, B_{i-1}, B_{i+1}, \ldots, B_k)$ to V_i and then either all verifier instances accept or all verifier instances fail, the first event occurring with probability at least $1/2$. If there are fewer than k instances the remaining C_i's can be defined as ciphertexts of zero.

Why the Example is Unsatisfactory. The example requires that the complexity of the verifier in each instance grows linearly with the number of instances. In other words, the example does not imply that k'-parallel repetition of the protocol for $k' > k$ does not lower the error probability.

This deficiency motivated Bellare et al. [2] to consider if there exists any analytical method, i.e, an error-reduction procedure, whereby one can show that the error probability is lowered by the parallel repetition of a protocol. They prove that there exists no such *black-box* error-reduction procedure. Although we agree that this result is a strong indication that there exists no error-reduction procedure at all, it does not preclude the possibility of a non-black-box error-reduction procedure.

4.1 Our Counter Example

The idea of our counter example is to reduce the complexity of the verifier by making the long messages submitted by the prover in Bellare et al's protocol implicit. More precisely, we let the prover choose k on the fly, and hand a hash value of the list of ciphertext $(C_1, \ldots, C_{k-1})$ instead of sending them explicitly. It also sends a hash value of $(b'_1, r'_1), \ldots, (b'_{k-1}, r'_{k-1})$ instead of sending them explicitly. The problem with this is of course that now the verifier can not perform the original verification. To solve this problem without increasing the complexity of any instance of the verifier the prover proves using universal arguments [1] that it knows correct preimages of the hash values. For technical reasons we replace addition modulo 2 by addition modulo 17. The reader may think of 17 as some constant to be defined in the proof such that the theorem holds.

We assume that there exists a cryptosystem that is chosen ciphertext secure in the sense of Rackoff and Simon [11] against adversaries in PT^*_ν where $\nu(\cdot)$ is a polynomially computable superpolynomial function (the reader can think of $\nu(n)$ as $n^{\log n}$). It should be possible to construct such a scheme from any family of trap-door permutations secure against adversaries in PT^*_ν following Dolev, Dwork, and Naor [5] or Sahai [13], but we are not aware of any explicit proof of this. We also assume the existence of a family of hash functions that is collision-free against adversaries in PT^*_ν.

Denote by $\mathcal{R}_h$ the relation consisting of pairs $((B, H, h, k), (C_1, \ldots, C_{k-1}))$ such that $h = H(C_1, \ldots, C_{k-1})$ and $B \neq C_i$ for $i = 1, \ldots, k - 1$. Denote by $\mathcal{R}_a$ the relation consisting of pairs $((pk, H, h, b, a, k), ((b'_1, r'_1), \ldots, (b'_{k-1}, r'_{k-1})))$ such that $b = - \sum_{i=1}^{k-1} b'_i \bmod 17$, $a = H((b'_1, r'_1), \ldots, (b'_{k-1}, r'_{k-1}))$, and

$$h = H(\mathsf{Enc}_{pk}(b'_1, r'_1), \ldots, \mathsf{Enc}_{pk}(b'_{k-1}, r'_{k-1})).$$

Denote by $M_{\mathcal{R}_h}$ a canonical Turing machine that decides $\mathcal{R}_h$ in polynomial time in n and k and correspondingly for $M_{\mathcal{R}_a}$.

Protocol 3 (Don't Do In Parallel)
Common input: Public key pk and collision-free hash function H.
Private input to prover: Private key sk.

1. *V chooses $b \in \mathbb{Z}_{17}$ and $r \in \{0, 1\}^n$ randomly, computes $B = \mathsf{Enc}_{pk}(b, r)$, and sends B to P.*
2. *P computes $b' = \mathsf{Dec}_{sk}(B)$. Then it chooses $r' \in \{0, 1\}^n$ randomly, computes $C = \mathsf{Enc}_{pk}(b', r')$ and $h = H(C)$, and hands (h, k, t_h) to V, where $k = 1$ and $t_h = T_{M_{\mathcal{R}_h}}((B, H, h, k), C)$.*
3. *If $k > \nu(n)$ or $t_h > \nu(n)$, then V outputs 0. Otherwise P and V execute a universal argument on common input $y_h = (M_{\mathcal{R}_h}, (B, H, h, k), t_h)$ and private input $w_h = C$ to the prover.*
4. *If V accepts the universal argument, then it hands (b, r) to P. Otherwise it outputs 0.*
5. *P computes $a = H(b', r')$ and $t_a = T_{M_{\mathcal{R}_a}}((pk, H, h, b, a, k), (b', r'))$ and hands (a, t_a) to V.*

6. *If $t_a > \nu(n)$, then V outputs 0. Otherwise P and V execute a universal argument on common input $y_a = (M_{\mathcal{R}_a}, (pk, H, h, b, a, k), t_a)$ and private input $w_a = (b', r')$.*
7. *If V accepts the universal argument it outputs 1 and otherwise 0.*

We stress that k, t_h, and t_a are encoded in binary. Thus, even though the adversary can choose t_h and t_a larger than any polynomial (as they only have to be smaller than the superpolynomial $\nu(n)$), the complexity of the verifier can still be bounded by some fixed polynomial in n as it is polynomial only in n and $\log(\nu(n))$. This means that also k can be larger than any polynomial. This freedom is needed since we do not want to put any fixed polynomial bound on the "width" of the parallel repetition. On the other hand this is what forces us to consider superpolynomial adversaries. The problem is that when reducing soundness of the protocol to breaking the cryptosystem or the collision-freeness of the hash function we need to extract the ciphertexts $C_1, \ldots, C_{k-1}$, but we can not guarantee that a polynomial time adversary can not use implicit such values, which could give a superpolynomial witness during extraction.

Proposition 4 (Single Instance). *The protocol is overwhelmingly complete and has 8 rounds. Let $\nu : \mathbb{N} \to \mathbb{N}$ be a fixed superpolynomial and polynomial-time computable function, let the hash function be PT^*_ν-collision-free, and let $\mathcal{CS}$ be PT^*_ν-CCA2-secure. Then for every prover $P^* \in \mathrm{PT}^*_\nu$ for all sufficiently large n: $\Pr_{(pk,sk),s}[\langle V_s(pk), P^*(pk)\rangle = 1] < \frac{3}{4}$.*

The relation between the constants $3/4$ and 17 is essentially that in the reduction we need to "split" the success probability of the adversary twice, giving a factor $1/8$, and we need to extract, giving a factor $(3/4)^2$. Thus, the resulting adversary has success probability at least $1/16$, which is bigger than $1/17$.

Before we prove the above theorem we show that its error probability does not decrease if repeated in parallel. We stress that each instance V_i of the verifier V^k has the same complexity both in terms of computation and communication independently of k.

Proposition 5 (Parallel Repetition). *For every polynomially bounded $k(\cdot)$ there is a prover $P^* \in \mathrm{PT}$ such that $\Pr_{(pk,sk),s}[\langle V_s^k(pk), P^*(pk)\rangle = 1] > \frac{1}{17} - \mathsf{negl}(n)$.*

Proof. We define the prover P^* interacting with V^k, i.e., the parallel repetition of k instances of V, as follows. Given the cryptotexts B_i from all V_i it defines $(C_{i,1}, \ldots, C_{i,k-1}) = (B_1, \ldots, B_{i-1}, B_{i+1} \ldots, B_k)$. Then it executes the first universal argument honestly. When it gets (b_i, r_i) from V_i it defines

$$((b'_{i,1}, r'_{i,1}), \ldots, (b'_{i,k-1}, r'_{i,k-1}))$$
$$= ((b_1, r_1), \ldots, (b_{i-1}, r_{i-1}), (b_{i+1}, r_{i+1}), \ldots, (b_k, r_k)) .$$

If $\sum_{i=1}^k b_i \neq 0 \bmod 17$ it fails and stops. Otherwise it executes the rest of the protocol honestly. With probability $\frac{1}{17}$ we have $\sum_{i=1}^k b_i = 0 \bmod 17$ and the probability that $B_i = B_j$ for some $i \neq j$ is negligible. Thus, it follows that the prover succeeds at least with probability $\frac{1}{17} - \mathsf{negl}(n)$. □

Proof (Proposition 4). Completeness follows by inspection. Although the naive implementation of the protocol has more than eight rounds, it is easy to see that one can combine the rounds of the universal argument with the main protocol and achieve eight rounds.

Suppose there exists a prover $P^* \in \text{PT}_\nu^*$ with $\text{Pr}_{(pk,sk),s}[\langle V_s(pk), P^*(pk)\rangle = 1] = \delta \geq \frac{3}{4}$ for n in some infinite index set $\mathcal{N}$. Consider the following experiment. The adversary is given a public key pk and a challenge ciphertext $B = \text{Enc}_{pk}(b, r)$ where b is chosen randomly in $\mathbb{Z}_{17}$. Then it may ask any decryption queries except B and then output a guess b' of b. A simple averaging argument implies that if $|\text{Pr}[b' = b] - 1/17|$ is non-negligible, then the cryptosystem is not CCA2-secure.

The CCA2-Adversary. We define an adversary $A \in \text{PT}_\nu^*$ against the above experiment run with the cryptosystem $\mathcal{CS}$ as follows. It accepts a public key pk and a challenge $B = \text{Enc}_{pk}(b, r)$, where b is chosen randomly in $\mathbb{Z}_{17}$. Then it generates a collision-free hash function H and simulates the honest verifier V except that it instructs it to use B instead of generating this ciphertext as in the protocol. If t_h is too large and V outputs 0, then A outputs 0. The simulation proceeds until the first universal argument has been executed. Then A invokes the knowledge extractors of the universal argument to extract $C_1, \ldots, C_{k-1}$ such that $((B, H, h, k), (C_1, \ldots, C_{k-1})) \in \mathcal{R}_h$. More precisely, it tries a random r and computes $(C_1, \ldots, C_{k-1}) = (E_r^{P^*}(y_h, 1), \ldots, E_r^{P^*}(y_h, k-1))$, where $y_h = (M_{\mathcal{R}_h}, (B, H, h, k), t_h)$ and $E_r^{P^*}$ is the extraction algorithm guaranteed by the weak proof of knowledge property of universal arguments. If $w_h = (C_1, \ldots, C_{k-1})$ does not satisfy $(y_h, w_h) \in \mathcal{R}_U$ it tries again with a fresh r. This procedure is repeated at most $g_h(n)$ times, where $g_h(n)$ is a polynomial to be determined in the analysis below. If extraction fails it outputs 0. Otherwise it asks its decryption oracle for $b'_i = \text{Dec}_{sk}(C_i)$ for $i = 1, \ldots, k-1$ and outputs as its guess of b the value $b' = -\sum_{i=1}^{k-1} b'_i \mod 17$.

We want to show that the CCA2-security of $\mathcal{CS}$ is broken by A, since this contradicts the security of $\mathcal{CS}$. To do that we must argue that extraction succeeds from the first universal argument, but this is not sufficient. The problem is that it is conceivable that the adversary uses one set of ciphertext as a preimage of h in the first universal argument and another set in the second. Intuitively, the collision-freeness of the hash function prohibits this, but we must prove that this is so.

Divide the randomness s used by the verifier into three parts: s_B is used to form B, s_h is used in the first universal argument, and s_a is used in the second universal argument. Denote by S_{good} the set of tuples (H, pk, sk, s_B) such that

$$\Pr_{s_h, s_a}[\langle V_{(s_B, s_h, s_a)}(H, pk), P^*(H, pk)\rangle = 1] \geq \delta/2 .$$

An averaging argument implies $\text{Pr}[(H, pk, sk, s_B) \in S_{good}] \geq \delta/2$. Note that the common input $y_h = (M_{\mathcal{R}_h}, (B, H, h, k), t_h)$ is defined by (H, pk, sk, s_B).

Claim 1. For every $f > 0$ there is a polynomial $g_h(n)$ such that the probability that A fails to extract w_h such that $(y_h, w_h) \in \mathcal{R}_U$ on a common input y_h induced by $(H, pk, sk, s_B) \in S_{good}$ is bounded by $\delta 2^{-f}$.

Proof. From the weak proof of knowledge property of a universal argument follows that there exists a positive polynomial $p'(\cdot)$ such that

$$\Pr_r[\exists w_h \cap \{0,1\}^t \ \forall i \in \{1,\dots,t\} : (y_h, w_h) \in \mathcal{R}_U \wedge E_r^{P_n^*}(y_h, i) = w_{h,i}] > \frac{1}{p'(|y_h|)} .$$

for common inputs y_h induced by $(H, pk, sk, s_B) \in S_{good}$. Thus, for such common inputs the expected number of repetitions needed to extract a witness is bounded by $p'(|y_h|)$. If we define $g_h(n) = (2^f/\delta)p'(|y_h|)$ it follows from Markov's inequality that extraction fails with probability bounded by $\delta 2^{-f}$ for such inputs.   □

We conclude from the union bound that the probability that $(H, pk, sk, s_B) \in S_{good}$ and A succeeds to extract w_h such that $(y_h, w_h) \in \mathcal{R}_U$ is at least $(1/2 - 2^{-f})\delta$. Then we set $c_1 = 1/2 - 2^{-f}$ and note that we by choosing $f > 0$ appropriately may set $c_1 < 1/2$ arbitrarily close to $1/2$.

A Hypothetical Machine. Unfortunately, the above claim says nothing about the probability that the negative sum (modulo 17) of the plaintexts of the extracted $C_1, \dots, C_{k-1}$ equal the plaintext of B. Intuitively, the problem is that the prover could use one H-preimage of h in the first universal argument and another one in the second, but this should of course never happen due to the collision-freeness of H.

Denote by A_C the machine that simulates A until $C_1, \dots, C_{k-1}$ are extracted from the first universal argument, or until it outputs 0. Then it chooses s_a randomly and continues the simulation of the interaction of V and P^* until P^* hands (a, t_a) to V. Then it repeatedly, at most $g_a(n)$ times, invokes the extractors of the second universal argument with fresh randomness in the hope to extract $w_a = ((b'_1, r'_1), \dots, (b'_{k-1}, r'_{k-1}))$ such that $(y_a, w_a) \in \mathcal{R}_U$, and then outputs (w_h, w_a). Otherwise it outputs 0.

Denote by S'_{good} the set of tuples (H, pk, sk, s_B, s_h) such that $(H, pk, sk, s_B) \in S_{good}$ and

$$\Pr_{s_a}[\langle V_{(s_B, s_h, s_a)}(H, pk), P^*(H, pk)\rangle = 1] \geq \delta/4 .$$

An averaging argument implies that

$$\Pr_{s_h}[(H, pk, sk, s_B, s_h) \in S'_{good} \mid (H, pk, sk, s_B) \in S_{good}] \geq \delta/4 .$$

Claim 2. For every $f' > 0$ there is a polynomial $g_a(n)$ such that the probability that A_C fails to extract w_a such that $(y_a, w_a) \in \mathcal{R}_U$ on a common input y_a induced by $(H, pk, sk, s_B, s_h) \in S'_{good}$ is bounded by $\delta 2^{-f'}$.

Proof. This follows mutatis mutandi from the proof of the previous claim.   □

We conclude that the probability that A_C succeeds to extract w_a where $(y_a, w_a) \in \mathcal{R}_U$ conditioned on $(H, pk, sk, s_B) \in S_{good}$ is at least $(1/4 - 2^{-f'})\delta$. We define $c_2 = 1/4 - 2^{-f'}$ and note that we by choosing $f' > 0$ appropriately can set $0 < c_2 < 1/4$ arbitrarily close to $1/4$.

Claim 3. *The probability that the output (w_h, w_a) contains a collision for H, i.e., it satisfies $(C_1, \ldots, C_{k-1}) \neq (\mathsf{Enc}_{pk}(b'_1, r'_1), \ldots, \mathsf{Enc}_{pk}(b'_{k-1}, r'_{k-1}))$, conditioned on $(H, pk, sk, s_B) \in S_{good}$ is negligible.*

Proof. If this was not the case we could define A'_C as the adversary that takes a description H of a hash function as input and simply simulates A_C and outputs $(C_1, \ldots, C_{k-1})$ and $(\mathsf{Enc}_{pk}(b'_1, r'_1), \ldots, \mathsf{Enc}_{pk}(b'_{k-1}, r'_{k-1}))$. It would break the collision-freeness of H with non-negligible probability. □

Conclusion of Proof of Proposition. From our claims follow that the probability that A_C outputs (w_h, w_a) such that

$$(C_1, \ldots, C_{k-1}) = (\mathsf{Enc}_{pk}(b'_1, r'_1), \ldots, \mathsf{Enc}_{pk}(b'_{k-1}, r'_{k-1}))$$

and $b = -\sum_{i=1}^{k-1} b'_i \bmod 17$ is at least $(c_1\delta)(c_2\delta) - \mathsf{negl}(n) \geq c_3\delta^2 > \frac{1}{16}$, where the constant $0 < c_3 < 1/8$ may be chosen arbitrarily close to $1/8$. This concludes the proof. □

5 Parallel Repetition Relative to a Generic Group

In the previous section we gave – under standard assumptions – an eight-round protocol with constant communication complexity where parallel repetitions does not decrease the error. In this section we give such a protocol with optimal four rounds relative to a generic group oracle.

5.1 The Model

A generic group is a group where the group elements are encoded by random strings. Access to the encoding and the group operation are provided by a public oracle $\mathcal{O}$. This model was put forward by Nechaev [10] and extended by Shoup [14] to prove lower bounds on the running time of the best generic algorithms to solve the discrete logarithm and related problems. An algorithm is called generic, if it does not use the representation of the group elements, for example the baby-step giant-step algorithm for the discrete logarithm problem is generic, but index-calculus is not. Damgård and Koprowski [3] extend this model to groups of unknown order, our model is very similar to theirs, the main difference is that our group oracle does not provide any efficient way to invert elements.[1] For ease of notation we write $N = 2^n$.

The distribution of the group oracle is defined as follows. A random prime p in the range $N < p < 2N$ and a random injection $\phi' : \mathbb{Z}_p \to [0, 2N - 1]$ are chosen.

---

[1] There is no efficient generic algorithm to find the inverse of an element if a large prime divides the (unknown) group order. In [3] the oracle explicitly provides the operation of inverting elements, the reason is that [3] wanted to prove lower bounds on the hardness of a problem in the RSA-group, where there exists an efficient (non-generic) algorithm for inversion (Extended Euclid).

Let $\phi(x) \overset{\text{def}}{=} \phi'(x \bmod p)$ denote the natural extension of ϕ' to the whole of $\mathbb{Z}$. To find the encoding of an element the oracle is called with a single argument, i.e., we define $\mathcal{O}(x) = \phi(x)$. In addition to providing encodings, the oracle can be called with two arguments from $\phi(\mathbb{Z})$ to find their product, i.e, we define $\mathcal{O}(X, Y) = \phi(\phi^{-1}(X) + \phi^{-1}(Y))$ if $X, Y \in \phi(\mathbb{Z})$ and $\bot$ otherwise. As mentioned above, unlike [3] our oracle does not provide the inverse operation $\phi(-x \bmod p)$ from $\phi(x)$, in fact, for our proof it is necessary that computing $\phi(-x \bmod p)$ given $\phi(x)$ is hard.

We will often have to sample a random element from the range of $\phi(\mathbb{Z})$, unfortunately we cannot efficiently sample a uniformly random one, as we do not know p. We thus use the following observation.

Observation 1. If x is uniformly distributed over $[0, N^2]$, then $\phi(x)$ is statistically close to the uniform distribution over $\phi(\mathbb{Z})$ for every $\mathcal{O}$ with $N < p < 2N$.

We use a polynomial time computable predicate $\tau : [0, 2N - 1] \to \{0, 1\}$ such that $|\Pr_{X \in \phi(\mathbb{Z})}[\tau(X) = 1] - 1/2|$ is negligible. A simple way[2] to construct such a predicate is to set $\tau(x) = 1 \iff x > N$. Due to the random choice of ϕ it is not hard to see that it has the required property with overwhelming probability over the choice of ϕ. Below we assume that $\Pr_{X \in \phi(\mathbb{Z})}[\tau(X) = 1] = 1/2$ to simplify the exposition.

5.2   Our Counter Example

We present a protocol which can be seen as an interactive proof that the prover P "knows" the group order p of the group oracle $\mathcal{O}$. If P indeed knows p, he can make the verifier V accept with probability 1.

Protocol 4 (Don't Do In Parallel (Generic Group))
Common input: A predicate τ.
Private input to prover: A predicate τ and a group order p.

1. *$V^{\mathcal{O}}$ chooses $x \in [0, N^2]$ randomly and sends $X = \phi(x)$ to $P^{\mathcal{O}}$.*
2. *$P^{\mathcal{O}}$ chooses any $y \in [0, 2N - 1]$ which satisfies $\tau(\phi(y)) = 1$, computes $Z = \phi(y - x)$, and sends Z to $V^{\mathcal{O}}$.*
3. *$V^{\mathcal{O}}$ sends x to $P^{\mathcal{O}}$.*
4. *$P^{\mathcal{O}}$ sends y to $V^{\mathcal{O}}$.*
5. *$V^{\mathcal{O}}$ accepts if and only if $\phi(y - x) = Z$ and $\tau(\phi(y)) = 1$.*

Note that if the prover computes the messages Z and y as shown in the protocol, then the verifier accepts. In Step 2 the prover can compute $\phi(-x \bmod p) = \phi((p - 1)x)$ from X in polynomial time using his knowledge of p.

---

[2] Here we are using the fact that the representation is random, i.e., our argument is not purely generic. A simple way to avoid this is to use the predicate $\tau'(x) = \tau(\mathrm{PRF}_s(x))$ for some pseudo-random function PRF and public seed s.

Proposition 6 (Single Instance). *The protocol is overwhelmingly complete and has 4 rounds. For every prover $P^{\mathcal{O},*} \in \mathrm{TM}^{\mathcal{O}}$ with total query complexity polynomially bounded in n we have $\Pr[\langle V^{\mathcal{O}}(\tau), P^{\mathcal{O},*}(\tau)\rangle = 1] < \frac{1}{2} + \mathsf{negl}(n)$, where the probability is taken over $\mathcal{O}$, τ, and the internal randomness of $V^{\mathcal{O}}$.*

Before we prove the proposition above we show that parallel repetition fails to reduce the error probability.

Proposition 7 (Parallel Repetition). *For every polynomially bounded $k(\cdot)$ there is a prover $P^{\mathcal{O},*} \in \mathrm{PT}^{\mathcal{O}}$ such that $\Pr[\langle (V^{\mathcal{O}})^k(\tau), P^{\mathcal{O},*}(\tau)\rangle = 1] > \frac{1}{2} - \mathsf{negl}(n)$, where the probability is taken over $\mathcal{O}$, τ, and the internal randomness of $V^{\mathcal{O}}$.*

Proof. The prover $P^{\mathcal{O},*}$ after receiving the messages $X_i = \phi(x_i), 1 \le i \le k$, simply computes $Z_i = \phi(\sum_{l \in \{1,\ldots,k\}\setminus\{i\}} x_l)$. Then when it receives $x_1, \ldots, x_k$ it computes $y_1 = \ldots = y_k = \sum_{l=1}^{k} x_l$. Note that Z_i can be computed by repeated queries to $\mathcal{O}$ using only $X_1, \ldots, X_k$. By construction we have $\phi(y_i - x_i) = \phi(\sum_{l=1}^{k} x_l - x_i) = \phi(\sum_{l \in \{1,\ldots,k\}\setminus\{i\}} x_l) = Z_i$ for $i = 1, \ldots, k$. The distribution of $\phi(y_1)$ is statistically close to uniform, and thus $\tau(\phi(y_1)) = 1$ with probability at least $1/2 - \mathsf{negl}(n)$. $\square$

Proof (Proposition 6). Let $Q_0 = X = \phi(x)$ and for $i > 0$ we denote by Q_i the answer to the ith oracle query $P^{\mathcal{O},*}$ makes to $\mathcal{O}$. We define $Q^i = \{Q_0, \ldots, Q_i\}$. Without loss of generality we assume that the replies received by $P^{\mathcal{O},*}$ are either of the form $Q_i = \mathcal{O}(q_i) = \phi(q_i)$ for some query $q_i \in \mathbb{Z}$ or $Q_i = \mathcal{O}(Q_j, Q_k) = \phi(\phi^{-1}(Q_j), \phi^{-1}(Q_k))$ for $j, k < i$. Note that then each reply Q_i is of the form $\phi(a_i + b_i x)$ where $P^{\mathcal{O},*}$ knows $a_i, b_i \in \mathbb{Z}$.[3] Denote by $\ell = \ell(n)$ the polynomial number of oracle queries made by the prover. Without loss we assume that $(a_i, b_i) \ne (a_j, b_j)$ for $i \ne j$, and that $Z \in Q^{\ell}$. The latter holds, since the probability that $\phi(y - x) = Z$ conditioned on $Z \notin Q^{\ell}$ is easily bounded by $1/(N - \ell)$. We now prove two claims from which the proposition follows.

Claim 4 (Hard to find multiple of p). *For any algorithm $M \in \mathrm{TM}^{\mathcal{O}}$ which makes at most $m - 1$ oracle queries, each of length at most m bits and where the output is of length at most m bits, we have $\Pr[M^{\mathcal{O}} = v \wedge p \mid v] \in O\left(m^2/N\right)$ (which is negligible for a polynomially bounded m).*

Proof. Denote by $\mathcal{P}(N)$ the set of primes in $[N, 2N]$, by the prime number theorem $|\mathcal{P}(N)| = \Theta(N/n)$.

The machine M can choose a sequence $t_1, t_2, \ldots, t_{m-1}$ of values in $\mathbb{Z}$ and ask the oracle for $T_1, T_2, \ldots, T_{m-1}$ where $T_i = \phi(t_i)$. Moreover we allow M an additional mth query which must be its output, i.e. $v = t_m$. The ith oracle query can be either of the form $T_i = \mathcal{O}(t_i)$ or $T_i = \mathcal{O}(T_j, T_k)$ for $j, k < i$ (then $t_i = t_j + t_k$). We can upper bound the size of any t_i as $\log(t_i) \le 2m$ as

---

[3] Here "knows" means that one can efficiently extract a_i, b_i given the queries that $P^{\mathcal{O},*}$ makes to $\mathcal{O}$.

follows: if the ith query is of the form $\mathcal{O}(t_i)$ then $\log(t_i) \leq m$ (as no query can be longer than m bits). If the query is of the form $\mathcal{O}(T_j, T_k)$, then $\log(t_i) \leq 1 + \max\{\log(t_j), \log(t_k)\}$, so for any $i \leq m, \log(t_i) \leq m + i \leq 2m$.

Let $t = \prod_{i=1}^{m} t_i$. Then we have $\log(t) \leq \sum_{i=1}^{m} \log(t_i) \leq 2m^2$. So at most $2m^2/n$ primes from $\mathcal{P}(N)$ divide t, and thus also $v = t_m$. The probability that p is one of those primes is at most $(2m^2/n)/|\mathcal{P}(N)| = \Theta(m^2/N)$.   □

The following claim is very similar to Theorem 1 in [14].

Claim 5 (x close to uniform). Let γ denote the view of the prover $P^{\mathcal{O}}$ after step 2. Then with overwhelming probability x is statistically close to uniform (over $[0, N^2]$) given γ.

Proof. The view γ contains, for some s, the oracle answers $Q_0, \ldots, Q_s$ and (a_i, b_i) for $i = 1, \ldots, s$. In fact, we prove the slightly stronger statement where p is also contained in γ. Recall that $Q_i = \phi(a_i + xb_i)$ and note that since $Q_0 = \phi(x)$ we have $a_0 = 0, b_0 = 1$. Let $a_i' = a_i \bmod p, b_i' = b_i \bmod p$.

For $(a_i', b_i') \neq (a_j', b_j')$ we have $\Pr[Q_i = Q_j] \leq 2/p$ as $a_i + b_i x = a_j + b_j x \bmod p$ for at most one x in each interval $t \leq x \leq t + p - 1$. Thus by the union bound, the probability that there is any nontrivial collision, i.e. $Q_i = Q_j$ for some $(a_i', b_i') \neq (a_j', b_j')$, is at most $\epsilon = s(s-1)/p$. So with overwhelming probability $1 - \epsilon$ there is no nontrivial collision, and conditioned on this event, x is uniformly random over at least a $1 - \epsilon$ fraction of $[0, N^2]$.   □

We can now conclude the proof of the proposition, for this we must show that

$$\Pr[\phi(y - x) = Z \wedge \tau(\phi(y)) = 1] - 1/2 < \mathsf{negl}(n) \ .$$

Let $Z = \phi(a_i + b_i x)$ for some a_i, b_i. Then $y = a_i + (b_i + 1)x \bmod p$ when $\phi(y - x) = Z$. By Claim 4 we can assume that $p \nmid (b_i + 1)$. By Claim 5 x is close to uniformly random for the prover at the point where he must choose a_i, b_i, thus $a_i + (b_i + 1)x \bmod p$ is close to uniformly random over $\mathbb{Z}_p$ (as $b_i + 1$ generates $\mathbb{Z}_p$ additively). This implies that $\Pr[\tau(\phi(y)) = 1] - 1/2$ is negligible, since $\Pr[\tau(\phi(u)) = 1]$ is negligibly close to $1/2$ if u is chosen randomly in $[0, N^2]$.   □

Acknowledgments

We thank Thomas Holenstein for fruitful discussions.

References

1. B. Barak and O. Goldreich. Universal arguments and their applications. *Electronic Colloquium on Computational Complexity (ECCC)*, (093), 2001.
2. M. Bellare, R. Impagliazzo, and M. Naor. Does parallel repetition lower the error in computationally sound protocols? In *38th IEEE Symposium on Foundations of Computer Science (FOCS)*, pages 374–383. IEEE Computer Society Press, 1997.

3. I. Damgård and M. Koprowski. Generic lower bounds for root extraction and signature schemes in general groups. In *Advances in Cryptology – Eurocrypt 2002*, volume 2332 of *Lecture Notes in Computer Science*, pages 256–271. Springer Verlag, 2002.
4. I. Damgård and B. Pfitzmann. Sequential iteration of interactive arguments and an efficient zero-knowledge argument for np. In *25th International Colloquium on Automata, Languages and Programming (ICALP)*, pages 772–783, 1998.
5. D. Dolev, C. Dwork, and M. Naor. Non-malleable cryptography. In *23rd ACM Symposium on the Theory of Computing (STOC)*, pages 542–552. ACM Press, 1991.
6. U. Feige. On the success probability of the two provers in one-round proof systems. In *Structure in Complexity Theory Conference*, pages 116–123, 1991.
7. U. Feige and O. Verbitsky. Error reduction by parallel repetition – a negative result. *Combinatorica*, 22(4):461–478, 2002.
8. O. Goldreich and H. Krawczyk. On the composition of zero-knowledge proof systems. *SIAM Journal on Computing*, 25(1):169–192, 1996.
9. Silvio Micali. Computationally sound proofs. *SIAM J. Comput.*, 30(4):1253–1298, 2000.
10. V.I. Nechaev. Complexity of a determinate algorithm for the discrete logarithm. *Mathematical Notes*, 55(2):165–172, 1994.
11. C. Rackoff and D. Simon. Non-interactive zero-knowledge proof of knowledge and chosen ciphertext attack. In *Advances in Cryptology – Crypto '91*, volume 576 of *Lecture Notes in Computer Science*, pages 433–444. Springer Verlag, 1991.
12. R. Raz. A parallel repetition theorem. *SIAM Journal on Computing*, 27(3):763–803, 1998.
13. A. Sahai. Non-malleable non-interactive zero-knowledge and adaptive chosen-ciphertext security. In *40th IEEE Symposium on Foundations of Computer Science (FOCS)*, pages 543–553. IEEE Computer Society Press, 1999.
14. V. Shoup. Lower bounds for discrete logarithms and related problems. In *Advances in Cryptology – Eurocrypt '97*, volume 1233 of *Lecture Notes in Computer Science*, pages 256–266. Springer Verlag, 1997.

A   Omitted Proofs

Proof (Proposition 1). Completeness is clear and the number of rounds follow by counting. Suppose the claim is false, i.e., there exists a prover P^* that succeeds with probability at least $1/2 + n^{-c}$ for n in some infinite index set $\mathcal{N}$. Denote by A the CCA2-adversary that proceeds as follows. It accepts a public key pk, hands the pair of messages $(0, 1)$ to the experiment, and waits for a challenge ciphertext B. Then it starts a simulation of the interaction between V and P^* on the common input pk and using B. If P^* sends $C \neq B$ to the verifier it invokes its decryption oracle to compute $c = \mathsf{Dec}_{sk}(C)$ and hands it back. Finally, it outputs the reply b' of P^* as its guess of the contents of B. It follows that A breaks the CCA2-security of $\mathcal{CS}$, since when the verifier accepts the guess b' equal the content of B. $\square$

Proof (Proposition 2). We use a subscript i with the elements in the ith sequential execution, i.e., we write $(B_i, C_i, c_i, b_i', b_i)$ for the values in the ith execution.

Denote by E_i the event that the verifier accepts in the ith instance of the protocol. Thus, we have $\Pr[E_i] = \Pr[b'_i = b_i] = \frac{1}{2} + \frac{1}{2}(\Pr[b'_i = 1 \mid b_i = 0] - \Pr[b'_i = 1 \mid b_i = 1])$.

Suppose there exists a constant c, an infinite index set $\mathcal{N}$, and a prover P^* such that $\Pr_{(pk,sk),s}[\langle kV_s(sk, pk), P^*(pk)\rangle = 1] \geq (\frac{1}{2})^k + n^{-c}$ for $n \in \mathcal{N}$ and fix such a security parameter n. Then we have

$$\Pr[E_1]\Pr[E_2 \mid E_1]\Pr[E_3 \mid E_2 \wedge E_1]\cdots\Pr[E_3 \mid \wedge_{i=1}^{k-1} E_i] \geq (1/2)^k + n^{-c} .$$

This implies that there exists a fixed l such that $\Pr[E_l \mid \bigwedge_{i=1}^{l-1} E_i] \geq \frac{1}{2} + n^{-c}$. In other words $|\Pr[b'_l = 1 \mid b_l = 0 \wedge \bigwedge_{i=1}^{l-1} E_i] - \Pr[b'_l = 1 \mid b_l = 1 \wedge \bigwedge_{i=1}^{l-1} E_i]| \geq n^{-c}/2$. We clearly also have $\Pr[\wedge_{i=1}^{l-1} E_i] \geq n^{-c}$. Denote by A the adversary that accepts a public key pk and hands the pair of messages $(0, 1)$ to the experiment, and waits for a challenge ciphertext B. Then it proceeds as follows:

1. It simulates the interaction between kV and P^* on common input pk. The verifier V_i for $i = 1, \ldots, l-1$ is simulated honestly except that it invokes the decryption oracle to compute $c_i = \mathsf{Dec}_{sk}(C_i)$ if necessary. If any event $\bar{E}_i$ occur for an $1 \leq i \leq l-1$ it halts with output 0.
2. Then it defines $B_l = B$, continues the simulation computing c_l using the decryption oracle if necessary, and outputs the final message b'_l of P^* in the lth instance of the protocol.

By construction A never queries its decryption oracle on $B_l = B$. Thus, it follows that the CCA2-security of $\mathcal{CS}$ is broken. $\square$

Lower Bounds for Non-interactive Zero-Knowledge

Hoeteck Wee*

Computer Science Division
University of California, Berkeley
hoeteck@cs.berkeley.edu

Abstract. We establish new lower bounds and impossibility results for non-interactive zero-knowledge proofs and arguments with set-up assumptions.

- For the common random string model, we exhibit a lower bound for the trade-off between hardness assumptions and the length of the random string for non-interactive zero-knowledge *proofs*. This generalizes a previous result ruling out non-interactive zero-knowledge proofs for non-trivial languages with a random string of length $O(\log n)$.
- In the registered public key model, we show that there does not exist a non-interactive zero-knowledge *proof* for a non-trivial language.
- In the bare public key model with fully nonuniform simulation wherein the size of the simulator is also allowed to depend on the size of the distinguisher and the distinguishing gap, there does not exist a non-interactive zero-knowledge *proof* for an NP-complete language, unless the polynomial hierarchy collapses. On the other hand, there is a non-interactive zero-knowledge *argument* for all of NP with a fully nonuniform simulator.

Our negative results complement upper bounds and feasibility results from previous work.

Keywords: Non-interactive zero-knowledge, set-up assumptions, lower bounds.

1 Introduction

The seminal notion of *zero-knowledge proofs*, namely proofs that yield no knowledge beyond the validity of the assertion proved, was introduced by Goldwasser, Micali and Rackoff [GMR89]. Formally, an interactive protocol is zero-knowledge if there exists a simulator that can simulate the behavior of every, possibly malicious, verifier without access to the prover, such that its output is indistinguishable from the output of the verifier after having interacted with the honest prover.

Minimizing the number of rounds is an important goal in design of zero-knowledge proof systems. A lower bound was established by Goldreich and Oren [GO94], who showed that at least three rounds of interaction are necessary to achieve auxiliary-input zero-knowledge. To understand and overcome this limitation, recent work has focused on both impossibility and feasibility results for weaker notions of zero-knowledge. The relaxations include limiting the power of malicious verifiers [BLV06], limiting prover resources [DS02], quasipolynomial-time simulation [P03, BP04], and witness indistinguishability [DN00, BOV03].

---

* Part of this work was completed while visiting IBM T.J. Watson Research Center and IPAM, Los Angeles.

S.P. Vadhan (Ed.): TCC 2007, LNCS 4392, pp. 103–117, 2007.
© International Association for Cryptologic Research 2007

1.1 Non-interactive Zero-Knowledge with Set-Up

A different way to bypass the lower bound on interaction is to introduce set-up assumptions. This approach was initiated by Blum, Feldman and Micali [BFM88], who showed how to realize a non-interactive zero-knowledge protocol for NP, comprising a single message from the prover to the verifier. In this work, we will focus on set-up assumptions with a "public key" flavor, presented in decreasing order in the amount of the trust the prover and verifier needs to put in the set-up:

- *Common random string (CRS) model.* This is the original model proposed by Blum et al., wherein both the prover and the verifier receive a truly random string from a trusted party. A slight relaxation of this model is the *common reference string model*, wherein both parties receive a random string chosen accordingly to some polynomial-time samplable distribution.
- *Registered public key model.* Barak et al. [BCNP04] introduced the registered public key model as a relaxation of the CRS model under which general multi-party computations can still be securely realized within the UC framework. In addition, they showed how to realize non-interactive zero-knowledge in this model.[1] We will restrict ourselves to the special case wherein only the verifier registers a "public key" with a "registration authority". An honest verifier upon registration receives a randomly generated public key, and does not need to keep the secret data used for key generation, whereas a cheating verifier may register any public key of its choice, but must provide the secret data associated with the key to the registration authority. Prior to participating in the protocol, the prover obtains the verifier's key from the registration authority.
- *Bare public key model.* In the bare public key model introduced by Canetti et al. [CGGM00], the verifier again has a public key that has been registered prior to interacting with the prover. Here, there is no trusted "registration authority" that verifies (and enforces) properties of the registered key. In particular, an honest verifier registers a randomly generated key, whereas a cheating verifier may register any arbitrary key, possibly even a malformed one.

We stress that in each of these models, the proofs are publicly verifiable - verification does not require verifier's secret key. Note that the bare public key model imposes the strongest requirements on the simulator (minimal trust requirements) whereas the common reference string model imposes the weakest requirement (maximal trust requirements: a non-interactive protocol that is zero-knowledge in the bare public key model is also zero-knowledge in the registered public key model, and a protocol satisfying the latter is zero-knowledge in the common reference string model. We refer the reader to [CGGM00, BCNP04] for further cryptographic motivations for these set-up assumptions.

1.2 Weak Nonuniform Zero Knowledge

A non-interactive zero-knowledge protocol in the bare public key model is essentially a 2-round zero-knowledge protocol without set-up assumptions, except the verifier's first

---

[1] Specifically, Barak et al. demonstrated a non-interactive protocol that realizes the UC zero-knowledge functionality, which implies zero-knowledge.

message must be independent of the instance. This means that the result of Goldreich and Oren [GO94] also rules out non-interactive zero-knowledge argument systems in the bare public key model for languages outside BPP. Therefore, we will relax the zero-knowledge requirement for the bare public key model in the following ways (as has previously been done for general zero-knowledge in [DNRS03]):

- We allow the simulator to depend nonuniformly on the cheating verifier (namely that for every nonuniform probabilistic polynomial-time cheating verifier, there is a nonuniform probabilistic polynomial-time simulator), with the additional guarantee of a polynomial relation between the size of the verifier and that of the simulator.[2] The main difference from auxiliary-input zero knowledge is the latter guarantees a single (uniform) polynomial-time algorithm that on input a description of the verifier, outputs a description of the simulator.
- Next, we allow the size of the simulator to depend on that of the distinguisher and the distinguishing gap.[3] In particular, this guarantee (by itself, without nonuniformity) implies quasipolynomial-time simulation [P03] (and therefore the security guarantee is stronger than that offered by quasipolynomial-time simulation).

We refer to this relaxation as *weak nonuniform zero knowledge*, and it is meaningful also in the standard model without set-up assumptions. Informally, weak nonuniform zero knowledge guarantees that an efficient verifier can approximate whatever he learns from interacting with the prover within any inverse polynomial factor at a price of a polynomial blow-up in the running time and a polynomial amount of help (corresponding to the nonuniformity). One can regard the simulator's nonuniformity as a measure of the *knowledge* leaked by an interaction with the honest prover.

As with the standard notion of zero knowledge, weak nonuniform zero knowledge implies witness indistinguishability (in the standard sense with a negligible distinguishing gap). More generally, weak nonuniform zero knowledge is sufficient in applications wherein the zero knowledge property is only used to construct an intermediate hybrid distribution in order to establish computational indistinguishability. This occurs for instance in the construction of non-malleable cryptographic schemes by Dolev, Dwork and Naor [DDN00] (as pointed out in [DNRS03]) and specifically, in their non-malleable bit commitment scheme and Sahai's CCA2-secure encryption scheme [S99]. We stress that nonuniform zero knowledge yields nonuniform reductions in the proof of security for these schemes.

An analogous relaxation for circuit obfuscation as formalized by Barak et al. [BGI+01] was previously presented in [W05].[4] Indeed, the only known positive results for obfuscation in the standard model [C97, W05] merely achieve this weaker

---

[2] Refer to [G01, Sec 4.3.3] for a discussion of this relaxation.

[3] Dwork et al. [DNRS03] had earlier considered $S(V, T, D)$ zero-knowledge, the simulator is allowed to depend on both the verifier V and the distinguisher T (whereas we allow a dependency on the size of T but not T itself). A dependency on the distinguishing gap was introduced by Dwork, Naor and Sahai in their work on concurrent zero-knowledge [DNS04].

[4] In the formalization proposed in [W05], the simulator is allowed to depend on the distinguisher, although it is easy to verify that the virtual black-box simulators for the constructions in [C97, W05] only need to depend on the size of the distinguisher and the distinguishing gap.

requirement. One of the motivations for this work is to understand whether this relaxation may also be meaningfully exploited in the context of zero-knowledge.

1.3 Our Contributions

We present lower bounds and impossibility results for non-interactive zero-knowledge proofs and arguments with set-up assumptions, along with matching upper bounds and feasibility results. Our main contributions are in the lower bounds; the protocols, apart from the one for the bare public key model, follow readily from previous work [KP98, P03, BCNP04]. We stress that understanding fundamental limitations do faciliate protocol design in narrowing down possible approaches. As a whole, our results complement known protocols to provide a clearer picture of the qualitative differences in the various set-up assumptions, as well as better insight into what can and cannot be realized in each of these models, and why.

Common random string model. We already know how to construct non-interactive zero-knowledge proof systems for NP in the common random string model [BFM88, FLS99]. We establish a lower bound on the trade-off between hardness assumptions and length of the common random string used in these constructions:

> **Informal Theorem.** [Lower bound]   In the common random string model, if there is a polynomial-time algorithm for CIRCUIT-SAT with $\ell(n)$ variables, then non-interactive zero-knowledge proof systems with a random string of length $\ell(n)$ only exist for languages in BPP.
>
> **Informal Theorem.** [Upper bound]   In the common random string model, under $\ell^{-1}(n)$-hardness assumptions for enhanced trapdoor permutations, there is a non-interactive zero-knowledge proof system for all of NP with a CRS of length $\mathrm{poly}(\ell(n))$.

The trade-off achieved in the upper bound is widely believed to be optimal (up to polynomial factors in the length of the CRS) but has not been formally stated; we provide and prove a formal statement to that effect. In the proof, we use probabilistic hashing techniques from [GS89] to address an issue related to randomness-efficient sampling [DI06]. We point out two special cases of our lower bound: to achieve a CRS of length $\mathrm{poly}(\log n)$, sub-exponential hardness assumptions for CIRCUIT-SAT are indeed necessary. Also, if the CRS has length $O(\log n)$, then the language is in BPP (since there is a trivial CIRCUIT-SAT with an exponential dependency on the number of variables). This special case (which extends to readily to arguments, unlike the general case) is folklore and was stated without proof in [DDP97].

Registered public key model. For the registered public key model, we establish a separation between proof systems and argument systems:

> **Informal Theorem.** [Impossibility]   In the registered public key model, non-interactive zero-knowledge proof systems only exist for languages in BPP.
>
> **Informal Theorem.** [Feasibility]   In the registered public key model, under sub-exponential hardness assumptions for enhanced trapdoor permutations, there is a non-interactive zero-knowledge argument system for all of NP.

In the registered public key model, the only advantage the simulator has for generating accepting transcripts for YES instances is the secret key corresponding to the public key. However, a computationally unbounded adversary can easily sample a secret key corresponding to the public key, and then run the simulator. This will yield an accepting transcript for YES instances, but not for NO instances if the protocol is a proof system. As such, we can have non-trivial non-interactive argument systems but not proof systems in the registered public key model.

Bare public key model. For the weak non-uniform non-interactive zero-knowledge in the bare public key model, we also establish a separation between proof systems and argument systems. Our feasibility result shows that weak nonuniform simulation can indeed be meaningful exploited in the context of zero-knowledge. Both of the following results refer to the weak non-uniform setting:

> **Informal Theorem.** [Impossibility]   In the bare public key model, non-interactive zero-knowledge proof systems only exist for languages in coNP/poly. In particular, there is no non-interactive zero-knowledge proof system for all of NP unless the polynomial hierarchy collapses [KL80].
> **Informal Theorem.** [Feasibility]   In the bare public key model, under sub-exponential hardness assumptions for enhanced trapdoor permutations, there is a non-interactive zero-knowledge argument system for all of NP.

We exploit derandomization via nonuniformity using the probabilistic method for both results.

We use the same protocol, namely a straight-forward adapation of Pass's 2-round public-coin zero-knowledge argument [P03], for the feasibility results in the registered and bare public key models, with somewhat different simulators. One can therefore view the weak non-uniform zero-knowledge as a fall-back guarantee provided by Pass's protocol in the registered public key model: even if the assumption about the verifier's key being well-formed is not satisfied, the protocol still guarantees non-uniform zero-knowledge, which implies witness indistinguishability.

1.4   Additional Related Work

Lower bounds for non-black-box zero-knowledge. On the whole, our lower bounds and impossibility results make use many of the insights and techniques from the work of Barak, Lindell and Vadhan [BLV06] on lower bounds for zero-knowledge, specifically, those for 2-round zero-knowledge protocols against uniform adversaries. The latter is an arguably less natural notion than non-interactive zero-knowledge protocols with set-up assumptions, while imposing the technical constraint of uniformity. Indeed, our results show that the ideas from [BLV06] are applicable to a more general setting.

Unconditional characterizations of NIZK. Another closely related work is that of Pass and Shelat [PS05] providing a systematic unconditional study of non-interactive proof systems in the common reference string model, and the secret parameter variant thereof, wherein the verifier also has a secret key corresponding to the public key. A natural next project would be to extend their work to the registered public key and bare public key models and to argument systems (after all, our lower bounds indicate that the limitation to proof systems may be too restrictive), and to extend our work to secret parameter variants of the different models.

2 NIZK with Set-Up

2.1 Non-interactive Protocols with Set-Up

We consider a set-up phase, parameterized by a (deterministic) function $f : \{0,1\}^* \to \{0,1\}^*$, that represents a method for computing a public set-up key given a secret (and supposedly random) key. An protocol (P, V) is a *non-interactive proof system with set-up* for a language L if there is a relation R such that $L = L_R$, a set-up function f such that the following holds:

COMPLETENESS. If $(x, w) \in R$,

$$\Pr[\sigma \leftarrow f(U_k); \ \pi \leftarrow P(x, w, \sigma) \ : \ V(x, \sigma, \pi) = 1] \geq 2/3$$

SOUNDNESS. If $x \notin L$, then for every P^*,

$$\Pr[\sigma \leftarrow f(U_k); \ \pi \leftarrow P^*(x, \sigma) \ : \ V(x, \sigma, \pi) = 1] \leq 1/3$$

We say that a protocol has *perfect completeness* if the expression $2/3$ is replaced by 1, and *negligible soundness* if the expression $1/3$ is replaced by $\text{neg}(|x|)$. We say that (P, V) is a *non-interactive argument system with set-up* if the soundness condition is replaced by:

COMPUTATIONAL SOUNDNESS. If $x \notin L$, then for every nonuniform PPT P^*,
$$\Pr[\sigma \leftarrow f(U_k); \ \pi \leftarrow P^*(x, \sigma) \ : \ V(x, \sigma, \pi) = 1] \leq 1/3$$

We note that our positive results satisfy the following stronger notion of soundness:

ADAPTIVE SOUNDNESS. For every P^*,

$$\Pr[\sigma \leftarrow f(U_k); \ (x, \pi) \leftarrow P^*(\sigma) \ : \ x \notin L \text{ and } V(x, \sigma, \pi) = 1] \leq 1/3$$

We emphasize that in the formulations of completeness and soundness, both parties receive a randomly generated public key, and the verifier does not receive the secret randomness used to generate the key.

2.2 Non-interactive Zero-Knowledge

Since our main contributions are the negative results, we present the weakest possible notion of security (in particular, we consider non-adaptive zero-knowledge). Establishing lower bounds for weaker notions makes our results stronger. We only present definitions for zero-knowledge in the non-interactive setting.

ZERO-KNOWLEDGE IN COMMON REFERENCE STRING MODEL. There exists a PPT simulator S such that the following distributions are nonuniformly computationally indistinguishable:

$$\{\sigma \leftarrow f(U_k); \ \pi \leftarrow P(x, w, \sigma) \ : \ (\sigma, \pi)\}_{(x,w) \in R}$$
$$\text{and } \{(\sigma, \pi) \leftarrow S(x) \ : \ (\sigma, \pi)\}_{(x,w) \in R}$$

We refer to the special case where f is the identity as the Common Random String model.

ZERO-KNOWLEDGE IN REGISTERED PUBLIC KEY MODEL. There exists a PPT simulator S such that the following distributions are nonuniformly computationally indistinguishable:

$$\{ P(x, w, f(r)) \}_{(x,w) \in R, r \in \{0,1\}^k} \text{ and } \{ S(x, r) \}_{(x,w) \in R, r \in \{0,1\}^k}$$

ZERO-KNOWLEDGE IN BARE PUBLIC KEY MODEL. There exists a PPT simulator S such that the following distributions are nonuniformly computationally indistinguishable:

$$\{ P(x, w, \sigma) \}_{(x,w) \in R, \sigma \in \{0,1\}^{\mathrm{poly}(k)}} \text{ and } \{ S(x, \sigma) \}_{(x,w) \in R, \sigma \in \{0,1\}^{\mathrm{poly}(k)}}$$

Note that zero-knowledge in the bare public key model implies zero-knowledge in the registered public key model, which in turn implies zero-knowledge in the common reference string model. Also, recall that in the definition of (auxiliary-input) zero-knowledge in the interactive setting, it suffices to consider deterministic cheating verifiers; for the same reason, once we have established the zero-knowledge property for a fixed public key in the registered public key and bare public key models, we derive the zero-knowledge property for any (adversarial) distribution over public keys.

3 Common Random String Model

3.1 Lower Bounds

Theorem 1. *If a language L has a non-interactive zero-knowledge proof system in the common string model with a CRS of length $\ell(n)$ (where n is the length of the instance) and there exists a probabilistic $\mathrm{poly}(\ell^{-1}(\#\mathrm{variables}), \mathrm{circuitsize})$ algorithm for the* CIRCUIT-SAT *Problem, then $L \in$ BPP.*

By zero-knowledge and soundness, the distribution of the simulated random strings is pseudorandom for YES instances and statistically far from uniform for NO instances. The idea is to use the CIRCUIT-SAT algorithm to design an efficient test that

- outputs 1 with high probability for samplable distributions over $\{0,1\}^{\ell(n)}$ that are statistically far from uniform.
- outputs 1 with small probability on the uniform distribution over $\{0,1\}^{\ell(n)}$.

The latter will correspond to YES instances and the former will correspond to NO instances. The difficulty lies in that the sampling algorithm may use $\text{poly}(n)$ bits of randomness, so we cannot directly test if the input lies in the support of the sampling distribution. To overcome this, we use pairwise independent sampling to reduce the randomness complexity of the sampling algorithm. This is inspired by the Goldwasser-Sipser protocol for proving lower bounds on set sizes [GS89]; the formal analysis is also very similar.

Proof (sketch). Suppose L has a non-interactive zero-knowledge proof system (P, V) in the CRS model with a CRS of length $\ell = \ell(n)$ and a simulator S. We modify the proof system to satisfy the following additional properties:

- The completeness and soundness errors are both at most $1/64$. This can be achieved using randomness-efficient error reduction while increasing the CRS by an additive $O(1)$ bits [DDP02], although naive parallel repetition with a $O(1)$ multiplicative increase is fine too.
- On every input x, the simulator S always outputs accepting transcripts, and the distinguishing error for YES instances is at most $1/32$.

Let r denote the number of random bits used by S, and let S_1 be S with the output truncated to just the simulated CRS. Consider the following algorithm M for deciding L: on input x,

1. Run $S(x)$ for n independent iterations to obtain transcripts (σ_i, π_i), $i = 1, 2, \ldots, n$. In addition, pick n independent pairwise-independent hash functions $h_i : \{0, 1\}^{\ell-2} \to \{0, 1\}^r$.
2. Reject if for the majority of $i = 1, 2, \ldots, n$, we have $(\sigma_i, x, h_i) \in L_{\text{aux}}$, where

$$L_{\text{aux}} = \left\{ (\sigma, x, h) \mid \exists\, u \in \{0, 1\}^{\ell-2} \text{ s.t. } S_1(x; h(u)) = \sigma \right\} \in \mathsf{BPP}$$

For each i, we show that $\Pr_{h_i, \sigma_i}[(\sigma_i, x, h_i) \in L_{\text{aux}}]$ is small for $x \in L$ and large for $x \notin L$:

- $x \in L$. By a union bound,

$$\Pr_{h_i, \sigma_i} \left[(\sigma_i, x, h_i) \in L_{\text{aux}} \right] \leq \Pr_{h_i}[(U_\ell, x, h_i) \in L_{\text{aux}}] + \frac{1}{32} \leq \frac{2^{\ell-2}}{2^\ell} + \frac{1}{32} < \frac{1}{3}$$

- $x \notin L$. By soundness, $|S_1(x; \{0,1\}^r)| \leq \frac{1}{64} \cdot 2^\ell$. Let Λ be the set of "low probability" strings in $S_1(x; \{0,1\}^r)$, that is,

$$\Lambda = \left\{ \sigma : \Pr[S_1(x; U_r) = \sigma] \leq \frac{1}{2^{\ell-4}} \right\}$$

A union bound yields

$$\Pr[\sigma_i \in \Lambda] \leq \frac{1}{64} \cdot 2^\ell \cdot \frac{1}{2^{\ell-4}} = \frac{1}{4}$$

On the other hand, for the "high probability" strings, a standard analysis via the Chebyshev inequality yields

$$\Pr_{h_i}\big[(\sigma_i, x, h_i) \notin L_{\mathsf{aux}} \mid \sigma_i \notin \Lambda\big] \leq \frac{1}{4}$$

Hence,

$$\Pr_{h_i, \sigma_i}\big[(\sigma_i, x, h_i) \in L_{\mathsf{aux}}\big] \geq \frac{1}{2}$$

Hence, M is a BPP algorithm for deciding L. $\qquad\qquad\square$

3.2 Upper Bounds

The following result follows from a variant of the Kilian-Petrank non-interactive zero-knowledge proof system for NP in the CRS model [KP98] (alluded to in [GOS06]) wherein the length on the random string depends polynomially on the security parameter (and not the length of the instance). The idea is to rewrite the input as a conjunction of a polynomial number of constant-sized statements and prove each of these statements using the same CRS (as in [FLS99]).

Proposition 1 ([KP98, GOS06]). *Suppose there exist enhanced trapdoor permutations secure against $\ell^{-1}(n)^{\omega(1)}$-size circuits. Then, there exists a non-interactive zero-knowledge proof system for* NP *in the common random string model wherein the CRS has length $O(\ell(n)^3)$ (where n is the length of the instance). In addition, the proof system has perfect completeness, negligible adaptive soundness error and an efficient prover.*

4 Registered Public Key Model

4.1 Impossibility Results

Theorem 2. *If a language L has a non-interactive zero-knowledge proof system in the registered public key model, then $L \in$* BPP.

Proof. Consider the following algorithm M for deciding L: on input x,

1. pick $r \leftarrow U_k$.
2. accept iff $V(x, f(r), S(x, r))$ accepts.

Completeness and zero-knowledge guarantees that for all $x \in L$, M accepts with probability at least $2/3 - \mathsf{neg}(|x|)$. Next, consider a (unbounded) cheating prover P^* that for all $x \notin L$ and all σ, outputs π such that $V(x, \sigma, \pi) = 1$ if such a π exists, and $\perp$ otherwise. Then, for all $x \notin L$ and $\sigma \in f(\{0,1\}^k)$,

$$\Pr[V(x, \sigma, P^*(x, \sigma)) = 1] \geq \Pr_{r: f(r) = \sigma}[V(x, \sigma, S(x, r)) = 1]$$

Averaging over σ, we obtain, for all $x \notin L$,

$$\begin{aligned}
\Pr[M(x) = 1] &= \Pr_r[V(x, f(r), S(x, r)) = 1] \\
&\leq \Pr_r[V(x, f(r), P^*(x, f(r))) = 1] \\
&\leq 1/3 \quad \text{(by soundness)}
\end{aligned}$$

Hence, M is a BPP algorithm for deciding L.   □

4.2 Feasibility Results

Indeed, by relaxing the soundness requirement to computational soundness, Barak et al. constructed a non-interactive UC zero-knowledge protocol in the registered public key model [BCNP04]. The protocol requires that the prover also has a public key in order to achieve additional guarantees required by universal composability. We observe that it is not necessary for the prover to register a key if zero-knowledge is our only goal (but paying the price of subexponential hardness assumptions); in particular, we may use the variant of Pass's protocol [P03] shown in Fig 1.

Proposition 2. *Suppose there exist enhanced trapdoor permutations secure against 2^{n^δ}-size circuits for some constant $\delta > 0$. Then, there exists a non-interactive zero-knowledge argument system for NP in the registered public key model. In addition, the argument system has perfect completeness, negligible adaptive soundness error and an efficient prover.*

Note that our negative results do not extend to the secret parameter model. There, Pass, Shelat, Vaikuntanathan [PSV06] constructed a non-interactive zero-knowledge proof system for all of NP assuming the existence of standard trapdoor permutations (or any semantically secure encryption scheme).

5 Bare Public Key Model

5.1 Weak Nonuniform Zero Knowledge

As noted in the introduction, the lower bound of Goldreich and Oren [GO94] also extends to the bare public key model:

Theorem 3 (implicit in [GO94]). *If a language L has a non-interactive zero-knowledge argument system in the bare public key model, then $L \in$ BPP.*

As such, we will focus on weak nonuniform zero-knowledge in the bare public model. We say that a nonuniform PPT A has size s if the running time and the length of the nonuniform advice for A is bounded by s. Two distributions A, B are (s, ϵ)-*indistinguishable* if for every nonuniform PPT D of size s, $|\Pr[D(A) = 1] - \Pr[D(B) = 1]| < \epsilon$. Unlike the uniform setting, we need to define zero-knowledge for distributions over public keys chosen by an adversarial verifier V^*.

WEAK NON-UNIFORM ZERO-KNOWLEDGE IN BARE PUBLIC KEY MODEL.
There exists a polynomial p such that for every function $s(n) = n^{O(1)}$ and
$\epsilon(n) = 1/n^{O(1)}$, and for every nonuniform PPT V^* of size s, there exists a
nonuniform PPT S of size $p(n, s, 1/\epsilon)$ such that for all sufficiently large n and
for all $(x, w) \in R$ with $|x| = n$, the following distributions

$$\{(\tau, \sigma) \leftarrow V^*(1^n); \ (\tau, \sigma, P(x, w, \sigma))\} \ \text{and} \ \{(\tau, \sigma, \pi) \leftarrow S(x); \ (\tau, \sigma, \pi)\}$$

are $(s(n), \epsilon(n))$-indistinguishable.

We stress once again that the definition allows for the size of the simulator to depend
on s, an upper bound on the sizes of the malicious verifier and the distinguisher,
and on ϵ, the distinguishing gap, although the dependency is determined by a *fixed*
polynomial p.

5.2   Impossibility Results

Theorem 4. *If a language L has a weak nonuniform non-interactive zero-knowledge
proof system in the bare public model, then $L \in \mathsf{P}/\mathsf{poly}$.*

Proof. The idea behind the proof is to use the probabilistic method to derandomize the
verifier in the NIZK proof system and obtain a polynomial number of deterministic
nonuniform verifiers with some randomness hardwired into it. We then use the
nonuniform simulators for these verifiers to decide the language.

   Fix an input length n, and by the probabilistic method, there exists a set $\Lambda \subseteq$
$\{0, 1\}^{\text{poly}(n)}$ of polynomial size satisfying the following properties:

– for all $x \in L \cap \{0, 1\}^n$ and a fixed witness w for each x,

$$\left| \Pr_{\alpha \in \{0,1\}^{\text{poly}(n)}} [V(x, f(\alpha), P(x, w, f(\alpha))) = 1] \right.$$

$$\left. - \Pr_{\alpha \in \Lambda} [V(x, f(\alpha), P(x, w, f(\alpha))) = 1] \right| < \tfrac{1}{12}$$

where the probabilities are also taken over the coin tosses of the prover.
– for all $x \in \{0, 1\}^n \setminus L$,

$$\left| \Pr_{\alpha \in \{0,1\}^{\text{poly}(n)}} [\exists \pi : V(x, f(\alpha), \pi) = 1] - \Pr_{\alpha \in \Lambda} [\exists \pi : V(x, f(\alpha), \pi) = 1] \right| < \tfrac{1}{12}$$

   Now, for each $r \in \Lambda$, consider the malicious verifier V_r^* with r hardwired into it and
sends $f(r)$ as its public key, and the class of distinguishers $\{D_{x,r} \mid x \in \{0, 1\}^n\}$ that
on input a transcript $(r', \sigma'\pi)$ accepts iff $r' = r$ and $V(x, \pi) = 1$. Let S_r denote the
nonuniform PPT simulator for V_r^* with distinguishing probability $\tfrac{1}{12}$ and which fools
$\{D_{x,r} \mid x \in L \cap \{0, 1\}^n\}$. Hence, for all $x \in \{0, 1\}^n$:

YES instance: $\Pr_{r \in \Lambda}[D_{x,r}(S_r(x)) = 1] > \tfrac{2}{3} - \tfrac{1}{12} - \tfrac{1}{12} = \tfrac{1}{2}$

NO instance: $\Pr_{r \in \Lambda}[D_{x,r}(S_r(x)) = 1] < \tfrac{1}{3} + \tfrac{1}{12} = \tfrac{5}{12}$

where the probabilities are also taken over the coin tosses of S_r. By hardwiring Λ and
$\{S_r | r \in \Lambda\}$ as nonuniform advice, we obtain $L \in \mathsf{BPP}/\mathsf{poly} = \mathsf{P}/\mathsf{poly}$. $\square$

Remark 1. The analogous result in [BLV06] requires that the proof system has either perfect completeness or an efficient prover.

5.3 Feasibility Results

The idea is to derandomize the adversary and the distinguisher and hardwire the trapdoor information about the public key into the simulator.

Theorem 5. *Suppose there exist enhanced trapdoor permutations secure against 2^{n^δ}-size circuits for some constant $\delta > 0$. Then, there exists a weak nonuniform non-interactive zero-knowledge argument system for NP in the bare public key model. Furthermore, the argument system has perfect completeness, negligible soundness error and an efficient prover.*

Proof. Let L be an NP-complete language for some relation R. Under the assumed trapdoor permutation family, we can construct the following primitives:

- a one-way permutation $\pi : \{0,1\}^n \to \{0,1\}^n$ secure against 2^{n^δ}-sized circuits;
- a non-interactive (perfectly binding, computationally hiding) commitment scheme Com that can be broken (that is, recover the plaintext from the commitment) in time $2^{n^{\delta/2}}$; and
- a zap system [DN00], namely a 2-round public-coin witness-indistinguishable proof system for NP. For simplicity and ease of presentation, we present the protocol and analysis assuming the existence of a 1-round zap (e.g. [BOV03]); for a 2-round zap, we include the first round message as part of the public key.

The argument system for L is shown in Fig 1. The completeness property of this protocol follows from that of the zap system. To prove computational soundness, consider a nonuniform PPT cheating prover P^* that convinces the honest verifier to accept some $x \notin L$ with non-negligible probability. By adaptive soundness of the zap system, the commitment sent by P^* must contain the value $\pi^{-1}(\sigma)$, which can be extracted in time $2^{O(n^{\delta/2})}$. Hence, we derive from P^* a nonuniform algorithm running in time $2^{O(n^{\delta/2})}$ and inverts π with non-negligible probability, a contradiction.

To prove weak nonuniform zero-knowledge, fix s, ϵ, a nonuniform PPT V^* and an input length n. Consider the following distributions for each $(x, w) \in R$ with $|x| = n$:

- Hybrid H_1. This is the distinguisher's view in an interaction with the honest prover.

$$\{(\tau, \sigma), \mathsf{Com}(0^n), P_{\mathrm{zap}}((x, \mathsf{Com}(0^n)), (w, \bot));\ (\tau, \sigma) \xleftarrow{\text{R}} V^*(U_s)\}$$

- Hybrid H_2. This is the distinguisher's view when the prover commits to $\pi^{-1}(\sigma)$ instead of 0^n.

$$\{(\tau, \sigma), \mathsf{Com}(\pi^{-1}(\sigma)), P_{\mathrm{zap}}((x, \mathsf{Com}(\pi^{-1}(\sigma))), (w, \bot));\ (\tau, \sigma) \xleftarrow{\text{R}} V^*(U_s)\}$$

- Hybrid H_3. We modify H_2 so that the prover uses $\pi^{-1}(\sigma)$ (and the private randomness used for the commitment) instead of w as the witness in the zap system.

$$\{(\tau, \sigma), \mathsf{Com}(\pi^{-1}(\sigma)), P_{\mathrm{zap}}((x, \mathsf{Com}(\pi^{-1}(\sigma))), (\bot, \pi^{-1}(\sigma)));\ (\tau, \sigma) \xleftarrow{\text{R}} V^*(U_s)\}$$

Set-up function: $f(r) = \pi(r)$, where π is a permutation.

Common input: An instance $x \in \{0,1\}^n$, public key σ.

Prover's private input: A witness $w \in \{0,1\}^{\text{poly}(n)}$.

$P \to V$: Send $z = \mathsf{Com}(0^n)$ and a zap proving the statement "$x \in L$ OR z is a commitment to $\pi^{-1}(\sigma)$" using witness w.

Fig. 1. Variant of Pass's protocol [P03] for an NP-complete language L

Note that H_1 and H_2 are $(s, \epsilon/4)$-indistinguishable by the hiding property of Com, and that H_2 and H_3 are $(s, \epsilon/4)$-indistinguishable by witness indistinguishability of the zap system. Observe that for a fixed choice of x and coin tosses for Com and P_{zap}, a sample from the distribution H_3 may be computed as a deterministic function of the choice of random coin tosses for V^*. Hence, by the probabilistic method, there exists a set $\Lambda \subseteq \{0,1\}^s$ of size $\Theta((s \log s + p'(n))/\epsilon^2)$ where p' is a fixed polynomial equal to $|x|$ plus the total randomness used by Com and P_{zap}, such that the following distribution H_4 is $(s, \epsilon/4)$-indistinguishable from H_3.

- Hybrid H_4. We modify H_3 so that we replace V^*'s coin tosses with a random sample from Λ, where $\Lambda \subseteq \{0,1\}^s$ is to be determined. We stress that Λ only depends on $|x|$, s, ϵ and not on x itself.

$$\{(\tau, \sigma), \mathsf{Com}(\pi^{-1}(\sigma)), P_{\text{zap}}((x, \mathsf{Com}(\pi^{-1}(\sigma))), (\bot, \pi^{-1}(\sigma)), \sigma);$$

$$(\tau, \sigma) = V^*(r'), r' \xleftarrow{\text{R}} \Lambda\}$$

By hardwiring Λ and $\{\pi^{-1}(\sigma) \mid r' \in \Lambda, (\tau, \sigma) = V^*(r')\}$ as nonuniform advice, we have a nonuniform PPT S of size $O(|\Lambda|n + s) = \text{poly}(n, s, 1/\epsilon)$ that on input x, computes the distribution H_4, which is (s, ϵ)-indistinguishable from H_1. □

Acknowledgements

I am very grateful towards Ran Canetti and Vinod Vaikuntanathan for many interesting discussions on non-interactive zero-knowledge, and in particular, for raising the question of NIZKs with short CRS. I would also like to thank Moni Naor, Alon Rosen, Luca Trevisan and Salil Vadhan for helpful discussions on an earlier version of this work, Joe Kilian for clarifications regarding [KP98], and the anonymous referees for various feedback and pointers.

References

[BCNP04] B. Barak, R. Canetti, J. B. Nielsen, and R. Pass. Universally composable protocols with relaxed set-up assumptions. In *Proc. 45th FOCS*, 2004.

[BFM88] M. Blum, P. Feldman, and S. Micali. Non-interactive zero-knowledge and its applications. In *Proc. 20th STOC*, 1988.

[BGI+01]   B. Barak, O. Goldreich, R. Impagliazzo, S. Rudich, A. Sahai, S. Vadhan, and K. Yang. On the (im)possibility of obfuscating programs. In *Proc. Crypto '01*, 2001.

[BLV06]   B. Barak, Y. Lindell, and S. Vadhan. Lower bounds for non-black-box zero knowledge. *JCSS*, 72(2):321–391, 2006.

[BOV03]   B. Barak, S. J. Ong, and S. Vadhan. Derandomization in cryptography. In *Proc. Crypto '03*, 2003.

[BP04]   B. Barak and R. Pass. On the possibility of one-message weak zero-knowledge. In *Proc. 1st TCC*, 2004.

[C97]   R. Canetti. Towards realizing random oracles: Hash functions that hide all partial information. In *Proc. Crypto '97*, 1997.

[CGGM00]   R. Canetti, O. Goldreich, S. Goldwasser, and S. Micali. Resettable zero-knowledge. In *Proc. 32nd STOC*, 2000.

[DI06]   B. Dubrov and Y. Ishai. On the randomness complexity of efficient sampling. In *Proc. 38th STOC*, 2006.

[DDN00]   D. Dolev, C. Dwork, and M. Naor. Non-malleable cryptography. *SIAM Journal on Computing*, 30(2):391–437, 2000.

[DDP97]   A. De Santis, G. Di Crescenzo, and P. Persiano. Randomness-efficient non-interactive zero knowledge. In *Proc. 24th ICALP*, 1997.

[DDP02]   A. De Santis, G. Di Crescenzo, and G. Persiano. Randomness-optimal characterization of two NP proof systems. In *Proc. Random '02*, 2002.

[DN00]   C. Dwork and M. Naor. Zaps and their applications. In *Proc. 41st FOCS*, 2000.

[DNRS03]   C. Dwork, M. Naor, O. Reingold, and L. Stockmeyer. Magic functions. *JACM*, 50(6):852–921, 2003.

[DNS04]   C. Dwork, M. Naor, and A. Sahai. Concurrent zero-knowledge. *JACM*, 51(6):851–898, 2004.

[DS02]   C. Dwork and L. Stockmeyer. 2-round zero knowledge and proof auditors. In *Proc. 34th STOC*, 2002.

[FLS99]   U. Feige, D. Lapidot, and A. Shamir. Multiple noninteractive zero knowledge proofs under general assumptions. *SICOMP*, 29(1):1–28, 1999.

[G01]   O. Goldreich. *Foundations of Cryptography: Basic Tools*. Cambridge University Press, 2001.

[GMR89]   S. Goldwasser, S. Micali, and C. Rackoff. The knowledge complexity of interactive proof systems. *SIAM Journal on Computing*, 18(1):186–208, 1989.

[GO94]   O. Goldreich and Y. Oren. Definitions and properties of zero-knowledge proof systems. *J. Cryptology*, 7(1):1–32, 1994.

[GOS06]   J. Groth, R. Ostrovsky, and A. Sahai. Perfect non-interactive zero knowledge for NP. In *Proc. Eurocrypt '06*, 2006.

[GS89]   S. Goldwasser and M. Sipser. Private coins versus public coins in interactive proof systems. *Advances in Computing Research*, 5:73–90, 1989.

[KL80]   R. M. Karp and R. J. Lipton. Some connections between nonuniform and uniform complexity classes. In *Proc. 12th STOC*, 1980.

[KP98]   J. Kilian and E. Petrank. An efficient noninteractive zero-knowledge proof system for NP with general assumptions. *J. Cryptology*, 11(1):1–27, 1998.

[P03]   R. Pass. Simulation in quasi-polynomial time and its application to protocol composition. In *Proc. Eurocrypt '03*, 2003.

[PS05]   R. Pass and A. Shelat. Unconditional characterizations of non-interactive zero-knowledge. In *Proc. Crypto '05*, 2005.

[PSV06]   R. Pass, A. Shelat, and V. Vaikuntanathan.   Construction of a non-malleable encryption scheme from any semantically secure one. In *Proc. Crypto '06*, 2006.

[S99]   A. Sahai.   Non-malleable non-interactive zero knowledge and adaptive chosen-ciphertext security. In *Proc. 40th FOCS*, 1999.

[W05]   H. Wee. On obfuscating point functions. In *Proc. 37th STOC*, 2005.

Perfect NIZK with Adaptive Soundness

Masayuki Abe[1] and Serge Fehr[2]

[1] Information Sharing Platform Laboratories
NTT Corporation, Japan
abe.masayuki@lab.ntt.co.jp
[2] CWI Amsterdam, The Netherlands
fehr@cwi.nl

Abstract. This paper presents a very simple and efficient adaptively-sound perfect NIZK argument system for any NP-language. In contrast to recently proposed schemes by Groth, Ostrovsky and Sahai, our scheme does not pose any restriction on the statements to be proven. Besides, it enjoys a number of desirable properties: it allows to re-use the common reference string (CRS), it can handle *arithmetic* circuits, and the CRS can be set-up very efficiently without the need for an honest party. We then show an application of our techniques in constructing *efficient* NIZK schemes for proving arithmetic relations among committed secrets, whereas previous methods required expensive generic NP-reductions.

The security of the proposed schemes is based on a strong non-standard assumption, an extended version of the so-called Knowledge-of-Exponent Assumption (KEA) over bilinear groups. We give some justification for using such an assumption by showing that the commonly-used approach for proving NIZK *arguments* sound does not allow for adaptively-sound statistical NIZK arguments (unless NP $\subset$ P/poly). Furthermore, we show that the assumption used in our construction holds with respect to generic adversaries that do not exploit the specific representation of the group elements. We also discuss how to avoid the non-standard assumption in a pre-processing model.

1 Introduction

1.1 Background

NON-INTERACTIVE ZERO-KNOWLEDGE (NIZK). The notion of NIZK captures the problem of proving a statement by just sending one message and without revealing any additional information besides the validity of the statement, provided that a common reference string (CRS) has been properly set up. Since its introduction by Blum, Feldman and Micali in 1988 [7], NIZK has been a fundamental cryptographic primitive used throughout modern cryptography in essential ways.

There is a considerable amount of literature dedicated to NIZK, in particular to the study of which languages allow for what flavor of NIZK proof. As in case of interactive ZK it is well known that there cannot be statistical NIZK proofs

S.P. Vadhan (Ed.): TCC 2007, LNCS 4392, pp. 118–136, 2007.

(i.e., both ZK and soundness are unconditional) for NP-complete languages unless the polynomial hierarchy collapses [24,3,32]. Hence, when considering general NP-languages, this only leaves room for a NIZK proof with *computational* ZK or *computational* soundness (where the proof is also called an *argument*), or both. However, in contrast to interactive ZK where it has long been known that both flavors can exist [9,8,25], all proposed NIZK proofs or arguments for general NP-languages have computational ZK (see e.g. [7,22,6,29,17]). Hence the construction of a statistically NIZK (NISZK) argument has remained an open problem (until very recently, see below). The question of the existence of NISZK arguments is in particular interesting in combination with a result by De Santis *et al.* [17], where they observe that for a strong notion of NIZK, called *same-string* NIZK, soundness can only be computational when considering NP-complete languages (assuming that one-way functions exist).

STATISTICAL NIZK ARGUMENTS. Recently, Groth, Ostrovsky and Sahai proposed an elegant construction for a perfect NIZK (NIPZK) argument for circuit-SAT [26] by using bilinear groups. This shows NIZK *can* come with perfect ZK for any NP-language. However, the scheme only provides security against a non-adaptive dishonest prover who chooses the target instance $x^* \notin L$ (for which it wants to fake a proof) *independent* of the CRS. In an application though, it is likely that the adversary first sees the CRS and then chooses the false statement on which he wants to cheat. Using a counting argument, they argue that under some strengthened assumption their scheme is secure against an adaptive dishonest prover if the size of the circuit to be proven is a-priori limited. However, the bound on the size of the circuit is so restrictive that the circuit must be smaller than sublinear in the bit size of the CRS (as discussed in Section 1.3). Groth *et al.* also proposed a perfect NIZK argument for SAT which is provably secure in Canetti's Universal Composability (UC) framework [10]. However, besides being much less efficient than their first construction, the scheme still does not guarantee unrestricted security against an adaptive dishonest prover who chooses the target instance $x^* \notin L$ depending on the CRS. For instance, the UC security does not exclude the possibility that a dishonest prover comes up with an accepting proof for the statement "the CRS is invalid or S is true" for an arbitrary false statement S. Since in a real-life execution the CRS is assumed to be valid, this is a convincing argument of the false statement S. Accordingly, the existence of an unrestricted statistical or perfect NIZK argument, which does not pose any restriction on the instances to be proven, is still an open problem.

THE KNOWLEDGE-OF-EXPONENT ASSUMPTION. Informally, the Knowledge-of-Exponent Assumption (KEA) says that for certain groups, given a pair g and $\hat{g} = g^x$ of group elements with unknown discrete-log x, the only way to efficiently come up with another pair A and $\hat{A}$ such that $\hat{A} = A^x$ (for the same x) is by raising g and $\hat{g}$ to some power a: $A = g^a$ and $\hat{A} = \hat{g}^a$. KEA was first introduced and used by Damgård in 1991 [14], and later, together with an extended version (KEA2), by Hada and Tanaka [27]. Recently, Bellare and Palacio [5] showed that KEA2 does not hold, and proposed a new extended version called KEA3 in order to save Hada and Tanaka's results. KEA3, which we call XKEA for eXtended KEA, says that given *two* pairs $(g, \hat{g})$ and $(h, \hat{h})$ with the same unknown discrete-log x, the only way to efficiently come up with another pair A and $\hat{A}$ such that $\hat{A} = A^x$ is by computing $A = g^a h^\alpha$ and $\hat{A} = \hat{g}^a \hat{h}^\alpha$. Assumptions like KEA and XKEA are widely criticized in particular because they do not appear to be "efficiently falsifiable", as Naor put it [30], though Bellare and Palacio showed that this is not necessarily the case.

1.2 Our Result

Based on XKEA over bilinear groups, we construct an adaptively-sound NIPZK argument for circuit-SAT without any restrictions on the instances to be proven. Besides being the first unrestricted adaptively-sound NISZK argument for any NP-language, the proposed scheme enjoys a number of additional desirable properties: It is *same-string* NIZK, which allows to re-use the CRS. It is very efficient: the CRS essentially consists of a few group elements, and a proof consists of a few group elements per multiplication gate; this is comparable (if not better) to the first scheme by Groth *et al.*, which is the most efficient general-purpose NIZK scheme known up to date (see the comparison in [26]). Furthermore, our scheme can also be applied to arithmetic circuits over $\mathbb{Z}_q$ for a large prime q whereas known schemes are tailored to binary circuits; this often allows a more compact representation of the statement to be proven. Finally, the CRS does not need to be set-up by a trusted party. It can efficiently be set-up jointly by the prover and the verifier. Furthermore, it can even be provided solely from a (possibly dishonest) verifier without any correctness proof if we view the proof system as a *zap* [21] rather than a NIZK. We are not aware of any other NIZK arguments or proofs that enjoy all these desirable properties.

Based on the techniques developed for the perfect NIZK argument for SAT, we also construct an *efficient* NIPZK argument for arithmetic relations among committed secrets over $\mathbb{Z}_q$ with large prime q. To the best of our knowledge, all known schemes only work for secrets from restricted domains such as $\mathbb{Z}_2$ and have to rely on generic inefficient reductions to NP-complete problems to handle larger secrets. Our approach in particular allows for additive and multiplicative relations among secrets committed to by standard Pedersen commitments.

We give two justifications for using such a strong non-standard assumption like XKEA. First, we prove that KEA and XKEA hold in the generic group model (even over bilinear groups). This suggests that if there exists an algorithm that breaks, say, KEA in a certain group, then this algorithm must use the specific representation of the elements of that group, and it is likely to fail when some other group (representation) is used. A similar result was independently developed by Dent [20] for non-bilinear groups. Second, we give some indication that a non-standard assumption is unavoidable for adaptively-sound NISZK arguments. We prove that the common approach for proving computational soundness, which has been used for all NIZK arguments (we are aware of), does not allow for statistical ZK unless $NP \subset P/poly$ (i.e. unless any NP-problem can be solved by an efficient non-uniform algorithm). Due to lack of space, this result is moved to the full version of this paper [1].

Finally, we discuss how to avoid XKEA in our NIZK arguments by allowing a pre-processing phase. Our scheme allows very efficient pre-processing where the prover only needs to make commitments for random values and prove their knowledge using efficient off-the-shelf zero-knowledge schemes.

1.3 Related Work

In order to make it easier for the reader to position our results, we would like to give a brief discussion about recently proposed NIPZK arguments. In [26] Groth *et al.* presented two schemes for proving circuit satisfiability, where the first one comes in two flavors. Let us name the resulting three schemes by the *non-adaptive*, the *adaptive* and the *UC* GOS scheme. These are the first (and so far only) NISZK arguments proposed in the literature. The non-adaptive GOS scheme is admitted by the authors to be *not* adaptively sound. The adaptive GOS scheme *is* adaptively sound, but it only allows for circuits that are limited in size, and the underlying computational assumption is somewhat non-standard in that it requires that some problem can only be solved with "sub-negligible" probability, like $2^{-\epsilon \kappa^\epsilon \log \kappa} negl(\kappa)$ where κ is the bit size of the problem instance. The more one relaxes the bound on the size of the circuits, the stronger the underlying assumption gets in terms of the assumed bound on the success probability of solving the problem; but in any case the size of the circuits are doomed to be sub-linear in the size of the CRS.

Concerning the UC GOS scheme, we first would like to point out that it is of theoretical interest, but it is very inefficient (though poly-time). Furthermore, it has some tricky weak soundness property in that if a dishonest prover should succeed in proving a false statement, then the statement cannot be distinguished from a true one. It is therefore claimed in [26] that the scheme "achieves a weaker, but sufficient, form of adaptive security." This is true but only if some care is taken with the kind of statements that the (dishonest) prover is allowed to prove; in particular, soundness is only guaranteed if the statement to be proven does not incorporate the CRS. Indeed, the same example that the authors use to reason that their first scheme is not adaptively sound can also be applied to the UC secure scheme: Consider a dishonest prover that comes up with an accepting proof for the statement "the CRS is invalid", or for a statement like "the CRS is invalid or S is true" where S is an arbitrary false statement. In real-life, where the CRS is guaranteed to be correct, this convinces the verifier of the truth of the false statement S. would expect such a dishonest prover to be ruled out. However, such a prover is *not* ruled out by the UC security: the simulator given in [26] does generate an *invalid* CRS so that the statement in fact becomes true; and thus the proof can obviously be simulated in the ideal-world (when given a corresponding witness, which the simulator has in case of the UC GOS scheme). We stress that this is not a flaw in the UC GOS scheme but it is the UC security definition that does not provide any security guarantees for statements that incorporate the CRS, essentially because in the ideal-life model there is *no* (guaranteed-to-be-correct) CRS.[1] This issue is addressed in a recent extension of the UC framework [11].

In conclusion, UC NIZK security provides good enough security under the condition that the statements to be proven do not incorporate the CRS. This is automatically guaranteed in a UC setting, where the statements to be proven must make sense in the ideal-world model, but not necessarily in other settings.

2   Preliminaries

2.1   Notation

We consider uniform probabilistic algorithms (i.e. Turing machines) which take as input (the unary encoding of) a security parameter $\kappa \in \mathbb{N}$ and possibly other inputs and run in deterministic poly-time in κ. We thus always implicitly require the size of the input to be bounded by some polynomial in κ. Adversarial behavior is modeled by *non-uniform* poly-time probabilistic algorithms, i.e., by algorithms which together with the security parameter κ also get some poly-size auxiliary input aux_κ. In order to simplify notation, we usually leave the dependency on κ (and on aux_κ) implicit. By $y \leftarrow \mathcal{A}(x)$, we mean that algorithm $\mathcal{A}$ is executed on input x (and the security parameter κ and, in the non-uniform case, aux_κ) and the output is assigned to y. Similarly, for any finite set S, we use the notation $y \leftarrow S$ to denote that y is sampled uniformly from S, and $y \leftarrow x$ means that the value x is assigned to y.

For two algorithms $\mathcal{A}$ and $\mathcal{B}$, $\mathcal{A}\|\mathcal{B}$ denotes the joint execution of $\mathcal{A}$ and $\mathcal{B}$ on the same input and the same random tape, and we write $(x; y) \leftarrow (\mathcal{A}\|\mathcal{B})(w)$ to express that in the joint execution on input w (and the same random tape) $\mathcal{A}$'s output is assigned to x and $\mathcal{B}$'s to y. Furthermore, $P[y = \mathcal{A}(x)]$ denotes the probability (taken over the uniformly distributed random tape) that $\mathcal{A}$ outputs y on input x, and we write $P[x \leftarrow \mathcal{B} : \mathcal{A}(x) = y]$ for the (average) probability that $\mathcal{A}$ outputs y on input x when x is output by $\mathcal{B}$: $P[x \leftarrow \mathcal{B} : \mathcal{A}(x) = y] = \sum_x P[y = \mathcal{A}(x)] P[x = \mathcal{B}]$. We also use natural self-explanatory extensions of this notation.

An *oracle* algorithm $\mathcal{A}$ is an algorithm in the above sense connected to an oracle in that it can write on its own tape an input for the oracle and tell the oracle to execute, and then, in a single step, the oracle processes its input in a prescribed way, and writes its output to the tape. We write $\mathcal{A}^{\mathcal{O}}$ when we consider $\mathcal{A}$ to be connected to the particular oracle $\mathcal{O}$.

---

[1] A minor flaw regarding the UC GOS scheme though is that Groth *et al.* claim the scheme to be *non-malleable*, and their UC NIZK functionality indeed does guarantee non-malleability in that a proof cannot be transformed into a different proof for the same instance without knowing a witness. But it is easy to see that the UC GOS scheme is *not* non-malleable, because the NIZK proof π generated at step 6. in Figure 4 (by using the non-adaptive GOS scheme) is malleable: it uses the NIZK proof from Figure 1 (with h of order n though) which is malleable by raising π_1 and π_3 to some power $s \in \mathbb{Z}_n^*$ and π_2 to power $s^{-1} \pmod{n}$.

As is common practice, a value $\nu(\kappa) \in \mathbb{R}$, which depends on the security parameter κ, is called *negligible*, denoted by $\nu(\kappa) \leq negl(\kappa)$ or $\nu \leq negl$, if $\forall c > 0 \; \exists \kappa_0 \in \mathbb{N} \; \forall \kappa \geq \kappa_0 : \nu(\kappa) < 1/\kappa^c$. Furthermore, $\nu(\kappa) \in \mathbb{R}$ is called *noticeable* if $\exists c > 0, \; \kappa_0 \in \mathbb{N} \; \forall \kappa \geq \kappa_0 : \nu(\kappa) \geq 1/\kappa^c$.

2.2 Definition

Let $L \subseteq \{0,1\}^*$ be an NP-language.

Definition 1. *Consider poly-time algorithms $\mathcal{G}$, $\mathcal{P}$ and $\mathcal{V}$ of the following form: $\mathcal{G}$ takes the security parameter κ (implicitly treated hereafter) and outputs a common reference string (CRS) Σ together with a trapdoor τ. $\mathcal{P}$ takes as input a CRS Σ and an instance $x \in L$ together with an NP-witness w and outputs a proof π. $\mathcal{V}$ takes as input a CRS Σ, an instance x and a proof π and outputs 1 or 0. The triple $(\mathcal{G}, \mathcal{P}, \mathcal{V})$ is a statistical/perfect NIZK argument for L if the following properties hold.*

Completeness: For any $x \in L$ with corresponding NP-witness w

$$P\big[(\Sigma, \tau) \leftarrow \mathcal{G}, \pi \leftarrow \mathcal{P}(\Sigma, x, w) : \mathcal{V}(\Sigma, x, \pi) = 0 \big] \leq negl \, .$$

Soundness: For any non-uniform poly-time adversary $\mathcal{P}^*$

$$P\big[(\Sigma, \tau) \leftarrow \mathcal{G}, (x^*, \pi^*) \leftarrow \mathcal{P}^*(\Sigma) : x^* \notin L \wedge \mathcal{V}(\Sigma, x^*, \pi^*) = 1 \big] \leq negl \, .$$

Statistical/Perfect Zero-Knowledge (ZK): There exists a poly-time simulator $\mathcal{S}$ such that for any instance $x \in L$ with NP-witness w, and for $(\Sigma, \tau) \leftarrow \mathcal{G}$, $\pi \leftarrow \mathcal{P}(\Sigma, x, w)$ and $\pi_{sim} \leftarrow \mathcal{S}(\Sigma, \tau, x)$, the joint distributions of (Σ, π) and (Σ, π_{sim}) are statistically/perfectly close.

Remark 2. The notion of soundness we use here guarantees security against an *adaptive* attacker, which may choose the instance x^* depending on the CRS. We sometimes emphasize this issue by using the term *adaptively-sound*. Note that this is a strictly stronger notion than when the adversary must choose x^* independent of the CRS.

Remark 3. In the notion of ZK we use here, $\mathcal{P}$ and $\mathcal{S}$ use the *same* CRS string. In [17], this is called *same-string* ZK. In the context of *statistical* ZK, this notion is *equivalent* (and not only sufficient) to *unbounded* ZK, which captures that the same CRS can be used an unbounded number of times. This is obviously much more desirable compared to the original notion of NIZK, where every proof requires a fresh CRS. In [17], it is shown that there cannot be a *same-string* NIZK *proof* with statistical soundness for a NP-complete language unless there exist no one-way functions. This makes it even more interesting to find out whether there exists a same-string NIZK argument with statistical security on at least one side, namely the ZK side.

2.3 Bilinear Groups and the Hardness Assumptions

We use the standard setting of bilinear groups. Let $\mathcal{BGG}$ be a *bilinear-group generator* that (takes as input the security parameter κ and) outputs $(\mathbb{G}, \mathbb{H}, q, g, e)$ where $\mathbb{G}$ and $\mathbb{H}$ is a pair of groups of prime order q, g is a generator of $\mathbb{G}$, and e is a non-degenerate bilinear map $e : \mathbb{G} \times \mathbb{G} \to \mathbb{H}$, meaning that $e(g^a, g^b) = e(g, g)^{ab}$ for any $a, b \in \mathbb{Z}_q$ and $e(g, g) \neq 1_{\mathbb{H}}$.

We assume the Discrete-Log Assumption, DLA, that for a random $h \in \mathbb{G}$ it is hard to compute $w \in \mathbb{Z}_q$ with $h = g^w$. In some cases, we also assume the Diffie-Hellman Inversion Assumption, DHIA, which states that, for a random $h = g^w \in \mathbb{G}$, it is hard to compute $g^{1/w}$. Formally, these assumptions for a bilinear-group generator $\mathcal{BGG}$ are stated as follows. In order to simplify notation, we abbreviate the output $(\mathbb{G}, \mathbb{H}, q, g, e)$ of $\mathcal{BGG}$ by *pub* (for "public parameters").

Assumption 4 (DLA). For every non-uniform poly-time algorithm $\mathcal{A}$

$$P\big[\, pub \leftarrow \mathcal{BGG}, \; h \leftarrow \mathbb{G}, \; w \leftarrow \mathcal{A}(pub, h) : \; g^w = h \,\big] \leq negl \,.$$

Assumption 5 (DHIA). For every non-uniform poly-time algorithm $\mathcal{A}$

$$P\big[\, pub \leftarrow \mathcal{BGG}, \; h \leftarrow \mathbb{G}, \; g^{1/w} \leftarrow \mathcal{A}(pub, h) : \; g^w = h \,\big] \leq negl \,.$$

Furthermore, we assume XKEA, a variant of the Knowledge-of-Exponent Assumption KEA, (referred to as KEA3 respectively KEA1 in [5]). KEA informally states that given $\hat{g} = g^x \in \mathbb{G}$ with unknown discrete-log x, the only way to efficiently come up with a pair $A, \hat{A} \in \mathbb{G}$ such that $\hat{A} = A^x$ for the same x is by choosing some $a \in \mathbb{Z}_q$ and computing $A = g^a$ and $\hat{A} = \hat{g}^a$. XKEA states that given $\hat{g} = g^x \in \mathbb{G}$ as well as another pair h and $\hat{h} = h^x$ with the same unknown discrete-log x, the only way to efficiently come up with a pair $A, \hat{A}$ such that $\hat{A} = A^x$ is by choosing $a, \alpha \in \mathbb{Z}_q$ and computing $A = g^a h^\alpha$ and $\hat{A} = \hat{g}^a \hat{h}^\alpha$. Formally, KEA and XKEA are phrased by assuming that for every algorithm which outputs A and $\hat{A}$ as required, there exists an *extractor* which outputs a (and α in case of XKEA) when given the same input and randomness.

Assumption 6 (KEA). For every non-uniform poly-time algorithm $\mathcal{A}$ there exists a non-uniform poly-time algorithm $\mathcal{X}_\mathcal{A}$, the *extractor*, such that

$$P\big[\, pub \leftarrow \mathcal{BGG}, \; x \leftarrow \mathbb{Z}_q, \; (A, \hat{A}; a) \leftarrow (\mathcal{A} \| \mathcal{X}_\mathcal{A})(pub, g^x) : \; \hat{A} = A^x \wedge A \neq g^a \,\big] \leq negl \,.$$

Recall that $(A, \hat{A}; a) \leftarrow (\mathcal{A} \| \mathcal{X}_\mathcal{A})(pub, g^x)$ means that $\mathcal{A}$ and $\mathcal{X}_\mathcal{A}$ are executed on the same input (pub, g^x) and the *same random tape*, and $\mathcal{A}$ outputs $(A, \hat{A})$ whereas $\mathcal{X}_\mathcal{A}$ outputs a.

Assumption 7 (XKEA). For every non-uniform poly-time algorithm $\mathcal{A}$ there exists a non-uniform poly-time algorithm $\mathcal{X}_\mathcal{A}$, the *extractor*, such that

$$P\left[\, \begin{array}{l} pub \leftarrow \mathcal{BGG}, \; x \leftarrow \mathbb{Z}_q, \; h \leftarrow \mathbb{G}, \\ (A, \hat{A}; a, \alpha) \leftarrow (\mathcal{A} \| \mathcal{X}_\mathcal{A})(pub, g^x, h, h^x) \end{array} : \; \hat{A} = A^x \wedge A \neq g^a h^\alpha \,\right] \leq negl \,.$$

It is well known that DLA holds provably with respect to generic algorithms (see e.g. [34]), which operate on the group elements only by applying the group operations (multiplication and inversion), but do not make use of the specific representation of the group elements. It is not so hard to see that this result extends to groups $\mathbb{G}$ that come with a bilinear pairing $e : \mathbb{G} \times \mathbb{G} \rightarrow \mathbb{H}$, i.e., to generic algorithms that are additionally allowed to apply the pairing and the group operations in $\mathbb{H}$. We prove in Section 5 that also KEA and XKEA hold with respect to generic algorithms.

We would also like to point out that we only depend on XKEA for "proof-technical" reasons: our perfect NIZK argument still *appears* to be secure even if XKEA should turn out to be false (for the particular generator $\mathcal{BGG}$ used), but we cannot prove it anymore formally. This is in contrast to how KEA and XKEA are used in [27] respectively [5] for 3-round ZK, where there seems to be no simulator anymore as soon as KEA is false.

3 A Perfect NIZK Argument for SAT

3.1 Handling Multiplication Gates

Let $(\mathbb{G}, \mathbb{H}, q, g, e)$ be generated by $\mathcal{BGG}$, as described in Section 2.3 above. Furthermore, let $h = g^w$ for a random $w \in \mathbb{Z}_q$ which is unknown to anybody. Consider a prover who announces an arithmetic circuit over $\mathbb{Z}_q$ and who wants to prove in NIZK that there is a satisfying input for it. Following a standard design principle, where the prover commits to every input value using Pedersen's commitment scheme with "basis" g and h as well as to every intermediate value of the circuit when evaluating it on the considered input, the problem reduces to proving the consistency of the multiplication gates in NIZK (whereas the addition gates come for free due to the homomorphic property of Pedersen's commitment scheme).

Concretely, though slightly informally, given commitments $A = g^a h^\alpha$, $B = g^b h^\beta$ and $C = g^c h^\gamma$ for values a, b and $c \in \mathbb{Z}_q$, respectively, the prover needs to prove in NIZK that $c = a \cdot b$. Note that

$$e(A, B) = e(g^a h^\alpha, g^b h^\beta) = e(g, g)^{ab} \, e(g, h)^{a\beta + \alpha b} \, e(h, h)^{\alpha\beta} \qquad \text{and}$$

$$e(C, g) = e(g^c h^\gamma, g) = e(g, g)^c \, e(g, h)^\gamma$$

and hence, if indeed $c = a \cdot b$, then

$$e(A, B)/e(C, g) = e(g, h)^{a\beta + \alpha b - \gamma} \, e(h, h)^{\alpha\beta} = e(g^{a\beta + \alpha b - \gamma} h^{\alpha\beta}, h) . \qquad (1)$$

Say that, in order to prove that $c = a \cdot b$, the prover announces $P = g^{a\beta + \alpha b - \gamma} h^{\alpha\beta}$ and the verifier accepts if and only if P is *satisfying* in that

$$e(A, B)/e(C, g) = e(P, h) .$$

Then, by the above observations it is immediate that an honest verifier accepts the correct proof of an honest prover. Also, it is quite obvious that a simulator which knows w can "enforce" $c = a \cdot b$ by "cheating" with the commitments, and thus perfectly simulate a satisfying P for the multiplication gate. Note that the simulator needs to know *some* opening of the commitments in order to simulate P; this though is good enough for our purpose. For completeness, though, we address this issue again in Section 4 and show a version which allows a full-fledged simulation. Finally, it appears to be hard to come up with a satisfying P unless one can indeed open A, B and C to a, b and c such that $c = a \cdot b$. Concretely, the following holds.

Lemma 8. *Given openings of A, B and C to a, b and c, respectively, with $c \neq a \cdot b$, and given an opening of a satisfying P, one can efficiently compute w.*

Proof. Let $P = g^\rho h^\varpi$ be the given opening of P. Then, inheriting the notation from above,

$$e(A, B)/e(C, g) = e(g^a h^\alpha, g^b h^\beta)/e(g^c h^\gamma, g) = e(g, g)^{ab-c} e(g, h)^{a\beta + \alpha b - \gamma} e(h, h)^{\alpha\beta}.$$

and

$$e(A, B)/e(C, g) = e(P, h) = e(g^\rho h^\varpi, h) = e(g, h)^\rho \, e(h, h)^\varpi$$

are two different representations of the same element in $\mathbb{H}$ with respect to the "basis" $e(g, g)$, $e(g, h) = e(g, g)^w$, $e(h, h) = e(g, g)^{w^2}$. This allows to compute w by solving a quadratic equation in $\mathbb{Z}_q$. □

The need for an opening of P can be circumvented by basing security on DHIA rather than DLA as stated in the following lemma.

Lemma 9. *Given openings of A, B and C to a, b and c, respectively, with $c \neq a \cdot b$, and given a satisfying P, one can efficiently compute $g^{1/w}$.*

Proof. For a satisfying P it holds that

$$e(P, h) = e(A, B)/e(C, g) = e(g, g)^{ab-c} \, e(g, h)^{a\beta + b\alpha - \gamma} \, e(h, h)^{\alpha\beta}$$

and thus, when $c \neq a \cdot b$ as assumed, the following equalities follow one after the other.

$$e(g, g) = e\big((P g^{-a\beta - b\alpha + \gamma} h^{-\alpha\beta})^{1/(ab-c)}, h\big)$$

$$e(g^{1/w}, g) = e\big((P g^{-a\beta - b\alpha + \gamma} h^{-\alpha\beta})^{1/(ab-c)}, g\big)$$

$$g^{1/w} = (P g^{-a\beta - b\alpha + \gamma} h^{-\alpha\beta})^{1/(ab-c)}$$

Thus, $g^{1/w}$ can be computed from the available information. □

It remains to argue that a (successful) prover can indeed open all the necessary commitments. This can be enforced as follows. Instead of committing to every value s by $S = g^s h^\sigma$, the prover has to commit to s by $S = g^s h^\sigma$ and $\hat{S} = \hat{g}^s \hat{h}^\sigma$, where $\hat{g} = g^x$ for a random $x \in \mathbb{Z}_q$ and $\hat{h} = h^x$ (with the same x). Note that the same randomness σ is used for computing S and $\hat{S}$, such that $\hat{S} = S^x$; this can be verified using the bilinear map: $e(\hat{S}, g) = e(S, \hat{g})$. XKEA now guarantees that for every correct double commitment $(S, \hat{S})$ produced by the prover, he knows (respectively there exists an algorithm that outputs) s and σ such that $S = g^s h^\sigma$.

Based on the above observations, we construct and prove secure an adaptively-sound perfect NIZK argument for circuit-SAT in the next section.

3.2   The Perfect NIZK Scheme

The NIZK scheme for circuit-SAT is given in Figure 1. Note that we assume an arithmetic circuit C over $\mathbb{Z}_q$ (rather than a binary circuit), but of course it is standard to "emulate" a binary circuit by an arithmetic one.

CRS Generator $\mathcal{G}(1^\kappa)$:

G-1. $(\mathbb{G}, \mathbb{H}, q, g, e) \leftarrow \mathcal{BGG}(1^\kappa)$, $w \leftarrow \mathbb{Z}_q$, $\hat{g} \leftarrow \mathbb{G}$, $h \leftarrow g^w$, $\hat{h} \leftarrow \hat{g}^w$,

G-2. output $\Sigma \leftarrow (\mathbb{G}, \mathbb{H}, q, g, h, \hat{g}, \hat{h}, e)$ and $\tau \leftarrow w$.

Prover $\mathcal{P}(\Sigma, \mathsf{C}, x = (x_1, \ldots, x_n))$:

P-1. Compute commitments for every input x_i by $X_i = g^{x_i} h^{\xi_i}$ and $\hat{X}_i = \hat{g}^{x_i} \hat{h}^{\xi_i}$.

P-2. Inductively, for every multiplication gate in C for which the two input values a and b are committed upon (either directly or indirectly via the homomorphic property) by $A = g^a h^\alpha$ and $\hat{A} = \hat{g}^a \hat{h}^\alpha$ respectively $B = g^b h^\beta$ and $\hat{B} = \hat{g}^b \hat{h}^\beta$, do the following. Compute a (double) commitment $C = g^c h^\gamma$ and $\hat{C} = \hat{g}^c \hat{h}^\gamma$ for the corresponding output value $c = a \cdot b$, and compute the (double) commitment $P = g^{a\beta+\alpha b - \gamma} h^{\alpha\beta}$ and $\hat{P} = \hat{g}^{a\beta+\alpha b - \gamma} \hat{h}^{\alpha\beta}$.

P-3. As proof π output all the commitments as well as the randomness η for the commitment $Y = g^{\mathsf{C}(x)} h^\eta$ for the output value $\mathsf{C}(x) = 1$.

Verifier $\mathcal{V}(\Sigma, \mathsf{C}, \pi)$:

Output 1 (i.e. "accept") if all of the following holds, otherwise output 0.

V-1. Every double commitment $(S; \hat{S})$ satisfies $e(\hat{S}, g) = e(S, \hat{g})$.

V-2. Every multiplication gate in C, with associated (double) commitments $(A, \hat{A})$, $(B, \hat{B})$, $(C, \hat{C})$ and $(P, \hat{P})$ for the two input values, the output value and the "multiplication proof", satisfies $e(A, B)/e(C, g) = e(P, h)$.

V-3. The commitment Y for the output value satisfies $Y = g^1 h^\eta$.

Fig. 1. Perfect NIZK argument for circuit-SAT

Theorem 10. $(\mathcal{G}, \mathcal{P}, \mathcal{V})$ *from Fig. 1 is an adaptively-sound perfect NIZK argument for circuit-SAT, assuming* XKEA *and* DLA.

Proof. Completeness is straightforward using observation (1). Also, perfect ZK is easy to see. Indeed, the simulator $\mathcal{S}$ can run $\mathcal{P}$ with a default input for x, say $o = (0, \ldots, 0)$, and then simply open the commitment Y for the output value

$y = \mathsf{C}(o)$ (which is likely to be different from 1) to 1 using the trapdoor w. Since Pedersen's commitment scheme is perfectly hiding, and since P and $\hat{P}$ computed in step P-2. for every multiplication gate are uniquely determined by A, B, and C, it is clear that this simulation is perfectly indistinguishable from a real execution of $\mathcal{P}$.

It remains to argue soundness. Assume there exists a dishonest poly-time prover $\mathcal{P}^*$, which on input the CRS Σ outputs a circuit C^* together with a proof π^* such that with non-negligible probability, C^* is not satisfiable but $\mathcal{V}(\Sigma, \mathsf{C}^*, \pi^*)$ outputs 1. By XKEA, there exists a poly-time extractor $\mathcal{X}_{\mathcal{P}^*}$ such that when run on the same CRS and the same random tape as $\mathcal{P}^*$, the extractor $\mathcal{X}_{\mathcal{P}^*}$ outputs the opening information for all (double) commitments in the proof with non-negligible probability.[2] Concretely, for every multiplication gate and the corresponding commitments A, B, C and P, the extractor $\mathcal{X}_{\mathcal{P}^*}$ outputs $a, \alpha, b, \beta, c, \gamma, \rho, \varpi$ such that $A = g^a h^\alpha$, $B = g^b h^\beta$, $C = g^c h^\gamma$ and $P = g^\rho h^\varpi$. If $\mathcal{P}^*$ succeeds in forging a proof for an unsatisfiable circuit, then there obviously must be an inconsistent multiplication gate with inputs a and b and output $c \neq a \cdot b$. (Note that since addition gates are processed using the homomorphic property, there cannot be an inconsistency in an addition gate.) But this contradicts DLA by Lemma 8. $\qquad\square$

Remark 11. The NIZK argument from Fig. 1 actually provides *adaptive ZK*, which is a stronger flavor of ZK than guaranteed by Definition 1. It guarantees that $\mathcal{S}$ cannot only perfectly simulate a proof π for any circuit C, but when later given a satisfying input x for C, it can also provide the randomness that explains how π could have been generated by running $\mathcal{P}$ on witness x.

Remark 12. It is reasonable to assume that one can efficiently verify that, for given $(\mathbb{G}, \mathbb{H}, q, g, e)$, $\mathbb{G}$ and $\mathbb{H}$ are groups of order q, g generates $\mathbb{G}$, and e is a non-degenerate bilinear map. Then, the CRS $\Sigma = (\mathbb{G}, \mathbb{H}, q, g, h, \hat{g}, \hat{h}, e)$ may be generated by the (possibly dishonest) verifier, together with an (interactive) ZK proof of the knowledge of w with $g^w = h$, which can be done very efficiently by using the 4-round ZK proof from [13] for instance. The prover additionally needs to check if $e(\hat{g}, h) = e(g, \hat{h})$. Hence, the set-up of the CRS requires no honest party nor any expensive 2-party (or multi-party) computation. If the proof of knowledge of w is omitted, so that the verifier only publishes the CRS Σ, then the argument is still *witness indistinguishable*. Thus, our scheme can also be appreciated as a (computationally sound) *zap* [21].

Remark 13. By omitting $\hat{P}$ (and the corresponding verifications), one can obtain a slightly more efficient protocol based on the possibly stronger assumption DHIA instead of DLA. The security can be proven in exactly the same way based on

---

[2] As a matter of fact, XKEA guarantees that for every double commitment there exists an extractor that outputs the opening *for that particular commitment* with non-negligible probability; but of course running all these extractors together allows to extract for all commitments simultaneously with non-negligible probability (since the size of C must be polynomially bounded).

Lemma 9 instead of Lemma 8. Furthermore, if one is willing to trade XKEA by a new assumption (though of similar flavor, but which can also be proven in the generic model) that the only way to come up with A, B, C and P such that $e(A, B)/e(C, g) = e(P, h)$ is by choosing A, B and C as commitments of a, b and $c = a \cdot b$, respectively, then one can get a NIZK scheme where (not counting the unavoidable commitments A, B and C) the proof for each multiplication gate consists of only 1 group element, P. Note that this requires less communication than using standard interactive ZK techniques in combination with the Fiat-Shamir heuristic [23].

4   Efficient Proof for Relations Among Commitments

We again consider the problem of proving that a Pedersen commitment $C = g^c h^\gamma$ "contains" the product $c = a \cdot b$ of a and b committed to by $A = g^a h^\alpha$ respectively $B = g^b h^\beta$. Recall that the multiplication proof discussed in Section 3.1, consisting of P such that $e(A, B)/e(C, g) = e(P, h)$ (and maybe the corresponding $\hat{P}$), can only be simulated if the simulator knows *some* openings of A, B and C. This was good enough for the application to NIZK for SAT, as in this case all commitments may be prepared by the simulator. However, for other applications, it might be desirable to have a similarly efficient non-interactive multiplication proof which allows a fully-fledged simulation, i.e., which can be simulated for any *given* A, B and C. We show in this section a modification of the multiplication proof of Section 3.1 which has this property.

The setting is the same as in the previous section; We assume that a CRS $\Sigma = (\mathbb{G}, \mathbb{H}, q, g, h, \hat{g}, \hat{h}, e)$ has been properly set up and is publicly available. Per default, the prover is required to provide $\hat{A}$, $\hat{B}$ and $\hat{C}$ for the commitments A, B and C in question, and the verifier should check if $e(\hat{g}, A) = e(g, \hat{A})$ etc., so that the opening of A, B and C can be extracted via XKEA. Note that such $\hat{A}$, $\hat{B}$ and $\hat{C}$ can be computed from A, B and C and $x = \log_g \hat{g} \in \mathbb{Z}_q$. Thus, the ZK simulator who knows x can simulate $\hat{A}$, $\hat{B}$ and $\hat{C}$ without knowing the openings of the original commitments. For ease of description, these "hatted" commitments and corresponding verifications are treated implicitly hereafter. We begin with a simple relation for proving that a commitment A can be opened to zero.

Open to Zero ($a = 0$): For $A = g^0 h^\alpha$, the prover publishes $P = g^\alpha$. The verifier accepts if $e(g, A) = e(h, P)$.

It is obvious that an honest verifier accepts the correct proof of an honest prover. ZK is straightforward: the simulator who knows w can compute $P = A^{1/w}$ (without knowing the opening of A). Finally, given an opening (a, α) of A and a satisfying P, i.e., such that $e(g, A) = e(h, P)$, if $a \neq 0$ then one can efficiently compute $g^{1/w}$. This follows from the following equalities:

$$e(g, g^a h^\alpha) = e(h, P)$$
$$e(g, g^a) = e(g, Pg^{-\alpha})^w$$
$$e(g, g^{1/w}) = e(g, (Pg^{-\alpha})^{1/a})$$

and thus $g^{1/w} = (Pg^{-\alpha})^{1/a}$.

The above protocol for opening to zero can be easily applied to show equality $(a = b)$ and addition $(a + b = c)$ by replacing A in the above protocol with A/B and AB/C, respectively.

We next show a proof system for multiplicative relation $a \cdot b = c$. Recall that the goal is to have a multiplication proof which allows a simulation for any A, B and C given to the simulator.

Multiplication $(ab = c)$: The prover publishes $P = (R, S, T)$ such that $R = h^r$, $S = g^r$ for random r and $T = g^{a\beta + ab - \gamma - ar} h^{\alpha\beta - \alpha r}$. The verifier accepts if $e(g, R) = e(h, S)$ and $e(A, B)/e(g, C) = e(A, R)e(h, T)$.

Completeness is verified by seeing that the first verification equation follows from $e(g, R) = e(g, h^r) = e(g^r, h) = e(h, S)$, and the second from $e(A, B)/e(g, C) = e(g, g)^{ab-c} e(g, h)^{a\beta + ab - \gamma} e(h, h)^{\alpha\beta}$ in combination with

$$e(A, R)e(h, T) = e(g^a h^\alpha, h^r)e(h, g^{a\beta + ab - \gamma - ar} h^{\alpha\beta - \alpha r})$$
$$= e(g^a, h^r)e(h^\alpha, h^r)e(h, g^{a\beta + ab - \gamma - ar})e(h, h^{\alpha\beta - \alpha r})$$
$$= e(g, h)^{a\beta + ab - \gamma} e(h, h)^{\alpha\beta},$$

which gives the desired equality if indeed $ab - c = 0$.

Fully-fledged ZK and soundness are captured by following Lemma 14 and 15, respectively.

Lemma 14. *Given A, B and C, one can efficiently simulate random R, S and T such that $e(g, R) = e(h, S)$ and $e(A, B)/e(g, C) = e(A, R)e(h, T)$.*

Proof. Given the trapdoor w, the simulator picks random u and sets $R = Bh^u$, $S = R^{1/w}$, and $T = A^{-u} C^{-1/w}$. As in the real proof, (S, R, T) is uniformly distributed subject to the verification equations:

$$e(h, S) = e(h, R^{1/w}) = e(g, R)$$

and

$$e(A, R)e(h, T) = e(A, Bh^u)e(h, A^{-u} C^{-1/w})$$
$$= e(A, B)e(A^u, h)e(h, A^{-u})e(g, C^{-1})$$
$$= e(A, B)/e(g, C).$$

Thus, the simulation is perfect. $\square$

Lemma 15. *Given openings of A, B and C to a, b and c, respectively, with $c \neq a \cdot b$, and given a satisfying $P = (R, S, T)$, one can efficiently compute $g^{1/w}$.*

Proof. We first observe that R and S constitute a proof of zero-opening. Hence we can say that $R = h^r$ for some r. Furthermore, we can extract such r by applying XKEA to R and S since they are in correct relation verified by $e(g, R) = e(h, S)$. Now, for a, α, b, β, c, γ and r, we see the following holds from the second verification equation:

$$e(A, B)/e(g, C) = e(A, R)\, e(h, T)$$

$$e(g^a h^\alpha, g^b h^\beta)\, e(g, g^{-c} h^{-\gamma}) = e(g^a h^\alpha, R)\, e(h, T)$$

$$e(g, g^{ab-c}) = e(h, g^{-a\beta - \alpha b + ra + \gamma} h^{-\alpha\beta} R^\alpha T)$$

$$e(g, g^{1/w}) = e(g, g^{-a\beta - \alpha b + ra + \gamma} h^{-\alpha\beta} R^\alpha T)^{1/(ab-c)}$$

Therefore, if $ab \neq c$, one can compute $g^{1/w} = (g^{-a\beta - \alpha b + ra + \gamma} h^{-\alpha\beta} R^\alpha T)^{1/(ab-c)}$, which contradicts to DHIA. $\qquad\square$

Now, we need to discuss what kind of NIZK arguments these protocols formally are. The crucial issue stems from the fact that Pedersen's commitment scheme is unconditionally hiding and thus the language of all triples (A, B, C) which allow an opening with $a \cdot b = c$ is trivial. Therefore, proving a statement among these commitments only makes sense in terms of *proof of knowledge*. By Lemma 15, the "knowledge soundness" can be proven by using the extractor of XKEA as knowledge extractor. Accordingly, the quality of the extractor of XKEA immediately affects to the quality of the knowledge extractor. Since XKEA only provides a non-black-box extractor, the best the protocol can achieve is a proof of knowledge characterized by a non-black-box knowledge extractor.

Let $\mathcal{R}$ be a binary poly-time relation, which we allow to depend on (κ and) Σ in order to capture schemes that prove something about commitments with "basis" g and h, which are part of the CRS. Let $L_\mathcal{R} = \{x \mid \exists w : (x, w) \in \mathcal{R}\}$ be the language characterized by $\mathcal{R}$.

Definition 16. *A NIZK proof of knowledge for $\mathcal{R}$ is a NIZK proof (or argument) for $L_\mathcal{R}$ such that additionally for every non-uniform poly-time prover $\mathcal{P}^*$ there exists a non-uniform poly-time extractor $\mathcal{E}_{\mathcal{P}^*}$ such that*

$$P\left[\begin{matrix} (\Sigma, \tau) \leftarrow \mathcal{G}, \\ (x^*, \pi^*; w^*) \leftarrow (\mathcal{P}^* \| \mathcal{E}_{\mathcal{P}^*})(\Sigma) \end{matrix} : (x^*, w^*) \notin \mathcal{R} \wedge \mathcal{V}(\Sigma, x^*, \pi^*) = 1\right] \leq negl.$$

Such NIZK proof of knowledge with non-black-box extractor might be weaker than the one with universal black-box extractor originally defined in [19]. This issue is analogue to black-box vs non-black-box ZK where both definitions are widely accepted. Although a stronger definition is in general favorable, a weaker definition has potential to capture nicer schemes with weaker assumptions or even schemes that are impossible otherwise, but still guarantees sufficient security.

The following now follows immediately from the above lemmas.

Theorem 17. *The above scheme gives a perfect NIZK proof of knowledge for*
$\mathcal{R}_{\text{mult}} = \{((A,B,C),(a,\alpha,b,\beta,\gamma)) \in \mathbb{G}^3 \times \mathbb{Z}_q^5 \mid A = g^a h^\alpha, B = g^b h^\beta, C = g^{ab} h^\gamma\}$
under XKEA *and* DHIA.

Finally, we note that all the statements in this section can be strengthened to
be based on DLA rather than DHIA by additionally providing $\hat{P} = (\hat{R}, \hat{S}, \hat{T})$,
similarly as for the proof of SAT in Section 3.

5   The Security of (X)KEA Against Generic Attacks

The notion of a generic algorithm is due to Nechaev [31] and Shoup [34], where
it was shown that the discrete-log problem is hard for generic algorithms. Informally, a *generic algorithm* for trying to solve some DL-related problem in a
group $\mathbb{G}$ is one that does not exploit and thus does not depend on the actual
encoding of the group elements, but only relies on the group structure (and
that the encoding is injective). In our context, where $\mathbb{G}$ allows a bilinear map
$e : \mathbb{G} \times \mathbb{G} \to \mathbb{H}$, we also allow the algorithm to make use of the bilinear map and
the group structure in $\mathbb{H}$.

   Formally, a generic algorithm for a bilinear group is an oracle algorithm $\mathcal{A}$
which takes as input a prime q, encodings of elements of $\mathbb{Z}_q$ with respect to a *random* injective encoding function $\sigma : \mathbb{Z}_q \to S$, and possibly encodings of elements
of $\mathbb{Z}_q$ with respect to another random encoding function $\tau : \mathbb{Z}_q \to T$ (with finite
$S, T \subset \{0,1\}^*$). Furthermore, $\mathcal{A}$ is allowed to make oracle queries in order to
compute on encoded group elements: upon a query $(\text{add\_in\_G}, sign, \sigma(y), \sigma(z))$
the oracle $\mathcal{O}$ replies by $\sigma(y + (-1)^{sign} z)$ and upon $(\text{add\_in\_H}, sign, \tau(y), \tau(z))$ by
$\tau(y + (-1)^{sign} z)$, and upon $(\text{pair}, \sigma(y), \sigma(z))$ the oracle replies by $\tau(y \cdot z)$. Note
that the **add**-queries model the group operations in $\mathbb{G}$ and $\mathbb{H}$, and the **pair**-query
models the pairing $e : \mathbb{G} \times \mathbb{G} \to \mathbb{H}$.

   Interestingly, in the literature a generic algorithm $\mathcal{A}$ is typically only allowed
to query the oracle on encodings that it has received either as input or as a reply
to one of the previous queries, but it is not allowed to take such an encoding
and, say, flip the last bit and query the oracle on that encoding. Sometimes
(but not always), this is argued by letting the set of encodings (here S and T)
be so large that essentially any such query would be invalid anyway. But this
also implies that $\mathcal{A}$ cannot sample random group elements without "knowing"
their discrete-log. We do not want to make such a restriction, in particular in the
context of KEA; even though such a step does not appear to be beneficial, we still
feel it should be taken care off in a rigorous analysis. In order to avoid having to
deal with invalid encodings, we assume that $S = T = \mathbb{Z}_q$ (actually, the natural
representation of $\mathbb{Z}_q$ as strings) and that $\mathcal{A}$ queries $\mathcal{O}$ only on valid encodings,
meaning strings in $\mathbb{Z}_q$. In some sense this models groups whose elements can be
efficiently recognized.

Theorem 18. *The assumptions* KEA *and* XKEA *over bilinear groups hold with
respect to generic algorithms (as long as $1/q$ is negligible).*

Note that the generic security of KEA in the standard (rather than the bilinear) group setting was concurrently and independently shown by Dent [20].

Proof. Let us first consider KEA. A generic algorithm $\mathcal{A}$ takes as input $\sigma(1)$ and $\sigma(x)$ for a random $x \in \mathbb{Z}_q$, and it should output $\sigma(a)$ and $\sigma(ax)$ for some $a \in \mathbb{Z}_q$. Let m be the (polynomial) number of oracle queries $\mathcal{A}$ makes. It is easy to see that any encoding that $\mathcal{A}$ might use (or receive) in an oracle query or that $\mathcal{A}$ might output is of the form $\sigma\big(P_k(x, u_1, \ldots, u_{2m})\big)$ respectively $\tau\big(Q_k(x, u_1, \ldots, u_{2m})\big)$ for multi-variate polynomials $P_k \in \mathbb{Z}_q[X, U_1, \ldots, U_{2m}]$ of total degree at most 1 respectively $Q_k \in \mathbb{Z}_q[X, U_1, \ldots, U_{2m}]$ of total degree at most 2, and for random (but fixed once and for all P_k and Q_k) pairwise-different $u_1, \ldots, u_{2m} \in \mathbb{Z}_q \setminus \{x\}$. Indeed, $\sigma(1)$ and $\sigma(x)$ correspond to the polynomials 1 and X, every encoding that $\mathcal{A}$ uses in a query which is fresh in that it has not been given to $\mathcal{A}$ in a reply (or as input) corresponds to a new variable U_i, and any reply given by the oracle corresponds to a P_k respectively Q_k that is inductively computed as $P_k = P_i \pm P_j$ respectively as $Q_k = Q_i \pm Q_j$ or $Q_k = P_i \cdot P_j$. In particular, it is easy to see that by observing $\mathcal{A}$'s oracle queries, one can keep track of these polynomials.

We now define the extractor $\mathcal{X}_{\mathcal{A}}$ as follows. $\mathcal{X}_{\mathcal{A}}$ runs $\mathcal{A}$ but keeps track of these polynomials P_k and Q_k; and if the two polynomials $P_{\text{out}_0}, P_{\text{out}_1} \in \{P_k\}_{k=1\ldots m}$ that correspond to the two encodings that $\mathcal{A}$ outputs are of the form $P_{\text{out}_0} = a$ and $P_{\text{out}_1} = aX$, then it outputs a and otherwise 0.

Let us analyze $\mathcal{X}_{\mathcal{A}}$. Obviously, if $\mathcal{X}_{\mathcal{A}}$ fails (in that $\mathcal{A}$ outputs $\sigma(a)$ and $\sigma(ax)$ but $\mathcal{X}_{\mathcal{A}}$ does not output a) then, by the restriction on the degree, $P_{\text{out}_1} \neq X \cdot P_{\text{out}_0}$, whereas P_{out_1} and $X \cdot P_{\text{out}_0}$ coincide when evaluated at $(x, u_1, \ldots, u_{2m})$. The event that $\mathcal{X}_{\mathcal{A}}$ fails is thus contained in the event $\mathcal{E}$ that at least two distinct polynomials in $\{P_{\text{out}_1}, X \cdot P_{\text{out}_0}\}$, in $\{P_k\}_{k=1\ldots m}$ or in $\{Q_k\}_{k=1\ldots m}$ evaluate to the same value when applied to $(x, u_1, \ldots, u_{2m})$. The standard argument for analyzing generic algorithms, using Schwartz' Lemma below, guarantees that the probability of $\mathcal{E}$ is upper bounded by $O(m^2/q)$; since m is polynomial in κ, this proves the claim (for KEA).[3]

The proof for XKEA uses exactly the same reasoning, the only difference is that $\mathcal{A}$ gets four inputs, encodings of $1, x, w, xw \in \mathbb{Z}_q$, which are associated with the polynomials $1, X, W, XW \in \mathbb{Z}_q[X, W, U_1, \ldots, U_{2m}]$, and $\mathcal{X}_{\mathcal{A}}$ outputs a, α if P_{out_0} is of the form $P_{\text{out}_0} = a + \alpha W$ (which is the only P_{out_0} which allows $P_{\text{out}_1} = X \cdot P_{\text{out}_0}$). As above we can argue that if $\mathcal{X}_{\mathcal{A}}$ fails then $P_{\text{out}_1}(x, w, u_1, \ldots, u_{2m}) = x \cdot P_{\text{out}_0}(x, w, u_1, \ldots, u_{2m})$ but $P_{\text{out}_1} \neq X \cdot P_{\text{out}_0}$. Rea-

---

[3] To make the argument rigorous, one has to consider a modified "game" where $\mathcal{A}$ is provided with random encodings as long as the corresponding polynomial (rather than its evaluation) is new, and then observe that one can define a joint probability distribution for the original and the modified game which leaves the individual (marginal) distributions intact, and such that $\mathcal{E}$ occurs in the original game if and only if it occurs in the modified one (and thus has the same probability in both cases). In the modified game, however, the polynomials are chosen completely independent of $(x, u_1, \ldots, u_{2m})$ and thus we can apply Schwartz' Lemma.

soning as above, the probability of this to happen can again be upper bounded by $O(m^2/q)$. $\qquad\square$

Lemma 19 (Schwartz [33]). *Let $q \in \mathbb{Z}$ be a prime. For any polynomial P in $\mathbb{Z}_q[X_1, \ldots, X_n]$ of total degree at most d, the probability that P vanishes on a uniformly distributed tuple $(x_1, \ldots, x_n) \in \mathbb{Z}_q^n$ is at most d/q.*

6 Eliminating XKEA by Pre-processing

In this section, we briefly discuss the possibility of circumventing XKEA, and to solely rely on standard assumptions, by allowing *pre-processing*. Note that in all of the above results, XKEA is only needed in order to *extract* openings of commitments that are prepared by the possibly dishonest player. There-fore, a possible way to circumvent XKEA is to have all players prepare in a pre-processing phase *random* commitments $U = g^u h^\nu$ in such a way that one can extract the openings (u, ν) of these commitments in the security proof: for instance in the 2-player setting by a standard interactive ZK proof of knowl-edge (e.g. the 4-round ZK scheme from [13]), or in the multi-player setting with dishonest minority by a simple Pedersen VSS sharing. Then, when the actual NIZK argument needs to be executed, instead of providing for every commit-ment $S = g^s h^\sigma$ its hatted version $\hat{S} = \hat{g}^s \hat{h}^\sigma$, for every commitment $S = g^s h^\sigma$ the opening $(s + u, \sigma + \nu)$ of SU is provided, where U is a fresh unused random commitment from the pre-processing phase. This then obviously also allows to extract s in the security proof as required. There are some feasibility results about statistical NIZK arguments in the pre-processing model, cf. [18,28,15], which rely only on general assumptions but require a complicated pre-processing stage.

Beaver's pre-processing techniques [4] can be applied in a straightforward way to yield similarly efficient schemes as we do. However, this approach requires the generation of random commitments *with multiplicative relations* in the pre-processing phase, whereas with our techniques purely random commitments, which are potentially easier to prepare, suffice. For instance in the multi-player setting, this is known as the *linear* pre-processing model [16], and when the number of players is small, using the techniques of [12], one can have a once-and-for-all pre-processing stage that allows to produce an *unbounded* number of pseudo-random commitments on the fly.

Acknowledgments

We would like to thank Alexander Dent for useful and interesting remarks re-garding our proof of the generic security of KEA and XKEA in the bilinear group setting (Section 5). Thanks also to the anonymous reviewers of TCC'07 for their invaluable comments.

References

1. M. Abe and S. Fehr. Perfect NIZK with adaptive soundness. Cryptology ePrint Archive, Report 2006/423, 2006. http://eprint.iacr.org.
2. L. M. Adleman. Two theorems on random polynomial time. In *19th Annual IEEE Symposium on Foundations of Computer Science (FOCS)*, 1978.
3. W. Aiello and J. Håstad. Perfect zero-knowledge languages can be recognized in two rounds. In *28th Annual IEEE Symposium on Foundations of Computer Science (FOCS)*, 1987.
4. D. Beaver. Efficient multiparty protocols using circuit randomization. In *Advances in Cryptology—CRYPTO '91*, volume 576 of *Lecture Notes in Computer Science*. Springer, 1991.
5. M. Bellare and A. Palacio. The knowledge-of-exponent assumptions and 3-round zero-knowledge protocols. In *Advances in Cryptology—CRYPTO '04*, volume 3152 of *Lecture Notes in Computer Science*. Springer, 2004.
6. M. Blum, A. De Santis, S. Micali, and G. Persiano. Non-interactive zero-knowledge proof systems. *SIAM Journal on Computing*, 20(6), 1991.
7. M. Blum, P. Feldman, and S. Micali. Non-interactive zero-knowledge and its applications. In *20th Annual ACM Symposium on Theory of Computing (STOC)*, 1988.
8. G. Brassard, D. Chaum, and C. Crépeau. Minimum disclosure proofs of knowledge. *Journal of Computer and System Science*, 37(2), 1988.
9. G. Brassard and C. Crépeau. Non-transitive transfer of confidence: A perfect zero-knowledge interactive protocol for SAT and beyond. In *28th Annual IEEE Symposium on Foundations of Computer Science (FOCS)*, 1987.
10. R. Canetti. Universally composable security: a new paradigm for cryptographic protocols. In *42nd Annual IEEE Symposium on Foundations of Computer Science (FOCS)*, 2001. Full version available from http://eprint.iacr.org/2000/067.
11. R. Canetti, Y. Dodis, R. Pass, and S. Walfish. Universally composable security with global setup. In *Theory of Cryptography Conference (TCC)*, *Lecture Notes in Computer Science*. Springer, 2007.
12. R. Cramer, I. B. Damgård, and Y. Ishai. Share conversion, pseudorandom secret-sharing and applications to secure computation. In *Theory of Cryptography Conference (TCC)*, volume 3378 of *Lecture Notes in Computer Science*. Springer, 2005.
13. R. Cramer, I. B. Damgård, and P. MacKenzie. Efficient zero-knowledge proofs of knowledge without intractability assumptions. In *Practice and Theory in Public Key Cryptography (PKC)*, volume 1751 of *Lecture Notes in Computer Science*. Springer, 2000.
14. I. B. Damgård. Towards practical public-key cryptosystems provably-secure against chosen ciphertext attacks. In *Advances in Cryptology—CRYPTO '91*, volume 576 of *Lecture Notes in Computer Science*. Springer, 1991.
15. I. B. Damgård. Non-interactive circuit based proofs and non-interactive perfect zero-knowledge with preprocessing. In *Advances in Cryptology—EUROCRYPT '92*, volume 658 of *Lecture Notes in Computer Science*. Springer, 1992.
16. I. B. Damgård and Y. Ishai. Constant-round multiparty computation using a black-box pseudorandom generator. In *Advances in Cryptology—CRYPTO '05*, volume 3621 of *Lecture Notes in Computer Science*. Springer, 2005.
17. A. De Santis, G. Di Crescenzo, R. Ostrovsky, G. Persiano, and A. Sahai. Robust non-interactive zero knowledge. In *Advances in Cryptology—CRYPTO '01*, volume 2139 of *Lecture Notes in Computer Science*. Springer, 2001.

18. A. De Santis, S. Micali, and G. Persiano. Non-interactive zero-knowledge with preprocessing. In *Advances in Cryptology—CRYPTO '88*, volume 403 of *Lecture Notes in Computer Science*. Springer, 1988.
19. A. De Santis and G. Persiano. Zero-knowledge proofs of knowledge without interaction. In *33rd Annual IEEE Symposium on Foundations of Computer Science (FOCS)*, 1992.
20. A. W. Dent. The hardness of the DHK problem in the generic group model. Cryptology ePrint Archive, Report 2006/156, 2006. http://eprint.iacr.org.
21. C. Dwork and M. Naor. Zaps and their applications. In *41st Annual IEEE Symposium on Foundations of Computer Science (FOCS)*, 2000.
22. U. Feige, D. Lapidot, and A. Shamir. Multiple non-interactive zero-knowledge proofs based on a single random string. In *31st Annual IEEE Symposium on Foundations of Computer Science (FOCS)*, 1990.
23. A. Fiat and A. Shamir. How to prove yourself: Practical solutions to identification and signature problems. In *Advances in Cryptology—CRYPTO '86*, volume 263 of *Lecture Notes in Computer Science*. Springer, 1986.
24. L. Fortnow. The complexity of perfect zero-knowledge. In *19th Annual ACM Symposium on Theory of Computing (STOC)*, 1987.
25. O. Goldreich, S. Micali, and A. Wigderson. Proofs that yield nothing but their validity or all languages in NP have zero-knowledge proof systems. *Journal of the ACM*, 38(3), 1991.
26. J. Groth, R. Ostrovsky, and A. Sahai. Perfect non-interactive zero knowledge for NP. In *Advances in Cryptology—EUROCRYPT '06*, volume 4004 of *Lecture Notes in Computer Science*. Springer, 2006.
27. S. Hada and T. Tanaka. On the existence of 3-round zero-knowledge protocols. In *Advances in Cryptology—CRYPTO '98*, volume 1462 of *Lecture Notes in Computer Science*. Springer, 1998. Full version available from http://eprint.iacr.org/1999/009.
28. J. Kilian, S. Micali, and C. Rackoff. Minimum resource zero-knowledge proofs. In *Advances in Cryptology—CRYPTO '89*, volume 435 of *Lecture Notes in Computer Science*. Springer, 1989.
29. J. Kilian and E. Petrank. An efficient noninteractive zero-knowledge proof system for NP with general assumptions. *Journal of Cryptology*, 11(1), 1998.
30. M. Naor. On cryptographic assumptions and challenges. In *Advances in Cryptology—CRYPTO '03*, volume 2729 of *Lecture Notes in Computer Science*. Springer, 2003.
31. V. I. Nechaev. Complexity of a determinate algorithm for the discrete logarithm. *Mathematical Notes*, 55(2), 1994.
32. R. Pass and A. Shelat. Unconditional characterizations of non-interactive zero-knowledge. In *Advances in Cryptology—CRYPTO '05*, volume 3621 of *Lecture Notes in Computer Science*. Springer, 2005.
33. J. T. Schwartz. Fast probabilistic algorithms for verification of polynomial identities. *Journal of the ACM*, 27(4), 1980.
34. V. Shoup. Lower bounds for discrete logarithms and related problems. In *Advances in Cryptology—EUROCRYPT '97*, volume 1233 of *Lecture Notes in Computer Science*. Springer, 1997.

Security Against Covert Adversaries: Efficient Protocols for Realistic Adversaries⋆

Yonatan Aumann and Yehuda Lindell

Department of Computer Science
Bar-Ilan University, Israel
{aumann,lindell}@cs.biu.ac.il

Abstract. In the setting of secure multiparty computation, a set of mutually distrustful parties wish to securely compute some joint function of their private inputs. The computation should be carried out in a secure way, meaning that no coalition of corrupted parties should be able to learn more than specified or somehow cause the result to be "incorrect". Typically, corrupted parties are either assumed to be semi-honest (meaning that they follow the protocol specification) or malicious (meaning that they may deviate arbitrarily from the protocol). However, in many settings, the assumption regarding semi-honest behavior does not suffice and security in the presence of malicious adversaries is excessive and expensive to achieve.

In this paper, we introduce the notion of *covert adversaries*, which we believe faithfully models the adversarial behavior in many commercial, political, and social settings. Covert adversaries have the property that they may deviate arbitrarily from the protocol specification in an attempt to cheat, but do not wish to be "caught" doing so. We provide a definition of security for covert adversaries and show that it is possible to obtain highly efficient protocols that are secure against such adversaries. We stress that in our definition, we quantify over all (possibly malicious) adversaries and do not assume that the adversary behaves in any particular way. Rather, we guarantee that if an adversary deviates from the protocol in a way that would enable it to "cheat", then the honest parties are guaranteed to detect this cheating with good probability. We argue that this level of security is sufficient in many settings.

1 Introduction

1.1 Background

In the setting of secure multiparty computation, a set of parties with private inputs wish to jointly compute some functionality of their inputs. Loosely speaking, the security requirements of such a computation are that (i) nothing is learned from the protocol other than the output (privacy), (ii) the output is distributed according to the prescribed functionality (correctness), and (iii) parties cannot

⋆ Work supported in part by an Infrastructures grant from the Ministry of Science, ISRAEL. The full version of this paper is available from the *Cryptology ePrint Archive*.

make their inputs depend on other parties' inputs. Secure multiparty computation forms the basis for a multitude of tasks, including those as simple as coin-tossing and agreement, and as complex as electronic voting, electronic auctions, electronic cash schemes, anonymous transactions, remote game playing (a.k.a. "mental poker"), and privacy-preserving data mining.

The security requirements in the setting of multiparty computation must hold even when some of the participating parties are adversarial. It has been shown that, with the aid of suitable cryptographic tools, *any* two-party or multiparty function can be securely computed [23,12,10,3,6], even in the presence of very strong adversarial behavior. However, the efficiency of the computation depends dramatically on the adversarial model considered. Classically, two main categories of adversaries have been considered:

1. *Malicious adversaries:* these adversaries may behave arbitrarily and are not bound in any way to following the instructions of the specified protocol. Protocols that are secure in the malicious model provide a very strong security guarantee, as honest parties are "protected" irrespective of the adversarial behavior of the corrupted parties.
2. *Semi-honest adversaries:* these adversaries correctly follow the protocol specification, yet may attempt to learn additional information by analyzing the transcript of messages received during the execution. Security in the presence of semi-honest adversaries provides only a weak security guarantee, and is not sufficient in many settings. Semi-honest adversarial behavior primarily models inadvertent leakage of information, and is suitable only where participating parties essentially trust each other, but may have other concerns.

Secure computation in the semi-honest adversary model can be carried out very efficiently, but, as mentioned, provides weak security guarantees. Regarding malicious adversaries, it has been shown that, under suitable cryptographic assumptions, *any* multiparty probabilistic polynomial-time functionality *can* be securely computed for any number of *malicious* corrupted parties [12,10]. However, this comes at a price. These feasibility results of secure computation typically do not yield protocols that are efficient enough to actually be implemented and used in practice (particularly if standard *simulation-based security* is required). Their importance is more in telling us that it is perhaps worthwhile searching for other efficient protocols, because we at least know that a solution exists in principle. However, the unfortunate state of affairs today – many years after these feasibility results were obtained – is that very few truly efficient protocols exist for the setting of malicious adversaries. Thus, we believe that some middle ground is called for: an adversary model that accurately models adversarial behavior in the real world, on the one hand, but for which efficient, secure protocols can be obtained, on the other.

1.2 Our Work – Covert Adversaries

In this work, we introduce a new adversary model that lies between the semi-honest and malicious models. The motivation behind the definition is that in many real-world settings, adversaries are willing to actively cheat (and as such are not semi-honest), but only if they are not caught (and as such they are not arbitrarily malicious). This, we believe, is the case in many business, financial, political and diplomatic settings, where honest behavior cannot be assumed, but where the companies, institutions and individuals involved cannot afford the embarrassment, loss of reputation, and negative press associated with being *caught* cheating. It is also the case, unfortunately, in many social settings, e.g. elections for a president of the country-club. Finally, in remote game playing, players may also be willing to actively cheat, but would try to avoid being caught, or else they may be thrown out of the game. In all, we believe that this type of *covert* adversarial behavior accurately models many real-world situations. Clearly, with such adversaries, it may be the case that the risk of being caught is weighed against the benefits of cheating, and it cannot be assumed that players would avoid being caught at any price and under all circumstances. Accordingly, our definition explicitly models the probability of catching adversarial behavior; a probability that can be tuned to the specific circumstances of the problem. In particular, we do not assume that adversaries are only willing to risk being caught with negligible probability, but rather allow for much higher probabilities.

The definition. Our definition of security is based on the classical *ideal/real simulation paradigm*. Loosely speaking, our definition provides the following guarantee. Let $0 < \epsilon \leq 1$ be a value (called the *deterrence factor*). Then, any attempt to cheat by an adversary is detected by the honest parties with probability at least ϵ. Thus, provided that ϵ is sufficiently large, an adversary that wishes not to be caught cheating, will refrain from *attempting* to cheat, lest it be caught doing so. Clearly, the higher the value of ϵ, the greater the probability that the adversary is caught and thus the greater the *deterrent* to cheat. We therefore call our notion security in the presence of covert adversaries with ϵ-deterrent. Note that the security guarantee does not preclude successful cheating. Indeed, if the adversary decides to cheat then it may gain access to the other parties' private information or bias the result of the computation. The only guarantee is that if it attempts to cheat, then there is a fair chance that it will be caught doing so. This is in contrast to standard definitions, where absolute privacy and security are guaranteed, for the given type of adversary. We remark that by setting $\epsilon = 1$, our definition can be used to capture a requirement that cheating parties are always caught.

When attempting to translate the above described basic approach into a formal definition, we obtain three different possible formulations, which form a hierarchy of security guarantees. In Section 3 we present the three formulations, and discuss the relationships between them and between the standard definitions of security for semi-honest and malicious adversaries. We also present *modular sequential composition* theorems (like that of [4]) for all of our definitions. Such composition theorems are important as security goals within themselves and as tools for proving the security of protocols.

Protocol constructions. As mentioned, the aim of this work is to provide a definition of security for which it is possible to construct highly efficient protocols. We demonstrate this fact by presenting a generic protocol for secure two-party computation that is only mildly less efficient than the protocol of Yao [23], which is secure only for semi-honest adversaries. The first step of our construction is a protocol for oblivious transfer that is based on homomorphic encryption schemes. Highly efficient protocols under this assumption are known [1,17]. However, these protocols do not achieve *simulation-based* security. Rather, only privacy is guaranteed (with the plus that privacy is preserved even in the presence of fully malicious adversaries). Having constructed an oblivious transfer protocol that meets our definition, we use it in the protocol of Yao [23]. We modify Yao's protocol so that two garbled circuits are sent, and then a random one is opened in order to check that it was constructed correctly. Our basic protocol achieves deterrent $\epsilon = 1/2$, but can be extended to greater values of ϵ at a moderate expense in efficiency. (For example, 10 copies of the circuit yields $\epsilon = 9/10$.)

Protocol efficiency. The protocol we present offers a great improvement in efficiency, when compared to the best known results for the malicious adversary model. The exact efficiency depends on the variant used in the definition of covert adversary security. For the weakest variant, our protocol requires only *twice* the amount of work and twice the bandwidth of the basic protocol of [23] for semi-honest adversaries. Specifically, it requires only a constant number of rounds, a single oblivious transfer for each input bit, and has communication complexity $O(n|C|)$ where n is the security parameter and $|C|$ is the size of the circuit being computed. For the intermediate variant, the complexity is slightly higher, requiring twice the number of oblivious transfers than in the weakest variant. For the strongest variant, the complexity increases to n oblivious transfers for each input bit. This is still much more efficient than any known protocol for the case of malicious adversaries. We view this as a "proof of concept" that highly efficient protocols are achievable in this model, and leave the construction of such protocols for specific tasks of interest for future work.

1.3   Related Work

The idea of allowing the adversary to cheat as long as it will be detected was first considered by [9] who defined a property called *t-detectability*; loosely speaking, a protocol fulfilling this property provides the guarantee that no coalition of t parties can cheat without being caught. The work of [9] differs to ours in that (a) they consider the setting of an honest majority, and (b) their definition is not simulation based. Another closely related work to ours is that of [5] that considers *honest-looking adversaries*. Such adversaries may deviate arbitrarily from the protocol specification, but only if this deviation cannot be detected. Our definition differs from that of [5] in a number of important ways. First, we quantify over *all* adversaries, and not only over adversaries that behave in a certain way. Second, our definition provides guarantees even for adversaries that may be willing to risk being caught cheating with non-negligible (or even constant) probability. Third, we place the onus of detecting any cheating by an adversary on the protocol, and not on the chance that the honest parties will analyze the distribution of the messages generated by the corrupted parties. (See Section 3 for more discussion on why these differences are important.) Finally, we remark that [5] considered a more stringent setting where all parties are either malicious or honest-looking. In contrast, we consider a *relaxation* of the adversary model (where parties are either fully honest or covert).

We remark that the idea of allowing an adversary to cheat with non-negligible probability as long as it will be caught with good probability has been mentioned many times in the literature; see [15,20] for just two examples. We stress, however, that none of these works formalized this idea. Furthermore, our experience in proving our protocol secure is that simple applications of cut-and-choose do not meet our definition (and there are actual attacks that can be carried out on the cut-and-choose technique used in [20], for example).

Our work studies a weaker definition of security than the standard one. Weaker definitions have been used before in order to construct efficient protocols for specific problems. However, in the past these relaxed definitions typically have not followed the simulation paradigm, but rather have considered privacy via indistinguishability (and sometimes correctness); see [7] for one example. Our work takes a completely different approach.

2 Secure Multiparty Computation – Standard Definition

In this section we briefly present the standard definition for secure multiparty computation and refer to [10, Chapter 7] for more details and motivating discussion. The following description and definition is based on [10], which in turn follows [13,21,2,4].

Multiparty computation. A multiparty protocol problem is cast by specifying a random process that maps sets of inputs to sets of outputs (one for each party). We refer to such a process as a functionality and denote it $f : (\{0,1\}^*)^m \to (\{0,1\}^*)^m$, where $f = (f_1, \ldots, f_m)$. That is, for every vector of inputs $\overline{x} = (x_1, \ldots, x_m)$, the output-vector is a random variable $\overline{y} = (f_1(\overline{x}), \ldots, f_m(\overline{x}))$ ranging over vectors of strings. The ith party P_i, with input x_i, wishes to obtain $f_i(\overline{x})$. We sometimes denote such a functionality by $(\overline{x}) \mapsto (f_1(\overline{x}), \ldots, f_m(\overline{x}))$. Thus, for example, the oblivious transfer functionality is denoted by $((x_0, x_1), \sigma) \mapsto (\lambda, x_\sigma)$, where (x_0, x_1) is the first party's input, σ is the second party's input, and λ denotes the empty string (meaning that the first party has no output).

Security of protocols (informal). The security of a protocol is analyzed by comparing what an adversary can do in a real protocol execution to what it can do in an ideal scenario that is secure by definition. This is formalized by considering an *ideal* computation involving an incorruptible *trusted third party* to whom the parties send their inputs. The trusted party computes the functionality on the inputs and returns to each party its respective output (in order to model the possibility of early aborting, the adversary receives its outputs first and then can decide if the honest parties also receive output). Loosely speaking, a protocol is secure if any adversary interacting in the real protocol (where no trusted third party exists) can do no more harm than if it was involved in the above-described ideal computation. We consider malicious adversaries and static corruptions in all of our definitions in this paper.

Execution in the ideal model. Let the set of parties be $P_1, \ldots, P_m$ and let $I \subseteq [m]$ denote the indices of the corrupted parties, controlled by an adversary $\mathcal{A}$. An ideal execution proceeds as follows:

Inputs: Each party obtains an input; the ith party's input is denoted x_i. The adversary $\mathcal{A}$ receives an auxiliary input denoted z (and we assume that it knows the length of all inputs).

Send inputs to trusted party: Any honest party P_j sends its received input x_j to the trusted party. The corrupted parties controlled by $\mathcal{A}$ may either abort, send their received input, or send some other input of the same length to the trusted party. This decision is made by $\mathcal{A}$ and may depend on the values x_i for $i \in I$ and its auxiliary input z. Denote the vector of inputs sent to the trusted party by $\overline{w}$ (note that $\overline{w}$ does not necessarily equal $\overline{x}$).

  If the trusted party does not receive m valid inputs (including the case that one of the inputs equals $\perp$), it replies to all parties with a special symbol $\perp$ and the ideal execution terminates. Otherwise, the execution proceeds to the next step.

Trusted party sends outputs to adversary: The trusted party computes $(f_1(\overline{w}), \ldots, f_m(\overline{w}))$ and sends $f_i(\overline{w})$ to party P_i, for all $i \in I$ (i.e., to all corrupted parties).

Adversary instructs trusted party to continue or halt: $\mathcal{A}$ sends either CONTINUE or HALT to the trusted party. If it sends CONTINUE, the trusted party sends $f_j(\overline{w})$ to party P_j, for all $j \notin I$ (i.e., to all honest parties). Otherwise, if it sends HALT, the trusted party sends $\perp$ to all parties P_j for $j \notin I$.

Outputs: An honest party always outputs the message it obtained from the trusted party. The corrupted parties output nothing. The adversary $\mathcal{A}$ outputs any arbitrary (probabilistic polynomial-time computable) function of the initial inputs $\{x_i\}_{i \in I}$ and the messages $\{f_i(\overline{w})\}_{i \in I}$ obtained from the trusted party.

Let $f : (\{0,1\}^*)^m \to (\{0,1\}^*)^m$ be an m-party functionality, where $f = (f_1, \ldots, f_m)$, let $\mathcal{A}$ be a non-uniform probabilistic polynomial-time machine, and let $I \subseteq [m]$ be the set of corrupted parties. Then, the ideal execution of f on inputs $\overline{x}$, auxiliary input z to $\mathcal{A}$ and security parameter n, denoted IDEAL$_{f,\mathcal{A}(z),I}(\overline{x}, n)$, is defined as the output vector of the honest parties and the adversary $\mathcal{A}$ from the above ideal execution.

Execution in the real model. We next consider the real model in which a real m-party protocol π is executed (and there exists no trusted third party). In this case, the adversary $\mathcal{A}$ sends all messages in place of the corrupted parties, and may follow an arbitrary polynomial-time strategy. In contrast, the honest parties follow the instructions of π.

Let f be as above and let π be an m-party protocol for computing f. Furthermore, let $\mathcal{A}$ be a non-uniform probabilistic polynomial-time machine and let I be the set of corrupted parties. Then, the real execution of π on inputs $\overline{x}$, auxiliary input z to $\mathcal{A}$ and security parameter n, denoted REAL$_{\pi,\mathcal{A}(z),I}(\overline{x}, n)$, is defined as the output vector of the honest parties and the adversary $\mathcal{A}$ from the real execution of π.

Security as emulation of a real execution in the ideal model. Having defined the ideal and real models, we can now define security of protocols. We will consider executions where all inputs are of the same length (see discussion in [10]), and will therefore say that a vector $\overline{x} = (x_1, \ldots, x_m)$ is balanced if for every i and j it holds that $|x_i| = |x_j|$.

Definition 1. (secure multiparty computation) *Let f and π be as above. Protocol π is said to securely compute f with abort in the presence of malicious adversaries if for every non-uniform probabilistic polynomial-time adversary $\mathcal{A}$ for the real model, there exists a non-uniform probabilistic polynomial-time adversary $\mathcal{S}$ for the ideal model, such that for every $I \subseteq [m]$, every balanced vector $\overline{x} \in (\{0,1\}^*)^m$, and every auxiliary input $z \in \{0,1\}^*$:*

$$\left\{ \text{IDEAL}_{f,\mathcal{S}(z),I}(\overline{x}, n) \right\}_{n \in \mathbb{N}} \stackrel{c}{\equiv} \left\{ \text{REAL}_{\pi,\mathcal{A}(z),I}(\overline{x}, n) \right\}_{n \in \mathbb{N}}$$

where $\stackrel{c}{\equiv}$ indicates computational indistinguishability.

3 Definitions – Security with Covert Adversaries

3.1 Motivation

The standard definition of security (see Definition 1) is such that all possible (polynomial-time) adversarial behavior is simulatable. In contrast, as we have mentioned, here we wish to model the situation that parties may cheat. However, if they do so, they are likely to be caught. There are a number of ways of defining this notion. In order to motivate ours, we begin with a somewhat naive implementation of the notion, and show its shortcoming.

First attempt. Define an adversary to be covert if the distribution over the messages that it sends during an execution is computationally indistinguishable from the distribution over the messages that an honest party would send. Then quantify over all covert adversaries $\mathcal{A}$ for the real world (rather than all adversaries).[1] A number of problems arise with this definition. First, the fact that the distribution generated by the adversary can be distinguished from the distribution generated by honest parties does not mean that the honest parties indeed detect this. This is due to the fact that the honest parties may not have an efficient distinguisher; it is only guaranteed that there exists one. Furthermore, in

---

[1] We remark that this is the conceptual approach taken by [5], and that there are important choices that arise when attempting to formalize the approach. In any case, as we have mentioned, the work of [5] differs greatly because their aim was to model all parties as somewhat adversarial.

order to guarantee that the honest parties detect the cheating, they would have to analyze all traffic during an execution. However, this analysis *cannot* be part of the protocol because then the distinguishers used by the honest parties would be known (and potentially bypassed). Another problem is that, as mentioned in the introduction, adversaries may be willing to risk being caught with more than negligible probability, say 10^{-6}. With such an adversary, the definition would provide no security guarantee. In particular, the adversary may be able to always learn all parties' inputs, and only risk being caught in one run in a million.

Second attempt. To solve the aforementioned problems, we first we require that the protocol itself be responsible for detecting cheating. Specifically, in the case that a party P_i attempts to cheat, the protocol may instruct the honest parties to output a message saying that "party P_i has cheated" (we require that this only happens if P_i indeed cheated). This solves the first problem. To solve the second problem, we explicitly quantify the probability that an adversary is caught cheating. Roughly, given a parameter ϵ, a protocol is said to be **secure against covert adversaries with ϵ-deterrent** if any cheating adversary will necessarily be caught with probability at least ϵ.

This definition captures the spirit of what we want, but is still problematic. To illustrate the problem, consider an adversary that plays honestly with probability 0.99, and cheats otherwise. Such an adversary can only ever be caught with probability 0.01 (because otherwise it is honest). If $\epsilon = 1/2$ for example, then such an adversary must be caught with probability 0.5, which is impossible. We therefore conclude that an *absolute* parameter cannot be used, and the probability of catching the adversary must be related to the probability that it cheats.

Final definition. We thus arrive at the following approach. First, as mentioned, we require that the protocol itself be responsible for detecting cheating. That is, if a party P_i successfully cheats, then with good probability (ϵ), the honest parties in the protocol will all receive a message that "P_i cheated". Second, we do not quantify only over adversaries that are covert (i.e., those that are not detected cheating by the protocol). Rather, we allow all possible adversaries, even completely malicious ones. Then, we require either that this malicious behavior can be successfully simulated (as in Definition 1), or that the honest parties will receive a message that cheating has been detected, and this happens with probability at least ϵ times the probability that successful cheating takes place. In other words, when an adversarial attack is carried out, we are guaranteed that one of the following two happens:

1. *The attack fails:* this event is represented by the fact that the adversary can simulate the interaction on its own, and so the attack cannot yield any more than what is possible in the ideal model.
2. *The attack succeeds:* in this case we are guaranteed that with good probability (and this probability is a parameter in the definition), the adversarial parties will be caught.

We stress that in the second case, the adversary may actually learn secret information or cause some other damage. However, since it is guaranteed that such a strategy will likely be caught, there is strong motivation to refrain from carrying it out.

As it turns out, the above intuition can be formalized in three different ways, which form a hierarchy of security guarantees. Since we view the definitional part of this work as of no less importance than the protocol constructions, we present all three formulations. In practice, the practitioner should choose the formulation that best suites her needs, and for which sufficiently efficient protocols exists. All three definitions are based on the ideal/real simulation paradigm, as presented in Section 2. We now present the definitions in order of security, starting with the weakest (least secure) one.

3.2   Version 1: Failed Simulation Formulation

The first formulation we present is based on allowing the simulator to fail sometimes, where by "fail" we mean that its output distribution is not indistinguishable from the real one. This corresponds to an event of successful cheating. However, we guarantee that the probability that the adversary is caught cheating is at least ϵ times the probability that the simulator fails. The details follow.

Recall that we call a vector balanced if all of its items are of the same length. In addition, we denote the output vector of the honest parties and adversary $\mathcal{A}$ in an ideal execution of f by $\text{IDEAL}_{f,\mathcal{A}(z),I}(\overline{x}, n)$, where $\overline{x}$ is the vector of inputs, z is the auxiliary input to $\mathcal{A}$, I is the set of corrupted parties, and n is the security parameter. Finally, we denote the analogous outputs in a real execution of π by $\text{REAL}_{\pi,\mathcal{A}(z),I}(\overline{x}, n)$. We begin by defining what it means to "detect cheating":

Definition 2. *Let π be an m-party protocol, let $\mathcal{A}$ be an adversary, and let I be the index set of the corrupted parties. A party P_j is said to* detect cheating *in π if its output in π is* corrupted$_i$*; this event is denoted* $\text{OUTPUT}_j(\text{REAL}_{\pi,\mathcal{A}(z),I}(\overline{x})) =$ corrupted$_i$*. The protocol π is called* detection accurate *if for every $j, k \notin I$, the probability that P_j outputs* corrupted$_k$ *is negligible.*

We require that all protocols be detection accurate (meaning that only corrupted parties can be "caught cheating"). This is crucial because otherwise a party that is detected cheating can just claim that it is due to a protocol anomaly and not because it really cheated. The definition follows:

Definition 3. (security – failed simulation formulation) *Let f and π be as in Definition 1, and let $\epsilon : \mathbb{N} \to [0,1]$ be a function. Protocol π is said to* securely compute f in the presence of covert adversaries with ϵ-deterrent *if it is* detection accurate *and if for every non-uniform probabilistic polynomial-time adversary $\mathcal{A}$ for the real model, there exists a non-uniform probabilistic polynomial-time adversary $\mathcal{S}$ for the ideal model such that for every $I \subseteq [m]$, every balanced vector $\overline{x} \in (\{0,1\}^*)^m$, every auxiliary input $z \in \{0,1\}^*$, and every non-uniform polynomial-time distinguisher D, there exists a negligible function $\mu(\cdot)$ such that,*

$$\Pr\left[\exists i \in I \; \forall j \notin I \; : \; \text{OUTPUT}_j(\text{REAL}_{\pi,\mathcal{A}(z),I}(\overline{x}, n)) = \text{corrupted}_i\right]$$

$$\geq \epsilon(n) \cdot \left|\Pr\left[D(\text{IDEAL}_{f,\mathcal{S}(z),I}(\overline{x}, n)) = 1\right] - \Pr\left[D(\text{REAL}_{\pi,\mathcal{A}(z),I}(\overline{x}, n)) = 1\right]\right| - \mu(n)$$

The parameter ϵ indicates the probability that successful adversarial behavior is detected (observe that when such a detection occurs, *all* honest parties must detect the same corrupted party). Clearly, the closer ϵ is to one, the higher the deterrence to cheat, and hence the level of security, assuming covert adversaries. Note that the adversary can decide to never be detected cheating, in which case the IDEAL and REAL distributions are guaranteed to be *computationally indistinguishable*, as in the standard definition of security. In contrast, it can choose to cheat with some noticeable probability, in which case the IDEAL and REAL output distribution may be distinguishable (while guaranteeing that the adversary is caught with good probability). This idea of allowing the ideal and real models to not be fully indistinguishable in order to model "allowed cheating" was used in [11].

We stress that the definition does not *require* the simulator to "fail" with some probability. Rather, it is *allowed* to fail with a probability that is at most $1/\epsilon$ times the probability that the adversary is caught cheating. As we shall see, this is what enables us to construct highly efficient protocols. We also remark that due to the required detection accuracy, the simulator cannot fail when the adversary behaves in a fully honest-looking manner (because in such a case, no honest party will output corrupted$_i$). Thus, security is always preserved in the presence of adversaries that are willing to cheat arbitrarily, as long as their cheating is not detected.

Cheating and aborting. It is important to note that according to the above definition, a party that halts mid-way through the computation may be considered a "cheat". Arguably, this may be undesirable due to the fact that an honest party's computer may crash (such unfortunate events may not even be that rare). Nevertheless, we argue that as a basic definition it suffices. This is due to the fact that it is possible for all parties to work by storing their input and random-tape on disk before they begin the execution. Then, before sending any message, the incoming messages that preceded it are also written to disk. The result of this is that if a party's machine crashes, it can easily reboot and return to its previous state. (In the worst case the party will need to request a retransmit of the last message if the crash occurred before it was written.) We therefore believe that honest parties cannot truly hide behind the excuse that their machine crashed (it would be highly suspicious that someone's machine crashed in an irreversible way that also destroyed their disk at the critical point of a secure protocol execution).

Despite the above, it is possible to modify the definition so that honest halting is never considered cheating. This modification only needs to be made to the notion of "detection accuracy" and uses the notion of a fail-stop party who acts semi-honestly, except that it may halt early.

Definition 4. *A protocol π is* non-halting detection accurate *if it is detection accurate as in Definition 2 and if for every honest party P_j and fail-stop party P_k, the probability that P_j outputs* corrupted$_k$ *is negligible.*

The definition of security in the presence of covert adversaries can then be modified by requiring non-halting detection accuracy. We remark that although this strengthening is highly desirable, it may also be prohibitive. For example, we are able to modify our main protocol so that it meets this stronger definition. However, in order to do so, we need to assume fully secure oblivious transfer, for which highly efficient (fully simulatable) protocols are not really known.

3.3   Version 2: Explicit Cheat Formulation

The drawback of Definition 3 is that it allows the adversary to decide whether to cheat as a function of the honest parties' inputs or of the output. This is undesirable since there may be honest parties' inputs for which it is more "worthwhile" for the adversary to risk being caught. We therefore wish to force the adversary to make its decision about whether to cheat *obliviously* of the honest parties' inputs. This brings us to an alternate definition, which is based on redefining the ideal functionality so as to explicitly include the option of cheating. Aside from overcoming the input dependency problem this alternate formulation has two additional advantages. First, it makes the security guarantees more explicit. Second, it makes it easy to prove a sequential composition theorem.

We modify the ideal model in the following way. Let $\epsilon : \mathbb{N} \rightarrow [0,1]$ be a function. Then, the ideal execution with ϵ proceeds as follows:

Inputs: Each party obtains an input; the ith party's input is denoted by x_i; we assume that all inputs are of the same length, denoted n. The adversary receives an auxiliary-input z.

Send inputs to trusted party: Any honest party P_j sends its received input x_j to the trusted party. The corrupted parties, controlled by $\mathcal{A}$, may either send their received input, or send some other input of the same length to the trusted party. This decision is made by $\mathcal{A}$ and may depend on the values x_i for $i \in I$ and the auxiliary input z. Denote the vector of inputs sent to the trusted party by $\overline{w}$.

Abort options: If a corrupted party sends $w_i = \mathsf{abort}_i$ to the trusted party as its input, then the trusted party sends abort_i to all of the honest parties and halts. If a corrupted party sends $w_i = \mathsf{corrupted}_i$ to the trusted party as its input, then the trusted party sends $\mathsf{corrupted}_i$ to all of the honest parties and halts.

Attempted cheat option: If a corrupted party sends $w_i = \mathsf{cheat}_i$ to the trusted party as its input, then the trusted party sends to the adversary all of the honest parties' inputs $\{x_j\}_{j \notin I}$. Furthermore, it asks the adversary for outputs $\{y_j\}_{j \notin I}$ for the honest parties. In addition,

  1. With probability ϵ, the trusted party sends $\mathsf{corrupted}_i$ to the adversary and all of the honest parties.
  2. With probability $1 - \epsilon$, the trusted party sends $\mathsf{undetected}$ to the adversary and the outputs $\{y_j\}_{j \notin I}$ to the honest parties (i.e., for every $j \notin I$, the trusted party sends y_j to P_j).

The ideal execution then ends at this point.

If no w_i equals abort$_i$, corrupted$_i$ or cheat$_i$, the ideal execution continues below.

Trusted party answers adversary: The trusted party computes $(f_1(\overline{w}), \ldots, f_m(\overline{w}))$ and sends $f_i(\overline{w})$ to $\mathcal{A}$, for all $i \in I$.

Trusted party answers honest parties: After receiving its outputs, the adversary sends either abort$_i$ for some $i \in I$, or continue to the trusted party. If the trusted party receives continue then it sends $f_j(\overline{w})$ to all honest parties P_j ($j \notin I$). Otherwise, if it receives abort$_i$ for some $i \in I$, it sends abort$_i$ to all honest parties.

Outputs: An honest party always outputs the message it obtained from the trusted party. The corrupted parties output nothing. The adversary $\mathcal{A}$ outputs any arbitrary (probabilistic polynomial-time computable) function of the initial inputs $\{x_i\}_{i \in I}$ and the messages obtained from the trusted party.

The output of the honest parties and the adversary in an execution of the above ideal model is denoted by $\mathrm{IDEALC}^{\epsilon}_{f, \mathcal{S}(z), I}(\overline{x}, n)$.

Notice that there are two types of "cheating" here. The first is the classic abort, except that unlike in Definition 1, the honest parties here are informed as to who caused the abort. Thus, although it is not possible to guarantee fairness here, we do achieve that an adversary who aborts after receiving its output is "punished" in the sense that its behavior is always detected.[2] The other type of cheating in this ideal model is more serious for two reasons: first, the ramifications of the cheat are greater (the adversary may learn all of the parties' inputs and may be able to determine their outputs), and second, the cheating is only guaranteed to be detected with probability ϵ. Nevertheless, if ϵ is high enough, this may serve as a deterrent. We stress that in the ideal model the adversary must decide whether to cheat obliviously of the honest-parties inputs and before it receives any output (and so it cannot use the output to help it decide whether or not it is "worthwhile" cheating). We define:

Definition 5. (security – explicit cheat formulation) *Let f, π and ϵ be as in Definition 3. Protocol π is said to* securely compute f in the presence of covert adversaries with ϵ-deterrent *if for every non-uniform probabilistic polynomial-time adversary $\mathcal{A}$ for the real model, there exists a non-uniform probabilistic polynomial-time adversary $\mathcal{S}$ for the ideal model such that for every $I \subseteq [m]$, every balanced vector $\overline{x} \in (\{0,1\}^*)^m$, and every auxiliary input $z \in \{0,1\}^*$:*

$$\left\{ \mathrm{IDEALC}^{\epsilon}_{f, \mathcal{S}(z), I}(\overline{x}, n) \right\}_{n \in \mathbb{N}} \stackrel{\mathrm{c}}{\equiv} \left\{ \mathrm{REAL}_{\pi, \mathcal{A}(z), I}(\overline{x}, n) \right\}_{n \in \mathbb{N}}$$

Definition 5 and detection accuracy. We note that in Definition 5 it is not necessary to explicitly require that π be detection accurate because this is taken

---

[2] Note also that there are two types of abort: in one the honest parties receive abort$_i$ and in the second they receive corrupted$_i$. This is included to model behavior by the real adversary that results in it being caught cheating with probability greater than ϵ (and not with probability exactly ϵ as when the ideal adversary sends a cheat$_i$ message).

care of in the ideal model (in an ideal execution, only a corrupted party can send a cheat$_i$ input). However, if *non-halting detection accuracy* is desired (as in Definition 4), then this should be explicitly added to the definition.

3.4 Version 3: Strong Explicit Cheat Formulation

The third, and strongest version follows the same structure and formulation of the previous version (Version 2). However, we make the following slight, but important change to the ideal model. In the case of an attempted cheat, if the trusted party sends corrupted$_i$ to the honest parties and the adversary (an event which happens with probability ϵ), then the adversary does *not* obtain the honest parties' inputs. Thus, if cheating is detected, the adversary does not learn anything and the result is essentially the same as a regular abort. This is in contrast to Version 2, where a detected cheat may still be successful. (We stress that in the "undetected" case here, the adversary still learns the honest parties' private inputs and can set their outputs.) We denote the resultant ideal model by IDEALSC$^\epsilon_{f,\mathcal{S}(z),I}(\overline{x}, n)$ and have the following definition:

Definition 6. (security – strong explicit cheat formulation): *Let f, π and ϵ be as in Definition 3. Protocol π is said to* securely compute f in the presence of covert adversaries with ϵ-deterrent *if for every non-uniform probabilistic polynomial-time adversary $\mathcal{A}$ for the real model, there exists a non-uniform probabilistic polynomial-time adversary $\mathcal{S}$ for the ideal model such that for every $I \subseteq [m]$, every balanced vector $\overline{x} \in (\{0,1\}^*)^m$, and every auxiliary input $z \in \{0,1\}^*$:*

$$\left\{ \text{IDEALSC}^\epsilon_{f,\mathcal{S}(z),I}(\overline{x}, n) \right\}_{n \in \mathbb{N}} \overset{c}{\equiv} \left\{ \text{REAL}_{\pi,\mathcal{A}(z),I}(\overline{x}, n) \right\}_{n \in \mathbb{N}}$$

The difference between the regular and strong explicit cheat formulations is perhaps best exemplified in the case that $\epsilon = 1$. In both versions, all potentially successful cheating attempt are detected. However, in the regular formulation, the adversary may learn the honest parties' private inputs (albeit, while being detected). In the strong formulation, in contrast, the adversary learns nothing when it is detected. Since it is always detected, this means that full security is achieved.

3.5 Relations Between Security Models

Relations between covert security definitions. It is not difficult to show that the three security definitions for covert adversaries constitute a strict hierarchy, with version 1 being strictly weaker than version 2, which is strictly weaker than version 3. We explicitly prove this in the full version of the paper.

Relation to the malicious and semi-honest models. As a sanity check regarding our definitions, we present two propositions that show the relation between security in the presence of covert adversaries and security in the presence of malicious and semi-honest adversaries.

Proposition 7. *Let π be a protocol that securely computes some functionality f with abort in the presence of malicious adversaries, as in Definition 1. Then, π securely computes f in the presence of covert adversaries with ϵ-deterrent, for any of the three formulations and for every $0 \le \epsilon \le 1$.*

This proposition follows from the simple observation that according to Definition 1, there exists a simulator that always succeeds in its simulation. Thus, Definition 3 holds even if the probability of detecting cheating is 0. Likewise, for Definitions 5 and 6 the same simulator works (there is simply no need to ever send a cheat input). Next, we consider the relation between covert and semi-honest adversaries.

Proposition 8. *Let π be a protocol that securely computes some functionality f in the presence of covert adversaries with ϵ-deterrent, for any of the three formulations and for $\epsilon \ge 1/\mathrm{poly}(n)$. Then, π securely computes f in the presence of semi-honest adversaries.*

This proposition follows from the fact that due to the requirement of detection accuracy, no party outputs corrupted$_i$ when the adversary is semi-honest. Since $\epsilon \ge 1/\mathrm{poly}(n)$ this implies that the REAL and IDEAL distributions can be distinguished with at most negligible probability, as is required for semi-honest security. We stress that if $\epsilon = 0$ (or is negligible) then the definition of covert adversaries requires nothing, and so the proposition does not hold for this case.

We conclude that, as one may expect, security in the presence of covert adversaries with ϵ-deterrent lies in between security in the presence of malicious adversaries and security in the presence of semi-honest adversaries.

Strong explicit cheat formulation and the malicious model. The following proposition shows that the strong explicit cheat formulation converges to the malicious model as ϵ approaches 1.

Proposition 9. *Let π be a protocol. Then π securely computes some functionality f in the presence of covert adversaries with $\epsilon = 1$ under Definition 6 if and only if it securely computes f with abort in the presence of malicious adversaries.*

This is true since, by definition, either the adversary does not attempt cheating, in which case the ideal execution is the same as in the regular ideal model, or it attempts cheating, in which case it is caught with probability 1 and the protocol is aborted. In both cases, the adversary gains no advantage, and the outcome can be simulated in the standard ideal model. (There is one technicality here relating to whether the output of an honest party due to an abort is $\bot$, or abort/corrupted. In order for the proposition to go through, we actually have to modify the basic ideal model so that abort$_i$ is received rather than $\bot$.)

3.6   Modular Sequential Composition

Sequential composition theorems for secure computation are important for two reasons. First, they constitute a security goal within themselves. Second, they are useful tools that help in writing proofs of security. As such, we believe that when presenting a new definition, it is of great importance to also prove an appropriate composition theorem for that definition. In our case, we obtain composition theorems that are analogous to that of [4] for all three of our definitions. The exact formulation of these theorems and the proofs appear in the full version.

4   Secure Two-Party Computation

In this section, we show how to securely compute any two-party functionality in the presence of covert adversaries. We have three different protocols, one for each of the three different security definitions. We first present the protocol for the strong explicit cheat formulation, which provides $\epsilon = 1/2$-deterrent. The variations for the other models are minor and will be presented later. In all cases, the deterrent can be boosted to $1 - 1/p(n)$ for any polynomial $p(\cdot)$, with an additional price in complexity, as will be explained later.

The protocol is based on Yao's protocol for semi-honest adversaries [23]. We will base our description on the write-up of [18] of this protocol, and due to lack of space we will assume familiarity with it. The protocol uses an oblivious transfer (OT) protocol that is secure in the presences of covert adversaries. In the full version, we prove the following theorem (via a highly efficient protocol):

Theorem 10. *Assume the existence of semantically secure homomorphic encryption schemes with errorless decryption. Then, for any $k = \text{poly}(n)$ there exists a secure protocol for computing the parallel string oblivious transfer functionality $((x_1^0, x_1^1), \ldots, (x_n^0, x_1^n), (\sigma_1, \ldots, \sigma_n)) \mapsto (\lambda, (x_1^{\sigma_1}, \ldots, x_n^{\sigma_n}))$ in the presence of covert adversaries with ϵ-deterrent for $\epsilon = 1 - \frac{1}{k}$, under any of the three security definitions.*

4.1   The Protocol

The original protocol of Yao is not secure when the parties may be malicious. Intuitively, there are two main reasons for this. First, the circuit constructor P_1 may send P_2 a garbled circuit that computes a completely different function. Second, the oblivious transfer protocol that is used when the parties can be malicious must be secure for this case. The latter problem is solved here by using the protocol guaranteed by Theorem 10. The first problem is solved by having P_1 send P_2 two garbled circuits. Then, P_2 asks P_1 to open one of the circuits at random, in order to check that it is correctly constructed. (This takes place before P_1 sends the keys corresponding to its input, so nothing is revealed by opening one of the circuits.) The protocol then proceeds similarly to the semi-honest case. The main point here is that if the unopened circuit is correct, then this will constitute a secure execution that can be simulated. However, if it is not correct, then with probability 1/2 party P_1 will have been caught cheating and so P_2 will output corrupted$_1$. While the above intuition forms the basis for our protocol, the actual construction of the appropriate simulator is somewhat delicate, and requires a careful construction of the protocol. We note some of these subtleties hereunder.

First, it is crucial that the oblivious transfers are run before the garbled circuit is sent by P_1 to P_2. This is due to the fact that the simulator sends a corrupted P_2 a fake garbled circuit that evaluates to the exact output received from the trusted party (and only this output), as described in [18]. However, in order for the simulator to receive the output from the trusted party, it must first send it the input used by the corrupted P_2. This is achieved by first running the oblivious transfers, from which the simulator is able to extract the corrupted P_2's input.

The second subtlety relates to an issue we believe may be a problem for many other implementations of Yao that use cut-and-choose. The problem is that the adversary can construct (at least in theory) a garbled circuit with two sets of keys, where one set of keys decrypt the circuit to the specified one and another set of keys decrypt the circuit to an incorrect one. This is a problem because the adversary can supply "correct keys" to the circuits that are opened and "incorrect keys" to the circuit (or circuits) that are computed. Such a strategy cannot be carried out without risk of detection for the keys that are associated with P_2's input because these keys are obtained by P_2 in the oblivious transfers *before* the garbled circuits are even sent (thus if incorrect keys are sent for one of the circuits, P_2 will detect this if that circuit is opened). However, it is possible for a corrupt P_1 to carry out this strategy for the input wires associated with its own input. We prevent this by having P_1 commit to these keys and send the commitments together with the garbled circuits. Then, instead of P_1 just sending the keys associated with its input, it sends the appropriate decommitments.

A third subtlety that arises is connected to the difference between Definitions 3 and 5 (where the latter is the stronger definition where the decision by the adversary to cheat is not allowed to depend on the honest parties' inputs or on the output). Consider a corrupted P_1 that behaves exactly like an honest P_1 except that in the oblivious transfers, it inputs an invalid key in the place of the key associated with 0 as the first bit of P_2. The result is that if the first bit of P_2's input is 1, then the protocol succeeds and no problem arises. However, if the first bit of P_2's input is 0, then the protocol will always fail and P_2 will always detect cheating. Thus, P_1's decision to cheat may depend on P_2's private input, something that is impossible in the ideal models of Definitions 5 and 6. In summary, this means that the protocol achieves Definition 3 (with $\epsilon = 1/2$) but not Definition 5. In order to solve this problem, we use a circuit that computes the function $g(x_1, x_2^1, \ldots, x_2^n) = f(x_1, \oplus_{i=1}^{n} x_2^i)$, instead of a circuit that directly computes f. Then, upon input x_2, party P_2 chooses random $x_2^1, \ldots, x_2^{n-1}$ and sets $x_2^n = (\oplus_{i=1}^{n-1} x_2^i) \oplus x_2$. This makes no difference to the result because $\oplus_{i=1}^{n} x_2^i = x_2$ and so $g(x_1, x_2^1, \ldots, x_2^n) = f(x_1, x_2)$. However, this modification makes every bit of P_2's input uniform when considering any proper subset of $x_2^1, \ldots, x_2^n$. This helps because as long as P_1 does not provide invalid keys for all n shares of x_2, the probability of failure is independent of P_2's actual input (because any set of $n-1$ shares is independent of x_2). If, on the other hand, P_2 attempts to provide invalid keys for all the n shares, then it is caught with probability almost 1. This method was previously used in [19]. We are now ready to describe the actual protocol.

Protocol 11 (two-party computation of a function f)

- **Inputs:** *Party P_1 has input x_1 and party P_2 has input x_2, where $|x_1| = |x_2|$. In addition, both parties have a security parameter n. For simplicity, we will assume that the lengths of the inputs are n.*
- **Auxiliary input:** *Both parties have the description of a circuit C for inputs of length n that computes the function f. The input wires associated with x_1 are $w_1, \ldots, w_n$ and the input wires associated with x_2 are $w_{n+1}, \ldots, w_{2n}$.*
- **The protocol**

  1. *Parties P_1 and P_2 define a new circuit C' that receives $n + 1$ inputs $x_1, x_2^1, \ldots, x_2^n$ each of length n, and computes the function $f(x_1, \oplus_{i=1}^n x_2^i)$. Note that C' has $n^2 + n$ input wires. Denote the input wires associated with x_1 by $w_1, \ldots, w_n$, and the input wires associated with x_2^i by $w_{in+1}, \ldots, w_{(i+1)n}$, for $i = 1, \ldots, n$.*

  2. *Party P_2 chooses $n - 1$ random strings $x_2^1, \ldots, x_2^{n-1} \in_R \{0,1\}^n$ and defines $x_2^n = (\oplus_{i=1}^{n-1} x_2^i) \oplus x_2$, where x_2 is P_2's original input (note that $\oplus_{i=1}^n x_2^i = x_2$). The value $z_2 \stackrel{\text{def}}{=} x_2^1, \ldots, x_2^n$ serves as P_2's new input of length n^2 to C'.*

  3. *Party P_1 chooses two sets of $2n^2$ random keys by running $G(1^n)$, the key generator for the encryption scheme:*

$$\hat{k}_{n+1}^0, \ldots, \hat{k}_{n^2+n}^0 \qquad \tilde{k}_{n+1}^0, \ldots, \tilde{k}_{n^2+n}^0$$
$$\hat{k}_{n+1}^1, \ldots, \hat{k}_{n^2+n}^1 \qquad \tilde{k}_{n+1}^1, \ldots, \tilde{k}_{n^2+n}^1$$

  4. *P_1 and P_2 run n^2 executions of an oblivious transfer protocol, as follows. In the ith execution, party P_1 inputs the pair $\left([\hat{k}_{n+i}^0, \tilde{k}_{n+i}^0], [\hat{k}_{n+i}^1, \tilde{k}_{n+i}^1] \right)$ and party P_2 inputs the bit z_2^i. (Note, P_2 receives for output the keys $\hat{k}_{n+i}^{z_2^i}$ and $\tilde{k}_{n+i}^{z_2^i}$.) The executions are run using a parallel oblivious transfer functionality, as in Theorem 10. If a party receives a corrupted$_i$ or abort$_i$ message as output from the oblivious transfer, it outputs it and halts.*

  5. *Party P_1 constructs two garbled circuits $G(C')_0$ and $G(C')_1$ using independent randomness. The keys to the input wires $w_{n+1}, \ldots, w_{n^2+n}$ in the garbled circuits are taken from above (i.e., in $G(C')_0$ they are $\hat{k}_{n+1}^0, \hat{k}_{n+1}^1, \ldots, \hat{k}_{n^2+n}^0, \hat{k}_{n^2+n}^1$, and in $G(C')_1$ they are $\tilde{k}_{n+1}^0, \tilde{k}_{n+1}^1, \ldots, \tilde{k}_{n^2+n}^0, \tilde{k}_{n^2+n}^1$). Let $\hat{k}_1^0, \hat{k}_1^1, \ldots, \hat{k}_n^0, \hat{k}_n^1$ be the keys associated with P_1's input in $G(C')_0$ and $\tilde{k}_1^0, \tilde{k}_1^1, \ldots, \tilde{k}_n^0, \tilde{k}_n^1$ the analogous keys in $G(C')_1$. Then, for every $i \in \{1, \ldots, n\}$ and $b \in \{0, 1\}$, party P_1 computes $\hat{c}_i^b = \mathsf{Com}(\hat{k}_i^b; \hat{r}_i^b)$ and $\tilde{c}_i^b = \mathsf{Com}(\tilde{k}_i^b; \tilde{r}_i^b)$, where Com is a perfectly-binding commitment scheme and $\mathsf{Com}(x; r)$ denotes a commitment to x using randomness r. P_1 sends the garbled circuits to P_2 together with all of the above commitments. The commitments are sent as two vectors of pairs; in the first vector the ith pair is $\{\hat{c}_i^0, \hat{c}_i^1\}$ in a random order, and in the second vector the ith pair is $\{\tilde{c}_i^0, \tilde{c}_i^1\}$ in a random order.*

  6. *Party P_2 chooses a random bit $b \in_R \{0, 1\}$ and sends b to P_1.*

7. P_1 sends P_2 all of the keys for the inputs wires $w_1, \ldots, w_{n^2+n}$ of the garbled circuit $G(C')_b$, together with the associated mappings and the decommitment values. (I.e. if $b = 0$, then party P_1 sends $(\hat{k}_1^0, 0), (\hat{k}_1^1, 1)$, $\ldots, (\hat{k}_{n^2+n}^0, 0), (\hat{k}_{n^2+n}^1, 1)$ and $\hat{r}_1^0, \hat{r}_1^1, \ldots, \hat{r}_n^0, \hat{r}_n^1$ for the circuit $G(C')_0$.)

8. P_2 checks the decommitments to the keys associated with $w_1, \ldots, w_n$, decrypts the entire circuit (using the keys and mappings that it received) and checks that it is exactly the circuit C' derived from the auxiliary input circuit C. In addition, it checks that the keys that it received in the oblivious transfers match the correct keys that it received in the opening (i.e., if it received $(\hat{k}, \tilde{k})$ in the ith oblivious transfer, then it checks that $\hat{k} = \hat{k}_{n+i}^{z_2^i}$ if $G(C')_0$ was opened, and $\tilde{k} = \tilde{k}_{n+i}^{z_2^i}$ if $G(C')_1$ was opened). If all the checks pass, it proceeds to the next step. If not, it outputs corrupted$_1$ and halts. In addition, if P_2 does not receive this message at all, it outputs corrupted$_1$.

9. P_1 sends decommitments to the input keys associated with its input for the unopened circuit. That is, if $b = 0$, then P_1 sends P_2 the keys and decommitment values $(\tilde{k}_1^{x_1^1}, \tilde{r}_1^{x_1^1}), \ldots, (\tilde{k}_n^{x_1^n}, \tilde{r}_n^{x_1^n})$ to P_2. Otherwise, if $b = 1$, then P_2 sends the keys $(\hat{k}_1^{x_1^1}, \hat{r}_1^{x_1^1}), \ldots, (\hat{k}_n^{x_1^n}, \hat{r}_n^{x_1^n})$.

10. P_2 checks that the values received are valid decommitments to the commitments received above. If not, it outputs abort$_1$. If yes, it uses the keys to compute $C'(x_1, z_2) = C'(x_1, x_2^1, \ldots, x_2^n) = C(x_1, x_2)$, and outputs the result. If the keys are not correct (and so it is not possible to compute the circuit), or if P_2 doesn't receive this message at all, it outputs abort$_1$.

Note that steps 7–10 are actually a single step of P_1 sending a message to P_2, followed by P_2 carrying out a computation.

If any party fails to receive a message as expected during the execution, it outputs abort$_i$ (where P_i is the party who failed to send the message). This holds unless the party is explicitly instructed above to output corrupted instead (as in Step 8).

We have the following theorem:

Theorem 12. *Let f be any probabilistic polynomial-time function. Assume that the encryption scheme used to generate the garbled circuits has indistinguishable encryptions under chosen-plaintext attacks (and has an elusive and efficiently verifiable range), and that the oblivious transfer protocol used is secure in the presence of covert adversaries with $1/2$-deterrent by Definition 6. Then, Protocol 11 securely computes f in the presence of covert adversaries with $1/2$-deterrent by Definition 6.*

The full proof of this theorem can be found in the full version.

4.2   Protocols for the Other Security Definitions

We present more efficient protocols for the two other security formulations (versions 1 and 2) which are more efficient. The protocols are essentially identical to the one described above, with the only difference being the number of shares used to split the inputs of P_2 in step 2:

- For the *failed-simulation formulation* (Version 1), we do not split the input of P_2 at all and use the original inputs (i.e., the original circuit C is used). This reduces the number of oblivious transfers from n^2 to n. The revised protocol provides security for covert adversaries in the failed simulation formulation with deterrence $1/2$.
- For the *explicit cheat formulation* (not strong) (Version 2), we split the input of P_2 into 2 shares, instead of n. Note again that this reduces the number of oblivious transfers from n^2 to $2n$. The revised protocol provides security for covert adversaries in the explicit cheat formulation with deterrence $1/4$.

4.3   Higher Deterrence Values

For all three versions, it is possible to boost the deterrence value to $1-1/\mathrm{poly}(n)$, with an increased price in performance. Let $p(\cdot)$ be a polynomial. Then, Protocol 11 can be modified so that a deterrent of $1 - 1/p(n)$ is obtained, as follows. First, we use an oblivious transfer protocol that is secure in the presence of covert adversaries with deterrent $\epsilon = 1 - 1/p(n)$. Then, Protocol 11 is modified by having P_1 send $p(n)$ garbled circuits to P_2 and then P_2 randomly asking P_1 to open all circuits except one. Note that when doing so it is not necessary to increase the number of oblivious transfers, because the same oblivious transfer can be used for all circuits. This is important since the number of oblivious transfers is a dominant factor in the complexity. The modification yields a deterrent $\epsilon = 1 - 1/p(n)$ and thus can be used to obtain a high deterrent factor. For example, using 10 circuits the deterrence is $9/10$.

4.4   Non-halting Detection Accuracy

It is possible to modify Protocol 11 so that it achieves *non-halting detection accuracy*; see Definition 4. Before describing how we do this, notice that the reason that we need to recognize a halting-abort as cheating in Protocol 11 is that if P_1 generates one faulty circuit, then it can always just refuse to continue (i.e., abort) in the case that P_2 asks it to open the faulty circuit. This means that if aborting is not considered cheating, then a corrupted P_1 can form a strategy whereby it is never detected cheating, but succeeds in actually cheating with probability $1/2$. In order to solve this problem, we construct a method whereby P_1 does not know if it will be caught or not. We do so by having P_2 receive the circuit opening via a fully secure oblivious transfer protocol, rather than having P_1 send it explicitly. This forces P_1 to either abort before learning anything, or to risk being caught with probability $1/2$. The details are provided in the full version. The price of this modification is that of one additional fully secure oblivious transfer and the replacement of all of the original oblivious transfer protocols with fully secure ones. (Of course, we could use an oblivious transfer protocol that is secure in the presence of covert adversaries with non-halting detection accuracy, but we do not know how to construct one.) Since fully-secure oblivious transfer is expensive, this is a considerable overhead.

References

1. W. Aiello, Y. Ishai and O. Reingold. Priced Oblivious Transfer: How to Sell Digital Goods. In *EUROCRYPT 2001*, Springer-Verlag (LNCS 2045), pages 119–135, 2001.
2. D. Beaver. Foundations of Secure Interactive Computing. In *CRYPTO'91*, Springer-Verlag (LNCS 576), pages 377–391, 1991.
3. M. Ben-Or, S. Goldwasser and A. Wigderson. Completeness Theorems for Non-Cryptographic Fault-Tolerant Distributed Computation. In *20th STOC,* pages 1–10, 1988.
4. R. Canetti. Security and Composition of Multiparty Cryptographic Protocols. *Journal of Cryptology*, 13(1):143–202, 2000.
5. R. Canetti and R. Ostrovsky. Secure Computation with Honest-Looking Parties: What If Nobody Is Truly Honest? In *31st STOC*, pages 255–264, 1999.
6. D. Chaum, C. Crépeau and I. Damgard. Multi-party Unconditionally Secure Protocols. In *20th STOC*, pages 11–19, 1988.
7. B. Chor, O. Goldreich, E. Kushilevitz and M. Sudan. Private Information Retrieval. *Journal of the ACM*, 45(6):965–981, 1998.
8. S. Even, O. Goldreich and A. Lempel. A Randomized Protocol for Signing Contracts. In *Communications of the ACM,* 28(6):637–647, 1985.
9. M.K. Franklin and M. Yung. Communication Complexity of Secure Computation. In *24th STOC*, 699–710, 1992.
10. O. Goldreich. *Foundations of Cryptography: Volume 2 – Basic Applications.* Cambridge University Press, 2004.
11. O. Goldreich and Y. Lindell. Session-Key Generation using Human Passwords Only. *Journal of Cryptology*, 19(3):241–340, 2006.
12. O. Goldreich, S. Micali and A. Wigderson. How to Play any Mental Game – A Completeness Theorem for Protocols with Honest Majority. In *19th STOC*, pages 218–229, 1987.
13. S. Goldwasser and L. Levin. Fair Computation of General Functions in Presence of Immoral Majority. In *CRYPTO'90*, Springer-Verlag (LNCS 537), 77–93, 1990.
14. S. Goldwasser and Y. Lindell. Secure Computation Without Agreement. *Journal of Cryptology*, 18(3):247–287, 2005.
15. Y. Ishai, J. Kilian, K. Nissim and E. Petrank. Extending Oblivious Transfers Efficiently. In *CRYPTO 2003,* Springer-Verlag (LNCS 2729), pp. 145–161, 2003
16. Y. Ishai, E. Kushilevitz, Y. Lindell and E. Petrank. Black-Box Constructions for Secure Computation. In *38th STOC*, pages 99–108, 2006.
17. Y.T. Kalai. Smooth Projective Hashing and Two-Message Oblivious Transfer. In *EUROCRYPT 2005*, Springer-Verlag (LNCS 3494) pages 78–95, 2005.
18. Y. Lindell and B. Pinkas. A Proof of Yao's Protocol for Secure Two-Party Computation. To appear in the *Journal of Cryptology. Cryptology ePrint Archive,* Report 2004/175, 2004.
19. Y. Lindell and B. Pinkas. An Efficient Protocol for Secure Two-Party Computation in the Presence of Malicious Adversaries. Manuscript, 2006.
20. D. Malkhi, N. Nisan, B. Pinkas and Y. Sella. Fairplay – A Secure Two-Party Computation System. In the *13th USENIX*, pages 287–302, 2004.
21. S. Micali and P. Rogaway. Secure Computation. Unpublished manuscript, 1992.
22. M. Rabin. How to Exchange Secrets by Oblivious Transfer. Tech. Memo TR-81, Aiken Computation Laboratory, Harvard U., 1981.
23. A. Yao. How to Generate and Exchange Secrets. In *27th FOCS*, pages 162–167, 1986.

On the Necessity of Rewinding in
Secure Multiparty Computation

Michael Backes[1], Jörn Müller-Quade[2], and Dominique Unruh[1]

[1] Saarland University, Saarbrücken, Germany
{backes,unruh}@cs.uni-sb.de
[2] Universität Karlsruhe, Germany
muellerq@ira.uka.de

Abstract. We investigate whether security of multiparty computation in the information-theoretic setting implies their security under concurrent composition. We show that security in the stand-alone model proven using black-box simulators in the information-theoretic setting does not imply security under concurrent composition, not even security under 2-bounded concurrent self-composition with an inefficient simulator and fixed inputs. This in particular refutes recently made claims on the equivalence of security in the stand-alone model and concurrent composition for perfect and statistical security (STOC'06). Our result strongly relies on the question whether every rewinding simulator can be transformed into an equivalent, potentially inefficient non-rewinding (straight-line) simulator. We answer this question in the negative by giving a protocol that can be proven secure using a rewinding simulator, yet that is not secure for any non-rewinding simulator.

1   Introduction

Multiparty computation allows a set of parties with private inputs to jointly compute a given function on their inputs such that the function evaluation does not reveal any information about the inputs of other parties except for what can already be deduced from the result of the evaluation. These properties should hold even in the presence of a malicious adversary which fully controls the network and which may control some subset of the parties that then may arbitrarily deviate from the protocol.

Defining the security of a multiparty computation via an ideal execution with an incorruptible trusted party has proven a salient technique in the past. More precisely, the trusted party receives the inputs of all parties, correctly evaluates the considered function and hands back the result. In this work, we consider multiparty computation in the information-theoretic setting, where the adversary is computationally unbounded, and where consequently no underlying complexity-theoretic assumptions are required.

Multiparty computation has been investigated for a variety of different security levels and execution scenarios. As far as security levels are concerned, *perfect security* means that the result obtained in the real protocol run with

S.P. Vadhan (Ed.): TCC 2007, LNCS 4392, pp. 157–173, 2007.

the real adversary is identical to the result obtained in an ideal protocol run with the simulator; *statistical security* is defined analogously but allows the real and ideal results to deviate from each other by a small amount. As far as different execution scenarios are concerned in which the protocol is executed in, we distinguish between security in the *stand-alone model*, security under *concurrent self-composition*, and security under *concurrent general composition*. The stand-alone model only considers a single execution of the protocol under consideration, and no other protocol is run concurrently. While this constituted the standard setting for analyzing distributed security protocols in the past, a common understanding arose that protocols have to be secure even when executed many times in parallel (concurrent self-composition), or even when run in an arbitrary network, where many different protocols may run concurrently (concurrent general composition).

A considerable amount of work has been dedicated to carrying over results obtained in the stand-alone model into the more realistic concurrent setting. In particular, it is highly desirable to analyze protocols in the stand-alone model with its much simpler execution scenario and restricted adversary capabilities, and to derive theorems that allow for subsequently carrying these analyses over into the more sophisticated models of concurrent composition. Our main result however constitutes a separation of these two notions, i.e., we show that stand-alone security and security under concurrent composition do not coincide in the information-theoretic setting, neither for perfect nor for statistical security, and not even for fixed inputs and 2-bounded concurrent self-composition (i.e., only two executions of the same protocol are executed concurrently). We believe that this helps to foster our understanding of the relationships of the respective security notions, thereby refuting some recently made claims, see below.

1.1   Related Work

Defining the security of a protocol by comparing it with an ideal specification has proven a salient technique in the past, see e.g. [7,8,2,16,3,17,18,4,1], since it entails strong compositionality properties. In recent years, several results have been obtained concerning the relation of concurrent composition of function evaluations to other security notions. First, [14] showed that a protocol that can be concurrently composed and used as a subprotocol of another protocol (concurrent general composition) is already secure with respect to specialised-simulator UC (a variant of the UC notion with another order of quantifiers). It was left open, however, whether concurrent general composition was also necessary for specialised-simulator UC. Then [15] showed that for a large class of functions (those which can be used to transfer a bit), the possibility to compose a protocol concurrently already implies the possibility to use that protocol as a subprotocol in arbitrary contexts, i.e., concurrent self-composition and concurrent general composition coincide for these functions. The relations left open by [14] were proven by [10,11] who showed that concurrent general composition is equivalent to specialised-simulator UC in the case of statistical and perfect security, and strictly stronger in the case of computational security.

All these results relate concurrent composition only to stronger notions, e.g., variants of the UC notion. To get feasibility results, it is necessary to look for relations to weaker security notions, e.g. variants of the stand-alone model. This approach was taken by [12], who could show that stand-alone security with a *non-rewinding* black-box simulator already implies concurrent self-composition (and in the perfect case even concurrent general composition). Unfortunately, they also showed that stand-alone security with a non-rewinding black-box simulator is not sufficient for concurrent general composition in the case of statistical or computational security. A similar approach had earlier successfully been pursued in [5] in a different security model based on [16]. They showed that perfect security with non-rewinding simulators allows for concurrent composition.

The central question left to solve consequently was how these results behave in the presence of a rewinding simulator, which arguably constitutes a crucial scenario in modern cryptography. It was thus investigated in [12] in which ways the requirement that the simulator has to be non-rewinding can be weakened such that the established implications remain valid. They gave theorems that every rewinding black-box simulator can be replaced by an equivalent, computationally unbounded non-rewinding simulator. A consequence of these theorems was that stand-alone security with a rewinding black-box simulator is already sufficient for concurrent self-composition in the statistical case and even for concurrent general composition in the perfect case. Our results however refute these claims.

In [9], it was shown that the task of performing a coin toss given a shorter coin toss as seed can be realised with respect to rewinding black-box simulators but not with respect to specialised-simulator UC. This resembles our results (see Section 5 for a discussion) but applies only to the hybrid model (i.e., with access to some ideal functionality) while the results in [12] were formulated in the bare model. Furthermore, in contrast to the examples given here, those in [9] do not cover the case of perfect security or of deterministic ideal functions, and they did not explicitly apply their examples to the problem of rewinding vs. non-rewinding simulators.

1.2   Our Results

We first show that rewinding constitutes a necessary ingredient for proving certain protocols secure:

Theorem 1 (Necessity of rewinding – informal). *There exist protocols that are secure in the information-theoretic stand-alone setting with a rewinding black-box simulator, and yet are not secure in this setting with any non-rewinding black-box simulator.*

This disproves the following claim from [12]: *Any black-box simulator for a perfect or statistically secure protocol can be transformed into a rewinding black-box simulator.*[1] However, it still leaves open the question if stand-alone security for

---

[1] The wording has been adapted to our notation.

protocols with rewinding black-box simulators implies security under at least concurrent self-composition (it only invalidates the existing proof chain). However, also this implication turns out not to hold, already if only two instances of the same protocol are run concurrently and if only fixed inputs are considered:

Theorem 2 (Separating stand-alone model and concurrent (self-)composition – informal). *There exist protocols that are secure in the stand-alone model with a rewinding black-box simulator, and yet are not secure under 2-bounded concurrent self-composition, not even with an inefficient simulator and fixed inputs. This holds for both the perfect and the statistical case.*

This refutes the following claim from [12]: *Every protocol that is perfectly/ statistically secure in the stand-alone model, and has a black-box simulator, is secure under concurrent self-composition with fixed inputs, with an inefficient simulator.*

The counterexample for proving this theorem exploits a specific protocol realization for a specific function. Thus one might still ask if there are other protocols that can securely implement the considered function while at the same time providing security under concurrent composition. If this was the case, the impact of Theorem 2 would be considerably weakened as one might identify the good protocol realizations for the troublesome function under consideration and then still achieve strong compositionality guarantees using those realizations.

However, we show that this is not the case in general, at least not for probabilistic functionalities, statistical security, and concurrent general composition: The task of extending coin toss (i.e., obtaining $k+1$ random coins from an ideal functionality which gives only k random bits) can be securely implemented with statistical security in the stand-alone model. However, there provably does not exist any protocol for coin toss extension with respect to statistical concurrent general composition.

Theorem 3 (A stronger separation – informal). *There exists a probabilistic function that can be securely implemented using a single instance of a probabilistic function in the stand-alone model with statistical security and an efficient rewinding black-box simulator, but that cannot be securely implemented by any protocol with a polynomial number of rounds with respect to statistical concurrent general composition.*

2   Notation and Definitions

The stand-alone model. In the *stand-alone model*, a protocol π *securely implements* an ideal function f if for every set of corrupted parties C and for every adversary $\mathcal{A}$ there is a simulator $\mathcal{S}$ such that the families of random variables $\mathrm{REAL}_{\pi,\mathcal{A},x}(k)$ and $\mathrm{IDEAL}_{f,\mathcal{S},x}(k)$ are indistinguishable in the security parameter k for all inputs $x = (x_1, \ldots, x_n)$. Here $\mathrm{REAL}_{\pi,\mathcal{A},x}$ is the output of the adversary and of the uncorrupted parties in the following interaction: The uncorrupted parties $i \notin C$ get input x_i. Then the parties interact as prescribed by the protocol π. The adversary controls the corrupted parties, i.e., he can send messages in the name of a party $i \in C$ and receives all messages for parties $i \in C$. Similarly, $\mathrm{REAL}_{f,\mathcal{S},x}$ consists of the output of the simulator $\mathcal{S}$ and of the results of the function f. Here the inputs of f corresponding to the uncorrupted parties are chosen according to x, and the inputs of the corrupted parties are entered by the simulator. The simulator can choose his output in dependence of the output of the function.[2]

In this paper, we distinguish two main flavors of the stand-alone model: *perfect* and *statistical* security. In the case of perfect security, $\mathrm{REAL}_{\pi,\mathcal{A},x}(k)$ and $\mathrm{IDEAL}_{f,\mathcal{S},x}(k)$ have to be identically distributed, while in the case of statistical security they must be statistically indistinguishable. Both cases do not impose any limitations on the adversary and the simulator. For completeness, we also mention *computational security* which requires the simulator and the adversary to be polynomially bounded and the two families of random variables to be computationally indistinguishable. (Sometimes, one also requires the simulator to be efficient in the case of statistical and perfect security. We address this case by explicitly stating whether the simulator is efficient or inefficient in the respective theorems.) A more detailed exposition of the stand-alone model can be found in [6, Chapter 7].

Concurrent self-composition. The stand-alone model does not a-priori guarantee that two or more concurrent executions of the same protocol are secure, even if a single instance is secure. Therefore one is interested in the notion of concurrent self-composition, which roughly says that several instances of a given protocol securely implement the same number of instances of the ideal function. In more detail, a protocol π securely implements a function f with respect to *g-bounded concurrent self-composition* if g instances of π (considered as a single protocol) securely implement g instances of the ideal function f in the stand-alone model. Here we distinguish two cases: either the inputs to the different instances of the protocol are all fixed in advance (i.e., each party i receives a vector $x_i = (x_{i,1}, \ldots, x_{i,g})$ of inputs and uses $x_{i,j}$ as input for the j-th instance), or the inputs to some instances can be chosen adaptively in dependence of messages sent in other instances. For us, only the first case is relevant, which is called *concurrent self-composition with fixed inputs*. More details on this definition can be found in [13]. The case of adaptive inputs is discussed in [15].

The special case, that g-bounded concurrent self-composition is given for any polynomial g we call *polynomially-bounded concurrent self-composition* or simply *concurrent self-composition*.

Concurrent general composition. The notion of *concurrent general composition* further extends the notion of concurrent self-composition. A protocol π securely implements an ideal function f with respect to *g-bounded concurrent general composition* if for any protocol σ that uses g copies of π as subprotocols,

---

[2] In case the function gives different output to different parties (i.e., are asymmetric), the situation gets slightly more complicated. However, all functions given in this paper are symmetric, so the issue does not arise.

σ securely implements σ^f in the stand-alone model (where σ^f denotes σ with all instances of π replaced by ideal evaluations of the function f). More details on this notion are found in [14].

Black-box simulators. A natural restriction on the simulators is to require *black-box simulators*, i.e., the simulator is not chosen in dependence of the adversary, but instead we require that there is an oracle Turing machine $\mathcal{S}$ (the black-box simulator) such that for every adversary $\mathcal{A}$ we have indistinguishability of $\text{REAL}_{\pi,\mathcal{A},x}(k)$ and $\text{IDEAL}_{f,\mathcal{S}^{\mathcal{A}},x}(k)$ where $\mathcal{S}^{\mathcal{A}}$ is $\mathcal{S}$ with black-box access to $\mathcal{A}$. A fine point in this definition is whether the simulator may rewind the adversary, i.e., whether the simulator may at some point in time make a snapshot of the state of the adversary and then return the adversary to that state. Normally, one permits this operation and speaks of *rewinding* black-box simulators. On the other hand, we may also require the simulator to be *non-rewinding*, so that it can perform only one execution of the black-box adversary. This is often also called a *straight-line simulator*.

3   The Necessity of Rewinding

We show that for certain functions and corresponding protocols, the ability to rewind a black-box simulator is a crucial and unavoidable ingredient for achieving simulation-based security proofs in secure multi-party computation. Throughout this section, we consider the multiplication function f_{mult} receiving two inputs from a simple domain.

Definition 4 (Function f_{mult}). *The function f_{mult} takes an input $a \in \{0,1\}$ from Alice and an input $b \in \{1,2\}$ from Bob and returns $a \cdot b$.*

The corresponding protocol π_{mult} that is intended to securely implement f_{mult} is defined as follows.

Definition 5 (Protocol π_{mult}). *Alice and Bob get inputs $a \in \{0,1\}$ and $b \in \{1,2\}$, respectively.*
  - *Alice sends a to Bob. If $a \notin \{0,1\}$, Bob assumes $a = 0$.*
  - *Bob sends $c := a \cdot b$ to Alice. If $a = 0$, but $c \neq 0$, Alice assumes $c = 0$. If $a = 1$, but $c \notin \{1,2\}$, Alice assumes $c = 1$.*
  - *Both parties output c.*

We first show that π_{mult} securely implements f_{mult} if rewinding of the black-box adversary is permitted. After that, we show that rewinding is also necessary, i.e., π_{mult} securely implements f_{mult} if and only if rewinding is permitted.

Lemma 6 (π_{mult} securely implements f_{mult}). *The protocol π_{mult} securely implements f_{mult} with perfect security in the stand-alone model with an efficient rewinding black-box simulator.*

We start with a short overview of the proof for the sake of illustration and subsequently delve into the details. First, consider the case that Bob is corrupted.

In this case, the simulator conducts a simulation of the real protocol by executing the real adversary in the role of Bob and choosing Alices input as 1. Since in this case the result of the function equals Bob's input, the simulator learns his input b as chosen by the adversary. Then the simulator enters this input b into the ideal function f_{mult}. From the result of the function f_{mult} one can deduce Alice's input a (the result is 0 if and only if a is zero). Then the simulator rewinds and restarts the adversary, this time choosing the true input a that Alice has input. Thus the simulator learns the output out the adversary gives when the input of Alice is a. Finally, the simulator outputs out. This constitutes a perfect simulation since the simulator enters the same input b into the function f_{mult} and produces the same output out as the adversary does in the real model.

Now consider the case that Alice is corrupted. In this case simulation is straightforward: Alice's input a is sent in the clear as the first message of the protocol (by the black-box adversary), so the simulator enters this input into the ideal function f_{mult}. The simulator finally has to simulate the message $c = a \cdot b$ sent by Bob. This is straightforward since the simulator knows the correct value of c from the output of f_{mult}. Finally, the simulator gives the same output as the simulated adversary, thus achieving a perfect simulation.

We now transform these intuitions into a rigorous proof.

Proof. In the case that no party is corrupted, the security (correctness) of the protocol is obvious.

Now first consider the case that Bob is corrupted. The simulator $\mathcal{S}$ for this case proceeds as follows:

- First fix the random tape of the adversary $\mathcal{A}$, which is given as a black-box.
- Then send the message $\hat{a} = 1$ to the adversary.
- Let $\hat{c}$ be the reply of the adversary. If $\hat{c} \notin \{1, 2\}$, set $\hat{c} := 1$ instead (as Alice would have done herself).
- Set $b := \hat{c}$ and use b as Bob's input to the function f_{mult}.
- Let res be the result of the function f_{mult}. Let $\tilde{a} := 0$ if $res = 0$ and $\tilde{a} := 1$ otherwise.
- Rewind the adversary (but use the same random tape) and send the message $\tilde{a}$ to the adversary.
- When the adversary outputs out, output out.

Let now an adversary $\mathcal{A}$ be given. Without loss of generality, assume $\mathcal{A}$ to be deterministic (this is indeed no restriction since the random tape for which the simulator is least successful can be hardwired into $\mathcal{A}$). Then the following values are defined:

- The message $\hat{c}_a$ sent by the adversary when receiving a message $a \in \{0, 1\}$.
- The output $\widehat{out}_a$ of the adversary when he receives a message $a \in \{0, 1\}$.

Without loss of generality again, assume $\hat{c}_0 = 0$ and $\hat{c}_1 \in \{1, 2\}$ (since Alice and the simulator replace other values by valid ones).

Given the values of $\hat{c}_a$ and out_a, and the input a, we can calculate the different values that occur during the run of the ideal protocol, in particular the values

$\mathrm{IDEAL}_{f_{mult}, \mathcal{S}^{\mathcal{A}}, a}(k) = (res, out)$. We summarise these values in the left table of Figure (1). Furthermore, we can also calculate the different values that occur during a real protocol run, i.e., c being the message sent by $\mathcal{A}$ to Alice, and $\mathrm{REAL}_{\pi_{mult}, \mathcal{A}, a}(k) = (res, out)$ the result of the function and the output of the real adversary. These values are summarized in the right table of Figure (1).

Ideal protocol:

a	$\hat{c}$	res	$\tilde{a}$	out
0	$\hat{c}_1$	0	0	out_0
1	$\hat{c}_1$	$\hat{c}_1$	1	out_1

Real protocol:

a	c	res	out
0	0	0	out_0
1	$\hat{c}_1$	$\hat{c}_1$	out_1

Fig. 1. Values occuring in the run of the ideal protocol (left side) and the real protocol (right side)

Since both out and res have the same values in real and ideal model (for all values of a), perfect security in the case that Bob is corrupted follows.

Now we consider the case that Alice is corrupted. In this case, the following simple simulator $\mathcal{S}$ achieves a perfect simulation:

- Query $\mathcal{A}$ for the first message a.
- If $a \notin \{0, 1\}$, set $a := 0$.
- Then a is passed to the function f_{mult} as Alice's input.
- The result $c := res = a \cdot b$ is given to the adversary as the answering message from Bob.
- Finally, output the simulated black-box adversary's output.

It is again straightforward to check that this constitutes a perfect simulation in the case of a corrupted Alice. □

The next lemma shows that considering only non-rewinding simulators is not sufficient to prove that π_{mult} securely implements f_{mult}.

Lemma 7 (π_{mult} needs rewinding). *The protocol π_{mult} does not securely implement f_{mult} in the stand-alone model with respect to perfect, statistical, or computational security, with any non-rewinding black-box simulator (not even with inefficient ones).*

We again start with a proof sketch. We consider the case that Bob is corrupted. To give the correct input to the ideal function f_{mult}, the simulator needs to interact with the black-box adversary before invoking f_{mult}. Furthermore, to get the correct value of Bob's input, the simulator has to choose Alice's input to be $a = 1$ in the interaction with the black-box adversary (this is exactly how the simulator in Lemma 6 was constructed). In addition to causing the result of the function to be correct, the simulator also needs to output what the black-box adversary would output in the same situation. The simulator already executed the adversary with $a = 1$; consequently if the true input turns out to be 0, the simulator cannot learn what the adversary would output in that situation unless he rewinds the adversary and executes it with Alice's input a set to 0. However, we assumed that rewinding is not permitted, and hence the simulation fails.

Proof. For contradiction, we assume that there is a non-rewinding black-box simulator $\mathcal{S}$ such that for all adversaries $\mathcal{A}$ corrupting Bob and all inputs $a \in \{0,1\}$ from Alice, we have

$$\text{REAL}_{\pi_{mult},\mathcal{A},a}(k) \approx \text{IDEAL}_{f_{mult},\mathcal{S}^{\mathcal{A}},a}(k). \tag{1}$$

For brevity, we write $\mathcal{S}$ instead of $\mathcal{S}^{\mathcal{A}}$.

Furthermore, we construct a family of adversaries $\mathcal{A} = \mathcal{A}(b,o)$ with $b \in \{1,2\}$ and $o \in \{0,1\}$. The adversary $\mathcal{A}$ corrupts Bob and behaves as follows: When receiving a message a from Alice, he sends $c := a \cdot b$ to Alice. Finally, he outputs o if $a = 0$ and $\bot$ otherwise.

To describe the real model, we use the following notation: Let a denote the input of Alice. Since we only consider the case that Bob is corrupted, a is also the message from Alice to Bob in the real model. Let c denote Bob's answer. Since Alice is uncorrupted, the result of the function evaluation in the real model is also c. When using the adversary $\mathcal{A}$ given above, it is $c = a \cdot b$. Finally let *out* denote $\mathcal{A}$'s output.

To prevent confusion, we add a swung dash ($\sim$) to the random variables in the ideal model. That is, $\tilde{c}$ is the result of the function f_{mult}, and $\tilde{b}$ is the input given by $\mathcal{S}$ in Bob's stead to the function f_{mult}. Further, let $\tilde{a}$ be the first message given by $\mathcal{S}$ to the black-box $\mathcal{A}$ (which corresponds to the message sent by Alice in the real protocol). Finally, let $\widetilde{out}$ denote the output of the simulator $\mathcal{S}$. The input of the uncorrupted Alice is still called a, since it is the same in real and ideal model (a, b and o are not random variables).

The simulator $\mathcal{S}$ has two possibilities: Either he queries the function f_{mult} (with some input $\tilde{b}$) before giving the message $\tilde{a}$ to the black-box adversary $\mathcal{A}$ (we call this event F), or he first sends the message $\tilde{a}$ to $\mathcal{A}$.

Assume that event F occurs with non-negligible probability $P(F)$. In that case, $\tilde{b}$ is chosen independently of b, so there exists a b, s.t. the probability that $b \neq \tilde{b}$ is at least $\frac{1}{2}P(F)$. Then, in the case $a = 1$ the probability that $a \cdot b \neq a \cdot \tilde{b} = \tilde{c}$ is also at least $\frac{1}{2}P(F)$. In the real model however, we have $c = a \cdot b$. Since c and $\tilde{c}$ denote the result of the function f_{mult} in the real and ideal model, this is a contradiction to Equation (1).

So event F happens only with negligible probability. Therefore, we can assume without loss of generality that the simulator $\mathcal{S}$ always first sends $\tilde{a}$ to the adversary, and only then inputs $\tilde{b}$ into f_{mult}. Assume now that the probability $P(\tilde{a} = 0)$ is non-negligible. Then consider the case $a = 1$. For $\tilde{a} = 0$ the adversary $\mathcal{A}$ answers with $\tilde{a} \cdot b = 0$, which is independent of b. In that case, $\tilde{b}$ is chosen by the simulator independently of b. Therefore, $P(\tilde{b} \neq b) \geq \frac{1}{2}P(\tilde{a} = 0)$ for some choice of b. Since $\tilde{c} = a \cdot \tilde{b}$, $P(\tilde{c} \neq a \cdot b)$ is non-negligible. But in the real model we have $c = a \cdot b$, in contradiction to Equation (1). Therefore $P(\tilde{a} = 0)$ is negligible, so we can assume without loss of generality that the simulator always sends $\tilde{a} = 1$ to $\mathcal{A}$ (before invoking f_{mult}).

By construction, when receiving $\tilde{a} = 1$, the adversary will give output $\bot$. Therefore, the simulator's output $\overline{out}$ is independent of o, so for some o we have $P(out \neq o) \geq \frac{1}{2}$. But in the case $a = 0$, the adversary outputs $out = o$ in the real model. This contradicts (1). So there cannot exist a non-rewinding simulator that fulfills Equation (1). □

The general statement induced by the previous results is summarized in the following corollary. Its proof is a direct consequence of Lemma 6 and 7.

Theorem 8. *Let f be a function and π a protocol that securely implements f with perfect, statistical, or computational security in the stand-alone model. Then π does not necessarily securely implement f with perfect, statistical, or computational security, respectively, with any non-rewinding black-box simulator.*

In the proofs we have assumed the following variant of the stand-alone model: After running the protocol, the *output* of the adversary should be indistinguishable in the real and the ideal model. Another popular variant of the model instead considers the *view* of the adversary. Our examples can be easily transferred to the latter setting. Instead of giving some output *out*, the adversary sends a protocol message containing *out* that is ignored by the protocol. Then our proofs directly carry over to that variant of the stand-alone model. This also applies to the proofs in the next section.

The example given in this section could also be used to show that stand-alone security does not imply concurrent self-composition: When two instances of π_{mult} run concurrently, a corrupted Bob can enforce the sum of the outputs to equal 2 (unless Alice inputs 0 in both cases), which is impossible given access to two copies of f_{mult}. However, instead of showing this in detail, we will show the separation between stand-alone security and concurrent self-composition in the next section using another example which we consider to be more instructive.

4   Perfect Stand-Alone Security Does Not Imply Concurrent Self-composition

In this section, we show that for certain functions and corresponding protocols, security in the stand-alone model is not necessarily sufficient for guaranteeing security under concurrent self-composition. Throughout this section, we consider the functions f_{min_x}, where x is a natural number that constitutes a parameter of the function. The function f_{min_x} outputs the minimum of its inputs.

Definition 9 (Function f_{min_x}). *Let $x \geq 2$ be an integer. The function f_{min_x} takes two inputs $a, b \in \{1, \ldots, x\}$ from Alice and Bob, where a is odd and b is even. The result $f_{min_x}(a, b)$ is the minimum of a and b.*

Definition 10 (Protocol π_{min_x}). *Let $x \geq 2$ be an integer. Alice gets an odd input $a \in \{1, \ldots, x\}$, Bob an even input $b \in \{1, \ldots, x\}$.*

- *The protocol π_{min_x} proceeds in at most $x - 2$ rounds $1, \ldots, x - 2$.*
- *In round r for an odd value r, Alice sends* no *if $r \neq a$, and* yes *if $r = a$.*

- *In round r for an even value r, Bob sends* **no** *if $r \neq b$, and* **yes** *if $r = b$.*
- *If any other message is sent, the message is assumed by the recipient to be* **no**.
- *As soon as a message* **yes** *has been sent (in some round r), the protocol terminates, and both parties output r.*
- *If no message* **yes** *is sent in any round, the output is $x - 1$.*

Lemma 11 (π_{min_x} securely implements f_{min_x}). *Let $x \geq 2$ be an integer. The protocol π_{min_x} securely implements f_{min_x} with perfect security in the stand-alone model and with an efficient rewinding black-box simulator.*

The proof can be sketched as follows; for the sake of readability, we only elaborate on the case that Bob was corrupted. The simulator starts a simulation of a real protocol where the (black-box) real adversary plays the role of Bob and Alice's input is set to its largest possible value a_{max}. Then the minimum of Alice's and Bob's input, which is returned by the ideal function f_{min_x}, allows us to calculate Bob's input b. This value b is then used as Bob's input in the ideal evaluation of f_{min_x}. So far, we have already guaranteed that the result of the function is identical in both the real and the ideal model. The simulator now has to learn what the adversary would output. Learning this output requires the simulator to perform a second simulation of the real protocol with the adversary using the correct value of Alice's input a (here we need the possibility to rewind). In the case that the result of the function is smaller than Bob's input b, this is an easy task since the input of Alice is equal to the result of the function. If the function result is however equal to the input of Bob, we can only deduce that Alice's input is larger than Bob's input. In this case, the simulator simply assumes the largest possible value a_{max} that Alice might have input. Since the protocol terminates in the round corresponding to Bob's input b, the adversary will never learn whether Alice used her maximum input or just some input greater than b. So in both cases, the simulator learns what output the adversary gives in the real model, and it can thus perform a perfect simulation.

Proof. Let a and b denote the inputs of Alice and Bob, respectively. Let a_{max} be the largest odd integer with $a_{max} \leq x$ (Alice's largest possible input), and b_{max} the largest even integer with $b_{max} \leq x$ (Bob's largest possible input).

If Alice and Bob are uncorrupted, it is easy to check, that the protocol indeed calculates the minimum of a and b. Hence correctness (security) in this case is clear.

Now consider the case that Bob is corrupted. In this case, the following simulator $\mathcal{S}$ achieves a perfect simulation:

- First fix the random tape of the adversary $\mathcal{A}$, which is given as a black-box.
- Simulate a protocol run of π_{min_x} with $\mathcal{A}$ where Bob is corrupted and Alice gets input a_{max}. Let $\widetilde{res}$ be Alice's output in this function evaluation.
- Let $\tilde{b} := \widetilde{res}$ if $\widetilde{res} < a_{max}$, and $\tilde{b} := b_{max}$ otherwise.
- Invoke the ideal function π_{min_x} using $\tilde{b}$ as Bob's input. Let *res* be the result of the function.

- Let $\tilde{a} := res$ if $res < \tilde{b}$, and let $\tilde{a} := a_{max}$ otherwise.
- Then simulate a protocol run of π_{min_x} with adversary $\mathcal{A}$ where Bob is corrupted and Alice gets input $\tilde{a}$. Let out be the adversary's output in that protocol run.
- Output out.

Now, consider an adversary $\mathcal{A}$ that corrupts Bob. Without loss of generality, assume $\mathcal{A}$ to be deterministic (cf. the proof of Lemma 6). Assume further that $\mathcal{A}$ only sends messages yes and no, and that these messages only occur in the appropriate (i.e., even) rounds.

Then we can associate the following values with this adversary $\mathcal{A}$:

- By $\hat{b}$ we denote the number of the first round in which the adversary answers with yes when Alice always sends no. If $\mathcal{A}$ never sends yes, let $\hat{b} := b_{max}$. (Intuitively, $\hat{b}$ denotes Bob's input as chosen by the adversary.)
- By out_a we denote the output made by the adversary in the execution of a real protocol when Alice has input a.

Note that both $\hat{b}$ and out_a are obtained deterministically since both $\mathcal{A}$ and the protocol π_{min_x} are deterministic. The following facts are easy to observe:

(i) In a protocol run of π_{min_x} with $\mathcal{A}$ where Alice gets input a, Alice's output (i.e., the result of the function) is always $\min(a, \hat{b})$. (This holds since in case $\hat{b} < a$, the adversary cannot distinguish Alice from an Alice with input a_{max}, and if $\hat{b} > a$, the protocol guarantees that a is output. The case $\hat{b} = a$ does not occur, since $\hat{b}$ is even and a is odd.)

(ii) For $a, a' > \hat{b}$ it is $out_a = out_{a'}$. (Because then the protocol terminates in round $\hat{b}$, so Alice behaves identically with inputs a and a'.)

We now show, that $\mathcal{S}$ entails a perfect simulation in the case that Bob is corrupted. Consider an ideal protocol run consisting of the simulator $\mathcal{S}$ (having black-box access to the adversary $\mathcal{A}$), and the ideal function f_{min_x} receiving some input a on Alice's side.

First, the simulator simulates a real protocol run with input a_{max} for Alice. By fact (i), Alice's output in that protocol run is $\min(a_{max}, \hat{b})$. Therefore it is $\widetilde{res} = \min(a_{max}, \hat{b})$.

If $\min(a_{max}, \hat{b}) = \widetilde{res} < a_{max}$, it follows $\widetilde{res} = \hat{b}$. If $\min(a_{max}, \hat{b}) = \widetilde{res} \geq a_{max}$, if follows $\hat{b} \geq a_{max}$, and therefore $\hat{b} = b_{max}$. In both cases the value $\tilde{b}$ calculated by the simulator equals $\hat{b}$.

Since the simulator enters $\tilde{b}$ as Bob's input into the function f_{min_x}, the result of the function is $res = \min(a, \tilde{b}) = \min(a, \hat{b})$.

Consider the case $\min(a, \tilde{b}) = res < \tilde{b}$. Then $a = res$ and the simulator sets $\tilde{a} := res$, so $out_a = out_{\tilde{a}}$. In the case $\min(a, \tilde{b}) = res \geq \tilde{b}$, it is $a > \tilde{b}$ (since a and $\tilde{b}$ cannot be equal, being of different parity). The simulator then chooses $\tilde{a} := a_{max} \geq a > \tilde{b}$, so $a, \tilde{a} > b$. By fact (ii), we then have $out_a = out_{\tilde{a}}$. So in both cases, the output of the simulator and the output of the adversary coincide, i.e., we have $out = out_a$.

In a nutshell, the result of the function (i.e., Alice's output) in an ideal protocol run is $res = \min(a, \hat{b})$, and the output of the simulator is $out = out_a$.

In the real model the result of the function is $\min(a, \hat{b})$ according to Fact (i). Moreover, the output of the adversary is out_a by definition.

Consequently, the output of the adversary and and the result of the function evaluation are identical in real and ideal model so that perfect security in the case that Bob is corrupted follows.

The case that Alice is corrupted is proven identically, except for exchanging the roles of Alice and Bob, the letters a and b and the words *odd* and *even*. $\square$

Lemma 12 (π_{min_x} **does not compose concurrently**). *Let $x \geq 4$ be an integer. The protocol π_{min_x} does not securely implement f_{min_x} with perfect, statistical, or computational 2-bounded concurrent self-composition with fixed inputs.*

This even holds if we allow unbounded non-black-box simulators that may adaptively query the two ideal instances of f_{min_x}.

The basic idea underlying the proof of the lemma can be given as follows. A corrupted Bob will start two parallel sessions of the real protocol π_{min_x} with Alice and subsequently forward all protocol messages from one protocol session into the other and vice versa. The output of both protocols will then be equal to the smaller input that Alice has made (± 1). This is impossible to achieve when concurrently interacting with two ideal functions: The simulator has to invoke one of the functions first. If it uses a large value for Bob's input, the simulation will fail if Alice gave a large input to that first function, and a small input to the second one. If it uses a small value for Bob's input, it will fail if Alice gave large inputs to both functions.

Actually, the proof gives a slightly stronger result than Lemma 12 since it shows that π_{min_x} does not even allow for *parallel* composition.

Proof. We construct an adversary $\mathcal{A}$ corrupting Bob and attacking two concurrently composed instances of π_{min_x} as follows:

- In each odd round, he receives messages $m_{A,1}, m_{A,2}$ from the two instances A_1, A_2 of Alice.
- In each even round, he sends $m_{A,1}$ to the second instance A_2 of Alice, and $m_{A,2}$ to the first instance A_1.

Let a_1 denote the input of the first instance of Alice, and a_2 the input of the second instance of Alice. Let further res_1 and res_2 denote the respective outputs.

Then if $a_1 < a_2$, one easily sees that $res_1 = a_1$ and $res_2 = a_1 + 1$ (since the forwarded **yes**-message reaches A_2 only in the $(a_1 + 1)$-st round). Similarly, if $a_1 > a_2$, it is $res_1 = a_2 + 1$ and $res_2 = a_2$. Finally, if $a_1 = a_2 < a_{max}$, we get $res_1 = res_2 = a_1 = a_2$, and if $a_1 = a_2 = a_{max}$ we finally have $res_1 = res_2 = x - 1$.

Now, consider an arbitrary simulator $\mathcal{S}$. This simulator has access to two instances $f_{min_x}^1, f_{min_x}^2$ of the function f_{min_x}, which receive inputs a_1 and a_2 from Alice, respectively. The simulator may now invoke the functions one after the other. Let $f_{min_x}^i$ denote the function invoked first (i.e., i is a random variable), and b_i the Bob-input given to $f^i_{min_x}$ by $\mathcal{S}$. Since both i and b_i cannot depend on the inputs a_1 and a_2 from Alice, at least one of the following two cases occurs with probability at least $\frac{1}{4}$ for suitable choices of a_1 and a_2: (i) It is $a_1 = a_2 = a_{max}$ and $b_i < x - 1$. (ii) It is $a_i = a_{max}$, $a_{3-i} = 1$ and $b_i \geq x - 1$.

In case (i) the result of $f^i_{min_x}$ will be $b_i < x - 1$. However, as we have shown above, in the real model, with inputs $a_1 = a_2 = a_{max}$ the result of $f^1_{min_x}$ and $f^2_{min_x}$ (i.e., the outputs of the Alice-instances) would be $x - 1$. Therefore the results differ in real and ideal model.

In case (ii) the result of $f^i_{min_x}$ will be greater or equal $x-1$ (since $a_{max} \geq x-1$). In the real model however, the results of the functions would be 1 and 2 (in some order), because $a_{3-i} = 1$. Since $x \geq 4$, both results are smaller than $x - 1$ and so the results differ in real and ideal model.

So with non-negligible probability, the function results in the ideal model do not match those in the real model. The insecurity of π_{min_x} under 2-bounded concurrent self-composition follows.   □

These lemmas yield the following theorem. Its proof is a direct consequence of Lemma 11 and 12.

Theorem 13. *Let f be a function and π a protocol that securely implements f with perfect, statistical, or computational security in the stand-alone model with an efficient rewinding black-box simulator. Then π does not necessarily securely implement f with perfect, statistical, or computational security under 2-bounded concurrent self-composition, not even with an inefficient non-black-box simulator and fixed inputs.*

5   A Stronger Separation

The results proven so far show that certain protocols can be secure in the stand-alone model, require rewinding, and do not allow composition. The natural question arising here is whether this issue of composition depends on a specific choice of the protocol while some other protocol for the same task might be composable. We show that, at least for the case of a probabilistic functionality, statistical security, and concurrent *general* composition, this is not the case: The task of extending coin toss (i.e., obtaining $k + 1$ random coins from an ideal funtionality which gives only k random bits) can be realised with statistical security in the stand-alone model. However, there does not exist a protocol for coin toss extension with respect to statistical concurrent general composition.

Corollary 14. *There exists a probabilistic function $\mathcal{F}$ that can be securely implemented using a single instance of a probabilistic function $\mathcal{G}$ in the stand-alone model with statistical security and an efficient rewinding black-box simulator, but that cannot be securely implemented using a single instance of $\mathcal{G}$ by any protocol with a polynomial number of rounds with respect to statistical concurrent general composition.*

Proof. Let $\mathcal{F}$ be the $(k+1)$-bit coin-toss functionality (i.e., the functionality that provides *one* uniformly chosen string of length $k+1$, cf. [9]). $\mathcal{F}$ is a function that ignores its inputs. Let $\mathcal{G}$ be the k-bit coin-toss functionality.

In [9] it is shown, that there is a protocol securely implementing $\mathcal{F}$ using $\mathcal{G}$ in the stand-alone model with statistical security (and an efficient rewinding black-box simulator).

Assume that there is a polynomial-round protocol π securely implementing $\mathcal{F}$ using a single instance of $\mathcal{G}$ with respect to statistical concurrent general composition. Then by the results in [14], π also securely implements $\mathcal{F}$ with respect to statistical specialised-simulator UC ([14] only shows the computational case, but the proof easily carries over to the statistical case). But in [9] it is shown that no polynomial-round protocol π using a single instance of $\mathcal{G}$ exists that securely implements $\mathcal{F}$ with respect to specialised-simulator UC (actually, [9] state the theorem for UC, but their proof also shows the case of specialised-simulator UC, since the environments constructed in their proof do not depend on the simulator). Thus we have a contradiction and the lemma follows. □

6   Conclusion and Open Questions

We have shown that in the information-theoretic setting, the existence of a rewinding simulator in the stand-alone model is not sufficient for the existence of a non-rewinding simulator, nor for achieving concurrent composition. In that light, the question naturally arises which additional constraints may be imposed on the black-box simulator so that it can be converted into a non-rewinding black-box simulator (which then in turn allows for concurrent composition, see [12]). A major problem in coming up with a constructive transformation of a rewinding simulator into a non-rewinding one seems to be the following: The original simulator's program may explicitly require several executions of the black-box adversary, e.g., the knowledge-extractor in most proofs of knowledge executes the adversary (i.e., the prover) twice or more often, and then uses the results of *several of the executions* to construct the required output. Such a knowledge-extractor cannot easily be transformed into a non-rewinding one, since then its program would suddenly find itself in the unexpected situation of terminating without having run the black-box adversary twice, yielding undefined results. The simulators in the counterexamples given in this work are of this form as well. On the other hand, many simulators found in the literature use rewinding only to backtrack from wrong choices. After having backtracked, they forget that they rewound the adversary and start anew, hoping to select the right choice this time. Protocol with simulators of this kind include, e.g., the well-known zero-knowledge proofs of graph-isomorphism and graph-3-colouring.

More formally we call a simulator *obliviously rewinding*, if it is an oracle Turing machine with the following extension: at any point during its execution, the simulator may set a *marker M*. Then a snapshot of the state of the simulator and of the black-box adversary is taken. Furthermore, the simulator may then at any other point of its execution choose to return to a marker M. If he chooses to

do so, the state of the black-box adversary *and of the simulator itself* are restored
to the snapshot that was taken when the marker M was set. The program of
an obliviously rewinding simulator does not run into an undefined situation
when the simulation suddenly goes through without any rewinding. Therefore,
it may be possible that such an obliviously rewinding simulator can indeed be
transformed into a non-rewinding one as proposed in [12]. We leave this as an
open question.

Acknowledgements. We thank the anonymous referees for helpful comments.

References

1. M. Backes, B. Pfitzmann, and M. Waidner. Secure asynchronous reactive systems.
   IACR Cryptology ePrint Archive 2004/082, Mar. 2004.
2. D. Beaver. Secure multiparty protocols and zero knowledge proof systems tolerating
   a faulty minority. *Journal of Cryptology*, 4(2):75–122, 1991.
3. R. Canetti. Security and composition of multiparty cryptographic protocols. *Journal of Cryptology*, 3(1):143–202, 2000.
4. R. Canetti. Universally composable security: A new paradigm for cryptographic
   protocols. In *Proc. 42nd IEEE Symposium on Foundations of Computer Science
   (FOCS)*, pages 136–145, 2001. Extended version in Cryptology ePrint Archive,
   Report 2000/67, http://eprint.iacr.org/.
5. Y. Dodis and S. Micali. Parallel reducibility for information-theoretically secure computation. In M. Bellare, editor, *Advances in Cryptology, Proceedings of
   CRYPTO '00*, volume 1880 of *Lecture Notes in Computer Science*, pages 74–92.
   Springer-Verlag, 2000.
6. O. Goldreich. *Foundations of Cryptography – Volume 2 (Basic Applications)*.
   Cambridge University Press, May 2004. Previous version online available at
   http://www.wisdom.weizmann.ac.il/~oded/frag.html.
7. O. Goldreich, S. Micali, and A. Wigderson. How to play any mental game – or – a
   completeness theorem for protocols with honest majority. In *Proc. 19th Annual
   ACM Symposium on Theory of Computing (STOC)*, pages 218–229, 1987.
8. S. Goldwasser and L. Levin. Fair computation of general functions in presence of
   immoral majority. In *Advances in Cryptology: CRYPTO '90*, volume 537 of *Lecture
   Notes in Computer Science*, pages 77–93. Springer, 1990.
9. D. Hofheinz, J. Müller-Quade, and D. Unruh. On the (im)possibility of extending coin toss. In S. Vaudenay, editor, *Advances in Cryptology, Proceedings of EUROCRYPT '06*, volume 4004 of *Lecture Notes in Computer Science*, pages 504–521. Springer-Verlag, 2006. Full version online available at
   http://eprint.iacr.org/2006/177.
10. D. Hofheinz and D. Unruh. Comparing two notions of simulatability. In J. Kilian,
    editor, *Theory of Cryptography, Proceedings of TCC 2005*, Lecture Notes in Com-
    puter Science, pages 86–103. Springer-Verlag, 2005. Online available at http://
    iaks-www.ira.uka.de/home/unruh/publications/hofheinz05comparing.html.
11. D. Hofheinz and D. Unruh. Simulatable security and polynomially bounded con-
    current composition. In *IEEE Symposium on Security and Privacy, Proceedings
    of SSP '06*, pages 169–182. IEEE Computer Society, 2006. Full version online
    available at http://eprint.iacr.org/2006/130.ps.

12. E. Kushilevitz, Y. Lindell, and T. Rabin. Information-theoretically secure protocols and security under composition. In *38th Annual ACM Symposium on Theory of Computing, Proceedings of STOC 2006*, pages 109–118. ACM Press, 2006. Online available at `http://www.cs.biu.ac.il/~lindell/abstracts/IT-composition_abs.html`.

13. Y. Lindell. Bounded-concurrent secure two-party computation without setup assumptions. In *35th Annual ACM Symposium on Theory of Computing, Proceedings of STOC 2003*, pages 683–692. ACM Press, 2003.

14. Y. Lindell. General composition and universal composability in secure multi-party computation. In *44th Annual Symposium on Foundations of Computer Science, Proceedings of FOCS 2003*, pages 394–403. IEEE Computer Society, 2003. Online available at `http://eprint.iacr.org/2003/141`.

15. Y. Lindell. Lower bounds for concurrent self composition. In M. Naor, editor, *Theory of Cryptography, Proceedings of TCC 2004*, volume 2951 of *Lecture Notes in Computer Science*, pages 203–222. Springer-Verlag, 2004.

16. S. Micali and P. Rogaway. Secure computation. In *Advances in Cryptology: CRYPTO '91*, volume 576 of *Lecture Notes in Computer Science*, pages 392–404. Springer, 1991.

17. B. Pfitzmann and M. Waidner. Composition and integrity preservation of secure reactive systems. In *Proc. 7th ACM Conference on Computer and Communications Security*, pages 245–254, 2000. Extended version (with Matthias Schunter) IBM Research Report RZ 3206, May 2000, `http://www.semper.org/sirene/publ/PfSW1_00ReactSimulIBM.ps.gz`.

18. B. Pfitzmann and M. Waidner. A model for asynchronous reactive systems and its application to secure message transmission. In *Proc. 22nd IEEE Symposium on Security & Privacy*, pages 184–200, 2001. Extended version of the model (with Michael Backes) IACR Cryptology ePrint Archive 2004/082, `http://eprint.iacr.org/`.

On Expected Probabilistic Polynomial-Time Adversaries: A Suggestion for Restricted Definitions and Their Benefits*

Oded Goldreich

Weizmann Institute of Science, Rehovot, Israel

Abstract. This paper concerns the possibility of developing a coherent theory of security when feasibility is associated with *expected* probabilistic polynomial-time (*expected* PPT). The source of difficulty is that the known definitions of *expected* PPT *strategies* (i.e., *expected* PPT *interactive* machines) do not support natural results of the type presented below. To overcome this difficulty, we suggest new definitions of *expected* PPT strategies, which are more restrictive than the known definitions (but nevertheless extend the notion of *expected* PPT *non-interactive* algorithms). We advocate the conceptual adequacy of these definitions, and point out their technical advantages. Specifically, identifying a natural subclass of black-box simulators, called *normal*, we prove the following two results:

1. Security proofs that refer to all *strict* PPT adversaries (and are proven via normal black-box simulators) extend to provide security with respect to all adversaries that satisfy the restricted definitions of *expected* PPT.
2. Security composition theorems of the type known for *strict* PPT hold for these restricted definitions of *expected* PPT, where security means simulation by normal black-box simulators.

Specifically, a normal black-box simulator is required to make an expected polynomial number of steps, when given oracle access to *any* strategy, where each oracle call is counted as a single step. This natural property is satisfies by most known simulators and is easy to verify.

1 An Opinionated Introduction

The title of this introduction and the use of first person singular are meant to indicate that this introduction is more opinionated than is customary in our field. Nevertheless, I will try to distinguish facts from my opinions by use of adequate phrases.

In my opinion, the first question that should be asked when suggesting and/or reviewing a definition is what is the purpose of the definition. When reviewing an existing definition, a good way to start is to look into the history of the

---

* This research was partially supported by the Israel Science Foundation (grant No. 460/05).

S.P. Vadhan (Ed.): TCC 2007, LNCS 4392, pp. 174–193, 2007.

definition, since the purpose may be more transparent in the initial works than in follow-up ones.

Before turning to the history and beyond, let me state that I assume that the reader is familiar with the notion of zero-knowledge and the underlying simulation paradigm (see, e.g., [G01, Sec. 4.3.1]). In fact, some familiarity with general secure multi-party computation (e.g., at the overview level of [G04, Sec. 7.1]) is also useful. Indeed, this paper is not intended for the novice: it deals with subtle issues that the novice may (or even should) ignore.

This is a trimmed version of my technical report [G06]. In particular, Sections 4-6 were omitted.

1.1 The History of Related Definitions

To the best of my recall, the first appearance in cryptography of the notion of expected (rather than strict) probabilistic polynomial-time was in the seminal work of Goldwasser, Micali, and Rackoff [GMR]. The reason was that the simulators presented in that paper (for the Quadratic Residuosity and the Quadratic Non-Residuosity interactive proofs) were only shown to run in *expected* probabilistic polynomial-time.[1] Recall that these simulators were used in order to simulate the interaction of arbitrary *strict* probabilistic polynomial-time (adversarial) verifiers with the honest prover.

At first, the discrepancy between the expected probabilistic polynomial-time allowed to the simulator and the restriction of the adversary to strict probabilistic polynomial-time did not bother anybody. One reason for this lack of concern seems to be that everybody was overwhelmed by the new fascinating notion of zero-knowledge proofs, its mere feasibility and its wide applicability (as demonstrated by [GMR, GMW]). But as time passed, some researchers became bothered by this discrepancy, which seemed to violate (at least to some extent) the intuition underlying the definition of zero-knowledge. Specifically, relating the complexity of the simulation to the complexity of the adversary is the essence of the simulation paradigm and the key to the conclusion that the adversary gains noting by the interaction (since it can obtain the same, essentially as easily, without any interaction). But *may we consider expected polynomial-time and strict (probabilistic) polynomial-time as being the same complexity?*

The original feeling was that the discrepancy between strict and expected polynomial-time is not very significant, and I do hold this view to this very day. After all, everybody seems quite happy with replacing one polynomial (bound

---

[1] Note that while a small definitional variation (cf. [G01, Sec. 4.3.1.1] versus [G01, Sec. 4.3.1.6]) suffices for obtaining a strict probabilistic polynomial-time (perfect) simulation for the QR protocol, this does not seem to be the case when the QNR protocol is concerned. The same dichotomy is manifested between the Graph Isomorphism and Graph 3-Colorability protocols (of [GMW]) on one hand and the constant-round zero-knowledge proof of [GK96] on the other hand. The dichotomy arises from two different simulation techniques; the first is tailored for "challenge-response" protocols, while the second refers to the use of "proofs-of-knowledge" (which may be implicit and trivial (as in [GK96])).

of the running time) by another, at least as a very first approximation of the intuitive notion of similar complexity.[2] Still, I cannot deny that there is something unpleasing about this discrepancy. Following [KL05], let me refer to this issue as an aesthetic consideration.

Jumping ahead in time, let me mention a more acute consideration articulated in [KL05]: A different handling of adversaries and simulations (e.g., the discrepancy between expected polynomial-time and strict probabilistic polynomial-time) raises technical difficulties and, in particular, stands in the way of various desired composition theorems (e.g., of the type presented in [GO94, C00]). But let me get back to the story.

Faced with the aforementioned aesthetic consideration, a few researchers suggested a simple solution: extending the treatment of adversaries to ones running in expected polynomial-time. This suggestion raised a few problems, the first being *how to define expected polynomial-time interactive machines?* (In addition, there are other problems, which I will discussed later.)

Feige's proposal [F90] was to consider the running-time of the adversary when it interacts with the honest party that it attacks, and require that the adversary runs in expected polynomial-time (in such a random interaction). My own proposal was to allow only adversaries that run in expected polynomial-time regardless with whom they interact; that is, the adversary is required to run in expected polynomial-time when interacting with any other strategy. Feige objected to my proposal saying that it unduly restricts the adversary, which is designed to attack a specific strategy and thus should be efficient only when attacking this strategy. My own feeling was that it is far more important to maintain a coherent theory by using a "stand-alone" notion of expected polynomial-time; that is, a notion that categorizes strategies regardless of their aim (e.g., without reference to whether or not these strategies model adversaries (and which strategies these adversaries attack)). The rationale underlying this feeling is discussed in Section 1.2. (Furthermore, Feige's definition also extends the standard definition of *strict* probabilistic polynomial-time adversaries by allowing adversaries that may not even halt when interacting with strategies other the those they were designed to attack (see proof of Proposition 5).)

In any case, a major problem regarding the suggestion of extending the treatment of adversaries to ones running in expected polynomial-time is *whether such an extension is at all possible*. One specific key question is *whether known simulators can handle expected polynomial-time adversaries*. As pointed out in [KL05], in some cases (e.g., the simulator of [GK96]), the answer is negative even if one uses the more restricted notion of expected polynomial-time adversaries (which refers to interaction with any possible strategy). Another important question is *whether composition theorems that are known to hold for strict probabilistic*

---

[2] It is telling that my advocacy of *knowledge tightness* [G01, Sec. 4.4.4.2], a notion aimed at quantitatively bounding the ratio of the running times of the simulator and adversary, has never gain much attention. (And yes, I am aware of the recent work of Micali and Pass [MP06] that introduces and advocates an even more refined notion.)

polynomial-time (strategies and simulators) *can be extended to the case of expected polynomial-time* (strategies and simulators).

Indeed, the "question of composition" became a major concern in the 1990's and motivated a re-examination of many aspects of the theory of cryptography. Here I refer specifically to the Sequential Composition Theorem of Canetti [C00], which supports modular construction of protocols, and to the Concurrent Composition Theorem of Canetti [C01], which is aimed at preserving security in settings where numerous executions of arbitrary protocols are taking place concurrently. These composition results were obtained when modeling adversaries as *strict* probabilistic polynomial-time strategies and allowing only *strict* probabilistic polynomial-time simulators. One consequence of the lack of analogous results for the case of *expected* polynomial-time was that the modular construction of secure protocol had to avoid protocols that were only known to be simulateable in *expected* polynomial-time.[3]

Recently, Katz and Lindell [KL05] initiated a study of the possibility of simulating *expected* polynomial-time adversaries and/or obtaining composition theorems (or sufficiently good alternatives) for the *expected* polynomial-time case. They showed that in some cases (e.g., when the simulator satisfies some additional properties and/or under some super-polynomial intractability assumptions) such partial results can be obtained.[4] These results do not provide a "free" transformation from the *strict* probabilistic polynomial-time model to the *expected* polynomial-time model, where "free" means without referring to additional assumptions. In my opinion, as long as this is the state of affairs, one better look for alternative directions.

1.2 Towards New Definitions

My starting point (or thesis) is that *we should not care about expected polynomial-time adversaries per se.* As hinted by my historical account, researchers were perfectly happy with strict probabilistic polynomial-time adversaries and would have probably remained so if it were not for the introduction of expected polynomial-time simulators. Indeed, at the end of the day, the user (especially a non-sophisticated one) should care about what an adversary can obtained within a specific time (or various possible amounts of work), where the term 'obtain' incorporates also a quantification of the success probability. I claim that our goal as researchers is to provide such statements (or rather techniques

---

[3] For example, relatively efficient proofs-of-knowledge (which only guarantee *expected* polynomial-time extraction) were avoided (e.g., in [G04, Sec. 7.4.1.3]) and strong proofs-of-knowledge (cf. [G01, Sec. 4.7.6]) were used instead.

[4] Roughly speaking, the two main results of [KL05] refers to versions of computational indistinguishability that are required to hold with respect to super-polynomial-time observers. This means that for obtaining (ordinary) computational security, somewhere along the way, one needs to make a super-polynomial-time intractability assumption. Also note that the simulators constructed in [KL05] use the corresponding adversaries in a "slightly non-black-box" manner in the sense that they terminate executions (of these adversaries) that exceed a specific number of steps.

for providing such statements), and that *expected polynomial-time machines may appear in the analysis only as intermediate steps* (or mental experiments).

My thesis is further enforced by the confusing and unintuitive nature of expected running-time especially when applied in the context of cryptography[5] and by numerous annoying phenomena related to expected-time complexity. In particular, note that, unlike strict polynomial-time, expected polynomial-time is a *highly non-robust notion that is not preserved under changes of computational model and standard algorithmic compositions.*[6] These "features" are an artifact of the "bad interaction" between the expectation operator and many non-linear operators: for example, for a random variable X, we cannot upper-bound $E[X^2]$ as a function of $E[X]$. Thus, if X is a random variable that represents the running-time of some process Π (where the probability space is that of the internal coin tosses of Π), then we cannot bound the expected running-time of various modest variants of Π (e.g., which square its running-time) in terms of the expected running-time of Π. (See Footnote 25, which refers to a natural case in which this problem arises.)

The foregoing reservations regarding expected polynomial-time are of lesser concern when expected running-time is only used as an intermediate step (rather than as a final statement). Taking this approach to its extreme, I claim that for this purpose (of an intermediate step) it is legitimate to use any (reasonable) definition of expected polynomial-time strategies, and that among such possibilities we better select a definition that supports the desired results (e.g., simulation of corresponding adversaries and composition theorems). Thus, we should seek a definition of expected polynomial-time strategies that enjoys the following properties:

1. The definition should include all strict probabilistic polynomial-time strategies (but should not extend "much beyond that"; e.g., super-polynomial-time computations may only occur with negligible probability).
2. When applied to non-interactive strategies (i.e., stand-alone algorithms) the definition of expected polynomial-time strategies should yield the standard notion of expected polynomial-time.

    This property is not only a matter of aesthetic considerations but is rather important for composition theorems (as desired in Property 3b). Furthermore, when applied to the context of zero-knowledge, the current property implies that expected polynomial-time simulators are deemed admissible by this definition.[7]

---

[5] Indeed, things become even worse if we bear in mind the need to keep track of both the running-time and the success probability (which should be calculated with respect to various strict time bounds). That is, I claim that providing only the expected running-time and the overall success probability is quite meaningless, since the success is likely to be correlated with the running-time.

[6] See analogous discussion of average-case complexity in [G97].

[7] In fact, we should strengthen Property 2 by requiring that also in the context of secure multi-party computation (where the simulators are themselves interactive machines) the known "expected polynomial-time" simulators (of strict probabilistic polynomial-time) are deemed admissible by the selected definition.

3. The definition should allow to derive the results that we seek:
   (a) Known simulators that handle strict probabilistic polynomial-time adversaries should also handle adversaries that satisfy the definition.[8]
   (b) The definition should support natural composition theorems (e.g., of the type proven by Canetti [C00]).

With the foregoing properties in mind, let me suggest a couple of new definitions of expected polynomial-time strategies. These definitions will be more restrictive than the existing definitions of this notion (which were reviewed in Section 1.1).

1.3   The New Definitions

Looking at the problem of simulating an "expected polynomial-time" adversary (cf. [KL05]), it becomes evident that the source of trouble is the fact that the bound on the running-time of the adversary (w.r.t any real interaction) is no longer guaranteed when the adversary is invoked by a simulator. The point being that the queries made by the simulator may have a different distribution than the messages sent in any real interaction (especially, since some of these queries may not appear in the transcript output by the simulator). Furthermore, the simulator is resetting the adversary, which may allow it to find queries that are correlated to the adversary's internal coin tosses in ways that are unlikely to happen in any real interaction (see examples in [KL05] and in the proof of Proposition 5). Such queries may cause the adversary to run for a number of steps that is not polynomial on the average. Indeed, this problem does not occur in the case of strict probabilistic polynomial-time adversaries because in that case we have an *absolute bound* on the number of steps taken by the adversary, regardless of which messages it receives.

Let me stress that assuming that the adversary runs in expected polynomial-time when interacting with any other party does not solve the problem, because the distribution of the simulator's queries may not correspond to the distribution of an interaction with any standard interactive machine. The simulator's queries correspond to a "reset attack" on the adversary, where reset attack are as defined in [CGGM] (except that here they are applied on the adversary's strategy rather than on the honest party's strategy). Specifically, in a reset attack, the internal coin tosses of the strategy are fixed (to a random value) and the attacker may interact several times with the resulting residual (deterministic) strategy.

The forgoing discussion suggests a simple fix to the problem. Just define expected polynomial-time strategies as ones that run in expected polynomial-time under any reset attack that interact with them for a polynomial number of times. Actually, we should allow attacks that interact with these strategies for an expected polynomial number of times.[9] (See Definition 3.)

---

[8] Actually, we may relax this condition by allowing a modification of the simulator but not of the protocol and/or the underlying intractability assumptions.

[9] When measuring the expected number of interactions, I refer to a variant of Feige's notion of expected complexity with respect to the designated machine. Indeed, this widens the class of possible (reset) attackers, which further limits the class of admissible strategies (i.e., those that are expected polynomial-time under such attackers).

It seems that any (black-box) simulator that handles strict probabilistic polynomial-time adversaries can also handle adversaries that run in expected polynomial-time under the foregoing definition. After all, this definition was designed to support such a result. However, I was not able to prove this result without further restricting the class of simulators (in a natural way). For details, see Section 1.4.

But before turning to the results, let me suggest an even more restricted notion of expected polynomial-time strategies. I suggest to consider strategies that run in expected polynomial-time when interacting with any ("magical") machine that receives the strategy's internal coin tosses as side information. Arguably, this is the most restricted (natural) notion of expected polynomial-time strategies (which, when applied to non-interactive machines, coincides with the standard definition of expected polynomial-time). Needless to say, this definition (which is more restrictive than the aforementioned resetting definition) also supports the extension of simulators that handle strict probabilistic polynomial-time adversaries to handle adversaries satisfying the current definition.

Clearly, both definitions satisfy the first two desirable properties stated in Section 1.2. As for the third desirable property, it at the focus of the next subsection.

1.4   The Main Results

The main results establish the third desirable property for both definitions, while assuming that the provided simulators (i.e., the simulators provided by the corresponding hypothesis) belong to a natural subclass of black-box simulators. Indeed, one could hope that these results would hold for all (universal) simulators or at least for all black-box simulators.[10]

The issue at hand is the definition of efficient black-box simulators. Since black-box simulators are typically given oracle access to an efficient strategy, some texts only refer to what happens in such a case (and mandate that the overall simulation be efficient, where one also accounts for the steps of the strategy). A more natural and robust definition mandates that the number of steps performed by the black-box simulator itself be feasible, when the simulator is given oracle access to *any* strategy. Specifically, I consider black-box simulators that, make an *expected* number of steps that is upper-bounded by a polynomial in the length of the input, *where each oracle call is counted as a single step*, and call such a simulator normal. Indeed, the known (black-box) simulations including those that run in *expected* polynomial-time (e.g., [GK96]) are normal. For further discussion see the beginning of Section 3.

As stated in Section 1.3, the new definitions (or actually the "resetting-based" one) were devised to support the first main result (stated in [G06, Thm. 10]). This result asserts that *any normal black-box simulator that handles strict prob-*

---

[10] Recall that a universal simulator is a universal machine that is given that the code of the adversary that it simulates. In contrast, a black-box simulator is only given oracle access to the corresponding strategy.

abilistic polynomial-time adversaries can also handle adversaries that run in expected polynomial-time under the new definition(s). In particular, it implies that normal black-box zero-knowledge protocols remain simulateable when attacked by adversaries that satisfy the new definition(s) of expected polynomial-time. This applies, in particular, to the proof system of [GK96], for which analogous ("free") results were not known under the previous definitions of expected polynomial-time.[11]

Note that the fact that the aforementioned (normal black-box) simulations run in *expected* polynomial-time also when given access to any *expected* polynomial-time adversary is quite obvious from the new definition(s). This follows from the fact that normal black-box simulators invoke the adversary strategy for an *expected* polynomial number of times, while the "resetting-based definition" upperbounds the total *expected* time consumed by the adversary in such invocations. What should be shown is that, also in this case, the corresponding simulation produces good output (i.e., indistinguishable from the real interaction). This can be shown by using a rather straightforward "truncation" argument.[12]

Let us now turn to the question of composition, starting with the sequential composition of zero-knowledge protocols. The known result (of [GO94]) refers to *strict* probabilistic polynomial-time adversaries (and holds both with respect to strict and expected polynomial-time simulation).[13] However, the known argument does not extend to *expected* polynomial-time adversaries. Recall that the said argument transforms any adversary that attacks the composed protocol into a residual adversary that attacks the basic protocol. The source of trouble is that the fact that the former adversary is expected polynomial-time (under any definition) does not imply that the latter adversary is expected polynomial-time (under this definition). See the proof of Theorem 9 for details. Fortunately, there is an alternative way: just note that the simulator obtained by [GO94], which refers to *strict* probabilistic polynomial-time adversaries, can handle *expected* polynomial-time adversaries (i.e., by invoking [G06, Thm. 10] (or rather its zero-knowledge version – Theorem 8)).

The foregoing idea can also be applied to the general setting of secure multiparty computation, but additional care is needed to deal with the extra com-

---

[11] Note that Katz and Lindell [KL05] showed that the simulator presented in [GK96] fails (w.r.t expected polynomial-time under the previous definitions). Their work implies that, if strongly hiding commitment schemes are used in the protocol, then an alternative simulator does work. In contrast, my result applies to the simulator presented in [GK96] and does not require strengthening the commitment scheme used in the protocol. Furthermore, the running-time is preserved also for no-instances (cf., in contrast, [KL05, Sec. 3.3]).

[12] Indeed, the running-time analysis relies on the hypothesis that the simulator is normal, whereas the analysis of its output only relies on the hypothesis that the simulator is black-box. In contrast, for the claim itself to make sense at all it suffices to have a universal simulator (as otherwise it is not clear what we mean by saying that a simulator that handles any $A \in \mathcal{C}$ can handle any $A' \in \mathcal{C}'$).

[13] The original proof (of [GO94]) refers to strict polynomial-time simulators, but it extends easily to expected polynomial-time simulators.

plexities of this setting (as described next). Specifically, the so-called *sequential composition theorem* of Canetti [C00] (see also [G04, Sec. 7.4.2]) refers to an oracle-aided (or "hybrid") protocol Π that uses oracle calls to a functionality[14] f, which can be securely computed by a protocol ρ. (Note that the corresponding oracle-aided protocol was not mentioned in the context of zero-knowledge, because it is trivial (i.e., it merely invokes the basic protocol several times).) The theorem asserts that the security of Π (with respect to a specific functionality unmentioned here) is preserved when Π uses subroutine calls to ρ rather than oracle calls to f. *This result refers to security with respect to strict probabilistic polynomial-time adversaries that is demonstrated by strict probabilistic polynomial-time simulators.* One point to notice is that the proof of security of the resulting protocol, denoted Π', proceeds by incorporating the simulator of ρ into an adversary for Π. Thus, if the simulator of ρ runs in *expected* polynomial-time then so does the resulting adversary (for Π), and thus the simulator for Π has to handle *expected* polynomial-time adversaries (even if we only care of *strict* polynomial-time adversaries attacking Π'). Indeed, having a simulator for Π that handles any *expected* polynomial-time adversaries suffices for a partial result that refers to *strict* probabilistic polynomial-time adversaries for the resulting protocol Π' and to *expected* polynomial-time simulators (for ρ, Π, and Π'). The general (sequential) composition theorem for the case of expected polynomial-time (which refers to expected polynomial-time adversaries and simulators) follows by applying [G06, Thm. 10].

An important corollary to the foregoing extendability and composition theorems (i.e., [G06, Thm. 10] and [G06, Thm. 11]) asserts that it is possible to compose secure protocols, *when security is demonstrated via expected polynomial-time simulators but refers only to strict probabilistic polynomial-time adversaries.* In such a case, the extendability theorem allows to use these simulators with respect to *expected* polynomial-time adversaries, whereas the composition theorem applies to the latter. Thus, one may freely use *expected* polynomial-time simulators, and be assured that the corresponding secure protocols can be composed (just as in the case that their security is demonstrated via *strict* polynomial-time simulators).

Turning to the concurrent composition theorem of Canetti [C01], recall that it evolves around the notion of environmental security (a.k.a UC-security [C01]). Specifically, Canetti proved that any protocol that is environmentally secure preserves security under arbitrary concurrent executions, where the adversaries, simulators, and environments are all modeled as *strict* probabilistic polynomial-time strategies (with non-uniform auxiliary inputs for the environments). He then suggested the methodology of establishing environmental-security as a way of obtaining security under concurrent composition. Consequently, an extension of Canetti's methodology to the *expected* polynomial-time setting requires (1) verifying that Canetti's proof extends to this setting, and (2) obtaining environmental security for *expected* polynomial-time adversaries and environ-

---

[14] A functionality is a randomized version of a multi-input multi-output function (cf. [G04, Sec. 7.2.1]).

ments. Using the new definitions of expected polynomial-time strategies, the first requirement follows analogously to the proof of the sequential composition theorem, while the second requirement follows by generalizing [G06, Thm. 10] (which may be viewed as referring to trivial environments).

The bottom-line is that, for normal black-box simulators, the new definitions of expected polynomial-time strategies provide a "free" transformation from the *strict* probabilistic polynomial-time model to the *expected* polynomial-time model. In particular, *normal black-box simulators that work in the strict model extend to the expected model, and the most famous composition theorems extend similarly.*

1.5 Why Deal with Expected Polynomial-Time at All?

In light of the difficulties discussed in Section 1.1, one may ask *why do we need this headache* (of dealing with expected polynomial-time) *at all?* This question is further motivated by my views (expressed in Section 1.2) by which *we should not care about expected polynomial-time adversaries per se.* The answer, as hinted in Section 1.1, is that *we do care about expected polynomial-time simulators.*

Specifically, some natural protocols are known to be secure (or zero-knowledge) only when the definition of security allows expected polynomial-time simulators. A notable example, already mentioned several times is the constant-round zero-knowledge proof system of [GK96]. Furthermore, as proved in [BL02], constant-round proof system for sets outside $\mathcal{BPP}$ do not have strict polynomial-time black-box simulators (although they do have such non-black-box simulators [B01], which are less preferable for reasons discussed below).

In general, expected polynomial-time simulators seem to allow more efficient protocols and/or tighter security analysis. Whereas various notions of protocol efficiency are well-understood, a few words about the tightness of various security analyses are in place. Loosely speaking, *security tightness*[15] refers to the ratio between the running-time of the adversary and the (expected) running-time of the simulator that handles it. The security tightness of a protocol is a lower-bound on this ratio that holds for every probabilistic polynomial-time adversary.[16] Indeed, in many cases (also when strict polynomial-time simulators exist), the expected running-time of the simulator provides a better bound than the worst-case running-time of the simulator.

In my opinion, security tightness should serve as a major consideration in the evaluation of alternative protocols, and claims about protocol efficiency are almost meaningless without referring to their security tightness. For example, in many cases, modest parallelization can be achieved at the cost of a deterioration

---

[15] In the special case of zero-knowledge, the corresponding notion is called *knowledge tightness* [G01, Sec. 4.4.4.2]. Note a minor technicality: here tightness is define as the reciprocal of the ratio in [G01, Sec. 4.4.4.2].

[16] Thus, if there exists a polynomial q such that, for every polynomial p, every p-time adversary is simulated in time $q \cdot p$ then the protocol has (noticeable) security tightness $1/q$. But if the simulation of p-time adversaries requires time p^3 then the protocol does not have a noticeable security tightness.

in the security tightness (cf. [G01, Sec. 4.4.4.2]). Let me stress that, by definition, black-box simulators always yield a *noticeable*[17] bound on the security tightness (and in some cases they offer a constant bound), whereas non-black-box simulators may fail to have such bound (e.g., indeed, that's the case with Barak's simulators [B01]).

Thus, I suggest the following methodology: When designing your protocol and proving its security, allow yourself expected polynomial-time simulations. To assist the design and analysis, use the "extendability results" (e.g., [G06, Thm. 10]) provided in this work as well as relevant composition theorems (e.g., [G06, Thm. 11]). Finally, when obtaining the desired protocol with a security analysis that refers to an expected polynomial-time simulator, you may interpret it as providing a trade-off between the simulation time and the corresponding deviation (from the real interaction). But actually, a final claim that refers to expected simulation time may be as appealing when stated in terms of security tightness (e.g., the effect of any strict polynomial-time adversary can be achieved by a simulation that is expected to run three times as long).

Indeed, my opinion is that *there is no contradiction between not caring about expected polynomial-time adversaries and providing security guarantees that refer to the expected simulation time*: Whereas (at least potentially) the adversary is a real entity, its simulation is (always) a mental experiment. Furthermore, I believe that the foregoing methodology may yield the best trade-offs between the efficiency of the protocol and the tightness of its security.

Finally, let me note that there are alternative ways of handling the problems that motivate the introduction of expected polynomial-time to Cryptography (i.e., the failure of strict polynomial-time simulation in some cases). These alternatives are based on different measures that are applicable to "varying" running-time (i.e., running-time that is expressed as a random variable). In each case, one should start with a definition that refers to standard algorithms, and extend it to a definition that refers to interactive machines. For details, see Section 5 in my technical report [G06]. Indeed, the issues arising in such extensions are the same as the ones discussed throughout the rest of this paper. It is my belief, however, that expected running-time (as treated in the rest of this paper) provides the best trade-offs between the efficiency of the protocol and the tightness of its security.

1.6   Organization

Section 2 provides formal statements of the aforementioned (old and new) definitions as well as a demonstration of a hierarchy among them. Since the special case of zero-knowledge protocols provides a good benchmark for the general case of secure protocols, the main results are first presented in that setting (see Section 3). This simplifies things, because in that special case the simulators are standard algorithms rather than interactive strategies (for the so-called "ideal-model"; see, e.g., [G04, Sec. 7.2]). Nevertheless, I believe that the main ideas are

---

[17] As usual, a noticeable function is one that decreases slower than the reciprocal of some positive polynomial.

already present in the zero-knowledge setting, and that this belief is supported by the treatment of general protocols (provided in Section 4 of my technical report [G06]). Section 5 of [G06] discusses the applicability of my approach to alternative notions of expected polynomial-time algorithms, while Section 6 contains conclusions and open problems.

2   The Definitions

We adopt the standard terminology of interactive machines, while occasionally identifying strategies (which specify the next message to be sent by an interactive machine given its view so far) with the interactive machines that activate them. We use the shorthand PPT for *probabilistic polynomial-time* whenever using the full term is too cumbersome; typically, we do so when contrasting strict PPT and expected PPT. For simplicity, we only consider the two-party case. We denote by x the common (part of the) input, and denote by y and z the corresponding private inputs of the two parties. The reader may ignore y and z, which model (possibly non-uniform) auxiliary information.

2.1   Known Definitions

We start by formulating the two known definitions that were mentioned in Section 1.1.

Definition 1 (Feige [F90]). *The strategy σ is* expected PPT w.r.t a specific interactive machine M_0 *if, for some polynomial p and every x, y, z, the expected number of steps taken by $\sigma(x, z)$ during an interaction with $M_0(x, y)$ is upperbounded by $p(|x|)$, where the expectation is taken over the internal coin tosses of both machines.*

We stress that σ may be expected PPT with respect to some interactive machines but not with respect to others.

Definition 2 (attributed to Goldreich, e.g., in [KL05]). *The strategy σ is* expected PPT w.r.t any interactive machine *if, for some polynomial p, every interactive machine M, and every x, y, z, the expected number of steps taken by $\sigma(x, z)$ during an interaction with $M(x, y)$ is upper-bounded by $p(|x|)$.*

Here we may assume, without loss of generality, that M (which is computationally unbounded) is deterministic, and thus the expectation is only taken over the internal coin tosses of σ. The same convention is applied also in Definition 4 (but not in Definition 3; see discussion there).

2.2   New Definitions

In the first new definition, we refer to the notion of a *reset attack* as put forward in [CGGM]. Such an attack proceeds as follows. First, we uniformly select and fix a sequence of internal coin tosses, denoted ω, for the attacked strategy σ, obtaining a residual deterministic strategy σ_ω. Next, we allow the attacker to interact with σ_ω numerous times (rather than a single time). Specifically, for each possible value of ω, the expected number of times that attacker interacts with σ_ω is upper-bounded by a polynomial.[18]

Note that the attacker is not given ω explicitly, but its ability to (sequentially) interact with the residual strategy σ_ω for several times provides it with additional power (beyond interacting with σ itself for several times, *where in each interaction σ uses a fresh sequence of coin tosses*). As shown in [CGGM], such an attack is equivalent to a single interaction in which the attacker may (repeatedly) "rewind" σ (or rather σ_ω) to any prior point in the interaction and ask to resume the interaction from that point. Indeed, such an attack is reminiscent of the way that a (black-box) simulator uses an adversary strategy.

Definition 3 (tailored for simulation). *A q-reset attack on σ is an attack that, for every x, y, z and ω, interacts with σ_ω for an expected number of times that is upper-bounded by $q(|x|)$.[19] The strategy σ is* expected PPT w.r.t any reset attack *if, for some polynomial p, every polynomial q, every q-reset attack on σ, and every x, y, z, the expected total number of steps taken by $\sigma(x, z)$ during this attack is upper-bounded by $q(|x|) \cdot p(|x|)$.[20]*

We stress that the number of invocations of σ (like the total number of steps taken by σ) is a random variable defined over the probability space consisting of all possible interactions of the attacker and σ. Here (unlike in Definition 2), allowing the potential attacker to be probabilistic increases its power (and thus adds restrictions on strategies satisfying the definition). The reason is that, for each fixed ω, the number of invocations of σ_ω is allowed to be an arbitrary random variable with a polynomially bounded expectation (rather than being strictly bounded by a polynomial).

In the next (and last) definition, we consider a "magical" attacker that is given the outcome of the strategy's internal coin tosses as side information. That is, such an attack proceeds as follows. First, we uniformly select and fix a

---

[18] Indeed, the restriction on the number of interactions is a hybrid of the spirit of Definitions 1 and 2. We are upper-bounding the (expected) number of interactions initiated by the attacker (rather than its running-time), but do so not with respect to the designated σ but rather with respect to each of the residual σ_ω. Note that a simplified version that refers to the expected number of interactions with σ (i.e., the expectation is taken also over the coins of σ) yield a "bad" definition. (For example, suppose that σ_ω sends ω and makes $2^{|\omega|}$ steps if $\omega = 1^{|\omega|}$ and halt immediately otherwise. Then, intuitively σ is expected PPT (and in fact it even satisfies Definition 4), but the reset attack that, upon receiving ω in the first interaction, invokes σ_ω for $2^{|\omega|}$ additional times if and only if $\omega = 1^{|\omega|}$, causes σ to make an expected exponential number of steps.)

[19] As in Definitions 1 and 2, such an attack is given x and y as its input.

[20] The upper-bounded of $q(|x|) \cdot p(|x|)$ seems natural; however, an upper-bounded of $p(|x| + q(|x|))$ would work just as well (for all results stated in this work), but would yield weaker quantitative bounds.

sequence of internal coin tosses, denoted ω, for the attacked strategy σ, obtaining a residual deterministic strategy σ_ω. Next, we provide the attacker with ω (as well as with z) and allow it a single interaction with σ_ω. We stress that this attacker is merely a mental experiment used for determining whether or not σ is expected polynomial-time (under the following definition).

Definition 4 (seemingly most restrictive). *The strategy σ is* expected PPT w.r.t any magical machine *if, for some polynomial p, every interactive machine M' that is provided with the internal coin tosses of σ as side information, and every x, y, z, the expected number of steps taken by $\sigma(x, z)$ during an interaction with M' is upper-bounded by $p(|x|)$. That is, for a randomly selected ω, the expected number of steps taken by $\sigma_\omega(x, z)$ during its interaction with $M'(x, y, z, \omega)$ is upper-bounded by $p(|x|)$.*[21]

Here as in Definition 2, we may assume, without loss of generality, that M' (which is computationally unbounded) is deterministic, and thus the expectation is only taken over the internal coin tosses of σ. Thus, Definition 4 refers to the expectation, taken uniformly over all choices of ω, of the number of steps taken by (the residual deterministic strategy) $\sigma_\omega(x, z)$ during an interaction with (the deterministic strategy) $M'(x, y, z, \omega)$. Indeed, a strategy σ that satisfies Definition 4 *runs in expected polynomial-time even if each of the incoming messages is selected to maximize its running-time, when this selection may depend on the internal coin tosses of σ* (and its auxiliary-input z). This formulation is closest in spirit to the standard definition of strict PPT strategies.

2.3 Relating the Definitions

It is easy to see that, for $i = 1, 2, 3$, Definition $i+1$ implies Definition i. In fact, it is not hard to see that the converses do not hold. That is:

Proposition 5. *For $i = 1, 2, 3$, the set of strategies that satisfy Definition $i+1$ is strictly contained in the set of the strategies that satisfy Definition i.*

Proof: The first two containments (i.e., for $i = 1, 2$) are plainly syntactic. Intuitively, the fact that Definition 4 implies Definition 3 follows by noting that a reset attack does not add power to a computationally unbounded machine that gets σ's internal coin tosses. (A rigorous proof of this implication is provided in our technical report [G06].)

To show that the foregoing containments are strict we present corresponding strategies that witness the separations. The following examples are rather minimal, but they can be augmented into strategies that make sense (even for natural protocols). For example, a strategy that halts immediately upon receiving the message 0 and runs forever upon receiving the message 1 witnesses the

---

[21] Note that, unlike in Definitions 1-3, the attacker is given σ's auxiliary input (i.e., z). This is most natural in the context of the current attack, which is also given σ's internal coin tosses (i.e., ω).

separation between Definition 1 and Definition 2. Note that this example has nothing to do with the issue of expected polynomial-time (although an example that does relate to the latter issue can be constructed similarly).

To separate Definition 3 from Definition 4 consider a strategy that uniformly selects an n-bit long string r, and upon receiving a message s halts immediately if $s \neq r$ and halts after making 2^n steps otherwise. Clearly, this strategy does not satisfy Definition 4, but it does satisfy Definition 3.

A small twist on the foregoing example can be used to separate Definition 2 from Definition 3: Suppose that upon receiving s, the strategy first sends r, and then halts immediately if $s \neq r$ and halts after making 2^n steps otherwise. In this case a 2-reset attack can cause this strategy to always run for 2^n steps, while no ordinary interactive machine can do so. ∎

Discussion. Consider a restriction of all four definitions such that each bound on an expectation is replaced by a corresponding strict bound. Then the resulting (strict) versions of Definition 2–4 coincide but remain separated from the (strict) version of Definition 1. We believe that this fact speaks against Definition 1.

3   Results for Zero-Knowledge

The setting of zero-knowledge provides a good warm-up for the general study of secure protocols. Recall that, in the context of zero-knowledge, simulators are used to establish the security of predetermined prover strategies with respect to attacks by adversarial verifiers. We start by showing that (normal black-box) simulators that handle strict PPT adversaries also handle adversaries that are expected PPT (under Definitions 3 and 4). We next turn to an expected PPT version of the standard sequential composition theorem. (In our technical report [G06], analogous results are proved for general secure protocols.) To shorthand the text, when we say that some quantity (referring to an interaction) is polynomial, we mean that it is polynomial in the length of the common input.

Since the notion of *normal black-box simulators* is pivotal to our results, let us start by briefly recalling the standard definition of *black-box simulators* (see, e.g., [G01, Def. 4.5.10]). Loosely speaking, a black-box simulator is a universal machine that is given oracle access to a deterministic strategy and provides a simulation of the interaction of this strategy with the party attacked by this strategy.[22] In extending this notion to randomized strategies, we refer to providing the simulator with oracle access to a residual (deterministic) strategy obtained by fixing random coin tosses to the given randomized strategy.

Typically, one considers the execution of black-box simulator when given oracle access to any (strict or expected) PPT adversary. In that case, one sometimes

---

[22] In typical use of a black-box simulator one refers to the quality of this simulation. Specifically, it is require that if the former strategy is efficient (in some adequate sense) then the simulation is computationally indistinguishable from the real corresponding interaction. Since the notion of efficiency will vary (i.e., from strict PPT to expected PPT), we shall not couple the operational aspect of the black-box simulator with the quality of the output that it produces, but rather separate the two.

states both the complexity and the quality of the simulation when referring only to the case that the oracle is a PPT strategy.[23] While the restriction of the quality requirement to the said case is often essential, this is typically not the case with respect to the complexity requirement. Indeed, it is more natural to formulate the complexity requirement when referring to any possible oracle. We adopt this convention below, but in order to avoid possible confusion (with different views) we refer to simulators that satisfy this convention as normal.

Definition 6 (normal black-box simulators). *A black-box simulator is called normal if, on any input and when given oracle access to any strategy, it make an* expected *number of steps that is upper-bounded by a polynomial in the length of the input,* where each oracle call is counted as a single step.

Although it is possible to construct black-box simulators that are not normal (e.g., they run forever if the black-box manages to solve a hard problem), the standard black-box simulators (e.g., the ones of [GMR, GMW, GK96]) are all normal. Furthermore, normality seems a very natural property and *it is easy to verify.* For example, if the running-time analysis of a simulator (unlike the analysis of the quality of its output) does not rely on any intractability assumptions, then it is probably the case that the simulator is normal.[24]

The total simulation time. We will often refer to the (total) simulation time of the combined simulator S^{V^*}, which consists of a normal black-box simulator S that is given oracle access to an adversarial verifier V^*. Needless to say, for any normal simulator S, if V^* is *strict* PPT then the *expected* (total) simulation time of S^{V^*} is polynomial. As observed by Katz and Lindell [KL05], this is not necessarily the case if V^* is *expected PPT w.r.t Definition 2.* The key observation, which motivates Definition 3, is that the desired bound on the *expected* (total) simulation time of S^{V^*} does hold if V^* is *expected PPT w.r.t any reset attack.*

Observation 7. *If S is a normal black-box simulator and V^* is expected polynomial-time w.r.t Definition 3 then the expected total simulation time of S^{V^*} is polynomial.*

The straightforward proof is provided in our technical report [G06].

3.1   Simulating Expected PPT Adversaries

Bearing in mind that (in the context of zero-knowledge) the simulator is a standard algorithm, it suffices to state the following result with respect to Definition 3, and its applicability to Definition 4 follows as a special case.

---

[23] See corresponding footnote in our technical report [G06].

[24] The word "probably" indicates that the said implication is not claimed as a fact but rather suggested as a conjecture regarding any natural case.

Theorem 8 (extendability of normal black-box simulators, the zero-knowledge case). *Let (P, V) be an interactive proof* (or argument) *system for a set L, and $\langle P, V^* \rangle(x)$ denote the output of the adversarial verifier strategy V^* on input x after interacting with the prescribed prover P. Let M be a normal black-box simulator that, on input in L and when given access to any strict PPT strategy V^*, produces output that is computational indistinguishable from $\langle P, V^* \rangle$. Then, when M is given oracle access to any strategy V^* that is* expected PPT w.r.t any reset attack, *the expected simulation time of M^{V^*} is polynomial and the output is computational indistinguishable from $\langle P, V^* \rangle$.*

Note that the hypothesis allows the simulator to run in expected PPT while simulating a strict PPT adversary. This makes the hypothesis weaker and the theorem stronger; that is, the theorem can be applied to a wider class of protocols (including protocols that are not known to have strict PPT simulators such as, e.g., the constant-round zero-knowledge proof of [GK96]).

Proof: Fixing any expected PPT w.r.t Definition 3 strategy V^*, we first note that (by Observation 7) the expected simulation time of M^{V^*} is polynomial. To analyze the quality of this simulation, suppose towards the contradiction that D distinguishes between the simulation and the real interaction, and let p be a polynomial such that the distinguishing gap of D for infinitely many $x \in L$ is at least $\epsilon(|x|) \stackrel{\text{def}}{=} 1/p(|x|)$. Let $t^*(x)$ denote the total (over all invocations) expected number of steps taken by V^* when invoked by M. Note that $t^*(x)$ is upper-bounded by a polynomial in $|x|$, and assume (without loss of generality) that $t^*(x)$ also upper-bounds the expected running time of V^* in the real interaction (with P). Now, consider a *strict* PPT V^{**} that emulates V^*, while truncating the emulation as soon as $3t^*/\epsilon$ steps are emulated. Then, the variation distance (a.k.a statistical difference) between $M^{V^*}(x)$ and $M^{V^{**}}(x)$ is at most $\epsilon(|x|)/3$, because $\epsilon/3$ upper-bounds the probability that the total number of steps taken by V^* during all invocations by M exceeds $3t^*/\epsilon$ (and otherwise V^{**} perfectly emulates all these invocations, since none exceeds $3t^*/\epsilon$ steps). Similarly, the variation distance between $\langle P, V^* \rangle(x)$ and $\langle P, V^{**} \rangle(x)$ is upper-bounded by $\epsilon(|x|)/3$. It follows that D distinguishes the simulation $M^{V^{**}}$ from the real interaction $\langle P, V^{**} \rangle$ with a gap that exceeds $\epsilon/3$, on infinitely many inputs in L, in contradiction to the hypothesis that M simulates all *strict* PPT verifiers. ∎

Discussion. We believe that the fact that the proof of Theorem 8 is rather straightforward should not be counted against Definition 3, but rather the other way around. That is, we believe that the claim that the simulation of strict PPT adversaries extends (without modifications) to expected PPT adversaries is natural, and as such a good definition of expected PPT adversaries should support it. It may be that Theorem 8 can be generalized also to arbitrary black-box simulators and even to arbitrary universal simulators, but the current proof fails to show this: the running-time analysis relies on the hypothesis that the simulator is normal, whereas the output-quality analysis relies on the hypothesis that the simulator is black-box.[25]

Note that the combined simulator resulting from Theorem 8 is trivially expected PPT under reset attacks (and also under Definition 4), because it is a non-interactive machine (which runs in expected polynomial-time). Things are not as simple when we move to the setting of secure protocols, where the simulator is an interactive strategy (which operates in a so-called ideal-model). See [G06, Sec. 4.1].

3.2 Sequential Composition

The following Theorem 9 is an expected PPT version of the standard result (of [GO94]) that refers to *strict* PPT adversaries and simulators (see also [G01, Lem. 4.3.11]). Note that the standard result does not require the simulator to be black-box (let alone normal). The reason for the extra requirement will become clear in the proof.

Theorem 9 (expected PPT version of sequential composition for zero-knowledge). *In this theorem zero-knowledge means the existence of a normal black-box simulator that handles any expected PPT w.r.t Definition 3 (resp., w.r.t Definition 4) adversarial verifier, where handling means that the corresponding combined simulator runs in expected PPT and produces output that is computationally indistinguishable from the real interaction. Suppose that (P, V) is a zero-knowledge protocol. Then, sequentially invoking (P, V) for a polynomial number of times yields a protocol, denoted (P', V'), that is zero-knowledge.*

Proof: The proof of the strict PPT version (see [G01, Sec. 4.3.4]) proceeds in two steps: First, any verifier V^* that attacks the composed protocol (or rather the prover P') is transformed into an verifier V^{**} that attacks the basic protocol (or actually the prover P). This transformation is quite straightforward; that is, V^{**} handles a single interaction with P (while receiving the transcript of previous interactions as auxiliary input). Let M denote a simulator for (P, V^{**}). Then, a simulator for the composed protocol (or rather for the attack of V^* on P') is obtained by invoking M for an adequate number of times (using a correspondingly adequate auxiliary input in each invocation).

---

[25] Recall that a universal simulator obtains the code of the adversary's strategy rather than a black-box access to it. Thus, it may be the case that such a simulator can distinguish the code of V^* from the code of V^{**} (i.e., the timed version of V^*), and produce bad output in the latter case. Indeed, a "natural" simulator will not do so, but we cannot rely on this. Turning to a more natural example, we note that the known non-black-box simulator of Barak [B01] (as well as its modification [BG02]) may fail to simulate expected PPT verifiers, because the random variable representing its simulation time is polynomially related (rather than linearly related) to the running-time of the verifier. Recall that it may be the case that $t(x)$ has expectation that is upper-bounded by a polynomial in $|x|$ while $t(x)^2$ has expectation that is lower-bounded by $\exp(|x|)$; for example, consider $t : \{0, 1\}^* \to \mathsf{N}$ such that $\Pr[t(x) = 2^{|x|}] = 2^{-|x|}$ and $\Pr[t(x) = |x|^2] = 1 - 2^{-|x|}$.

Wishing to pursue the foregoing route, we merely need to check that any verifier V^* that is expected PPT w.r.t Definition 3 (resp., Definition 4) is transformed into a verifier V^{**} that is expected PPT w.r.t Definition 3 (resp., Definition 4). Unfortunately, this is not necessarily the case. Indeed, the expected running-time of V^{**} when given a *random* auxiliary input (i.e., one produced at random by prior interactions) is polynomial, but this does not mean that the expected running-time of V^{**} *on each possible value* of the auxiliary input is polynomial. For example, it may be the case that, with probability $2^{-|x|}$ over the history of prior interactions, the current interaction of V^* (i.e., V^{**} with the corresponding auxiliary input) runs for $2^{|x|}$ steps. The bottom-line is that V^{**} may not be expected PPT w.r.t any reasonable definition (let alone w.r.t Definition 3 or Definition 4).

In view of the forgoing, we take an alternative route. We only use the hypothesis that some normal black-box simulator M can handle all *strict* PPT verifiers that attack the basic prover P. Next, we observe that the proof of [G01, Lem. 4.3.11] (i.e., the strict PPT version) can be extended to the case that the simulation of the basic protocol (w.r.t *strict* PPT adversaries) runs in *expected* PPT. The key observation is that in this case V^{**} is strict PPT, although it will be fed with auxiliary inputs that are produced in expected PPT (by the simulation of prior interactions of V^{**} with P). Thus, we obtain an *expected* PPT simulation that handles any *strict* PPT attack on P'. Furthermore, the simulation amounts to invoking M for a polynomial number of times (while providing it with black-box access to V^{**}, which in turn is implemented by a black-box access to V^*). It follows that the simulation of (P', V^*) is performed by a normal black-box simulator (because M is normal). Hence, we have obtained a *normal black-box simulator that can handle any strict PPT attack on the composed protocol* (or rather on the prover P'). The current theorem follows by applying Theorem 8 to the latter simulator.   ∎

Discussion. The proof of Theorem 9 is somewhat disappointing because it does not use the hypothesis that P is zero-knowledge w.r.t *expected* PPT verifiers. Instead, Theorem 8 is used to bridge the gap between strict and expected PPT verifiers. A similar (but not identical) phenomenon will occur in the sequential composition theorem for general protocols, presented in [G06, Sec. 4.2].

Acknowledgments

I am grateful to Salil Vadhan for a discussion that inspired this work (and in particular Definition 3). I should be equally grateful to Yehuda Lindell for a discussion that inspired Definition 4, but I only understood this in retrospect. In addition, I wish to thank Salil and Yehuda for many insightful discussions and helpful comments on earlier drafts of this write-up. Finally, I wish to thank the reviewers of TCC'07 for their comments, although I disagree with most comments.

References

[B01]   B. Barak. How to Go Beyond the Black-Box Simulation Barrier. In *42nd FOCS*, pages 106–115, 2001.

[BG02]   B. Barak and O. Goldreich, Universal arguments and their applications. In the *17th Conf. on Comput. Complex.*, pages 194–203, 2002.

[BL02]   B. Barak and Y. Lindell. Strict Polynomial-time in Simulation and Extraction. In *34th STOC*, pages 484–493, 2002.

[C00]   R. Canetti. Security and Composition of Multi-party Cryptographic Protocols. *J. of Crypto.*, Vol. 13, No. 1, pages 143–202, 2000.

[C01]   R. Canetti. Universally Composable Security: A New Paradigm for Cryptographic Protocols. In *42nd FOCS*, pages 136–145, 2001. Full version is available from the author.

[CGGM]   R. Canetti, O. Goldreich, S. Goldwasser, and S. Micali. Resettable Zero-Knowledge. In *32nd STOC*, pages 235–244, 2000.

[DNS]   C. Dwork, M. Naor, and A. Sahai. Concurrent Zero-Knowledge. In *30th STOC*, pages 409–418, 1998.

[F90]   U. Feige. *Alternative Models for Zero-Knowledge Interactive Proofs*. Ph.D Thesis, Weizmann Institute of Science, 1990.

[G97]   O. Goldreich. Notes on Levin's Theory of Average-Case Complexity. *ECCC*, TR97-058, Dec. 1997.

[G01]   O. Goldreich. *Foundation of Cryptography: Basic Tools*. Cambridge University Press, 2001.

[G04]   O. Goldreich. *Foundation of Cryptography: Basic Applications*. Cambridge University Press, 2004.

[G06]   O. Goldreich. On Expected Probabilistic Polynomial-Time Adversaries: A suggestion for restricted definitions and their benefits. *ECCC*, TR06-099, Aug. 2006. See revision, Nov. 2006.

[GK96]   O. Goldreich and A. Kahan. How to Construct Constant-Round Zero-Knowledge Proof Systems for NP. *J. of Crypto.*, Vol. 9, No. 2, pages 167–189, 1996. Preliminary versions date to 1988.

[GL06]   O. Goldreich and Y. Lindell. Session-Key Generation using Human Passwords Only. *J. of Crypto.*, Vol. 91, No. 3, pages 241–340, July 2006.

[GMW]   O. Goldreich, S. Micali and A. Wigderson. Proofs that Yield Nothing but their Validity or All Languages in NP Have Zero-Knowledge Proof Systems. *JACM*, Vol. 38, No. 1, pages 691–729, 1991. Preliminary version in *27th FOCS*, 1986.

[GO94]   O. Goldreich and Y. Oren. Definitions and Properties of Zero-Knowledge Proof Systems. *J. of Crypto.*, Vol. 7, No. 1, pages 1–32, 1994.

[GMR]   S. Goldwasser, S. Micali and C. Rackoff. The Knowledge Complexity of Interactive Proof Systems. *SIAM J. on Comput.*, Vol. 18, pages 186–208, 1989. Preliminary version in *17th STOC*, 1985.

[KL05]   J. Katz and Y. Lindell. Handling Expected Polynomial-Time Strategies in Simulation-Based Security Proofs. In *2nd TCC*, 2005. To appear in *J. of Crypto.*.

[L86]   L.A. Levin. Average Case Complete Problems. *SIAM J. on Comput.*, Vol. 15, pages 285–286, 1986.

[L03]   Y. Lindell. General Composition and Universal Composability in Secure Multi-Party Computation. In *44th FOCS*, pages 384–393, 2003.

[MP06]   S. Micali and R. Pass. Local Zero-Knowledge. In *38th STOC*, 2006.

On Best-Possible Obfuscation

Shafi Goldwasser[1,2,*] and Guy N. Rothblum[2,**]

[1] Weizmann Institute of Science, Rehovot 76100, Israel
[2] CSAIL, MIT, Cambridge MA 02139, USA
{shafi,rothblum}@theory.csail.mit.edu

Abstract. An obfuscator is a compiler that transforms any program (which we will view in this work as a boolean circuit) into an obfuscated program (also a circuit) that has the same input-output functionality as the original program, but is "unintelligible". Obfuscation has applications for cryptography and for software protection.

Barak *et al.* initiated a theoretical study of obfuscation, which focused on *black-box obfuscation*, where the obfuscated circuit should leak no information except for its (black-box) input-output functionality. A family of functionalities that cannot be obfuscated was demonstrated. Subsequent research has showed further negative results as well as positive results for obfuscating very specific families of circuits, all with respect to black box obfuscation.

This work is a study of a new notion of obfuscation, which we call *best-possible obfuscation*. Best possible obfuscation makes the relaxed requirement that the obfuscated program leaks as little information as *any other program* with the same functionality (and of similar size). In particular, this definition allows the program to leak non black-box information. Best-possible obfuscation guarantees that *any* information that is not hidden by the obfuscated program is also not hidden by *any other* similar-size program computing the same functionality, and thus the obfuscation is (literally) the best possible. In this work we study best-possible obfuscation and its relationship to previously studied definitions. Our main results are:

1. A separation between black-box and best-possible obfuscation. We show a natural obfuscation task that can be achieved under the best-possible definition, but cannot be achieved under the black-box definition.
2. A hardness result for best-possible obfuscation, showing that strong (information-theoretic) best-possible obfuscation implies a collapse in the polynomial hierarchy.
3. An impossibility result for efficient best-possible (and black-box) obfuscation in the presence of random oracles. This impossibility result uses a random oracle to construct hard-to-obfuscate circuits, and thus it does *not* imply impossibility in the standard model.

* Supported by NSF grant CNS-0430450, NSF grant CFF-0635297 and a Cymerman-Jakubskind award.
** Supported by NSF grant CNS-0430450 and NSF grant CFF-0635297.

S.P. Vadhan (Ed.): TCC 2007, LNCS 4392, pp. 194–213, 2007.

1   Introduction

An open question in computer security is whether computer programs can be *obfuscated*; whether code can be made unintelligible while preserving its functionality. This question is important as obfuscation has wide-ranging applications, both for software protection and for cryptography. Beyond its theoretical importance, the question of obfuscation is of great practical importance. Numerous ad-hoc heuristical techniques are used every day by practitioners to obfuscate their code, even though many of these techniques do not supply any provable notion of security.

A theoretical study of obfuscation was initiated by Barak, Goldreich, Impagliazzo, Rudich, Sahai, Vadhan and Yang [2]. They studied several notions of obfuscation, primarily focusing on *black-box obfuscation*, in which an obfuscator is viewed as a compiler that, given *any* input program or circuit, outputs a program with the same functionality from which it is hard to find any deterministic information on the input program. Formally, black-box obfuscation requires that anything that can be efficiently computed from the obfuscated program, can also be computed efficiently from *black-box* (i.e. input-output) access to the program. Their main result was that this (strong) notion of obfuscation cannot always be achieved, as they were able to present an explicit family of circuits that provably cannot be black-box obfuscated.

Barak *et al.* [2] considered also an alternative notion of obfuscation called *indistinguishability obfuscation* that sidesteps the black-box paradigm. An indistinguishability obfuscator guarantees that if two circuits compute the same function, then their obfuscations are indistinguishable in probabilistic polynomial time. This definition avoids the black-box paradigm, and also avoids the impossibility results shown for the black-box obfuscation notion. Indeed, Barak *et al.* showed that it is simple to build *inefficient* indistinguishability obfuscators. One main disadvantage of indistinguishability obfuscation is that it does not give an intuitive guarantee that the circuit "hides information". This is apparent in their proposed construction of an inefficient indistinguishability obfuscator, where a small circuit is revealed which is equivalent to the original circuit. For some functionalities, this is a great deal of information to give away.

This Work. We propose a new notion of obfuscation, *best-possible obfuscation*, that avoids the black-box paradigm, and also gives the appealing intuitive guarantee that the obfuscated circuit leaks less information than *any other* circuit (of a similar size) computing the same function. This work is a study of this new notion of best-possible obfuscation.

Instead of requiring that an obfuscator strip a program of *any* non black-box information, we require only that the (best-possible) obfuscated program leak as little information as possible. Namely, the obfuscated program should be "as private as" any other program computing the same functionality (and of a certain size). A *best-possible* obfuscator should transform any program so that anything that can be computed given access to the obfuscated program should also be computable from *any other* equivalent program (of some related size).

A best-possible obfuscation may leak non black-box information (e.g. the code of a hard-to-learn function), as long as whatever it leaks is efficiently learnable from any other similar-size circuit computing the same functionality.

While this relaxed notion of obfuscation gives no absolute guarantee about what information is hidden in the obfuscated program, it does guarantee (literally) that the obfuscated code is the best possible. It is thus a meaningful notion of obfuscation, especially when we consider that programs are obfuscated every day in the real world without *any* provable security guarantee.

In this work we initiate a study of *best-possible* obfuscation. We explore its possibilities and limitations, as well as its relationship with other definitions of obfuscation that have been suggested. We formalize the best-possible requirement in Definition 5, by requiring that for every *efficient learner* who tries to extract information from an obfuscated circuit, there exists an *efficient simulator* that extracts similar information from any other circuit with the same functionality and of the same size. We consider both *computationally* best-possible obfuscation, where the outputs of the learner and simulator are indistinguishable with respect to *efficient* distinguishers, and *information theoretically* best-possible obfuscation (perfect or statistical), where even an *unbounded* distinguisher cannot tell the difference between the two. We emphasize that statistically or perfectly best-possible obfuscation refer to the *distinguisher*, whereas we only consider information that can be learned *efficiently* given the obfuscated circuit. This strengthens negative results. Our positive result on perfectly best-possible obfuscation applies also to unbounded *learners*.

Relationship with Previous Definitions. We study how best-possible obfuscation relates to black-box obfuscation, and present a separation between the two notions of obfuscation. The proof of this result also gives the first known separation between black-box and *indistinguishability* obfuscation. The separation result considers the complexity class of languages computable by polynomial sized ordered decision diagrams or POBDDs; these are log-space programs that can only read their input tape once, from left to right (see Section 3). We observe that *any* POBDD can be best-possible obfuscated *as a POBDD* (Proposition 2), whereas there are many natural functions computable by POBDDs that provably cannot be black-box obfuscated *as any POBDD* (Proposition 3). These two results give new possibility results (for best-possible and indistinguishability obfuscation), and simple natural impossibility results (for black-box obfuscation). Note that the impossibility result for black-box obfuscation only applies when we restrict the representation of the obfuscator's output to be a POBDD itself.

We also compare the notions of best-possible and indistinguishability obfuscation. Proposition 4 shows that any best-possible obfuscator is also an indistinguishability obfuscator. For *efficient* obfuscators the definitions are equivalent (Proposition 5). For inefficient obfuscation, the difference between the two definitions is sharp, as inefficient information-theoretic indistinguishability obfuscators are easy to construct (see [2]), but the existence of inefficient statistically best-possible obfuscators even for the class of languages recognizable by 3-CNF circuits (a sub-class of AC^0) implies that the polynomial hierarchy collapses to its second level.

We believe that the equivalence of these two definitions for efficient obfuscation motivates further research on both, as the "best-possible" definition gives a strong intuitive security guarantee, and the indistinguishability definition may sometimes be technically easier to work with.

Impossibility Results. We explore the limits of best-possible obfuscation. As noted above, we begin by considering information-theoretically (statistically) best-possible obfuscation. In Theorem 1 we show that if there exist (not necessarily efficient) statistically secure best-possible obfuscators for the simple circuit family of 3-CNF circuits (a sub-class of AC^0), then the polynomial hierarchy collapses to its second level. Corollary 1 of this theorem states that also if there exists an *efficient* statistically secure indistinguishability obfuscator for the same simple circuit family, then the polynomial hierarchy collapses to its second level. This is the first impossibility result for indistinguishability obfuscation in the standard model.

We also consider best-possible obfuscation in the (programmable) random oracle model. In this model, circuits can be built using special random oracle gates that compute a completely random function. Previously, this model was considered by Lynn, Prabhakaran and Sahai [17] as a promising setting for presenting positive results for obfuscation. We show that the random oracle can also be used to prove strong *negative results* for obfuscation. In Theorem 2 we present a simple family of circuits with access to the random oracle, that are provably hard to best-possible obfuscate efficiently. This impossibility results extends to the black-box and indistinguishability obfuscation notions. We note that using random oracles for obfuscation was originally motivated by the hope that giving circuits access to an idealized "box" computing a random function would make it easier to obfuscate more functionalities (and eventually perhaps the properties of the "box" could be realized by a software implementation). We, on the other hand, show that the existence of such boxes (or a software implementation with the idealized properties) could actually allow the construction of circuits that are impossible to obfuscate. Although this negative result does not rule out that every circuit without random oracle gates *can* be best-possible obfuscated, we believe it is illuminating for two reasons. First, as a warning sign when considering obfuscation in the random oracle model, and secondly as its proof hints that achieving general purpose best-possible obfuscation *in the standard model* would require a significant leap (a discussion of this point appears at the end of Section 4).

1.1   Related Work

Negative Results. Barak *et al.* showed that black-box obfuscation cannot always be achieved. They showed this by presenting families of circuits that cannot be black-box obfuscated: there exists a predicate that cannot be computed from black-box access to a random circuit in the family, but can be computed from (non black-box access to) any circuit in the family. Thus they showed that there *exist* circuits that cannot be obfuscated, but it remained possible that almost any natural circuit could be obfuscated. Goldwasser and Kalai [12], showed that if the definition of obfuscation is strengthened even further with a requirement that the obfuscation leak no more information than black-box access even in the presence of auxiliary input, then a large class of more natural circuits cannot be obfuscated.

Positive Results. The functionalities for which obfuscation was ruled out in [2] and [12] are somewhat complex. An interesting open question is whether obfuscation can be achieved for simpler classes of functionalities and circuits. Lynn, Prabhakaran and Sahai [17] were the first to explicitly explore this question. They suggested working in the random oracle model and focused on obfuscating access control functionalities (note that impossibility results of [2] and [12] extend to the random oracle model). At the heart of their construction is the obfuscation of a *point function*. A point function $I_p(x)$ is defined to be 1 if $x = p$, or 0 otherwise, and they observed that in the random oracle model point functions can be obfuscated, leading to obfuscation algorithms for more complex access control functionalities. Under cryptographic assumptions, it is also known how to obfuscate point functions without a random oracle. Canetti [6] showed (implicitly) how to obfuscate point functions (even under a strong auxiliary-input definition), using a strong variant of the Decisional Diffie-Hellman assumption. Wee [21] presented a point function obfuscator based on the existence of one-way permutations that are hard to invert on a very strong sense.

Other solutions for obfuscating point functions are known if the obfuscator doesn't need to work for *every* point, but rather for a point selected at random from a distribution with some min-entropy. For this relaxed requirement Canetti, Micciancio and Reingold [8] presented a scheme that uses more general assumptions than those used by [6] (their solution is not, however, secure in the presence of auxiliary inputs). Dodis and Smith [9] were able to obfuscate proximity queries in this framework.

The Random Oracle Model. The random oracle model is an idealization, in which it is assumed that all parties have oracle access to a truly random function $\mathcal{R}$. The parties can access this function by querying the random oracle at different points. The Random oracle methodology is a heuristic methodology, in which the random oracle is used for building provably secure cryptographic objects, but then, to implement the cryptographic object in the real world, the random oracle is replaced by some real function with a succinct representation. This methodology was introduced by Fiat and Shamir [15], and later formalized by Bellare and Rogaway [3].

A clear question raised by this methodology is whether the security of the cryptographic objects in an ideal world with a random oracle can be translated into security for the real-world implementation. In principle, this was answered negatively by Canetti, Goldreich and Halevi [7], who showed that there exist cryptographic schemes that are secure in the presence of a random oracle, but cannot be secure in the real world, regardless of the implementation of the random oracle. Their work left open the possibility that the random oracle methodology could still work for "natural" cryptographic practices. This was ruled out by Goldwasser and Kalai [11] for the Fiat-Shamir method [15], which uses a random oracle for obtaining digital signatures from identification schemes. The method was shown to lead to insecure signature schemes regardless of the possible implementation of the random oracle.

In the context of obfuscation, Lynn, Prabhakaran and Sahai [17] explored which circuits could be obfuscated in the (programmable) random oracle model, where the view generated by the black-box simulator is indistinguishable when taken over a randomly selected oracle. This work considers the same model. They used the random oracle $\mathcal{R}$ to obfuscate a point function I_p (when p is given to the obfuscator) using the value $\mathcal{R}(p)$. On input x the obfuscated circuit outputs 1 if and only if $\mathcal{R}(x) = \mathcal{R}(p)$. The only information about p in the obfuscated circuit is the value $\mathcal{R}(p)$, and this ensures that the obfuscation does not leak any non black-box information about I_p. They then proceeded to show how to obfuscate point functions with more general outputs (on input $x = p$ the function outputs some value, and otherwise it outputs $\perp$), multi-point functions and other more complex access control circuits. Narayanan and Shmatikov [16] gave a positive result for obfuscating databases in the random oracle model. In this work we explore whether indeed the random oracle model is a promising setting for further work on obfuscation.

1.2   Organization

We begin by presenting notation and formal definitions in **Section 2**. We compare our new definition of obfuscation with previous definitions in **Section 3**. In **Section 4** we present impossibility results for statistically best-possible obfuscation, and for best-possible obfuscation in the random oracle model. We conclude with discussions and extensions in **Section 5**.

2   Definitions and Discussion

2.1   Notation and Preliminaries

Notation. Let $[n]$ be the set $\{1, 2, \ldots n\}$. For $x \in \{0,1\}^n$, where $x = x_1 x_2 \ldots x_n$, and an index subset $M \subseteq [n]$, where $M = \{i_1, i_2, \ldots i_m\}$, we use $x|_M$ to denote the restriction of x to the indices in M. I.e. $x|_M = x_{i_1} x_{i_2} \ldots x_{i_m}$. For a (discrete) distribution D over a set X we denote by $x \sim D$ the experiment of selecting $x \in X$ by the distribution D. A function $f(n)$ is *negligible* if it smaller than any (inverse) polynomial: for any polynomial $p(n)$, there exists some n_0 such that for all $n \geq n_0$ we get that $f(n) < p(n)$.

Distributions, Ensembles and Indistinguishability. An ensemble $D = \{D_n\}_{n \in N}$ is a sequence of random variables, each ranging over $\{0,1\}^{\ell(n)}$, we consider only ensembles where $\ell(n)$ is polynomial in n (we occasionally abuse notation and use D in place of D_n). An ensemble D is polynomial time constructible if there exists a probabilistic polynomial time Turing Machine (PPTM) $\mathcal{M}$ such that $D_n = \mathcal{M}(1^n)$.

Definition 1. *The* statistical distance *between two distributions X and Y over $\{0,1\}^\ell$, which we denote by $\Delta(X,Y)$, is defined as:*

$$\Delta(X,Y) = \frac{1}{2} \sum_{\alpha \in \{0,1\}^\ell} |Pr[X = \alpha] - Pr[Y = \alpha]|$$

Definition 2. Computational Indistinguishability (Goldwasser Micali [13], Yao [22]) *Two probability ensembles D and F are computationally indistinguishable if for any PPTM $\mathcal{M}$, that receives 1^n and one sample s from D_n or F_n, and outputs 0 or 1, there exists a negligible function neg, such that for all n's:*

$$|Pr_{s \sim D_n}[\mathcal{M}(1^n, s) = 1] - Pr_{s \sim F_n}[\mathcal{M}(1^n, s) = 1]| \leq neg(n)$$

The Random Oracle Model. In the random oracle model we assume that all parties (the circuits, obfuscator, adversary etc.) have access to a random oracle and can make oracle queries. All oracle queries are answered by a single function $\mathcal{R}$, that is selected uniformly and at random from the set of all functions. Specifically, for each input length n, $\mathcal{R}$ will be a function from $\{0,1\}^n$ to $\{0,1\}^{p(n)}$ for some polynomial p. For simplicity, we will assume throughout this work that for all n's the function $\mathcal{R}$ is a random *permutation* [1] on $\{0,1\}^n$. Circuits access the random oracle by making oracles queries using a special *oracle gate*. It is important that we assume that calls to these oracle gates are clearly visible when running the circuit.

2.2 Definitions of Obfuscation

In the subsequent definitions, we consider a family $\mathcal{C}$ of probabilistic polynomial size circuits to be obfuscated. For a length parameter n let $\mathcal{C}_n$ be the circuits in $\mathcal{C}$ with input length n. The size of the circuits in $\mathcal{C}_n$ is polynomial in n. If the obfuscator $\mathcal{O}$ is a polynomial-size circuit, then we say it *efficiently obfuscates* the family $\mathcal{C}$, and that $\mathcal{C}$ is *efficiently obfuscatable*. Note that when considering obfuscation in the random oracle model, all circuits are allowed oracle access (*including the circuits to be obfuscated*), and all probabilities are taken over the selection of a random oracle. Whenever we refer to obfuscation, we will mean (efficient) black-box obfuscation unless explicitly noted otherwise.

Definition 3 (Black-Box Obfuscation [2]). *An algorithm $\mathcal{O}$, which takes as input a circuit in $\mathcal{C}$ and outputs a new circuit, is said to be a* black-box obfuscator *for the family $\mathcal{C}$, if it has the following properties:*

---

[1] Note that all our results hold for random function oracles (as long as the function's range is significantly larger than its domain, say at least twice as large.)

– *Preserving Functionality:*
  There exists a negligible function neg(n), such that for any input length n, for any $C \in \mathcal{C}_n$:

$$Pr[\exists x \in \{0,1\}^n : \mathcal{O}(C)(x) \neq C(x)] \leq neg(n)$$

  The probability is over the random oracle and $\mathcal{O}$'s coins.
– *Polynomial Slowdown:*
  There exists a polynomial p(n) such that for all but finitely many input lengths, for any $C \in \mathcal{C}_n$, the obfuscator $\mathcal{O}$ only enlarges C by a factor of p: $|\mathcal{O}(C)| \leq p(|C|)$.
– *Virtual Black-box:*
  For any polynomial size circuit adversary $\mathcal{A}$, there exists a polynomial size simulator circuit $\mathcal{S}$ and a negligible function neg(n) such that for every input length n and every $C \in \mathcal{C}_n$:

$$|Pr[\mathcal{A}(\mathcal{O}(C)) = 1] - Pr[|\mathcal{S}^C(1^n) = 1]| \leq neg(n)$$

  Where the probability is over the coins of the adversary, the simulator and the obfuscator. In the presence of a random oracle, the probability is also taken over the random oracle.

Definition 4 (Indistinguishability Obfuscation [2]). *An algorithm $\mathcal{O}$, that takes as input a circuit in $\mathcal{C}$ and outputs a new circuit, is said to be a (computational/statistical/perfect) indistinguishability obfuscator for the family $\mathcal{C}$, if it has the preserving functionality and polynomial slowdown properties as above, and also has the following property (instead of the virtual black-box property).*

– *Computationally/Statistically/Perfectly Indistinguishable Obfuscation:*
  For all large enough input lengths, for any circuit $C_1 \in \mathcal{C}_n$ and for any circuit $C_2 \in \mathcal{C}_n$ that computes the same function as C_1 and such that $|C_1| = |C_2|$, the two distributions $\mathcal{O}(\mathbf{C_1})$ and $\mathcal{O}(\mathbf{C_2})$ are (respectively) computationally/statistically/perfectly indistinguishable.

Definition 5 (Best-Possible Obfuscation). *An algorithm $\mathcal{O}$, which takes as input a circuit in $\mathcal{C}$ and outputs a new circuit, is said to be a (computationally/statistically/perfectly) best-possible obfuscator for the family $\mathcal{C}$, if it has the preserving functionality and polynomial slowdown properties as above, and also has the following property (instead of the virtual black-box property).*

– *Computational/Statistical/Perfect Best-Possible Obfuscation:*
  For any polynomial size learner $\mathcal{L}$, there exists a polynomial size simulator $\mathcal{S}$ such that for every large enough input length n, for any circuit $C_1 \in \mathcal{C}_n$ and for any circuit $C_2 \in \mathcal{C}_n$ that computes the same function as C_1 and such that $|C_1| = |C_2|$, the two distributions $\mathcal{L}(\mathcal{O}(\mathbf{C_1}))$ and $\mathcal{S}(\mathbf{C_2})$ are (respectively) computationally/statistically/perfectly indistinguishable.

Informally, this definition guarantees that anything that can be learned efficiently from the obfuscated $\mathcal{O}(C_1)$, can also be extracted efficiently (*simulated*) from *any program C_2 of similar size for the same function*. Thus, any information that is exposed by $\mathcal{O}(C_1)$ is exposed by *every other* equivalent circuit of a similar size, and we conclude that $\mathcal{O}(C_1)$ is a better obfuscation than any of these other circuits.

When dealing with best-possible obfuscators, we often refer to the *"empty" learner*; this is the learner that simply outputs whatever obfuscation it gets as input. It is simple to see that if there exists an efficient simulator $\mathcal{M}$ for the "empty" learner, then there exists an efficient simulator $\mathcal{M}'$ for every efficient learner $\mathcal{L}$: $\mathcal{M}'$ on input C_2 simply computes $\mathcal{M}(C_2)$ and outputs the result of $\mathcal{L}(\mathcal{M}(C_2))$. Thus, an equivalent definition to 'Best Possible' can do away with the leaner and only require the existence of an *efficient* simulator, i.e., a simulator $\mathcal{S}$ such that for circuits C_1, C_2 of identical size and identical functionality the distributions $\mathcal{O}(C_1)$ and $\mathcal{S}(C_2)$ are indistinguishable.

Note that when we refer to best-possible or indistinguishability obfuscators we always mean *efficient* and *computational* obfuscators unless we explicitly note otherwise. By perfect indistinguishability, we mean that the distributions are identical (statistical distance 0). For statistical indistinguishability, unless noted otherwise, we only assume that the distinguisher's advantage (the statistical distance) is smaller than a (specific) constant.[2] This strengthens negative results.

3   Comparison with Prior Definitions

In this section we compare the new definition of best-possible obfuscation to the black-box and indistinguishability definitions proposed by Barak *et al.* [2].

3.1   Best-Possible vs. Black-Box Obfuscation

Best-possible obfuscation is a relaxed requirement that departs from the black-box paradigm of previous work. We first observe that any black-box obfuscator is also a best-possible obfuscator.

Proposition 1. *If $\mathcal{O}$ is an efficient black-box obfuscator for circuit family C, then $\mathcal{O}$ is also an efficient (computationally) best-possible obfuscator for C.*

Proof. Assume for a contradiction that $\mathcal{O}$ is not a best-possible obfuscator for C. This implies that there is no best-possible simulator for the "empty" learner that just outputs the obfuscated circuit it gets. In particular, $\mathcal{O}$ itself is not a good simulator. Thus there exists a polynomial p and a distinguisher $\mathcal{D}$, such

---

[2] The existence of an *inefficient* perfectly best-possible obfuscator, implies the existence of an efficient one that uses the simulator to obfuscate. A similar argument also applies to statistically best-possible obfuscation, unless the statistical distance guarantee is very weak.

that for infinitely many input lengths n, there exist two circuits $C_1, C_2 \in \mathcal{C}_n$, such that $|C_1| = |C_2|$ and C_1 and C_2 are equivalent, but:

$$|Pr[\mathcal{D}(\mathcal{O}(C_1)) = 1] - Pr[\mathcal{D}(\mathcal{O}(C_2)) = 1]| \geq p(n)$$

Now consider $\mathcal{D}$ as a predicate adversary for the black-box obfuscator $\mathcal{O}$. The black-box simulator $\mathcal{S}$ for $\mathcal{D}$ clearly behaves identically on C_1 and C_2 (because they have the same functionality), but $\mathcal{D}$'s behavior on $\mathcal{O}(C_1)$ and $\mathcal{O}(C_2)$ is non-negligibly different. Thus (for infinitely many input lengths) $\mathcal{S}$ is not a black-box simulator for $\mathcal{D}$, a contradiction.

Next, we provide a (weak) separation result. We exhibit a natural (low) complexity class, that of languages computable by polynomial size ordered binary decision diagrams (POBDDs), such that best-possible obfuscation *within* the class is achievable, but there are simple functionalities that are provably impossible to black-box obfuscate *within the class*.

Ordered Binary Decision Diagrams (OBDDs). The computational model of ordered binary decision diagrams was introduced by Bryant [5]. An ordered binary decision diagram is a rooted directed acyclic graph with a vertex set V containing non-terminal vertices, each with two children, and terminal vertices (without children), each labeled 0 or 1. Each edge e in the graph is marked with an input literal ℓ_e (e.g. ℓ_e could be x_1, $\overline{x_8}$ etc.). For every non-terminal vertex, the labels of its (two) outgoing edges should be negations of each other (e.g. x_3 and $\overline{x_3}$). An input $x \in \{0,1\}^n$ is accepted by an OBDD if and only if after removing every edge e for which $\ell_e = 0$ there exists a path from the root node to a terminal node labeled by 1. In addition, in an OBDD, on *every* path from the root vertex to a terminal vertex, the indices of the literals of edges on the path must be strictly increasing. We will focus on polynomial-size OBDDs, or POBDDs. We note that another way to view POBDDs is as logarithmic-space deterministic Turing Machines whose input tape head can only move in one direction (from the input's first bit to its last).

Bryant [5] showed that OBDDs have a simple canonical representation. For any function, there exists a *unique* smallest OBDD that is its canonical representation. Moreover, for polynomial-size OBDDs, this canonical representation is efficiently computable.

Note that we defined obfuscation for *circuits*, not OBDDs, but for every OBDD, there exists a boolean circuit (that computes the same functionality) from which it is easy to extract the OBDD. When we refer to obfuscating the family of OBDDs, we are implicitly referring to obfuscating the underlying family of circuits representing OBDDs.

We begin by observing that POBDDs can be perfectly best-possible obfuscated *as POBDDs* (namely the output of the obfuscator is a POBDD itself). This is a corollary of POBDDs having efficiently computable *canonical representations*.

Proposition 2. *There exists an efficient perfectly best-possible (and perfectly indistinguishable) obfuscator for POBBDs.*

Proof. The best-possible obfuscator for a POBDD P simply takes P, computes (efficiently) its canonical representation, and outputs that program as the best-possible obfuscation. The canonical representation has the same functionality as P, is no larger than P, and (most significantly) is unique, depending only on the functionality of P. The simulator gets a POBDD P' and also efficiently computes its canonical representation. The canonical representations of P and P' are identical if and only if P and P' compute the same functionality. Thus the obfuscator is indeed a perfectly best-possible obfuscator for the family of POBDDs.

We next show that there exists a family of languages computable by POBDDs, that cannot be black-box obfuscated (efficiently or inefficiently) as POBDDs (i.e the resulting program itself being represented as a POBDD). This gives a (weak) separation between best-possible and black-box obfuscation. The weakness is that it remains possible that any input POBDD *can* be black-box obfuscated such that the output circuit is no longer a POBDD but is in some higher complexity class.

Proposition 3. *There exists a family of languages computable by POBDDs, that cannot be black-box obfuscated as POBDDs.*

Proof (Sketch). Intuitively, because POBDDs have a simple efficiently computable canonical representation, non black-box information can be extracted from a POBDD by reducing it to its "nice" canonical form, and then extracting information from this canonical form.

More formally, consider (for example) the simple family of point functions $\{I_p\}_{p \in \{0,1\}^n}$, where the function I_p outputs 1 on input the point p and 0 everywhere else. Note that point functions are computable by POBDDs. Now observe that any POBDD computing a point function for a point p can be reduced to its canonical representation, from which p is easily extracted. Thus for any supposed obfuscator that obfuscates point functions as POBDDs there exists an adversary that (for *every point*) can extract all the bits of the point from the "obfuscated" POBDD. Clearly, no black-box simulator can successfully extract even a single bit of the point for a non-negligible fraction of point functions. Thus there exists no black-box obfuscator that obfuscates POBDDS computing point functions *as POBDDs*.

We note that many other natural languages computable by POBDDs cannot be black-box obfuscated as POBDDs. Black-box obfuscation of POBDDs as more complex circuits remains an intriguing open question.

3.2   Best-Possible vs. Indistinguishability Obfuscation

As mentioned above, the notions of best-possible obfuscation and indistinguishability obfuscation are related, though the guarantees given by these two types of obfuscation are different. In this section we will show that any best-possible obfuscator is also an indistinguishability obfuscator. Furthermore, for *efficient* obfuscation, the two notions are equivalent. For *inefficient* obfuscation (which is still interesting), however, the notions are *not* equivalent unless the polynomial hierarchy collapses. In fact, inefficient indistinguishability obfuscators exist unconditionally (see [2]). On the other hand, building even inefficient best-possible obfuscators remains an interesting open question. We begin by showing that best-possible obfuscation is in fact at least as strong as indistinguishability obfuscation.

Proposition 4. *If $\mathcal{O}$ is a perfectly/statistically/computationally best-possible obfuscator for circuit family $\mathcal{C}$, then $\mathcal{O}$ is also a (respectively) perfect/statistical/computational indistinguishability obfuscator for $\mathcal{C}$.*

Proof (Sketch). To prove the claim, consider the "empty" learner $\mathcal{L}$ that just outputs whatever obfuscation it is given, and its simulator $\mathcal{S}$. Let δ be the computational or statistical distinguishability in the (computational or perfect/statistical) guarantee of the obfuscator. We get that for any two circuits C_1 and C_2 that are of the same size and compute the same functionality:

$$\delta(\mathcal{L}(\mathcal{O}(C_1)), \mathcal{S}(C_2)) = \delta(\mathcal{O}(C_1), \mathcal{S}(C_2)) \leq \varepsilon$$

$$\delta(\mathcal{L}(\mathcal{O}(C_2), \mathcal{S}(C_2))) = \delta(\mathcal{O}(C_2), \mathcal{S}(C_2)) \leq \varepsilon$$

Thus (since computational and statistical distinguishabilities are transitive):

$$\delta(\mathcal{O}(C_1), \mathcal{O}(C_2)) \leq 2\varepsilon$$

Note that the perfect/statistical/computational guarantee is preserved.

As noted above, if we restrict our attention to efficient obfuscators, indistinguishability obfuscators are also best-possible obfuscators.

Proposition 5. *If $\mathcal{O}$ is an* efficient *perfect/statistical/computational indistinguishability obfuscator for a circuit family $\mathcal{C}$, then $\mathcal{O}$ is also an efficient (respectively) perfectly/statistically/computationally best-possible obfuscator for $\mathcal{C}$.*

Proof. Let $\mathcal{O}$ be an efficient indistinguishability obfuscator. Then for any learner $\mathcal{L}$, let $\mathcal{S}$ be the (efficient) simulator that gets a circuit C_2, runs $\mathcal{O}(C_2)$, and then activates $\mathcal{L}$ on $\mathcal{O}(C_2)$. We get that if $\mathcal{O}$ is a perfect/statistical/computational indistinguishability obfuscator, then for any two circuits C_1 and C_2 that are of the same size and compute the same functions, the two distributions $\mathcal{L}(\mathcal{O}(C_1))$ and $\mathcal{S}(C_2) = \mathcal{L}(\mathcal{O}(C_2))$ are perfectly/statistically/computationally indistinguishable (because $\mathcal{O}$ is an indistinguishability obfuscator). Thus $\mathcal{O}$ is also an efficient best-possible obfuscator.

Note that the efficiency of the indistinguishability obfuscator is *essential* to guarantee the efficiency of the simulator, without which the obfuscator does not meet the best-possible definition.

It is important to note that there is no reason to believe that the two notions of obfuscation are equivalent for *inefficient* obfuscation. In fact, whereas [2] design exponential-time indistinguishability obfuscators, there is no known construction for inefficient *best-possible* obfuscators. We believe that even constructing *inefficient* best-possible obfuscators is interesting.

We end this subsection by observing that if $P = NP$ then it is not hard to construct efficient perfect best-possible obfuscators (and indistinguishability obfuscators) for every polynomial-size circuit. In fact this complexity assumption is almost tight. We will show in Theorem 1, that if statistically best-possible obfuscators can be built even for very simple circuits, then the polynomial hierarchy collapses to its second level.

Proposition 6. *If P=NP then the family of polynomial-sized circuits can be efficiently perfectly best-possible obfuscated.*

Proof. Assume $P = NP$. For any circuit C, it is possible to *efficiently* extract the smallest lexicographically first circuit C_{min} that is equivalent to C (this problem is solvable using a language in the second level of the polynomial hierarchy). As Barak *et al.* [2] note, such an extraction procedure is a perfectly indistinguishable obfuscation of C, and thus there exists an efficient perfect indistinguishability obfuscator for the family of polynomial-size circuits. By Proposition 5 it is also an efficient perfect best-possible obfuscator for the family of polynomial-size circuits. Note that even if $P \neq NP$ then we get an (inefficient) indistinguishability obfuscator. It remains, however, unclear whether we can get an inefficient best-possible obfuscator, as the (always efficient!) simulator can no longer run the "circuit minimization" procedure.

4   Impossibility Results for Best-Possible Obfuscation

4.1   Statistically Best-Possible Obfuscation

In this section we present a hardness result for statistically best-possible obfuscation. In Section 3 it was shown that if $P = NP$ then every polynomial-sized circuit can be perfectly best-possible obfuscated, thus we cannot hope for an *unconditional* impossibility result. We show that the condition $P = NP$ is (nearly) tight, and in fact the existence of statistically best-possible obfuscators even for the class of languages recognizable by 3-CNF circuits (a sub-class of AC^0) implies that the polynomial hierarchy collapses to its second level. This result shows the impossibility of statistically best-possible obfuscation for any class that contains 3-CNF formulas (and in particular also for the class of general polynomial sized circuits).

Theorem 1. *If the family of 3-CNF formulas can be statistically best-possible obfuscated (not necessarily efficiently), then the polynomial hierarchy collapses to its second level.*

Proof (Intuition). We begin by considering the case that the family of 3-CNF formulas can be *perfectly* best-possible obfuscated (not necessarily efficiently) while *perfectly* preserving functionality (i.e. the obfuscated circuit never errs). We can use the Simulator $\mathcal{S}$ for the "empty" learner, to construct an NP proof for Co-SAT (a Co-NP -complete problem). To see this, consider an input 3-CNF formula φ of size $|\varphi|$. We would like to find a witness for non-satisfiability of φ. Towards this end, we first construct an unsatisfiable formula ψ of size $|\varphi|$. A witness for the non-satisfiability of φ is a pair of random strings (r, r') such that the output of the simulator $\mathcal{S}$ on φ with randomness r is equal to its output on ψ with randomness r'. This proof system is indeed in NP:

- Efficiently Verifiable. The simulator is efficient, and thus the witness is efficiently verifiable.
- Complete. If φ is unsatisfiable, then φ and ψ compute the same function (the constant 0 function) and are of the same size. We know that $\mathcal{O}$ is a perfect best-possible obfuscator and thus the distributions $\mathcal{O}(\varphi)$, $\mathcal{S}(\varphi)$, $\mathcal{S}(\psi)$, $\mathcal{O}(\psi)$ are all identical. This implies that there must exist (r, r') such that $\mathcal{S}(\varphi, r) = \mathcal{S}(\psi, r')$.
- Sound. If φ is satisfiable, then because the obfuscator perfectly preserves functionality, the distributions $\mathcal{O}(\varphi)$, $\mathcal{O}(\psi)$ are disjoint (they are distributions of circuits with different functionalities). Thus the distributions $\mathcal{S}(\varphi)$, $\mathcal{S}(\psi)$ of the (perfect) simulator's output are also disjoint, and there exist no (r, r') such that $\mathcal{S}(\varphi, r) = \mathcal{S}(\psi, r')$.

The full proof for the case of statistically best-possible obfuscation follows along similar lines, giving a reduction from a Co-NP -complete problem (circuit equivalence) to a problem in AM.[3] By the results of Fortnow [10], Aiello and Håstad [1], and Boppana, Håstad and Zachos [4] (see also Feigenbaum and Fortnow [14]), this collapses the polynomial hierarchy to its second level. The full proof is omitted from this extended abstract.

Proposition 2 and Theorem 1 give examples of circuit classes can and cannot be statistically best-possible obfuscated. The proofs give characterizations of circuit classes that can be statistically best-possible obfuscated. A *sufficient* condition for statistically best-possible obfuscation of a class of circuits is having an efficiently computable canonical representation, a *necessary* condition is having a statistical zero knowledge proof for the equivalence problem.

Finally, a corollary of this theorem is that the same class of 3-CNF formulas cannot be statistically *indistinguishability* obfuscated in polynomial time unless the polynomial hierarchy collapses. This is the first impossibility result for indistinguishability obfuscation in the standard model.

Corollary 1. *If the family of 3-CNF formulas can be* efficiently *statistically indistinguishability obfuscated, then the polynomial hierarchy collapses to its second level.*

---

[3] Actually, this is a problem in statistical zero knowledge: the complement of the Statistical Difference Problem, introduced by Sahai and Vadhan [20].

Proof. By Proposition 5, if there exists an *efficient* statistical indistinguishability obfuscator for the family of 3-CNFs, then there also exists an efficient
statistically best-possible obfuscator for the same family. This, in turn, implies
(by Theorem 1) that the polynomial hierarchy collapses to its second level.

4.2   Computationally Best-Possible Obfuscation

In this section we present an impossibility result for (efficient) computationally
best-possible obfuscation in the (programmable) random oracle model. We show
how to use a random oracle to build circuits for point functions that cannot
be best-possible obfuscated. We note that the use of the random oracle both
strengthens and weakens this result. The result is strengthened because a random
oracle could conceivably help obfuscation (*a la* [17]), but weakened because the
random oracle is used to build a circuit that cannot be obfuscated. Moreover, in
the proof we need to assume that a distinguisher can see the obfuscated circuit's
oracle calls and that it can access the random oracle itself. It is still possible
that circuits that do not use the random oracle can be best-possible obfuscated.

We show that a specific family of circuits for computing point functions cannot be obfuscated in the presence of a random oracle $\mathcal{R}$. A point function I_p
is the function that outputs 1 on input p and 0 on all other inputs. We begin by presenting the family of point function circuits for which we will show
impossibility of obfuscation.

Definition 6 (The circuit family $\{C_p^M\}$). *For any input length n, the family
of circuits $\{C_p^M\}_n$ defines a set of circuits on inputs of length n. Each circuit
C_p^M computes the point function I_p on the point $p \in \{0,1\}^n$, and is defined by
the point p and an index subset $M \subseteq [n]$ (all index subsets in this section are of
size $\frac{n}{2}$). The information that the circuit C_p^M gives about p is:*

- *The index subset M is included in C_p^M "in the clear".*
- *The bits of p that aren't in the index subset M ($p|_{[n]-M}$) are also given in
  the clear.*
- *The bits of p that are in M ($p|_M$) are "hidden", the only information given
  about them is $\mathcal{R}(p|_M)$.*

*For an input x, to compute the point function I_p, the circuit C_p^M outputs 1 if
and only if x is equal to p in the indices that aren't in M ($x|_{[n]-M} = p|_{[n]-M}$), and
the random oracle gives the same values on x and p restricted to M ($\mathcal{R}(x|_M) =
\mathcal{R}(p|_M)$). Thus $C_p^M(x) = 1$ if and only if $x = p$, otherwise the circuit outputs 0.*

We also take the family $\{I_p\}$ to be the family of point function circuits that
contain their point in the clear, and are padded to be of the same size as the
circuits $\{C_p^M\}$ on each input length. We claim that the family of point function
circuits $\{C_p^M\} \cup \{I_p\}$ cannot be best-possible obfuscated.

Theorem 2. *The circuit family $\{C_p^M\} \cup \{I_p\}$ cannot be efficiently computationally best-possible obfuscated.*

Proof (Proof Intuition). Observe that any obfuscator $\mathcal{O}$ must preserve the functionality of a circuit C_p^M. Furthermore, the only information the obfuscator has about the indices of the point p that are in the subset M is the value $\mathcal{R}(p|_M)$. To preserve functionality, for any input x, the obfuscated circuit $\mathcal{O}(C_p^M)$ needs to find out whether $x = p$. Now because the only information available to the obfuscator and the obfuscated circuit about $p|_M$ is the value $\mathcal{R}(p|_M)$, for most inputs x, the obfuscated circuit must ask the random oracle for the value $\mathcal{R}(x|_M)$. Thus for many x's one of the (polynomially many) oracle calls of $\mathcal{O}(C_p^M)$ should be to $\mathcal{R}(x|_M)$.

In the proof we construct a distinguisher between obfuscated circuits and the output of the "empty" learner's simulator. For an index subset $T \subseteq [n]$ and input $x \in \{0,1\}^n$, we examine the distinguisher $\mathcal{D}_{T,x}$ that activates the obfuscated circuit it was given, $\mathcal{O}(C_p^M)$ (for some index subset M and point p), on the input x, and tries to guess whether T was the subset used in the underlying circuit that was obfuscated (i.e. whether $M = T$). To do this, the distinguisher runs $\mathcal{O}(C_p^M)$ on the input x and outputs 1 if and only if the obfuscated circuit queried the random oracle on the input $x|_T$. Recall that we concluded above that if $M = T$, then we expect the obfuscated circuit to query the random oracle on the input $x|_T$. Thus, when the distinguisher $\mathcal{D}_{T,x}$ gets $\mathcal{O}(C_p^M)(x)$, it has an advantage in deciding whether $M = T$ or not. This advantage disappears when the distinguisher is activated on the output of a simulator that was given the circuit I_p: the simulator was given no information about M, so its output cannot help the distinguisher determine whether or not $M = T$. The full proof is omitted from this extended abstract.

The family of circuits that we show cannot be obfuscated is a family that computes point functions. This may seem contradictory, as Lynn, Prabhakaran and Sahai [17] showed that a class of circuits computing point functions *can* be obfuscated in the random oracle model. The source of this disparity is that they (as well as all other known positive results on obfuscating point functions) only consider obfuscators that get the point *in the clear*, whereas the family of point function circuits that we present ($\{C_p^M\}$) hides information about the point. Malkin [18], was the first to ask whether *any* point function implementation can be black-box obfuscated.

Thus Theorem 2 shows impossibility for simpler and more natural functionalities than those considered in previous results, but does so using circuits with random oracle gates.

Extensions. We note that this impossibility result applies also to black-box obfuscation (the proof is omitted from this extended abstract, but note that the distinguisher in the theorem can be viewed as a predicate adversary). One possible objection to this impossibility result, is that the information revealed by obfuscation of circuits in the family $\{C_p^M\}$ (namely the subset M) is not necessarily information related to the point p. We note, however, that unless an obfuscator guarantees that *no* non black-box information is revealed by the obfuscation, for circuits for which the point p *is* related to the subset M, the obfuscated circuit may leak non black-box information *about the point p.*

Implications for a world without random oracles. We conclude with a discussion of the ways in which our proof uses the random oracle model, and how one could hope to remove this assumption. Our construction uses the random oracle $\mathcal{R}$ in two ways. First, $\mathcal{R}$ is used to hide information about p in the circuit family $\{C_p^M\}$. Essentially, we use $\mathcal{R}$ to obfuscate a point function (where the point is $p|_M$). Intuitively, since we know how to (black-box) obfuscate point functions without using random oracles, we could use (strong) cryptographic assumptions in place of the random oracle for this.

The second place in our proof where we use the properties of random oracles is when we assume a distinguisher can see the points on which the obfuscated circuit queries the random oracle. If we want to get rid of the random oracles, this is a more troubling assumption. The issue is that even if we could use some other method to hide information about the point p in the standard model, there is no reason to assume we could identify any internal computation of the obfuscated circuit. For example, consider using Canetti's point function obfuscation and giving the obfuscator a circuit C that hides some information on p by exposing only $(r, r^{p|_M})$. Even if on any input x the obfuscated circuit *always* computes $(r, r^{x|_M})$, there is no guarantee that a distinguisher can identify these computations! Thus $\mathcal{O}(C)$ may not expose any information on M. We note, however, that to *prove* that an obfuscator can obfuscate *any* circuit computing a point function, one would have to construct an obfuscator that indeed hides internal computations. Thus it seems that even for achieving the (seemingly modest) goal of best-possible obfuscation for polynomial-size point-function circuits, one would have to present a method for hiding complex internal computations of a circuit. Such a method, in and of itself, would likely have interesting implications.

5   Concluding Remarks and Discussions

We conclude with a discussion of best-possible obfuscation and issues raised in this work.

Input/Output Representation. Several of our results highlight the issue of the *representation* of an obfuscator's input and output. At times (in Section 3) we restrict the *representation* of both the obfuscator's input and output functionality to be "simple" circuits representing POBDDs. At other times (in the proof of Theorem 2), we construct complex circuits that hide information about their functionality from the obfuscators. In general, restricting the input representation makes the task of obfuscation easier (see discussion in section 4.2), whereas restricting the output representation makes the task of obfuscation harder, and we use this in Proposition 3 to show that point functions cannot be black-box obfuscated as POBDDs. Previous positive results on obfuscation considered obfuscators that get a particular representation of the functionality (e.g. the point

p for the point function I_p). Future work on black-box (and non black-box) obfuscation should consider the question of which *representations* of the desired functionality are obfuscated.

This issue was also raised by Malkin [18], who asked whether *any* point function implementation can be black-box obfuscated in the standard model. An relaxed (but related[4]) formulation of this question is whether the family of polynomial-size circuits computing point functions can be best-possible obfuscated. The proof of Theorem 2 answers this question negatively in the presence of random oracles, but either an impossibility proof or a provably secure obfuscator would likely have interesting consequences.

Circuit Sizes. In our definition of best-possible obfuscation (Definition 5) we compare the obfuscated circuit $\mathcal{O}(C_1)$ with circuits C_2 of the same size as C_1 (and computing the same functionality). This definition requires that the obfuscation of C_1 leak as little information as any equivalent circuit of a specific (polynomially) smaller size. We could make stronger requirements, such as leaking less information than an equivalent circuit C_2 that is as large as $\mathcal{O}(C_1)$, twice as large as C_1, etc. (all results would still hold). In general, the larger the circuit used as a benchmark (C_2), the stronger the definition. The important point is guaranteeing that $\mathcal{O}(C_1)$ leaks as little information as any other functionally equivalent circuit of a related size.

Auxiliary Input. Goldwasser and Kalai [12] augment the virtual black-box requirement of obfuscation to hold in the presence of auxiliary input. They note that this is an important requirement for any obfuscation that is used in practice, as auxiliary input comes into play in the real world. Following this argument, we could extend the best-possible obfuscation requirement to hold in the presence of auxiliary input. This is a strengthening of the definition, and thus all negative results clearly still hold. The positive result of Proposition 2 (obfuscating POBDDs) also holds even in the presence of (dependent) auxiliary input.

Weaker Variants. In light of the negative results of Theorems 1 and 2 it is interesting to consider *weaker* variants of best-possible obfuscation (Definition 5). While the variants below lose some of the appealing intuitive "garbling" guarantee of Definition 5, meeting any of them would all give at least *some* indication that the obfuscator truly garbles circuits.

– *Hiding Less Information.* One natural approach is to follow in the footsteps of Barak *et al.* [2], and consider best-possible *predicate* obfuscators: an obfuscation is predicate best-possible if any *predicate* of the original circuit that can be learned from the obfuscation, could also be learned from any other circuit of a similar size computing the same functionality. While this definition is weaker than computationally best-possible obfuscation, the proof of

---

[4] This formulation is equivalent to the original question raised by Malkin under the assumption that point functions can indeed be obfuscated when the point is given in the clear. In this case, a best-possible obfuscation leaks as little information as the black-box obfuscated point function circuits, and is thus also a black-box obfuscation.

Theorem 2 rules out even general-purpose predicate best-possible obfuscation in the random oracle model (and gives some intuition that this type of obfuscation would be hard to achieve int he standard model).
- *Weaker Indistinguishability.* Canetti [6] and Wee [21] relax the virtual blackbox requirement, requiring only *polynomially* small indistinguishability between the output of an adversary and its simulator. Moreover, they allow the simulator's size to depend (polynomially) on this indistinguishability parameter. We note that negative results in this work (Theorems 1 and 2) hold even if we require only polynomially small indistinguishability and allow the simulator's size to depend (polynomially) on the indistinguishability parameter.
- *Weaker Functionality.* Definition 5 requires that with all but negligible probability, the obfuscated circuit *perfectly* preserves the functionality of the original circuit. We could relax this, and require only that for every input, with all but a small constant error probability, the obfuscated circuit outputs the same output as the original circuit. Our negative results apply even under this weakened preserving functionality requirement. The positive result on best-possible obfuscation of POBDDs (Proposition 2) gives an obfuscator that *perfectly* preserves the functionality of the circuit it obfuscates.

Acknowledgements

We thank Yael Tauman Kalai and Tali Kaufman for helpful and enjoyable discussions. Thanks also to the anonymous reviewers for their insightful comments which much improved (or so we hope) the presentation.

References

1. William Aiello, Johan Håstad. *Statistical Zero-Knowledge Languages can be Recognized in Two Rounds.* Journal of Computer and System Sciences 42(3): 327-345 (1991)
2. Boaz Barak, Oded Goldreich, Russell Impagliazzo, Steven Rudich, Amit Sahai, Salil P. Vadhan, Ke Yang. *On the (Im)possibility of Obfuscating Programs.* CRYPTO 2001: 1-18
3. Mihir Bellare, Phillip Rogaway. *Random Oracles are Practical: A Paradigm for Designing Efficient Protocols.* ACM Conference on Computer and Communications Security 1993: 62-73
4. Ravi B. Boppana, Johan Håstad, Stathis Zachos. *Does co-NP Have Short Interactive Proofs?* Information Processing Letters 25(2): 127-132 (1987)
5. Randal E. Bryant. *Graph-Based Algorithms for Boolean Function Manipulation.* IEEE Transactions on Computers, C(35), No. 8, August, 1986: 677-691
6. Ran Canetti. *Towards Realizing Random Oracles: Hash Functions That Hide All Partial Information.* CRYPTO 1997: 455-469
7. Ran Canetti, Oded Goldreich, Shai Halevi. *The random oracle methodology, revisited.* Journal of the ACM 51(4): 557-594 (2004)
8. Ran Canetti, Daniele Micciancio, Omer Reingold. *Perfectly One-Way Probabilistic Hash Functions (Preliminary Version).* STOC 1998: 131-140

9. Yevgeniy Dodis, Adam Smith. *Correcting errors without leaking partial information.* STOC 2005: 654-663
10. Lance Fortnow. *The complexity of perfect zero-knowledge.* In S. Micali, editor, Advances in Computing Research, 5: 327-343. JAI Press, Greenwich, 1989
11. Shafi Goldwasser, Yael Tauman Kalai. *On the (In)security of the Fiat-Shamir Paradigm.* FOCS 2003: 102-113
12. Shafi Goldwasser, Yael Tauman Kalai. *On the Impossibility of Obfuscation with Auxiliary Input.* FOCS 2005: 553-562
13. Shafi Goldwasser and Silvio Micali. *Probabilistic Encryption and How to Play Mental Poker Keeping Secret All Partial Information.* STOC 1982: 365-377
14. Joan Feigenbaum, Lance Fortnow. *Random-Self-Reducibility of Complete Sets.* SIAM Journal on Computing 22(5): 994-1005 (1993)
15. Amos Fiat, Adi Shamir. *How to Prove Yourself: Practical Solutions to Identification and Signature Problems.* CRYPTO 1986: 186-194
16. Arvind Narayanan, Vitaly Shmatikov. *Obfuscated databases and group privacy.* ACM Conference on Computer and Communications Security 2005: 102-111
17. Ben Lynn, Manoj Prabhakaran, Amit Sahai. *Positive Results and Techniques for Obfuscation.* EUROCRYPT 2004: 20-39
18. Tal Malkin. *Personal Communication* (2006).
19. Tatsuaki Okamoto. *On Relationships between Statistical Zero-Knowledge Proofs.* Journal of Computer and System Sciences 60(1): 47-108 (2000)
20. Amit Sahai, Salil P. Vadhan. *A complete problem for statistical zero knowledge.* Journal of the ACM 50(2): 196-249 (2003)
21. Hoeteck Wee. *On obfuscating point functions.* STOC 2005: 523-532
22. Andrew Chi-Chih Yao. *Theory and Applications of Trapdoor Functions (Extended Abstract).* FOCS 1982: 80-91

Obfuscation for Cryptographic Purposes

Dennis Hofheinz[1], John Malone-Lee[2], and Martijn Stam[3]

[1] CWI, Amsterdam
Dennis.Hofheinz@cwi.nl
[2] University of Bristol
malone@cs.bris.ac.uk
[3] EPFL, Lausanne
martijn.stam@epfl.ch

Abstract. An obfuscation $\mathcal{O}$ of a function F should satisfy two requirements: firstly, using $\mathcal{O}$ it should be possible to evaluate F; secondly, $\mathcal{O}$
should not reveal anything about F that cannot be learnt from oracle
access to F. Several definitions for obfuscation exist. However, most of
them are either too weak for or incompatible with cryptographic applications, or have been shown impossible to achieve, or both.

    We give a new definition of obfuscation and argue for its reasonability
and usefulness. In particular, we show that it is strong enough for cryptographic applications, yet we show that it has the potential for interesting
positive results. We illustrate this with the following two results:

1. If the encryption algorithm of a secure secret-key encryption scheme
   can be obfuscated according to our definition, then the result is a
   secure public-key encryption scheme.
2. A uniformly random point function can be easily obfuscated according to our definition, by simply applying a one-way permutation.
   Previous obfuscators for point functions, under varying notions of
   security, are either probabilistic or in the random oracle model (but
   work for arbitrary distributions on the point function).

On the negative side, we show that

1. Following Hada [12] and Wee [25], any family of deterministic functions that can be obfuscated according to our definition must already
   be "approximately learnable." Thus, many deterministic functions
   cannot be obfuscated. However, a probabilistic functionality such as
   a probabilistic secret-key encryption scheme can potentially be obfuscated. In particular, this is possible for a public-key encryption
   scheme when viewed as a secret-key scheme.
2. There exists a secure probabilistic secret-key encryption scheme that
   cannot be obfuscated according to our definition. Thus, we cannot
   hope for a general-purpose cryptographic obfuscator for encryption
   schemes.

Keywords: obfuscation, point functions.

1 Introduction

The obfuscation of a function (or, more generally, a program) should provide
nothing more than the possibility of evaluating that function. In particular,

S.P. Vadhan (Ed.): TCC 2007, LNCS 4392, pp. 214–232, 2007.

from an obfuscation of a function, one should not be able to learn more than one can learn from oracle access to that function.

History and Related Work. Practical yet informal approaches to code obfuscation were considered in [13,16]. The first theoretical contributions were made by Hada [12] who gives a security definition for obfuscations and relates it to zero-knowledge proof systems.

   In their seminal paper [1], Barak *et al.* define a hierarchy of obfuscation definitions, the weakest of which is predicate-based and the strongest of which is simulation-based. They show that there are languages that cannot be obfuscated, even under the weakest definition that they proposed. Specifically, they show that there are (contrived) sets of programs such that no single obfuscation algorithm can work for all of them (and output secure obfuscations of the given input program). Hence, Barak *et al.* rule out the possibility of *generic* obfuscation (and the proof argument they give also applies to our notion), yet they leave room for the possibility of obfuscators for *specific* families of programs.

   Goldwasser and Tauman Kalai present obfuscation definitions which model several types of auxiliary information available to an adversary [11]. They show general impossibility results for these definitions using *filtered functions* (functions whose output is forced to zero if the input is not of a special form). They also show that secure (without auxiliary information) obfuscations of *point functions* (functions that are 0 everywhere except at one point, see Definition 5) are automatically secure with respect to independent auxiliary information.

   Even before a precise definition of obfuscation was formulated, positive obfuscation results could be given implicitly and in a different context for a special class of functions. Namely, Canetti [3] and Canetti *et al.* [4] essentially obfuscate point functions. The construction from Canetti [3] works for (almost) arbitrary function distributions and hence requires a very strong computational assumption. On the other hand, one construction from Canetti *et al.* [4] requires only a standard computational assumption, but is also proven only a for uniform function distribution. Both of these constructions are probabilistic and technically very sophisticated.

   Positive results for the predicate-based definition of Barak *et al.* [1] were demonstrated by Lynn *et al.* [17] who show how to employ a random oracle to efficiently obfuscate the control flow of programs, which includes point functions.

   Subsequently Wee showed how to obfuscate point functions in the standard model [25] (still predicate-based). Yet he only does this under very strong computational assumptions and for a weakened definition of obfuscation. Wee also shows that, at least under one of the original obfuscation definitions of Barak *et al.* [1], point functions can only be obfuscated under strong computational assumptions.

   Finally, a generalisation of one-way functions can be found in the work of Dodis and Smith [6] who show how to obfuscate a proximity function.

Our Work. We concentrate on a new definition that is a variant of the simulation-based definition of Barak *et al.* [1]. We deviate from the notion of [1] only in that we consider probabilistic functions and also pick the function to be obfuscated according to a distribution and hence demand only "good obfuscations on average." (This is similar to Canetti's "oracle hashing" definition from [3], or to the "approximate functionality with respect to a particular input distribution" variant [1].) However, we stress that the impossibility results from [1] also carry over to such a weakened notion in a meaningful sense. In fact, a similar notion is already informally considered in [1, Discussion after Theorem 4.5]. That means that also for our notion, there can be no general-purpose obfuscators. Yet our goal is to consider obfuscations for specific applications such as obfuscating secret-key encryption schemes. We stress that there *are* secret-key encryption schemes that are unobfuscatable (see Remark 1), so general-purpose obfuscators cannot exist; then again, there are also schemes that can be obfuscated (e.g., a public-key encryption scheme when viewed as secret-key). We are interested in *specific* obfuscations, not in general-purpose obfuscation.

Intuitively, our variation makes sense in many cryptographic applications where the function to be obfuscated is chosen at random by a trusted party. This could for instance model the situation where an issuer of smartcards selects a signing key and then hardwires it on a smartcard.

To show the usefulness of our notion, we demonstrate that by obfuscating the encryption algorithm of an IND-CPA secure symmetric encryption scheme, we obtain an IND-CPA secure asymmetric scheme. As a sidenote, we prove that, surprisingly, the analogous result does *not* hold for IND-CCA schemes. The latter is not a consequence of our relaxed definition, but is inherent in all existing definitions of obfuscation. We also prove similar results (both positive and negative) concerning the construction of signature schemes from MACs using obfuscation.

Although we are not yet able to give a concrete (interesting) example of a public-key encryption scheme or a signature scheme produced using our new definition of obfuscation, we provide evidence that the definition is satisfiable in other contexts. In particular, we show that, given a one-way permutation, it is possible to obfuscate a point function *deterministically* and in a quite straightforward way in the standard model. Previous obfuscations of point functions (under different definitions) were either probabilistic and quite involved or in the random oracle model; this owes to the fact that they were geared to work for *arbitrary* distributions on the point function to be obfuscated.

Our definition does not overcome the impossibility results of Barak *et al.* [1]. Actually, following Hada [12] and Wee [25], we remark that any family of deterministic functions must be approximately learnable to be obfuscatable. We prove that in particular, it is not possible to obfuscate pseudorandom functions under our definition.

2   Previous Definitions

Barak *et al.* [1] discuss the obfuscation of an arbitrary Turing machine or circuit. This leads to the same notation for the description of a circuit and the function it evaluates. Moreover, from an adversary's point of view, the security does not depend on the particular way some function is implemented prior to obfuscation.

Under our definitions, the implementation or representation of a function prior to obfuscation is relevant for the security of the obfuscation. To emphasize this, we will make an explicit distinction between keys of a function on the one hand, and keyed functions on the other.

Let $\mathcal{F} = \{\mathcal{F}_k\}_{k \in \mathbb{N}}$ be a class of probabilistic functions $\mathcal{F}_k = \{F_K\}_{K \in \mathcal{K}_k}$ where all $F_K \in \mathcal{F}_k$ have an identical domain X_k. We call K the key and we assume there is some probabilistic polynomial time (in k) algorithm F that, on input of $K \in \mathcal{K}_k$ and $x \in X_k$, samples from $F_K(x)$.

Formally, we regard an obfuscator as a combination of two algorithms: a key transformation algorithm $\mathcal{O} : \bigcup_k \mathcal{K}_k \rightarrow \bigcup_k \mathcal{K}'_k$ that takes a key $K \in \mathcal{K}_k$ and returns the obfuscated key $K' \in \mathcal{K}'_k$; and a probabilistic polynomial time algorithm G that, on input a key $K' \in \mathcal{K}'_k$ and $x \in X_k$, samples from $F_K(x)$. Depending on the context, we will not always make explicit mention of G.

We are now ready to rephrase Barak et al.'s definition of obfuscation in the setting that we consider. We only give a uniform model and, for the moment, concentrate on obfuscating deterministic functions. Note that our model is slightly more restrictive, as a result of which polynomial slowdown is automatically satisfied. (This refers to the slowdown between the function and its obfuscation.)

Definition 1 (Universal Obfuscation). *A PPT algorithm $\mathcal{O}$ is a universal obfuscator for a class $\mathcal{F}$ of deterministic functions if the following holds.*

- *Approximate Functionality: For all $k \in \mathbb{N}$, for all $K \in \mathcal{K}_k$ and all $x \in X_k$ it holds that $F_K(x) = G_{\mathcal{O}(K)}(x)$ with overwhelming probability over the choices of $\mathcal{O}$.*
- *Virtual Black-Box: Loosely speaking, given access to the obfuscated key $\mathcal{O}(K)$ of a function F_K, an adversary cannot learn anything about the original function F_K that it could not learn from oracle access to F_K. Formal definitions follow.*

Note that we call Definition 1 "universal" to indicate that—in contrast to our own upcoming refinement of this definition—security for each individual function F_K to be obfuscated is required.

Barak et al. [1] give several ways to formulate the notion of not learning anything, two of which we recall below.

Predicate-Based Obfuscation. This is based on *computing a predicate*. In this case the task of an adversary given the obfuscation $\mathcal{O}(K)$ is to compute any boolean predicate on K. That is to say, for any adversary and any boolean predicate π, the probability that an adversary computes $\pi(K)$ given $\mathcal{O}(K)$ is no greater than the probability that a simulator, given only oracle access to F_K, computes $\pi(K)$. This notion is formally defined by a slightly simpler, but equivalent, notion.

Definition 2 (Predicate-Based Universal Black-Box). *A probabilistic algorithm $\mathcal{O}$ for a family $\mathcal{F}$ of functions satisfies the weak universal black-box* *property if for all PPT D, there exist a PPT S and a negligible function ν such that for all k and all $K \in \mathcal{K}_k$,*

$$\left| \Pr\big[K' \leftarrow \mathcal{O}(K) \ : \ D(K') = 1 \big] - \Pr\big[S^{F_K}(1^k) = 1 \big] \right| \leq \nu(k) \ .$$

Simulation-Based Obfuscation. This is based on *computational indistinguishability*. Under this formulation one does not restrict the nature of what an adversary must compute: it says that for any adversary, given $\mathcal{O}(K)$, it is possible to simulate the output of the adversary given only oracle access to F_K. The outputs of the adversary and the simulator must be computationally indistinguishable. It is easy to see that in fact it is necessary and sufficient to simulate the output of the obfuscator (thus removing the need to quantify over all adversaries). This equivalence has also been observed by Wee [25] who gives the following formulation.

Definition 3 (Simulation-Based Universal Black-Box). *A probabilistic algorithm $\mathcal{O}$ for a class $\mathcal{F}$ of functions satisfies the strong universal black-box property if there exists a PPT S such that for all PPT D there exists a negligible function ν such that for all k and all $K \in \mathcal{K}_k$*

$$\left| \Pr\big[K' \leftarrow \mathcal{O}(K) \ : \ D(K') = 1 \big] - \Pr\big[\tilde{K}' \leftarrow S^{F_K}(1^k) \ : \ D(\tilde{K}') = 1 \big] \right| \leq \nu(k) \ .$$

3   Our Definition

Inspired by existing impossibility results, we endeavour to find a definition of cryptographic obfuscation that both allows meaningful applications such as transforming secret-key cryptosystems into public-key systems, and at the same time is satisfiable. Recall that previous work on obfuscation uses a universal quantifier for the functions $F_K \in \mathcal{F}_k$ to be obfuscated. In contrast, we will assume a key distribution on the keys K and hence on the functions F_K that have to be obfuscated. For simplicity we will assume a uniform distribution on the keys.

First, we will define and discuss a new notion of obfuscation that is a relaxation of the simulation-based definition of Section 2. In Section 4 we will examine some applications and limitations of our new definition. In the following, we will put emphasis on the virtual black-box property. (However, we include a short discussion of an adaptation of the approximate functionality requirement.)

3.1   The Definition

Simulation-based obfuscation refers to computational indistinguishability: given only oracle access one can produce something indistinguishable from the obfuscator's output. Note that the most straightforward analogue of Definition 1 with a randomized key distribution is not very meaningful. Since the distinguisher does not know the key K, a simulator can make up a different key and obfuscate it, so trivial obfuscation would be possible. To prevent this (and end up with a more sensible definition), we additionally give the distinguisher, similarly to the simulator, *oracle access* to the function.

Definition 4 (Simulation-Based Virtual Black-Box Property). *An obfuscation $\mathcal{O}$ for a class of functions $\mathcal{F}$ has the simulation-based virtual black-box property iff for all PPT distinguishers D there is a PPT simulator S such that the following quantity is negligible in k.*

$$\left| \Pr\left[K \leftarrow \mathcal{K}_k : D^{F_K}(1^k, \mathcal{O}(K)) = 1\right] - \Pr\left[K \leftarrow \mathcal{K}_k : D^{F_K}(1^k, S^{F_K}(1^k)) = 1\right] \right|.$$

3.2 On the Approximate Functionality Requirement

The natural analogue of the "approximate functionality" requirement from Definition 1 for the case of function *distributions* would be the following. For all keys K and inputs x, with overwhelming probability over the random choices of $\mathcal{O}$, the obfuscation $G_{\mathcal{O}(K)}(x)$ has *exactly* the same distribution as $F_K(x)$. This is a very strong requirement, and we can actually relax this a little.

Definition 3.1. *An obfuscator $\mathcal{O}$ satisfies the* statistical functionality requirement *for $\mathcal{F}$ iff there exists a negligible function ν such that for all k, the following holds:*

$$\sum_{K,K'} \Pr[K, K' : K' \leftarrow \mathcal{O}(K), K \leftarrow \mathcal{K}_k] \max_x(\sigma(G_{K'}(x), F_K(x))) \le \nu(k).$$

Here, σ is used to denote the statistical distance.

For deterministic functions, the requirement reduces to the statement that, with overwhelming probability over the choice of the key generation and the obfuscator $\mathcal{O}$, F_K and $G_{\mathcal{O}(K)}$ should be the same functions. This is similar to the approximate functionality of the universal definition [1, Definition 4.3], with the caveat that we take our probability over F_K as well. We note that all the results to follow do not depend on the choice of functionality requirement.

3.3 Comparison to Previous Definitions

The Definitions of Barak et al. Definition 4 differs in several aspects from [1, Definition 2.1]. First, Definition 4 requires security w.r.t. a randomly chosen key from a given set, whereas [1, Definition 2.1] demands security for every key in that set. In that sense, Definition 4 is a relaxation of [1, Definition 2.1] (although this does not protect Definition 4 from impossibility results for general-purpose obfuscation; see below).

On the other hand, Definition 4 requires a multi-bit output from the simulator, whereas [1, Definition 2.1] restricts adversary and simulator to a one-bit output. As [1, Definition 2.1] demands security for all keys in a set, this one-bit output can be seen as an approximation of a predicate on the key. In fact, when directly relaxing [1, Definition 2.1] by randomizing the distribution on the key, approximating a predicate on the key leads to a more sensible definition than simply comparing the probabilities for 1-output of adversary and simulator. Such a predicate-based formulation, even with randomly chosen key and a distinguisher with oracle access, is incomparable to our definition (since the predicate could be not computable in polynomial time).

Perfect One-Way Hashing and Point Functions. We note that a distribution on the keys (or, on the function to obfuscate) was already considered in other definitions, e.g., in the security definition for perfect one-way hashing (that is actually an obfuscation of a point function) from [3]. In the case of [3], security could be achieved as long as the distribution on the functions was *well-spread*, which basically means that a brute-force search for the function has only negligible success. Our results from Section 4.3 (that also concern an obfuscation of a point function) are formulated with a *uniform* distribution on the key.

In contrast to the very sophisticated construction from [3], our construction is quite simple: an obfuscation of a point function P_x, is $\Pi(x)$ for a one-way permutation Π. However, there *can* be well-spread distributions (different from the uniform one) on the key for which our point function obfuscation becomes insecure. (Imagine a one-way permutation that leaks the upper half of the preimage, and a distribution that keeps the lower half of the preimage constant.) In other words, the price to pay for the simplicity of our construction is the dependency on a *uniform* distribution of the key.

Also, the construction from [3] is "semantically secure" in the sense that any predicate on the hashed value (i.e., the key of the point function to be obfuscated) is hidden. Our construction from Section 4.3 does not guarantee this; just like the one-way permutation that is employed, our construction only hides the key in its entirety. This may have the disadvantage that in some applications, this might not be sufficient, and in particular not a meaningful "idealization" of a point function. However, in other settings (such as a password query), this may be exactly the idealization one is interested in.

Other Similar Definitions. Technically, Definition 4 is quite similar to [12, Definition 10] (the latter definition which is also formulated with a distribution on the keys). Essentially, the only difference is that [12, Definition 10] equips the distinguisher with an extra copy of the obfuscation instead of oracle access to the function. As argued by Hada [12], this leads to a very strong definition (that is in particular strictly more restrictive than ours).

Finally, the definitions from Wee [25, Section 5.2] are technically similar to ours, in that they allow the adversary a multi-bit output. These definitions suffer from strong impossibility results (in particular, a function must be *exactly* learnable for obfuscation); this is partly due to the fact that these definitions demand security for *all* keys in a given set. In our case, a function must be *approximately* learnable for obfuscation, and this enables, e.g., the obfuscation of point functions (see Sections 4.3 and 4.4).

3.4   Specific vs. General-Purpose Obfuscation

Impossibility of General-Purpose Obfuscation. As indicated, also Definition 4 suffers from certain impossibility results. First, the argument from [1, Section 3] works also for the case of a randomized key distribution, and hence there *are* certain (albeit constructed) examples of unobfuscatable function families.

There are even less constructed examples, as we will show in Remarks 1 and 2, and in Section 4.4. In other words: there can be no general-purpose obfuscation.[1]

Specific Obfuscators. What we advocate here is to consider *specific* obfuscators for *specific* function families. For example, we will show (in Section 4.1) that obfuscating the encryption algorithm of a secret-key encryption scheme yields a public-key encryption scheme, and that such obfuscations (in principle at least) exist. However, our example that such obfuscations exist assumes a public-key encryption scheme in the first place. Plugging this example into the secret-key→public-key transformation gives (nearly) the same public-key encryption scheme one started with. So the following question arises:

What is Gained? First, the secret-key→public-key transformation can be seen, similarly to [5], as a technical paradigm to realize public-key encryption in the first place. In that context, a formalization of obfuscation can provide an interface and a technical guideline of what to exactly achieve.

Second, the mentioned impossibility results does not exclude that a sensible formulation of *what* can be obfuscated exists. In other words, there may be a large and easily describable class of functions which *can* be obfuscated. Universal, general-purpose obfuscators for this class may exist and provide solutions for applications which correspond to functions inside this class.

4   Results Concerning the Simulation-Based Virtual Black-Box Property

In this section we discuss two applications of Definition 4: transforming secret-key encryption into public-key encryption and transforming MACs into signature schemes. Although we are not yet able to give an example of such a construction we provide evidence that our definition is satisfiable by demonstrating how to obfuscate a point function using a one-way permutation. Finally we present an impossibility result concerning Definition 4, thereby demonstrating that obfuscation definitions should be tailored to the context in which one wishes to use them.

4.1   Transforming Secret-Key Encryption

When the idea of public-key cryptography was first proposed by Diffie and Hellman [5], they suggested that one way to produce a public-key encryption scheme was to obfuscate a secret-key scheme. This application of obfuscation was also suggested by Barak *et al.* [1].

A secret-key encryption scheme SKE consists of the following three algorithms.

---

[1] It is actually worse: there are function families that cannot be obfuscated even with very specific, case-tailored obfuscators.

- A PPT *key generation algorithm* SKE.KeyGen that takes as input 1^k for $k \in \mathbf{Z}_{\geq 0}$. It outputs a key K.
- A polynomial time *encryption algorithm* SKE.Enc that takes as input 1^k for $k \in \mathbf{Z}_{\geq 0}$, a secret key K, and a message $m \in \{0,1\}^*$. It outputs a ciphertext c. Algorithm SKE.Enc may be probabilistic or deterministic.
- A polynomial time *decryption algorithm* SKE.Dec that takes as input 1^k for $k \in \mathbf{Z}_{\geq 0}$, a key K, and a ciphertext c. It outputs a message m.

Functionality of the scheme requires that for all keys, encryption followed by decryption under the same key is the identity function (slight relaxations of this statement are possible).

A secret-key cryptosystem is IND-CPA secure if no adversary with access to an encryption oracle can pick two messages, of equal length, such that it can distinguish (still having encryption-oracle access) between encryptions of the two. This is the notion called *find-then-guess* CPA (FTG-CPA) security by Bellare *et al.* [2].

A public-key cryptosystem consists of the same three algorithms, but with the difference that the key generation algorithm now outputs two keys: one private and one public. The public key is used for encryption, the private key for decryption. The scheme is IND-CPA secure if no adversary with access to the public key (and hence an encryption oracle) can pick two messages, of equal length, the encryptions of which it can distinguish.

A secret-key encryption scheme is turned into a public-key encryption scheme by releasing as the public key an obfuscation $\mathcal{O}(K)$ of SKE.Enc$(1^k, K, \cdot)$, the private key encryption algorithm using key K.

Note that the correctness requirement of an obfuscation may not guarantee that the public-key scheme obtained in this way functions correctly in terms of decryption of encryption being the identity function. In fact, this is guaranteed only with overwhelming probability. However, we ignore this issue here as one can always demand perfect correctness from the obfuscation (which would result in a public-key encryption with perfect functionality), or one can weaken the functionality requirement for public-key encryption schemes.

Remark 1 (On the obfuscatability of secret-key encryption). In the following, we simply assume a secret-key encryption scheme with obfuscatable encryption algorithm. One may wonder how realistic that assumption is. First, there *are* unobfuscatable secret-key cryptosystems; any scheme that enjoys *ciphertext integrity* [14] in a "publicly verifiable way" cannot be obfuscated. That is, imagine that a keypair of a digital signature scheme is made part of the secret key, and any ciphertext is signed using the signing key, while the verification key is included in every ciphertext. Then by the functionality requirement, an obfuscation must be able to sign messages (i.e., ciphertexts) under this signing key (note that the "real" verification key can be obtained by oracle access to encryption, so the obfuscator cannot make up a different signing key). However, by unforgeability of the signature scheme, no simulator can do so with oracle access to encryption only.

But, on the other hand, *specific* secret-key encryption schemes *can* be obfuscated: imagine a public-key encryption scheme where the public key is part of every ciphertext.[2] The ability to encrypt arbitrary messages can then be acquired with one black-box query to the encryption oracle, hence if we view such a scheme as secret-key encryption, its encryption algorithm can be obfuscated.

So we find ourselves in a situation where we cannot hope for an *all-purpose* obfuscation (for secret-key encryption). In contrast, we hope for efficient obfuscations of *specific* schemes. We are unfortunately not able to give concrete examples here; instead, we simply assume such obfuscations and see how we could benefit:

Theorem 4.1. *If a secret-key encryption scheme SKE that is IND-CPA is turned into a public-key encryption scheme using the method above with an obfuscator satisfying Definition 4, then the resulting scheme is an IND-CPA secure public-key encryption scheme.*

Proof. For the sake of brevity, we will write $\mathsf{E}_K(\cdot)$ for $\mathsf{SKE.Enc}(1^k, K, \cdot)$ and O_{E_K} for an obfuscation thereof. Let a PPT adversary $A = (A_1, A_2)$ be an adversary of the public-key scheme whose advantage is

$$\mathsf{Adv}^{\mathsf{IND-CPA}} = \Big| \Pr[K \leftarrow \mathsf{SKE.KeyGen}(1^k), O_{\mathsf{E}_K} \leftarrow \mathcal{O}(K),$$

$$(m_0, m_1, h) \leftarrow A_1(O_{\mathsf{E}_K}), b \leftarrow \{0,1\} : A_2(h, O_{\mathsf{E}_K}(m_b)) = b] - \frac{1}{2} \Big|.$$

(In the above we have split the adversary in two and use h to denote state information it might wish to relay.) We must show that this advantage is negligible. By approximation of obfuscation, we have

$$\mathsf{Adv}^{\mathsf{IND-CPA}} \approx \Big| \Pr[K \leftarrow \mathsf{SKE.KeyGen}(1^k), O_{\mathsf{E}_K} \leftarrow \mathcal{O}(K),$$

$$(m_0, m_1, h) \leftarrow A_1(O_{\mathsf{E}_K}), b \leftarrow \{0,1\} : A_2(h, \mathsf{E}_K(m_b)) = b] - \frac{1}{2} \Big|,$$

where $X \approx Y$ denotes that $|X - Y|$ is negligible.

If we view A as a distinguisher against the obfuscation (that chooses b on its own and uses its oracle access to $E_K(\cdot)$ to obtain $E_K(m_b)$), then Definition 4 guarantees the following. There must be a simulator S that, given only oracle access to E_K, produces an output O'_{E_K} indistinguishable from O_{E_K} from A's point of view, yielding

$$\mathsf{Adv}^{\mathsf{IND-CPA}} \approx \Big| \Pr[K \leftarrow \mathsf{SKE.KeyGen}(1^k), O'_{\mathsf{E}_K} \leftarrow S^{E_K},$$

$$(m_0, m_1, h) \leftarrow A_1(O'_{\mathsf{E}_K}), b \leftarrow \{0,1\} : A_2(h, \mathsf{E}_K(m_b)) = b] - \frac{1}{2} \Big|.$$

---

[2] This trick was suggested by a TCC referee.

Now consider the adversary (A_1', A_2) of the symmetric scheme SKE that runs S^{E_K} to obtain O_{E_K}' and then runs $A = (A_1, A_2)$, replacing O_{E_K} with $O_{E_K'}'$.

From the above it follows that

$$\mathsf{Adv}^{\mathsf{IND-CPA}} \approx \Big| \Pr[K \leftarrow \mathsf{SKE.KeyGen}(1^k),$$

$$(m_0, m_1, h) \leftarrow A_1'^{E_K}(1^k),\ b \leftarrow \{0,1\} : A_2(h, \mathsf{E}_K(m_b)) = b] - \frac{1}{2} \Big|,$$

and since the term on the right hand side is negligible by assumption, it follows that A's advantage against the public-key scheme is negligible as well.   □

A stronger security requirement for secret-key and public-key encryption schemes is indistinguishability of ciphertexts under adaptive chosen-ciphertext attacks (IND-CCA, see [23]; for secret-key schemes, this is also called FTG-CCA in [2]). This notion is very similar to IND-CPA security, only an attacker (who tries to distinguish encryptions of two self-chosen plaintexts) is also given access to a decryption oracle. (Obviously, that access is limited in the sense that decryption of the ciphertext the adversary should distinguish is not allowed.)

It is natural to ask whether an IND-CCA secure secret-key scheme can be directly converted to an IND-CCA secure public-key scheme by obfuscating the encryption function. The next theorem shows that the answer is unfortunately negative:

Theorem 4.2. *Assuming that there is an IND-CCA secure secret-key encryption scheme* SKE *with obfuscatable encryption algorithm. Then, there is also another obfuscatable, IND-CCA secure secret-key encryption scheme* SKE' *and an obfuscator* O' *for* SKE' *such that, after obfuscating the encryption function* O', *the result is not an IND-CCA secure public-key encryption scheme.*

Proof. Assume an IND-CCA secure secret-key encryption scheme SKE that is obfuscatable in the sense of Definition 4. Modify SKE into a scheme SKE' as follows: the modified key generation outputs (K, r) for a key K as produced by SKE.KeyGen and a uniformly chosen random k-bit string r. A message m is encrypted under key (K, r) to (c, d), where $c \leftarrow \mathsf{SKE.Enc}(1^k, K, m)$ and d is the empty bitstring. A ciphertext (c, d) is decrypted under secret-key (K, r) to $m \leftarrow \mathsf{SKE.Dec}(1^k, K, c)$ if d is the empty bitstring, to K is $d = r$, and to $\bot$ otherwise.

The IND-CCA security of SKE' can be reduced to that of SKE: say that A' is an IND-CCA adversary on SKE'. A corresponding adversary A on SKE can internally simulate A' and only needs to translate oracle calls accordingly. Namely, A appends to each SKE-encryption an empty component d so as to make it an SKE'-encryption; decryption queries (c, d) are answered depending on d: if d is empty, the query is relayed to A's own decryption oracle, otherwise A answers the query on its own with $\bot$. (Note that this ensures that A never asks for decryption of the challenge ciphertext, since by assumption A' does not do so.) Since A' can have only negligible probability in guessing r, this provides A'

with a view at most negligibly away from its own SKE'-IND-CCA game. Hence A is successful iff A' is, and thus, SKE' is IND-CCA secure because SKE is.

Consider an obfuscator $\mathcal{O}$ for the encryption algorithm of SKE that satisfies Definition 4. Modify $\mathcal{O}$ into $\mathcal{O}'$, such that obfuscations produced by $\mathcal{O}'$ append an empty bitstring d to encryptions. Furthermore, make $\mathcal{O}'$ include r in the obfuscation. Since only the encryption algorithm is considered for obfuscation, $\mathcal{O}'$ still satisfies Definition 4. (Specifically, a simulator S' for $\mathcal{O}'$ can be obtained from a simulator S for $\mathcal{O}$ by simply appending a uniformly selected k-bit string r to the output of S.)

However, applying $\mathcal{O}'$ to SKE.Enc yields a public key encryption scheme in which r is part of the public key. Any query of the form (c, r) to the decryption oracle can be used by an IND-CCA attacker to obtain K and thus break the scheme.   □

Note that the precondition in Theorem 4.2—namely, an *obfuscatable* IND-CCA secure secret-key encryption scheme—is satisfiable by an IND-CCA *public-key* encryption scheme with a the argument from Remark 1.

Although, by Theorem 4.2, a direct construction of an IND-CCA secure public-key scheme is not possible using obfuscation, this does not mean that Definition 4 is not useful in this context: using Theorem 4.1 combined with a generic conversion from, say, IND-CPA to NM-CPA such as [22], one still obtains at least a non-malleable public-key scheme.

4.2   Transforming Message Authentication Codes

Another obvious application of obfuscation is to transform message authentication codes (in which a secret key is used for authenticating *and* verifying a message) into signature schemes (in which a secret key is used for signing, *i.e.*, authenticating, a message, but in which the verification key is public). Intuitively, this could be done by obfuscating the verification algorithm of the message authentication code. To begin with we will introduce the necessary concepts.

A message authentication code MAC consists of the three PPT algorithms MAC.Key, MAC.Sign and MAC.Verify, where

- MAC.Key(1^k) outputs a secret key K,
- MAC.Sign($1^k, K, m$) signs a message $m \in \{0, 1\}^*$ under key K and outputs a signature μ,
- MAC.Verify($1^k, K, m, \mu$) verifies a signature μ to the message m.

As a functionality (or, correctness) requirement, one demands that under any possible key K, verifying a legitimately generated signature should always succeed. Again, relaxations are possible.

We demand for security that it should be hard for a PPT adversary to come up with a valid message/signature pair without knowing the secret key. Here, the adversary may request signatures of messages of its choice from a signing oracle. (Of course, signatures which are obtained through that oracle do not count as successful forgeries.) Since we explicitly allow that the signing algorithm is probabilistic, there may be multiple signatures for a single message. Therefore, we also equip the adversary with an oracle for verifying messages. Formally, a message authentication code MAC is *secure under adaptive chosen-message attacks*, or *MAC-CMA secure*, if and only if, for all PPT adversaries A, the following probability is negligible.

$$\Pr\big[K \leftarrow \mathsf{MAC.KeyGen}(1^k), (m, \mu) \leftarrow A^{\mathsf{MAC.Verify}(1^k, K, \cdot, \cdot), \mathsf{MAC.Sign}(1^k, K, \cdot)}(1^k) :$$
$$\mathsf{MAC.Verify}(1^k, K, m, \mu) = 1\big]$$

Analogously, a weaker notion of security, namely *security under verify-only attacks*, or *MAC-VOA*, can be derived by omitting the signing oracle from the above definition. Note that we have restricted here to PPT adversaries; actually, there are even schemes that achieve *unconditional security* [24].

The natural public-key analogue of message authentication codes are digital signature schemes. These are identical to message authentication codes, only the key generation algorithm outputs two keys: one public key that is used for verifying signatures, and a private key that is used for signing messages. Correctness is defined as for message authentication codes. The security notions SIG-CMA and SIG-VOA are defined exactly analogously to their message authentication code counterparts (only the adversary is given the public verification key in addition to 1^k).

So, in a digital signature scheme, verification is public, whereas in a message authentication code it requires a secret key. Thus, the verification algorithm of a message authentication code is a natural candidate for obfuscation. In fact, by obfuscating the verification algorithm of a message authentication code, we syntactically obtain a digital signature scheme. (As in the case of secret/public-key encryption, we ignore the *perfect* functionality requirement; here either the functionality requirement on the obfuscation must be perfect, or the functionality definition for digital signature schemes must be weakened.)

Technically, this means that the key generation SIG.KeyGen of the transformed scheme outputs a secret key K obtained from MAC.KeyGen along with an obfuscation of the verification function, $\mathcal{O}(\mathsf{MAC.Verify}(1^k, K, \cdot, \cdot))$, (with hard-coded secret key) as the public key. (Signing is unchanged, and verification simply means using the algorithm given by the public key.)

Remark 2 (On the obfuscatability of message authentication). We will simply assume a message authentication code with obfuscatable verification algorithm. Again, one may wonder how realistic that assumption is. First, there are certain (artificial) MACs whose verification algorithm cannot be obfuscated. (In particular, there can be no general-purpose MAC authenticator.) Since this is less straightforward to see than the existence of unobfuscatable secret-key encryption schemes, we now sketch such a MAC. (This is basically the general construction from [1, Section 3] adapted to the MAC interface.)

Let MAC be a MAC-VOA secure MAC. Define MAC$'$ through

- MAC$'$.KeyGen(1^k) := (K, α, β) for $K \leftarrow$ MAC.KeyGen(1^k) and uniformly chosen $\alpha, \beta \in \{0, 1\}^k$.
- MAC$'$.Sign($1^k, K', m$) $= (0, \mu)$ for $\mu \leftarrow$ MAC.Sign($1^k, K, m$), where K' is parsed as (K, α, β).
- MAC$'$.Verify($1^k, K', m, (0, \mu)$) = MAC.Verify($1^k, K, \mu$)
- MAC$'$.Verify($1^k, K', m, (1, \mu)$) = 1 iff $\mu(\alpha) = \beta$, where μ is interpreted as an algorithm description.[3]
- MAC$'$.Verify($1^k, K', m, (2, \mu, i)$) = "the i-th bit of β", but only if $\mu = \alpha$ (otherwise, MAC$'$.Verify returns 0).

First, it is easy to see that with oracle access to MAC$'$.Verify, no valid (in the sense of MAC$'$.Verify) "signatures" of the form $(1, \mu)$ or $(2, \mu)$ can be generated. Hence, MAC$'$ inherits MAC's MAC-VOA security.

But now consider a distinguisher D who, on input an obfuscation O of MAC$'$.Verify($1^k, K', \cdot, \cdot$), returns $O(0^k, (1, O'))$, where the algorithm description O' is constructed from O such that

$$O'(x) = O(0^k, (2, x, 1)) || \ldots || O(0^k, (2, x, k)).$$

Then, functionality of an obfuscation dictates that $O'(\alpha) = \beta$ with overwhelming probability, and hence, $\Pr[D(O) = 1]$ must be overwhelming. On the other hand, no simulator can (with non-negligible probability, and from oracle access to MAC$'$.Verify alone) produce a fake obfuscation $\tilde{O}$ that satisfies $\tilde{O}'(\alpha) = \beta$, so $\Pr\left[D(\tilde{O}) = 1\right]$ is negligible. Hence, MAC$'$ cannot be obfuscated in the sense of Definition 4.

On the other hand, there also *are* obfuscatable MACs. Similarly to the encryption setting, any MAC-VOA secure digital signature scheme can be converted into a MAC-VOA secure MAC that is obfuscatable: simply declare the verification key part of the (secret) MAC key K. The obfuscation of the verification algorithm is simply the verification key. To achieve simulation of this obfuscation in the sense of Definition 4, we cannot use the trick from the encryption setting, where the public key was part of every encryption. In our setting, the verification algorithm outputs only one bit, and we must take care not to make a trivial signature forgery possible. However, a simulation of obfuscation is possible by simply outputting a *fresh* verification key randomly. No distinguisher can, with oracle access to the "right" verification routine, distinguish the "right" verification key from an independently generated one; to do so, it would need to forge a signature, which would contradict the MAC-VOA security of the digital signature scheme.

Analogously to the results in the previous section, we can make two statements about the security of the digital signature schemes that are obtained by obfuscating the verification algorithm of a message authentication code:

---

[3] Here and in the further analysis, we ignore complexity issues; techniques similar to those from [1] can and must be used to make MAC$'$.Verify PPT.

Theorem 4.3. *Say that a message authentication code* MAC *is MAC-VOA. If* MAC *is turned into a digital signature scheme by obfuscating using the method above and using an obfuscator satisfying Definition 4, then the resulting scheme is a SIG-VOA secure digital signature scheme.*

Proof. The proof is very similar to the proof of Theorem 4.1 (in particular, the idea is to first use the functionality and then the simulatability of obfuscation), so we omit it.   □

Theorem 4.4. *Assuming that there is a MAC-CMA secure message authentication code* MAC *with obfuscatable verification algorithm. Then there is also a MAC-CMA secure message authentication code* MAC′ *and an obfuscator (in the sense of Definition 4)* $\mathcal{O}′$ *for* MAC′*'s verification function, such that the result if not SIG-CMA secure as a digital signature scheme.*

Proof. The proof is analogous to the proof of Theorem 4.2. First, assume an obfuscatable MAC-CMA secure message authentication code MAC. Modify MAC into a code MAC′ by including a uniformly selected $r \in \{0,1\}^k$ to the secret key during key generation. Signing and verification take places as before, except that if $m = r$ is to be signed, the signing algorithm appends K to the signature. This authentication code is still MAC-CMA, since an attacker has negligible probability of guessing r.

Any obfuscation $\mathcal{O}$ of the verification function can be modified into another one $\mathcal{O}′$ that includes the second part r of the secret key. If $\mathcal{O}$ satisfies Definition 4 for MAC, then so does $\mathcal{O}′$ for MAC′, since a simulator that is to simulate an obfuscation of the verification function can simply choose r by itself. (It cannot be detected in doing so by only looking at the verification algorithm.)

However, applying $\mathcal{O}′$ to the verification algorithm of MAC′ obviously leads to a digital signature scheme that is *not* SIG-CMA secure: an attacker gets r as part of the public key and simply needs to submit it to the signing oracle to obtain the signing key K.   □

4.3   Deterministic Obfuscation of Point Functions

In this section we prove a concrete feasibility result for Definition 4 by showing how to obfuscate a point function, as defined below.

Definition 5 (Point Functions). *For* $k \in \mathbf{Z}_{\geq 0}$ *and* $x \in \{0,1\}^k$*, the point function* $P_x : \{0,1\}^k \to \{0,1\}$ *is defined by* $P_x(y) = 1$ *if* $y = x$ *and 0 otherwise. Define* $\mathcal{P}_k = \{P_x : x \in \{0,1\}^k\}$.

We now show how to obfuscate point functions under Definition 4. Note that the requirement that the obfuscation has the same functionality as the original function follows directly from the construction.

Theorem 4.5. *Let* Π *be a one-way permutation on* $\{0,1\}^k$ *with respect to the uniform distribution. Then the obfuscation* $\mathcal{O}(x) = \Pi(x)$ *satisfies Definition 4 with respect to the uniform distribution on* $\mathcal{P}_k$ *and where the obfuscated function on input* y *and* $\mathcal{O}(x)$ *outputs 1 iff* $\Pi(y) = \mathcal{O}(x)$.

Proof. Consider the simulator S that outputs a uniformly sampled $y \in \{0,1\}^k$. We need to show that the difference from Definition 4 is negligible for any PPT distinguisher D. This is done by

$$\Pr\big[x \leftarrow \{0,1\}^k : D^{P_x}(1^k, \Pi(x)) = 1\big]$$
$$= \Pr\big[y \leftarrow \{0,1\}^k : D^{P_y}(1^k, \Pi(y)) = 1\big]$$
$$\overset{(*)}{\approx} \Pr\big[x,y \leftarrow \{0,1\}^k : D^{P_x}(1^k, \Pi(y)) = 1\big]$$
$$= \Pr\big[x \leftarrow \{0,1\}^k : D^{P_x}(1^k, S(1^k)) = 1\big],$$

where $X \approx Y$ means that $|X - Y|$ is negligible in k. Here, $(*)$ can be shown by a reduction to the one-way property of Π. If there was a D for which $(*)$ does not hold, this D can be used to invert Π with non-negligible probability. A probabilistic Π-inverter D' then internally simulates D and works as follows. If D makes in any case at most, say, $p(k)$ oracle queries, D' chooses $i \in \{1, \ldots, p(k)\}$ uniformly and answers all queries up to the i-th with 0 and outputs the i-th query as a guess for a preimage. $\qquad\square$

Note 1. Very recently, [20] investigated to what extent point function obfuscations can be used to bootstrap other obfuscations. They did this under a definition of obfuscation in which adversaries are bounded *only* in their number of oracle queries, but not in the number of their computation steps. With respect to this definition, [20] shows that there are circuits which can be obfuscated with a random oracle, but not with just an oracle to a point function.

They also improve an upper bound on the concurrent self-composability of Wee's construction for point function obfuscation. Note that our construction, under our definition, is self-composable, which follows easily from the obfuscator being deterministic.

4.4 An Infeasibility Result

We conclude our work on Definition 4 by considering impossibility results on previous notions of obfuscation. The two notions that come closest to the new notion are simulation-based universal black box (Definition 3) and obfuscation with respect to independent auxiliary input [11, Definition 4].

Wee [25] shows that in the standard model obfuscation of deterministic functions with respect to Definition 3 is possible if and only if the functions are efficiently and exactly learnable (meaning that with a polynomial number of queries and effort one can construct a circuit that computes the function exactly). Since point functions are not efficiently and exactly learnable, it is clear from our positive result in the preceding section that Definition 4 is indeed a relaxation.

However, for a deterministic function one possible distinguisher simply samples random inputs and checks whether the obfuscated function (or simulated one) gives the same output as the real function (to which the distinguisher has oracle access). Consequently, the simulated function needs to correspond to the

real function on all inputs, except for a small (and hard to find) fraction. Hence
a function should be efficiently *approximately* learnable, that is, with a polynomial number of queries and effort one can construct a circuit that computes the
function except on a small (and hard to find) fraction of the inputs. This in particular rules out the obfuscation of deterministic signature schemes, public-key
decryption and pseudorandom functions.

To give a taste of a formal proof of the above, we give an explicit theorem and
proof for the impossibility to obfuscate a pseudorandom function [8], as defined
below.

Definition 6. *[8] A family of functions* $\mathcal{F} = \{f_k\}$, $f_k : \{0,1\}^k \to \{0,1\}^k$ *is*
pseudorandom *if any probabilistic polynomial time algorithm D the advantage*

$$\left| [f \leftarrow f_k \; : \; D^f(1^k) = 1] - \Pr[r \leftarrow r_k \; : \; D^r(1^k)] \right|$$

is negligible in k where r_k is the set of all functions from $\{0,1\}^k$ to $\{0,1\}^k$.

Theorem 4.6. *It is impossible to obfuscate a pseudorandom function under
Definition 4.*

Proof. Suppose for contradiction that there is an obfuscator $\mathcal{O}$ that satisfies
Definition 4 when applied to a pseudorandom function family $\mathcal{F} = \{f_k\}$. For
$f \leftarrow f_k$, consider the distinguisher D that, on input a supposed obfuscation g
and with oracle access to f, chooses $x \in \{0,1\}^k$ and compares $f(x)$ with $g(x)$.
By functionality of the obfuscation, in case $g = \mathcal{O}(f)$, we may assume that
$f(x) = g(x)$ with overwhelming probability.

Now by assumption, there is a simulator S for this distinguisher that satisfies
Definition 4. This S must thus be able to produce a function g with $f(x) = g(x)$,
but it has only negligible probability of guessing x and thus, with overwhelming
probability, does not query its f-oracle at x.

Thus S can be used to predict $f(x)$ with overwhelming probability without
explicitly querying f. Hence S can be modified into a distinguisher that distinguishes f from a truly random function r as in Definition 6, which contradicts
the pseudorandomness of $\mathcal{F}$.   □

To conclude, in addition to potential applications, the results of this section
demonstrate that while it is satisfiable, Definition 4 is not appropriate for all
application scenarios.

5   Conclusion

We have presented a simulation-based definition that, on the one hand, allows for
obfuscating point functions, yet at the same time is strong enough for converting
secret-key cryptography into public-key cryptography.

We would like to stress again that we do *not* rule out unobfuscatability results.
In fact, we have shown certain scenarios in which obfuscation is not possible.

On the other hand, our positive results (in particular the simplicity of our point function obfuscation) leave hope that obfuscations in interesting cryptographic scenarios are possible. We have given toy examples for the case of secret-key encryption or message authentication.

Acknowledgements

We are indebted to the Crypto 2006 and TCC 2007 referees who gave very valuable comments that helped to improve the paper. Specifically, the construction from Section 4.4 was suggested by one referee, and in particular one TCC referee had very constructive comments concerning presentation of our results. We also thank Alex Dent for motivating discussions.

References

1. B. Barak, O. Goldreich, R. Impagliazzo, S. Rudich, A. Sahai, S. Vadhan, and K. Yang. On the (im)possibility of obfuscating programs. In *Advances in Cryptology - CRYPTO 2001*, volume 2139 of *Lecture Notes in Computer Science*, pages 1–18. Springer-Verlag, 2001. Full version available at http://eprint.iacr.org/2001/069/.
2. M. Bellare, A. Desai, E. Jokipii, and P. Rogaway. A concrete security treatment of symmetric encryption. In 38th *Annual Symposium on Foundations of Computer Science*, pages 394–403. IEEE Computer Science Press, 1997.
3. R. Canetti. Towards realizing random oracles: Hash functions that hide all partial information. In *Advances in Cryptology - CRYPTO '97*, volume 1294 of *Lecture Notes in Computer Science*, pages 455–469. Springer-Verlag, 1997.
4. R. Canetti, D. Micciancio, and O. Reingold. Perfectly one-way probabilistic hash functions. In 30th *ACM Symposium on Theory of Computing*, pages 131–140. ACM Press, 1998.
5. W. Diffie and M. E. Hellman. New directions in cryptography. *IEEE Transactions on Information Theory*, 22(6):644–654, 1976.
6. Y. Dodis and A. Smith. Correcting errors without leaking partial information. In 37th *ACM Symposium on Theory of Computing*, pages 654–663. ACM Press, 2005.
7. R. Gennaro, A. Lysyanskaya, T. Malkin, S. Micali, and T. Rabin. Algorithmic tamper-proof (ATP) security: Theoretical foundations for security against hardware tampering. In *Theory of Cryptography, TCC 2004*, volume 2951 of *Lecture Notes in Computer Science*, pages 258–277. Springer-Verlag, 2004.
8. O. Goldreich, S. Goldwasser and S. Micali. How to construct random functions. *Journal of the ACM*, 33(4), pages 210-217, 1986.
9. O. Goldreich and L. Levin. A hard-core predicate to any one-way function. In 21st *ACM Symposium on Theory of Computing*, pages 25–32. ACM Press, 1989.
10. S. Goldwasser, S. Micali, and R. Rivest. A digital signature scheme secure against adaptive chosen-message attacks. *SIAM Journal on Computing*, 17(2):281–308, 1988.
11. S. Goldwasser and Y. Tauman Kalai. On the impossibility of obfuscation with auxiliary input. In 46th *IEEE Symposium on Foundations of Computer Science*, pages 553–562. IEEE Computer Society, 2005.

12. S. Hada. Zero-knowledge and code obfuscation. In *Advances in Cryptology - ASI-ACRYPT 2000*, volume 1976 of *Lecture Notes in Computer Science*, pages 443–457. Springer-Verlag, 2000.
13. R. Jaeschke. Encrypting C source for distribution. *Journal of C Language Translation*, 2(1), 1990.
14. J. Katz and M. Yung. Unforgeable Encryption and Chosen Ciphertext Secure Modes of Operation. In *Fast Software Encryption - FSE 2000*, volume 1978 of *Lecture Notes in Computer Science*, pages 284–299. Springer-Verlag, 2001.
15. P. Kocher, J. Jaffe, and B. Jun. Differential power analysis. In *Advances in Cryptology - CRYPTO '99*, volume 1666 of *Lecture Notes in Computer Science*, pages 388–397. Springer-Verlag, 1999.
16. C. Linn and S. Debray. Obfuscation of executable code to improve resistance to static disassembly. In 10th *ACM Conference on Computer and Communications Security*, pages 290 – 299. ACM Press, 2003.
17. B. Lynn, M. Prabhakaran, and A. Sahai. Positive results and techniques for obfuscation. In *Advances in Cryptology - EUROCRYPT 2004*, volume 3027 of *Lecture Notes in Computer Science*, pages 20–39. Springer-Verlag, 2004.
18. S. Micali and L. Reyzin. Physically observable cryptography (extended abstract). In *Theory of Cryptography, TCC 2004*, volume 2951 of *Lecture Notes in Computer Science*, pages 278–296. Springer-Verlag, 2004. Full version available at http://eprint.iacr.org/2003/120/.
19. M. Naor and M. Yung. Public-key cryptosystems provably secure against chosen ciphertext attack. In 22nd *ACM Symposium on Theory of Computing*, pages 427–437. ACM Press, 1990.
20. Arvind Narayanan and Vitaly Shmatikov. On the Limits of Point Function Obfuscation. IACR ePrint Archive, May 2006. Online available at http://eprint.iacr.org/2006/182.ps.
21. National Institute of Standards and Technology. *Data Encryption Standard (DES)*, 1993. FIPS Publication 46-2.
22. R. Pass and a. shelat and V. Vaikuntanathan. Construction of a Non-Malleable Encryption Scheme From Any Semantically Secure One. In *Advances in Cryptology - CRYPTO '06*, volume 4116 of *Lecture Notes in Computer Science*. Springer-Verlag, 2006.
23. C. Rackoff and D. Simon. Non-interactive zero-knowledge proof of knowledge and chosen ciphertext attack. In *Advances in Cryptology - CRYPTO '91*, volume 576 of *Lecture Notes in Computer Science*, pages 433–444. Springer-Verlag, 1992.
24. D. R. Stinson. *Cryptography: Theory and Practice*. CRC Press, 1995.
25. H. Wee. On obfuscating point functions. In 37th *ACM Symposium on Theory of Computing*, pages 523–532. ACM Press, 2005.
26. A. C. Yao. Theory and applications of trapdoor functions (extended abstract). In 23rd *Annual Symposium on Foundations of Computer Science*, pages 80–91. IEEE Computer Science Press, 1982.

Securely Obfuscating Re-encryption

Susan Hohenberger[1,2], Guy N. Rothblum[3,*],
abhi shelat[2], and Vinod Vaikuntanathan[3,**]

[1] Johns Hopkins University
susan@cs.jhu.edu
[2] IBM Zurich Research
{sus,abs}@zurich.ibm.com
[3] MIT CSAIL
{rothblum,vinodv}@mit.edu

Abstract. We present the first positive obfuscation result for a traditional cryptographic functionality. This positive result stands in contrast to well-known negative impossibility results [BGI+01] for *general obfuscation* and recent negative impossibility and improbability [GK05] results for obfuscation of many cryptographic functionalities.

Whereas other positive obfuscation results in the standard model apply to very simple point functions, our obfuscation result applies to the significantly more complicated and widely-used *re-encryption functionality*. This functionality takes a ciphertext for message m encrypted under Alice's public key and transforms it into a ciphertext for the same message m under Bob's public key.

To overcome impossibility results and to make our results meaningful for cryptographic functionalities, we use a new definition of obfuscation. This new definition incorporates more security-aware provisions.

1 Introduction

A recent line of research in theoretical cryptography aims to understand whether it is possible to *obfuscate* programs so that a program's code becomes unintelligible while its functionality remains unchanged. A general method for obfuscating programs would lead to the solution of many open problems in cryptography.

Unfortunately, Barak, Goldreich, Impagliazzo, Rudich, Sahai, Vadhan and Yang [BGI+01] show that for many notions of obfuscation, a general program obfuscator does not exist—i.e., they exhibit a class of circuits which cannot be obfuscated. A subsequent work of Goldwasser and Kalai [GK05] shows the impossibility and improbability of obfuscating more natural functionalities.

In spite of these negative results for general-purpose obfuscation, there are a few positive obfuscation results for simple functionalities such as point functions. A point function I_x returns 1 on input x and 0 on all other inputs. Canetti [Can97] shows that under a very strong Diffie-Hellman assumption point

* Research supported by NSFgrant CNS-0430450 and NSF grant CFF-0635297.
** Research supported by NSF grant CNS-0430450.

functions can be obfuscated. Further work of Canetti, Micciancio and Reingold [CMR98], Wee [Wee05] and Dodis and Smith [DS05] relaxes the assumptions required for obfuscation and considers other (related) functionalities. Despite these positive results, obfuscators for traditional cryptographic functionalities (such as those that deal with encryption) have remained elusive.

Our Results. In this work, we present the first obfuscator for a more traditional cryptographic functionality. Namely, we show that:

Main Theorem 1 (Informal). *Under reasonable bilinear complexity assumptions, there exists an efficient program obfuscator for a family of circuits implementing re-encryption.*

A *re-encryption program* for Alice and Bob takes a ciphertext for a message m encrypted under Alice's public key, and transforms it into a ciphertext for the same message m under Bob's public key. Re-encryption programs have many practical applications such as the iTunes DRM system (albeit, with symmetric keys [Smi05]), secure distributed file servers [AFGH06] and secure email forwarding.

The straightforward method to implement re-encryption is to write a program P which decrypts the input ciphertext using Alice's secret key and then encrypts the resulting message with Bob's public key. When P is run by Alice, this is a good solution.

In the practical applications noted above, however, the re-encryption program is executed by a third-party. When this is the case, the straightforward implementation has serious security problems since P's code may reveal Alice's secret key to the third party. A better solution is to design an obfuscator for the re-encryption program P. That is, we would like that:

> *A third party who has a re-encryption program* learns *no more from the re-encryption program than it does from interaction with a black-box oracle that provides the same functionality.*

As we discuss later in §1.2, several re-encryption schemes have been proposed before [BS97, BBS98, DI03, AFGH06], but none of these prior works satisfy the strong obfuscation requirement informally stated above. Our main technical contribution is the construction of a novel re-encryption scheme which meets this strong notion while remaining surprisingly practical. As a side note, in our construction, ciphertexts that are re-encrypted from Alice to Bob cannot be further re-encrypted from Bob to Carol. This may be a limitation in some scenarios, but it is nonetheless sufficient for the important practical applications noted above.

Our main conceptual contribution is a definition of obfuscation that both sidesteps impossibility results by considering *randomized* functionalities and is more meaningful for cryptographic applications than previous definitions of obfuscation. Let us briefly explain.

1.1   Notion of Secure Obfuscation

The work of [BGI+01] views an obfuscator as a compiler which takes a program (i.e., boolean circuit) P and turns it into an equivalent program that satisfies the *predicate black-box* property: any *predicate* that is computable from the obfuscated program should also be computable from black-box access to the program (see Definition 1).

Secure Obfuscation. Unfortunately, the predicate definition [BGI+01] and subsequent work does not provide a meaningful security guarantee when the obfuscated program is used as part of a larger cryptographic system. Intuitively, while the predicate black-box property gives a quantifiable guarantee that *some* information (namely, predicates) about the program is hidden by the obfuscated circuit, other "non-black-box information" may still leak. Moreover, this leaked information might compromise the security of a cryptographic scheme which uses the obfuscated circuit. For instance, it is completely possible that an obfuscated program for "delegating signatures" both meets the predicate black-box definition and is unforgeable under black-box access to a signature oracle, yet allows an adversary who has the obfuscated program code to forge a signature!

Since many potential applications of obfuscation use obfuscated circuits in larger cryptographic schemes, the definition of obfuscation *should* guarantee that the security of cryptographic schemes is preserved in the following sense.

> *If a cryptographic scheme is "secure" when the adversary is given black-box access to a program, then it remains "secure" when the adversary is given the obfuscated program.*

The most important feature of our new definition of obfuscation is that it preserves security in the above sense, and thus we refer to it as *secure obfuscation*. Informally, our definition requires that if there exists a *non black-box adversary* with access to an obfuscated program who can break the security of a cryptographic scheme, then there exists a *black-box simulator* which breaks the scheme with similar probability using only black-box access to the program. Thus, if the scheme is secure against black-box adversaries, then it is also secure against adversaries with access to obfuscated programs. The definition we give in this work gives the above guarantee for any cryptographic scheme with a *distinguishable attack* property; any scheme where a distinguisher with public information and black-box access to the obfuscated functionality can distinguish whether or not an attacker has broken the scheme. Semantically secure encryption and re-encryption are examples of such schemes.

This new definition of obfuscation can play an important role in the design of cryptographic schemes that use obfuscation. With secure obfuscation, the design of such schemes proceeds in two stages:

1. Specify the functionality of a program (or program family), and prove security of the cryptographic scheme against an adversary given *black-box* access to the program.
2. Construct a secure obfuscator for the program (or program family).

The combination of these two steps guarantees security of the scheme against an adversary that has full access to the obfuscated program. Indeed, for our scheme, we show step (1) in Theorem 3 and step (2) in Theorem 4.

Average-Case Obfuscation. Our new definition only requires obfuscation for a *random* circuit in a family of circuits (as in [Can97, CMR98, Had00, DS05, GK05]). This relaxed requirement remains meaningful for the many cryptographic applications of obfuscation in which the circuit to be obfuscated *is* chosen at random. Normally the random choice of a circuit corresponds to the random selection of cryptographic keys. We call the new definition of security *average-case secure obfuscation.* Combining more security-aware provisions (i.e. giving a distinguisher oracle access to the functionality) with this average-case relaxation was originally suggested by Pass [Pas06].

Obfuscating Probabilistic Circuits. It is not hard to see that *deterministic* functionalities that are not learnable cannot be obfuscated under our definition (see §2.2). In fact, security-preserving definitions considered in previous works ([BGI⁺01, Wee05]) are only achievable for learnable deterministic functions. An important conceptual contribution of our definition and of this work is showing that these impossibility results disappear when considering obfuscation of *probabilistic* circuits. Furthermore, obfuscation of probabilistic circuits is important because most interesting cryptographic functionalities are probabilistic.

Other work on Obfuscation. Recently, Ostrovsky and Skeith [OS05] consider a different notion of *public-key obfuscation* focused on keyword search. A public-key obfuscator does not maintain the functionality of a program, but rather ensures that the outputs of a public-key obfuscated program are *encryptions* of the original program's outputs. Adida and Wikström [AW05] use a variation of this definition for public mixing. Both of these works differ from our notion of obfuscation in that our notion preserves functionality and explicitly considers black-box versus non-black-box access to a program.

1.2   The Obfuscated Re-encryption Scheme

Comparison with Prior Work. Mambo and Okamoto [MO97] noted the popularity of re-encryption programs in practical applications and suggested efficiency improvements over the decrypt-and-encrypt approach. Blaze, Bleumer, and Strauss introduced the notion of *proxy re-encryption* [BS97, BBS98] in which the re-encryption program is executed by a third-party *proxy.* In their security notion, the proxy cannot read the messages of either Alice or Bob. The Blaze et al. construction is *bidirectional* (i.e., a program to translate ciphertexts from Alice to Bob can also be used to translate from Bob to Alice) and can be repeatedly applied (i.e., a ciphertext can be re-encrypted from Alice to Bob to Carol, etc.). Ateniese, Fu, Green, and Hohenberger [AFGH06] presented a semantic-security style definition for proxy re-encryption and designed the first *unidirectional* scheme, although their scheme can only be applied once. Ateniese et al. also built a secure distributed storage system using their algorithms.

While these prior works are secure under specialized definitions, they cannot be considered as obfuscations for re-encryption since they leak subtle non-black-box information. On the other hand, the re-encryption definitions of Ateniese et al. [AFGH06] provide some security guarantee with respect to *dependent* auxiliary inputs, which we will not consider in this work. For example, they show that even when Alice has the re-encryption program from Alice to Bob, Bob's semantic security still holds. (Although our definition does not require this, the scheme we present here satisfies this property.)

Overview of the Construction. We now provide intuition behind our construction of an obfuscator for re-encryption (see §4 for the full construction). In a series of attempts, we develop a cryptosystem and an obfuscated re-encryption program which translates ciphertexts under pk_1 to ciphertexts under pk_2. Our starting point is a suitable public key cryptosystem.

Recall the semantically-secure encryption scheme due to Boneh, Boyen, and Shacham [BBS04] as instantiated in a group $\mathbb{G}$ of order q equipped with a bilinear map. The keys in this scheme are generated by selecting a random $h \xleftarrow{r} \mathbb{G}$ and $a, b \xleftarrow{r} \mathbb{Z}_q$, and setting $sk = (a, b, h)$ and $pk = (h^a, h^b, h)$. To encrypt a message $m \in \mathbb{G}$, select two random values $r, s \xleftarrow{r} \mathbb{Z}_q$ and output the ciphertext $C = [h^{ar}, h^{bs}, h^{r+s} \cdot m]$. To decrypt a ciphertext $C = [W, X, Y]$, compute the plaintext $Y/(W^{1/a} \cdot X^{1/b})$. Let $pk_1 = (g^{a_1}, g^{b_1}, g)$ and $pk_2 = (h^{a_2}, h^{b_2}, h)$ be two public keys for this cryptosystem.

The basic (naive) re-encryption program from pk_1 to pk_2 contains (sk_1, pk_2). The program simply decrypts the input using sk_1 and encrypts the resulting message with pk_2. Clearly this program exposes both sk_1 and the underlying plaintext to any third-party executing the re-encryption program.

As a first attempt to obfuscate the basic program, consider the re-encryption program that contains $Z_1 = a_2/a_1$ and $Z_2 = b_2/b_1$ and re-encrypts the ciphertext $[W, X, Y]$ by computing $[W^{Z_1}, X^{Z_2}, Y]$ for pk_2. (On a different cryptosystem, a similar approach was suggested by Blaze et al. [BBS98].) Unfortunately, this re-encryption program leaks non-black-box information (i.e., does not satisfy the virtual black-box property in Def. 2). For example, the program containing (Z_1, Z_2) which translates ciphertexts from Alice to Bob can be transformed into a new program containing (Z_1^{-1}, Z_2^{-1}) which translates ciphertexts from Bob to Alice—a feat which black-box access does not allow.

As a second attempt, consider the re-encryption program containing $Z_1 = h^{a_2/a_1}$ and $Z_2 = h^{b_2/b_1}$. Alice, with $sk_1 = (a_1, b_1, g)$, can compute this program given Bob's public key $pk_2 = (h^{a_2}, h^{b_2}, h)$. (On a different cryptosystem, a similar approach was suggested by Ateniese et al. [AFGH06].) The re-encryption program works as follows: on input a ciphertext $[W, X, Y] = [g^{a_1 r}, g^{b_1 s}, g^{r+s} \cdot m]$ under pk_1, output the ciphertext $[\mathbf{e}(W, Z_1), \mathbf{e}(X, Z_2), \mathbf{e}(Y, h)] = [E, F, G]$ under pk_2. To decrypt $[E, F, G]$, the holder of sk_2 would first compute $Q = G/(E^{1/a_2} \cdot F^{1/b_2})$ and then find and output the message m_i in the message space M such that $\mathbf{e}(m_i, h) = Q$. Of course, to ensure efficient decryption, this limits the size of the message space M to be a polynomial. Notice the encryption scheme now support two "forms" of ciphertexts—an original form and a re-encrypted one, each containing elements from different groups. As a result, a re-encrypted ciphertext cannot be further re-encrypted. The question, though, is whether or not such a program is any closer to being an obfuscation.

To be a secure obfuscation according to our Def. 2, the output of an adversary who is given the obfuscated program must be indistinguishable—even to a distinguisher with oracle access to the re-encryption program—from the output of a simulator given only black-box access to the program. Unfortunately, in the second attempt, knowledge of the public keys $pk_1 = (g^{a_1}, g^{b_1}, g)$ and $pk_2 = (h^{a_2}, h^{b_2}, h)$ easily allows a distinguisher to test whether a program containing (Z_1, Z_2) is a valid re-encryption program for these keys by checking that $\mathbf{e}(g^{a_1}, Z_1) = \mathbf{e}(g, h^{a_2})$ and $\mathbf{e}(g^{b_1}, Z_2) = \mathbf{e}(g, h^{b_2})$. We do not know how to construct a simulator that can output a program which also passes this test.

To bypass this problem, we design our re-encryption program to be a probabilistic function of the keys. More specifically, consider the program containing $(y^{a_2/a_1}, y^{b_2/b_1}, y) = (Z_1, Z_2, Z_3)$ for a randomly selected $y \in \mathbb{G}$. (In the context of point functions, a similar approach was suggested by Canetti [Can97].) Alice can still generate this re-encryption program using only Bob's public key. The re-encryption program becomes: on input $[W, X, Y] = [g^{a_1 r}, g^{b_1 s}, g^{r+s} \cdot m]$ under pk_1, output the ciphertext under pk_2 as $[\mathbf{e}(W, Z_1), \mathbf{e}(X, Z_2), \mathbf{e}(Y, Z_3), Z_3] = [E, F, G, H]$. Decryption works as follows: first compute $Q = G/(E^{1/a_2} \cdot F^{1/b_2})$ and then output message m_i in the message space M such that $\mathbf{e}(m_i, H) = Q$.

This solution has one subtle problem because *all* ciphertexts produced by the obfuscated re-encryption program include $H = y$ as the fourth component, whereas ciphertexts produced by the decrypt-and-encrypt approach contain a fresh random value in that position. Thus, the obfuscated program does not "preserve the functionality" of the original one. This is easily fixed by having the obfuscated program re-randomize its output by choosing $z \xleftarrow{r} \mathbb{Z}_q$ and outputting $[E^z, F^z, G^z, H^z]$. (Note, it is not sufficient that we choose y randomly, since this choice is only made once for all re-encrypted ciphertexts, whereas z is chosen freshly for each re-encryption.)

Even this, however, falls short, because we do not know how to prove this construction is secure. In particular, since the distinguisher has access to a re-encryption oracle, it can query the oracle on the values contained in the obfuscated program! Indeed, in the above scheme, there is a specific (complicated) property of valid obfuscated programs that a distinguisher can test for, and we do not know how to construct a simulator that also passes this test.

In order to overcome this final hurdle, our program re-randomizes the input ciphertext before applying the transformation above. If the public key is (g^a, g^b, g), and the input ciphertext is $C = [W, X, Y]$, our program re-randomizes C by sampling r', s' and computing the ciphertext $[W \cdot (g^a)^{r'}, X \cdot (g^b)^{s'}, Y \cdot g^{r'+s'}]$. Finally, we are able to show this construction meets our obfuscation definition under two reasonable complexity assumptions.

As a final point about our complexity assumptions, because our obfuscation definition only requires average-case obfuscation, we do not have to make the strong complexity assumptions necessary in the constructions of Canetti [Can97] and Wee [Wee05]. Thus, our scheme simultaneously meets a strong theoretical definition while retaining the sensibility associated with standard assumptions and efficient algorithms.

2 Definitions

Barak et al. [BGI+01] required that an obfuscator strip programs of non-black-box information. They formalized this by requiring that any predicate computable from the obfuscated program is also computable from black-box access to it. Goldwasser and Kalai [GK05] gave a stronger definition, guaranteeing security in the presence of (dependent) auxiliary input. A formal definition, which we call *predicate black-box obfuscation* (or predicate obfuscation, for short), follows.

For a family $\mathbf{C}$ of polynomial-size circuits, for a length parameter n let $\mathbf{C}_n$ be the circuits in $\mathbf{C}$ with input length n (i.e. $\mathbf{C} = \{\mathbf{C}_n\}$).

Definition 1 (Predicate Obfuscation [BGI+01, GK05]). *An efficient algorithm* Obf *is a* predicate obfuscator *for the family* $\mathbf{C} = \{\mathbf{C}_n\}$, *if it has the following properties:*

- *Preserving Functionality: There exists a negligible function* $neg(n)$, *s.t. for all input lengths* n, *for any* $C \in \mathbf{C}_n$:

$$\Pr[\exists x \in \{0,1\}^n : (\mathsf{Obf}(C))(x) \neq C(x)] \leq neg(n)$$

  The probability is taken over Obf*'s random coins.*
- *Polynomial Slowdown: There exists a polynomial* $p(n)$ *such that for sufficiently large input lengths* n, *for any* $C \in \mathbf{C}_n$, *the obfuscator* Obf *only enlarges* C *by a factor of* p: $|\mathsf{Obf}(C)| \leq p(|C|)$.
- *Predicate Virtual Black-box: For every polynomial sized adversary circuit* $\mathcal{A}$, *there exists a polynomial size simulator circuit* Sim *and a negligible function* $neg(n)$, *such that for every input length* n, *for every* $C \in \mathbf{C}_n$, *for every predicate* π, *for every auxiliary input* $z \in \{0,1\}^{q(n)}$:

$$\left| \Pr[\mathcal{A}(\mathsf{Obf}(C), z) = \pi(C, z)] - \Pr[\mathsf{Sim}^C(1^n, z) = \pi(C, z)] \right| \leq neg(n)$$

  The probability is over the coins of the adversary, the simulator and the obfuscator.

As discussed in §1, the predicate black-box definition does not guarantee *security* when obfuscated circuits are used in cryptographic settings. To address this, we introduce the new notion of *average-case secure obfuscation*:

Definition 2 (Average-Case Secure Obfuscation). *An efficient algorithm* Obf *that takes as input a (probabilistic) circuit and outputs a new (probabilistic)* *circuit, is an* average-case secure obfuscator *for the family* $\mathbf{C} = \{\mathbf{C}_n\}$, *if it satisfies the following properties:*

- *Preserving Functionality:* "With overwhelming probability Obf(C) behaves almost *identically to* C *on all inputs*". *There exists a negligible function* $neg(n)$, *such that for any input length* n, *for any* $C \in \mathbf{C}_n$:

$$\Pr_{\text{coins of Obf}} [\exists x \in \{0,1\}^n : \Delta((\text{Obf}(C))(x), C(x)) \geq neg(n)] \leq neg(n)$$

The distributions $(\text{Obf}(C))(x)$ *and* $C(x)$ *are taken over* Obf(C)'s *and* C's *random coins respectively.* Δ *denotes statistical (L1) distance between distributions.*

- *Polynomial Slowdown: (identical to Definition 1)*
- *Average-Case Secure Virtual Black-Box: For any efficient adversary* $\mathcal{A}$, *there exists an efficient simulator* Sim *and a negligible function* $neg(n)$, *such that for every efficient distinguisher* D, *for every input length* n *and for every polynomial-size auxiliary input* z:

$$\left| \begin{array}{l} \Pr[C \xleftarrow{r} \mathbf{C}_n : D^C(\mathcal{A}(\text{Obf}(C), z), z) = 1] \\ - \Pr[C \xleftarrow{r} \mathbf{C}_n : D^C(\text{Sim}^C(1^n, z), z) = 1] \end{array} \right| \leq neg(n)$$

The probability is over the selection of a random *circuit* C *from* $\mathbf{C}_n$, *and the coins of the distinguisher, the simulator, the oracle and the obfuscator. Note that entities with black-box access to* C *cannot set* C's *random tape.*

Note that without loss of generality it is sufficient to require the existence of a simulator for the "dummy" adversary that just outputs its input. This would give an equivalent definition, but it loses some intuitive appeal.

Discussion. Intuitively, Definition 2 guarantees that any attack that a non-black box adversary can mount using the obfuscated circuit, can also be mounted by a black-box simulator with (oracle) access to the functionality. This new definition differs from the predicate definition in several ways. It considers obfuscation of a *random* circuit from a family, and furthermore, the circuit families considered can be *probabilistic* (this allows us to side-step impossibility results, see §2.2). We also follow [GK05] in requiring that the obfuscation be secure in the presence of (independent) auxiliary input, where the auxiliary input is selected first, and then a random circuit is chosen from the family. Note that the average-case secure virtual black-box requirement of our new definition is incomparable to the predicate black-box requirement of [BGI+01]; the latter is weaker in that it only requires that the obfuscator hides *predicates*, but is stronger in that it provides the predicate distinguisher with the actual program (whereas our definition only gives our predicate distinguisher black-box access).

Finally, we emphasize that in this new definition there are *two* important sources of randomness. The first source of randomness is in the circuits being obfuscated, which are probabilistic. The second, more subtle, source of randomness is in the selection of a random circuit C from the family $\mathbf{C}_n$. The average-case secure virtual black-box requirement guarantees security when a circuit is selected from the family by a *specific* distribution (i.e., the uniform distribution—one should think of this as uniformly choosing random keys for a cryptographic scheme). The predicate black-box definition, on the other hand, guarantees security for *every* circuit in the family, or (equivalently) for *every* *distribution* on circuits. Other work [CMR98, DS05] guarantees security for a large class of distributions on circuits from a family, such as *every* distribution with at least super-logarithmic min-entropy. Our notion of secure obfuscation can be generalized to give security against more general classes of distributions. For clarity, we choose to present the less general definition above.

2.1   Meaningfulness for Security

This section serves as an *informal* discussion of the security guarantee provided by average-case secure obfuscation. As mentioned in §1, the definition of obfuscation should be security-preserving in the following sense: *"If a cryptographic scheme is secure when the adversary is given black-box access to a program, then it remains secure when the adversary is given the obfuscated program."* We claim that for a large class of applications (including re-encryption), average-case secure obfuscation indeed gives this guarantee.

To see this, consider any cryptographic scheme, for which a *distinguisher*, that has only public information (e.g. public keys) and *black-box* access to an obfuscated program, can test whether a given adversary can break a scheme (we call this the *distinguishable attack* property). Many standard cryptographic schemes, such as semantically secure encryption and re-encryption, have this property. For such schemes Definition 2 indeed guarantees that for every adversary that mounts an attack using an obfuscated circuit, there exists a *black-box* simulator that can mount an attack with a similar success probability. Thus, if the scheme is secure against black-box adversaries, it is also secure against non black-box adversaries that are given the obfuscated program.

To illustrate the meaningfulness of the notion of average-case secure obfuscation, we propose to use the following informal argument as a methodology for constructing secure cryptographic schemes:

If a cryptographic scheme has the following three properties:

1. The scheme is secure against black-box adversaries with oracle access to functionality X selected randomly from a family F
2. A distinguisher D with oracle access to X can test whether an adversary $\mathcal{A}$ can break the security guarantee of the scheme (we call this property the *distinguishable attack* property)
3. There exists an average-case secure obfuscator for a family C_F of circuits implementing the functionalities in F,

Then the cryptographic scheme is also secure against adversaries who are given an obfuscation of a circuit selected at random from the family C_F.

As a case study, consider semantically-secure re-encryption (see Def. 3). An attacker is given two relevant public keys and black-box access to a re-encryption oracle. The attacker is successful if it can distinguish the encryptions of two different messages (of its choice) under one of the public keys. As with many cryptographic schemes, re-encryption schemes have Property (1). For Property (2), a distinguisher who is given public keys and oracle access to the re-encryption functionality can indeed *test* whether an adversary has a noticeable chance of mounting a successful attack.[1] Thus, for any re-encryption functionality, assuming that Property (3) holds (i.e. there exists an average-case secure obfuscator for some circuit family computing re-encryption), we conclude that the scheme is also secure against adversaries who are given an obfuscated re-encryption circuit. The predicate definition would *not* let us make such a conclusion.

2.2   Obfuscating Probabilistic Programs

In this section we discuss an impossibility result for average-case secure obfuscation of *deterministic* circuits, and explain how we side-step this impossibility by considering probabilistic circuits. Wee [Wee05] observes that the only deterministic circuits that can be obfuscated under strong security-preserving notions of obfuscation are those that are *learnable*. This result also applies to obfuscating *deterministic* circuits under Definition 2. To see the intuition behind this result, consider a circuit family **C**, the "empty" adversary who simply outputs the obfuscated circuit $\mathsf{Obf}(C)$ it gets, and a distinguisher (with black-box access to C) that outputs 1 only if whatever circuit it gets agrees with C for random inputs.[2] Because Obf preserves functionality, the above adversary that outputs $\mathsf{Obf}(C)$ will get the distinguisher to accept with all but negligible probability. To make the distinguisher accept with similar probability, the simulator must learn, from black-box access, a circuit that is (at the very least) very close to C on random inputs. Thus random circuits from **C** must be learnable from black-box access. In particular, deterministic circuit families that are not learnable cannot be obfuscated under Definition 2.

This impossibility disappears when we consider probabilistic circuit families. This is because the (efficient) distinguisher with black-box access to a probabilistic C and non black-box access to $\mathsf{Obf}(C)$ cannot necessarily distinguish whether the *distributions* that C and $\mathsf{Obf}(C)$ output on a particular input are *statistically* close or far. This is similar to the case of encryption (see Goldwasser and Micali [GM84]), where only randomness can prevent an adversary from recognizing whether two ciphertexts are encryptions of the same bit. Our obfuscation of re-encryption programs uses this observation. In fact, our simulator outputs a "dummy circuit" that has little to do with the circuit being obfuscated, but is still indistinguishable from the true obfuscated circuit.

---

[1] To do this, the distinguisher simply runs the adversary with the public keys, answering the adversary's re-encryption requests using the re-encryption oracle.

[2] Wee considers different distinguishers that check different inputs.

3   Algebraic Setting and Assumptions

Bilinear Groups. Let BMsetup be an algorithm that, on input the security param-
eter 1^k, outputs the parameters for a bilinear map as $(q, g, \mathbb{G}, \mathbb{G}_T, \mathbf{e})$, where $\mathbb{G}, \mathbb{G}_T$
are groups of prime order $q \in \Theta(2^k)$. The efficient mapping $\mathbf{e} : \mathbb{G} \times \mathbb{G} \to \mathbb{G}_T$
is both *bilinear*, i.e., for all $g \in \mathbb{G}$ and $a, b \in \mathbb{Z}_q$, $\mathbf{e}(g^a, g^b) = \mathbf{e}(g, g)^{ab}$, and
non-degenerate, i.e., if g generates $\mathbb{G}$, then $\mathbf{e}(g, g) \neq 1$.

For simplicity, we present our solution using bilinear maps of the form $\mathbf{e} :
\mathbb{G} \times \mathbb{G} \to \mathbb{G}_T$. Our scheme can also be implemented in the more general setting
where $\mathbf{e} : \mathbb{G}_1 \times \mathbb{G}_2 \to \mathbb{G}_T$ and isomorphisms between $\mathbb{G}_1$ and $\mathbb{G}_2$ may not be
efficiently computable. Galbraith, Paterson, and Smart [GPS06] provide more
information on various implementation options.

Complexity Assumptions. In this paper, we make the following two complexity
assumptions in bilinear groups. When we say two distributions are computationally indistinguishable, we mean with respect to a distinguisher with auxiliary
information (which is selected independently of the instance).

Assumption 1 (Strong Diffie Hellman Indistinguishability). *Let $\mathbb{G}$ be a
group of order q where q is a k-bit prime, $g \xleftarrow{r} \mathbb{G}$ and $a, b, c, d \xleftarrow{r} \mathbb{Z}_q$. Then the
following two distributions are computationally indistinguishable:*

$$\left\{g, g^a, g^b, g^c, g^{abc}\right\}_k \stackrel{c}{\approx} \left\{g, g^a, g^b, g^c, g^d\right\}_k$$

This assumption has not been proposed before, but it is implied by the *Decision
3-party Diffie-Hellman* assumption proposed in [BSW06].

Assumption 2 (Decision Linear [BBS04]). *Let $\mathbb{G}$ be a group of order q
where q is a k-bit prime, $f, g, h \xleftarrow{r} \mathbb{G}$ and $a, b, c \xleftarrow{r} \mathbb{Z}_q$. Then the following two
distributions are computationally indistinguishable:*

$$\left\{f, g, h, f^a, g^b, h^{a+b}\right\}_k \stackrel{c}{\approx} \left\{f, g, h, f^a, g^b, h^c\right\}_k$$

4   A Special Encryption Scheme and Re-encryption
   Functionality

In this section, we describe a special encryption scheme and a *re-encryption
functionality* for which we later present a secure obfuscation scheme.

4.1   A Special Encryption Scheme Π

Our special encryption scheme Π is described in Fig.1. The encryption algorithm
supports two forms of ciphertexts and takes an additional input $\beta \in \{0, 1\}$ to
choose between them. For the first form, encryption and decryption work as
per the Boneh et al. [BBS04] construction. For the second form, the encryption
and decryption are novel and relevant for re-encryption. Note that this encryption system also requires the message space M to be a subset of $\mathbb{G}$ which is
of size polynomial in k. The semantic security of this scheme will be proven in
Thm. 3.

Common: For a security parameter 1^k, let $(q, g, \mathbb{G}, \mathbb{G}_T, \mathbf{e}) \leftarrow \mathsf{BMsetup}(1^k)$ be a common parameter and let $M \subset \mathbb{G}$ where $|M| = O(\mathrm{poly}(k))$ be the message space.

$\mathsf{KeyGen}(1^k, (q, g, \mathbb{G}, \mathbb{G}_T, \mathbf{e}))$:

    1. Randomly select a new generator $h \xleftarrow{r} \mathbb{G}$ and random $a, b \xleftarrow{r} \mathbb{Z}_q$.

    2. Output $pk = (h^a, h^b, h)$ and $sk = (a, b, h)$.

$\mathsf{Enc}(pk, \beta, m)$:

    1. Parse $pk = (h^a, h^b, h)$.

    2. Choose random $r, s \xleftarrow{r} \mathbb{Z}_q$.

    3. If $\beta = 0$, output the ciphertext $[0, (h^a)^r, (h^b)^s, h^{r+s} \cdot m, 0]$.

    4. If $\beta = 1$, then choose a random $t \xleftarrow{r} \mathbb{G}$, and output the ciphertext $[1, \mathbf{e}((h^a)^r, t), \mathbf{e}((h^b)^s, t), \mathbf{e}(h^{r+s} \cdot m, t), t]$.

$\mathsf{Dec}(sk, [s, W, X, Y, Z])$:

    1. Parse $sk = (a, b, h)$.

    2. If $s = 0$, then output $Y/(W^{1/a} \cdot X^{1/b})$.

    3. If $s = 1$, then

        (a) Compute $Q = Y/(W^{1/a} \cdot X^{1/b})$.

        (b) For each $m \in M$, test if $\mathbf{e}(m, Z) = Q$. If so, output m and halt.

Fig. 1. Encryption Scheme Π

4.2 Re-encryption Functionality

Recall that obfuscation is with respect to a class of circuits. We now define a special class of re-encryption circuits for the encryption scheme Π which can be easily analyzed.

Let (pk_1, sk_1) and (pk_2, sk_2) be two keys pairs which were generated by running KeyGen on independent random tapes. When given an honestly-generated ciphertext encrypted under pk_1, a *re-encryption circuit* decrypts the ciphertext and then re-encrypts the resulting message under a second public key pk_2. For technical reasons, we also require the circuit to produce the pairs of public keys for which it transforms ciphertexts.

More formally, the re-encryption circuit F_{sk_1, pk_2}, when run on input $c_1 = \mathsf{Enc}(pk_1, 0, m)$ for any message $m \in M$, computes $m \leftarrow \mathsf{Dec}(sk_1, c_1)$, then computes $c_2 \leftarrow \mathsf{Enc}(pk_2, 1, m)$, and finally outputs c_2. When F_{sk_1, pk_2} is run on input the special symbol **keys**, it outputs the ordered pair of public keys (pk_1, pk_2). For ciphertexts corresponding to messages in not in M, the circuit returns a randomized ciphertext of the second form for the same message. For other ill-formed inputs, it returns $\bot$.

Furthermore, let C_{sk_1, pk_2} be the same as F_{sk_1, pk_2} with the exception that the values sk_1 and pk_2 can be read from the circuit description. This property is easy to achieve by adding a "message" section to the circuit which does not affect the circuit's output, but encodes a message, with say, AND gates encoding a 1 and OR gates encoding a 0. We now define the family of circuits:

$$C_k = \left\{ C_{sk_1, pk_2} \mid (pk_1, sk_1) \leftarrow \mathsf{KeyGen}(1^k), (pk_2, sk_2) \leftarrow \mathsf{KeyGen}(1^k) \right\}$$

4.3   Security for Re-encryption

We generalize the standard notion of indistinguishability [GM84] for encryption schemes by allowing the adversary to have access to a re-encryption oracle. In particular, the following definition captures the notion that "given a ciphertext y and black-box access to a re-encryption circuit, an adversary does not learn any information about the plaintext corresponding to y."

Definition 3 (IND-security with Oracle C_{sk_1, pk_2}). *Let Π be an encryption scheme and let the random variable $\mathsf{IND}_b(\Pi, \mathcal{A}, k)$ where $b \in \{0, 1\}$, $\mathcal{A} = (\mathcal{A}_1, \mathcal{A}_2)$ and $k \in \mathbb{N}$ denote the result of the following probabilistic experiment:*

$$\mathsf{IND}_b(\Pi, \mathcal{A}, k)$$
$$(pk_1, sk_1) \leftarrow \mathsf{KeyGen}(1^k), \ (pk_2, sk_2) \leftarrow \mathsf{KeyGen}(1^k)$$
$$(m_0, m_1, i, \beta, z) \leftarrow \mathcal{A}_1^{C_{sk_1, pk_2}}(1^k)$$
$$y \leftarrow \mathsf{Enc}(pk_i, \beta, m_b)$$
$$B \leftarrow \mathcal{A}_2^{C_{sk_1, pk_2}}(y, z)$$
$$Output \ B$$

Scheme Π is indistinguishable under a chosen-plaintext attack if $\forall$ p.p.t. algorithms $\mathcal{A}$ the following two ensembles are computationally indistinguishable:

$$\left\{ \mathsf{IND}_0(\Pi, \mathcal{A}, k) \right\}_k \ \overset{c}{\approx} \ \left\{ \mathsf{IND}_1(\Pi, \mathcal{A}, k) \right\}_k$$

Remark 1. For simplicity, we allow the adversary to pick the key pk_i under which the challenge is encrypted and the form β of the encryption. By a standard hybrid argument, the above definition is equivalent to one in which the adversary is given four encryptions of the challenge message—one per key and per form.

Theorem 3. *The encryption scheme Π (in Fig. 1) is an indistinguishable-secure encryption scheme with respect to oracle C_{sk_1, pk_2} under the Decision Linear assumption in $\mathbb{G}$.*

The proof sketch is given in Appendix A.

5   The Obfuscator for Re-encryption

In Fig. 2, we describe an obfuscator Obf for the class of re-encryption circuits C_k relative to the encryption scheme Π defined in the previous section.

5.1   Main Result

Theorem 4. *The obfuscator in Fig. 2 is a secure obfuscator for family C_k.*

Proof sketch. Let $pk_1 = (g^{a_1}, g^{b_1}, g)$ and $pk_2 = (h^{a_2}, h^{b_2}, h)$ with appropriately defined secret keys. Let C denote the re-encryption circuit C_{sk_1, pk_2}, and let $R \leftarrow \mathsf{Obf}(C)$ be an obfuscated version of C.

Algorithm Obf, on input a circuit $C_{sk_1,pk_2} \in C_k$,

1. Reads $sk_1 = (a_1, b_1, g)$ and $pk_2 = (h^{a_2}, h^{b_2}, h)$ from the description of C_{sk_1,pk_2}.
2. Selects a random integer $z \xleftarrow{r} \mathbb{Z}_q^*$ and compute the re-encryption tuple $(Z_1, Z_2, Z_3) = ((h^{a_2})^{z/a_1}, (h^{b_2})^{z/b_1}, h^z)$.
3. Constructs and outputs an obfuscated circuit R_{sk_1,pk_2} that contains the values $pk_1, pk_2, Z_1, Z_2, Z_3$ and does the following:
   - On input **keys**, output $pk_1 = (g^{a_1}, g^{b_1}, g)$ and $pk_2 = (h^{a_2}, h^{b_2}, h)$.
   - On input a 5-tuple $[0, W, X, Y, 0]$ where $W, X, Y \in \mathbb{G}$, then:
     (a) Select input re-randomization values $r, s \xleftarrow{r} \mathbb{Z}_q^*$.
     (b) Re-randomize the input as $W' \leftarrow W \cdot (g^{a_1})^r$, $X' \leftarrow X \cdot (g^{b_1})^s$, and $Y' \leftarrow Y \cdot g^{r+s}$.
     (c) Compute $E \leftarrow \mathbf{e}(W', Z_1)$.
     (d) Compute $F \leftarrow \mathbf{e}(X', Z_2)$.
     (e) Compute $G \leftarrow \mathbf{e}(Y', Z_3)$.
     (f) Select an output re-randomization value $y \xleftarrow{r} \mathbb{Z}_q^*$.
     (g) Output the ciphertext $[1, E^y, F^y, G^y, Z_3^y]$.
   - Otherwise return $\perp$.

Fig. 2. Obfuscator Obf for Re-encryption circuits for Π

Preserving Functionality. Consider any $C \in C_k$ and let circuit $R \leftarrow \mathsf{Obf}(C)$. We claim that the output distributions of C and R are statistically close (in fact, identical). To see this, we must consider three classes of inputs. First, for any message $m \in M$, observe that

$$\mathsf{Enc}(pk_1, 0, m) = [0, \ g^{a_1 r}, \ g^{b_1 s}, g^{r+s} \cdot m, \ 0]$$

for a randomly chosen $r, s \xleftarrow{r} \mathbb{Z}_q^*$. When such a ciphertext is fed as input to R, the circuit outputs

$$[1, \ \mathbf{e}(g^{a_1(r+r')}, h^{a_2 z/a_1})^y, \ \mathbf{e}(g^{b_1(s+s')}, h^{b_2 z/b_1})^y, \ \mathbf{e}(g^{r+s+r'+s'} \cdot m, h^z)^y, \ h^{zy}]$$

for randomly chosen $r', s', y \xleftarrow{r} \mathbb{Z}_q^*$. Substituting $\bar{r} = \frac{r+r'}{\ell}, \bar{s} = \frac{s+s'}{\ell}, \bar{t} = h^{yz}$, and ℓ is such that[3] $g^\ell = h$, this 5-tuple can be re-written as

$$[1, \ \mathbf{e}(h^{a_2 \bar{r}}, \bar{t}), \ \mathbf{e}(h^{b_2 \bar{r}}, \bar{t}), \ \mathbf{e}(h^{\bar{r}+\bar{s}} \cdot m, \bar{t}), \ \bar{t}]$$

which is identically distributed to the output of $\mathsf{Enc}(pk_2, 1, m)$. Second, the same holds for all $m \in \mathbb{G} \backslash M$. Lastly, for **keys** and junk input, the outputs are identical.

Polynomial slowdown. This property follows by inspection because the obfuscated circuit computes a few bilinear maps and exponentiations.

---

[3] We do not need to compute ℓ explicitly.

Virtual Blackbox. In order to satisfy the virtual black-box property, it suffices to only consider the "dummy" adversary. Thus, we must construct a simulator $\mathsf{Sim}^C(1^k, z)$ such that for all distinguishers D^C which take as input an obfuscated circuit R and auxiliary input z,

$$\left| \Pr[D^C(\mathsf{Obf}(C), z) = 1] - \Pr[D^C(\mathsf{Sim}^C(1^k, z), z) = 1] \right| < neg(k).$$

Let us define the simulator $\mathsf{Sim}^C(1^k, z)$ as follows:

1. Query the oracle C on keys to get pk_1, pk_2.
2. Sample $Z'_1, Z'_2, Z'_3 \xleftarrow{r} \mathbb{G}$.
3. As in Step (3) of the Obf algorithm, create and output a circuit R' using the values $(pk_1, pk_2, Z'_1, Z'_2, Z'_3)$.

Notice that Sim^C produces a circuit which does not correctly compute the re-encryption function. However, we now show that under appropriate complexity assumptions, no p.p.t. distinguisher D^C will notice.

Towards this goal, notice that the output of $D^C(\mathsf{Obf}(C), z)$ is distributed identically to $\mathsf{Nice}(D^C, k, z)$ and the output of $D^C(\mathsf{Sim}^C(1^k, z))$ is distributed identically to $\mathsf{Junk}(D^C, k, z)$ where

$\mathsf{Nice}(D^C, k, z)$

    $q, \mathbb{G} \leftarrow \mathsf{BMsetup}(1^k)$

    $g, h, r \xleftarrow{r} \mathbb{G}$

    $a_1, a_2, b_1, b_2 \xleftarrow{r} \mathbb{Z}_q$

    $pk_1 \leftarrow (g^{a_1}, g^{b_1}, g)$

    $pk_2 \leftarrow (h^{a_2}, h^{b_2}, h)$

    $Z_1 \leftarrow r^{a_2/a_1}; \; Z_2 \leftarrow r^{b_2/b_1}$

    $b \leftarrow D^C(pk_1, pk_2, Z_1, Z_2, r, z)$

    Output b

$\mathsf{Junk}(D^C, k, z)$

    $q, \mathbb{G} \leftarrow \mathsf{BMsetup}(1^k)$

    $g, h, r \xleftarrow{r} \mathbb{G}$

    $a_1, a_2, b_1, b_2 \xleftarrow{r} \mathbb{Z}_q$

    $pk_1 \leftarrow (g^{a_1}, g^{b_1}, g)$

    $pk_2 \leftarrow (h^{a_2}, h^{b_2}, h)$

    $Z'_1, Z'_2 \xleftarrow{r} \mathbb{G}$

    $b \leftarrow D^C(pk_1, pk_2, Z'_1, Z'_2, r, z)$

    Output b

In the above experiments, the oracle C represents the re-encryption oracle for the public keys pk_1 to pk_2 which are chosen in the experiment. There is a slight abuse of notation here; when we write $\mathsf{expt}(D^C, k, z)$ we mean that the distinguisher D has oracle access to C_{sk_1, pk_2} for the keys sk_1, pk_2 chosen in the experiment. The virtual blackbox property follows immediately from the following lemma. $\square$

Lemma 1. *Under the SDHI and Decision Linear assumptions, for all p.p.t. distinguishers D and auxiliary information z, the following two distributions are statistically close.*

$$\left\{ \mathsf{Nice}(D^C, k, z) \right\}_k \quad and \quad \left\{ \mathsf{Junk}(D^C, k, z) \right\}_k$$

Proof Outline. We prove this lemma in a series of incremental steps. We begin with a simple indistinguishability problem and incrementally add elements and provide access to various oracles until the experiments are equivalent to their final form in Lemma 1. Let us now start with a claim which relates the SDHI problem to a simple indistinguishability problem: (In all of the following experiments, we implicitly generate q, $\mathbb{G} \leftarrow$ BMsetup(1^k) and each experiment is indexed by k and z although we omit this extra notation when the context is clear.)

Proposition 1. *Under the SDHI assumption,* $\text{Nice}_{k,z}^{(1)} \overset{c}{\approx} \text{Junk}_{k,z}^{(1)}$ *where*

   $\text{Nice}^{(1)}$: *Proceeds as* Nice *except that the output is* $(g^{a_1}, g, h^{a_2}, h, Z_1, r, z)$.
   $\text{Junk}^{(1)}$: *Proceeds as* Junk *except that the output is* $(g^{a_1}, g, h^{a_2}, h, Z_1', r, z)$.

If there exists a distinguisher D which distinguishes $\text{Nice}^{(1)}$ *from* $\text{Junk}^{(1)}$ *with advantage ε, then there exists an distinguisher D' which solves the SDHI problem with the same advantage (in roughly the same time).*

Proof sketch. The algorithm $D'(g, g^a, g^b, g^c, Q, z)$ works as follows:

1. D' chooses a random $w \overset{r}{\leftarrow} \mathbb{Z}_q$.
2. D' runs $D(g^w, (g^b)^w, g^a, g, Q, g^c, z)$ and echoes the response.

   Consider $a_1 = 1/b$, $a_2 = a$ and $r = g^c$. Thus, if $Q = g^{abc}$, then we have $Q = r^{ab} = r^{a_2/a_1}$ in which case the input to D is identically distributed to $\text{Nice}^{(1)}$. Otherwise, Q is equal to r^t for some random t and the input to D is identically distributed to $\text{Junk}^{(1)}$.   □

We now extend Proposition 1 to include more input values.

Proposition 2. *Under the SDHI assumption,* $\text{Nice}_{k,z}^{(2)} \overset{c}{\approx} \text{Junk}_{k,z}^{(2)}$ *where*

   $\text{Nice}^{(2)}$: *Same as* Nice *except that the output is* $(pk_1, pk_2, Z_1, Z_2, r, z)$.
   $\text{Junk}^{(2)}$: *Same as* Junk *except that the output is* $(pk_1, pk_2, Z_1', Z_2', r, z)$.

Proof sketch. Consider the hybrid distribution $T^{(2)}$ which is the same as $\text{Nice}^{(2)}$ except that $Z_2' \overset{r}{\leftarrow} \mathbb{G}$ and the output is $(pk_1, pk_2, Z_1, Z_2', r, z)$. If $\text{Nice}^{(2)}$ and $\text{Junk}^{(2)}$ are distinguishable with advantage ε, then either $\text{Nice}^{(2)}$ and $T^{(2)}$ or $T^{(2)}$ and $\text{Junk}^{(2)}$ are distinguishable by algorithm D with advantage $\varepsilon/2$. Either case implies a distinguisher for $\text{Nice}^{(1)}$ from $\text{Junk}^{(1)}$. In the later case, this involves picking $b_1, b_2 \in \mathbb{Z}_q$ to form public keys, picking Z_2' randomly, and using the input instance from $\text{Nice}^{(1)}$ (or $\text{Junk}^{(1)}$) to simulate the input distribution for D. The former case does the same, but swaps the role of a_i and b_i.   □

Towards the proof of our main theorem, we now extend Prop. 2 by providing the distinguisher with an oracle which returns a five-tuple of random values which works as follows. On input $[0, W, X, Y, 0]$, where $W, X, Y \in \mathbb{G}$, $\mathcal{R}$ selects three random values $E, F, G \overset{r}{\leftarrow} \mathbb{G}_T$ and a random value $H \overset{r}{\leftarrow} \mathbb{G}$ and returns $[1, E, F, G, H]$. Otherwise, $\mathcal{R}$ returns $\bot$. Intuitively, oracle $\mathcal{R}$ outputs only random values and thus should not help any distinguisher.

Proposition 3. *Under the SDHI assumption,* $\text{Nice}_{k,z}^{(3)} \overset{s}{\approx} \text{Junk}_{k,z}^{(3)}$ *where*

   $\text{Nice}^{(3)}$: *Same as* Nice($D^{\mathcal{R}}, k, z$).
   $\text{Junk}^{(3)}$: *Same as* Junk($D^{\mathcal{R}}, k, z$).
   (That is, the distinguishers have oracle access to $\mathcal{R}$ instead of C and unlike the $^{(2)}$-experiments which output a tuple, these experiments output a bit.)

Proof sketch. The oracle $\mathcal{R}$ can be perfectly simulated without any auxiliary information. Thus, for any D^R, there exists another non-oracle distinguisher D' (which internally runs D while perfectly simulating $\mathcal{R}$ to D) whose output distribution is identical to D. Proposition 2 therefore implies that for all distinguishers D^R, $\mathsf{Nice}^{(2)} \overset{c}{\approx} \mathsf{Junk}^{(2)}$ which implies $\mathsf{Nice}^{(3)} \overset{s}{\approx} \mathsf{Junk}^{(3)}$ (since the later experiment outputs a bit) □

We now return to the first experiments in which the distinguisher has oracle access to the re-encryption circuit C.

Proposition 4. *For any p.p.t. distinguisher D, let*

$$\alpha(k, z) = \mathsf{Adv}\left(\mathsf{Nice}(D^C, k, z), \ \mathsf{Junk}(D^C, k, z)\right)$$
$$\beta(k, z) = \mathsf{Adv}\left(\mathsf{Nice}(D^R, k, z), \ \mathsf{Junk}(D^R, k, z)\right)$$

be the advantage[4] that D has in distinguishing Nice from Junk given either a re-encrypting oracle C or a random oracle $\mathcal{R}$ respectively. There exists a p.p.t. algorithm $\mathcal{A}$ which decides the Decision Linear problem with probability at least $\frac{1}{2} + \frac{1}{4}(\alpha(k, z) - \beta(k, z))$.

Proof. Without loss of generality, assume that $\alpha > \beta$. (If not, then we flip the way $\mathcal{A}$ guesses in its final step.) The algorithm $\mathcal{A}$ takes as input, a Decision Linear instance $\Gamma = (h_1, h_2, h, h_1^x, h_2^y, Q)$ and auxiliary information z, and:

1. $\mathcal{A}$ samples a challenge bit $c \overset{r}{\leftarrow} \{0, 1\}$ to pick whether to run Nice or Junk.
2. $\mathcal{A}$ samples integers $a, b, u \overset{r}{\leftarrow} \mathbb{Z}_q$ and group elements $g, Z_1', Z_2', Z_3' \overset{r}{\leftarrow} \mathbb{G}$.
3. $\mathcal{A}$ sets $pk_1 = (g^a, g^b, g)$ and $pk_2 = (h_1, h_2, h)$ and computes a valid re-encryption tuple (Z_1, Z_2, Z_3) by $Z_1 \leftarrow h_1^{u/a}$, $Z_2 \leftarrow h_2^{u/b}$, and $Z_3 \leftarrow h^u$.
4. If $c = 1$, then $\mathcal{A}$ runs $D^{\mathcal{O}}(pk_1, pk_2, Z_1, Z_2, Z_3, z)$ where $\mathcal{O}$ is defined below. If $c = 0$, then $\mathcal{A}$ runs $D^{\mathcal{O}}(pk_1, pk_2, Z_1', Z_2', Z_3', z)$.
   When D queries the oracle $\mathcal{O}$ on input $[0, W, X, Y, 0]$, $\mathcal{A}$ responds as follows:
   (a) Sample input re-randomization values $r, s, t \overset{r}{\leftarrow} \mathbb{Z}_q$.
   (b) Re-randomize the input as $W' \leftarrow W \cdot g^{ar}$, $X' \leftarrow X \cdot g^{bs}$, and $Y' \leftarrow Y \cdot g^{r+s}$.
   (c) Compute $E \leftarrow \mathbf{e}(W', Z_1) \cdot \mathbf{e}(g, h_1^{tx})$.
   (d) Compute $F \leftarrow \mathbf{e}(X', Z_2) \cdot \mathbf{e}(g, h_2^{ty})$.
   (e) Compute $G \leftarrow \mathbf{e}(Y', Z_3) \cdot \mathbf{e}(g, Q^t)$.
   (f) Sample output re-randomization value $\ell \overset{r}{\leftarrow} \mathbb{Z}_q$.
   (g) Respond with the ciphertext $[1, E^\ell, F^\ell, G^\ell, Z_3^\ell]$.

   Whenever D queries its oracle on input keys, $\mathcal{A}$ responds with pk_1 and pk_2, and on all other queries, $\mathcal{A}$ responds with $\perp$.
5. Eventually D outputs $c' \in \{0, 1\}$. If $c = c'$, $\mathcal{A}$ outputs 1 (i.e., it guesses that $Q = h^{x+y}$). Else if $c \neq c'$, then $\mathcal{A}$ outputs 0 (i.e., it guesses that $Q \neq h^{x+y}$).

---

[4] By advantage, we mean the following. Suppose D_0, D_1 are two probability distributions. Then for any adversary $\mathcal{A}$, the *advantage* in distinguishing D_0 from D_1 is defined as: $\mathsf{Adv}_{\mathcal{A}}(D_0, D_1) = |\Pr[x_0 \overset{r}{\leftarrow} D_0 : \mathcal{A}(x_0) = 1] - \Pr[x_1 \overset{r}{\leftarrow} D_1 : \mathcal{A}(x_1) = 1]|$.

Note that $\mathcal{A}$ almost mimics the real obfuscated program. The difference is that when computing (4c)-(4e), additional terms are multiplied in to the ciphertext. When the Γ instance is a decision linear tuple, then these operations simply contribute to additional re-randomization of the ciphertext (this does not change the ciphertext distribution). However, if Γ is not a decision linear instance, then these operations make E, F, G a random 3-tuple that is also independent of Z_3. This proof step is essential.

Claim: If Γ is a decision linear instance, then $\Pr[\mathcal{A}(\Gamma) = 1] = \frac{1}{2} + \alpha(k, z)/2$.
Proof of Claim: When $Q = h^{x+y}$, then $\mathcal{A}$ perfectly simulates Nice^C or Junk^C towards the algorithm D. The key point is to recognize that (h_1, h_2, h) can be interpreted as a randomly generated public key since h_1, h_2 can be rewritten as $h_1 = h^{e_1}$ and $h_2 = h^{e_2}$ for some (unknown) e_1, e_2. Since the re-encryption tuple Z_1, Z_2, Z_3 is also a valid re-encryption tuple for $pk_1 \to pk_2$, the input parameters to D in step 4 are identically distributed to the inputs to D in either experiment Nice or Junk. Moreover, the response to an oracle query on **keys** is also identically distributed. All that remains is to show that the responses $\mathcal{A}$ provides to oracle queries on $[0, W, X, Y, 0]$ are also identically distributed. This last point follows by inspection because $Q = h^{x+y}$ and Z_1, Z_2, Z_3 are a valid re-encryption tuple. A simple probability analysis completes the result:

$$\Pr[\mathcal{A}(\Gamma) = 1 | \Gamma \in DL] = \frac{1}{2} \left(\Pr[\mathsf{Nice}(D^C) = 1] + \Pr[\mathsf{Junk}(D^C) = 0] \right)$$

$$= \frac{1}{2} \left(\Pr[\mathsf{Nice}(D^C) = 1] + 1 - \Pr[\mathsf{Junk}(D^C) = 1] \right)$$

$$= \frac{1}{2} + \frac{\mathsf{Adv}(\mathsf{Nice}(D^C), \ \mathsf{Junk}(D^C))}{2}$$

$$= \frac{1}{2} + \frac{\alpha}{2}$$

Claim: If Γ is not a decision linear instance, then $\Pr[\mathcal{A}(\Gamma) = 1] = \frac{1}{2} + \beta(k, z)/2$.
Proof of Claim: This proof is almost identical to the previous one. The only difference is we must show that responses to the oracle queries return four randomly selected group elements. Let us denote by ω, χ, γ, v the values such that $W = g^\omega, X = g^\chi, Y = g^\gamma$ and $Q = h^v$, and by e_1, e_2 the values such that $h_1 = h^{e_1}$ and $h_2 = h^{e_2}$. Observe that the elements returned by the oracle are

$$E = [\mathbf{e}(W \cdot g^{ar}, h_1^{u/a}) \cdot \mathbf{e}(g, h_1^{tx})]^\ell = \mathbf{e}(g, h)^{\ell e_1 [\omega u/a + tx] + rz\ell e_1}$$

$$F = [\mathbf{e}(X \cdot g^{bs}, h_2^{u/b}) \cdot \mathbf{e}(g, h_2^{ty})]^\ell = \mathbf{e}(g, h)^{\ell e_2 [\chi u/b + ty] + sz\ell e_2}$$

$$G = [\mathbf{e}(Y \cdot g^{r+s}, h^u) \cdot \mathbf{e}(g, Q^t)]^\ell = \mathbf{e}(g, h)^{\ell[(\gamma + r + s)u] + tv\ell}$$

$$H = h^{u\ell}$$

Since r, s, t, ℓ are fresh independently selected values, then E, F, G, H will also be independent on every invocation of the oracle.

Proof of Lemma 1. By the decision linear assumption and Prop. 4, it follows that $|\alpha(k, z) - \beta(k, z)|$ is negligible. By Prop. 3, $\beta(k, z)$ must be a negligible function, and therefore, so too must $\alpha(k, z)$. This establishes the lemma.

Acknowledgments

We would like to thank Ran Canetti for suggesting the problem and the TCC 2007 anonymous reviewers for their helpful comments.

References

[AFGH06]   Giuseppe Ateniese, Kevin Fu, Matthew Green, and Susan Hohenberger. Improved Proxy Re-encryption Schemes with Applications to Secure Distributed Storage.   *ACM Trans. on Information and System Security*, 9(1):1–30, February 2006. Previously, in *NDSS*, pages 29-43, 2005.

[AW05]   Ben Adida and Douglas Wikström. How to shuffle in public. Cryptology ePrint Archive, Report 2005/394, 2005. http://eprint.iacr.org/.

[BBS98]   Matt Blaze, Gerrit Bleumer, and Martin Strauss. Divertible protocols and atomic proxy cryptography. In *EUROCRYPT '98*, volume 1403 of LNCS, pages 127–144, 1998.

[BBS04]   Dan Boneh, Xavier Boyen, and Hovav Shacham. Short Group Signatures Using Strong Diffie Hellman.   In *CRYPTO*, volume 3152, pages 41–55, 2004.

[BGI+01]   Boaz Barak, Oded Goldreich, Russell Impagliazzo, Steven Rudich, Amit Sahai, Salil P. Vadhan, and Ke Yang. On the (im)possibility of obfuscating programs. In *CRYPTO '01*, volume 2139 of LNCS, pages 1–18, 2001.

[BS97]   Matt Blaze and Martin Strauss. Atomic proxy cryptography. Technical report, AT&T Research, 1997.

[BSW06]   Dan Boneh, Amit Sahai, and Brent Waters. Fully collusion resistant traitor tracing with short ciphertexts and private keys. In *EUROCRYPT '06*, volume 4004 of LNCS, pages 573–592, 2006.

[Can97]   Ran Canetti. Towards realizing random oracles: Hash functions that hide all partial information. In *CRYPTO*, volume 1294, pages 455–469, 1997.

[CMR98]   Ran Canetti, Daniele Micciancio, and Omer Reingold. Perfectly one-way probabilistic hash functions (preliminary version). In *STOC*, pages 131–140, 1998.

[DI03]   Yevgeniy Dodis and Anca Ivan. Proxy cryptography revisited. In *NDSS*, 2003.

[DS05]   Yevgeniy Dodis and Adam Smith. Correcting errors without leaking partial information. In *STOC '05*, pages 654–663, 2005.

[GK05]   Shafi Goldwasser and Yael Tauman Kalai. On the impossibility of obfuscation with auxiliary input. In *FOCS '05*, pages 553–562, 2005.

[GM84]   Shafi Goldwasser and Silvio Micali. Probabilistic encryption. *Journal of Computer and System Sciences*, 28(2):270–299, 1984. Previously, in *STOC*, pages 365-377, 1982.

[GPS06]   Steven D. Galbraith, Kenneth G. Paterson, and Nigel P. Smart. Pairings for cryptographers, 2006. Cryptology ePrint Archive: Report 2006/165.

[Had00]   Satoshi Hada. Zero-knowledge and code obfuscation. In *ASIACRYPT '00*, volume 1976 of LNCS, pages 443–457, 2000.

[MO97]   Masahiro Mambo and Eiji Okamoto. Proxy Cryptosystems: Delegation of the Power to Decrypt Ciphertexts. *IEICE Trans. Fund. Electronics Communications and Computer Science*, E80-A/1:54–63, 1997.

[OS05]   Rafail Ostrovsky and William E. Skeith III. Private searching on streaming data. In *CRYPTO*, volume 3621 of LNCS, pages 223–240, 2005.

[Pas06]   Rafael Pass, 2006. Personal Communication.

[Smi05]   Tony Smith. DVD Jon: buy DRM-less Tracks from Apple iTunes, March 18, 2005. http://www.theregister.co.uk/2005/03/18/itunes_pymusique.

[Wee05]   Hoeteck Wee. On obfuscating point functions. In *STOC*, pages 523–532, 2005.

A   Proof of Security for Encryption Scheme

Proof sketch.[of Thm. 3] Let us first argue that Π is an encryption scheme, i.e., it is perfectly complete. When $\beta = 0$, this follows from the BBS scheme. For the second form of ciphertexts, on input $[1, E, F, G, H]$, the decryption algorithm first computes $Q = \frac{G}{E^{1/a_2} \cdot F^{1/b_2}}$, which by inspection is equal to $\mathbf{e}(m, H)$. The decryption algorithm loops over each (of the polynomially many) $m_i \in M$ and tests whether $\mathbf{e}(m_i, H) = Q$ and therefore eventually recovers m as required.

To argue that the scheme meets the security definition, suppose adversary $\mathcal{A} = (\mathcal{A}_1, \mathcal{A}_2)$ and distinguisher D has advantage ε in distinguishing $\mathsf{IND}_0(\cdots)$ from $\mathsf{IND}_1(\cdots)$. Then, we construct an adversary $\mathcal{A}'$ that decides the Decision Linear problem with advantage $\varepsilon/4$ as follows. Let $\Gamma = (h_1, h_2, h, h_1^x, h_2^y, Q)$ be a DL instance; $\mathcal{A}'$ works as follows:

1. Sample $a, b, c \xleftarrow{r} \mathbb{Z}_q$.
2. Set $pk_1 = (h_1, h_2, h)$ and $pk_2 = (h_1^{ac}, h_2^{bc}, h^c)$.
3. Run $\mathcal{A}_1^{\mathcal{O}}(1^k)$ to produce a tuple (m_0, m_1, i, β, z).
   When $\mathcal{A}$ queries $[s, W, X, Y, Z]$ to its oracle, respond as follows:
   (a) Return $\perp$ if $s \neq 0$ or $Z \neq 0$, or if $W, X, Y \notin \mathbb{G}$, etc.
   (b) Sample $r \in \mathbb{G}$ and use the valid re-encryption program $Z_1 \leftarrow r^a, Z_2 \leftarrow r^b$ and $Z_3 \leftarrow r$ to compute a response.
4. Sample a bit $t \xleftarrow{r} \{0, 1\}$.
5. Set y to be the ciphertext $[0, h_1^x, h_2^y, Q \cdot m_t, 0]$ if $i = 0$ and $[0, (h_1^x)^{ac}, (h_2^y)^{bc}, Q^c \cdot m_t, 0]$ if $i = 1$. Furthermore, if $\beta = 1$, transform y into a second-form cipher-text (this can be done with public information).
6. Run $B \leftarrow \mathcal{A}_2^{\mathcal{O}}(y, z)$
7. Run $t' \leftarrow D(B)$ and output 1 if $t' = t$ (guess that Γ is a DLA instance) and otherwise output 0.

We argue that when Γ is a DL instance, $\mathcal{A}'$ perfectly simulates the experiment IND_t. When Γ is not an instance, then the encryption y is independent of the message m_t and so the probability that $t' = t$ is exactly $1/2$. The proof of the theorem follows by standard probability manipulation of these two facts. $\square$

Weakly-Private Secret Sharing Schemes*

Amos Beimel[1] and Matthew Franklin[2]

[1] Department of Computer Science, Ben-Gurion University
[2] Department of Computer Science, University of California, Davis

Abstract. Secret-sharing schemes are an important tool in cryptography that is used in the construction of many secure protocols. However, the shares' size in the best known secret-sharing schemes realizing general access structures is exponential in the number of parties in the access structure, making them impractical. On the other hand, the best lower bound known for sharing of an ℓ-bit secret with respect to an access structure with n parties is $\Omega(\ell n / \log n)$ (Csirmaz, EUROCRYPT 94). No major progress on closing this gap has been obtained in the last decade.

Faced by our lack of understanding of the share complexity of secret sharing schemes, we investigate a weaker notion of privacy in secrets sharing schemes where each unauthorized set can never rule out any secret (rather than not learn any "probabilistic" information on the secret). Such schemes were used previously to prove lower bounds on the shares' size of perfect secret-sharing schemes. Our main results is somewhat surprising upper-bounds on the shares' size in weakly-private schemes.

- For every access structure, we construct a scheme for sharing an ℓ-bit secret with $(\ell + c)$-bit shares, where c is a constant depending on the access structure (alas, c can be exponential in n). Thus, our schemes become more efficient as ℓ – the secret size – grows. For example, for the above mentioned access structure of Csirmaz, we construct a scheme with shares' size $\ell + n \log n$.
- We construct efficient weakly-private schemes for threshold access structures for sharing a one bit secret. Most impressively, for the 2-out-of-n threshold access structure, we construct a scheme with 2-bit shares (compared to $\Omega(\log n)$ in any perfect secret sharing scheme).

1 Introduction

Secret-sharing schemes are a tool used in many cryptographic protocols. A secret-sharing scheme involves a dealer who has a secret, a finite set of n participants, and a collection $\mathcal{A}$ of subsets of the set of participants called the access structure. A perfect secret-sharing scheme for $\mathcal{A}$ is a method by which the dealer distributes shares to the parties such that: (1) any subset in $\mathcal{A}$ can reconstruct the secret from its shares, and (2) any subset not in $\mathcal{A}$ can never reveal any partial information on the secret (in the information theoretic sense). Secret-sharing schemes were first introduced by Blakley [10] and Shamir [44] for the threshold case, that

---

* The work of the first author was done while on sabbatical at the University of California, Davis, partially supported by the Packard Foundation. The second author is partially supported by the NSF and the Packard Foundation.

is, for the case where the subsets that can reconstruct the secret are all the sets whose cardinality is at least a certain threshold. Secret-sharing schemes for general access structures were introduced by Ito, Saito, and Nishizeki [28]. More efficient schemes were presented in, e.g., [9,45,15,30,46,25]. Originally motivated by the problem of secure information storage, secret-sharing schemes have found numerous other applications in cryptography and distributed computing, e.g., Byzantine agreement [42], secure multiparty computations [8,18,19], threshold cryptography [23], access control [40], and attribute based encryption [26].

A major problem with secret-sharing schemes is that the shares' size in the best known secret-sharing schemes realizing general access structures is exponential in the number of parties in the access structure (e.g., in the schemes based on monotone span programs [30] presented in 1993). Thus, the known constructions for general access structures are impractical. This is true even for explicit access structures (e.g., access structures whose characteristic function can be computed by a small uniform circuit). On the other hand, the best known lower bounds on the shares' size for sharing a secret with respect to an access structure (e.g., in [31,9,17,12,24,20,21,11,41]) are far from the above upper bounds. The best lower bound was proved by Csirmaz [20] in 1994, proving that, for every n, there is an access structure with n parties such that sharing an ℓ-bit secrets requires shares of length $\Omega(\ell n / \log n)$. The question if there exist more efficient schemes, or if there exists an access structure that does not have (space) efficient schemes remains open. The following widely believed conjecture was made by the first author in 1996 [3]:

Conjecture 1. There exists an $\epsilon > 0$ such that for every positive integer n there is an access structure with n parties, for which every secret sharing scheme distributes shares of length exponential in the number of parties n, that is, $2^{\epsilon n}$.

Proving (or disproving) this conjecture is one of the most important open questions concerning secret sharing. No major progress on proving or disproving this conjecture has been obtained in the last decade.

Faced by our lack of understanding of the share complexity of secret sharing schemes, we investigate a weaker notion of privacy of secrets sharing schemes where each unauthorized set can never rule out any secret (rather than not learn any "probabilistic" information on the secret). Our belief is that studying these schemes will shed light on perfect secret-sharing schemes and the techniques needed to prove lower bounds and upper bounds for them. Our main results is somewhat surprising upper-bounds on the shares' size in weakly-private secret-sharing schemes.

Weakly-private scheme were studied implicitly and explicitly in previous papers. They were first studied in [16], where it is proved that ideal weakly-private secret-sharing schemes are perfect (a scheme is ideal if the domain of shares of each party is the same as the domain of shares). Thus, relaxing the privacy requirement does not help for ideal schemes. The relation between perfect secret sharing and weakly-private secret sharing was further discussed in [29]. Lower bounds for secret-sharing schemes were proved in [35] using combinatorial arguments; their results actually apply to weakly-private schemes. In particular, they show that the size of the share of each (non-redundant) party in a weakly-private scheme is at least the size of the secret (such result was proved for perfect schemes in [31]). Weakly-private secret-sharing schemes were used in [43,7] to prove lower bounds on the shares' size of perfect secret-sharing schemes of a certain (matroidial) access structure.

Our main motivation studying weakly-private secret-sharing schemes is to understand what makes them hard (if they are hard). The strongest lower bounds for secret-sharing schemes [17,12,24,20,21] consider the shares as random variables and use entropy arguments to prove the lower bounds. In particular, the proofs rely on the perfectness (or near perfectness) of the schemes. We raise the question if this requirement is essential for proving lower bounds for secret-sharing schemes. This can help in understanding what techniques can be used to prove such lower bounds. While more direct combinatorial methods used to prove lower bounds for weakly-private secret-sharing schemes (e.g., in [43,35,7]) are more intuitive, they might not be strong enough to prove super-polynomial lower-bounds.

To understand this question, let us consider two additional cryptographic protocols. Blundo et al. [13] proved a lower bound on the size of the shares in perfectly private key distribution schemes using entropy arguments. Beimel and Chor [5] showed that the same lower bound holds even for weakly-private key distribution schemes. A similar phenomenon is true for 2-party secure computation in the honest-but-curious model. Kushilevitz [36] characterizes the functions that can be computed privately in this model; in particular, a function can be computed in the honest-but-curious 2-party model with weak privacy if and only if it can be computed with perfect privacy.[1] As we have seen that weak privacy suffices for proving lower bounds and impossibility results for some cryptographic tasks, it is natural to ask if this is the case for secret-sharing schemes.

1.1   Our Results

Our main results in this paper are somewhat surprising upper-bounds on the shares' size in weakly-private secret-sharing schemes. In addition we prove some lowers bounds.

A generic construction of weakly-private schemes. For every access structure, we construct a scheme for sharing an ℓ-bit secret with $(\ell + c)$-bit shares, where c is a constant depending on the access structure (alas, c can be exponential in n – the number of the parties in the access structure). For comparison, in the best known constructions of perfect secret-sharing schemes realizing an arbitrary access structure, the size of the shares is $\ell c'$, where c' is a constant (which can also be exponential in n).

Let us consider a few examples. Capocelli et al. [17] proved that there is an access structures with 4 parties such that in every perfect secret-sharing scheme realizing it with ℓ-bit secrets, the shares of at least one party is at least 1.5ℓ-bit strings. In contrast, we show how to realize this access structures by a weakly-private scheme with $(\ell + 2)$-bit shares. Csirmaz [20] proved that for every $n \in \mathbb{N}$

---

[1] The notion of weak privacy in [36] is different than ours; however, the impossibility result for our definition of weak privacy follows from the proof in that paper. See treatment in [4].

there is an access structures $\mathcal{A}_n$ with n parties such that in every perfect secret-sharing scheme realizing $\mathcal{A}_n$ with ℓ-bit secrets, the shares of at least one party are $\Omega((n/\log n)\ell)$-bit strings. In contrast, we show how to realize this access structures by a weakly-private scheme in which the shares are $(\ell + n \log n)$-bit strings. In particular, if we take $\ell = n \log n$, then in any perfect scheme the shares are $\Omega(\ell^2/\log^2 \ell)$-bit strings, while in the weakly-private schemes we construct the shares are 2ℓ-bit strings.

As discussed above, one of the motivations for weakly-private secret-sharing schemes is for proving lower bounds on perfect schemes. For example, Kurosawa and Okada [35] have used combinatorial arguments to prove an inferior version of the above mentioned result of [17]. However, their proof applies to weakly-private schemes and our results show that using weakly-private schemes one cannot hope to prove a lower bound of $\ell + \omega(1)$. Beimel and Livne [7] (improving on Seymour [43]) proved lower bounds of the shares' size in a matroidial access structure $\mathcal{M}$ with 7 parties. On one hand, the shares in the best known perfect secret-sharing scheme realizing $\mathcal{M}$ with with ℓ-bit secrets are 1.5ℓ-bit strings [38]. On the other hand, by our result, there is secret-sharing scheme realizing $\mathcal{M}$ with with ℓ-bit secrets and $(\ell + 16)$-bit shares. Thus, if the lower bound for perfect scheme realizing $\mathcal{M}$ can be improved to $\ell + \omega(1)$, then such proof must use the fact that the scheme is perfect (e.g., generalize the combinatorial proof of [7] to use some additional ideas).

In addition, we present a construction, due to Yuval Ishai [27], giving efficient weakly-private secret-sharing schemes for a doubly exponential number of access structures. Specifically, for every $n \in \mathbb{N}$, there are 2^{2^n} access structures with $2n$ parties that have a weakly-private scheme for sharing a 1-bit secret using shares of length $O(n^3)$. This should be contrasted with perfect secret sharing schemes where efficient schemes for sharing a 1-bit secret are known only for exponentially many access structures.

Weakly-private threshold schemes for sharing one bit. The most important secret-sharing schemes are threshold secret-sharing schemes. Shamir [44] shows that there are very efficient perfect t-out-of-n secret-sharing schemes for sharing ℓ-bit secrets when $\ell \geq \log n$, namely the shares are ℓ-bit strings as well. However, the best known perfect t-out-of-n schemes for sharing a 1-bit secret (when $2 \leq t \leq n-1$) use $\log n$-bit shares (e.g., in Shamir's scheme).[2] Kilian and Nisan [32] proved that this is unavoidable when $t \leq \alpha n$ for some constant $\alpha < 1$; they prove that the shares are at least $\log(n - t + 2)$-bit strings.

In contrast, we construct efficient weakly-private schemes for threshold access structures for sharing a 1-bit secret. Our most efficient construction is a simple weakly-private 2-out-of-n secret-sharing scheme with 1-bit secrets and 2-bit shares. For larger values of t, we construct weakly-private t-out-of-n schemes for sharing 1-bit secrets with $O(t)$-bit shares. In particular, our scheme improves the share size when $t \leq \log n - 2 \log \log n$. These schemes have the additional nice property that they are anonymous, that is, the reconstruction of the secret does not depend on the identity of the authorized set. Anonymous secret-sharing schemes were introduced by [47], and were further studied in [14,33,39].

---

[2] For $t = 2$ and $t = n - 1$, we can use the formula-based scheme of [9].

We present an additional construction of weakly-private threshold scheme that is efficient for big thresholds. When n is a prime-power and $n > t/2$, we construct a weakly-private t-out-of-n scheme that is better than the *known* perfect schemes, that is, our scheme uses a domain of shares of size $n - 1$ when $t \approx n/2$ and a domain of size $3n/4$ when $t = n - 1$ (the size of the domain of shares in the best known perfect secret-sharing scheme is at least n). We remark that the size of the shares in the optimal perfect $(n - 1)$-out-of-n schemes for sharing a 1-bit secret is unknown as the lower bound of [32] for this case on the size of the domain of shares is 3, and the upper bound is n.

Our last result is a lower bound on the size of shares in weakly-private t-out-of-n schemes for sharing a 1-bit secret. We prove that in this case the secrets are taken from a domain of size $\min\left\{t, \Omega(\frac{\log\log(n-t)}{\log\log\log(n-t)})\right\}$. For anonymous weakly-private t-out-of-n schemes for sharing a 1-bit secret we prove a much stronger lower bound of $\min\left\{2t, \sqrt{(n-t)/2}\right\}$. This should be compared to the lower bound of $n - t + 2$ for perfect t-out-of-n schemes for sharing a 1-bit secret.

Are weakly-private schemes suitable for proving lower bounds? Our results suggest that weakly-private schemes are indeed weaker than perfect schemes. The ideas used in constructing our weakly-private schemes guarantee the weak privacy, but they are far from providing perfect privacy or statistical privacy. We conclude that weakly-private secret-sharing schemes are not useful for proving lower bounds for large domains of secrets (e.g., for proving that the information rate of an access structure is bounded from 1). The situation is less clear for secret sharing of a 1-bit secret. In this case the share complexity of weakly-private secret schemes is still open; weakly-private secret-sharing schemes might be useful for proving lower bounds for perfect scheme for sharing a 1-bit secret. The efficient weakly-private schemes for the doubly exponential family and the efficient weakly-private threshold schemes might discourage such belief.

Alternative notions of "weaker" secret sharing. In this work we discuss weakly-private secret-sharing schemes as a relaxation of perfect secret-sharing schemes. Below we mention a few other relaxations of perfect secret-sharing schemes; all these relaxations are incomparable to weakly-private secret-sharing schemes. A notion that is related is statistical secret-sharing schemes, considered in, e.g., [6,22]. In these schemes the privacy and possibly also the correctness are only statistical. Another related notion is computational secret-sharing schemes, considered in [49,34,2,48]. In these schemes, unauthorized sets of parties cannot distinguish in polynomial time between the different secrets.

Organization. In Section 2 we define perfect and weakly-private secret-sharing schemes. In Section 3 we present the construction of the generic weakly-private secret-sharing scheme for arbitrary access structures, and in Section 4 we describe efficient weakly-private secret-sharing schemes for doubly exponential number of access structures. In Section 5 we construct weakly-private threshold schemes for sharing 1-bit secrets, and in Section 6 we prove lower bounds for them.

2   Definitions and Notations

In this section we define perfect secret sharing and weakly-private secret sharing. We start by defining an access structure – the collection of sets that should be able to reconstruct the secret.

Definition 1 (Access Structure). *Let $U = \{P_1, \ldots, P_n\}$ be a set of parties. A collection $\mathcal{A} \subseteq 2^U$ is monotone if $B \in \mathcal{A}$ and $B \subseteq C$ imply that $C \in \mathcal{A}$. An access structure is a monotone collection $\mathcal{A} \subseteq 2^U$ of non-empty subsets of U. Sets in $\mathcal{A}$ are called* authorized, *and sets not in $\mathcal{A}$ are called* unauthorized.

Definition 2 (Perfect Secret-Sharing Schemes). *Let S be a finite set of secrets, where $|S| \geq 2$, and R be a set of random strings. An n-party secret-sharing scheme Π with domain of secrets S is a mapping from $S \times R$ to a set of n-tuples $S_1 \times \cdots \times S_n$, where S_i is called the* share-domain *of P_i. A dealer shares a secret $s \in S$ among the n parties according to Π by first sampling a random string $r \in R$ (according to some given distribution), computing the vector of shares $\Pi(s,r) = \langle s_1, \ldots, s_n \rangle$, and then privately communicating each share s_i to the party P_i. We say that Π realizes an access structure $\mathcal{A} \subseteq 2^U$ if the following two requirements hold:*

CORRECTNESS. *The secret s can be reconstructed by any authorized set of parties. That is, for any set $B \in \mathcal{A}$ (where $B = \{P_{i_1}, \ldots, P_{i_{|B|}}\}$), there exists a* reconstruction function $\text{RECON}_B : S_{i_1} \times \cdots \times S_{i_{|B|}} \to S$ such that $\text{RECON}_B(\Pi_B(s,r)) = s$ for every $s \in S$, every $r \in R$, and every possible value of $\Pi_B(s,r)$, the restriction of $\Pi(s,r)$ to its B-entries.

PRIVACY. *Every unauthorized set can never learn anything about the secret (in the information theoretic sense) from their shares. Formally, for any set $C \notin \mathcal{A}$, for every two secrets $a, b \in S$, and for every possible $|C|$-tuple of shares $\langle s_i \rangle_{P_i \in C}$: $\Pr[\, \Pi_C(a,r) = \langle s_i \rangle_{P_i \in C} \,] = \Pr[\, \Pi_C(b,r) = \langle s_i \rangle_{P_i \in C} \,]$.*

In this work we concentrate on weakly-private secret-sharing schemes, where an unauthorized set can never rule out any secret.

Definition 3 (Weakly-Private Secret-Sharing Schemes). *We say that a secret-sharing scheme Π weakly realizes an access structure $\mathcal{A} \subseteq 2^U$ if it satisfies the correctness requirement of Definition 2 and it satisfies the following weak privacy requirement:*

WEAK PRIVACY. *Every unauthorized set can never rule out any secret from its shares. Formally, for any set $C \notin \mathcal{A}$, for every two secrets $a, b \in S$, and for every possible $|C|$-tuple of shares $\langle s_i \rangle_{P_i \in C}$: $\Pr[\, \Pi_C(a,r) = \langle s_i \rangle_{P_i \in C} \,] > 0$ if and only if $\Pr[\, \Pi_C(b,r) = \langle s_i \rangle_{P_i \in C} \,] > 0$.*

In this work we measure the share complexity of a scheme either as the length of the strings representing the shares or as the size of the domain of shares. The latter is used mainly when we discuss threshold schemes.

Definition 4 (Possible Vectors of Shares). *Let Π be a secret-sharing scheme and A be a set of parties. We say that a vector of shares $\langle s_i \rangle_{P_i \in A}$ is possible with secret a for A in Π if $\Pr[\, \Pi_A(a,r) = \langle s_i \rangle_{P_i \in A} \,] > 0$.*

The most important secret-sharing schemes are threshold schemes, where the authorized sets are all sets whose size is at least some given threshold.

Definition 5 (*t*-out-of-*n* Secret Sharing). *A secret-sharing scheme Π is a t-out-of-n secret-sharing scheme if it realizes the access structure $\mathcal{A}_{t,n} \stackrel{def}{=} \{A \subseteq \{P_1, \ldots, P_n\} : |A| \geq t\}$. We say that a secret-sharing scheme Π is a weakly-private t-out-of-n secret-sharing scheme if it weakly realizes $\mathcal{A}_{t,n}$.*

In the definition of secret-sharing schemes we say that for every set B there is a reconstruction function RECON_B that takes the shares of the parties of B and reconstructs the secret. That is, the reconstruction function can use the identities of the parties of B. For example, in Shamir's scheme the parties in every set B of size t reconstruct the secret by applying a linear function to their shares; the coefficients in this linear function depend on the set B. A scheme is anonymous if the reconstruction is done as a function of the shares without knowing the identities of the parties in B. The following definition, which is equivalent to the definition of [14], captures this intuition by requiring that if a vector of shares is possible given a secret s, then every possible permutation in the order of the coordinates in this vector is possible given s.

Definition 6 (Anonymous *t*-out-of-*n* Secret Sharing). *We say that a perfect or weakly-private t-out-of-n secret-sharing scheme is anonymous if for every $s \in S$, every vector of shares $\langle s_1, s_2, \ldots, s_n \rangle$, and every permutation $\pi : \{1, \ldots, n\} \rightarrow \{1, \ldots, n\}$, the vector $\langle s_1, s_2, \ldots, s_n \rangle$ is possible with secret s for $\{P_1, \ldots, P_n\}$ iff the vector $\langle s_{\pi(1)}, s_{\pi(2)}, \ldots, s_{\pi(n)} \rangle$ is possible with secret s for $\{P_1, \ldots, P_n\}$.*

Notation. For a set Σ, let $\binom{\Sigma}{<t}$ be the collections of subsets of Σ of size less than t and let $\binom{\Sigma}{t}$ be the collections of subsets of Σ of size exactly t. For an integer $n \in \mathbb{N}$, let $[n] \stackrel{def}{=} \{1, \ldots, n\}$.

3   A Generic Construction of Weakly-Private Secret-Sharing Schemes

In this section we show that weakly-private schemes can be more efficient than perfect schemes. We construct for every access structure $\mathcal{A}$ a weakly-private secret-sharing scheme realizing $\mathcal{A}$ with shares whose size is linear in the size of the domain of secrets (but possibly exponential in the number of parties).

Theorem 1. *For every access structure $\mathcal{A}$ with n parties there is some constant c such that for every $\ell \in \mathbb{N}$ there exists a weakly-private secret-sharing scheme realizing $\mathcal{A}$ with ℓ-bit secrets and $(\ell + c)$-bit shares for each party (however, c may be exponential in n).*

The theorem is proven in Lemma 1. For comparison, the size of the shares in the best known constructions of perfect secret-sharing schemes realizing an arbitrary access structure, the size of the shares is $\ell \cdot c'$, where c' is a constant that can also be exponential in n.

Define the following sets of vectors of shares for a secret $s \in \{0,1\}^{\ell}$ are:

{P_1}-**Vectors:** $\langle\langle a,r\rangle, \langle \overline{a},s\rangle\rangle$, for every $r \in \{0,1\}^{\ell}$ and every $a \in \{0,1\}$.
{P_2}-**Vectors:** $\langle\langle a,s\rangle, \langle a,r\rangle\rangle$ for every $r \in \{0,1\}^{\ell}$ and every $a \in \{0,1\}$.

To share a secret s, choose at random $i \in \{1,2\}$ and choose a random vector from the {P_i}-vectors.

Fig. 1. The weakly-private scheme realizing Γ

A Warmup. Let Γ be the access structure with two participants P_1 and P_2 and one authorized set $\{P_1, P_2\}$. As a warm up, we describe a weakly-private scheme realizing Γ. The scheme we describe is inferior to the best perfect scheme realizing Γ. The main purpose of describing this scheme is to introduce the ideas of the general scheme.

In the scheme we construct, the secret is an ℓ-bit string and the shares are $(\ell + 1)$-bit strings. The scheme is described in Fig. 1. In this scheme, in each vector of shares exactly one party holds the secret and the other party holds a random element. Only both parties together know which party holds the secret, thus they can reconstruct the secret and each individual party can never rule out any secret. The vectors of shares are divided to two sets, {P_1}-vectors and {P_2}-vectors. The {P_i}-vectors, where P_i holds a random element, disable {P_i} from ruling out any secret (where $i \in \{0,1\}$).

We first explain how P_1 and P_2, holding shares $\langle a_1, b_2 \rangle$ and $\langle a_2, b_2 \rangle$ respectively (where a_i is a bit and b_i is either the secret or a random string), reconstruct the secret: If $a_1 = a_2$, then the secret is b_1, otherwise the secret is b_2. To argue that the scheme is weakly private, note that for every secret $s \in \{0,1\}^{\ell}$, Party P_i can get every share in $\{0,1\} \times \{0,1\}^{\ell}$ in the {P_i}-vectors.

The construction of the weakly-private schemes. Let $\mathcal{A}$ be an access structure with n parties and $\ell \in \mathbb{N}$. We first describe a very simple scheme with ℓ-bit secrets and ℓ-bit shares that has useful properties. To share a secret s, there is a set of vectors of shares for every maximal unauthorized set $C \notin \mathcal{A}$, called the C-vectors, which prevent C from ruling out any secret. In the C-vectors, the share of every $P_i \notin C$ is the secret s, while the share of each party in C ranges over all the possible shares in $\{0,1\}^{\ell}$. Thus, the number of C-vectors, for a given secret s, is $2^{\ell|C|}$.

Clearly, weak privacy holds in the above scheme (that is, every unauthorized set can never rule out any secret). We next argue that every authorized set B can reconstruct the secret with probability at least $1/|B| \geq 1/n$ (even when $\ell \gg n$). Let $B \in \mathcal{A}$ be any authorized set holding a vector of shares v. This vector is a sub-vector of a C-vector for some $C \notin \mathcal{A}$. Since B is authorized and C is unauthorized, there must be some $P_i \in B \setminus C$, thus P_i holds s in v. The parties in B, which do not know C, choose $P_i \in B$ at random and output its share as the secret.

In the previous scheme, the authorized set B could not know the set C. However, to reconstruct the secret with certainty, the set B needs to know C

Let $s \in \{0,1\}^\ell$ be the secret.

1. Choose at random a maximal unauthorized set $C \notin \mathcal{A}$.
2. Share the n-bit string representing C using a weakly-private scheme Π_{weak} realizing $\mathcal{A}$. Let $a_1, \ldots, a_n$ be the generated shares.
3. Choose a random $b_i \in \{0,1\}^\ell$ for every $P_i \in C$ and set $b_i = s$ for every $P_i \notin C$.
4. The share of P_i is (a_i, b_i).

Fig. 2. A generic weakly-private scheme Π_{generic} realizing an access structure $\mathcal{A}$

(or at least some $P_i \in B \setminus C$). Thus, we represent C as an n-bit string and share this string using a weakly-private scheme realizing $\mathcal{A}$. That is, we reduced the question of sharing a secret taken from a big domain to sharing a secret from a domain of size 2^n. Such (perfect) schemes, with $2^{O(n)}$-bit shares, exist for every access structure (e.g., [28,9,30]). The formal description of the scheme Π_{generic} appears in Fig. 2. The possible vectors of shares generated in Π_{generic} when the maximal unauthorized set chosen in Step (1) of the scheme is C are called the C-vectors.

Lemma 1. *The generic weakly-private scheme Π_{generic}, described in Fig. 2, weakly realizes the access structure $\mathcal{A}$. Furthermore, if Π_{generic} uses a weakly-private scheme Π_{weak} with n-bit secrets and c-bit shares, then, to share ℓ-bit secrets, Π_{generic} distributes $(\ell + c)$-bit shares.*

Proof. To prove that Π_{generic} weakly realizes $\mathcal{A}$, we prove the correctness and weak privacy of the scheme. To reconstruct the secret, an authorized set B, holding shares $\langle(a_i, b_i)\rangle_{P_i \in B}$, reconstructs the set C from the shares $\langle a_i \rangle_{P_i \in B}$, finds some $P_i \in B \setminus C$, and returns b_i.

To argue that the scheme is weakly private, consider a maximal unauthorized set C holding shares $\langle(a_i, b_i)\rangle_{P_i \in C}$ that are possible with some secret s_0. These shares are possible given any secret s: First, the shares $\langle a_i \rangle_{P_i \in B}$ are possible in Π_{weak} for the set C. Thus, by the definition of the C-vectors, the shares $\langle(a_i, b_i)\rangle_{P_i \in B}$ are a restriction of a C-vector that is possible for the secret s. $\square$

Example 1. Csirmaz [20] proved that for every $n \in \mathbb{N}$ there is an access structures $\mathcal{A}_n$ with n parties such that in every perfect secret-sharing scheme realizing $\mathcal{A}_n$ with ℓ-bit secrets, the shares of at least one party are $\Omega((n/\log n)\ell)$-bit strings. The description of the access structure $\mathcal{A}_n$ is somewhat technical. The only property we need is that in $\mathcal{A}_n$ each party is contained in at most n minimal authorized sets. Thus, by [28], there is a perfect scheme realizing $\mathcal{A}_n$ for sharing n-bit secrets using $O(n^2)$ bit shares. By Lemma 1, there is a scheme weakly realizing $\mathcal{A}_n$ with ℓ-bit secrets and $(\ell + n^2)$-bit shares. If we use Lemma 2 and Lemma 3 (proved in section 3.1), we get a scheme weakly realizing $\mathcal{A}_n$ with ℓ-bit secrets and $(\ell + n \log n)$-bit shares. In particular, if we take $\ell = n \log n$, then in perfect scheme shares are $\Omega(\ell^2/\log^2 \ell)$-bit strings, while in the weakly-private schemes we construct the shares are 2ℓ-bit strings.

3.1   Improvements of the Generic Scheme

In the generic scheme Π_{generic}, presented in Fig. 2, the shares are $(\ell + c)$-bit strings, where c can be large, that is, it is the size of the shares in a scheme Π_{weak} realizing $\mathcal{A}$ with n-bit secrets. In this section we try to reduce the constant c. We observe that in the proof of Lemma 1, the properties required from the secret-sharing scheme Π_{weak} are the following:

- Every authorized set B can compute the identity of a party $P_i \in B \setminus C$, and
- Every unauthorized set C can never rule out that the shared set is C.

Next we formally define schemes satisfying these conditions.

Definition 7 (Weakly-Private Sharing of Unauthorized Sets). *Let S be the set of maximal unauthorized sets in $\mathcal{A}$. We say that a secret-sharing scheme Π with domain of secrets S weakly shares the unauthorized sets of an access structure $\mathcal{A}$ if it satisfies the following two requirements:*

- *For any set $B \in \mathcal{A}$ (where $B = \{P_{i_1}, \ldots, P_{i_{|B|}}\}$), there exists a reconstruction function $\text{RECON}_B : S_{i_1} \times \cdots \times S_{i_{|B|}} \to S$ such that for every maximal $C \in S$, for every $r \in R$, and for every possible value of $\Pi_B(C, r)$,*

$$\text{RECON}_B(\Pi_B(C, r)) = P_i \text{ such that } P_i \in B \setminus C.$$

- *Every unauthorized set can never rule out itself from its shares. Formally, for any maximal unauthorized set $C \notin \mathcal{A}$, for every possible $|C|$-tuple of shares $\langle s_i \rangle_{P_i \in C}$: If there is some maximal unauthorized set $C_0 \notin \mathcal{A}$ such that $\Pr[\Pi_C(C_0, r) = \langle s_i \rangle_{P_i \in C}] > 0$ then $\Pr[\Pi_C(C, r) = \langle s_i \rangle_{P_i \in C}] > 0$.*

In Π_{generic}, if we use a scheme that weakly shares the unauthorized sets of $\mathcal{A}$, then the proof of Lemma 1 remains valid.

Lemma 2. *Assume that there is a scheme Π_{set} that weakly shares the unauthorized sets of $\mathcal{A}$ with c_{set}-bit shares. To share ℓ-bit secrets, the generic weakly-private scheme Π_{generic}, when using Π_{set} instead of Π_{weak}, weakly realizes the access structure $\mathcal{A}$ distributing $(\ell + c_{\text{set}})$-bit shares.*

We next give an example of weakly-private schemes for sharing unauthorized sets. We first use ideas similar to Ito, Saito, and Nishizeki [28]. They proved that if every party is contained in at most d minimal sets of an access structure $\mathcal{A}$, then there is a scheme perfectly realizing $\mathcal{A}$ with ℓ-bit secrets and ℓd-bit shares.

Lemma 3. *Assume $\mathcal{A}$ is an access structure such that every party is contained in at most d minimal authorized sets of $\mathcal{A}$. Then, there is a scheme for weakly sharing the unauthorized sets of $\mathcal{A}$ distributing $d\lceil \log n \rceil$-bit shares.*

Proof. To share a maximal unauthorized set C, for every minimal authorized set B, choose a random party $P_{j_B} \in B \setminus C$, choose $|B|$ random elements $\langle s_{i,B} \rangle_{P_i \in B}$ such that $s_{i,B} \in \{0, \ldots, n-1\}$ and $\sum_{\{i : P_i \in A\}} s_{i,B} \equiv j_B \pmod{n}$. The share of P_i is $\langle s_{i,B} : P_i \in B, B \in \mathcal{A}$ is a minimal authorized set$\rangle$. Clearly, this scheme is correct. Furthermore, each maximal unauthorized set can never rule out itself as the parties in C cannot rule out any j_B for a minimal authorized set B in $\mathcal{A}$. $\qquad\square$

4   Upper Bounds for Efficient Weakly-Private Sharing of Double Exponential Number of Access Structures

In this section we present a construction due to Yuval Ishai [27] giving an efficient weak secret-sharing schemes with a 1-bit secret for a family of access structures of a doubly exponential size. We first define this family.

Definition 8 (The Access Structure $\mathcal{A}_C$). *For every n and every $C \subseteq \{0,1\}^n$, we define an access structure $\mathcal{A}_C$ with $2n$ parties denoted $P_1^0, P_1^1, \ldots, P_n^0, P_n^1$. For every $c = \langle c_1, \ldots, c_n \rangle \in \{0,1\}^n$ define a set $Q_c \stackrel{def}{=} \{P_1^{c_1}, P_2^{c_2}, \ldots, P_n^{c_n}\}$. The minimal authorized sets in $\mathcal{A}_C$ are $\{Q_c : c \in C\} \cup \{\{P_j^0, P_j^1\} : j \in [n]\}$.*

Theorem 2. *For every $C \subseteq \{0,1\}^n$ there is a weakly-private secret-sharing scheme realizing $\mathcal{A}_C$ with domain of secrets $\{0,1\}$ and $O(n^3)$-bit shares.*

Proof. The idea, again, is that for every unauthorized set we construct a set of vectors that prevent the set from ruling out a secret. Towards this goal, we define the following function: For $a, b \in \{0,1\}$ and $x, y \in \{0,1\}^n$, let $f(a, b, x, y)$ be the function which outputs a if $x \neq y$ and outputs b otherwise. Informally, the input a of f is the secret we want to share, the input b is a random input, and if we set $x = y = z$, we will prevent the set Q_z from ruling out the secret b. To construct the scheme, we use the randomized encodings of Applebaum, Ishai, and Kushilevitz [1]. Specifically, the function f can be efficiently encoded by a function $f'((a, b, x, y), r)$ such that:

1. The output distribution of f' induced by a random choice of r reveals the output of f and no additional information about a, b, x, y, that is, there are two distributions D_0, D_1 such that
   (a) If $f(a, b, x, y) = 0$ then $f'((a, b, x, y), r)$ is distributed according to D_0 and if $f(a, b, x, y) = 1$ then $f'((a, b, x, y), r)$ is distributed according to D_1, and
   (b) The distributions D_0 and D_1 have a disjoint support.
2. The length of the output of f' is $O(n^3)$, and
3. Each output bit of f' depends on at most a single bit of (a, b, x, y).

In particular, if the ith bit of f' depends on x_j and we fix r, then we can compute the ith bit of f' from r and x_j without knowing the other bits of x (or knowing a, b, y).

For any subset $C \subseteq \{0,1\}^n$, we describe in Fig. 3 a weakly-private scheme realizing $\mathcal{A}_C$. First note that every pair $\{P_j^0, P_j^1\}$ can reconstruct the secret using the shares given in Step (1) of the scheme. Second, consider a set that contains at most one party from every pair $\{P_j^0, P_j^1\}$ and for some $j \in [n]$ does not contain neither P_j^0 nor P_j^1. Such set can never rule out any value of s_0, hence can never rule out any value of s. Thus, it remains to prove that a set Q_c can reconstruct the secret if and only if $c \in C$.

If $c \in C$, then in Step (4) of the scheme a $w \neq c$ is chosen. The parties of Q_c together hold the bits of $f'((s_1, b, c, w), r)$, which is an element of D_{s_1}, hence they can also compute $f(s_1, b, c, w) = s_1$ (since the support of D_0 and the support

To share a secret $s \in \{0,1\}$:

1. For every j choose $r_j \in \{0,1\}$ at random, and send to P_j^0 the bit r_j and to P_j^1 the bit $r_j \oplus s$,
2. Choose $s_0 \in \{0,1\}$ at random, define $s_1 \leftarrow s \oplus s_0$,
3. For every $j \in [n-1]$ choose $q_j \in \{0,1\}$ at random, set $q_n = s_0 \oplus \bigoplus_{j=1}^{n-1} q_j$, and send to P_j^0 and P_j^1 the bit q_j.
4. Choose $w \notin C$ at random, choose $b \in \{0,1\}$ at random, and choose a random r.
5. Send to player P_j^d, for $j \in [n]$ and $d \in \{0,1\}$, the value of output bits of $f'((s_1, b, x, w), r)$ that depend on x_j assuming that $x_j = d$.
6. All bits of $f'((s_1, b, x, w), r)$ that do not depend on bits of x are sent to all parties.

Fig. 3. A weakly-private scheme realizing $\mathcal{A}_C$

of D_1 are disjoint). Furthermore, they hold $q_1, \ldots, q_n$, hence, they can compute $s = s_0 \oplus \bigoplus_{j=1}^{n} q_j$.

For any $z \notin C$, the set of n players Q_z can never rule out any value of s_1: When $w = z$ and $b = 0$ are chosen in Step (4) of the scheme, the parties of Q_z can compute a random element of D_0 and when $w = z$ and $b = 1$ are chosen in Step (4) of the scheme they can compute a random element of D_1. Thus, the parties do not know if $w \neq z$ and they got an element of D_{s_1} or $w = z$ and $b = \overline{s_1}$ and they got an element of $D_{\overline{s_1}}$. □

5 Upper Bounds for Weakly-Private Threshold Sharing of One Bit

In this section we construct weakly-private t-out-of-n secret-sharing schemes for sharing one bit. We first present a simple weakly-private 2-out-of-n scheme in which the size of the domain of shares of each party is 4. Generalizing the ideas of this scheme we present a 3-out-of-n scheme in which the size of the domain of shares of each party is 6, and a t-out-of-n scheme in which the size of the domain of shares of each party is $\tilde{O}(2^t)$. Finally, we present a different scheme, based on Shamir's scheme, in which the size of domain of shares is roughly $n - t/(2(n-t+1))$ (when n is a prime-power). The best known perfect t-out-of-n schemes use domain of shares of size n. By a lower bound of [32], the size of the domain of shares in every perfect t-out-of-n schemes is at least $n - t$. Thus, our weakly-private t-out-of-n secret-sharing schemes are more efficient than every perfect t-out-of-n secret-sharing schemes when $t < \log n - 2 \log \log n$ and more efficient than known schemes when $t > n/2$.

5.1 The Weakly-Private Scheme for $t = 2$

Lemma 4. *There exists an anonymous weakly-private 2-out-of-n secret-sharing scheme with domain of secrets $\{0,1\}$ in which the size of the domain of shares of every party is 4.*

Proof. To prove the claim we describe a scheme with domain of shares $\{0, 1, 2, 3\}$ for each party.

- To share the secret 0, choose a random index $i \in [n]$ and choose a random $\sigma \in \{2, 3\}$. The share of P_i is σ. The share of P_j, for $j \neq i$, is 0 if $\sigma = 2$ and 1 if $\sigma = 3$.
- To share the secret 1, choose a random index $i \in [n]$ and choose a random $\sigma \in \{0, 1\}$. The share of P_i is σ. The share of P_j, for $j \neq i$, is 3 if $\sigma = 0$ and 2 if $\sigma = 1$.

The 2-out-4 scheme is explicitly described in Example 2.

On one hand, the reconstruction of the secret by any two parties is simple: If the shares are $\{0, 0\}$, $\{0, 2\}$, $\{1, 1\}$, or $\{1, 3\}$, then the secret is 0. Otherwise the secret is 1. On the other hand, each value is possible for each coordinate for each secret, thus, the scheme is weakly private.   □

Example 2. We explicitly describe the weakly-private 2-out-4 anonymous secret-sharing scheme. The shares for the secret 0 are randomly chosen from $\langle 0, 0, 0, 2 \rangle$, $\langle 0, 0, 2, 0 \rangle$, $\langle 0, 2, 0, 0 \rangle$, $\langle 2, 0, 0, 0 \rangle$ and $\langle 1, 1, 1, 3 \rangle$, $\langle 1, 1, 3, 1 \rangle$, $\langle 1, 3, 1, 1 \rangle$, $\langle 3, 1, 1, 1 \rangle$. The shares for the secret 1 are randomly chosen from $\langle 2, 2, 2, 1 \rangle$, $\langle 2, 2, 1, 2 \rangle$, $\langle 2, 1, 2, 2 \rangle$, $\langle 1, 2, 2, 2 \rangle$ and $\langle 3, 3, 3, 0 \rangle$, $\langle 3, 3, 0, 3 \rangle$, $\langle 3, 0, 3, 3 \rangle$, $\langle 0, 3, 3, 3 \rangle$.

5.2   Weakly-Private Schemes for $t < \log n$

We now describe a generalization of the above scheme for larger thresholds. Specifically, in the scheme we design: (1) the scheme is anonymous (as defined in Definition 6), (2) in each vector of shares all but at most $t - 1$ coordinates are equal, and (3) every vector of values in Σ^{t-1} is possible for every $t - 1$ parties for every secret (where Σ is the domain of shares of each party).

We will first describe a generic way to construct a weakly-private t-out-of-n scheme based on the existence of two functions f_0, f_1 with certain properties. Roughly speaking, these functions take an arbitrary vector of shares of length $t - 1$ and stretch it to a vector of shares of length n. The exact properties we require from these functions are sufficient for proving the correctness of the scheme (however, weaker conditions may also be sufficient for proving correctness). We show a simple construction of f_0, f_1 satisfying these properties for $t = 3$ with domain of size 6. We then show that certain combinatorial structure can be used to construct such functions f_0 and f_1, and show that such structures exist implying a t-out-of-n scheme with domain of shares of size $O(t^2 2^t)$.

Lemma 5. *Let t be an integer, Σ be a finite domain, and Σ_0 and Σ_1 be a partition of Σ. Assume there are two functions f_0, f_1, where $f_s : \binom{\Sigma_{\bar{s}}}{< t} \to \Sigma_s$ for $s \in \{0, 1\}$ satisfying*

$$\forall_{A_0 \subseteq \Sigma_0, A_1 \subseteq \Sigma_1 \text{such that } |A_0| + |A_1| \leq t} \; f_0(A_1) \notin A_0 \vee f_1(A_0) \notin A_1. \tag{1}$$

Then, for every $n \geq t$ there is an anonymous weakly-private t-out-of-n scheme with domain of secrets $\{0, 1\}$ and domain of shares Σ for each party.

Proof. We describe the scheme using the given functions f_0 and f_1. To share the secret $s \in \{0, 1\}$, do the following:

1. Choose $t - 1$ random distinct indices $i_1, \ldots, i_{t-1} \in [n]$ and choose $t - 1$ random values $\sigma_1, \ldots, \sigma_{t-1}$ for the parties $P_{i_1}, \ldots, P_{i_{t-1}}$ respectively.
2. Let $A_{\bar{s}}$ be the set of elements of $\Sigma_{\bar{s}}$ in $\sigma_1, \ldots, \sigma_{t-1}$. For every $\ell \notin \{i_1, \ldots, i_{t-1}\}$, the share of P_ℓ is $f_s(A_{\bar{s}})$.

The privacy is guaranteed since every $t - 1$ parties can be chosen in Step 1. We next argue that Property (1) implies the correctness of the scheme. That is, every vector of t shares is possible for at most one secret. Assume towards contradiction that $\boldsymbol{b} = \langle b_1, \ldots, b_t \rangle$ is possible both for the secret 0 and for the secret 1. For $s \in \{0, 1\}$, let B_s be the set of elements of Σ_s in the vector $\boldsymbol{b}$ (without repetition). As $\boldsymbol{b}$ is possible for a secret $s \in \{0, 1\}$, in Step 1 of the scheme some vector $\boldsymbol{\sigma} = \langle \sigma_1, \ldots, \sigma_{t-1} \rangle$ could have been chosen, where $A_{\bar{s}}$ are the elements of $\Sigma_{\bar{s}}$ in this vector (without repetition). The vector $\boldsymbol{b}$ is obtained by taking a sub-vector of $\boldsymbol{\sigma}$ and completing it to a vector of length t with the value $f_s(A_{\bar{s}})$ (possibly with repetitions). Therefore, the following conditions must hold:

1. $B_{\bar{s}} \subseteq A_{\bar{s}}$,
2. Let n^s be the number of times that $f_s(A_{\bar{s}})$ appears in $\boldsymbol{b}$. Thus,

$$n^s \geq |A_{\bar{s}}| - |B_{\bar{s}}| + 1 \geq 1. \tag{2}$$

In particular, $f_s(A_{\bar{s}})$ appears at least once in $\boldsymbol{b}$.

Thus, $f_0(A_1) \in B_0 \subseteq A_0$ and $f_1(A_0) \in B_1 \subseteq A_1$. Furthermore, $|B_0| + |B_1| \leq t - n^0 - n^1 + 2$ (since $f_0(A_1)$ appears n^0 times in $\boldsymbol{b}$ and $f_1(A_0)$ appears n^1 times in $\boldsymbol{b}$), thus $|A_0| + |A_1| \leq t$ (by (2)). This contradicts Property (1), and thus the scheme is correct. □

We next reformulate Lemma 5 using only one function f_0.

Lemma 6. *Let t be an integer, Σ be a finite domain, and Σ_0 and Σ_1 be a partition of Σ. Assume there is a function $f_0 : \binom{\Sigma_1}{<t} \to \Sigma_0$ such that for every $A_0 \subseteq \Sigma_0$, where $|A_0| < t$,*

$$\bigcup \{A_1 \subseteq \Sigma_1 : |A_0| + |A_1| \leq t \text{ and } f_0(A_1) \in A_0\} \subsetneq \Sigma_1. \tag{3}$$

Then, for every $n \geq t$ there is an anonymous weakly-private t-out-of-n scheme with domain of secrets $\{0, 1\}$ and domain of shares Σ for each party.

Proof. We show that there is a function f_1 such that f_0, f_1 satisfy Property (1) of Lemma 5. For every $A_0 \subseteq \Sigma_0$ define $f_1(A_0)$ as any element σ in

$$\Sigma_1 \setminus \left(\bigcup \{A_1 \subseteq \Sigma_1 : |A_0| + |A_1| \leq t \text{ and } f_0(A_1) \in A_0\} \right).$$

Now, if $f_0(A_1) \in A_0$, then $\sigma \notin A_1$, thus, f_0, f_1 satisfy Property (1). □

Specific implementation for $t = 3$

Lemma 7. *There exists an anonymous weakly-private 3-out-of-n secret-sharing scheme with domain of secrets $\{0, 1\}$ in which the size of the domain of shares of every party is 6.*

Proof. We show how to implement the functions f_0, f_1 satisfying Property (1) of Lemma 5 with $\Sigma_0 = \{0, 1, 2\}$, $\Sigma_1 = \{3, 4, 5\}$, and $\Sigma = \Sigma_0 \cup \Sigma_1$. Define f_0 and f_1 as follows:

A_1	$f_0(A_1)$
$\emptyset$	0
$\{3\}$	0
$\{4\}$	1
$\{5\}$	2
$\{3, 4\}$	1
$\{3, 5\}$	0
$\{4, 5\}$	2

A_0	A_1 s.t. $\lvert A_0 \rvert + \lvert A_1 \rvert \le t$ and $f_0(A_1) \in A_0$	$f_1(A_0)$
$\emptyset$	–	3
$\{0\}$	$\emptyset, \{3\}, \{3, 5\}$	4
$\{1\}$	$\{4\}, \{3, 4\}$	5
$\{2\}$	$\{5\}, \{4, 5\}$	3
$\{0, 1\}$	$\emptyset, \{3\}, \{4\}$	5
$\{0, 2\}$	$\{3\}, \{5\}$	4
$\{1, 2\}$	$\{4\}, \{5\}$	3

As indicated by the table, Property (3) holds for f_0; the function f_1 is constructed using Lemma 6. □

Remark 1. We next explain why we need a share domain of size six in the above 3-out-of-n scheme. Assume, f_0, f_1 satisfy Property (1). Thus, for example, if $f_1(\{0, 1\}) = \sigma$ we require $f_0(\{\sigma\}) \ne 0, 1$, and $\lvert \Sigma_0 \rvert \ge 3$. Similarly, $\lvert \Sigma_1 \rvert \ge 3$.

Generic implementation using set-systems

To construct the weakly-private t-out-of-n secret-sharing schemes for larger values of t we use a set-system with specific properties. The existence of such set-system is basically equivalent to the existence of a function f_0 satisfying Property (3) in Lemma 6. The definition of the set-system we use is similar to the definition used in [37] and the construction we present is the same as theirs.

Definition 9. *Let $\mathcal{C} = \{C_1, \ldots, C_m\}$ be a collection of m sets and $B = \bigcup_{i=1}^{m} C_i$. We say that $\mathcal{C}$ is an (ℓ, m, b) set-system if the following three requirements hold:*

1. *$\lvert \mathcal{C} \rvert = m$ and $\lvert B \rvert \le b$,*
2. *The union of every ℓ sets in $\mathcal{C}$ is properly contained in B. That is, for every $A_0 \subset [m]$, where $\lvert A_0 \rvert = \ell$,*

$$\bigcup_{i \in A_0} C_i \subsetneq B,$$

3. *Every subset of $A_1 \subseteq B$ of size ℓ is contained in at least one C_i, that is, $A_1 \subseteq C_i$.*

It is easy to satisfy one of the above Conditions 2 and 3. For example, to satisfy Conditions 2 we can partition B to $\ell + 1$ disjoint non-empty sets. To satisfy Conditions 3 we can take $\mathcal{C} = \{B\}$. The difficulty is to satisfy the two conditions simultaneously.

Example 3. Let $B = [\ell^2 + 1]$ and $\mathcal{C}$ be the collection of all subsets of size ℓ of B. Then, $\mathcal{C}$ is an $(\ell, m, \ell^2 + 1)$ set-system, where $m \stackrel{\text{def}}{=} \binom{\ell^2+1}{\ell} = 2^{O(\ell \log \ell)}$. Clearly, Items 1 and 3 of Definition 9 hold. To prove that Item 2 holds, notice that the size of B is $\ell^2 + 1$ and the size of each set in $\mathcal{C}$ is ℓ, thus the size of the union of ℓ subsets is at most ℓ^2, that is, there exists at least one element of B that is not in the union of the ℓ sets.

Lemma 8. *If there is a $(t - 1, m, b)$ set-system, then there is an anonymous weakly-private t-out-of-n secret-sharing scheme with domain of secrets $\{0, 1\}$ and domain of shares of size $m + b$.*

Proof. Let $C_1, \ldots, C_m$ be a $(t - 1, m, b)$ set-system and $B = \bigcup_{i=1}^{m} C_i$. Without loss of generality, assume that $B \cap [m] = \emptyset$. Let $\Sigma_0 = [m]$ and $\Sigma_1 = B$. We define $f_0 : \binom{[\Sigma_1]}{<t} \to \Sigma_0$ satisfying the condition of Lemma 6: For every $A_1 \subset B$ of size at most $t - 1$, we define $f_0(A_1)$ as the smallest i such that $A_1 \subseteq C_i$. By Item 3 such i exists.

We prove that a stronger condition that Property (3) of Lemma 6 holds, namely, we prove that for every $A_0 \subseteq \Sigma_0$

$$\bigcup \{A_1 \subseteq \Sigma_1 : |A_1| \leq t - 1 \text{ and } f_0(A_1) \in A_0\} \subsetneq \Sigma_1. \tag{4}$$

Notice that

$$\bigcup \{A_1 \subseteq \Sigma_1 : |A_1| \leq t - 1 \text{ and } f_0(A_1) \in A_0\}$$
$$= \bigcup_{i \in A_0} \left(\bigcup \{A_1 \subseteq \Sigma_1 : |A_1| \leq t - 1 \text{ and } f_0(A_1) = i\} \right).$$

However, $f_0(A_1) = i$ implies that $A_1 \subseteq C_i$. Thus,

$$\bigcup \{A_1 \subseteq \Sigma_1 : |A_1| \leq t - 1 \text{ and } f_0(A_1) \in A_0\} \subseteq \bigcup_{i \in A_0} C_i \subsetneq \Sigma_1$$

(by Item 2 of Definition 9). By Lemma 6, there is a secret-sharing scheme with the parameters promised in the lemma.   □

We show the existence of an (ℓ, m, m) set-system using a probabilistic proof provided that $\ell = O(\log m)$. The construction is simple; we choose m subsets independently with uniform distribution.

Lemma 9. *Let $m = 2^{\ell+1} \ell^2$. There exists an (ℓ, m, m) set-system.*

Proof. We show the existence using a probabilistic proof. Define $B = [m]$. Pick m sets $C_1, \ldots, C_m \subset B$ where each set is chosen independently with uniform distribution (in particular, $\Pr[j \in C_i] = 1/2$ for every i and j).

We prove that with positive probability Conditions 2 and 3 hold, thus, there exists a "good" choice such that $\{C_1, \ldots, C_m\}$ is an (ℓ, m, m) set-system.

We first prove that Condition 2 holds with probability greater than 0.5. First fix a set $A_0 \in [m]$ of size ℓ. For every index $j \in [m]$, the probability that for at least one $i \in A_0$ the index j is in C_i is $1 - 2^{-\ell}$. The probability that

To share a secret $s \in \{0, 1\}$ using a domain of shares $\Sigma \subseteq GF(q)$ of size $\lfloor q - (q-1)(2(n-t+1)) \rfloor + 1$, where $q \geq n$ is a prime-power:

1. Pick random $s_1, \ldots, s_{t-1} \in \Sigma$, and let $a \leftarrow s$.
2. Compute the unique polynomial Q_a of degree at most $t-1$ such that
   - $Q_a(i) = s_i$ for every $1 \leq i \leq t-1$.
   - The coefficient of x^{t-1} in Q_a is a.
3. If $Q_a(i) \notin \Sigma$ for some $t \leq i \leq n$, then $a \leftarrow a+2$; Goto Step 2.
4. (* We found an a such $s \equiv a \pmod 2$ and $Q_a(i) \in \Sigma$ for $1 \leq i \leq n$ *)
   The share of P_i is $Q_a(i)$.

Fig. 4. A t-out-of-n secret-sharing scheme with domain of shares of size $\lfloor q - (q-1)(2(n-t+1)) \rfloor + 1$

$\cup_{i \in A_0} C_i = B$ is the probability that for every $j \in [m]$ for at least one $i \in A_0$ the index j is in C_i. This probability is $(1 - 2^{-\ell})^m \leq e^{-m/2^\ell}$. Thus, by the union bound, the probability that there exists a set A_0 violating Condition 2 is at most $\binom{m}{\ell} e^{-m/2^\ell} < e^{\ell \ln m - m/2^\ell}$. By our choice of m, this probability is less than half.

The same calculations show that Condition 3 holds with probability greater than 0.5. First fix a set $A_1 \subset B$ of size ℓ. The probability that $A_1 \subseteq C_i$ for a fixed i is $2^{-\ell}$. Thus, the probability that $A_1 \nsubseteq C_i$ for every $i \in [m]$ is $(1 - 2^{-\ell})^m \leq e^{-m/2^\ell}$. By the union bound, the probability that there exists a set A_1 violating Condition 3 is at most $\binom{m}{\ell} e^{-m/2^\ell} < e^{\ell \ln m - m/2^\ell} < 1/2$. $\qquad \square$

Theorem 3. *There is an anonymous weakly-private t-out-of-n secret-sharing scheme with domain of secrets $\{0, 1\}$ in which the size of the domain of shares of each party is $2(t-1)^2 2^t$.*

In the full version of this paper, we discuss the restriction that we used in the construction of the above scheme. In particular, we prove that in every t-out-of-n scheme implementing Lemma 5 the size of the domain is $2^{\Omega(t)}$, thus our implementation in Theorem 3 is almost optimal.

5.3 Weakly-Private Schemes for $t \geq n/2$

We next present weakly-private t-out-of-n secret-sharing schemes for large values of t. For example, when n is a prime-power, we construct an $(n-1)$-out-of-n scheme with share domain of size roughly $0.75n$ for every party. For $t \approx n/2$, we construct a scheme with domain of shares of size $n-1$. In our scheme, we restrict the domain of shares in a variant of Shamir's scheme [44] to a subset of the field. In this variant of Shamir's scheme, the secret is the coefficient of x^{t-1} in the polynomial (compared to x^0 in Shamir's scheme). The advantage of this variant is that it reduces the size of the field by 1 (yielding the best known perfect t-out-of-n scheme for sharing 1-bit secrets). Unlike the previous schemes for $t < \log n$, the scheme we present in this section is not anonymous.

Theorem 4. *Let $n \in \mathbb{N}$ be integer and $q \geq n$ be a prime-power. For every $1 < t < n$ there is a weakly-private t-out-of-n secret-sharing scheme in which the size of the domain of shares of each party is $\left\lfloor q - \frac{q-1}{2(n-t+1)} \right\rfloor + 1$.*

Proof. We describe the scheme in Fig. 4. All arithmetic in the scheme is in $GF(q)$. To simplify the notations, we assume that the elements of $GF(q)$ are $\{0,\ldots,q-1\}$. In the proof below of the weak privacy, we prove that for every $s_1,\ldots,s_{t-1} \in \Sigma$ there exists at least one value a satisfying the conditions of Step 4, thus the scheme terminates. We say that a polynomial Q passes through a share s_i of P_i if $Q(i) = s_i$.

The reconstruction of the secret by t parties is done as in Shamir's scheme: the parties compute the unique polynomial Q of degree $t-1$ that passes through their shares, compute the coefficient a of x^{t-1} in Q, and output $a \bmod 2$. We next prove the weak privacy of the scheme, that is, every $t-1$ parties are unable to rule out either secret. Fix any set C of $t-1$ parties, fix any $t-1$ values $\langle s_i \rangle_{P_i \in C}$ in Σ as the shares of C, and fix a secret $s \in \{0,1\}$. There are at least $(q-1)/2$ values a such that $a \equiv s \bmod 2$. If for one such a the unique polynomial Q of degree $t-1$ with coefficient a of x^{t-1} that passes through the shares of C satisfies $Q(i) \in \Sigma$ for every $i \notin C$, then the shares $\langle s_i \rangle_{P_i \in C}$ are possible for C given s. We will show that every party $P_i \notin C$ eliminates at most $q - |\Sigma| < \frac{q-1}{2(n-t+1)}$ values of a and there are $n - t + 1$ parties not in C. Thus, since $(n - t + 1)\frac{q-1}{2(n-t+1)} \leq \frac{q-1}{2}$, there is at least one a that survives.

To complete the proof, we fix $P_i \notin C$, and prove that P_i eliminates at most $q - |\Sigma|$ values of a. For each value $s_i \in \{0,\ldots,q-1\} \setminus \Sigma$, there is a unique polynomial of degree $t - 1$ that passes through the shares of $C \cup \{P_i\}$. Thus, such value s_i only eliminates the coefficient of x^{t-1} in this polynomial. $\qquad\square$

6 Lower Bounds for Weakly-Private Threshold Schemes

We state lower bounds on the size of domain of shares in weakly-private t-out-of-n schemes. The proofs of these results appear in the full version of this paper.

Lemma 10. *Let $n \geq 9$. In every weakly-private 2-out-of-n secret-sharing scheme with domain of secrets $\{0,1\}$, the size of the domain of shares of at least one party is at least 4.*

Theorem 5. *In every anonymous weakly-private t-out-of-n secret-sharing scheme with domain of secrets $\{0,1\}$, the size of the domain of shares of at least one party is at least $\min\left\{2t, \sqrt{(n-t)/2}\right\}$.*

Theorem 6. *In every weakly-private t-out-of-n secret-sharing scheme with domain of secrets $\{0,1\}$, the size of the domain of shares of at least one party is at least $\min\left\{t, \frac{\log\log(n-t)}{2\log\log\log(n-t)}\right\}$. Furthermore, if $n > t-1+(t-1)(2t-1)^{2\left((2t-1)^{t-1}\right)}$, in every weakly-private t-out-of-n secret-sharing scheme with domain of secrets $\{0,1\}$, the size of the domain of shares of at least one party is at least $2t$.*

Acknowledgments. We thank Benny Chor, Eyal Kushilevitz, Noam Livne, and Enav Weinreb for helpful discussions on this subject. We thank Yuval Ishai for allowing us to include the results described in Section 4.

References

1. B. Applebaum, Y. Ishai, and E. Kushilevitz. Cryptography in NC^0. In *Proc. of the 45th Symp. on Foundations of Computer Science*, pages 166–175, 2004.
2. P. Beguin and A. Cresti. General short computational secret sharing schemes. In *EUROCRYPT '95*, vol. 921 of *LNCS*, pages 194–208. 1995.
3. A. Beimel. *Secure Schemes for Secret Sharing and Key Distribution*. PhD thesis, Technion – Israel Institute of Technology, 1996.
4. A. Beimel. On private computation in incomplete networks. *Distributed Computing*, 2006.
5. A. Beimel and B. Chor. Communication in key distribution schemes. *IEEE Trans. on Information Theory*, 42(1):19–28, 1996.
6. A. Beimel and Y. Ishai. On the power of nonlinear secret-sharing. *SIAM J. on Discrete Mathematics*, 19(1):258–280, 2005.
7. A. Beimel and N. Livne. On matroids and non-ideal secret sharing. In *TCC 2006*, vol. 3876 of *LNCS*, pages 482–501, 2006.
8. M. Ben-Or, S. Goldwasser, and A. Wigderson. Completeness theorems for non-cryptographic fault-tolerant distributed computations. In *Proc. of the 20th STOC*, pages 1–10, 1988.
9. J. Benaloh and J. Leichter. Generalized secret sharing and monotone functions. In *CRYPTO '88*, vol. 403 of *LNCS*, pages 27–35. 1990.
10. G. R. Blakley. Safeguarding cryptographic keys. In *Proc. of the 1979 AFIPS National Computer Conference*, pages 313–317. 1979.
11. C. Blundo, A. De Santis, R. de Simone, and U. Vaccaro. Tight bounds on the information rate of secret sharing schemes. *Designs, Codes and Cryptography*, 11(2):107–122, 1997.
12. C. Blundo, A. De Santis, A. Giorgio Gaggia, and U. Vaccaro. New bounds on the information rate of secret sharing schemes. *IEEE Trans. on Information Theory*, 41(2):549–553, 1995.
13. C. Blundo, A. De Santis, A. Herzberg, S. Kutten, U. Vaccaro, and M. Yung. Perfectly secure key distribution for dynamic conferences. *Info. and Comput.*, 146(1):1–23, 1998.
14. C. Blundo and D. R. Stinson. Anonymous secret sharing schemes. *Discrete Applied Math. and Combin. Operations Research and Comp. Sci.*, 77:13–28, 1997.
15. E. F. Brickell. Some ideal secret sharing schemes. *Journal of Combin. Math. and Combin. Comput.*, 6:105–113, 1989.
16. E. F. Brickell and D. M. Davenport. On the classification of ideal secret sharing schemes. *J. of Cryptology*, 4(73):123–134, 1991.
17. R. M. Capocelli, A. De Santis, L. Gargano, and U. Vaccaro. On the size of shares for secret sharing schemes. *J. of Cryptology*, 6(3):157–168, 1993.
18. D. Chaum, C. Crépeau, and I. Damgård. Multiparty unconditionally secure protocols. In *Proc. of the 20th STOC*, pages 11–19, 1988.
19. R. Cramer, I. Damgård, and U. Maurer. General secure multi-party computation from any linear secret-sharing scheme. In *EUROCRYPT 2000*, vol. 1807 of *LNCS*, pages 316–334. 2000.
20. L. Csirmaz. The size of a share must be large. In *EUROCRYPT '94*, vol. 950 of *LNCS*, pages 13–22. 1995. Also in: *J. of Cryptology*, 10(4):223–231, 1997.
21. L. Csirmaz. The dealer's random bits in perfect secret sharing schemes. *Studia Sci. Math. Hungar.*, 32(3–4):429–437, 1996.
22. I. Damgård and R. Thorbek. Linear integer secret sharing and distributed exponentiation. In *PKC 2006*, vol. 3958 of *LNCS*, pages 75 – 90. 2006.
23. Y. Desmedt and Y. Frankel. Shared generation of authenticators and signatures. In *CRYPTO '91*, vol. 576 of *LNCS*, pages 457–469. 1992.

24. M. van Dijk. On the information rate of perfect secret sharing schemes. *Designs, Codes and Cryptography*, 6:143–169, 1995.

25. M. van Dijk. A linear construction of secret sharing schemes. *Designs, Codes and Cryptography*, 12(2):161–201, 1997.

26. V. Goyal, O. Pandey, A. Sahai, and B. Waters. Attribute based encryption for fine-grained access control of encrypted data. In *CCS 2006*, 2006.

27. Y. Ishai. Personal communication. 2006.

28. M. Ito, A. Saito, and T. Nishizeki. Secret sharing schemes realizing general access structure. In *Proc. of Globecom 87*, pages 99–102, 1987.

29. W.-A Jackson and K. M. Martin. Combinatorial models for perfect secret sharing schemes. *J. of Comb. Mathematics and Comb. Computing*, 28:249–265, 1998.

30. M. Karchmer and A. Wigderson. On span programs. In *Proc. of the 8th Structure in Complexity Theory*, pages 102–111, 1993.

31. E. D. Karnin, J. W. Greene, and M. E. Hellman. On secret sharing systems. *IEEE Trans. on Information Theory*, 29(1):35–41, 1983.

32. J. Kilian and N. Nisan. Private communication, 1990.

33. W. Kishimoto, K. Okada, K. Kurosawa, and W. Ogata. On the bound for anonymous secret sharing schemes. *Discrete Appl. Math.*, 121(1-3):193–202, 2002.

34. H. Krawczyk. Secret sharing made short. In *CRYPTO '93*, vol. 773 of *LNCS*, pages 136–146. 1994.

35. K. Kurosawa and K. Okada. Combinatorial lower bounds for secret sharing schemes. *Inform. Process. Lett.*, 60(6):301–304, 1996.

36. E. Kushilevitz. Privacy and communication complexity. *SIAM J. on Discrete Mathematics*, 5(2):273–284, 1992.

37. C. Lund and M. Yannakakis. On the hardness of approximating minimization problems. *J. of the ACM*, 41(5):960–981, 1994.

38. J. Martí-Farré and C. Padró. On secret sharing schemes, matroids and polymatroids. Technical Report 2006/077, Cryptology ePrint Archive, 2006.

39. Y. Miao. A combinatorial characterization of regular anonymous perfect threshold schemes. *Inform. Process. Lett.*, 85(3):131–135, 2003.

40. M. Naor and A. Wool. Access control and signatures via quorum secret sharing. *IEEE Transactions on Parallel and Distributed Systems*, 9(1):909–922, 1998.

41. C. Padró and G. Sáez. Secret sharing schemes with bipartite access structure. *IEEE Trans. on Information Theory*, 46:2596–2605, 2000.

42. M. O. Rabin. Randomized Byzantine generals. In *Proc. of the 24th IEEE Symp. on Foundations of Computer Science*, pages 403–409, 1983.

43. P. D. Seymour. On secret-sharing matroids. *J. of Combinatorial Theory, Series B*, 56:69–73, 1992.

44. A. Shamir. How to share a secret. *Communications of the ACM*, 22:612–613, 1979.

45. G. J. Simmons, W. Jackson, and K. M. Martin. The geometry of shared secret schemes. *Bulletin of the ICA*, 1:71–88, 1991.

46. D. R. Stinson. Decomposition construction for secret sharing schemes. *IEEE Trans. on Information Theory*, 40(1):118–125, 1994.

47. D. R. Stinson and S. A. Vanstone. A combinatorial approach to threshold schemes. *SIAM J. on Discrete Mathematics*, 1(2):230–236, 1988.

48. V. Vinod, A. Narayanan, K. Srinathan, C. Pandu Rangan, and K. Kim. On the power of computational secret sharing. In *Indocrypt 2003*, vol. 2904 of *LNCS*, pages 162–176. 2003.

49. A. C. Yao. Unpublished manuscript, 1989. Presented at Oberwolfach and DIMACS workshops.

On Secret Sharing Schemes, Matroids and Polymatroids*

Jaume Martí-Farré and Carles Padró

Universitat Politècnica de Catalunya, Barcelona, Catalonia, Spain
{jaumem,cpadro}@ma4.upc.edu

Abstract. One of the main open problems in secret sharing is the characterization of the access structures of ideal secret sharing schemes. As a consequence of the results by Brickell and Davenport, every one of those access structures is related in a certain way to a unique matroid.

Matroid ports are combinatorial objects that are almost equivalent to matroid-related access structures. They were introduced by Lehman in 1964 and a forbidden minor characterization was given by Seymour in 1976. These and other subsequent works on that topic have not been noticed until now by the researchers interested on secret sharing.

By combining those results with some techniques in secret sharing, we obtain new characterizations of matroid-related access structures. As a consequence, we generalize the result by Brickell and Davenport by proving that, if the information rate of a secret sharing scheme is greater than 2/3, then its access structure is matroid-related. This generalizes several results that were obtained for particular families of access structures.

In addition, we study the use of polymatroids for obtaining upper bounds on the optimal information rate of access structures. We prove that every bound that is obtained by this technique for an access structure applies to its dual structure as well.

Finally, we present lower bounds on the optimal information rate of the access structures that are related to two matroids that are not associated with any ideal secret sharing scheme: the Vamos matroid and the non-Desargues matroid.

Keywords: Secret sharing, Information rate, Ideal secret sharing schemes, Ideal access structures, Matroids, Polymatroids.

1   Introduction

1.1   The Problems

A *secret sharing scheme* is a method to distribute a *secret value* into *shares* in such a way that only some *qualified subsets* of *participants* are able to recover the

---

* This work was partially supported by the Spanish Ministry of Education and Science under project TIC 2003-00866. This work was done while the second author was in a sabbatical stay at CWI, Amsterdam. This stay was funded by the *Secretaría de Estado de Educación y Universidades* of the Spanish Ministry of Education.

S.P. Vadhan (Ed.): TCC 2007, LNCS 4392, pp. 273–290, 2007.

secret from their shares. Secret sharing schemes were independently introduced by Shamir [34] and Blakley [5]. Only *unconditionally secure perfect secret sharing schemes* will be considered in this paper. That is, the shares of the participants in a non-qualified subset must not contain any information about the secret value.

The family of the qualified subsets is the *access structure* of the scheme, which is supposed to be *monotone increasing*, that is, every subset containing a qualified subset must be qualified. Then an access structure is determined by its *minimal qualified subsets*.

The complexity of a secret sharing scheme can be measured by the length of the shares. In all secret sharing schemes, the length of every share is greater than or equal to the length of the secret [20]. A secret sharing scheme is said to be *ideal* if all shares have the same length as the secret.

The qualified subsets of a *threshold access structure* are those having at least a fixed number of participants. Shamir's construction [34] provides an ideal scheme for every threshold access structure. Even though there exists a secret sharing scheme for every access structure [18], in general some shares must be much larger than the secret [12,13].

This paper deals with the optimization of the complexity of secret sharing schemes for general access structures.

The characterization of the *ideal access structures*, that is, the access structures of ideal secret sharing schemes, is one of the main open problems in that direction. Brickell and Davenport [10] discovered important connections of this problem with matroid theory. The main definitions and basic facts about secret sharing schemes, matroids, and polymatroids are presented in Section 2. Table 1, at the end of the paper, may be helpful to the readers that are not familiar with the concepts that are discussed here.

A necessary condition for an access structure to be ideal is obtained from the results by Brickell and Davenport [10]. They proved that every ideal secret sharing scheme on a set P of participants univocally determines a matroid $\mathcal{M}$ on the set $Q = P \cup \{D\}$, where $D \notin P$ is a special participant, usually called *dealer*. In addition, the access structure Γ of the ideal scheme is determined by this matroid. Specifically, the minimal qualified subsets of Γ are

$$\min \Gamma = \{A \subseteq P \,:\, A \cup \{D\} \text{ is a circuit of } \mathcal{M}\}.$$

Therefore, every ideal access structure is *matroid-related*, that is, it can be defined in this way from a matroid. This necessary condition is not sufficient, because there exist matroids that cannot be defined from any ideal secret sharing scheme [27,33], and hence the access structures that are related to these matroids are not ideal.

The matroids that are obtained from ideal secret sharing schemes are generally called *secret sharing matroids*, but we prefer to call them *ideal secret sharing representable matroids*, or *iss-representable matroids* for short. This is due to the fact that an ideal secret sharing scheme can be seen as a representation of its associated matroid.

Brickell [9] proposed a special class of ideal schemes, the *vector space secret sharing schemes*. The matroids that are associated with these ideal schemes are precisely the linearly representable ones. Therefore, all linearly representable matroids are iss-representable. This implies that the representation by ideal secret sharing schemes is a generalization of the linear representation of matroids. In addition, every access structure that is related to a linearly representable matroid is ideal. These access structures are called *vector space access structures*. This sufficient condition is not necessary, because there exist iss-representable matroids that are not linearly representable [35].

As a consequence of the results by Brickell and Davenport [10] the open problem of characterizing the access structures of ideal secret sharing schemes can be splitted into the following two open problems.

Problem 1. Characterize the matroid-related access structures.

Problem 2. Characterize the ideal secret sharing representable matroids.

Surprisingly enough, almost all authors interested on secret sharing, including the ones of this paper, have been unaware that matroid-related access structures were studied before secret sharing was invented. Of course, a different name was used: *matroid ports*.

A *clutter* on a set P is a family Λ of subsets of P such that there do not exist two different subsets $A, B \in \Lambda$ with $A \subset B$. A clutter Λ on P is a *matroid port* if there exists a matroid $\mathcal{M}$ on $Q = P \cup \{D\}$, where $D \notin P$, such that

$$\Lambda = \{A \subseteq P : A \cup \{D\} \text{ is a circuit of } \mathcal{M}\}.$$

Therefore, an access structure is matroid-related if an only if the clutter formed by its minimal qualified subsets is a matroid port. Matroid ports were introduced by Lehman [21] in 1964 to solve the Shannon switching game. Seymour [32] presented in 1976 a characterization of matroid ports by excluded minors that is based on a previous characterization of matroid ports due to Lehman [22]. As a consequence, an answer to Problem 1 is obtained.

A more general open problem in secret sharing is to determine the complexity of the best secret sharing scheme for any given access structure. For instance, we can try to maximize the *information rate*, which is the ratio between the length in bits of the secret and the maximum length of the shares. The *optimal information rate* of an access structure Γ, which is denoted by $\rho(\Gamma)$, is defined as the supremum of the information rates of all secret sharing schemes with access structure Γ. Clearly, $0 < \rho(\Gamma) \leq 1$, and $\rho(\Gamma) = 1$ if Γ is ideal.

Problem 3. Determine the value of $\rho(\Gamma)$ or, at least, improve the known bounds on this function.

Duality has been defined for matroids, for linear codes, and for access structures. It plays an important role in the considered problems. For instance, if an access structure is related to a matroid, its dual is related to the dual matroid. One can consider the dual of a linear secret sharing scheme by identifying it with a linear code. A linear scheme with the same information rate for the dual access structure is obtained in this way. Nevertheless, it is not known whether the dual of an ideal access structure is ideal as well. In addition, the relation between the optimal information rates of an access structure and its dual is equally an open problem.

1.2   Our Results

Because of their important implications to the problems we are considering here, one of the main goals of this paper is to point out the results by Lehman [21,22] and Seymour [32] on matroid ports to researchers interested on secret sharing. We think that they will be very useful to obtain new general results on the problems we are considering here as well as to solve them for particular families of access structures.

One of our main results, Theorem 17, is a new characterization of matroid-related access structures in terms of the existence of *independent sequences*. These sequences are combinatorial configurations that were introduced in [6,30] to obtain upper bounds on the optimal information rate. Our characterization is obtained by combining Seymour's characterization of matroid ports [32] with the fact that the Shannon entropy defines a polymatroid over a set of random variables [15,16]. As a corollary of Theorem 17 we obtain a generalization of the result by Brickell and Davenport [10]. Namely, they proved that the access structure of every ideal secret sharing scheme is matroid-related, and we prove that this is so for every secret sharing scheme with information rate greater than 2/3. This is the main result in this paper.

Theorem 4. *The access structure of every secret sharing scheme with information rate greater than 2/3 is matroid-related.*

Our proof for this theorem, as well as the ones for the results we apply in it, do not rely on the result by Brickell and Davenport [10]. Moreover, except for the relation between entropy and polymatroids, those proofs use only combinatorial techniques. Therefore, we can say that we present here a new, almost purely combinatorial proof for that important result.

Theorem 4 explains a gap property that has been observed in some particular classes of access structures that have been previously studied, in which every access structure is either ideal or has optimal information at most 2/3. So, there is no access structure Γ with $2/3 < \rho(\Gamma) < 1$ in these families. Specifically, this has been proved for the access structures on sets of four [36] and five [19] participants, the ones defined by graphs [7,10,12], the bipartite ones [30], the ones with three or four minimal qualified subsets [24], the ones with intersection number equal to one [26], and for a special class of homogeneous structures with rank three [23]. This fact was proved by methods that seemed to be specific to every one of those families, and hence it was not clear to which extent this result could be generalized. Since in all those families every matroid-related access structure is ideal, this gap property is a direct consequence of Theorem 4, which implies that $\rho(\Gamma) \leq 2/3$ if Γ is not matroid-related. Therefore, we generalize and explain a phenomenon that had been observed in several particular situations.

Moreover, our result can be applied to other families that have been studied previously as, for instance, the weighted threshold access structures [3] and the access structures with rank three [25].

In addition, we present in Section 3 a new result about the use of polymatroids to obtain upper bounds on the information rate, a technique that was introduced by Csirmaz [13]. Specifically, we prove that every bound on the optimal information rate of a given access structure that can be obtained by using polymatroids applies also to the dual access structure. In order to do that, we define in a suitable way the *dual* of a polymatroid. The interest of this result is that, for the first time, we present a connection between the complexities of the secret sharing schemes for an access structure and the ones for its dual that is not restricted to linear schemes.

Finally, Section 5 is devoted to present lower bounds on the optimal information rate of the access structures related to the Vamos matroid and the non-Desargues matroid. Since these matroids are not iss-representable, the related access structures are not ideal. We prove that the optimal information rate of the access structures related to the Vamos matroid is at least 2/3, while this parameter is at least 3/4 for the structures related to the non-Desargues matroid. The only previously known results on the optimal information rate of non-ideal matroid-related access structures have been presented in a recent work by Beimel and Livne [2]. They give lower bounds on the length of the shares in secret sharing schemes for the access structures related to the Vamos matroid.

1.3   Related Work

As a sequel of the results by Brickell and Davenport [10], there is a number of works dealing with Problem 2. The Vamos matroid was the first matroid that was proved to be non-iss-representable. This was done by Seymour [33] and different proofs were given later by Simonis and Ashikhmin [35] and Beimel and Livne [2]. An infinite family of non-iss-representable matroids was given by Matúš [27]. As we said before, all linearly representable matroids are iss-representable [9]. The first example of an iss-representable matroid that is not linearly representable, the non-Pappus matroid, was presented in [35].

A number of important results and interesting ideas for future research on Problem 2 can be found in the works by Simonis and Ashikhmin [35] and Matúš [27]. The first one deals with the geometric structure that lies behind iss-representations of matroids. The second one analyzes the algebraic properties that the matroid induces in all its iss-representations. These properties make it possible to find some restrictions on the iss-representations of a given matroid and, in some cases, to exclude the existence of such representations. By using these tools, Matúš [27] presented an infinite family of non-iss-representable matroids with rank three.

One of the most important results on the optimization of the complexity of secret sharing schemes for general access structures is the fact that nonlinear secret sharing schemes are in general more efficient than the linear ones. By using the results and techniques in [1,17], Beimel and Weinreb [4] presented families of access structures for which there exist nonlinear secret sharing schemes whose complexity is polynomial on the number of participants while the complexity of the best linear schemes is not polynomial.

Lower bounds on the optimal information rate of wide families of access structures can be found by applying the different techniques to construct secret sharing schemes with high information rate given in [8,11,31,37,38]. Upper bounds on this parameter have been found by using Information Theory [6,7,12]. In particular, Capocelli, De Santis, Gargano, and Vaccaro [12] presented for the first time bounds smaller than 1 on the optimal information rate. Specifically, they showed access structures whose optimal information rates are at most 2/3. Csirmaz [13] proved that every secret sharing scheme defines a polymatroid that is related to the access structure and he observed that those upper bounds on the optimal information rate could be derived from this fact. A general combinatorial method to find upper bounds, the *independent sequence method*, was given in [6] and was improved in [30]. However, there exists a wide gap between the best known upper and lower bounds on the optimal information rate for most access structures.

2   Basics on Secret Sharing, Matroids, and Polymatroids

The reader is referred to [36] for an introduction to secret sharing and to [29,39] for general references on matroid theory. The book by Welsh [39] contains a chapter about polymatroids. Table 1 summarizes the connections between some of the concepts that are introduced here.

Let Q be a finite set of *participants* and $D \in Q$ a special participant called *dealer*. Consider a finite set E with a probability distribution on it. For every $i \in Q$, consider a finite set E_i and a surjective mapping $\pi_i \colon E \to E_i$. Those mappings induce random variables on the sets E_i. Let $H(E_i)$ denote the Shannon entropy of one of these random variables. For a subset $A = \{i_1, \ldots, i_r\} \subseteq Q$, we write $H(A)$ for the joint entropy $H(E_{i_1} \ldots E_{i_r})$, and a similar convention is used for conditional entropies as, for instance, in $H(E_j|A) = H(E_j|E_{i_1} \ldots E_{i_r})$.

The mappings π_i define a *secret sharing scheme* Σ with *access structure* Γ on the set $P = Q - \{D\}$ of participants if $H(E_D) > 0$ and $H(E_D|A) = 0$ if $A \in \Gamma$ while $H(E_D|A) = H(E_D)$ if $A \notin \Gamma$. In this situation, every random choice of an element $\mathbf{x} \in E$, according to the given probability distribution, results in a *distribution of shares* $((s_i)_{i \in P}, s)$, where $s_i = \pi_i(\mathbf{x}) \in E_i$ is the *share* of the participant $i \in P$ and $s = \pi_D(\mathbf{x}) \in E_D$ is the *shared secret value*.

A participant is said to be *redundant* in an access structure if there is no minimal qualified set containing it. An access structure is *connected* if there is not any redundant participant in it.

The ratio $\rho(\Sigma) = H(E_D)/\max_{i \in P} H(E_i)$ is called the *information rate* of the scheme Σ, and the *optimal information rate* $\rho(\Gamma)$ of the access structure Γ is the supremum of the information rates of all secret sharing schemes with access structure Γ. It is not difficult to check that $H(E_i) \geq H(E_D)$ for every non-redundant participant $i \in P$, and hence $\rho(\Sigma) \leq 1$. Secret sharing schemes with $\rho(\Sigma) = 1$ are said to be *ideal* and their access structures are called *ideal* as well. Of course, $\rho(\Gamma) = 1$ for every ideal access structure Γ.

If the sets E and E_i are vector spaces over some finite field $\mathbb{K}$, the mappings π_i are linear mappings, and the uniform probability distribution is considered in E, we say that Σ is a $\mathbb{K}$-*linear secret sharing scheme*. The linear schemes in which $E_i = \mathbb{K}$ for every $i \in Q$ are ideal and they are called $\mathbb{K}$-*vector space secret sharing schemes*. Their access structures are called $\mathbb{K}$-*vector space access structures*. Observe that there exist ideal linear schemes that are not vector space secret sharing schemes. In such schemes, $\dim E_i = \dim E_D > 1$ for every $i \in P$.

We notate $\mathcal{P}(Q)$ for the power set of Q. Given a secret sharing scheme Σ on the set $P = Q - \{D\}$, consider the mapping $h \colon \mathcal{P}(Q) \to \mathbb{R}$ defined by $h(X) = H(X)/H(E_D)$. This mapping satisfies the following properties [13].

1. $h(\emptyset) = 0$, and
2. h is *monotone increasing*: if $X \subseteq Y \subseteq Q$, then $h(X) \leq h(Y)$, and
3. h is *submodular*: if $X, Y \subseteq Q$, then $h(X \cup Y) + h(X \cap Y) \leq h(X) + h(Y)$, and
4. for every $X \subseteq Q$, either $h(X \cup \{D\}) = h(X) + 1$ or $h(X \cup \{D\}) = h(X)$.

A *polymatroid* is any pair $\mathcal{S} = (Q, h)$ satisfying the first three properties. Polymatroids satisfying the fourth property as well will be called here *D-secret sharing polymatroids*, or *D-ss-polymatroids* for short. Therefore, every secret sharing scheme Σ defines a *D*-ss-polymatroid $\mathcal{S} = \mathcal{S}(\Sigma) = (Q, h)$. Nevertheless, there exist *D*-ss-polymatroids that are not associated with any secret sharing scheme.

For a *D*-ss-polymatroid $\mathcal{S} = (Q, h)$, we consider the access structure

$$\Gamma_D(\mathcal{S}) = \{A \subseteq P : h(A \cup \{D\}) = h(A)\}.$$

Clearly, the access structure of a secret sharing scheme Σ is the one defined in this way by the associated polymatroid $\mathcal{S}(\Sigma)$. Since there exists a secret sharing scheme for every access structure Γ, all access structures are of the form $\Gamma_D(\mathcal{S})$ for some *D*-ss-polymatroid $\mathcal{S}$. Nevertheless, different *D*-ss-polymatroids can define the same access structure.

A *matroid* can be defined as a polymatroid $\mathcal{M} = (Q, h)$ with the following additional property.

4'. $h(X) \in \mathbb{Z}$ and $0 \leq h(X) \leq |X|$ for every $X \subseteq Q$, or, equivalently, for every $X \subseteq Q$ and $x \in Q$, either $h(X \cup \{x\}) = h(X) + 1$ or $h(X \cup \{x\}) = h(X)$.

We need to recall now some terminology and basic facts about matroids. For a matroid $\mathcal{M} = (Q, r)$ (we change from h to r because this is the usual notation for matroids), the set Q and the mapping r are called, respectively, the *ground set* and the *rank function* of the matroid $\mathcal{M}$. The value $r(X)$ is called the *rank* of the subset X while the *rank of the matroid* $\mathcal{M}$ is defined to be $r(\mathcal{M}) = r(Q)$. A subset $X \subseteq Q$ is said to be *independent* if $r(X) = |X|$. The *dependent* subsets are those that are not independent. A *circuit* is a minimally dependent subset while a *basis* is a maximally independent subset. All bases have the same number of elements, which coincide with the rank of the matroid.

As a consequence of the results by Brickell and Davenport [10], if Σ is an ideal scheme, then the polymatroid $\mathcal{S} = \mathcal{S}(\Sigma)$ is a matroid and, hence, $\mathcal{S}$ is a j-ss-polymatroid for every $j \in Q$. Moreover, by considering $(\pi_i(\mathbf{x}))_{i \in Q - \{j\}}$ as shares of the secret value $\pi_j(\mathbf{x})$, the scheme Σ defines an ideal secret sharing scheme with access structure $\Gamma_j(\mathcal{S})$ on the set of participants $Q - \{j\}$. We say that Γ is a *matroid-related* access structure if $\Gamma = \Gamma_D(\mathcal{M})$ for some matroid $\mathcal{M}$. It is not difficult to check that this definition is equivalent to the one we gave in the Introduction. Observe that the results by Brickell and Davenport [10] imply that all ideal access structures are matroid-related.

Let $\mathbb{K}$ be a finite field and let M be a $r_0 \times n$ matrix with entries in $\mathbb{K}$. If $|Q| = n$ and the points in Q are put in a one-to-one correspondence with the columns of M, a matroid $\mathcal{M}$ on the set Q is obtained by considering that the rank of a subset $X \subseteq Q$ is equal to the rank of the corresponding columns of M. In this situation, we say that the matrix M is a $\mathbb{K}$-*representation* of the matroid $\mathcal{M}$. The matroids that can be defined in this way are called *linearly representable*. Observe that linearly representable matroids coincide with the ones that are obtained from vector space secret sharing schemes and their related access structures are precisely the vector space access structures. The matroids that are associated with an ideal linear secret sharing scheme are called *multilinearly representable*, a class that contains the linearly representable matroids. The non-Pappus matroid is not linearly representable [29], but it was proved to be multilinearly representable in [35]. The existence of iss-representable matroids that are not multilinearly representable is an open problem.

The matroid $\mathcal{M}$ is said to be *connected* if, for every two different points $i, j \in Q$, there exists a circuit C with $i, j \in C$. As a consequence of [29, Proposition 4.1.2], the matroid $\mathcal{M}$ is connected if and only if the access structure $\Gamma_D(\mathcal{M})$ is connected. A connected matroid is determined by the circuits that contain some given point [21]. Therefore, if Γ is a matroid-related connected access structure, there exists a unique matroid $\mathcal{M}$ with $\Gamma = \Gamma_D(\mathcal{M})$.

3   Polymatroids and Optimal Information Rate

Most of the upper bounds on the optimal information rate that have been given until now were obtained by information-theoretical arguments. Specifically, by using basic properties of the Shannon entropy function. Csirmaz [13] pointed out that all those results are based solely on the so-called *Shannon inequalities* on the entropy of subsets of variables and, hence, they can be deduced from the fact that every secret sharing scheme defines a D-ss-polymatroid related to the access structure.

If $\mathcal{S} = (Q, h)$ is a polymatroid, we define $\sigma(\mathcal{S}) = \max\{h(\{x\}) : x \in Q\}$. For every access structure Γ, we consider the value $\kappa(\Gamma) = \inf \sigma(\mathcal{S})$, where the infimum is taken over all D-ss-polymatroids $\mathcal{S}$ with $\Gamma = \Gamma_D(\mathcal{S})$. The upper bounds on the optimal information rate that can be obtained by using polymatroids (that is, by using Shannon inequalities) are based on the following proposition.

Proposition 5. *The optimal information rate of every access structure Γ is upper bounded by $\rho(\Gamma) \leq 1/\kappa(\Gamma)$.*

Proof. Let Σ be a secret sharing scheme with access structure Γ and let $\mathcal{S}$ be the D-ss-polymatroid defined by Σ. Then $\rho(\Sigma) = 1/\sigma(\mathcal{S}) \leq 1/\kappa(\Gamma)$.   □

Therefore, upper bounds on $\rho(\Gamma)$ can be found by deriving lower bounds on $\kappa(\Gamma)$ from combinatorial properties of the access structure. Actually, $1/\kappa(\Gamma)$ is the best upper bound that can be obtained by this technique. Since $\kappa(\Gamma)$ deals only with the properties of the D-ss-polymatroids $\mathcal{S}$ such that $\Gamma = \Gamma_D(\mathcal{S})$, and some of these polymatroids may not be associated with any secret sharing scheme, there can exist access structures Γ such that $\rho(\Gamma) < 1/\kappa(\Gamma)$. As far as we know, no examples of access structures in this situation are known, but Theorem 8 gives some intuition supporting their existence.

Since $\kappa(\Gamma) = 1$ if Γ is matroid-related, it is clear that no upper bounds on the optimal information rate of matroid-related access structures can be found by using this method.

As far as we know, the only known upper bounds that do not fit this pattern are the one given by Gál [17], which was improved in [28], and the one presented by Beimel and Livne [2]. The first one applies only to linear secret sharing schemes and it is the basis for proving the separation between the complexities of linear and nonlinear schemes [1,4]. The second one applies to the access structures related to the Vamos matroid.

As an example of the kind of results that are obtained by using polymatroids, we present the *independent sequence method*, which was introduced in [6] and was improved in [30]. Let Γ be an access structure on a set of participants P. Consider $A \subseteq P$ and an increasing sequence of subsets $B_1 \subseteq \cdots \subseteq B_m \subseteq P$. We say that $(B_1, \ldots, B_m \,|\, A)$ is an *independent sequence* in Γ with *length* m and *size* s if $|A| = s$ and, for every $i = 1, \ldots, m$, there exists $X_i \subseteq A$ such that $B_i \cup X_i \in \Gamma$, while $B_m \notin \Gamma$ and $B_{i-1} \cup X_i \notin \Gamma$ if $i \geq 2$. The independent sequence method is based on the following result. We notice that this theorem was not stated in [6,30] in terms of polymatroids, but in terms of the entropy function. The proof in [6] is easily adapted to this new statement.

Theorem 6. ([6,30]) *Let Γ be an access structure on the set P. Let $\mathcal{S} = (Q, h)$ be a D-ss-polymatroid such that $\Gamma = \Gamma_D(\mathcal{S})$. If there exists in Γ an independent sequence $(B_1, \ldots, B_m \,|\, A)$ with length m and size s, then $h(A) \geq m$. As a consequence, $\kappa(\Gamma) \geq m/s$ and $\rho(\Gamma) \leq s/m$.*

The following corollary of that theorem points out that independent sequences can be used in the characterization of matroid-related access structures. Actually, the converse of this result will be proved in Section 4.

Corollary 7. *An access structure is not matroid-related if it admits an independent sequence with length m and size $s < m$.*

The next result by Csirmaz [13] points out the limitations of the use of polymatroids to find upper bounds on the optimal information rate.

Theorem 8. ([13]) *If Γ is an access structure on a set P of participants with $|P| = n$, then $\kappa(\Gamma) \le n$.*

Proof. It is not difficult to prove that there exists a D-ss-polymatroid $\mathcal{S} = (Q, h)$ with $\Gamma = \Gamma_D(\mathcal{S})$ such that $h(X) = n + (n - 1) + \cdots + (n - (k - 1))$ for every subset of participants $X \subseteq P$ with $|X| = k$. $\qquad\square$

By taking into account the known methods to construct secret sharing schemes, it is against intuition to suppose that there can exist, for every access structure, a secret sharing scheme such that the length of the shares is around n times the length of the secret. Therefore, as a consequence of Theorem 8, it seems that the optimal information rate of an access structure will be in general much smaller than $1/\kappa(\Gamma)$, the best upper bound that can be obtained by using polymatroids. Nevertheless, besides the Shannon inequalities, the properties of the entropy function imply other inequalities, the so-called *non-Shannon inequalities*. Thus, it might be possible to find better upper bounds on the optimal information rate than the ones derived from Proposition 5 by using information theory. This may be the case for matroid-related access structures as well.

Anyway, the polymatroid technique has proved to be very useful when studying some particular families of access structures. In some cases the obtained upper bounds are tight or, at least, close to the best known lower bounds. In the following we prove a positive result for the polymatroid technique. Namely, we prove in Theorem 12 that the bounds that are obtained by this technique for an access structure apply also to its dual.

Before presenting our result, we recall some facts about dual access structures and dual matroids. The *dual* of the access structure Γ on the set P is defined as the access structure $\Gamma^* = \{A \subseteq P : P - A \notin \Gamma\}$. If $\mathcal{M} = (Q, r)$ is a matroid, the mapping $r^* \colon \mathcal{P}(Q) \to \mathbb{Z}$ defined by $r^*(X) = |X| - r(Q) + r(Q - X)$ is the rank function of a matroid $\mathcal{M}^* = (Q, r^*)$, which is called the *dual* of the matroid $\mathcal{M}$. Since $\Gamma_D(\mathcal{M}^*) = (\Gamma_D(\mathcal{M}))^*$, the dual of a matroid-related access structure is matroid-related. If Σ is an ideal secret sharing scheme with access structure Γ, then there exists a linear scheme Σ^* with access structure Γ^* and information rate $\rho(\Sigma^*) = \rho(\Sigma)$ [14]. Actually, Σ can be seen as a linear code, and the linear scheme Σ^* is the one constructed from the dual code. As a consequence, if a matroid is linearly or multilinearly representable, the same applies to the dual matroid. Nevertheless, it is not known whether the dual of an iss-representable matroid is iss-representable, and the relation between $\rho(\Gamma)$ and $\rho(\Gamma^*)$ is an open problem too. Our result, Theorem 12, deals with this open problem. Specifically, we prove that the upper bounds for $\rho(\Gamma)$ that are obtained by the polymatroid technique apply also to $\rho(\Gamma^*)$.

There exist several inequivalent ways to define the dual of a polymatroid [39] and we have to choose the suitable one to prove our result. Specifically, if $\mathcal{S} = (Q, h)$ is a polymatroid, we consider the *dual polymatroid* $\mathcal{S}^* = (Q, h^*)$, where $h^* \colon \mathcal{P}(Q) \to \mathbb{R}$ is defined by $h^*(X) = \sum_{x \in X} h(\{x\}) - h(Q) + h(Q - X)$. This definition generalizes the duality that is usually considered for matroids. Clearly, if $\mathcal{M} = (Q, r)$ is a *loopless* matroid, that is, with $r(\{x\}) = 1$ for every $x \in Q$, then the dual matroid of $\mathcal{M}$ coincides with the dual polymatroid. We prove in the next lemma that $\mathcal{S}^*$ is actually a polymatroid, and we describe in Lemma 10 the relation between the dual of a D-ss-polymatroid and the dual of the corresponding access structure.

Lemma 9. $\mathcal{S}^* = (Q, h^*)$ *is a polymatroid.*

Proof. Obviously, $h^*(\emptyset) = 0$. Take a subset $X \subseteq Q$ and a point $y \notin X$. Since $h(\{y\}) + h(Q - (X \cup \{y\})) \geq h(Q - X)$, we get that $h^*(X \cup \{y\}) \geq h^*(X)$. Therefore, h^* is monotone increasing. Finally, consider two arbitrary subsets $X, Y \subseteq Q$. Then from the definition of h^* and the submodularity of h,

$$h^*(X) + h^*(Y) - h^*(X \cup Y) - h^*(X \cap Y) =$$

$$= h(Q - X) + h(Q - Y) - h(Q - (X \cup Y)) - h(Q - (X \cap Y)) \geq 0.$$

This proves that h^* is submodular.   □

Lemma 10. *Let $\mathcal{S} = (Q, h)$ be a D-ss-polymatroid. Assume that $\Gamma_D(\mathcal{S}) \neq \emptyset$ and $\emptyset \notin \Gamma_D(\mathcal{S})$. Then $\mathcal{S}^* = (Q, h^*)$ is also a D-ss-polymatroid and $\Gamma_D(\mathcal{S}^*) = (\Gamma_D(\mathcal{S}))^*$.*

Proof. Let $\Gamma = \Gamma_D(\mathcal{S})$. Since $\emptyset \notin \Gamma$ and $P = Q - \{D\} \in \Gamma$, we have that $h(\{D\}) = 1$ and $h(P) = h(Q)$, and hence $h^*(\{D\}) = 1$. Consider a subset $X \subseteq P$. Then $h^*(X \cup \{D\}) = h(\{D\}) + \sum_{x \in X} h(\{x\}) - h(Q) + h(P - X)$. If $X \in \Gamma^*$, then $P - X \notin \Gamma$ and $h(P - X) = h(Q - X) - 1$. In this case, $h^*(X \cup \{D\}) = h^*(X)$. Analogously, if $X \notin \Gamma^*$ then $h(P - X) = h(Q - X)$, and hence $h^*(X \cup \{D\}) = h^*(X) + 1$.   □

To be precise, the polymatroid $\mathcal{S}^*$ is properly a dual of $\mathcal{S}$, in the sense that $\mathcal{S}^{**} = \mathcal{S}$, if and only if $h(Q - \{x\}) = h(Q)$ for every $x \in Q$. The polymatroids satisfying this property will be said to be *normalized*. In addition, we need some technical results that are given in the next lemma, whose proof is an easy exercise.

Lemma 11. *Let $\mathcal{S} = (Q, h)$ be a polymatroid. Then the following properties hold.*

1. *The polymatroid $\mathcal{S}^* = (Q, h^*)$ is normalized.*
2. *$h^{**}(X) \leq h(X)$ for every $X \subseteq Q$.*
3. *$\mathcal{S}$ is normalized if and only if $\mathcal{S}^{**} = \mathcal{S}$.*
4. *If $\mathcal{S}$ is normalized, then $h^*(\{x\}) = h(\{x\})$ for every $x \in Q$.*

Theorem 12. *Let Γ be an access structure with $\Gamma \neq \emptyset$ and $\emptyset \notin \Gamma$, and let Γ^* be its dual. Then $\kappa(\Gamma) = \kappa(\Gamma^*)$.*

Proof. Let Γ be an access structure. Consider the sets of real numbers $\Omega(\Gamma) = \{\sigma(\mathcal{S}) : \Gamma = \Gamma_D(\mathcal{S})\}$ and $\widehat{\Omega}(\Gamma) = \{\sigma(\mathcal{S}) : \mathcal{S} \text{ is normalized}, \Gamma = \Gamma_D(\mathcal{S})\}$. If $\mathcal{S}$ is a D-ss-polymatroid such that $\Gamma = \Gamma_D(\mathcal{S})$, then $\mathcal{S}^{**}$ is normalized, $\Gamma = \Gamma_D(\mathcal{S}^{**})$ and $\sigma(\mathcal{S}^{**}) \leq \sigma(\mathcal{S})$. Therefore, $\kappa(\Gamma) = \inf \Omega(\Gamma) = \inf \widehat{\Omega}(\Gamma)$. The proof is concluded by taking into account that $\widehat{\Omega}(\Gamma) = \widehat{\Omega}(\Gamma^*)$.   □

4   On the Characterization of Matroid-Related Access Structures

4.1   A Theorem by Seymour

Let Γ be an access structure on a set P and take a subset $Z \subseteq P$. We define the access structures $\Gamma \backslash Z$ and Γ / Z on the set $P - Z$ by $\Gamma \backslash Z = \{A \subseteq P - Z : A \in \Gamma\}$ and $\Gamma / Z = \{A \subseteq P - Z : A \cup Z \in \Gamma\}$. Every access structure that can be obtained from Γ by repeatedly applying the operations $\backslash$ and $/$ is called a *minor of the access structure* Γ. If Z_1 and Z_2 are disjoint subsets then $(\Gamma \backslash Z_1)/Z_2 = (\Gamma/Z_2) \backslash Z_1$, and $(\Gamma \backslash Z_1) \backslash Z_2 = \Gamma \backslash (Z_1 \cup Z_2)$, and $(\Gamma/Z_1)/Z_2 = \Gamma/(Z_1 \cup Z_2)$. Therefore, every minor of Γ is of the form $(\Gamma \backslash Z_1)/Z_2$ for some disjoint subsets $Z_1, Z_2 \subseteq P$. In addition, $(\Gamma \backslash Z)^* = \Gamma^*/Z$ and $(\Gamma/Z)^* = \Gamma^* \backslash Z$.

We can consider as well *minors* of matroids and polymatroids. Let $\mathcal{S} = (Q, h)$ be a polymatroid. Given a subset $Z \subseteq Q$, we define the polymatroids $\mathcal{S} \backslash Z = (Q - Z, h_{\backslash Z})$ and $\mathcal{S}/Z = (Q - Z, h_{/Z})$, where $h_{\backslash Z}(X) = h(X)$ and $h_{/Z}(X) = h(X \cup Z) - h(Z)$ for every $X \subseteq Q - Z$. It is not difficult to prove that, if $\mathcal{S}$ is a D-ss-polymatroid and $\Gamma = \Gamma_D(\mathcal{S})$, then for every $Z \subseteq P$, both $\mathcal{S} \backslash Z$ and $\mathcal{S}/Z$ are D-ss-polymatroids and $\Gamma \backslash Z = \Gamma_D(\mathcal{S} \backslash Z)$ and $\Gamma/Z = \Gamma_D(\mathcal{S}/Z)$. Moreover, if $\mathcal{M} = (Q, r)$ is a matroid, then $\mathcal{M} \backslash Z$ and $\mathcal{M}/Z$ are matroids as well. The following proposition is a direct consequence of all these considerations.

Proposition 13. *Every minor of a matroid-related access structure is matroid-related.*

We introduce now the forbidden minors in the characterization by Seymour. The set of participants of the access structures Φ and $\widehat{\Phi}$ is $P = \{p_1, p_2, p_3, p_4\}$. The minimal qualified subsets of Φ are $\{p_1, p_2\}$, $\{p_2, p_3\}$ and $\{p_3, p_4\}$, while the minimal qualified subsets $\widehat{\Phi}$ are $\{p_1, p_2\}$, $\{p_2, p_3\}$, $\{p_2, p_4\}$ and $\{p_3, p_4\}$. For every $s \geq 3$, the set of participants of the access structure Ψ_s is $P = \{p_1, \dots, p_s, p_{s+1}\}$ and its minimal qualified subsets are $\{p_1, \dots, p_s\}$ and $\{p_i, p_{s+1}\}$ for every $i = 1, \dots, s$. Observe that $\Phi^* \cong \Phi$ and $\Psi_s^* = \Psi_s$. The minimal qualified subsets of $\widehat{\Phi}^*$ are $\{p_1, p_3, p_4\}$, $\{p_2, p_3\}$ and $\{p_2, p_4\}$.

The forbidden minor characterization of matroid ports by Seymour is stated here in our terminology.

Theorem 14. (Seymour [32]) *An access structure is matroid-related if and only if it has no minor isomorphic to Φ, $\widehat{\Phi}$, $\widehat{\Phi}^*$ or Ψ_s with $s \geq 3$.*

4.2   Generalizing the Result by Brickell and Davenport

New characterizations of matroid-related access structures are given in Theorem 17. They are obtained by combining Theorem 14 with the results in Section 3. As a consequence we obtain Theorem 4, a generalization of the result by Brickell and Davenport [10].

We need to introduce two technical results that are used in the proof of Theorem 17. First, the independent sequence method we have described in Section 3 has a good behavior with respect to minors, and second, all the forbidden minors in Seymour's characterization admit an independent sequence with length $m = 3$ and size $s = 2$.

Lemma 15. *Let Γ' be a minor of an access structure Γ. If there exists in Γ' an independent sequences with length m and size s, then the same occurs for Γ.*

Proof. Consider disjoint subsets $Z_1, Z_2 \subseteq P$ such that $\Gamma' = (\Gamma \backslash Z_1)/Z_2$. Suppose that $(B_1, \ldots, B_m \mid A)$ is an independent sequence with length m and size $s = |A|$ in Γ'. Then $(B_1 \cup Z_2, \ldots, B_m \cup Z_2 \mid A)$ is an independent sequence in Γ. $\square$

Proposition 16. *Every one of the access structures Φ, $\widehat{\Phi}$, $\widehat{\Phi}^*$, and Ψ_s with $s \geq 3$ admits an independent sequence with length $m = 3$ and size $s = 2$.*

Proof. We are going to consider sequences $(B_1, B_2, B_3 \mid a_1 a_2)$ with $B_1 \subseteq B_2 \subseteq B_3 \subseteq P$ and $a_1, a_2 \in P$. Such a sequence will be independent in the access structure Γ if the subsets $B_1 \cup \{a_1, a_2\}$, $B_2 \cup \{a_1\}$ and $B_3 \cup \{a_2\}$ are in Γ while $B_1 \cup \{a_1\}$, $B_2 \cup \{a_2\}$ and B_3 are not in Γ. The sequence $(\emptyset, \{p_1\}, \{p_1, p_4\} \mid p_2 p_3)$ is independent for both Φ and $\widehat{\Phi}$, while an independent sequence for $\widehat{\Phi}^*$ is $(\emptyset, \{p_4\}, \{p_1, p_4\} \mid p_2 p_3)$. Finally, $(\emptyset, \{p_s\}, \{p_2, \ldots, p_s\} \mid p_{s+1} p_1)$ is an independent sequence in Ψ_s. $\square$

Theorem 17. *Let Γ be an access structure. Then the following statements are equivalent.*

1. *Γ is matroid-related.*
2. *There does not exist in Γ any independent sequence with length m and size $s < m$.*
3. *There does not exist in Γ any independent sequence with length $m = 3$ and size $s = 2$.*
4. *$\kappa(\Gamma) < 3/2$.*

Proof. If Γ is matroid-related, then $\kappa(\Gamma) = 1$ and, by Corollary 7, there does not exist in Γ any independent sequence with length m and size $s < m$. In addition, by Theorem 6, there does not exist in Γ any independent sequence with length $m = 3$ and size $s = 2$ if $\kappa(\Gamma) < 3/2$. Finally, if Γ is not matroid-related, then there exists a minor Γ' of Γ that is isomorphic to one of the forbidden minors in Theorem 14. From Proposition 16, Γ' admits an independent sequence with length $m = 3$ and size $s = 2$ and, by Lemma 15, the same occurs with Γ. $\square$

Two direct consequences of Theorem 17 are stated in Corollary 18. Our main result, Theorem 4, is proved from the second one. As we said before, we have obtained in this way a generalization of the important result by Brickell and Davenport [10], who proved that the access structure of every ideal secret sharing scheme is matroid-related. Moreover, since the result by Brickell and Davenport has not been used in the proof of Theorem 17, we have presented here an alternative proof for it.

Corollary 18. *Let Γ be an access structure. Then the following statements hold.*

1. *Γ is matroid-related if and only if $\kappa(\Gamma) = 1$.*
2. *If Γ is not matroid-related, then $\kappa(\Gamma) \geq 3/2$, and hence $\rho(\Gamma) \leq 2/3$.*

This result implies a gap in the values of $\kappa(\Gamma)$. Namely, there does not exist any access structure Γ with $1 < \kappa(\Gamma) < 3/2$. This gap does not mean that the corresponding gap appears in the values of the optimal information rate $\rho(\Gamma)$. Specifically, the existence of non-ideal matroid-related access structures Γ with $2/3 < \rho(\Gamma) < 1$ is an open problem.

5   On Non-ideal Matroid-Related Access Structures

Since there exist matroids that are not iss-representable, there are matroid-related access structures that are not ideal. Very little is known about the optimal information rate of these structures. We cannot find upper bounds by the techniques in Section 3 because $\kappa(\Gamma) = 1$ if Γ is matroid-related. By using other techniques, upper bounds have been given by Beimel and Livne [2]. We present here some lower bounds on the optimal information rate of the access structures related to the Vamos matroid and the non-Desargues matroid.

The *Vamos matroid* $\mathcal{V}$ is the matroid on the set $Q_1 = \{v_1, \ldots, v_8\}$ such that its bases are all sets with cardinality 4 except the following five: $\{v_1, v_2, v_3, v_4\}$, $\{v_1, v_2, v_5, v_6\}$, $\{v_3, v_4, v_5, v_6\}$, $\{v_3, v_4, v_7, v_8\}$ and $\{v_5, v_6, v_7, v_8\}$. The Vamos matroid is not iss-representable [33] and, hence, the access structures related to it are not ideal. In a recent work, Beimel and Livne [2] prove that, for every secret sharing scheme realizing one of these access structures with domain of the secrets of size k, the size of the domain of the shares is at least $k + \Omega(\sqrt{k})$. Observe that this bound does not exclude that the optimal information rate of these structures may be equal to one, because $\rho(\Gamma)$ is the *supremum* of the information rates of the schemes realizing Γ.

The *non-Desargues matroid* $\mathcal{N}$ is the matroid with rank 3 on a set with 10 points determined by a non-Desargues configuration on a projective plane. That is, take three different lines L_1, L_2, L_3 that meet in a point p_0 and, on the line L_i, two different points $q_i, r_i \neq p_0$. Finally, consider the points s_{12}, s_{23}, and s_{31}, where s_{ij} is the intersection of the lines $q_i q_j$ and $r_i r_j$. If such a configuration has been taken on a projective plane over a field, the points s_{12}, s_{23} and s_{31} must be collinear by the Desargues' Theorem. The non-Desargues matroid is defined by this configuration but considering that the three points s_{ij} are not collinear. That is, the set of points of $\mathcal{N}$ is $Q_2 = \{p_0, q_1, q_2, q_3, r_1, r_2, r_3, s_{12}, s_{23}, s_{31}\}$, and the bases are all subsets with three points that are not supposed to be collinear. As a consequence of the Desargues' Theorem, this matroid is not linearly representable. Moreover, Matúš [27] proved that it is not iss-representable.

Lower bounds on the optimal information rate of the access structures related to those matroids are given in the next theorem. We just present here a sketch of the proof. All details will be discussed in the full version.

Theorem 19. *Consider two arbitrary points $D_1 \in Q_1$ and $D_2 \in Q_2$ and the access structures $\Gamma_1 = \Gamma_{D_1}(\mathcal{V})$ and $\Gamma_2 = \Gamma_{D_2}(\mathcal{N})$. Then $\rho(\Gamma_1) \geq 2/3$ and $\rho(\Gamma_2) \geq 3/4$.*

Proof. Suppose that $D_1 = v_1$. For every $2 \leq i < j \leq 8$, let $\Gamma^{(i,j)}$ be the access structure on P_1 whose minimal qualified subsets are the minimal qualified subsets A of Γ_1 such that $\{v_i, v_j\} \not\subseteq A$. It can be proved that $\Gamma^{(3,4)}$, $\Gamma^{(5,6)}$ and $\Gamma^{(7,8)}$ are $\mathbb{K}$-vector space access structures for some finite field $\mathbb{K}$. By applying the λ-decomposition technique to these substructures, we get that $\rho(\Gamma_1) \geq 2/3$. A similar construction can be obtained for other values of $D_1 \in Q_1$.

There exists a finite field $\mathbb{K}$ such that, for every $x \in P_2 = Q_2 - \{D_2\}$, the matroid $\mathcal{N} \backslash \{x\}$ is $\mathbb{K}$-representable and, hence, $\Gamma_2 \backslash \{x\}$ is a $\mathbb{K}$-vector space access structure. Therefore, we can apply the λ-decomposition technique by Stinson [38] to the nine access structures $\{\Gamma_2 \backslash \{x\}\}_{x \in P_2}$. By doing that, a secret sharing scheme for Γ_2 with information rate equal to $3/4$ is obtained. $\square$

Table 1

Access structures of. . .		Access structures related to. . .
SSS with $\rho > 2/3$	$\Longrightarrow$ [here] $\Longleftarrow$?	Matroids
Ideal SSS	$\Longleftrightarrow$ [10]	⇑ ⇓̸ [27,33] Iss-representable matroids
Ideal linear SSS	$\Longleftrightarrow$	⇑ ⇓? Multilinearly representable matroids
Vector space SSS	$\Longleftrightarrow$ [9]	⇑ ⇓̸ [35] Linearly representable matroids

6   Open Problems

The known results about the connection between secret sharing and matroids, including our main result, are summarized in Table 1. Equally, some open problems appear there. The following open problem was posed in [24,26].

Problem 20. Is there any access structure Γ with $2/3 < \rho(\Gamma) < 1$?

From Theorem 4, if such an access structure exists, it must be matroid-related. We proved before that there exist non-ideal matroid-related access structures Γ with $\rho(\Gamma) \geq 3/4$. Nevertheless, it is possible that $\rho(\Gamma) = 1$ even if Γ is not ideal. Observe that the results in [2] about the length of the shares for the access structures related to the Vamos matroid do not imply an affirmative answer to Problem 20. Actually, very little is known about the optimal information rate of non-ideal matroid-related access structures.

Problem 21. Is there any matroid-related access structure Γ with $\rho(\Gamma) < 1$? And with $\rho(\Gamma) \leq 2/3$?

The existence of ideal access structures that are not realized by any ideal linear secret sharing scheme is another unsolved question, which is equivalent to the following open problem.

Problem 22. Is there any iss-representable matroid that is not multilinearly representable?

Even though the existence of access structures Γ with $\rho(\Gamma) < 1/\kappa(\Gamma)$ is quite natural from Theorem 8, no actual example is known.

Problem 23. Present an access structure Γ with $\rho(\Gamma) < 1/\kappa(\Gamma)$.

Acknowledgments

The second author thanks Ronald Cramer, Bert Gerards, Robbert de Haan and Lex Schrijver for useful discussions, comments, and suggestions. Thanks to Lex Schrijver, who pointed out the existence of the paper by P.D. Seymour on matroid ports [32], and to Bert Gerards, who independently found the construction in Theorem 8.

References

1. A. Beimel, Y. Ishai. On the power of nonlinear secret sharing schemes. *SIAM J. Discrete Math.* **19** (2005) 258–280.
2. A. Beimel, N. Livne. On Matroids and Non-ideal Secret Sharing. *Third Theory of Cryptography Conference, TCC 2006. Lecture Notes in Comput. Sci.* **3876** (2006) 482–501.
3. A. Beimel, T. Tassa, E. Weinreb. Characterizing Ideal Weighted Threshold Secret Sharing. *Second Theory of Cryptography Conference, TCC 2005. Lecture Notes in Comput. Sci.* **3378** (2005) 600–619.
4. A. Beimel, E. Weinreb. Separating the power of monotone span programs over different fields. *SIAM J. Comput.* **34** (2005) 1196–1215.
5. G.R. Blakley. Safeguarding cryptographic keys. *AFIPS Conference Proceedings.* **48** (1979) 313–317.
6. C. Blundo, A. De Santis, R. De Simone, U. Vaccaro. Tight bounds on the information rate of secret sharing schemes. *Des. Codes Cryptogr.* **11** (1997) 107–122.
7. C. Blundo, A. De Santis, L. Gargano, U. Vaccaro. On the information rate of secret sharing schemes. *Advances in Cryptology - CRYPTO'92. Lecture Notes in Comput. Sci.* **740** (1993) 148–167.
8. C. Blundo, A. De Santis, D.R. Stinson, U. Vaccaro. Graph decompositions and secret sharing schemes. *J. Cryptology* **8** (1995) 39–64.
9. E.F. Brickell. Some ideal secret sharing schemes. *J. Combin. Math. and Combin. Comput.* **9** (1989) 105–113.
10. E.F. Brickell, D.M. Davenport. On the classification of ideal secret sharing schemes. *J. Cryptology* **4** (1991) 123–134.
11. E.F. Brickell, D.R. Stinson. Some improved bounds on the information rate of perfect secret sharing schemes. *J. Cryptology* **5** (1992) 153–166.

12. R.M. Capocelli, A. De Santis, L. Gargano, U. Vaccaro. On the size of shares of secret sharing schemes. *J. Cryptology* **6** (1993) 157–168.
13. L. Csirmaz. The size of a share must be large. *J. Cryptology* **10** (1997) 223–231.
14. S. Fehr. Efficient Construction of the Dual Span Program. Manuscript.
15. S. Fujishige. Entropy functions and polymatroids—combinatorial structures in information theory. *Electron. Comm. Japan* **61** (1978) 14–18.
16. S. Fujishige. Polymatroidal Dependence Structure of a Set of Random Variables. *Information and Control* **39** (1978) 55–72.
17. A. Gál. A characterization of span program size and improved lower bounds for monotone span programs. *Proceedings of 30th ACM Symposium on the Theory of Computing, STOC 1998* (1998) 429–437.
18. M. Ito, A. Saito, T. Nishizeki. Secret sharing scheme realizing any access structure. *Proc. IEEE Globecom'87* (1987) 99–102.
19. W.-A. Jackson, K.M. Martin. Perfect secret sharing schemes on five participants. *Des. Codes Cryptogr.* **9** (1996) 267–286.
20. E.D. Karnin, J.W. Greene, M.E. Hellman. On secret sharing systems. *IEEE Trans. Inform. Theory* **29** (1983) 35–41.
21. A. Lehman. A solution of the Shannon switching game. *J. Soc. Indust. Appl. Math.* **12** (1964) 687–725.
22. A. Lehman. Matroids and Ports. *Notices Amer. Math. Soc.* **12** (1965) 356–360.
23. J.Martí-Farré, C. Padró. Secret sharing schemes on sparse homogeneous access structures with rank three. *Electronic Journal of Combinatorics* **11(1)** (2004) Research Paper 72, 16 pp. (electronic).
24. J. Martí-Farré, C. Padró. Secret sharing schemes with three or four minimal qualified subsets. *Des. Codes Cryptogr.* **34** (2005) 17–34.
25. J.Martí-Farré, C. Padró. Ideal secret sharing schemes whose minimal qualified subsets have at most three participants. *Fifth Conference on Security and Cryptography for Networks, SCN 2006. Lecture Notes in Comput. Sci.* **4116** (2006) 201–215.
26. J. Martí-Farré, C. Padró. Secret sharing schemes on access structures with intersection number equal to one. *Discrete Applied Mathematics* **154** (2006) 552–563.
27. F. Matúš. Matroid representations by partitions. *Discrete Math.* **203** (1999) 169–194.
28. V. Nikov, S. Nikova, B. Preneel. On the Size of Monotone Span Programs. *Fourth Conference on Security in Communication Networks - SCN 2004. Lecture Notes in Comput. Sci.* **3352** (2004) 252–265.
29. J.G. Oxley. *Matroid theory*. Oxford Science Publications. The Clarendon Press, Oxford University Press, New York, 1992.
30. C. Padró, G. Sáez. Secret sharing schemes with bipartite access structure. *IEEE Trans. Inform. Theory* **46** (2000) 2596–2604.
31. C. Padró, G. Sáez. Lower bounds on the information rate of secret sharing schemes with homogeneous access structure. *Inform. Process. Lett.* **83** (2002) 345–351. Quoting Marc Heymann Pignolo ¡mheymann@ma4.upc.edu¿:
32. P.D. Seymour. A forbidden minor characterization of matroid ports. *Quart. J. Math. Oxford Ser.* **27** (1976) 407–413.
33. P.D. Seymour. On secret-sharing matroids. *J. Combin. Theory Ser. B* **56** (1992) 69–73.
34. A. Shamir. How to share a secret. *Commun. of the ACM* **22** (1979) 612–613.
35. J. Simonis, A. Ashikhmin. Almost affine codes. *Des. Codes Cryptogr.* **14** (1998) 179–197.

36. D.R. Stinson. An explication of secret sharing schemes. *Des. Codes Cryptogr.* **2** (1992) 357–390.
37. D.R. Stinson. New general lower bounds on the information rate of secret sharing schemes. *Advances in Cryptology - CRYPTO'92. Lecture Notes in Comput. Sci.* **740** (1993) 168-182.
38. D.R. Stinson. Decomposition constructions for secret-sharing schemes. *IEEE Trans. Inform. Theory* **40** (1994) 118–125.
39. D.J.A. Welsh. *Matroid Theory*. Academic Press, London, 1976.

Secure Linear Algebra Using Linearly Recurrent Sequences

Eike Kiltz[1], Payman Mohassel[2], Enav Weinreb[3], and Matthew Franklin[4]

[1] CWI Amsterdam, The Netherlands
kiltz@cwi.nl
[2] Department of Computer Science, University of California, Davis CA 95616
mohassel@cs.ucdavis.edu
[3] Dept. of Computer Science, Technion, Haifa, Israel
weinreb@cs.technion.ac.il
[4] Department of Computer Science, University of California, Davis CA 95616
franklin@cs.ucdavis.edu

Abstract. In this work we present secure two-party protocols for various core problems in linear algebra. Our main result is a protocol to obliviously decide singularity of an encrypted matrix: Bob holds an $n \times n$ matrix, encrypted with Alice's secret key, and wants to learn whether or not the matrix is singular (while leaking nothing further). We give an interactive protocol between Alice and Bob that solves the above problem in $O(\log n)$ communication rounds and with overall communication complexity of roughly $O(n^2)$ (note that the input size is n^2). Our techniques exploit certain nice mathematical properties of linearly recurrent sequences and their relation to the minimal and characteristic polynomial of the input matrix, following [Wiedemann, 1986]. With our new techniques we are able to improve the round complexity of the communication efficient solution of [Nissim and Weinreb, 2006] from $O(n^{0.275})$ to $O(\log n)$.

At the core of our results we use a protocol that securely computes the minimal polynomial of an encrypted matrix. Based on this protocol we exploit certain algebraic reductions to further extend our results to the problems of securely computing rank and determinant, and to solving systems of linear equations (again with low round and communication complexity).

Keywords: Secure Linear Algebra, Linearly Recurrent Sequences, Wiedemann's Algorithm.

1 Introduction

Linear algebra plays a central role in computer science in general and in cryptography in particular. Numerous cryptographic applications such as private information retrieval, secret sharing schemes, and multi-party secure computation make use of linear algebra. In particular, the ability to solve a set of linear equations is an important algorithmic and cryptographic tool. In this work we

S.P. Vadhan (Ed.): TCC 2007, LNCS 4392, pp. 291–310, 2007.

design efficient and secure protocols for various linear algebra problems. Our protocols enjoy both low communication and round complexity.

The secure computation of many linear algebra tasks efficiently reduces to the following problem. Alice holds the private key of a public-key homomorphic encryption scheme, and Bob holds a square matrix A whose entries are encrypted under Alice's public key. Alice and Bob wish to decide whether A is singular while leaking no other information on A. Our protocol is based on an algorithm by Wiedemann for "black-box linear algebra" [24] which is highly efficient when applied to sparse matrices. This algorithm uses *linearly recurrent sequences* and their relation to the *greatest common divisor* problem for polynomials (see Section 3). Somewhat surprisingly, we design a secure protocol based on this algorithm which is applicable to *general* matrices. Previous secure protocols for linear algebra problems used basic linear algebra techniques such as Gaussian Elimination. Our protocols exploit more advanced properties of linear systems to achieve improved complexity bounds.

Cramer and Damgård initiated the study of secure protocols for solving various linear algebra problems [6]. Their work was done in the information theoretic multi-party setting, with the main focus on achieving constant round complexity. The communication complexity of their protocols is $\Omega(n^3)$ while the size of the inputs is just $O(n^2)$. A generic approach for designing secure protocols is to apply the garbled circuit method of Yao [25], for which the communication complexity is related to the Boolean circuit complexity of the underlying function. However, these linear algebra functions are strongly related to the problem of matrix multiplication [4], with essentially the same circuit complexity. The best known upper bound for matrix multiplication is $O(n^\omega)$ [5] for $\omega \cong 2.38$, which is still larger than the input size. In a recent paper, Nissim and Weinreb [19] introduced an oblivious singularity protocol with communication complexity of roughly $O(n^2)$. However, their protocol, which relies on the Gaussian elimination procedure, has round complexity $\Omega(n^{0.275})$, which is considered relatively high. The need for low round complexity is motivated by the fact that in most practical systems the time spent on sending and receiving messages is large compared to local computation time.

Our Results. We design a secure protocol for deciding singularity of a matrix, which gets the best of previous results, both in terms of communication and round complexity, up to a logarithmic factor. We achieve communication complexity of roughly $O(n^2)$ and $O(\log n)$ round complexity. Our constructions are secure, assuming the existence of a *homomorphic* public-key encryption scheme and a secure instantiation of Yao's garbled circuit protocol. The latter can be constructed using an appropriate symmetric key encryption and an oblivious transfer protocol which is secure against semi-honest adversaries. Using the protocol for deciding singularity, we design a secure protocol for solving a linear system $Ax = y$ based on an algorithm by Kaltofen and Saunders [14]. The technical difficulty in applying this algorithm is that it depends on the rank of the matrix A. Computing the rank of A in the clear would compromise the privacy of the protocol. We overcome this problem by designing a protocol for computing an *encryption* of the rank of an encrypted matrix. As the rank of a matrix is a basic concept in linear algebra, this protocol is of independent interest. The above techniques also yield communication and round efficient secure protocols for computing the minimal polynomial and the determinant of an encrypted matrix. Our results give rise to communication and round efficient secure protocols for problems that are reducible to linear algebra, e.g., perfect matching and problems with low span program complexity [15]. We summarize our main protocols in Table 1. Note that the outputs of our protocols are always encrypted, which in particular enables composition of our protocols. Thus, our protocols may be conveniently used as sub-protocols in other secure protocols.

Our protocols are designed under the assumption that Bob holds an encrypted version of the input and Alice holds the decryption key. In practice, secure linear algebra is often needed when the inputs of Alice and Bob are in the clear. However, applying simple reductions, we are able to give improved secure protocols for many natural problems of this kind. For example, consider the linear subspace intersection problem, in which each of Alice and Bob holds a subspace of $\mathbb{F}^n$ and they wish to securely decide whether there is a non-zero vector in the intersection of their input subspaces. Even for *insecure* computation, it is shown in [2] that the deterministic communication complexity of the problem is $\Omega(n^2)$. This result agrees with ours up to a logarithmic factor.[1] Another natural problem that we can compute securely and efficiently is solving a shared system of linear equations. Here Alice and Bob both hold independent systems of linear equations in the same variables. They jointly want to compute a solution vector that satisfies both sets of equations, without revealing anything about their secret inputs.

Table 1. Basic linear algebra protocols with $O(n^2 \log n \cdot \log |\mathbb{F}|)$ communication complexity and $O(\log n)$ rounds. Here $A \in \mathbb{F}^{n \times n}$ is a matrix and $x \in \mathbb{F}^n$ is a vector.

Protocol name	INPUT		OUTPUT	
	Bob	Alice	Bob	Alice
MINPOLY	$\mathrm{Enc}(A)$	SK	$\mathrm{Enc}(m_A)$	—
SINGULAR	$\mathrm{Enc}(A)$	SK	$\mathrm{Enc}(\det(A) = 0?)$	—
RANK	$\mathrm{Enc}(A)$	SK	$\mathrm{Enc}(\mathrm{rank}(A))$	—
DET	$\mathrm{Enc}(A)$	SK	$\mathrm{Enc}(\det(A))$	—
LINEAR SOLVE	$\mathrm{Enc}(A), \mathrm{Enc}(x)$	SK	$\mathrm{Enc}(y)$ (y random s.t. $Ax = y$)	—

Techniques. Our protocols rely on random reductions from computing linear algebra properties of a matrix $A \in \mathbb{F}^{n \times n}$ to computing the *minimal polynomial* m_A of a certain matrix related to A [24,14,11]. In particular, the singularity of A is related to the constant coefficient of this minimal polynomial, and the rank of A to its degree.

---

[1] Although determining the *randomized* communication complexity of subspace intersection is an open problem, it serves as an evidence that our upper bound may be tight.

Since no efficient secure protocol for computing the minimal polynomial of a shared matrix is known, we exploit another probabilistic reduction from this problem to computing the minimal polynomial of a particular *linearly recurrent sequence*. A sequence of field elements $\mathbf{a} = (a_i)_{i\in\mathbb{N}} \in \mathbb{F}^{\mathbb{N}}$ is linearly recurrent of order $n \in \mathbb{N}$ if there exist $f_0,\ldots,f_n \in \mathbb{F}$ with $f_n \neq 0$ such that $\sum_{j=0}^n f_j a_{i+j} = 0$, for all $i \in \mathbb{N}$. The field elements $f_0,\ldots,f_{2n-1}$ completely characterize the sequence $\mathbf{a}$ and are, roughly speaking, related to its minimal polynomial (see Section 3). Picking u, v uniformly in $\mathbb{F}^n$, the minimal polynomial of the linearly recurrent sequence $\mathbf{a} = (\mathbf{u}^\top A^i \mathbf{v})_{i\in\mathbb{N}}$ of order n coincides, with high probability, with the minimal polynomial of matrix A [11].

To securely compute the minimal polynomial of the above sequence, we first show how to compute the first $2n$ elements of $\mathbf{a}$ in low round and communication complexity. Then we use the Berlekamp/Massey algorithm [17] to reduce the problem to computing the *extended greatest common divisor* of two polynomials derived from the first $2n$ elements of sequence $\mathbf{a}$. Finally, we exploit the fact the the Boolean circuit complexity of the extended GCD algorithm is significantly smaller than that of the original linear algebraic function to apply Yao's garbled circuit method here. Moreover, we show a general technique to apply Yao's garbled circuit method from a starting point where Bob holds an encrypted input and Alice holds the decryption key. As we discussed earlier in the introduction, trying to apply Yao's construction directly to the original linear algebra problems would result in $\Omega(n^\omega)$ communication complexity.

Organization. In Section 2 we discuss the setting and some basic building blocks. In Section 3 we define linearly recurrent sequences and discuss their basic properties. Then, in Section 4, we show how to compute the minimal polynomial of an encrypted matrix. We design protocols for deciding singularity, computing rank and determinant of an encrypted matrix, and solving an encrypted linear system in Section 5. The appendices contain some additional details and applications of our secure protocols.

2   General Framework

Homomorphic encryption schemes. As a first step in our protocols, we reduce the original linear algebra problems to a state where Bob holds data encrypted by a public key homomorphic encryption scheme, and Alice holds the private decryption key. Our constructions use semantically-secure public-key encryption schemes that allow for simple computations on encrypted data. In particular, we use encryption schemes where given two encryptions $\mathsf{Enc}(m_1)$ and $\mathsf{Enc}(m_2)$, we can efficiently compute a random encryption $\mathsf{Enc}(m_1 + m_2)$. Note that this implies that given an encryption $\mathsf{Enc}(m)$ and $c \in \mathbb{F}$, we can efficiently compute a random encryption $\mathsf{Enc}(cm)$. We will be working with encryption of elements in a finite field. Paillier's [20] cryptosystem is an appropriate choice for this purpose. One minor issue is that the domain of Paillier's cryptosystem is the ring Z_n, where n is the product of two large and secret primes. Note that Z_n has all of the properties of a finite field except that some of the non-zero elements in Z_n are not invertible. We assume that all the non-zero values used by our protocols are invertible elements of Z_n. This assumption is reasonable since otherwise one could use our protocols to factor n. Particularly, an extended GCD algorithm on any element x used by our protocols and n, would either find the inverse of $x \bmod n$, or find a non-trivial factor of n. So in the context of our paper, we can describe computations in Z_n as if it was a finite field. Several other constructions of homomorphic encryption schemes are known, each with their particular properties (see e.g. [22,13,10,21,23,3,18,7]).

We view our protocols as algorithms that Bob executes on his encrypted input. As mentioned above, the homomorphic encryption allows Bob to locally perform several simple computations on his input. However, other computations require the help of Alice. As a simple example of a protocol where Bob uses Alice's help, consider the following (folklore) protocol MATRIX MULT for encrypted matrix multiplication.

Bob holds the encryptions $\mathsf{Enc}(A)$ and $\mathsf{Enc}(B)$ of two matrices $A \in \mathbb{F}^{n \times \ell}$ and $B \in \mathbb{F}^{\ell \times m}$. Alice holds the private decryption key. At the end of the protocol Bob should hold the encryption $\mathsf{Enc}(AB)$ of the product matrix $AB \in \mathbb{F}^{n \times m}$. Bob chooses two random matrices $R_A \in \mathbb{F}^{n \times \ell}$ and $R_B \in \mathbb{F}^{\ell \times m}$ and sends Alice the two matrices $\mathsf{Enc}(A + R_A)$ and $\mathsf{Enc}(B + R_B)$, which he can locally compute using the homomorphic properties of $\mathsf{Enc}(\cdot)$. Alice decrypts these matrices and returns $\mathsf{Enc}((A + R_A) \cdot (B + R_B))$ to Bob. Finally Bob locally computes $\mathsf{Enc}(AB) = \mathsf{Enc}((A + R_A)(B + R_B)) - \mathsf{Enc}(AR_B) - \mathsf{Enc}(R_A B) - \mathsf{Enc}(R_A R_B)$. The protocol runs in two rounds and the communication complexity of this protocol is $n\ell + \ell m + nm$. The security proof for this protocol is straightforward.

Notation. We denote by $\mathbf{neg}(x)$ a function that is negligible in x, i.e., $\mathbf{neg}(x) = x^{-\omega(1)}$. Let $\mathbb{F}$ be a finite field with p elements, and denote $k = \log p$. To make our complexity statements simpler, we make the assumption that the size of the field $\mathbb{F}$ is not too big[2] with respect to the dimensions of the matrix, i.e. $\log |\mathbb{F}| = k = O(n)$. Our protocols usually work with error probability of about $n/|\mathbb{F}|$. That is, we also assume the field size to be super-polynomial in n. If the field size is too small, we can always work over an extension field of appropriate size. This may add a small multiplicative factor polylog(n) to the communication complexity of the protocol. For example, for the case of $\mathbb{F}_2$ we could view the elements as if they were from $(\mathbb{F}_2)^\alpha$ for $\alpha = (\log n)^{1+\epsilon}$. This would add a factor of $(\log n)^{1+\epsilon}$ to the communication complexity, and reduce the error probability to $\mathbf{neg}(n)$.

For an encryption scheme, we denote by λ its security parameter. We assume that the result of encrypting a field element is of length $O(\lambda + k)$. As a con-

---

[2] For bigger fields the complexity of our protocols grows at most by an additional factor of $\log k$.

vention, the complexities of our protocols count the number of encrypted field elements that are communicated during the protocol.

We view a vector $\mathbf{v} \in \mathbb{F}^n$ as a column vector. To denote a row vector we use $\mathbf{v}^\top$. For a vector $\mathbf{v} \in \mathbb{F}^n$, we denote by $\mathsf{Enc}(\mathbf{v})$ the coordinate-wise encryption of $\mathbf{v}$. That is, if $\mathbf{v} = \langle a_1, \ldots, a_n \rangle$ where $a_1, \ldots, a_n \in \mathbb{F}$, then $\mathsf{Enc}(\mathbf{v}) = \langle \mathsf{Enc}(a_1), \ldots, \mathsf{Enc}(a_n) \rangle$. Similarly, for a matrix $A \in \mathbb{F}^{m \times n}$, we denote by $\mathsf{Enc}(A)$ the $m \times n$ matrix such that $\mathsf{Enc}(A)[i, j] = \mathsf{Enc}(A[i, j])$. An immediate consequence of the above properties of homomorphic encryption schemes is the ability to perform the following operations without knowledge of the secret key: (i) Given encryptions of two vectors $\mathsf{Enc}(\mathbf{v_1})$ and $\mathsf{Enc}(\mathbf{v_2})$, we can efficiently compute $\mathsf{Enc}(\mathbf{v_1} + \mathbf{v_2})$, and similarly with matrices. (ii) Given an encryption of a vector $\mathsf{Enc}(\mathbf{v})$ and a constant $c \in \mathbb{F}$, we can efficiently compute $\mathsf{Enc}(c\mathbf{v})$. (iii) Given an encryption of a matrix $\mathsf{Enc}(A)$ and a matrix A' of the appropriate dimensions, we can efficiently compute $\mathsf{Enc}(AA')$ and $\mathsf{Enc}(A'A)$, as any entry in the resulting matrix is a linear combination of some encrypted matrix entries.

Adversary model. Our protocols are constructed for the two-party semi-honest adversary model.[3] Roughly speaking, both parties are assumed to act in accordance with their prescribed actions in the protocol. Each party may, however, collect any information he/she encounters during the protocol run, and try to gain some information about the other party's input. We will compose our protocols in a modular manner and will argue about their privacy using well-known sequential composition theorems [12] in the semi-honest adversary model. Designing communication and round efficient secure protocols for linear algebraic problems in the malicious model remains an open problem.

Complexity Measures. Any interaction between Alice and Bob in the protocol is called a *round* of communication. The total number of such interactions consists the *round complexity* of the protocol. In each round some data is sent from Bob to Alice or from Alice to Bob. The size of all the data (i.e. the total number of bits) that is communicated between Alice and Bob during the whole execution of the protocol is called the *communication complexity* of the protocol. We make the convention to count the communication complexity of our protocols in terms of the number of encrypted values $\mathsf{Enc}(\cdot)$ exchanged between Alice and Bob.

2.1 Applying Yao's Garbled Circuit Method

In Yao's garbled circuit method [25] Alice and Bob hold private binary inputs x and y, respectively, and wish to jointly compute a functionality $f(x, y)$, such that Alice learns $f(x, y)$ and Bob learns nothing. Let f be a functionality with

---

[3] Getting the same results in the multi-party information theoretic setting remains an open problem. In particular, our protocols reduce the linear algebra problems into a variant of the extended GCD problem for polynomials. Unfortunately, a communication and round efficient protocol for this problem is not known in the multi-party information theoretic setting.

m' inputs and ℓ' outputs, which can be computed by a Boolean circuit of size G. Then the construction of Yao results in a protocol that runs in a constant number of rounds and communication complexity $O(G + m' + \ell')$.[4]

In our (homomorphic encryption) setting, we typically get to a state where Bob holds $\mathsf{Enc}(y)$ and Alice holds a private decryption key, and they wish for Bob to learn $\mathsf{Enc}(f(y))$ while Alice learns nothing, for some function f computed by a given circuit. In our protocols, we sometimes need to switch from this "homomorphic encryption setting" to the setting of Yao's garbled circuit to perform some tasks more efficiently. Then, we change from this setting to the "homomorphic encryption setting" and continue. For completeness, next we explain a simple way of doing so securely and efficiently.

From Homomorphic Encryption to Yao's. Assume that Bob is holding $\mathsf{Enc}(a)$ where $a \in \mathbb{F}$. Parties want to switch to a circuit C_f that computes the function $f(a, ...)$ on a and other inputs, without revealing the value of a.

Bob generates a random $r \in \mathbb{F}$ and sends $\mathsf{Enc}(a + r)$ to Alice. Alice decrypts to get $a + r$. Now, parties create a circuit C'_f such that Bob feeds r to C'_f as his part of the input, and Alice feeds $a + r$ to C'_f as her part of the input. They also add the additional circuitry that subtracts $(a + r) - r = a$, and use the output of this circuitry in the same way that a would be used in C_f. Everything else will stay the same as it was in C_f. The circuit for subtraction requires $O(k)$ gates. This does not affect the overall complexity of the circuit.

From Yao's to Homomorphic Encryption. Assume that Bob and Alice want to apply Yao's garbled circuit method to compute the function f, and C_f is an appropriate circuit for this task. Lets denote the output of f by $o \in \mathbb{F}$. Then, parties want to have Bob hold $\mathsf{Enc}(o)$ without revealing o itself. In what follows, we assume that Bob creates the circuit and Alice evaluates it (for more information on Yao's protocol see [16]).

Bob generates a random value $r \in F$. Parties create a circuit C'_f such that Bob feeds r to C'_f as part of his input. C'_f is the same as C_f except that parties add the additional circuity to the end of the circuit to add r to o and output $o + r$ instead of o. Note that only Alice receives the output. She encrypts and sends $\mathsf{Enc}(o + r)$ to Bob. Bob computes $\mathsf{Enc}(o) = \mathsf{Enc}(o + r) - \mathsf{Enc}(r)$ on his own.

The circuit for addition requires $O(k)$ gates and does not affect the overall complexity of the circuit. Parties can use the above two transformation on the same circuit if the goal is to change back and forth between the two different settings.

---

[4] Here we make the (simplifying but reasonable) assumption that the primitives used in [25] (i.e., the 1-out-of-2 oblivious transfer protocol and sending one garbled gate of the circuit which is usually done by sending the output of a pseudorandom bit generator) have a communication complexity $O(\lambda)$ (where $\lambda = |\mathsf{Enc}(\cdot)|$) for each execution.

3   Linearly Recurrent Sequences

We reduce our various problems from linear algebra to computing the minimal polynomial of a certain *linearly recurrent sequence*. In this section we formally define linearly recurrent sequences and discuss some of their basic properties. We follow the exposition given in [11].

Let $\mathbb{F}$ be field and V be a vector space over $\mathbb{F}$. An infinite sequence $\mathbf{a} = (a_i)_{i \in \mathbb{N}} \in V^{\mathbb{N}}$ is linearly recurrent (over $\mathbb{F}$) if there exists $n \in \mathbb{N}$ and $f_0, \dots, f_n \in \mathbb{F}$ with $f_n \neq 0$ such that $\sum_{j=0}^n f_j a_{i+j} = 0$, for all $i \in \mathbb{N}$. The polynomial $f = \sum_{j=0}^n f_j x^j$ of degree n is called a *characteristic polynomial* of $\mathbf{a}$.

We now define a multiplication of a sequence by a polynomial. For $f = \sum_{j=0}^n f_j x^j \in \mathbb{F}[x]$ and $\mathbf{a} = (a_i)_{i \in \mathbb{N}} \in V^{\mathbb{N}}$, we set

$$f \bullet \mathbf{a} = (\sum_{j=0}^n f_j a_{i+j})_{i \in \mathbb{N}} \in V^{\mathbb{N}}.$$

This makes $\mathbb{F}^{\mathbb{N}}$, together with $\bullet$, into an $\mathbb{F}[x]$-module.[5]

The property of being a characteristic polynomial can be expressed in terms of the operation $\bullet$. A polynomial $f \in \mathbb{F}[x] \setminus \{0\}$ is a characteristic polynomial of $\mathbf{a} \in \mathbb{F}^{\mathbb{N}}$ if and only if $f \bullet \mathbf{a} = \mathbf{0}$ where $\mathbf{0}$ is the all-0 sequence. The set of all characteristic polynomials of a sequence $\mathbf{a} \in \mathbb{F}^{\mathbb{N}}$, together with the zero polynomial form an ideal in $\mathbb{F}[x]$. This ideal is called the *annihilator* of $\mathbf{a}$ and denoted by $\mathrm{Ann}(\mathbf{a})$. Since any ideal in $\mathbb{F}[x]$ is generated by a single polynomial, either $\mathrm{Ann}(\mathbf{a}) = \{0\}$ or there is a unique monic polynomial $m \in \mathrm{Ann}(\mathbf{a})$ of least degree such that $\langle m \rangle = \{rm : r \in \mathbb{F}[x]\} = \mathrm{Ann}(\mathbf{a})$. This polynomial is called the *minimal polynomial* of $\mathbf{a}$ and divides any other characteristic polynomial of $\mathbf{a}$. We denote the minimal polynomial of $\mathbf{a}$ by $m_{\mathbf{a}}$. The degree of $m_{\mathbf{a}}$ is called the *recursion order* of $\mathbf{a}$.

Let $A \in \mathbb{F}^{n \times n}$ be a matrix, and $\mathbf{u}, \mathbf{v} \in \mathbb{F}^n$ be vectors. We will be interested in the following three sequences:

- $\mathbf{A} = \mathbf{A}_A = (A^i)_{i \in \mathbb{N}}$ where the sequence elements are from $V = \mathbb{F}^{n \times n}$.
- $\mathbf{a} = \mathbf{a}_{A,\mathbf{v}} = (A^i \mathbf{v})_{i \in \mathbb{N}}$ where the sequence elements are from $V = \mathbb{F}^n$.
- $\mathbf{a}' = \mathbf{a}'_{A,\mathbf{u},\mathbf{v}} = (\mathbf{u}^\top A^i \mathbf{v})_{i \in \mathbb{N}}$ where the sequence elements are from $V = \mathbb{F}$.

Definition 1. *The minimal polynomial of a matrix $A \in \mathbb{F}^{n \times n}$ is defined as $m_A = m_{\mathbf{A}}$, i.e. as the minimal polynomial of the sequence $\mathbf{A} = (A^i)_{i \in \mathbb{N}}$.*

By our definition of the minimal polynomial of a sequence the minimal polynomial of A can alternatively be characterized as the unique monic polynomial $p(x)$ over $\mathbb{F}$ of least degree such that $p(A) = 0$.

We denote by $f_A = \det(x I_n - A) = \sum_{i=0}^n f_j x^j$ the characteristic polynomial of matrix $A \in \mathbb{F}^{n \times n}$. Note that f_A is monic.

---

[5] Roughly speaking, a module is something similar to a vector space, with the only difference that the "scalars" may be elements of an arbitrary ring instead of a field. A formal definition can be found in many linear algebra textbooks (e.g., [11]).

Lemma 1. *Consider $m_{a'}, m_a, m_A$, the minimal polynomials of the sequences* a', a, A *respectively. Then* $m_{a'}|m_a|m_A|f_A$.

Proof. We first show $m_A|f_A$. By the Cayley-Hamilton Theorem $f_A(A) = 0$. Consequently,

$$f_A \bullet \mathbf{A} = (\sum_{j=0}^{n} f_j A^{i+j})_{i \in \mathbb{N}} = (A^i f_A(A))_{i \in \mathbb{N}} = \mathbf{0},$$

and $f_A(A)$ is a characteristic polynomial of $\mathbf{A}$. Therefore m_A, the minimal polynomial of $\mathbf{A}$, divides f_A.

Next, to prove $m_a|m_A$, write $m_A = \sum_{i=0}^{n} a_i x^i$. As $m_A \bullet \mathbf{A} = \mathbf{0}$, we get that $(\sum_{j=0}^{n} a_j A^{i+j})_{i \in \mathbb{N}} = \mathbf{0}$. Hence,

$$m_A \bullet \mathbf{a} = (\sum_{j=0}^{n} a_j (A^{i+j} \cdot \mathbf{v}))_{i \in \mathbb{N}} = ((\sum_{j=0}^{n} a_j A^{i+j} \cdot)\mathbf{v})_{i \in \mathbb{N}} = (0 \cdot \mathbf{v})_{i \in \mathbb{N}} = \mathbf{0}.$$

Therefore m_A is a characteristic polynomial of $\mathbf{a}$ as well, thus $m_a|m_A$. The proof of $m_{a'}|m_a$ is similar.

Corollary 1. *The sequences* $\mathbf{a}, \mathbf{a}', \mathbf{A}$ *are linearly recurrent of order at most n.*

We will use the following useful result (e.g. see [8, page 92]).

Lemma 2. *The minimal polynomial $m_A(x)$ of A divides the characteristic polynomial $f_A(x)$ of A, and both polynomials have the same irreducible factors.*

Since $f_A(0) = \det(-A) = (-1)^n \det(A)$, we obtain:

Corollary 2. $m_A(0) = 0$ *if and only if* $\det(A) = 0$.

Corollary 3. *If f_A is square-free, then $m_A = f_A$, which implies that $m_A(0) = f_A(0) = (-1)^n \cdot \det(A)$.*

4   Computing the Minimal Polynomial of a Matrix

In this section we consider the following problem: Bob holds an $n \times n$ dimensional matrix $\mathsf{Enc}(A)$ over a finite field $\mathbb{F}$, encrypted under a public-key homomorphic encryption scheme. Alice holds the private decryption key. We design a secure two-party protocol such that in the end Bob holds an encryption of m_A, the minimal polynomial of A. Computing the minimal polynomial of matrix A can be reduced to computing the minimal polynomial of the linearly recurrent sequence of field elements $\mathbf{a}' = (\mathbf{u}^\top A^i \mathbf{v})_{i \in \mathbb{N}}$. The correctness of the reduction is proved in Exercise 12.15 in [11].

Lemma 3. *Let $A \in \mathbb{F}^{n \times n}$ and let m_A be the minimal polynomial of matrix A. For $\mathbf{u}, \mathbf{v} \in \mathbb{F}^n$ chosen uniformly at random, we have $m_A = m_{\mathbf{a}'}$ with probability at least $1 - 2 \deg(m_A)/|\mathbb{F}|$.*

To compute $m_{\mathbf{a}'}$, the minimal polynomial of the sequence $\mathbf{a}'$, we first need to compute a prefix of the sequence itself. As we will later see, the $2n$ first entries of the sequence will suffice. As the communication complexity of the sub-protocol for matrix multiplication is linear in the matrix size, we are interested in computing $(\mathsf{Enc}(\mathbf{u}^\top A^i \mathbf{v}))_{0 \leq i \leq 2n-1}$ using the least number of matrix multiplication operations.

We now show how to compute the sequence using $2 \log n$ matrix multiplication operations. First compute $\mathsf{Enc}(A^{2^j})$ for $0 \leq j \leq \log n$. This can be easily done in $\log n$ sequential matrix multiplications. For two matrices X and Y of matching size let $X|Y$ be the matrix obtained by concatenating X with Y. Then compute the following using a sequence of $\log n$ matrix multiplications: (Note that all the matrices are of dimensions at most $n \times n$.)

$$
\begin{aligned}
\mathsf{Enc}(A\mathbf{v}) &= \mathsf{Enc}(A) \cdot \mathbf{v} \\
\mathsf{Enc}(A^3\mathbf{v}|A^2\mathbf{v}) &= \mathsf{Enc}(A^2) \cdot \mathsf{Enc}(A\mathbf{v}|\mathbf{v}) \\
\mathsf{Enc}(A^7\mathbf{v}|A^6\mathbf{v}|A^5\mathbf{v}|A^4\mathbf{v}) &= \mathsf{Enc}(A^4) \cdot \mathsf{Enc}(A^3\mathbf{v}|A^2\mathbf{v}|A\mathbf{v}|\mathbf{v}) \\
&\vdots \\
\mathsf{Enc}(A^{2n-1}\mathbf{v}|A^{2n-2}\mathbf{v}|\ldots|A^n\mathbf{v}) &= \mathsf{Enc}(A^n) \cdot \mathsf{Enc}(A^{n-1}\mathbf{v}|A^{n-2}\mathbf{v}|\ldots|A\mathbf{v}|\mathbf{v})
\end{aligned}
$$

Finally, multiply each vector $\mathsf{Enc}(A^i\mathbf{v})$ from the left by $\mathbf{u}^\top$ to get $\mathsf{Enc}(\mathbf{u}^\top A^i\mathbf{v})$ for $0 \leq i \leq 2n-1$.

Our next step is to compute the minimal polynomial. By Corollary 1, the order of the sequence $\mathbf{a}'$ is at most n. To compute the minimal polynomial of the sequence $\mathbf{a}'$ given the encryption of its first $2n$ elements, we use the following sub-protocol. Using the well-known Berlekamp/Massey algorithm [17] there exists an algebraic circuit of size $O(n^2)$ that computes the minimal polynomial from a sequence $\mathbf{a}' = (a_i')_{i \in \mathbb{N}}$ of maximal recursion order n. Further efficiency improvement can be obtained by noting that computing the minimal polynomial can actually be reduced to computing the greatest common division (GCD) of two polynomial of degree $2n$. For completeness we give further details in Appendix A.2. Using the fast Extended Euclidean algorithm [11, Chapter 11] the latter one can be carried out using an algebraic circuit of size $O(n \log n)$. By implementing each algebraic operation over $\mathbb{F}$ with a binary circuit of size $O(k \log k \log \log k)$ we get a binary circuit of size $O(nk \log n \log k \log \log k)$ for computing the minimal polynomial. We will use the fact that the size of this circuit is $O(n^2 k \log n)$, and so it will not be the dominate part in the overall complexity of our protocol (since we assume $|\mathbb{F}| = 2^{O(n)}$ and thus $k = \log |\mathbb{F}| = O(n)$). Using the techniques from Section 2.1 we now apply Yao's protocol to this circuit and obtain the following result.

Lemma 4. *Suppose Bob holds a sequence $\mathsf{Enc}(\mathbf{a}') = (\mathsf{Enc}(a_0'), \ldots, \mathsf{Enc}(a_{2n-1}'))$, where $\mathbf{a}' = (a_i')_{i \in \mathbb{N}}$ is a linearly recurrent sequence of order at most n. There exists a secure two-party protocol that runs in constant rounds and $O(n^2 k \log n)$ communication complexity that returns the encrypted minimal polynomial $\mathsf{Enc}(m_{\mathbf{a}'})$ of $\mathbf{a}'$ to Bob.*

The following protocol computes the minimal polynomial of matrix A.

Protocol MINPOLY

Input: $\mathsf{Enc}(A)$ where $A \in \mathbb{F}^{n \times n}$
Output: $\mathsf{Enc}(m_A)$.

1. Pick random vectors $\mathbf{u}, \mathbf{v} \in_R \mathbb{F}^n$.
   Compute $\mathsf{Enc}(\mathbf{a}')$, i.e. for $i = 0, \dots, 2n-1$ compute the values $\mathsf{Enc}(a'_i) = \mathsf{Enc}(\mathbf{u}^\top A^i \mathbf{v})$ using $2 \log n$ executions of the matrix multiplication protocol.
2. Compute $\mathsf{Enc}(m_{\mathbf{a}'})$, an encryption of the minimal polynomial of the sequence $\mathbf{a}' = (a'_i)_{0 \le i \le 2n-1}$ using Yao's Protocol.
3. Return $\mathsf{Enc}(m_{\mathbf{a}'})$ as encryption of the minimal polynomial of matrix A.

The following theorem summarizes the properties of Protocol MINPOLY.

Theorem 1. *Let* $\mathsf{Enc}(A)$ *be an encrypted* $n \times n$ *matrix over a finite field* $\mathbb{F}$. *Then protocol* MINPOLY *securely computes* $\mathsf{Enc}(m_A)$ *with probability* $1 - 2n/|\mathbb{F}|$, *communication complexity* $O(n^2 k \log n)$ *and round complexity* $O(\log n)$, *where* $k = \log |\mathbb{F}|$.

5   Singularity, Rank, Determinant, and Linear Equations

In this section, we present our main basic linear algebra protocols from Table 1 for testing if a matrix is singular, computing the rank of a matrix, computing the determinant of a matrix, and solving a system of linear equations.

5.1   Testing Matrix Singularity

One possible implementation of a protocol to securely test matrix singularity is based on Corollary 2 stating that $m_A(0) = 0$ if and only if $\det(A) = 0$. Hence, testing singularity can be reduced to computing the minimal polynomial of the matrix A and checking if its constant term equals zero. By Theorem 1 its success probability is bounded by $1 - 2n/|\mathbb{F}|$ and a secure implementation is given by protocol MINPOLY from Section 4. We now present an alternative protocol achieving a slightly improved error bound by exploiting certain algebraic properties of the minimal polynomial of sequence $\mathbf{a}$.

Again we reduce matrix singularity to computing the minimal polynomial of $\mathbf{a}'$. Our reduction works in three steps. Our first step is to reduce the problem of deciding whether $\det(A) = 0$ to deciding whether the linear system $A\mathbf{x} = \mathbf{v}$ is solvable for some random vector $\mathbf{v} \in \mathbb{F}^n$. If A is non-singular then, obviously, the linear system must be solvable. On the other hand, if $\det(A) = 0$, then with probability at least $1 - 1/|\mathbb{F}|$, the linear system has no solution.

In the second step we reduce the problem of deciding whether the linear system $A\mathbf{x} = \mathbf{v}$ is solvable to computing $m_{\mathbf{a}}$, the minimum polynomial of the recurrent sequence of vectors $\mathbf{a} = (A^i \mathbf{v})_{i \in \mathbb{N}}$.

Lemma 5. *If $m_\mathbf{a}(0) \neq 0$ then the system $A\mathbf{x} = \mathbf{v}$ is solvable.*

Proof. Since by Corollary 1 the order of $\mathbf{a}$ is at most n, we can write $m_\mathbf{a} = \sum_{i=0}^{n} m_i x^i$. As $m_\mathbf{a}$ is the minimal polynomial of $\mathbf{a}$, we get that

$$m_n A^n \mathbf{v} + m_{n-1} A^{n-1} \mathbf{v} + \ldots + m_1 A\mathbf{v} + m_0 I\mathbf{v} = 0.$$

Since $m_0 = m_\mathbf{a}(0)$ is non-zero, we get

$$-m_0^{-1}(m_n A^n \mathbf{v} + m_{n-1} A^{n-1} \mathbf{v} + \ldots + m_1 A\mathbf{v}) = \mathbf{v}$$

and hence

$$A(-m_0^{-1}(m_n A^{n-1} \mathbf{v} + m_{n-1} A^{n-2} \mathbf{v} + \ldots + m_1 I\mathbf{v})) = \mathbf{v}.$$

Therefore, the system $A\mathbf{x} = \mathbf{v}$ is solvable.

In the third step we reduce computing the minimal polynomial of sequence $\mathbf{a} = (A^i \mathbf{v})_{i \in \mathbb{N}}$ to computing the minimal polynomial of $\mathbf{a}' = (\mathbf{u}^\top A^i \mathbf{v})_{i \in \mathbb{N}}$, where $\mathbf{u} \in \mathbb{F}^n$ is a random vector. The correctness of the reduction is proved in Lemma 12.17 in [11].

Lemma 6. *Let $A \in \mathbb{F}^{n \times n}$, $\mathbf{v} \in \mathbb{F}^n$, $m_\mathbf{a}$ the minimal polynomial of the sequence $\mathbf{a} = (A^i \mathbf{v})_{i \in \mathbb{N}}$. For a $\mathbf{u} \in \mathbb{F}^n$ chosen uniformly at random we have that $m_\mathbf{a}$ is the minimal polynomial of the sequence $\mathbf{a}' = (\mathbf{u}^\top A^i \mathbf{v})_{i \in \mathbb{N}}$ with probability at least $1 - \deg(m_\mathbf{a})/|\mathbb{F}|$.*

Protocol SINGULAR

Input: $\mathsf{Enc}(A)$ where $A \in \mathbb{F}^{n \times n}$
Output: $\mathsf{Enc}(0)$ if $\det(A) = 0$ and $\mathsf{Enc}(1)$ otherwise.

1. Pick random vectors $\mathbf{u}, \mathbf{v} \in_R \mathbb{F}^n$.
   For $i = 0 \ldots 2n - 1$ compute the values $a_i' = \mathsf{Enc}(\mathbf{u}^\top A^i \mathbf{v})$ using $2 \log n$ executions of the matrix multiplication protocol.
2. Compute $\mathsf{Enc}(m_{\mathbf{a}'})$, an encryption of the minimal polynomial of the sequence $\mathbf{a}' = (a_i')_{0 \leq i \leq 2n-1}$ except that in the last step a circuit is used that returns 0 if $m_{\mathbf{a}'}(0) = 0$ and 1 otherwise using Yao's Protocol.

The following theorem summarizes the properties of Protocol SINGULAR.

Theorem 2. *Let $\mathsf{Enc}(A)$ be an encrypted $n \times n$ matrix over a finite field $\mathbb{F}$. Then Protocol SINGULAR securely checks if A is singular with probability $1 - (n + 1)/|\mathbb{F}|$, communication complexity $O(n^2 k \log n)$ and round complexity $O(\log n)$, where $k = \log |\mathbb{F}|$.*

Proof. We first prove that if $\det(A) \neq 0$ then the output of the protocol is $\mathsf{Enc}(1)$. If $m_{\mathbf{a}'}(0) = 0$, this means that the constant coefficient of $m_{\mathbf{a}'}$ is 0, thus $x | m_{\mathbf{a}'}$. By Lemma 1, $m_{\mathbf{a}'} | f_A$, where f_A is the characteristic polynomial of the matrix A. Hence, the constant coefficient of f_A is 0, which implies $\det(A) = 0$. Hence if A is non-singular, the output of the entire protocol must be $\mathsf{Enc}(1)$.

On the other hand, if $\det(A) = 0$ then, by Lemma 5, if the following two events happen, the output of the protocol is $\mathsf{Enc}(0)$: (i) The system $Ax = v$ is not solvable. (ii) $m_{\mathbf{a}'} = m_{\mathbf{a}}$. The probability of event (i) is at least $(1 - 1/|\mathbb{F}|)$. The probability of event (ii), by Lemma 6, is at least $1 - \deg(m_{\mathbf{a}})/|\mathbb{F}| \geq 1 - n/|\mathbb{F}|$. Therefore, with probability at least $1 - (n+1)/|\mathbb{F}|$ the output is $\mathsf{Enc}(0)$. Security of the protocol follows by security of the sub-protocols used. Round and communication complexity of the protocol is easy to verify.

5.2 Computing the Rank

In this section we show how to compute $\mathsf{Enc}(\mathrm{rank}(A))$ given an encryption $\mathsf{Enc}(A)$ of a matrix $A \in \mathbb{F}^{n \times n}$. To compute the rank of a matrix A, we use the following two results which are proved in [14].

Lemma 7. *Let A be a matrix in $\mathbb{F}^{n \times n}$ of (unknown) rank r. Let U and L be randomly chosen unit upper triangular and lower triangular Toeplitz matrices in $\mathbb{F}^{n \times n}$, and let $B = UAL$. Lets denote the $i \times i$ leading principal of B by B_i. The probability that $\det(B_i) \neq 0$ for all $1 \leq i \leq r$ is greater than $1 - n^2/|\mathbb{F}|$.*

Lemma 8. *Let matrix $B \in \mathbb{F}^{n \times n}$ have leading invertible principals up to B_r where r is the (unknown) rank of B. Let X be a randomly chosen diagonal matrix in $\mathbb{F}^{n \times n}$. Then, $r = \deg(m_{XB}) - 1$ with probability greater than $1 - n^2/|\mathbb{F}|$.*

The above two results lead to the following protocol for computing the rank of a matrix.

Protocol RANK

Input: $\mathsf{Enc}(A)$ where $A \in \mathbb{F}^{n \times n}$.
Output: $\mathsf{Enc}(r)$ where r is rank of A.

1. Generate random unit upper and lower triangular Toeplitz matrices $U, L \in \mathbb{F}^{n \times n}$ and a random diagonal matrix $X \in \mathbb{F}^{n \times n}$.
2. Compute $\mathsf{Enc}(M) = XU \cdot \mathsf{Enc}(A) \cdot L$.
3. Run the protocol MINPOLY on M except that in the last step, use a circuit that only outputs the degree of the minimal polynomial minus 1, and not the polynomial itself.

The following theorem is implied by the above two lemmas and summarizes the properties of our RANK protocol.

Theorem 3. *Let* $\mathsf{Enc}(A)$ *be an encrypted* $n \times n$ *matrix over a finite field* $\mathbb{F}$. *Then Protocol* RANK *securely outputs the encrypted rank of* A *with probability at least* $1 - 2n^2/|\mathbb{F}|$, *communication complexity* $O(n^2 k \log n)$, *and round complexity* $O(\log n)$, *where* $k = \log |\mathbb{F}|$.

5.3   Computing the Determinant

In this section we show how to compute $\mathsf{Enc}(\det(A))$, given an encryption $\mathsf{Enc}(A)$ of a matrix $A \in \mathbb{F}^{n \times n}$. The protocol uses the following fact from linear algebra [24].

Lemma 9. *Let* B *be an* $n \times n$ *matrix over* $\mathbb{F}$ *where all the leading principal submatrices of* B, *including* B *itself are nonsingular, and let* X *be a uniformly chosen diagonal matrix in* $\mathbb{F}^{n \times n}$. *Then,* f_{XB} *is square-free with probability greater than* $1 - n/|\mathbb{F}|$.

Protocol DET

Input: $\mathsf{Enc}(A)$ where $A \in \mathbb{F}^{n \times n}$.
Output: $\mathsf{Enc}(d)$ where d is the determinant of A.

1. Generate random unit upper and lower triangular Toeplitz matrices $U, L \in \mathbb{F}^{n \times n}$ and a random diagonal matrix $X \in \mathbb{F}^{n \times n}$.
2. Computes $\mathsf{Enc}(Z) = XU \cdot \mathsf{Enc}(A) \cdot L$.
3. Run the protocol MINPOLY on Z except that in the last step, use a circuit that computes $(-1)^n m_Z(0)/\det(X)$ instead of m_Z (note that Bob knows $\det(X)$, and feeds it to the circuit as part of his input.).

The following theorem summarizes the properties of our DET protocol.

Theorem 4. *Let* $\mathsf{Enc}(A)$ *be an encrypted* $n \times n$ *matrix over a finite field* $\mathbb{F}$. *Then Protocol* DET *securely outputs the encryption of determinant of* A *with probability at least* $1 - 3n^2/|\mathbb{F}|$, *communication complexity* $O(n^2 k \log n)$, *and round complexity* $O(\log n)$, *where* $k = \log |\mathbb{F}|$.

Proof. If A is singular, $Z = XUAL$ is also singular. Therefore, based on Corollary 2, $m_Z(0) = 0$. Hence, the protocol correctly returns $\mathsf{Enc}(0)$ as the answer. On the other hand, if A is non-singular, Z is also non-singular. Note that from given the determinant of Z, Bob can easily derive the determinant of A, as he has all the other matrices in the clear.

Based on Corollary 3, if Z is square-free, computing the constant coefficient of m_Z is sufficient to compute the $det(Z)$. We now show that the probability that Z is square-free is high. By Lemma 7, the probability that all the leading principals of the matrix UAL are of full rank is $1 - n^2/|\mathbb{F}|$. Conditioned on the latter, Lemma 9 implies that with probability $1 - n/|\mathbb{F}|$ the matrix $Z = XUAL$ is square-free. Hence, with probability greater than $1 - 2n^2/|\mathbb{F}|$, the matrix Z

is indeed square free. We also need the minimal polynomial protocol to succeed, which happens with probability $1 - 2n/|\mathbb{F}|$. Hence, the overall success probability of the protocol is at least $1 - 3n^2/|\mathbb{F}|$. The round and communication complexity of the protocol is easy to verify.

5.4   Solving Linear Equations

In this section we discuss the problem of solving a system of linear equations. Given encryptions $\mathsf{Enc}(M)$ and $\mathsf{Enc}(\mathbf{y})$, where $M \in \mathbb{F}^{m \times n}$ and $\mathbf{y} \in \mathbb{F}^m$, we are interested in outputting an encryption $\mathsf{Enc}(\mathbf{x})$ of a random solution to the linear system $M\mathbf{x} = \mathbf{y}$, if the system is solvable.

The easy case is where M is a non-singular square matrix. In this case it is enough to compute $\mathsf{Enc}(M^{-1})$ and then execute Protocol MATRIX MULT once to compute $\mathsf{Enc}(M^{-1})\mathsf{Enc}(\mathbf{y}) = \mathsf{Enc}(M^{-1}\mathbf{y})$, which is the unique solution to the system (and hence is also a random solution). To compute $\mathsf{Enc}(M^{-1})$ from $\mathsf{Enc}(M)$ we use Protocol MATRIX INVERT which assume the encrypted input matrix M to be invertible. see Appendix A.1 for an implementation of protocol MATRIX INVERT based on [1].

To reduce the general case to the non-singular case, we adapt an algorithm of Kaltofen and Saunders [14]. Their algorithm solves $M\mathbf{x} = \mathbf{y}$ in the following way: (i) Perturb the linear system $M\mathbf{x} = \mathbf{y}$ to get a system $M'\mathbf{x} = \mathbf{y}'$ with the same solution space. The perturbation has the property that, with high probability, if M is of rank r, then M'_r, the top-left $r \times r$ sub-matrix of M', is non-singular. (ii) Pick a random vector $\mathbf{u} \in \mathbb{F}^n$ and set $\mathbf{y}'_r$ to be the upper r coordinates of the vector $\mathbf{y}' + M'\mathbf{u}$. (iii) Solve the linear system $M'_r\mathbf{x}_r = \mathbf{y}'_r$, and denote the solution by $\mathbf{u}_r$. (iv) Let $\mathbf{u}^* \in \mathbb{F}^n$ be a vector with upper part u_r and lower part 0^{n-r}. It can be shown that $\mathbf{x} = \mathbf{u}^* - \mathbf{u}$ is a uniform random solution to the system $M'\mathbf{x} = \mathbf{y}'$ and thus is a uniform random solution to the original system. The correctness proof for this algorithm may be found in [14, Theorem 4]. Note that this algorithm is correct assuming that the system $M\mathbf{x} = \mathbf{y}$ is solvable. An implementation of the first step relies on the following simple linear algebraic lemma.

Lemma 10. *Let M be a matrix in $\mathbb{F}^{m \times n}$ of (unknown) rank r. Let $P \in \mathbb{F}^{m \times m}$ and $Q \in \mathbb{F}^{n \times n}$ randomly chosen full rank matrices, and let $M' = PMQ$. Denote the $r \times r$ leading principal of M' by M'_r. The probability that $\det(M'_r) \neq 0$ is greater than $1 - 2n/|\mathbb{F}|$.*

Implementing Kaltofen-Saunders algorithm in a secure protocol is not straight-forward. On one hand, we need to compute r, the rank of M, in order to invert the top-left sub-matrix of M. On the other hand, computing r violates the privacy of the protocol, as r cannot be extracted from a random solution to the linear system. We overcome this problem by showing how to implement the Kaltofen-Saunders algorithm using only an encryption of r (computed using Protocol RANK from Section 5.1). The key idea is that we can work with the $r \times r$ top-left sub-matrix of the perturbed matrix M', without knowing the value of r in the clear.

Protocol LINEAR SOLVE

Input: $\mathsf{Enc}(M)$ where $M \in \mathbb{F}^{m \times n}$ and $n \leq m$, and $\mathsf{Enc}(\mathbf{y})$ where $\mathbf{y} \in \mathbb{F}^m$.
This protocol assumes the system $M\mathbf{x} = \mathbf{y}$ is solvable.
Output: $\mathsf{Enc}(\mathbf{x})$, where $\mathbf{x} \in \mathbb{F}^n$ is a random solution to the system $M\mathbf{x} = \mathbf{y}$.

1. Execute Protocol RANK on $\mathsf{Enc}(M)$ to compute $\mathsf{Enc}(r)$ where $r = \mathrm{rank}(M)$.
2. Locally compute $\mathsf{Enc}(M') = P \cdot \mathsf{Enc}(M) \cdot Q$ and $\mathsf{Enc}(\mathbf{y}') = P \cdot \mathsf{Enc}(\mathbf{y})$ where P and Q are random non-singular $m \times m$ and $n \times n$ matrices respectively.
3. Compute the encrypted matrix $\mathsf{Enc}(N')$, where $N' \in \mathbb{F}^{n \times n}$ consists of the r by r leading principal of M' in the top-left corner and of the unit matrix in the bottom-right corner.
4. Compute $\mathsf{Enc}(N'^{-1})$ using protocol MATRIX INVERT.
5. Pick a random vector $\mathbf{u} \in \mathbb{F}^n$ and set $\mathsf{Enc}(\mathbf{y}'_r)$ for $\mathbf{y}'_r \in \mathbb{F}^n$ to be a vector whose upper r coordinates are the upper r coordinates of $\mathsf{Enc}(\mathbf{y}')$ + $\mathsf{Enc}(M')\mathbf{u}$ and lower $n - r$ coordinates are $\mathsf{Enc}(0)$.
6. Compute $\mathsf{Enc}(\mathbf{u}_r) = \mathsf{Enc}(N'^{-1}) \cdot \mathsf{Enc}(\mathbf{y}'_r)$ and output $\mathsf{Enc}(\mathbf{x}) = Q^{-1} \cdot \mathsf{Enc}(\mathbf{u} - \mathbf{u}_r)$.

Some remarks are in place. First, note that the protocol is valid only for solvable linear system. To check if a system is solvable, it is sufficient to compare the rank of the matrices M and $M|\mathbf{y}$ where $|$ stands for concatenation. The encryption of the rank of these matrices can be computed using Protocol RANK, while the comparison can be easily done using Yao's garbled circuit method.

In Step 3 to compute $\mathsf{Enc}(N')$ from $\mathsf{Enc}(M')$ and $\mathsf{Enc}(r)$ one proceeds as follows. First $\mathsf{Enc}(r)$ is converted into unary representation (i.e., $(\mathsf{Enc}(\delta_1), \ldots, \mathsf{Enc}(\delta_n))$ with $\delta_i = 1$ if $i \leq r$ and $\delta_i = 0$ otherwise) using Yao's garbled circuit method. Then create $\mathsf{Enc}(\Delta)$, where Δ is the $n \times n$ matrix $\Delta = \mathrm{diag}(\delta_1, \delta_2, \ldots, \delta_n)$. Then $\mathsf{Enc}(N')$ is computed as $\mathsf{Enc}(N') = \mathsf{Enc}(M')\mathsf{Enc}(\Delta) + I_n - \mathsf{Enc}(\Delta)$, where I_n is the $n \times n$ identity matrix.

As a final note, we stress that the requirement that $n \leq m$ is made only for simplicity of presentation. Otherwise, N' would have been of dimension $\min(m, n) \times \min(m, n)$ instead of $n \times n$, and the changes needed in the rest of the protocol are minor. The following Theorem concludes the properties of Protocol LINEAR SOLVE.

Theorem 5. *Let $\mathsf{Enc}(M)$ be an encrypted $m \times n$ matrix over a finite field $\mathbb{F}$, and let $\mathsf{Enc}(\mathbf{y})$ be an encrypted vector $\mathbf{y} \in \mathbb{F}^m$. Protocol LINEAR SOLVE securely computes $\mathsf{Enc}(\mathbf{x})$, where $\mathbf{x} \in \mathbb{F}^n$ is a random solution of $M\mathbf{x} = \mathbf{y}$, with probability $1 - 3n^2/|\mathbb{F}|$, communication complexity $O(n^2 k \log n)$ and round complexity $O(\log n)$, where $k = \log |\mathbb{F}|$.*

We now discuss the success probability of Protocol LINEAR SOLVE. In Step 1, we compute an encryption of the rank of M which is by Theorem 3 correct with probability $1 - 2n^2/|\mathbb{F}|$. In Step 2, we multiply the matrix M from the right and from the left by random non-singular matrices to get the matrix M'. By Lemma 10, the top left $r \times r$ sub-matrix of M' is of rank r with probability $1 - 2n/|\mathbb{F}|$. If this is the case, then the rest of the protocol follows the Kaltofen-Saunders algorithm, and thus its correctness is implied by [14, Theorem 4].

Acknowledgments. We thank Amos Beimel for valuable comments on a previous version of this paper. The first author was supported by the research program Sentinels (http://www.sentinels.nl). Sentinels is being financed by Technology Foundation STW, the Netherlands Organization for Scientific Research (NWO), and the Dutch Ministry of Economic Affairs.

References

1. J. Bar-Ilan and D. Beaver. Non-cryptographic fault-tolerant computing in constant number of rounds of interaction. In *PODC '89: Proceedings of the eighth annual ACM Symposium on Principles of distributed computing*, pages 201–209, New York, NY, USA, 1989. ACM Press.
2. A. Beimel and E. Weinreb. Separating the power of monotone span programs over different fields. In *Proc. of the 44th IEEE Symp. on Foundations of Computer Science*, pages 428–437, 2003.
3. D. Boneh, E. Goh, and K. Nissim. Evaluating 2-DNF formulas on ciphertexts. In *the Second Theory of Cryptography Conference – TCC 2005*, pages 325–341, 2005.
4. P. Bürgisser, M. Clausen, and M. A. Shokrollahi. *Algebraic complexity theory.* Springer-Verlag, Berlin, 1997.
5. D. Coppersmith and S. Winograd. Matrix multiplication via arithmetic progressions. In *STOC '87: Proceedings of the nineteenth annual ACM conference on Theory of computing*, pages 1–6. ACM Press, 1987.
6. R. Cramer and I. Damgaard. Secure distributed linear algebra in a constant number of rounds. In *CRYPTO '01: Proceedings of the 21st Annual International Cryptology Conference on Advances in Cryptology*, pages 119–136. Springer-Verlag, 2001.
7. R. Cramer, R. Gennaro, and B. Schoenmakers. A secure and optimally efficient multi-authority election scheme. In *Advances in Cryptology – EUROCRYPT '97*, Lecture Notes in Computer Science, pages 103–118. Springer-Verlag, 1997.
8. M. Curtis. *Abstract Linear Algebra.* Springer-Verlag, 1990.
9. J. L. Dornstetter. On the equivalence between Berlekamp's and Euclid's algorithms. *IEEE Trans. Inf. Theory*, it-33(3):428–431, 1987.
10. T. El Gamal. A public key cryptosystem and a signature scheme based on discrete logarithms. In *Proceedings of CRYPTO 84 on Advances in cryptology*, pages 10–18, New York, NY, USA, 1985. Springer-Verlag New York, Inc.
11. J. von zur Gathen and J. Gerhard. *Modern computer algebra.* Cambridge University Press, New York, 1999.
12. O. Goldreich. *Foundations of Cryptography, Voume II Basic Applications.* Cambridge University Press, 2004.
13. S. Goldwasser and S. Micali. Probabilistic encryption & how to play mental poker keeping secret all partial information. In *STOC '82: Proceedings of the fourteenth annual ACM symposium on Theory of computing*, pages 365–377, New York, NY, USA, 1982. ACM Press.

14. E. Kaltofen and D. Saunders. On Wiedemann's method of solving sparse linear systems. In *AAECC-9: Proceedings of the 9th International Symposium, on Applied Algebra, Algebraic Algorithms and Error-Correcting Codes*, pages 29–38, London, UK, 1991. Springer-Verlag.

15. M. Karchmer and A. Wigderson. On span programs. In *Proc. of the 8th IEEE Structure in Complexity Theory*, pages 102–111, 1993.

16. Y. Lindell and B. Pinkas. A proof of yao's protocol for secure two-party computation. *eprint archive*, 2004.

17. J. L. Massey. Shift-register synthesis and BCH decoding. *IEEE Trans. Inf. Theory*, it-15:122–127, 1969.

18. D. Naccache and J. Stern. A new public-key cryptosystem based on higher residues. In *ACM CCS 98*, pages 59–66, 1998.

19. K. Nissim and E. Weinreb. Communication efficient secure linear algebra. In *the Third Theory of Cryptography Conference – TCC 2006*, pages 522–541, 2006.

20. P. Pallier. Public-key cryptosystems based on composite degree residuosity classes. In *Advances in Cryptology – EUROCRYPT '99*, pages 223–238, 1999.

21. T. P. Pedersen. A threshold cryptosystem without a trusted party. In *Advances in Cryptology – EUROCRYPT '91*, pages 522–526, 1991.

22. R. L. Rivest, A. Shamir, and L. Adleman. A method for obtaining digital signatures and public-key cryptosystems. *Commun. ACM*, 21(2):120–126, 1978.

23. T. Sander, A. Young, and M. Yung. Non-interactive cryptocomputing for NC1. In *FOCS '99: Proceedings of the 40th Annual Symposium on Foundations of Computer Science*, page 554, Washington, DC, USA, 1999. IEEE Computer Society.

24. D. H. Wiedemann. Solving sparse linear equations over finite fields. *IEEE Trans. Inf. Theor.*, 32(1):54–62, 1986.

25. A. C. Yao. How to generate and exchange secrets. In *Proc. of the 27th IEEE Symp. on Foundations of Computer Science*, pages 162–167, 1986.

A   More Protocols

A.1   Matrix Inversion

Bob holds an encrypted matrix $\mathsf{Enc}(M)$ such that $M \in \mathbb{F}^{n \times n}$ is guaranteed to be invertible. Alice holds the private decryption key. Based on the shared field inversion protocol from Bar-Ilan and Beaver [1] we design a protocol for computing $\mathsf{Enc}(M^{-1})$.

Protocol MATRIX INVERT

Input: $\mathsf{Enc}(M)$ where $M \in \mathbb{F}^{n \times n}$.
Output: $\mathsf{Enc}(M^{-1})$

1. Bob picks an $n \times n$ random non-singular matrix Q.
2. Bob computes the encrypted matrix $\mathsf{Enc}(QM)$ by multiplying $\mathsf{Enc}(M)$ from the left by the matrix Q, and sends $\mathsf{Enc}(QM)$ to Alice.
3. Alice decrypts $\mathsf{Enc}(QM)$ and compute $(QM)^{-1} = M^{-1}Q^{-1}$. Alice encrypts $M^{-1}Q^{-1}$ and sends Bob $\mathsf{Enc}(M^{-1}Q^{-1})$.
4. Bob computes $\mathsf{Enc}(M^{-1}) = \mathsf{Enc}(M^{-1}Q^{-1})Q$.
5. Bob locally outputs $\mathsf{Enc}(M^{-1})$.

It is easy to see that Alice gets a random non-singular matrix QM, and thus learns no information in the protocol. Since Bob only learns encrypted values from the protocol, he gets no information on the value of M.

A.2 Minimal Polynomial

We demonstrate an algorithm from [9] how to efficiently compute the minimal polynomial of a sequence $\mathbf{a} = (a_i)_{i \in \mathbb{N}}$ of recursion order n using the Extended Euclidean Algorithm on polynomials. By the definition from Section 3 the minimal polynomial $m_{\mathbf{a}}$ of the sequence $\mathbf{a}$ is the unique monic polynomial $m_{\mathbf{a}}(x) = m(x)$ of least degree $\leq n$ for which $m(x) \bullet \mathbf{a} = \mathbf{0}$. By division with remainder we can rewrite this as

$$m_{\mathbf{a}} \cdot (a_1 + a_2 x + \ldots + a_{2n} x^{2n-1}) - q(x) \cdot x^{2n} = r(x), \tag{1}$$

where $r(x)$ is a remainder polynomial of degree $< n$, and $q(x)$ is a quotient polynomial. Denote by $a(x)$ the sum $\sum_{i=1}^{2n} a_i x^{i-1}$. If we apply the extended GCD algorithm to the two polynomials $a(x)$ and x^{2n}, keeping track of remainders, we get two sequences $p_i(x), q_i(x)$ such that the $r_i := p_i(x) \cdot a(x) - q_i(x) \cdot x^{2n}$ form a series of polynomials whose degree is strictly decreasing. As soon as the degree of r_i is less than n, we have the required polynomials from (1) with $m_{\mathbf{a}}(x) = p_i(x)$, $q(x) = q_i$, and $r(x) = r_i(x)$.

B Applications

B.1 Linear Subspace Intersection

Let $\mathbb{F}$ be a finite field and n be a positive integer. Alice holds a subspace $V_A \subseteq \mathbb{F}^n$ of dimension $n_a \leq n$. The subspace V_A is represented by an $n_a \times n$ matrix A, where the rows of A span V_A. Similarly, Bob's input is a subspace $V_B \subseteq \mathbb{F}^n$ of dimension n_b, represented by an $n_b \times n$ matrix B. Letting $V_I = V_A \cap V_B$, Alice and Bob wish to securely study different properties of V_I.

In [19], constant round $O(n^2)$ protocols were designed for securely *computing* the subspace V_I, and for securely computing the rank of the subspace V_I. However, it turned out that the problem of securely *deciding* whether the subspace V_I is the trivial zero subspace seems harder to solve. Ignoring security issues, computing the intersection of the input subspaces is at least as hard as deciding whether they have a non trivial intersection. However, constructing a *secure* protocol for the latter turns to be somewhat harder as the players gain less information from its output.

The following lemma from [19] reduces the problem of deciding subspace intersection, to computing whether a matrix is of full rank:

Lemma 11 ([19]). *Define* $M = AB^{\top}$. *Then* $V_I \neq \{0\}$ *if and only if the matrix* M *is not of full row rank.*

This gives rise to the following protocol:

Protocol INTERSECTION DECIDE

Input: Alice (resp. Bob) holds a $n_a \times n$ (resp. $n_b \times n$) matrix A (resp. B) over a finite field $\mathbb{F}$ representing a subspace $V_A \subseteq \mathbb{F}^n$ (resp. $V_B \subseteq \mathbb{F}^n$). Let $B^\top$ be a $n \times n_b'$ matrix that represents the subspace $V_B^\top$, where $n_b' \stackrel{\text{def}}{=} n - n_b$.
Output: If V_I is the trivial zero subspace, Alice outputs 1. Else, Alice outputs 0.

1. Alice generates keys for a homomorphic public key encryption system, and sends Bob $\mathsf{Enc}(A)$ and the public key.
2. Bob locally computes $\mathsf{Enc}(M)$, where $M \stackrel{\text{def}}{=} AB^\top$. Note that M is a $n_a \times n_b'$ matrix.
3. Alice and Bob execute Protocol RANK on $\mathsf{Enc}(M)$. Denote by $\mathsf{Enc}(r)$ the output of the protocol held by Bob.
4. Alice and Bob execute protocol EQUAL on $\min n_a, n_b'$ and $\mathsf{Enc}(r)$. Bob sends the encrypted output to Alice who decrypts and outputs it.

This protocol has the same communication complexity as of the protocol designed in [19]. However, the round complexity of this protocol, which is $O(\log n)$ is substantially better than the round complexity of [19], which is $\Omega(n^{0.275})$. We note that the techniques in our paper are very different from those of [19].

B.2   Solving a Common Linear Equation System

Let $\mathbb{F}$ be a finite field and n be a positive integer. Alice holds an $n_a \times n$ matrix M_A and a vector $\mathbf{v}_a \in F^{n_a}$. Similarly, Bob's input is an $n_a \times n$ matrix M_B and a vector $\mathbf{v}_b \in F^{n_b}$. Alice and Bob wish to securely compute a random vector $\mathbf{x} \in \mathbb{F}^n$ such that both $M_A \mathbf{x} = \mathbf{v}_a$ and $M_B \mathbf{x} = \mathbf{v}_b$.

This problem can be viewed as computing a random vector from the intersection of the affine subspaces representing the solutions to the systems $M_A \mathbf{x} = \mathbf{v}_a$ and $M_B \mathbf{x} = \mathbf{v}_b$. This problem was considered in [19], who designed a protocol of communication complexity $O(n^2 k \log n)$ and round complexity $\Omega(n^{0.275})$. We show a protocol which improves the round complexity to $O(\log n)$ while keeping the communication complexity roughly $O(n^2)$.

The protocol is simple: Alice generates keys for a homomorphic public key encryption system, and sends Bob $\mathsf{Enc}(M_A)$, $\mathsf{Enc}(v_a)$ and the public key. Bob encrypts his input to get the encrypted linear system.

$$\begin{pmatrix} \mathsf{Enc}(M_A) \\ \mathsf{Enc}(M_B) \end{pmatrix} \mathbf{x} = \begin{pmatrix} \mathsf{Enc}(v_a) \\ \mathsf{Enc}(v_b) \end{pmatrix}$$

Alice and Bob then execute Protocol LINEAR SOLVE after which Bob holds $\mathsf{Enc}(\mathbf{x})$ where $\mathbf{x}$ is a random solution to the common system. Finally, bob sends $\mathsf{Enc}(\mathbf{x})$ to Alice, which decrypts and outputs $\mathbf{x}$.

Towards Optimal and Efficient
Perfectly Secure Message Transmission

Matthias Fitzi[1,*], Matthew Franklin[2], Juan Garay[3], and S. Harsha Vardhan[4,**]

[1] Department of Computer Science, ETH Zürich, Switzerland
fitzi@inf.ethz.ch
[2] Department of Computer Science, UC Davis, CA 95016
franklin@cs.ucdavis.edu
[3] Bell Labs, 600 Mountain Ave., Murray Hill, NJ 07974
garay@research.bell-labs.com
[4] Department of Computer Science and Engineering, IIT Madras, India
harshas@cse.iitm.ernet.in

Abstract. Perfectly secure message transmission (PSMT), a problem formulated by Dolev, Dwork, Waarts and Yung, involves a sender S and a recipient R who are connected by n synchronous channels of which up to t may be corrupted by an active adversary. The goal is to transmit, with perfect security, a message from S to R. PSMT is achievable if and only if $n > 2t$.

For the case $n > 2t$, the lower bound on the number of communication rounds between S and R required for PSMT is 2, and the only known efficient (i.e., polynomial in n) two-round protocol involves a communication complexity of $O(n^3 \ell)$ bits, where ℓ is the length of the message. A recent solution by Agarwal, Cramer and de Haan is provably communication-optimal by achieving an asymptotic communication complexity of $O(n\ell)$ bits; however, it requires the messages to be exponentially large, i.e., $\ell = \Omega(2^n)$.

In this paper we present an efficient communication-optimal two-round PSMT protocol for messages of length polynomial in n that is almost optimally resilient in that it requires a number of channels $n \geq (2 + \varepsilon)t$, for any arbitrarily small constant $\varepsilon > 0$. In this case, optimal communication complexity is $O(\ell)$ bits.

1 Introduction

In the problem of *perfectly secure message transmission* (PSMT) a sender S and a recipient R are connected by n distinct, synchronous communication channels. Of these channels, an active adversary may be corrupting any selection of up to t. The goal is to have S transmit a message to R perfectly securely, i.e., in such a way that (1) the adversary gets no information about the message, and (2) that R receives the correct message with probability 1. In general, a protocol for PSMT requires multiple communication exchanges—*rounds*—between S and R,

---

* Work partly done while at Aarhus University.
** Work partly done at Bell Labs Research, Bangalore, India.

S.P. Vadhan (Ed.): TCC 2007, LNCS 4392, pp. 311–322, 2007.

for example, to first agree on a one-time pad before having the padded message transmitted from S to R.

PSMT was introduced by Dolev, Dwork, Waarts and Yung in [8]. Their main result is that PSMT is achievable if and only if $n > 2t$. For this particular bound, they also showed that two communication rounds are necessary and sufficient in order to achieve PSMT (i.e., a communication flow from R to S, and then a flow from S to R). However, their protocol to achieve this bound is inefficient as it involves an exponential (in n) computation and communication overhead. In [17], Sayeed and Abu-Amara gave polynomial-time two-round protocol that requires a communication complexity of $O(n^3\ell)$ bits, where ℓ is the length of the message to be transmitted. More recently, Srinathan, Narayanan and Rangan [18] showed that, in order to achieve two-round PSMT, $\Omega(n\ell)$ bits must be communicated. This lower bound has been matched by the protocol by Agarwal, Cramer and de Haan [1], at the price, however, of requiring messages of length exponential in n. In [16], Patra, Choudhary, Srinathan and Rangan show that by using one additional round (i.e., three rounds in total), this communication bound can be achieved with polynomial message length.

Our contributions. In this paper, we present an efficient two-round protocol for PSMT with optimal communication complexity that works for messages of length polynomial in n. The protocol works for any parameterization of $n \geq (2+\varepsilon)t$, where $\varepsilon > 0$ is a fixed but arbitrarily small constant—i.e., the protocol is almost optimally resilient. Note, however, that our protocol is optimally resilient with respect to the communication complexity we achieve: $O(\ell)$, where ℓ is the length of the message—as it follows from the lower bound in [18] that $n = 2t + \Omega(t)$ is necessary in order to achieve communication complexity $O(\ell)$ (in contrast to $\Omega(n\ell)$ for the general case $n > 2t$).

Our protocol is derived from a modification of the communication-optimal one-round PSMT protocol for $n > 3t$ in [17], and by applying a technique that we call *player virtualization*, which can be viewed as a very simple and constructive instantiation of so-called *Bracha assignments* [6], which are used to "amplify" the resilience of a distributed computation protocol while preserving some of its other properties. (We describe this technique in more detail below.)

Additionally, we also show a tight bound on the communication complexity of one-round PSMT for $n > 3t$.

The "player virtualization" technique. The idea of creating *virtual players* whose behavior is simulated by the actions of groups of real players was introduced by Bracha in [6] in the context of Byzantine agreement [14], in order to prove the existence of a randomized protocol for the problem for any $n > (3 + \delta)t$, where n is the total number of players, t is the number of faulty players, and $\delta > 0$ is an arbitrary constant, running in expected $O(\log n)$ rounds. The goal was to simulate Ben-Or's randomized distributed coin-flipping protocol [2], which required, for good performance, that the number of faulty players be at most $O(\sqrt{n})$—i.e., the effect of the simulation is to obtain a set of virtual players with a lower corruption rate than in the original player set. While Bracha was able to prove the existence of such a protocol, the result is non-constructive.

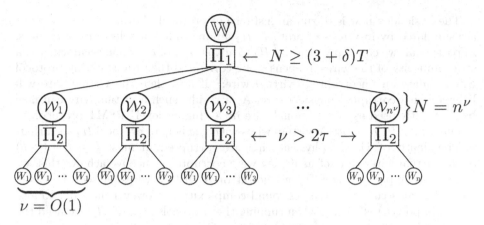

Fig. 1. The wire virtualization scheme for PSMT

A similar—but perhaps simpler—idea, also applied in the context of Byzantine agreement, is to partition the player set into smaller, non-overlapping "committees" (e.g., [5,7]), with the goal of obtaining at least one out of the several subsets of players that maintains the global corruption ratio (t/n). This approach, however, typically has the converse effect of the set of committees having a higher corruption rate than the original player set.

In the context of secure multi-party computation [19,10], Hirt and Maurer [11] essentially applied player virtualization in order to reduce a generalized-adversary computation to threshold-adversary computations of a small size. Their construction, however, generally yields protocols with exponential (in n) computation and communication complexities.

Constructive Bracha assignments have also been used for the leader election problem in the full information model [15,20], and recently in order to reduce the communication (to polylogarithmic in n) required for the task [13] ([13] also studies "almost-everywhere agreement" [9] under reduced communication). At a high level, these constructions are based on expander graphs, and typically carry a probability of error. We elaborate more on this type of approach in Section 5.

We now give a high-level description of how we apply player (more precisely, "wire") virtualization to PSMT. Recall that we are given $\mathcal{S}$ and $\mathcal{R}$ who are connected by n wires of which t might be corrupted by the adversary. We first observe the following facts about PSMT:

1. For any $N \geq 3(T + \delta)$, where $\delta > 0$ is a constant and N denotes the total number of wires and T the number of possibly corrupted wires, there is a one-round PSMT protocol Π_1 with constant communication overhead. Such a protocol is described in Section 3.1.

2. For any $\nu > 2\tau$, where ν denotes the total number of wires and τ the number of possibly corrupted wires, there is a two-round PSMT protocol Π_2 that is communication-optimal but requires messages of exponential size in ν. This is the protocol in [1].

The basic idea now is to run an instance of protocol Π_1 wherein the N wires are simulated by instances of protocol Π_2 among different selections of ν wires. In particular, we can apply protocol Π_2 to any subset of $\nu < n$ physical wires. If a strict minority of the wires happens to be corrupted then the resulting protocol will simulate an uncorrupted "virtual wire;" if not, then the virtual wire will behave like a corrupted physical wire. As a result, such a virtual wire can now be abstractly used as an additional wire by a "higher-level" PSMT protocol.

Our goal is to generate N virtual wires with the help of protocol Π_2 such that, independently of which t physical wires are corrupted, at most $T \leq N/(3 + \delta)$ of the virtual wires can act as if they were corrupted. Once we achieve this, we can simply apply protocol Π_1 on the set of N virtual wires. As can be easily seen, this construction preserves round complexity 2. However, in order to also maintain poly(n) efficiency when running the protocols Π_1 and Π_2, we need the additional constraints $N = \text{poly}(n)$ and $\nu = O(\log n)$.

We meet these constraints by choosing $\nu = O(1)$ and having each possible set of ν physical wires (including repetitions) simulate a different virtual wire, resulting in $N = n^{\nu}$. The approach is depicted in Figure 1. As we show in the sequel, it turns out that this construction works for any parameterization of $n \geq (2 + \varepsilon)t$, where $\varepsilon = \Omega(1)$; i.e., round-optimal, bit-optimal and efficient PSMT with almost optimal resiliency can be achieved in this way.

Organization of the paper. In the next section we present the model and the definition of the PSMT problem. We dedicate Section 3 to the treatment of the one-round case. We first present an efficient PSMT protocol for $n > 3t$ wires which, as we also show, has optimal communication overhead. Design and analysis of the virtualization construction yielding our main result are presented in Section 4. We conclude in Section 5 with some optimization considerations and final remarks.

2   Model and Definitions

Sender S and recipient R are connected by n distinct synchronous channels ("wires") $W_1, W_2, \ldots, W_n$. An adversary A may select up to t of the n wires and corrupt them actively, i.e., A may eavesdrop on the selected wires as well as change the messages being sent on them. The adversary is assumed to be computationally unbounded. Furthermore, the adversary is assumed to be *adaptive*, i.e., it can adaptively decide on which further wires to corrupt at any point during the protocol — but "non-mobile," i.e., the adversary is not allowed to have corrupted any more than t different wires by the end of the protocol, overall.

Definition 1. *A protocol between S and R, based on local computation and communication via the network described above, achieves* perfectly secure message transmission (PSMT) *if it transmits a message from S to R such that the following two conditions are satisfied:*

PRIVACY: *A does not get any information about the message being transmitted.*

Protocol 1-PSMT(n, t, m)

- Given a message $m = [m_1 m_2 \ldots m_k]$ $(k = n - 3t)$, the sender $\mathcal{S}$ randomly forms a polynomial $f(x)$ of degree at most $d = (n - 2t - 1)$ by choosing its coefficients as follows:

$$\text{coeff}(x^i) = \begin{cases} m_{i+1}, & \text{if } 0 \leq i < k, \\ c_{i-k}, & \text{if } k \leq i < (k+t), \end{cases}$$

where the c_{i-k}'s are chosen uniformly at random from $\mathbb{F}$.
- On wire W_j, $1 \leq j \leq n$, the sender $\mathcal{S}$ sends the share $r_j = f(\alpha^{j-1})$, where α is a generator of the multiplicative group of $\mathbb{F}$.
- The recipient $\mathcal{R}$ uses the Welch-Berlekamp decoding algorithm [4] on the received values in order to obtain the message.

Fig. 2. One-round PSMT with low communication overhead

CORRECTNESS: *$\mathcal{R}$ gets full information about the message transmitted by $\mathcal{S}$; i.e., $\mathcal{R}$ learns the message with probability 1.*   ◇

In the sequel, and without loss of generality, we assume that the messages are taken from a finite field $\mathbb{F}$ with $|\mathbb{F}| > n$.

We define the *bit-communication complexity* (or, *communication complexity*, for short) of a PSMT protocol to be the total number of bits being communicated between $\mathcal{S}$ and $\mathcal{R}$. For convenience, we also define the *communication overhead*, Λ, as the total number of bits communicated by the protocol divided by the length of the message. The *round complexity* of a PSMT protocol is its number of subsequent communication rounds between $\mathcal{S}$ and $\mathcal{R}$. In particular, a one-round PSMT protocol consists of a synchronous flow of communication on the wires from $\mathcal{S}$ to $\mathcal{R}$, and a two-round PSMT protocol has a synchronous flow from $\mathcal{R}$ to $\mathcal{S}$ followed by a synchronous flow from $\mathcal{S}$ to $\mathcal{R}$.

3   One-Round PSMT with Low Communication Overhead

In this section, we extend the one-round PSMT protocol in [8,17] for $n = 3t+1$ to handle any $n > 3t$ with low communication overhead—in fact, exactly $\Lambda = \frac{n}{n-3t}$, which, as we also show, is optimal for this case.

3.1   Protocol 1-PSMT

At a high level, the PSMT protocols in [8,17] hide the message to be transmitted using the approach in [3] of verifiable secret sharing (VSS) over a finite field $\mathbb{F}$ using Reed-Solomon codes. In contrast to their solutions, instead of hiding the message in one single coefficient of the polynomial, we split the message into "pieces" and assign each piece to a separate coefficient, and correspondingly increase the degree of the polynomial. Effectively, this allows us to hide $n - 3t$ different field elements in one VSS instance.

In more detail, assuming an adequate field size[1], the message is interpreted as a sequence of $k = n - 3t$ field elements, and transmitted using the protocol of Figure 2. We are able to show:

Theorem 1. *Protocol 1-PSMT(n, t, m) is a one-round PSMT protocol for any $n > 3t$ with communication overhead $\Lambda = \frac{n}{n-3t}$.*

Proof (sketch).

CORRECTNESS: Since $n > d + 2t$, $\mathcal{R}$ can decode the complete polynomial, compute the low-degree coefficients m_i, $1 \leq i \leq k$, and extract the full message m.

PRIVACY: Since $f(x)$ is of degree $d = t + (k-1)$, any t shares of the form $f(\alpha^j)$ are independent from the k coefficients m_i. Thus, $\mathcal{A}$ gets no information about m.

COMMUNICATION OVERHEAD: The protocol communicates n field elements in order to transmit a secret message consisting of $k = n - 3t$ field elements. Thus, the communication overhead of the protocol is $\Lambda = \frac{n}{n-3t}$.   □

The following corollary will be useful for our main virtualization result in Section 4.

Corollary 1. *One-round PSMT with* constant *communication overhead is possible for $n = (3 + \delta)t$, for any constant $\delta > 0$.*

As we now show, the communication overhead of our one-round PSMT protocol is in fact optimal. The reader intrigued by the use of 1-PSMT in our virtualization scheme is invited to proceed directly to Section 4.

3.2 Communication Lower Bound for One-Round PSMT

In [18], Srinathan, Narayanan and Rangan established a lower bound on the communication overhead (of $\Lambda \geq \frac{n}{n-2t}$) for two-round PSMT. In this section we show a lower bound of $\Lambda \geq \frac{n}{n-3t}$ for one-round PSMT when $n > 3t$. Note that one-round PSMT is impossible if $n \leq 3t$.

Theorem 2. *Any one-round PSMT protocol for $n > 3t$ wires requires communication overhead $\Lambda \geq \frac{n}{n-3t}$.*

Proof. Let $\mathcal{M}$ be the message space from where the sender $\mathcal{S}$'s message is drawn. Let $\mathbf{T}_i^m$ denote the set of all possible transmissions that can occur on wire $W_i \in \{W_1, \ldots, W_n\}$ when $\mathcal{S}$ transmits message m. Furthermore, for $j \geq i$, let

---

[1] Alternatively, we would first split the message into blocks, and then transmit each block separately.

$\mathbf{M}_{i,j}^m \subseteq \mathbf{T}_i^m \times \mathbf{T}_{i+1}^m \times \cdots \times \mathbf{T}_j^m$ denote the set of all possible transmissions that can occur on the wires in $\{W_i, W_{i+1}, \ldots, W_j\}$ when $\mathcal{S}$ transmits message m. Finally, let $\mathbf{M}_{2t+1,n} = \bigcup_{m \in \mathcal{M}} \mathbf{M}_{2t+1,n}^m$, and $\mathbf{T}_i = \bigcup_{m \in \mathcal{M}} \mathbf{T}_i^m$, and let us call $\mathbf{T}_i$ the *capacity of wire* W_i and $\mathbf{M}_{k,\ell}$ the *capacity of the set of wires* $\{W_k, W_{k+1}, \ldots, W_\ell\}$.

Consider any one-round PSMT protocol for $n > 3t$. Perfect privacy requires that the transmissions on any t wires be independent of the message. Thus, for any two messages $m_1, m_2 \in \mathcal{M}$ it must hold that

$$\mathbf{M}_{2t+1,3t}^{m_1} = \mathbf{M}_{2t+1,3t}^{m_2} .$$

(The above must hold for any selection of t wires; we focus on the set $\{W_{2t+1}, \ldots, W_{3t}\}$ for simplicity.) Furthermore, perfect correctness implies that the (uncorrupted) transmissions on any $n - 2t$ wires must uniquely determine the message. Thus, it must also hold that

$$\mathbf{M}_{2t+1,n}^{m_1} \cap \mathbf{M}_{2t+1,n}^{m_2} = \emptyset .$$

Since $\mathbf{M}_{2t+1,3t}^m$ may be the same for every message m, it follows that

$$\prod_{i=3t+1}^{n} |\mathbf{T}_i| \geq |\mathbf{M}_{3t+1,n}| \geq |\mathcal{M}| .$$

Let $d = n - 3t$. More generally, the above inequality holds for any selection of d wires $\mathcal{D} \subset \{W_1, W_2, \ldots, W_n\}$, $|\mathcal{D}| = d$, i.e., $\prod_{W_i \in \mathcal{D}} |\mathbf{T}_i| \geq |\mathcal{M}|$, and in particular it holds for every selection $\mathcal{D}_k = \{W_{(kd+1) \mod n}, W_{(kd+2) \mod n}, \ldots, W_{(kd+d) \mod n}\}$, with $k \in \{0, 1, \ldots, n-1\}$.

If we consider all sets $\mathcal{D}_k$ separately, then each wire is accounted for exactly d times. Thus, the product of the capacities of all $\mathcal{D}_k$ yields the capacity of the full wire set to the d-th power, and since each $\mathcal{D}_k$ has capacity at least $|\mathcal{M}|$, we get

$$|\mathcal{M}|^n \leq \prod_{k=0}^{n-1} \prod_{W_j \in \mathcal{D}_k} |\mathbf{T}_j| = \left(\prod_{i=1}^{n} |\mathbf{T}_i| \right)^d ,$$

and therefore

$$\Lambda \geq \frac{\sum_{i=1}^{n} log|\mathbf{T}_i|}{log|\mathcal{M}|} \geq \frac{n}{d} = \frac{n}{n - 3t} .$$

$\square$

4   Communication-Optimal Two-Round PSMT for $n \geq (2 + \varepsilon)t$

In this section, we use wire virtualization and protocol 1-PSMT from the previous section to construct our new two-round PSMT protocol.

4.1   The Wire Virtualization Construction

Let $n \geq (2+\varepsilon)t$ for some $\varepsilon > 0$. Let Π_2 be the communication-optimal (but inefficient) two-round PSMT protocol in [1] (or even the communication-suboptimal protocol in [17]) for ν wires tolerating $\tau = \lfloor \frac{\nu-1}{2} \rfloor$ corrupted wires, where $\nu = O(1)$ (ν will be quantified later, based on the analysis below). Choosing $\nu = O(1)$ implies that protocol Π_2's communication overhead is constant, and thus that Π_2 is communication-optimal.

Further, let Π_1 be 1-PMST, the communication-optimal one-round protocol from Section 3.1 for N wires tolerating $T \leq \frac{N}{3+\delta}$ corrupted wires for some fixed constant $\delta > 0$ (where $N = n^\nu$; see below).

We start by forming all $N = n^\nu$ possible virtual wires $\mathcal{W}_1, \ldots, \mathcal{W}_{n^\nu}$ involving ν wires from the set of real wires $W = \{W_1, \ldots, W_n\}$, allowing repetitions. We call this collection of virtual wires $\mathcal{W}$, $\mathcal{W} = \{\mathcal{W}_1, \ldots, \mathcal{W}_{n^\nu}\}$. We can apply protocol Π_2 to any element of $\mathcal{W}$ with the effect that it will achieve PSMT as long as at most $\tau = \lfloor \frac{\nu-1}{2} \rfloor$ of the involved real wires are actually corrupted. We thus call a virtual wire *correct* when it involves at most τ corrupted real wires and *corrupted* otherwise. Let T be the number of corrupted virtual wires in $\mathcal{W}$.

Our goal now is to find a constant ν such that of all $N = n^\nu$ possible virtual wires out of $\mathcal{W}$, at most $T = \frac{N}{3+\delta}$ are corrupted. This will then allow us to apply one-round protocol Π_1 to the N virtual wires where, in turn, every virtual wire is simulated by the two-round protocol Π_2 (see Figure 1). The analysis in the next section will yield constant ν.

4.2   Virtualization Analysis

We consider the following random experiment in order to give a (deterministic) estimation on the ratio of corrupted virtual wires.

Let ν be fixed. Let p be the probability that, picking one of the $N = n^\nu$ possible ν-tuples of n real wires uniformly at random, the respective virtual wire is corrupted. If this probability is at most $\frac{T}{N} = \frac{1}{3+\delta}$ then, clearly, at most $T = \frac{N}{3+\delta}$ virtual wires are corrupted — which is tolerated by protocol Π_1.

For this, we consider random variable $X \in \{0, \ldots, \nu\}$ denoting the number of corrupted wires in the selection. Let P be the probability distribution induced by the following random experiment: pick a wire out of W uniformly at random, repeat this ν times, and let the resulting selection of wires form a tuple of size ν (i.e., a virtual wire).

Our goal is to show that there is a constant ν such that $p = \Pr(X \geq \nu/2) \leq \frac{1}{3+\delta}$, and thus, that the number of actual corrupted virtual wires in Π_1 is at most $T = \frac{N}{3+\delta}$. We achieve this with help of the Chernoff bound (see Appendix A).

According to the process associated with P, let X_i be the 0-1 distributed random variable describing whether the i-th chosen wire is corrupted. Then $X = \sum_{i=1}^{\nu} X_i$. We demand

$$\Pr\left(X \geq \frac{\nu}{2}\right) \leq \frac{1}{3+\delta} .$$

Since, clearly, the random variables $X_1, \ldots, X_\nu$ are independent, we can estimate this probability by the Chernoff bound (Equation 2) as

$$\Pr\left(X \geq \frac{\nu}{2} = \lambda \mu \nu = \lambda \frac{\nu}{2+\varepsilon}\right) \leq e^{-\frac{\nu}{2(2+\varepsilon)}(\lambda-1)^2} \quad \text{where} \quad \lambda = \frac{2+\varepsilon}{2},$$

and demand

$$e^{-\frac{\nu}{2(2+\varepsilon)}(\lambda-1)^2} \overset{!}{\leq} \frac{1}{3+\delta}.$$

We thus require that

$$\frac{\nu}{2(2+\varepsilon)}(\lambda-1)^2 = \frac{\nu}{8(2+\varepsilon)}\varepsilon^2 \geq \ln(3+\delta),$$

which yields

$$\nu \geq \left\lceil \frac{8\ln(3+\delta)(2+\varepsilon)}{\varepsilon^2} \right\rceil, \tag{1}$$

obtaining a lower-bound estimation on ν depending on constants ε and δ, where ε is an input parameter and δ is any positive constant of free choice.

Theorem 3. *The construction described above is a two-round PSMT protocol for any $n \geq (2+\varepsilon)t$, $\varepsilon > 0$, and has constant communication overhead, which is optimal.*

Proof (sketch).

CORRECTNESS AND PRIVACY. Correctness and privacy of the protocol follow from the above quantitative analysis and from the respective properties of protocols Π_1 and Π_2.

NUMBER OF ROUNDS. The top-level protocol Π_1 is one-round and operates on virtual wires. Every virtual wire can be independently simulated in parallel by the two-round protocol Π_2. Thus, the resulting protocol involves two communication rounds.

COMMUNICATION OVERHEAD. Protocol Π_2 operates on ν real wires. Since $\nu = O(1)$, the protocol has constant communication overhead. Protocol Π_1 operates on $N = n^\nu$ virtual wires and also has constant communication overhead since we have $T = \frac{N}{3+\delta}$. Thus Π_1 involves N messages of size $\frac{\ell}{N} \cdot O(1)$ which are each transmitted by an instance of protocol Π_2 with constant communication overhead, resulting in the total communication of $N \cdot \frac{\ell}{N} \cdot O(1) = O(\ell)$ bits — or communication overhead $\Lambda = O(1)$ — matching the lower bound for two-round PSMT established in [18]. $\qquad\square$

5 Conclusions

In this paper, we presented a communication-optimal two-round PSMT protocol for $n \geq (2+\varepsilon)t$ where $\varepsilon > 0$ is an arbitrary, small constant. For the protocol to be communication-optimal, messages of length only polynomial in n are required. The communication complexity of the protocol is $O(\ell)$.

As it follows from the lower bound in [18], communication complexity $O(\ell)$ can only be achieved if $n = 2t + \Omega(t)$. Thus, our protocol is optimally resilient under the constraint of communication complexity $O(\ell)$. Our protocol is constructed along the lines of Bracha's player-virtualization technique, systematically extending the player set in order to amplify the resilience of a lower-level protocol.

We also obtained a tight bound on the communication complexity of one-round PSMT for $n > 3t$.

Regarding optimizations to our construction, note that our estimation on ν is rather conservative since it is based on a rough Chernoff-bound estimation. Experiments computing minimal values ν for particular values of ε show that much better results can be achieved. However, depending on the particular value of ε, our construction may still demand the message size to be a polynomial in n of high degree. For example, $\varepsilon = .6$ yields $\nu = 3$, $\varepsilon = .3$ yields $\nu = 11$, while $\varepsilon = .1$ yields $\nu = 83$.

In some cases, variations of the given construction achieve better results — for example, by setting $\nu = 3$ and applying virtualization recursively. Another possibility, at least in order to non-constructively prove the existence of protocols for smaller message sizes, is to have $\nu = \Theta(\log n)$ and proceed along the lines of Bracha [6]. We note that in this case constant communication overhead can still be achieved while requiring low-level protocol Π_2 to be of lower-than-optimal resilience, i.e., $\nu \geq (2 + \alpha)\tau$, where $\alpha > 0$ is a constant. Yet another direction worth investigating in order to achieve a lower number of virtual wires, as suggested by one of the reviewers, would be a "de-randomized" choice of sets obtained from short walks on low-degree expander graphs.

Acknowledgements

We thank the anonymous reviewers for *TCC* for their many useful comments. The work of Matthias Fitzi and Matt Franklin was partly supported by a David and Lucile Packard Fellowship for Science and Engineering.

References

1. S. Agarwal, R. Cramer, and R. de Haan. Asymptotically optimal two-round perfectly secure message transmission. In *Advances in Cryptology: CRYPTO '06*. Springer-Verlag, 2006.
2. M. Ben-Or. Another advantage of free choice: Completely asynchronous agreement protocols. In *Proceedings of the 2nd ACM Symposium on Principles of Distributed Computing (PODC '83)*, pages 17–19. ACM, 1983.
3. M. Ben-Or, S. Goldwasser, and A. Wigderson. Completeness theorems for non-cryptographic fault-tolerant distributed computation. In *Proceedings of the 20th Annual ACM Symposium on Theory of Computing (STOC '88)*, pages 1–10. Springer-Verlag, 1988.

4. E. Berlekamp and L. Welch. Error correction of algebraic block codes. US Patent 4,633,470.
5. P. Berman, J. A. Garay, and K. J. Perry. Bit optimal distributed consensus. In *Computer Science Research*, pages 313–322. Plenum Publishing Corporation, 1992.
6. G. Bracha. An $O(\log n)$ expected rounds randomized Byzantine generals protocol. *Journal of the Association for Computing Machinery*, 34(4):910–920, Oct. 1987.
7. B. A. Coan and J. L. Welch. Modular construction of a Byzantine agreement protocol with optimal message bit complexity. *Information and Computation*, 97(1):61–85, Mar. 1992.
8. D. Dolev, C. Dwork, O. Waarts, and M. Yung. Perfectly secure message transmission. *Journal of the ACM*, 40(1):17–47, Jan. 1993.
9. C. Dwork, D. Peleg, N. Pippinger, and E. Upfal. Fault tolerance in networks of bounded degree. In *Proceedings of the 18th Annual ACM Symposium on Theory of Computing (STOC '86)*, pages 370–379, 1986.
10. O. Goldreich, S. Micali, and A. Wigderson. How to play any mental game. In *Proceedings of the 19th Annual ACM Symposium on Theory of Computing (STOC '87)*, pages 218–229, 1987.
11. M. Hirt and U. Maurer. Player simulation and general adversary structures in perfect multiparty computation. *Journal of Cryptology*, 13(1):31–60, Winter 2000.
12. W. Hoeffding. Probability inequalities for sums of bounded random variables. *Journal of the American Statistical Association*, 58(301):13–30, Mar. 1963.
13. V. King, J. Saia, V. Sanwalani, and E. Vee. Towards secure and scalable computation in Peer-to-Peer networks. In *Proceedings of the 47th Annual IEEE Symposium on Foundations of Computer Science (FOCS '06)*, 2006.
14. L. Lamport, R. Shostak, and M. Pease. The Byzantine generals problem. *ACM Trans. Prog. Lang. Syst.*, 4(3):382–401, July 1982.
15. R. Ostrovsky, S. Rajagopalan, and U. Vazirani. Simple and efficient leader election in the full information model. In *Proceedings of the 26th Annual ACM Symposium on Theory of Computing (STOC '94)*, pages 234–242, 1994.
16. A. Patra, A. Choudhary, K. Srinathan, and C. Pandu Rangan. Constant phase bit optimal protocols for perfectly secure message transmission. In *Indocrypt '06*, 2006.
17. H. Sayeed and H. Abu-Amara. Efficient perfectly secure message transmission in synchronous networks. *Information and Communication*, 126(1):53–61, 1996.
18. K. Srinathan, A. Narayanan, and C. Pandu Rangan. Optimal perfectly secure message transmission. In *Advances in Cryptology: CRYPTO '04*, volume 3152 of *Lecture Notes in Computer Science*, pages 545–561. Springer-Verlag, 2004.
19. A. C. Yao. Protocols for secure computations. In *Proceedings of the 23rd Annual IEEE Symposium on Foundations of Computer Science (FOCS '82)*, pages 160–164. IEEE, 1982.
20. D. Zuckerman. Randomness-optimal sampling, extractors, and constructive leader election. In *Proceedings of the 28th Annual ACM Symposium on Theory of Computing (STOC '96)*, pages 286–295, 1996.

A   Chernoff Bounds

Chernoff bounds [12] give bounds on the probability that of n independent Bernoulli trials the outcome deviates form the expected value by a given fraction. Here we present the "upper tail" version.

Let X_i $(1 \leq i \leq n)$ be a sequence of independent 0-1 distributed random variables with expected value μ. By $\mathcal{C}(\mu, n, \lambda)$ $(\lambda > 1)$ we denote the probability that, out of n trials, the outcome exceeds the expected value $n\mu$ by a given factor depending on λ. The following inequality, which holds for $1 < \lambda < 2e$, bounds this probability.

$$\mathcal{C}(\mu, n, \lambda) = \Pr\left(\sum_{i=1}^{n} X_i \geq \lambda\mu n\right) \leq e^{-\frac{\mu n(\lambda-1)^2}{2}} \tag{2}$$

Concurrently-Secure Blind Signatures Without Random Oracles or Setup Assumptions*

Carmit Hazay[1], Jonathan Katz[2], Chiu-Yuen Koo[2], and Yehuda Lindell[1]

[1] Bar-Ilan University
{harelc,lindell}@cs.biu.ac.il
[2] University of Maryland
{jkatz,cykoo}@cs.umd.edu

Abstract. We show a new protocol for blind signatures in which security is preserved even under arbitrarily-many concurrent executions. The protocol can be based on standard cryptographic assumptions and is the first to be proven secure in a concurrent setting (under *any* assumptions) without random oracles or a trusted setup assumption such as a common reference string. Along the way, we also introduce new definitions of security for blind signature schemes.

1 Introduction

Blind signature schemes, introduced by Chaum [11], are a fascinating primitive that (roughly speaking) enable a user to interact with a signer and obtain a signature on a message m without revealing anything about m to the signer. Blind signature schemes are a crucial component of many systems in which certain values need to be *certified*, yet *anonymity* should be ensured: classical examples include e-cash (where a bank signs 'e-coins' that are withdrawn by customers) and e-voting (where an authority signs public keys for voters to use when they later cast their votes).

Definitions of security for blind signature schemes were first proposed by Pointcheval and Stern [29], though many refinements and extensions of their original definitions have since been suggested. At a high level, all existing definitions impose two basic requirements: *blindness* (or anonymity) and *unforgeability*. Blindness formalizes the notion that a malicious signer should be unable to 'link' any message/signature pair with a particular execution of the signing protocol. Unforgeability for blind signatures is the analogue of the notion of unforgeability for standard signature schemes: informally, a malicious user should be unable to output a valid signature on any message other than those whose signatures were explicitly requested from the signer. A subtlety in the case of blind signatures is that a malicious user's execution of the protocol with the

---

* This research was supported by US-Israel Binational Science Foundation grant #2004240. Work of the first author was also supported by an Eshkol fellowship from the Israel Ministry of Science and Technology. Work of the second author was also supported by NSF CAREER award #0447075.

S.P. Vadhan (Ed.): TCC 2007, LNCS 4392, pp. 323–341, 2007.

signer may not result in any well-defined message whose signature is being requested. Because of this, the formal definition requires that for any polynomial ℓ and any user executing the protocol ℓ times with the signer, the user should be unable to output $\ell + 1$ valid signatures on $\ell + 1$ distinct messages.

When defining blindness and unforgeability it is necessary to distinguish whether security requires different executions of the protocol to be carried out *sequentially* (i.e., waiting for one execution to finish before beginning the next), or whether security holds even when multiple executions are performed *concurrently* (i.e., in an arbitrarily-interleaved manner). (One can also consider the intermediate case in which executions are run *in parallel*.) Concurrency in the context of blindness has received little attention, both because the 'standard' definition of blindness considers only two executions of the protocol and also, perhaps, because many known constructions of blind signature schemes achieve perfect blindness. In contrast, handling concurrency in the context of unforgeability has received much attention (surveyed below), and it is not hard to see that — assuming there exist blind signature schemes at all — there exist schemes that are unforgeable in the sequential setting but *not* in a concurrent setting.

1.1   Previous Constructions

Chaum [11] proposed a candidate blind signature scheme without any proof of security (though his scheme was later proven secure in the random oracle model under a somewhat non-standard cryptographic assumption [5]). Since then, numerous works have aimed to design secure schemes. We review these here, with particular attention to the type of unforgeability proved.

Schemes in the random oracle model. Initial constructions of blind signature schemes were in the random oracle model [6], and, in fact, until relatively recently all efficient constructions relied on random oracles. Pointcheval and Stern [28] showed the first secure blind signature schemes, though they prove unforgeability (in the parallel setting) only for a user who requests *logarithmically*-many signatures. This was improved in later work by Pointcheval [27], who showed schemes that are unforgeable (in a restricted variant of the parallel setting) for polynomially-many signatures. Abe [1] gave a protocol with improved round complexity, and also proved unforgeability in the concurrent setting. Bellare, et al. [5] and Boldyreva [8] present 2-round blind signature schemes; note that 2-round protocols (which consist of a single message from the user and a response by the signer) are automatically secure in a concurrent setting.

Schemes in the standard model. Relatively early, it was suggested [12] that blind signatures might be constructed using protocols for generic secure 2-party computation. Juels, Luby, and Ostrovsky [19] point out that the naïve way of implementing this approach does not work, but show how to adapt and extend this idea so as to achieve a secure solution. Although they claim security in the concurrent setting, no details of the proof in this case are provided; as best as we can tell, their solution is secure in the sequential setting only. Indeed, security of their protocol in the concurrent setting seems to require a *concurrently-secure* protocol for 2-party computation, but constructing such protocols without random oracles or setup assumptions is currently a major open question. The work of [4] could be used here, but then security would require sub-exponential hardness assumptions (something avoided in our work).

Camenisch, et al. [9] show the first *efficient* protocol secure in the standard model, proven unforgeable only for the case of sequential attacks.

Lindell [23] has shown the impossibility of concurrently-secure blind signatures if simulation-based definitions of security are used.[1] In an effort to overcome the limitations of the above protocols, as well as Lindell's impossibility result, much recent work focused on proving security for blind signature schemes in the concurrent setting by assuming a *common reference string* [26,21,16]. However, although Lindell's impossibility result was used as justification for relying on a common reference string in these works, Lindell's results do not apply if *game-based* security definitions (rather than *simulation-based* security definitions) are used. Indeed, this serves as the starting point for our work.[2]

1.2 Our Contributions

As hinted at earlier, the standard definition of blindness considers only the interaction of a malicious signer with two users; furthermore, the definition does not seem to reasonably extend for the case of multiple users (the issue is how to deal with a signer who may abort some sessions). We propose a new definition here which extends seamlessly to the multi-user setting, and (in retrospect) seems to capture better the security requirements of a blind signature scheme.

As our main contribution, we present the first concurrently-secure blind signature scheme that does not rely on random oracles or any setup assumptions such as a common reference string. In order to 'bypass' the impossibility result of Lindell [23], we prove security using game-based definitions that have anyway been standard in almost all prior work in this area. Our protocol relies on standard cryptographic assumptions (e.g., trapdoor permutations and the decisional Diffie-Hellman assumption), and we prove security with respect to game-based definitions that are stronger than others that have appeared in the literature.

Besides being interesting in its own right, our construction serves as yet another illustration that known impossibility results for concurrently-secure 2-party computation [23,24] might be overcome for *specific* functionalities of interest by considering relaxed (yet still meaningful) definitions of security. In this sense, our work exemplifies what we see as a viable alternative to the approaches to concurrently-secure computation taken by, e.g., [10,23,31,3,20,4,25], who focus on staying within the simulation paradigm (in part, because they are striving for a generic result) but are thus forced to impose additional assumptions (e.g., a common reference string [10] or a bound on network delay [20]) or to settle for alternate definitional relaxations (e.g., bounded concurrency [23] or super-polynomial-time simulation [4]).

---

[1] Technically, he only rules out *black-box* proofs of security.

[2] We do not formally define what it means for a definition to be 'simulation-based' or 'game-based,' but instead appeal to the reader's intuition regarding such matters.

1.3 Outline

In Section 2 we discuss definitions of security for blind signature schemes, and present a new set of definitions that are stronger than any to have previously appeared in the literature. We also propose, for the first time, a definition of blindness for the case of a signer interacting with an arbitrary number of users.

We then build up to our main result in stages: in Section 3.1 we describe the recent blind signature scheme of Fischlin [16] which is used as a building block in our work, and then in Section 3.2 we construct a blind signature scheme that can be proven concurrently-secure using *complexity leveraging*.[3] Our main result (which does not rely on complexity leveraging) appears in Section 4, along with proof sketches of the blindness and unforgeability properties. Due to space limitations, complete proofs are omitted but will appear in the full version.

2   Definitions

A standard signature scheme is a tuple of PPT algorithms (Gen, Sign, Vrfy), where the *key generation algorithm* Gen takes as input a security parameter 1^k and outputs a pair of keys (pk, sk) with the security parameter implicit in both; the *signing algorithm* Sign takes as input a message m and a secret key sk and outputs a signature σ; and the *verification algorithm* Vrfy takes as input a public key pk, a message m, and a candidate signature σ and outputs a decision bit. Correctness requires that if (pk, sk) is output by Gen(1^k) then $\mathsf{Vrfy}_{pk}(m, \mathsf{Sign}_{sk}(m)) = 1$ for all m. We use the standard definition of existential unforgeability under adaptive chosen-message attacks [18].

We assume signature schemes that are *length-regular*: i.e., there exists a polynomial $p(\cdot)$ such that if (pk, sk) are output by Gen(1^k) then for any m in the message space (1) $\mathsf{Sign}_{sk}(m) \in \{0,1\}^{p(|m|)}$ and (2) $\mathsf{Vrfy}_{pk}(m, \sigma) = 0$ if $\sigma \notin \{0,1\}^{p(|m|)}$. We will not write this explicitly in the rest of the paper.

We now define a *blind* signature scheme.

Definition 1. *A* blind signature scheme *consists of* PPT *algorithms* Gen, Vrfy *along with interactive* PPT *algorithms* $\mathcal{S}, \mathcal{U}$ *such that:*

- Gen, *on input* 1^k, *outputs a key pair* (PK, SK) *with k implicit in both.*
- *The joint execution of $\mathcal{S}$, holding input* SK, *and $\mathcal{U}$, holding inputs* PK, m, *results in an output σ for $\mathcal{U}$, assuming neither $\mathcal{S}$ nor $\mathcal{U}$ abort. We write this as $\sigma \leftarrow \langle \mathcal{S}_{\mathsf{SK}}, \mathcal{U}_{\mathsf{PK}}(m) \rangle$. If $\mathcal{U}$ aborts, its output is $\bot$ (which is never a valid signature) and we assume that it notifies $\mathcal{S}$.*
- Vrfy, *on input* PK, m, σ, *outputs a decision bit.*

Correctness requires that for all (PK, SK) *output by* Gen(1^k) *and all m, if $\sigma \leftarrow \langle \mathcal{S}_{\mathsf{SK}}, \mathcal{U}_{\mathsf{PK}}(m) \rangle$ then $\mathsf{Vrfy}_{\mathsf{PK}}(m, \sigma) = 1$.*

---

[3] Roughly speaking, this means we assume primitives A and B such that A cannot be broken in polynomial time but can be broken in time $T(k)$ for some super-polynomial function T, while B cannot be broken in time $T(k)$.

We now define *unforgeability* and *blindness*. In both definitions, the adversary maintains state throughout its execution.

Definition 2. *Blind signature scheme* (Gen, $\mathcal{S}, \mathcal{U}$, Vrfy) *is* unforgeable *if for any polynomial ℓ, the success probability of any* PPT *algorithm $\hat{\mathcal{U}}$ in the following game is negligible:*

- Gen(1^k) *outputs keys* (PK, SK), *and $\hat{\mathcal{U}}$ is given* PK.
- $\hat{\mathcal{U}}$(PK) *interacts concurrently with $\ell = \ell(k)$ instances $\mathcal{S}_{SK}^1, \ldots, \mathcal{S}_{SK}^\ell$.*
- $\hat{\mathcal{U}}$ *outputs* $(m_1, \sigma_1, \ldots, m_{\ell+1}, \sigma_{\ell+1})$.

$\hat{\mathcal{U}}$ *succeeds if the $\{m_i\}$ are distinct and* Vrfy$_{PK}(m_i, \sigma_i) = 1$ *for all i.*

We next turn to defining blindness. We begin with a (strong) variant of the standard definition of blindness, which only considers the execution of the signer with two users. This is followed by some discussion of how the definition might be extended for the case of multiple users.

Definition 3. *Blind signature scheme* (Gen, $\mathcal{S}, \mathcal{U}$, Vrfy) *satisfies* blindness *if the advantage of any* PPT *algorithm $\hat{\mathcal{S}}$ in the following game is negligible:*

1. *$\hat{\mathcal{S}}(1^k)$ outputs an arbitrary public key* PK *along with equal-length messages m_0, m_1.*

2. *A random bit b is chosen, and $\hat{\mathcal{S}}$ interacts concurrently with $\mathcal{U}_b \stackrel{def}{=} \mathcal{U}_{PK}(m_b)$ and $\mathcal{U}_{\bar{b}} \stackrel{def}{=} \mathcal{U}_{PK}(m_{\bar{b}})$. When $\mathcal{U}_b, \mathcal{U}_{\bar{b}}$ have completed their execution, σ_0, σ_1 are defined as follows:*

    - *If either $\mathcal{U}_b$ or $\mathcal{U}_{\bar{b}}$ abort, then $(\sigma_0, \sigma_1) := (\perp, \perp)$.*
    - *Otherwise, let σ_0 (resp, σ_1) be the output of $\mathcal{U}_0$ (resp., $\mathcal{U}_1$).*

    $\hat{\mathcal{S}}$ is given (σ_0, σ_1).

3. *Finally, $\hat{\mathcal{S}}$ outputs a bit b'.*

$\hat{\mathcal{S}}$ *succeeds (denoted Succ) if $b' = b$. The advantage of $\hat{\mathcal{S}}$ is $\left| \Pr[Succ] - \frac{1}{2} \right|$.*

For the definition to be meaningful, we cannot give $\hat{\mathcal{S}}$ the signature output by one user in case the other aborts: if we did, $\hat{\mathcal{S}}$ could simply abort the execution with its 'left' oracle and then, depending on whether it is given a signature on m_0 or m_1, easily determine b. On the other hand, in contrast to [16], we allow the game to continue if either user aborts (this only strengthens the definition). Note also that $\hat{\mathcal{S}}$ may generate PK in an arbitrary manner, not necessarily using Gen. It seems perfectly natural to us to allow this possibility, though it appears to have been formally considered only relatively recently [2,26,16].

In extending the above definition to the case of a signer interacting with an arbitrary number of users, an obvious approach is to allow the signer to output two *vectors* m_0, m_1 containing the same messages $m_1, \ldots, m_\ell$ (possibly allowing repeats) in permuted order. A difficulty that arises is how to deal with a signer who aborts some of the sessions. Some natural ways of dealing with this are (1) if the signer aborts any session, it receives no signatures; or (2) say $m_0 = (m_1^0, \ldots, m_\ell^0)$ and $m_1 = (m_1^1, \ldots, m_\ell^1)$. Then if the signer aborts the ith

session, it is given neither the signature on m_i^0 nor the signature on m_i^1. The first option seems (to us) to be too weak. The second option seems a bit arbitrary, though reasonable; an aesthetic drawback is that it is not clear that it is implied by Definition 3. In the full version we sketch a third possibility, intermediate in strength between the above two, which *is* implied by Definition 3.

In any case, all the above ways of dealing with abort (even in the original case with two users) seem a bit arbitrary even though for technical reasons they are necessary to make the definitions non-trivial. We therefore propose a new definition which, in our opinion, handles the issue of abort in a cleaner way. Though it allows some 'attacks' which are ruled out by Definition 3, we believe it models the security desired of typical proposed applications of blind signatures (such as e-cash or e-voting). Further discussion follows the definition.

Definition 4. *Blind signature scheme* (Gen, $\mathcal{S}, \mathcal{U}$, Vrfy) *satisfies a posteriori* blindness *if for any polynomial ℓ, any ℓ' such that $1 \leq \ell'(k) \leq \ell(k)$ for all k, and any* PPT *algorithm $\hat{S}$, the advantage of $\hat{S}$ in the following game is at most a negligible quantity:*

1. *$\hat{S}(1^k)$ outputs an arbitrary public key* PK *and a message distribution[4] $\mathcal{M}$ sampleable in polynomial time.*
2. *Messages $m_1, \ldots, m_\ell$ are sampled according to $\mathcal{M}$, and $\hat{S}$ interacts concurrently with $\mathcal{U}_{\text{PK}}(m_1), \ldots, \mathcal{U}_{\text{PK}}(m_\ell)$. The game ends if the number of non-aborted sessions is not equal to ℓ'. Otherwise, we say event* NA(ℓ') *occurs and the game continues.*
3. *Let $i_1, \ldots, i_{\ell'}$ denote the indices of the non-aborted sessions and let π be a random one-to-one function mapping $\{1, \ldots, \ell'\}$ to these indices. $\hat{S}$ is given $(m_{\pi(1)}, \sigma_{\pi(1)}), \ldots, (m_{\pi(\ell')}, \sigma_{\pi(\ell')})$.*
4. *Finally, $\hat{S}$ outputs (i, i').*

$\hat{S}$ succeeds (this event is denoted by Succ*) if $\pi(i) = i'$. The advantage of $\hat{S}$ is* $\Pr[\text{Succ}] - \frac{1}{\ell'} \Pr[\text{NA}(\ell')]$.

Note that allowing the signer to choose the message distribution is stronger than quantifying over all sampleable distributions, since it allows the signer to choose a distribution that depends on the (maliciously-chosen) public key.

The intent of the above definition is to model the scenario where (honest) users anyway choose the 'messages' to be signed from some known distribution. For example, in the case of e-cash the message might be a random string; in the case of e-voting the message might be an honestly-generated public key; finally, a scenario similar (but not identical) to that of Definition 3 can be achieved if $\mathcal{M}$ is the uniform distribution over $\{m_0, m_1\}$. After interacting with users who choose their messages according to this distribution, the signer is given all message/signature pairs (in a randomly-permuted order) from the non-aborted sessions; this corresponds to the scenario when the users in the non-aborted sessions reveal their message/signature pairs (e.g., by spending an e-coin or

---

[4] This could be specified, e.g., by a circuit whose output (on uniform input) defines the distribution.

casting a vote). Informally, the signer 'wins' if it can link some message/signature pair to its corresponding session with probability better than randomly guessing a non-aborted session.

The nice thing about the above definition is that it models exactly what the signer actually 'sees' in the real world, without imposing any artificial (though necessary) restrictions. We remark also that Definition 4 in the special case $\ell = 2$ implies the general case.

We stress, however, that Definition 4 guarantees no 'blindness' whatsoever in the aborted sessions. In particular, a scheme in which the user reveals m (and aborts) if the signer sends an improper first message could still potentially be secure with respect to Definition 4 though it would not satisfy Definition 3. We do not view this as a problem since we view 'messages' as having no inherent secrecy requirement (indeed, the user eventually reveals its message anyway); rather, the goal is to prevent the linking of a particular message (that is later used) to a particular session. In this sense, schemes satisfying a posteriori blindness are analogous to commitment schemes with a posteriori secrecy (cf. [17, Section 4.8.2.5]). For this reason, schemes satisfying this notion may not be appropriate for all possible applications of blind signatures.[5]

3   A Warm-Up for Our Main Result

Our blind signature scheme builds on an elegant construction due to Fischlin [16] that relies on a common reference string. We review Fischlin's scheme and then, as a step toward our main result, present a blind signature scheme that can be proven concurrently-secure using complexity leveraging (cf. footnote 3).

3.1   Fischlin's Blind Signature Scheme

We describe a simplified[6] version of Fischlin's scheme that satisfies our definitions of blindness and unforgeability in the *common reference string* (CRS) model. Let $\Pi' = (\mathsf{Gen}', \mathsf{Sign}', \mathsf{Vrfy}')$ be a standard signature scheme, and let Com be a perfectly-binding commitment scheme. Fischlin's scheme is defined as follows (see also Figure 1):

Setup: The CRS contains a public key pk_E for a semantically-secure public-key encryption scheme, and a string ρ used as a CRS for a non-interactive zero-knowledge (NIZK) proof system. $\mathcal{E}_{pk_E}(\cdot)$ denotes encryption using pk_E.

Key generation: $\mathsf{Gen}(1^k)$ runs $\mathsf{Gen}'(1^k)$ to obtain keys $(\mathsf{pk}', \mathsf{sk}')$ and outputs these keys.

Signing: The protocol for a user $\mathcal{U}$ to obtain a signature on a message m is as follows:

---

[5] However, we conjecture that any scheme satisfying Def. 4 can be converted to one satisfying Def. 3 by using a commitment to the message in the signing protocol.

[6] The scheme presented by Fischlin includes some additional complications that are used to achieve *strong* unforgeability, which we do not consider here.

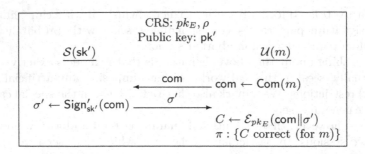

Fig. 1. Fischlin's protocol

- $\mathcal{U}$ computes $\mathsf{com} \leftarrow \mathsf{Com}(m)$ and sends com to the signer.
- $\mathcal{S}$ computes $\sigma' \leftarrow \mathsf{Sign}'_{\mathsf{sk}'}(\mathsf{com})$ and sends σ' to $\mathcal{U}$.
- $\mathcal{U}$ verifies the signature sent in the previous step, and aborts if it is invalid. Otherwise, the user computes $C \leftarrow \mathcal{E}_{pk_E}(\mathsf{com}\|\sigma')$ and computes an NIZK proof π (using ρ) that $(m, C, pk_E, \mathsf{pk}') \in L$ where L is defined as the set of tuples $(m, C, pk_E, \mathsf{pk}')$ for which there exists $\omega_1, \omega_2, \mathsf{com}, \sigma'$ such that

$$\mathsf{com} := \mathsf{Com}(m; \omega_1) \bigwedge C := \mathcal{E}_{pk_E}(\mathsf{com}\|\sigma'; \omega_2) \bigwedge \mathsf{Vrfy}'_{\mathsf{pk}'}(\mathsf{com}, \sigma') = 1.$$

(Note that L is an $\mathcal{NP}$ language.) The signature is (C, π).

Verification: To verify signature (C, π) on m with respect to public key pk' and CRS (pk_E, ρ), verify that π is a valid proof (with respect to ρ) that $(m, C, pk_E, \mathsf{pk}') \in L$.

We now sketch the proofs of blindness and unforgeability. For blindness, note that the signer observes only a commitment to m, an encryption of this commitment, and an NIZK proof π; it is not too hard to see that none of these leaks information about m, nor allows the signer to correlate a particular execution of the protocol with a particular signature output by $\mathcal{U}$.

For unforgeability, an adversary $\hat{\mathcal{U}}$ that forges a signature in the sense of Definition 2 can be used to construct a forger $\mathcal{F}$ for standard signature scheme Π': given public key pk' of an instance of Π', forger $\mathcal{F}$ generates pk_E on its own (along with the corresponding secret key sk_E), generates ρ at random, and runs $\hat{\mathcal{U}}$ in the natural way. $\mathcal{F}$ can easily execute the protocol with $\hat{\mathcal{U}}$ using its own signing oracle. Finally, if $\hat{\mathcal{U}}$ outputs $\ell + 1$ distinct messages $\{m_i\}$ with valid signatures $\{(C_i, \pi_i)\}$, then with all but negligible probability (by soundness of the NIZK proof system and perfect binding of the commitment scheme) each C_i is a valid encryption of a *distinct* commitment com_i and a valid signature σ'_i (with respect to Π') on this commitment. Given this, $\mathcal{F}$ can recover all the $\{(\mathsf{com}_i, \sigma'_i)\}$ by decrypting all the ciphertexts using sk_E; since $\mathcal{F}$ accessed its signing oracle exactly ℓ times, at least one $(\mathsf{com}_i, \sigma'_i)$ leads to a valid forgery for Π'.

3.2   Concurrently-Secure Blind Signatures: A Partial Solution

If we try to adapt Fischlin's scheme so as to avoid the CRS, we encounter two main obstacles. We describe these now, along with our solutions.

Removing ρ: If the signer generates ρ, the proof π may leak information about the underlying m, com, or σ' (which would violate blindness); on the other hand, the user clearly cannot generate ρ itself since then soundness may no longer apply and forgery would be possible.

We can resolve this by relying on ZAPs [14] rather than NIZK, and having the signer include the first message ρ for a ZAP as part of its public key. (A ZAP is a two-round witness-indistinguishable proof system; see Appendix A). Since a ZAP is witness indistinguishable but not zero knowledge, however, the protocol must be changed so as to provide an alternate witness that will be available to a simulator (for proving blindness) but not to a malicious user (or else forgery becomes possible). We provide such a witness by having the signer include $y_0 = f(x_0)$ and $y_1 = f(x_1)$ in its public key, where f is a one-way function, and then having the signer give a witness-indistinguishable proof of knowledge of either x_0 or x_1 as part of the signing protocol [15]. When constructing the signature (after execution of the signing protocol), the user $\mathcal{U}$ computes C as in Fischlin's protocol and then gives a witness-indistinguishable proof π that (essentially) it either constructed C appropriately or it knows one of x_0 or x_1.

Removing pk_E: If the signer generates pk_E then it is trivial for a malicious signer to violate the blindness property; if the user generates pk_E on its own, then the reduction in the proof of unforgeability given in the previous section no longer works since $\mathcal{F}$ can no longer recover a forgery for Π' from a forgery for the blind signature scheme (since it cannot decrypt C).

If we are willing to rely on complexity leveraging, we can overcome this difficulty by using a *commitment scheme* Com^* to construct C rather than an *encryption scheme*. If Com^* is secure against PPT adversaries, blindness still holds. If, however, Com^* can be broken in time $T(k)$ for some super-polynomial function $T(\cdot)$, then (referring to the proof of unforgeability in the previous section) we can construct a forger $\mathcal{F}$ running in time $O(T(k))$ who extracts a valid signature for Π'. If we further assume that Π' is secure *even against adversaries running in time* $O(T(k))$, this still yields a contradiction and is enough to prove unforgeability of the blind signature scheme.

This gives the main intuition. We now give a more complete description of the protocol, along with sketches of the proofs of blindness and unforgeability. We take the liberty of being somewhat informal, as this protocol is meant mainly as a 'stepping stone' toward our main result (which does not use complexity leveraging).

Let $\Pi' = (\mathsf{Gen}', \mathsf{Sign}', \mathsf{Vrfy}')$ be a standard signature scheme, and let f be a one-way function. We assume these are secure (in the appropriate sense) for adversaries running in time $O(T(k))$, where $T(\cdot)$ is a super-polynomial function. Let $\mathsf{Com}, \mathsf{Com}^*$ be perfectly-binding commitment schemes, where Com^* is such

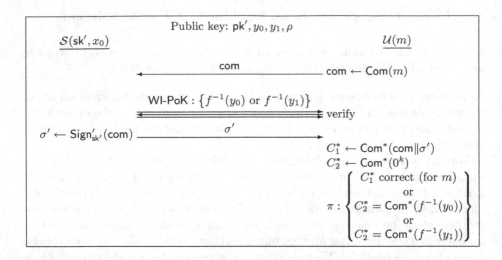

Fig. 2. A partial solution using complexity leveraging

that given $C^* \leftarrow \mathsf{Com}^*(m)$ it is possible to recover m in time $T(k)$. (However, Com^* is still hiding for PPT adversaries.) Our protocol is defined as follows:

Key generation: $\mathsf{Gen}(1^k)$ runs $\mathsf{Gen}'(1^k)$ to obtain keys $(\mathsf{pk}', \mathsf{sk}')$. It also chooses $x_0, x_1 \leftarrow \{0,1\}^k$ and sets $y_0 := f(x_0)$ and $y_1 := f(x_1)$. Finally, it computes ρ as the verifier's initial message in a ZAP. The public key is $\mathsf{PK} := (\mathsf{pk}', y_0, y_1, \rho)$ and the secret key is $\mathsf{SK} := (\mathsf{sk}', x_0)$.

Signing: The protocol for $\mathcal{U}$ to obtain a signature on message m is as follows:

- $\mathcal{U}$ computes $\mathsf{com} \leftarrow \mathsf{Com}(m)$ and sends com to the signer.
- $\mathcal{S}$ and $\mathcal{U}$ execute a witness-indistinguishable proof of knowledge (WI-PoK) in which $\mathcal{S}$ proves knowledge of either $f^{-1}(y_0)$ or $f^{-1}(y_1)$. (This should be witness indistinguishable even against adversaries running in $O(T(k))$ time.) If this proof fails, $\mathcal{U}$ aborts.
- $\mathcal{S}$ computes $\sigma' \leftarrow \mathsf{Sign}'_{\mathsf{sk}'}(\mathsf{com})$ and sends σ' to $\mathcal{U}$.
- $\mathcal{U}$ verifies the signature sent in the previous step, and aborts if it is invalid. Otherwise, the user computes $C_1^* \leftarrow \mathsf{Com}^*(\mathsf{com} \| \sigma')$ and $C_2^* \leftarrow \mathsf{Com}^*(0^k)$. It then computes a ZAP π (with respect to ρ) that $(m, C_1^*, C_2^*, \mathsf{pk}', y_0, y_1) \in L$, where L contains tuples for which there exist $\omega_1, \omega_2, \mathsf{com}, x, \sigma'$ such that:

$$\mathsf{com} := \mathsf{Com}(m; \omega_1) \bigwedge C_1^* := \mathsf{Com}^*(\mathsf{com} \| \sigma'; \omega_2) \bigwedge \mathsf{Vrfy}'_{\mathsf{pk}'}(\mathsf{com}, \sigma') = 1$$

$$or$$

$$C_2^* := \mathsf{Com}^*(x; \omega_2) \bigwedge f(x) \in \{y_0, y_1\}$$

(Note that $L \in \mathcal{NP}$.) The signature is (C_1^*, C_2^*, π).

Verification: To verify signature (C_1^*, C_2^*, π) on message m, verify that π is a valid proof (with respect to ρ) that $(m, C_1^*, C_2^*, \mathsf{pk}', y_0, y_1) \in L$.

We now sketch the proofs of blindness and unforgeability. Again, these are informal because they are mostly intended to provide the reader with some intuition toward our main result that appears in the following section.

Proof sketch (blindness). Given a malicious signer $\hat{S}$ we will consider a sequence of hybrid experiments, and argue that the success probability of $\hat{S}$ (in the sense of Definition 3) cannot change by more than a negligible amount in going from one experiment to the next. The first experiment is the original game of Definition 3, and in the final experiment the success probability of $\hat{S}$ will be exactly $1/2$. We conclude that the success probability of $\hat{S}$ in the original experiment is negligibly-close to $1/2$, thus proving blindness.

In the initial experiment H_0 the signer $\hat{S}$ outputs a public key $\mathsf{PK} = (\mathsf{pk}', y_0, y_1, \rho)$ and two equal-length messages m_0, m_1. A random bit b is chosen and $\hat{S}$ interacts with $\mathcal{U}_b \stackrel{\mathrm{def}}{=} \mathcal{U}_{\mathsf{PK}}(m_b)$ and $\mathcal{U}_{\bar{b}} \stackrel{\mathrm{def}}{=} \mathcal{U}_{\mathsf{PK}}(m_{\bar{b}})$. If neither of these users aborts, then $\hat{S}$ is given the signatures output by these users. Finally, $\hat{S}$ outputs a bit b', and succeeds if $b' = b$.

In the first hybrid experiment H_1, whenever $\mathcal{U}_0$ does not abort we extract from the WI-PoK (given by $\hat{S}$ to $\mathcal{U}_0$) a value x such that $f(x) \in \{y_0, y_1\}$. If $\hat{S}$ gives a valid WI-PoK but extraction fails, b' is chosen at random; otherwise, b' is computed as in H_0. Clearly, the success probabilities in games H_0 and H_1 differ by only a negligible amount. We remark that extraction here is only required from *one* of the proofs given by $\hat{S}$, and furthermore if the WI-PoK given to $\mathcal{U}_0$ fails then no signatures need be provided to $\hat{S}$ (even if the WI-PoK given to $\mathcal{U}_1$ succeeds). Thus, no difficulties arise due to the concurrent execution of two WI-PoKs by $\hat{S}$.

In H_2, the signatures output by $\mathcal{U}_0, \mathcal{U}_1$ are both computed using the witness x that was extracted (this is only done if neither user aborts and extraction is successful, as otherwise either $\hat{S}$ is given $(\bot, \bot)$ or else extraction failed and b' is chosen at random). Specifically, each user computes C_1^* as before but now sets $C_2^* := \mathsf{Com}^*(x; \omega)$; the proof π is constructed using (ω, x) as the witness. Hiding of Com^* (for PPT adversaries) and witness-indistinguishability of the ZAP imply that the success probabilities of $\hat{S}$ in experiments H_1 and H_2 differ by only a negligible amount.

In the final experiment H_3, the first component C_1^* of the signature generated by each user is computed as a commitment to 'garbage', i.e., an all-0s string of the appropriate length. Also, the commitments com sent by each of the users during their execution of the protocol are replaced with commitments to garbage as well. Hiding of Com and Com^* (against PPT adversaries) again implies that the success probabilities in experiments H_2 and H_3 differ by only a negligible amount.

In H_3, both protocol executions are distributed identically and both signatures are independent of these executions; thus, the probability of success is exactly $1/2$. This concludes the proof. ∎

Proof sketch (unforgeability). As in the analysis of the Fischlin scheme, an adversary $\hat{\mathcal{U}}$ that, with non-negligible probability, forges a signature with respect to the blind signature scheme can be used as a sub-routine of an algorithm that 'breaks' another cryptographic assumption. Here, however, there are two main differences:

1. First, the resulting algorithm must be able to extract the underlying messages being committed to in C_1^* and/or C_2^*; this can be done in time $T(k)$ (but *not* in polynomial time) and so we obtain an algorithm running in $O(T(k))$ time rather than in polynomial time.

2. Second, the algorithm is only ensured to extract (with non-negligible probability) *either* $\ell + 1$ distinct commitments $\{\text{com}_i\}$ along with $\ell + 1$ valid signatures $\{\sigma_i'\}$, *or* a value x with $f(x) \in \{y_0, y_1\}$ (in the proof for the Fischlin scheme only the first of these could occur). The first event immediately leads to a forgery on Π'. The second event leads to an algorithm $\mathcal{I}$ inverting f with non-negligible probability (using the technique of Feige and Shamir [15]).

If the signature scheme Π' and the one-way function f are secure even against adversaries running in time $O(T(k))$, the above leads to a contradiction. Hence, we conclude that the blind signature scheme is unforgeable. ∎

4   A Concurrently-Secure Blind Signature Scheme

In this section, we describe our main result: a concurrently-secure blind signature scheme based on standard cryptographic assumptions. In addition to a standard signature scheme, our construction also relies on a perfectly-binding commitment scheme and a ZAP, reviewed in Appendix A. We also use a special type of commitment scheme, described below, and a particular concurrent zero-knowledge protocol, discussed in detail in the following section.

For our protocol we will require a special type of commitment scheme that we call *ambiguous*. In such a scheme, commitment depends on a key pk_c which can be generated in one of two ways: either by a 'normal' key-generation procedure ComGen, or by an 'alternate' key-generation procedure ExtGen which outputs some additional trapdoor information td along with pk_c. If pk_c is generated by ComGen, the scheme is perfectly hiding. On the other hand, if pk_c is generated by ExtGen then td enables extraction of the committed value. Formally:

Definition 5. *An* ambiguous commitment scheme *is a tuple of* PPT *algorithms* (ComGen, ExtGen, Com, Extract) *such that:*

Functionality: ComGen(1^k) *outputs a key* pk_c. ExtGen(1^k) *outputs a key* pk_c *and a trapdoor* td.

Indistinguishability: *The keys output by* ComGen *and* ExtGen *are computationally indistinguishable; that is:*

$$\left\{ pk_c \leftarrow \mathsf{ComGen}(1^k) : pk_c \right\} \stackrel{c}{\approx} \left\{ (pk_c, \mathsf{td}) \leftarrow \mathsf{ExtGen}(1^k) : pk_c \right\}.$$

Perfect hiding: *If* pk_c *is output by* ComGen, *then (with probability 1)* $\mathsf{Com}_{pk_c}(\cdot)$ *is a perfectly-hiding commitment scheme.*

Extraction: *If* (pk_c, td) *is output by* ExtGen, *then* $\mathsf{Extract}_{\mathsf{td}}(\mathsf{Com}_{pk_c}(m)) = m$ *with probability 1. (This implies that if pk_c is output by* ExtGen, *then* $\mathsf{Com}_{pk_c}(\cdot)$ *is perfectly binding.)*

The last two requirements imply that the ranges of ComGen and ExtGen are disjoint.

Commitment schemes with the above functionality (satisfying also some additional requirements) were shown previously by Damgård and Nielsen [13] based on a variety of number-theoretic assumptions. The following construction is easily seen to satisfy Definition 5 under the decisional Diffie-Hellman assumption:

- $\mathsf{ComGen}(1^k)$ first generates a group $\mathbb{G}$ of prime order q, along with generators $g, h \in \mathbb{G}$. It then chooses $r_1, r_2 \leftarrow \mathbb{Z}_q$. If $r_1 \neq r_2$ it outputs $pk_c = (\mathbb{G}, q, g, h, g^{r_1}, h^{r_2})$, and otherwise[7] it outputs $pk_c = (\mathbb{G}, q, g, h, g^0, h^1)$.
- $\mathsf{ExtGen}(1^k)$ generates $\mathbb{G}, q, g, h$ exactly as ComGen. It then chooses $r \leftarrow \mathbb{Z}_q$ and outputs $pk_c = (\mathbb{G}, q, g, h, g^r, h^r)$ and $\mathsf{td} = r$.
- $\mathsf{Com}^*_{pk_c}(m)$, where $m \in \mathbb{G}$ and $pk_c = (\mathbb{G}, q, g, h, g_1, h_1)$, chooses random $x, y \leftarrow \mathbb{Z}_q$ and outputs $A = g^x h^y$ and $B = g_1^x h_1^y \cdot m$.
- $\mathsf{Extract}_r(A, B)$ outputs B/A^r.

4.1 The PRS Concurrent Zero-Knowledge Protocol

As part of our blind signature scheme, we rely on a concurrent zero-knowledge protocol adapted from work of Prabhakaran, Rosen, and Sahai [30,32] and described in Figure 3; we will refer to this protocol as cZK. Protocol cZK is almost identical to the protocol shown in [32, Section 4.8.2], with one difference being that we are satisfied with an *argument* system[8] rather than a *proof* system. The first step of the second stage of cZK is also added specifically for the proof of security of our blind signature scheme. Finally, cZK is also a (stand-alone) *argument of knowledge*, something we need for our protocol.

We do not offer a proof that cZK satisfies the definition of concurrent zero-knowledge, appealing instead to the analysis in [32] which extends without significant modification to our protocol. Actually, for the proof of security of our blind signature scheme we do not rely on the concurrent zero-knowledge property of cZK as a 'black-box,' but instead rely on the properties of the *specific* zero-knowledge simulator shown by Prabhakaran, et al. We therefore briefly describe their simulation strategy at a high level.

The keys to the simulation strategy of [30] are that (1) second-stage messages can be simulated (without knowing a witness) in a straight-line manner as long as the simulator learns in advance the value α that the verifier committed to in the first phase; and (2) the value α can be extracted if the verifier ever answers correctly for two different values of s^j. Correspondingly, the simulation

---

[7] We explicitly check whether $r_1 \neq r_2$, even though this occurs with negligible probability, since perfect hiding for keys output by ComGen must hold with probability 1.

[8] Recall that in a proof system soundness must hold unconditionally, while in an argument system soundness need only hold against a PPT cheating prover.

Inputs: The prover and verifier reduce their common input to a graph $G = (V, E)$. From its witness, the prover computes (as private input) a Hamiltonian cycle $C \subseteq E$. Let k be the security parameter.

First stage: Let $r = \log^2(k)$.

1. The verifier uniformly selects $\alpha \in \{0,1\}^r$, and then chooses values $\{\alpha_{i,j}^0\}_{i,j=1}^r$ and $\{\alpha_{i,j}^1\}_{i,j=1}^r$ at random subject to the constraint that $\alpha_{i,j}^0 \oplus \alpha_{i,j}^1 = \alpha$ for all i, j. The verifier sets $\mathsf{com} \leftarrow \mathsf{Com}(\alpha)$ and $\mathsf{com}_{i,j}^b \leftarrow \mathsf{Com}(\alpha_{i,j}^b)$, and sends all these commitments to the prover.

2. For $j = 1, \ldots, r$:

   1. The prover selects a random $s^j \in \{0,1\}^r$ and sends it to the verifier.

   2. Let $s^j = s_1^j \cdots s_r^j$. The verifier sends $\{\alpha_{i,j}^{s_i^j}\}_{i=1}^r$ along with the randomness used in generating $\{\mathsf{com}_{i,j}^{s_i^j}\}_{i=1}^r$. The prover verifies that these match the corresponding initial commitments sent by the verifier, and aborts if this is not the case.

Second stage: The prover and verifier run r parallel executions of (a modified version of) Blum's Hamiltonicity protocol [7]:

1. The verifier and prover execute a (standard) zero-knowledge proof in which the verifier proves that its commitments (sent in step 1 of the first phase) are 'consistent': namely, that there exist values α and $\{\alpha_{i,j}^0, \alpha_{i,j}^1\}_{i,j=1}^r$ such that (1) com is a commitment to α; (2) $\mathsf{com}_{i,j}^b$ is a commitment to $\alpha_{i,j}^b$ for all i, j, b; and (3) $\alpha_{i,j}^0 \oplus \alpha_{i,j}^1 = \alpha$ for all i, j. If the verifier's proof fails, the prover aborts.

2. The prover selects r random permutations $\pi_1, \ldots, \pi_r$ of the vertices V, and sends perfectly-binding commitments to the entries of the adjacency matrices of the resulting permuted graphs.

3. The verifier sends α, and the verifier and prover execute a (standard) zero-knowledge proof in which the verifier proves that com is a commitment to α. If the verifier's proof fails, the prover aborts.

4. For $j = 1, \ldots, r$ do: if $\alpha_j = 1$ send π_j and open all the commitments in the jth adjacency matrix. If $\alpha_j = 0$ open only the commitments to entries corresponding to the (permuted) cycle C.

5. The verifier checks the values sent by the prover in the standard way.

Fig. 3. A concurrent zero-knowledge argument of knowledge

used in [30,32] can be separated, both conceptually and functionally, into two parts: a 'look-ahead' sub-routine (whose goal is to extract α for all existing sessions) and a 'straight-line simulation' sub-routine (which actually generates the transcript that is output by the simulator). The look-ahead sub-routine dynamically updates a table containing (roughly speaking) all the α-values that have been extracted thus far; if the straight-line simulation sub-routine is reached and a corresponding value of α (needed to continue the simulation) is not in the table, the simulator aborts with output $\bot$.

Another important feature of the simulation strategy is that control alternates between the two sub-routines according to a *fixed* schedule that does not depend on the actions of the particular verifier under consideration. This, in turn, means that we can distinguish in advance the portion of the simulator's random coins that are used for 'look-aheads' and those that are used for straight-line simulation. We will exploit this feature in the unforegablility proof of our protocol. We remark also that the transcript generated by the 'straight-line simulation' sub-routine is built up incrementally, message-by-message, but once a message is placed in this transcript it is never removed.

4.2   Our Construction: An Overview

We begin with some intuition motivating our construction. Recalling the scheme presented in Section 3.2, we see that the use of complexity leveraging there is due to the need to extract from the commitments of $\hat{\mathcal{U}}$ in the proof of unforgeability (which requires super-polynomial time). A first thought is to let Com^* in that protocol be an *ambiguous* commitment scheme, with the public key pk_c for the commitment included in the signer's public key and generated using ComGen. Then, in the proof of unforgeability, we can generate pk_c using ExtGen (instead of ComGen) and thus extract the necessary values from the signature forgeries output by $\hat{\mathcal{U}}$.

An immediate problem is that a malicious signer could then easily violate blindness by generating pk_c using ExtGen. To prevent this, we have the signer provide a proof[9] that pk_c was correctly generated as part the signing protocol. Because we will want to replace pk_c with an incorrectly-generated key in the proof of unforgeability, this proof will need to be *(concurrent) zero knowledge* (witness indistinguishability does not help us here). Because we will again want to provide an 'alternate' witness in the proof of blindness, it will also be a proof of knowledge. We remark that once we introduce this change, we no longer need the values y_0, y_1 in the signer's public key

This almost completes the description of our protocol. However, a difficulty arises if we try to prove unforgeability of the construction as described to this point. Roughly speaking, for the construction thus far it is possible to prove the following:

> Given $\hat{\mathcal{U}}$ who interacts with ℓ instances of $\mathcal{S}$ and outputs $\ell + 1$ valid signatures on distinct messages with non-negligible probability (cf. Definition 2), we can construct an adversarial forger $\mathcal{F}$ who interacts with a signing oracle for (standard signature scheme) Π' and outputs $\ell + 1$ valid signatures on distinct messages with non-negligible probability.

The problem is that $\mathcal{F}$ *makes more than ℓ queries to its signing oracle*, and it is therefore not clear that the $\ell + 1$ signatures output by $\mathcal{F}$ yield a valid forgery! To see why, note that although $\hat{\mathcal{U}}$ invokes only ℓ instances of $\mathcal{S}$, simulation of the zero-knowledge proof by $\mathcal{F}$ requires rewinding of $\hat{\mathcal{U}}$, and many more than ℓ signatures will have to be generated as part of this rewinding. (In the protocol of Section 3.2 no rewinding was needed and so $\mathcal{F}$ made exactly ℓ queries to its

---

[9] Actually, we use an argument system but this does not affect the intuition.

338   C. Hazay et al.

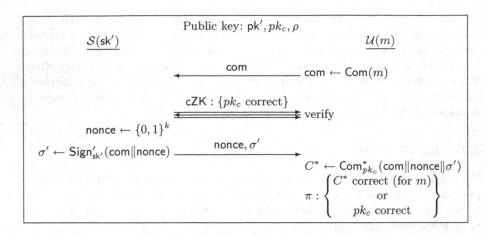

Fig. 4. A high-level overview of our protocol

signing oracle there.) Dealing with this issue is the most difficult and technically-involved aspect of our construction.

We resolve the issue in the following way: instead of having the signer generate a (standard) signature on the commitment com sent by the user in the first round, we have the signer choose a random string nonce $\in \{0,1\}^k$ and sign com$\|$nonce (computation of the final signature by $\mathcal{U}$ is changed in the obvious way). In the proof of unforgeability, we still construct a forger $\mathcal{F}$ who outputs $\ell + 1$ valid signatures on distinct messages $\{(\text{com}_i\|\text{nonce}_i)\}$, but requests *more* than ℓ signatures from its signing oracle. Now, however, we can show that these $\ell + 1$ messages are (in a certain sense) *independent* of the random nonces used during the rewinding done by $\mathcal{F}$. (Here, in particular, we rely on the fact that in step 1 of the second phase of cZK the verifier proves consistency of its commitments, and therefore it does not matter in which iteration the simulator extracted α.) Since the nonces used during rewinding are chosen at random, this means that with overwhelming probability at least one of the messages com$_i\|$nonce$_i$ will be different from any query made by $\mathcal{F}$ to its signing oracle, in which case $\mathcal{F}$ can output a forgery for Π'.

We remark that in proving the above we rely on the specific concurrent zero-knowledge protocol cZK, as well as a particular simulation strategy for this protocol, rather than relying on concurrent zero-knowledge in a 'black-box' way. Indeed, we do not know how to prove unforgeability of our construction when instantiated with an arbitrary concurrent zero-knowledge protocol.

4.3 Our Construction

We now give the details of our construction. Let $\Pi' = (\text{Gen}', \text{Sign}', \text{Vrfy}')$ be a standard signature scheme, let cZK be the protocol of Figure 3, and let

(ComGen, ExtGen, Com*, Extract) be an ambiguous commitment scheme. Our protocol is constructed as follows (see Figure 4):

Key generation: First, $\mathsf{Gen}'(1^k)$ is run to obtain keys $(\mathsf{pk}', \mathsf{sk}')$ and $\mathsf{ComGen}(1^k)$ is run to obtain pk_c. The signer also computes ρ as the verifier's initial message in a ZAP. The public key is $\mathsf{PK} := (\mathsf{pk}', pk_c, \rho)$ and the secret key is sk' along with the randomness used to generate pk_c.

Signing: The protocol for a user $\mathcal{U}$ to obtain a signature on a message m is as follows:

- $\mathcal{U}$ computes $\mathsf{com} \leftarrow \mathsf{Com}(m)$ and sends com to the signer.
- $\mathcal{S}$ and $\mathcal{U}$ execute protocol cZK by which $\mathcal{S}$ proves that pk_c was generated correctly. Formally, it proves that $pk_c \in L_{\mathsf{ComGen}}$, where

$$L_{\mathsf{ComGen}} \overset{\text{def}}{=} \left\{ pk_c : \exists \omega \text{ s.t. } pk_c := \mathsf{ComGen}(1^k; \omega) \right\}.$$

  If this proof fails, $\mathcal{U}$ aborts. If $\mathcal{S}$ aborts in cZK (because it detects that $\mathcal{U}$ is cheating), then $\mathcal{S}$ aborts the entire signing protocol.
- $\mathcal{S}$ chooses nonce $\leftarrow \{0,1\}^k$, computes $\sigma' \leftarrow \mathsf{Sign}'_{\mathsf{sk}'}(\mathsf{com}\|\mathsf{nonce})$, and sends nonce, σ' to $\mathcal{U}$.
- $\mathcal{U}$ verifies the signature sent in the previous step, and aborts if it is invalid. Otherwise, the user computes $C^* \leftarrow \mathsf{Com}^*_{pk_c}(\mathsf{com}\|\mathsf{nonce}\|\sigma')$. It then computes a ZAP π (with respect to ρ) that $(m, C^*, \mathsf{pk}', pk_c) \in L_2$, where L_2 contains tuples such that there exist $\omega_1, \omega_2, \mathsf{com}, \mathsf{nonce}, \sigma'$ with:

$$\Big(\mathsf{com} := \mathsf{Com}(m; \omega_1) \wedge$$
$$C^* := \mathsf{Com}^*_{pk_c}(\mathsf{com}\|\mathsf{nonce}\|\sigma'; \omega_2) \wedge \mathsf{Vrfy}'_{\mathsf{pk}'}(\mathsf{com}\|\mathsf{nonce}, \sigma') = 1 \Big)$$
$$or$$
$$pk_c := \mathsf{ComGen}(1^k; \omega_1)$$

  The signature is (C^*, π).

Verification: To verify signature (C^*, π) on message m, verify that π is a valid proof (with respect to initial message ρ) that $(m, C^*, \mathsf{pk}', pk_c) \in L_2$.

We claim the following about the above scheme:

Theorem 1. *Assuming that (1)* Com *is computationally hiding; (2)* (ComGen, ExtGen, Com*, Extract) *is an ambiguous commitment scheme; (3)* cZK *is an argument of knowledge with negligible knowledge error; and (4) the ZAP being used is witness indistinguishable, the blind signature scheme above satisfies blindness.*

Theorem 2. *Assuming that (1)* Com *is perfectly binding; (2)* (ComGen, ExtGen, Com*, Extract) *is an ambiguous commitment scheme; (3) the ZAP being used has negligible soundness error; and (4)* $\Pi' = (\mathsf{Gen}', \mathsf{Sign}', \mathsf{Vrfy}')$ *is existentially unforgeable under adaptive chosen-message attacks, the blind signature scheme above satisfies unforgeability.*

The proof of blindness (in the sense of Definition 3) follows the general structure of the proof of blindness sketched in Section 3.2. (The above scheme also satisfies all definitions of blindness mentioned in Section 2 and in particular Definition 4.) The proof of unforgeability was sketched in Section 4.2. Complete proofs of all the above will appear in the full version.

References

1. M. Abe. A Secure Three-Move Blind Signature Scheme for Polynomially-Many Signatures. Eurocrypt 2001.
2. M. Abdalla, C. Namprempre, and G. Neven. On the (Im)possibility of Blind Message Authentication Codes. CT-RSA 2006.
3. B. Barak, R. Canetti, J.B. Nielsen, and R. Pass. Universally Composable Protocols with Relaxed Set-Up Assumptions. FOCS 2004.
4. B. Barak and A. Sahai. How To Play Almost Any Mental Game Over The Net — Concurrent Composition via Super-Polynomial Simulation. FOCS 2005.
5. M. Bellare, C. Namprempre, D. Pointcheval, and M. Semanko. The One-More-RSA-Inversion Problems and the Security of Chaum's Blind Signature Scheme. *J. Cryptology* 16(3): 185–215 (2003).
6. M. Bellare and P. Rogaway. Random Oracles are Practical: A Paradigm for Designing Efficient Protocols. ACM CCCS '93.
7. M. Blum. How to Prove a Theorem so No One Else Can Claim It. *Proceedings of the International Congress of Mathematicians*, pp. 1444–1451, 1986.
8. A. Boldyreva. Efficient Threshold Signatures, Multisignatures, and Blind Signatures Based on the Gap-Diffie-Hellman-Group Signature Scheme. PKC 2003.
9. J. Camenisch, M. Koprowski, and B. Warinschi. Efficient Blind Signatures without Random Oracles. SCN 2004.
10. R. Canetti, Y. Lindell, R. Ostrovsky, and A. Sahai. Universally Composable Two-Party and Multi-Party Secure Computation. STOC 2002.
11. D. Chaum. Blind Signatures for Untraceable Payments. Crypto '82.
12. I. Damgård. Payment Systems and Credential Mechanisms with Provable Security against Abuse by Individuals. Crypto '88.
13. I. Damgård and J.B. Nielsen. Perfect Hiding and Perfect Binding Universally Composable Commitment Schemes with Constant Expansion Factor. Crypto 2002.
14. C. Dwork and M. Naor. Zaps and Their Applications. FOCS 2000.
15. U. Feige and A. Shamir. Zero-Knowledge Proofs of Knowledge in Two Rounds. Crypto '89.
16. M. Fischlin. Round-Optimal Composable Blind Signatures in the Common Reference String Model. Crypto 2006.
17. O. Goldreich. *Foundations of Cryptography, vol. 1: Basic Tools.* Cambridge University Press, 2001.
18. S. Goldwasser, S. Micali, and R. Rivest. A Digital Signature Scheme Secure against Adaptive Chosen-Message Attacks. *SIAM J. Computing* 17(2): 281–308 (1988).
19. A. Juels, M. Luby, and R. Ostrovsky. Security of Blind Digital Signatures. Crypto '97.
20. Y. Kalai, Y. Lindell, and M. Prabhakaran. Concurrent General Composition of Secure Protocols in the Timing Model. STOC 2005.
21. A. Kiayias and H.-S. Zhou. Two-Round Concurrent Blind Signatures without Random Oracles. SCN 2006.

22. Y. Lindell. Parallel Coin-Tossing and Constant-Round Secure Two-Party Computation. *J. Cryptology* 16(3): 143–184 (2003).
23. Y. Lindell. Bounded-Concurrent Secure Two-Party Computation without Setup Assumptions. STOC 2003.
24. Y. Lindell. Lower Bounds for Concurrent Self-Composition. TCC 2004.
25. S. Micali, R. Pass, and A. Rosen. Input-Indistinguishable Computation. FOCS 2006.
26. T. Okamoto. Efficient Blind and Partially Blind Signatures without Random Oracles. TCC 2006.
27. D. Pointcheval. Strengthened Security for Blind Signatures. Eurocrypt '98.
28. D. Pointcheval and J. Stern. Provably Secure Blind Signature Schemes. Asiacrypt '96.
29. D. Pointcheval and J. Stern. Security Arguments for Digital Signatures and Blind Signatures. *J. Cryptology* 13(3): 361–396 (2000).
30. M. Prabhakaran, A. Rosen, and A. Sahai. Concurrent Zero Knowledge with Logarithmic Round-Complexity. FOCS 2002.
31. M. Prabhakaran and A. Sahai. New Notions of Security: Achieving Universal Composability without Trusted Setup. STOC 2004.
32. A. Rosen. *Concurrent Zero-Knowledge*. Springer, 2006.

A ZAPs

A ZAP is a 2-round witness-indistinguishable proof (with negligible soundness error) for some $\mathcal{NP}$-language L with associated relation R_L. Formally, let $L_{p(k)} \stackrel{\text{def}}{=} L \cap \{0,1\}^{\leq p(k)}$. A ZAP consists of two polynomial-time interactive algorithms $\mathcal{P}, \mathcal{V}$ along with a polynomial p such that:

- On input 1^k, the verifier $\mathcal{V}$ outputs an initial message ρ.
- On input ρ, a statement $x \in L_{p(k)}$, and a witness w such that $(x, w) \in R_L$, the prover $\mathcal{P}$ outputs a proof π.
- Given ρ, x, and π, the verifier $\mathcal{V}$ outputs a decision bit.

For any k and (x, w) as above, $\mathcal{V}(\rho, x, \mathcal{P}(\rho, x, w)) = 1$ with probability 1. A ZAP also satisfies *adaptive soundness* even against all-powerful cheating provers; that is, for arbitrary $\mathcal{P}^*$ the following is negligible:

$$\Pr\left[\rho \leftarrow \mathcal{V}(1^k); (x, \pi) \leftarrow \mathcal{P}^*(\rho) : \mathcal{V}(\rho, x, \pi) = 1 \bigwedge x \notin L\right].$$

We define witness indistinguishability by requiring that the advantage of any PPT adversary $\mathcal{A}$ in the following game is negligible:

1. $\mathcal{A}(1^k)$ outputs a string ρ, a sequence $x_1, \ldots, x_\ell \in L_{p(k)}$, and two sequences $w_1^0, \ldots, w_\ell^0$ and $w_1^1, \ldots, w_\ell^1$. It is required that $(x_i, w_i^0), (x_i, w_i^1) \in R_L$ for all i.
2. A random bit b is chosen.
3. Compute $\pi_i \leftarrow \mathcal{P}(\rho, x_i, w_i^b)$ for all i, and give these to $\mathcal{A}$.
4. $\mathcal{A}$ outputs a bit b'. The advantage of $\mathcal{A}$ is $\left|\Pr[b' = b] - \frac{1}{2}\right|$.

ZAPs can be constructed based on trapdoor permutations [14].

Designated Confirmer Signatures Revisited*

Douglas Wikström

ETH Zürich, Department of Computer Science
douglas@inf.ethz.ch

Abstract. Previous definitions of designated confirmer signatures in the literature are incomplete, and the proposed security definitions fail to capture key security properties, such as unforgeability against malicious confirmers and non-transferability. We propose new definitions.

Previous schemes rely on the random oracle model or set-up assumptions, or are secure with respect to relaxed security definitions. We construct a practical scheme that is provably secure with respect to our security definition under the strong RSA-assumption, the decision composite residuosity assumption, and the decision Diffie-Hellman assumption.

To achieve our results we introduce several new relaxations of standard notions. We expect these techniques to be useful in the construction and analysis of other efficient cryptographic schemes.

1 Introduction

In a digital signature scheme, as introduced by Diffie and Hellman [10], a signer computes a signature of a message using its secret key, and then anybody holding the public key can verify the signature. This means that the receiver of a signature can show the signature to anybody. If the signer does not want the signer to transfer the signature it can use undeniable signatures [5] or designated verifier signatures [17], but then the holder of the signature no longer holds any indisputable evidence of a signature. Chaum [4] proposed designated confirmer signatures (DC-signatures) as a means to get the best of both worlds at the price of the introduction of a semi-trusted third party called the confirmer.

An example application for DC-signatures is a job offer scenario. Alice is offered a job by Bob and wishes to receive a formal signed offer at some point, but Bob wants to avoid that Alice shows this offer to his competitor Eve. To solve the problem Carol comes to the rescue. Bob computes a DC-signature using his own secret key and Carols public key. Then he proves to Alice that he formed the signature in this way. The DC-signature is special in that it can only be verified directly by Carol, and its distribution is indistinguishable from a distribution that can be computed using only the public keys of Carol and Bob. Furthermore, given a valid/invalid DC-signature, Carol can either convert it into a valid/invalid ordinary signature of Bob that can be verified by anybody, or she

---

* This is an extended abstract. The full paper [23] is available at the Cryptology ePrint Archive, http://eprint.iacr.org.

can prove that she has the ability to do this. Bob can assume that nobody can forge a signature for his public key, and that as long as Carol is honest nobody learns that he signed an offer. Alice can safely assume that Bob can not fool her, and that if Bob denies having signed an offer and Carol is honest, then Carol can prove to anybody that Bob is lying.

1.1 Previous Work

The first formal model of DC-signatures was given by Okamoto [19], but he did not consider the problem of signer coercion. Thus, a signer could be coerced into confirming/denying a signature without the randomness of the signature computation. This problem was considered by Camenisch and Michels [2], who provided stronger definitions. They also proposed both a scheme based on general primitives and more practically oriented schemes, and sketched a security proof for the general construction. In their work on verifiable encryption of discrete logarithms Camenisch and Shoup [3] give a very brief sketch of a DC-signature scheme where most interactive protocols use Schnorr-style techniques.

Goldwasser and Waisbard [16] proposed a relaxed security definition to allow the proofs of knowledge to be strong witness hiding instead of zero-knowledge, and thus allow concurrency. They give a transformation that converts an ordinary signature scheme into a DC-signature scheme secure according to their relaxed definition. They use no random oracles, but the disavowal protocol is based on general zero-knowledge techniques, and the other protocols are based on cut-and-choose techniques.

Gentry, Molnar, and Ramzan [13] considered another relaxation based on an observation originally made by Michels and Stadler [18]. Instead of computing a signature of the message directly, the signer computes a "confirmer commitment" of the message, and then sign the commitment. The constructions in [13] are efficient and do not rely on the random oracle model, but they require the existence of trusted RSA-parameters.

1.2 Our Contributions

Firstly, we take a careful look at existing definitions of DC-signatures. It turns out that two protocols that are not mentioned in previous works, are needed for successful deployment: a proof of correct conversion of a signature, and a proof that a public key is "well formed". We also observe that the definitions of security proposed by Camenisch and Michels [2], Goldwasser and Waisbard [16], and Gentry et al. [13] respectively do not ensure unforgeability when the confirmer is malicious. Furthermore, we note that the relaxed definition in [16] does not prevent transferability, which is arguably a key property of DC-signatures, and the definition in [2] is flawed and can not be satisfied at all. Thus, previous definitions do not capture the notion of DC-signatures correctly. We propose new definitions that correct these deficiencies.

Secondly, we consider how to construct a secure DC-signature scheme. We prove the security of a generic construction with respect to the new security definition. We then describe an instantiation of the generic construction that is secure under the strong RSA-assumption, the decision composite residuosity assumption, and the decision Diffie-Hellman assumption. In contrast to the scheme briefly sketched by Camenisch and Shoup [3] our scheme does not rely on the random oracle model, and it satisfies stronger security requirements than the schemes proposed in [16] and [13]. Furthermore, the setting we consider is stricter in that we do not assume the existence of a trusted key generator as is done in [3]. Despite this our scheme is practical.

Thirdly, our approach to the problem of constructing DC-signatures is different from previous in that instead of relaxing the security definitions of DC-signatures, we relax the security definitions of the primitives used to construct them and prove that weaker primitives suffice. The relaxed notions we introduce and our techniques are of general interest in the construction of efficient and provably secure cryptographic schemes.

Because of space requirements we only sketch most of our results and proofs in this extended abstract, and focus on the new ideas presented. For details and proofs of all claims we refer the reader to the full version [23] of this paper.

1.3 Notation

We consider security with respect to *uniform* algorithms and our assumptions are also uniform in nature, but our results are easily translated to their non-uniform analogs. We use PT, PPT and EPPT to denote the set of uniform, uniform and probabilistic, and uniform expected, polynomial time Turing machines respectively. We let κ be the main security parameter. We write $\langle V(x), P(y) \rangle(z)$ to denote the output of V with private input x when it interacts with P with private input y on common input z. We write $A[S(x)]$ to denote that A has "oracle access" to an interactive Turing machine S with private input x. Formally, we assume that A has a separate pair of communication tapes over which it communicates with S. Whenever we write an expression of the form $\langle V(x), A[S(y)](z) \rangle(w)$ it is assumed that communication takes place in two phases. Before any message is communicated between V and A, A and S may communicate freely. Then some messages are communicated between V and A. When some message is again communicated between A and S, communication is no longer possible between V and A. Finally, when A chooses part of the common input on which it interacts with V, we write $\langle V, A[S(x)](y) \rangle(z, \cdot)$. We abuse notation and say that a protocol is an interactive proof if it is overwhelmingly sound and we say that a protocol is a proof of knowledge only if it is also an interactive proof.

We use 1 and 0, and logical true and false interchangeably. We denote the natural numbers by $\mathbb{N}$, the integers by $\mathbb{Z}$, the integers modulo n by $\mathbb{Z}_n$, the multiplicative group modulo n by $\mathbb{Z}_n^*$ and the subgroup of squares modulo n by SQ_n. We call a prime integer p safe if $(p-1)/2$ is prime.

We use the variation of the strong RSA-assumption which says that given a product N of two random safe primes of the same bit-size and a random $g \in SQ_N$, it is infeasible to compute (b, η) such that $b^\eta = g \bmod N$ and $\eta \neq \pm 1$.

We use the variation of the decision composite residuosity assumption (DCR), which says that given a product n of two random safe primes of the same bit-size, it is infeasible to distinguish the uniform distribution on elements in $\mathbb{Z}_{n^2}^*$ from the uniform distribution on nth residues in $\mathbb{Z}_{n^2}^*$. We use the decision Diffie-Hellman assumption for the subgroup G_Q of squares of $\mathbb{Z}_P^*$, where $P = 2Q + 1$ is a safe prime. This says that if g generates G_Q and $\alpha, \beta, \gamma \in \mathbb{Z}_Q$ are random, then the distributions of $(g^\alpha, g^\beta, g^{\alpha\beta})$ and $(g^\alpha, g^\beta, g^\gamma)$ are indistinguishable.

2   Definition of Designated Confirmer Signature Schemes

A DC-signature scheme consists of algorithms and interactive protocols. There are two key generation algorithms Kg_s^{dc} and Kg_c^{dc} for the signer and confirmer respectively. There is a single signature algorithm Sig^{dc} that given a secret signature key, a message, and a confirmer public key outputs a signature. The signature can not be verified directly, but the confirmer can use a conversion algorithm Con^{dc} with his secret key and the signer public key to transform it into a signature that can be verified by anybody holding the public key of the signer using the verification algorithm Vf^{dc}. The protocols π_{wf} and π_c are used by the confirmer to prove that it formed its key correctly and the correctness of a conversion. The protocols π_v and π_e are used by the signer and confirmer respectively, to convince a verifier that a given DC-signature is valid and valid/invalid respectively. The protocols π_{wf} and π_c are not present in previous formalizations.

Definition 1 (DC Signature Scheme). *A designated confirmer signature scheme $\mathcal{DCS}$ consists of algorithms $\mathsf{Kg}_s^{dc}, \mathsf{Kg}_c^{dc}, \mathsf{Sig}^{dc} \in \mathrm{PPT}$ and $\mathsf{Con}^{dc}, \mathsf{Vf}^{dc} \in \mathrm{PT}$, and interactive protocols $\pi_{wf} = (P_{wf}, V_{wf})$, $\pi_c = (P_c, V_c)$, $\pi_v = (P_v, V_v)$, and $\pi_e = (P_e, V_e)$ with the following completeness properties. For every $\kappa \in \mathbb{N}$, for every possible outputs (ssk, spk) of $\mathsf{Kg}_s^{dc}(1^\kappa)$ and (sk, pk) of $\mathsf{Kg}_c^{dc}(1^\kappa)$ respectively, for every $m \in \{0,1\}^*$, and for every $r, \sigma_0 \in \{0,1\}^*$, and with $\sigma_1 = \mathsf{Sig}_{ssk,r}^{dc}(m, pk)$*

1. $\mathsf{Vf}_{spk}^{dc}(m, \mathsf{Con}_{sk}^{dc}(\sigma_1, spk)) = 1$,
2. $\Pr[\langle V_{wf}, P_{wf}(sk)\rangle(pk) = 1]$ *is overwhelming,*
3. $\Pr[\langle V_c, P_c(sk)\rangle(\sigma_0, \mathsf{Con}_{sk}^{dc}(\sigma_0, spk), pk) = 1]$ *is overwhelming,*
4. $\Pr[\langle V_e, P_e(sk)\rangle(m, \sigma_0, \mathsf{Vf}_{spk}^{dc}(m, \mathsf{Con}_{sk}^{dc}(\sigma_0, spk)), pk, spk) = 1]$ *is overwhelming, and*
5. $\Pr[\langle V_v, P_v(ssk, r)\rangle(m, \sigma_1, pk, spk) = 1]$ *is overwhelming.*

2.1   Well-Formed Keys and Signatures

We introduce the notion of well-formed confirmer keys to formalize the set of strings which behave as keys functionally, and we introduce the notion of well-formed signatures as a generalization of the set of honestly generated signatures.

Definition 2 (Well-Formed Keys). *Let* $\mathcal{DCS}$ *be a DC-signature scheme. We say that the tuple* $(\mathsf{Kg}^{dc}_{c,1}, \mathsf{Kg}^{dc}_{c,2}, \mathsf{Kg}^{dc}_{c,3})$ *splits* Kg^{dc}_c *if*

1. $\mathsf{Kg}^{dc}_{c,1}$ *and* $\mathsf{Kg}^{dc}_{c,2}$ *are probabilistic and* $\mathsf{Kg}^{dc}_{c,3}$ *is a deterministic polynomial time (in their first parameters) algorithms,*
2. *on input* 1^κ, Kg^{dc}_c *computes* $pk_1 = \mathsf{Kg}^{dc}_{c,1}(1^\kappa)$, $sk = \mathsf{Kg}^{dc}_{c,2}(pk_1)$, *and* $pk_2 = \mathsf{Kg}^{dc}_{c,3}(1^\kappa, pk_1, sk)$, *and outputs* $(pk, sk) = ((pk_1, pk_2), sk)$,
3. *for every* $\kappa \in \mathbb{N}$ *and* $pk_1, pk_2 \in \{0,1\}^*$ *there exists at most one* $sk \in \{0,1\}^*$ *such that* $pk_2 = \mathsf{Kg}^{dc}_{c,3}(1^\kappa, pk_1, sk)$, *and*
4. *if* $pk_2 = \mathsf{Kg}^{dc}_{c,3}(1^\kappa, pk_1, sk)$, *then for every output* (spk, ssk) *of* $\mathsf{Kg}^{dc}_s(1^\kappa)$ *and* $m, r \in \{0,1\}^*$: $\mathsf{Vf}^{dc}_{spk}(m, \mathsf{Con}^{dc}_{sk}(\mathsf{Sig}^{dc}_{ssk,r}(m, pk), spk)) = 1$.

We say that (pk_1, pk_2) *is well-formed with respect to the splitting if* $pk_2 = \mathsf{Kg}^{dc}_{c,3}(pk_1, sk)$ *for some* sk. *We say that* $((pk_1, pk_2), sk)$ *is well-formed for such a secret key* sk.

If the signer or the confirmer proves to the verifier that a DC-signature is valid/invalid relative a well-formed confirmer public key, then the verifier is confident that if converted, the result is also valid/invalid in a consistent way. Every key generator can be trivially split, but we are interested in splittings that given (pk_1, pk_2) allow a simple proof of knowledge of sk such that $((pk_1, pk_2), sk)$ is well-formed.

Definition 3 (Well-Formed Signature). *Let* $\mathcal{DCS}$ *be a DC-signature scheme and let* $wf : \{0,1\}^* \times \{0,1\}^* \times \{0,1\}^* \to \{0,1\}$ *be a polynomially computable function. We say that* wf *is a well-formedness function with respect to* $\mathcal{DCS}$ *and a splitting of* Kg^{dc}_c *if for every well-formed* (pk, sk), *every output* (spk, ssk) *of* $\mathsf{Kg}^{dc}_s(1^\kappa)$, *and every possible output* $s = \mathsf{Con}^{dc}_{sk}(\mathsf{Sig}^{dc}_{ssk}(m, pk), spk)$, *we have* $wf(s, pk, spk) = 1$.

All honestly generated signatures are well formed, but some valid signatures may not be. It is trivial to see that there exists a well-formedness function for every DC-signature scheme, but we are interested in well-formedness functions that simplify the construction of our protocols.

2.2 Definition of Security

Assume that some splitting and well-formedness functions are fixed and define the following relations.

Definition 4 (Relations)

1. WELL-FORMED CONFIRMER KEYS. *Denote by* $\mathcal{R}_{wf}$ *the set of all well-formed key pairs* $((pk_1, pk_2), sk)$.
2. CORRECT CONVERSION. *Denote by* $\mathcal{R}_c$ *the set of pairs* $((\sigma, s, pk, spk), sk)$ *such that* $(pk, sk) \in \mathcal{R}_{wf}$ *and* $s = \mathsf{Con}^{dc}_{sk}(\sigma, spk)$.

3. CORRECT EVALUATION. *Denote by* $\mathcal{R}_e$ *the set of pairs* $((m, \sigma, c, pk, spk), sk)$ *such that* $(pk, sk) \in \mathcal{R}_{wf}$, *and* $s = \mathsf{Con}^{\mathsf{dc}}_{sk}(\sigma, spk)$ *for some* s *such that* $\mathsf{Vf}^{\mathsf{dc}}_{spk}(m, s) = c$ *and* $wf(s, pk, spk) = 1$.

4. PROOF OF VALIDITY. *Denote by* $\mathcal{R}_v$ *the set of pairs* $((m, \sigma, pk, spk), w)$, *where* w *is a witness that* $\mathsf{Vf}^{\mathsf{dc}}_{spk}(m, \mathsf{Con}^{\mathsf{dc}}_{sk}(\sigma, spk)) = 1$ *for every* sk *such that* $(pk, sk) \in \mathcal{R}_{wf}$.

The relation $\mathcal{R}_e$ only considers *well-formed* signatures. If a signature is not well-formed, it was by definition computed by a corrupt signer. In this case the confirmer need not protect the signer and can simply convert the signature and prove that it did so correctly. The relation $\mathcal{R}_v$ does *not* capture the set of valid signatures. It captures the set of signatures that are valid *given that the public confirmer key is well-formed*.

We now consider what each honest party or group of honest parties in a DC-signature scheme might expect from a secure implementation.

Honest Verifier. An honest verifier naturally expects that it is infeasible to convince it of a false statement. Furthermore, it seems reasonable that the verifier only accepts a proof of well-formedness of a public key if it shows that the prover knows the secret key, since otherwise it can not be confident that the confirmer actually is able to convert a signature.

Definition 5 (Soundness). *A DC-signature scheme DCS is sound if* π_c, π_e, *and* π_v *are interactive proofs for the relations* $\mathcal{R}_c$, $\mathcal{R}_e$, *and* $\mathcal{R}_v$ *respectively, and* π_{wf} *is a proof of knowledge for the relation* $\mathcal{R}_{wf}$.

Note that if a confirmer public key pk is well-formed and σ is a candidate signature for a signer public key spk, then well-formedness implies that a signature can be converted in only one way. Thus, the confirmer can not choose if a signature should be considered valid or not. Well-formedness also implies that if a signer proves the validity of σ, it can not be converted into an invalid one.

It may be dangerous for a signer to use a confirmer public key that is not well-formed. Thus, we assume that any signer (or somebody the signer trusts) executes π_{wf} with the confirmer before using its public key.

Honest Signer. It must be infeasible for the adversary to convince anybody that the honest signer has signed a message m unless this is the case. This must hold even when the adversary can ask arbitrary signature queries and execute the proof of validity protocol π_v. In other words we need a slight generalization of security against chosen message attacks [15].

We formalize the honest signer S to be a probabilistic interactive Turing machine that accepts as input a key pair (spk, ssk). Whenever it receives a message with prefix π_{wf}, $\mathsf{Sig}^{\mathsf{dc}}$, or π_v on its communication tape it halts the execution of any executing interactive protocol and proceeds as follows. Given a message (π_{wf}, pk) it executes the verifier V_{wf} of the protocol π_{wf} on common input

pk. If V_{wf} accepts the proof, then S stores pk. Given a message $(\mathsf{Sig}^{dc}, m, pk)$ it checks if it has stored pk. If not, it returns $\perp$, and otherwise it computes $\sigma = \mathsf{Sig}^{dc}_{ssk,r}(m, pk)$ and writes σ on the communication tape. Given a message (π_v, m, σ, pk), where it previously computed $\sigma = \mathsf{Sig}^{dc}_{ssk,r}(m, pk)$ it executes the prover of the protocol π_v on common input (m, σ, pk, spk) and private input some witness w that $((m, \sigma, pk, spk), w) \in \mathcal{R}_v$. When we use several copies of S below we index them for easy reference.

Experiment 1 (CMA-Security, $\mathsf{Exp}^{cma}_{\mathcal{DCS}, A}(\kappa)$)

$$(spk, ssk) \leftarrow \mathsf{Kg}^{dc}_s(1^\kappa)$$
$$(m, s) \leftarrow A[S(spk, ssk)](spk)$$

If $\mathsf{Vf}^{dc}_{spk}(m, s) = 0$ or if S signed m return 0 and otherwise 1.

Definition 6 (CMA-Security). *A DC-signature scheme $\mathcal{DCS}$ is secure against chosen message attacks (CMA-secure) if for every $A \in$ PPT:* $\Pr[\mathsf{Exp}^{cma}_{\mathcal{DCS}, A}(\kappa) = 1]$ *is negligible.*

In the definitions of Camenisch and Michels [2], Goldwasser and Waisbard [16], and Gentry et al. [13] security hold only with respect to *honestly* generated confirmer keys, i.e., their definitions do not ensure any form of CMA-security for the signer when the confirmer is corrupted.

The definition does not say that an adversary can not form a bit-string σ and then convince an honest verifier using π_v or π_e that this is a valid designated signature of some message m not signed by S, but we prove in the full paper that this follows from soundness and CMA-security.

Honest Confirmer. Nobody except the confirmer should be able to play the role of the prover in the protocols π_{wf}, π_c, and π_e using the honest confirmers public key as common input, even after interacting with the real confirmer.

We formalize the honest confirmer C to be a probabilistic interactive Turing machine that accepts as input a key pair (pk, sk). Whenever it receives a message with prefix π_{wf}, Con^{dc}, or π_e on its input communication tape it halts the execution of any interactive protocol it is executing and proceeds as follows. Given a message (π_{wf}) it executes the prover of protocol π_{wf} on common input pk and private input sk. Given a message $(\mathsf{Con}^{dc}, \sigma, spk)$ it computes $s = \mathsf{Con}^{dc}_{sk}(\sigma, spk)$, writes s on its output communication tape, and executes the prover of protocol π_c on common input (σ, s, pk, spk) and private input sk. On input (π_e, m, σ, spk) it computes $s = \mathsf{Con}^{dc}_{sk}(\sigma, spk)$. If $wf(s, pk, spk) = 1$, i.e., s is well-formed, then C computes $c = \mathsf{Vf}^{dc}_{spk}(m, s)$, writes c on its output communication tape, and executes the prover of protocol π_e on common input (m, σ, c, pk, spk) and private input sk. Otherwise C writes $(\mathsf{malformed}, s)$ on its output communication tape and executes the prover of protocol π_c on common input (σ, s, pk, spk) and private input sk.

Experiment 2 (Impersonation-Resistance, $\mathsf{Exp}_{\mathcal{DCS},A}^{imp-res}(\kappa)$)

$$(pk, sk) \leftarrow \mathsf{Kg}_{\mathsf{c}}^{\mathsf{dc}}(1^{\kappa})$$
$$d_1 \leftarrow \langle V_{wf}, A[C(pk, sk)](pk)\rangle(pk)$$
$$d_2 \leftarrow \langle V_c, A[C(pk, sk)](pk)\rangle(\cdot, \cdot, pk, \cdot)$$
$$d_3 \leftarrow \langle V_e, A[C(pk, sk)](pk)\rangle(\cdot, \cdot, \cdot, pk, \cdot)$$

Return $d_1 \vee d_2 \vee d_3$.

Definition 7 (Impersonation Resistance). *A designated confirmer signa-ture scheme* $\mathcal{DCS}$ *is impersonation-resistant if for every* $A \in$ PPT: $\Pr[\mathsf{Exp}_{\mathcal{DCS},A}^{imp-res}(\kappa) = 1]$ *is negligible.*

Note that C never invokes P_e without executing $\mathsf{Con}^{\mathsf{dc}}$. This is without loss of generality, since $\mathsf{Con}^{\mathsf{dc}}$ is deterministic and the common input contains the extracted signature anyway. Note that this differs from the signature case, where a signer could potentially want to prove the correctness of a particular signature to several receivers.

Remark 1. A stronger definition would allow the adversary to interact with the confirmer and the verifier concurrently on other inputs. Unfortunately, such a definition requires the protocols π_c and π_e to be non-malleable [11] with respect to each other and themselves. General methods such as [21] can be used to construct non-malleable zero-knowledge protocols, but currently these techniques are far from practical. Thus, we do not follow this definitional path.

Honest Signer and Honest Confirmer. To start with we observe that from the point of view of an honest signer and honest verifier, or from the point of view of an honest verifier and honest confirmer, no additional requirements are natural to impose.

When the signer and the confirmer are honest we require that knowledge that the signer signed a particular message can not be transfered. Note that this is needed in the job offer scenario. Non-transferability can clearly only hold until a DC-signature has been converted.

We formalize this as follows. Let SC be the machine that simulates both S and C on inputs (pk, sk) and (spk, ssk) except for the following modifications. Given a message $(\mathsf{Sig}^{\mathsf{dc}}, m, pk')$ with $pk' = pk$ it waits for a message (m, σ), stores this, and writes σ on its communication tape. If later invoked on (π_v, m, σ, pk) it returns $\perp$ instead of invoking P_v. For $pk' \neq pk$ it behaves as S. Given a message $(\mathsf{Con}^{\mathsf{dc}}, \sigma, spk')$ such that $spk' = spk$ and (m, σ) is stored it checks if (m, σ, s) is stored for some s. If not, then it computes $s = \mathsf{Con}_{sk}^{\mathsf{dc}}(\mathsf{Sig}_{ssk}^{\mathsf{dc}}(m, pk), spk)$ and stores (m, σ, s). Finally, it writes s on its communication tape. It does not invoke the prover of π_c. If $spk' \neq spk$ or (m, σ) is not stored it behaves as C. Finally, given a message (π_e, m, σ, spk), where (m', σ) is stored for some m' it returns 0 if $m \neq m'$ and 1 otherwise. It does not execute the prover of π_e.

Intuitively, SC *delays* the computation of every DC-signature using the public key pk until it is converted. We want to say that if there is an adversary A that interacts with S and C, there is another adversary A' that interacts with SC such that its output is indistinguishable from that of A, despite that all its signature queries are "delayed". For this to make sense the order of messages sent to S, C, and SC must be given to the distinguisher as well. We say that an adversary is scheduled if whenever it writes a message with prefix Sig^{dc}, π_v, π_{wf}, Con^{dc}, and π_e the message and any return value (excluding the messages exchanged by the protocols that may be invoked) are stored on a special write only scheduling tape. Furthermore, when the adversary halts its output is prefixed by its scheduling tape. The additional input pk to S below is stored as a well-formed public key, and this is done in the simulation of SC as well.

Experiment 3 (Non-Transferability, $\mathsf{Exp}^{non-trans-b}_{\mathcal{DCS},A,V}(\kappa)$)

$$((pk, sk), (spk, ssk)) \leftarrow (\mathsf{Kg}^{dc}_{\mathsf{c}}(1^{\kappa}), \mathsf{Kg}^{dc}_{\mathsf{s}}(1^{\kappa}))$$

$$d \leftarrow \begin{cases} D(A(pk, spk, ssk)[SC(pk, sk, spk, ssk)]) & \text{if } b=0 \\ D(A(pk, spk, ssk)[S(pk, spk, ssk), C(pk, sk)]) & \text{if } b=1 \end{cases}$$

Definition 8 (Non-Transferability). *A DC-signature scheme $\mathcal{DCS}$ is non-transferable if for every scheduled $A_1 \in$ EPPT there exists a scheduled $A_0 \in$ EPPT such that for every distinguisher $D \in$ EPPT:*
$|\Pr[\mathsf{Exp}^{non-trans-1}_{\mathcal{DCS},A_1,D}(\kappa) = 1] - \Pr[\mathsf{Exp}^{non-trans-0}_{\mathcal{DCS},A_0,D}(\kappa) = 1]|$ *is negligible.*

Remark 2. Our definition is similar to the "liberal" definition of zero-knowledge in that the simulator is allowed to run in *expected* polynomial time.

Remark 3. Again, our experiment is not completely realistic. A stronger definition would allow the adversary to interact concurrently with S and C on other common inputs when trying to convince a verifier. Unfortunately, such definitions imply that the protocols π_{wf}, π_c, π_e, and π_v are non-malleable in a very strong sense. We are not aware of any general methods to achieve this.

Informally, a DC-signature scheme is coercion-free if a signer can reveal its secret signing key and still claim that it did not compute a particular DC-signature. Naturally, this can only hold as long as the DC-signature is not converted, or proved to be valid. Note that this is already captured in our definition of non-transferability.

The definition of non-transferability in Camenisch and Michels [2] can not be satisfied, since it requires the existence of a straight-line zero-knowledge simulator for an interactive proof without set-up assumptions. The definition of Goldwasser and Waisbard [16] only prevents the adversary from transferring confidence of validity of a signature using the confirmation protocol of the scheme. It says nothing about the possibility of using another confirmation protocol. The relaxed definition of Gentry et al. [13] explicitly allows some forms of transferability.

Most previous definitions require some form of indistinguishability of signatures computed by different signers, but this is unnecessarily strong. In any claim about a signature, the holder of the signature would disclose the identity of the claimed signer anyway, and our definition implies that anybody can generate something indistinguishable from a valid signature of any such signer.

Definition of Security. We now define security of a DC-signature scheme in the natural way.

Definition 9. *A designated confirmer signature scheme* $\mathcal{DCS}$ *is secure if it is sound, CMA-secure, impersonation-resistant, and non-transferable.*

On Concurrency. In our definitions the "oracle access" to the honest signer S and the honest confirmer C are sequential. Stronger definitions similar to those of Camenisch and Michels [2], where the adversary is given concurrent "oracle access", follow by giving the adversary access to several copies of S and C, each executing on the same input key pair.

3   Theoretical Tools

3.1   A Relaxation of Zero-Knowledge

The definition of zero-knowledge is very strong in that the simulation property must hold with respect to every verifier and *every instance* (x, w) in the relation $\mathcal{R}$ under consideration. As pointed out by Goldreich [14] it is quite natural to consider a uniform definition that only requires that it is *infeasible to find an instance* on which a verifier can gain knowledge.

In many cryptographic settings the instance can not be chosen completely freely by the adversary, e.g., the adversary may ask an honest party to prove that it performed a decryption correctly, where the keys to the cryptosystem are generated honestly. Furthermore, in some security proofs the simulator can be allowed an additional advice string *dependent* on the instance, e.g., if a decryption oracle is present in the environment where the simulator is invoked we may give the simulator decryptions of some ciphertexts. The following definition allows for both these settings.

Experiment 4 (Zero-Knowledge, $\mathsf{Exp}^{(T,F)-\mathsf{zk}-b}_{\pi,\mathcal{R},I,V^*,D}(\kappa)$)

$$t \leftarrow T(1^\kappa)$$
$$(i, z) \leftarrow I(1^\kappa, t)$$
$$(x, w, a) \leftarrow F(t, i)$$
$$d \leftarrow \begin{cases} D(x, z, a, \langle V^*(z), P(w)\rangle(x)) & \text{if } b=0 \\ D(x, z, a, M(z, a, x)) & \text{if } b=1 \end{cases}$$

Return 0 if $\mathcal{R}(x, w) = 0$ *and* d *otherwise.*

Definition 10. *Let* $\pi = (P, V)$ *be an interactive protocol, let* $T \in \mathrm{PPT}$ *and* $F : \{0,1\}^* \to \{0,1\}^*$, *and let* $\mathcal{R}$ *be a relation. We say that* π *is* (T, F)-*zero-knowledge for* $\mathcal{R}$ *if for every verifier* $V^* \in \mathrm{EPPT}$ *there exists a simulator* $M \in \mathrm{EPPT}$ *such that for every instance chooser* $I \in \mathrm{EPPT}$ *and every distinguisher* $D \in \mathrm{EPPT}$: $|\Pr[\mathsf{Exp}^{(T,F)-\mathsf{zk}-0}_{\pi,\mathcal{R},I,V^*,D}(\kappa) = 1] - \Pr[\mathsf{Exp}^{(T,F)-\mathsf{zk}-1}_{\pi,\mathcal{R},I,V^*,D}(\kappa) = 1]|$ *is negligible.*

Remark 4. We do not require that F is polynomial time in the definition, but the concrete protocols we present are (T, F)-zero-knowledge with efficiently computable functions F. This seems also essential to allow sequential composition.

Remark 5. The definition makes sense for non-uniform adversaries as well. Furthermore, the definition can be both generalized and relaxed. One natural relaxation is to give only part of the sample t to the instance finder. Note that a probabilistic F is captured by this relaxation. A related definition gives the instance finder access to some specific oracle, i.e., we would talk about (T, F, O)-zero-knowledge for some specific oracle O. This seems to make most sense when the oracle is efficiently computable using the sample t (of which not all is given to the instance finder).

Choose some canonical interpretation of strings such that the output of I is always of the form $((x_1, w_1), z)$. Then when the output of T is always of the form $t = (x_2, w_2)$, and $F(t, i) = ((x_1, x_2), (w_1, w_2), \emptyset)$, we simply say that the protocol is T-zero-knowledge.

   We show in the full paper that if a simulator satisfies the definition, then it can be used instead of a real protocol execution polynomially many times sequentially as long as F is polynomial time. This extension is necessary in our analysis.

3.2   Cryptosystems with Labels and Δ-CCA2-Security

Our starting point is the generalization of CCA2-security for cryptosystems with labels introduced by Shoup and Gennaro [22].

   In such a scheme the encryption algorithm Enc takes as input a label L in addition to a public key pk and a message m. The decryption algorithm Dec takes as input a label L in addition to a secret key sk and a ciphertext c. CCA2-security is then defined as usual except that the adversary must choose a label L in addition to the two challenge ciphertexts m_0 and m_1, and it may not ask the decryption query (L, c), where c is the challenge ciphertext.

   The definition of CCA2-security is strict in that the indistinguishability property of ciphertexts holds for *any* two messages. In our setting a weaker property suffices, namely that any two encrypted signatures from the *same* signer are indistinguishable. Thus, we introduce the following relaxed definition.

Experiment 5 (Δ-CCA2-Security, $\mathrm{Exp}_{CS,A}^{\Delta-\mathrm{cca2}-b}(\kappa)$)

$$(pk, sk) \leftarrow \mathsf{CSKg}(1^\kappa)$$
$$(r, m_0, m_1, \mathrm{state}) \leftarrow A^{\mathsf{Dec}_{sk}(\cdot,\cdot)}(\mathrm{choose}, pk)$$
$$c \leftarrow \mathsf{Enc}_{pk}(\Delta(r, m_b))$$
$$d \leftarrow A^{\mathsf{Dec}_{sk}(\cdot,\cdot)}(\mathrm{guess}, \mathrm{state}, c)$$

Interpret $\Delta(r, m_b)$ as a pair (L, m_b'). The experiment returns 0 if $\mathsf{Dec}_{sk}(\cdot, \cdot)$ was queried on (L, c), and otherwise d.

Definition 11 (Δ-CCA2-Security). *Let $\Delta \in$ PPT. A public key cryptosystem CS with labels is said to be Δ-CCA2-secure if for every adversary $A \in$ PPT: $|\Pr[\mathrm{Exp}_{CS,A}^{\Delta-\mathrm{cca2}-0}(\kappa) = 1] - \Pr[\mathrm{Exp}_{CS,A}^{\Delta-\mathrm{cca2}-1}(\kappa) = 1]|$ is negligible in κ.*

3.3 Collision-Free Signature Schemes

We say that a signature scheme is collision-free if it is infeasible to find two distinct messages and a signature such that the signature is a valid with respect to both messages, even if the adversary is given the honestly generated secret key and the public key.

4 A Generic Construction of Designated Confirmer Signatures

It is natural to construct DC-signatures from a CMA-secure signature scheme and a CCA2-secure cryptosystem. A signer holds a secret key for the signature scheme and the confirmer holds a secret key of the cryptosystem. A DC-signature is simply an ordinary signature encrypted with the cryptosystem, conversion corresponds to decryption, and zero-knowledge proofs of knowledge are used to instantiate the protocols. The theorem below implies that this is secure, but for most signature schemes and cryptosystems it is prohibitively inefficient.

As a first step in the construction of an *efficient* DC-signature scheme we prove that weaker primitives suffice to construct a secure DC-signature scheme, but the basic idea is the same. Let $\mathcal{CS} = (\mathsf{CSKg}, \mathsf{Enc}, \mathsf{Dec})$ be a cryptosystem with labels and let $\mathcal{SS} = (\mathsf{Kg}, \mathsf{Sig}, \mathsf{Vf})$ be a signature scheme with fixed size signatures (this is easy to ensure by padding) that fit in the plaintext space of $\mathcal{CS}$. Define Kg_s^{dc} to compute $(spk, ssk) = \mathsf{Kg}(1^\kappa)$ and $s_\perp = \mathsf{Sig}_{ssk}(\perp)$, where $\perp$ is a special symbol, and output $((spk, s_\perp), ssk)$ (we drop $s_\perp$ from our notation when convenient). Define $\mathsf{Kg}_c^{dc}(1^\kappa)$ to output $\mathsf{CSKg}(1^\kappa)$, and define Vf^{dc} to be Vf except that $\mathsf{Vf}_{spk}^{dc}(\perp, \cdot) = 0$. On input ssk, m, and pk the DC-signature algorithm Sig^{dc} is defined to compute $s = \mathsf{Sig}_{ssk}(m)$ and $\sigma = \mathsf{Enc}_{pk}(spk, s)$, and output σ. On input sk, σ, and spk the conversion algorithm Con^{dc} is defined to output $s = \mathsf{Dec}_{sk}(spk, \sigma)$. In other words, the public signature key spk is used as a label. Let $(\mathsf{Kg}_{c,1}^{dc}, \mathsf{Kg}_{c,2}^{dc}, \mathsf{Kg}_{c,3}^{dc})$ be a splitting of Kg_c^{dc} and let wf be some well-formedness function with respect to $\mathcal{DCS}$. Let π_{wf}, π_c, π_e, and π_v be interactive protocols, complete with respect to the relations $\mathcal{R}_{wf}$, $\mathcal{R}_c$, $\mathcal{R}_e$, and $\mathcal{R}_v$. It is easy to see that $\mathcal{DCS} = (\mathsf{Kg}_s^{dc}, \mathsf{Kg}_c^{dc}, \mathsf{Con}^{dc}, \mathsf{Vf}^{dc}, \pi_{wf}, \pi_c, \pi_e, \pi_v)$ is a DC-signature scheme.

The algorithm $\Delta(r, (r', m))$ first computes $(spk, ssk) = \mathsf{Kg}_r(1^\kappa)$ and $s = \mathsf{Sig}_{ssk,r'}(m)$, and then outputs (spk, s). Define $T_{hs}(1^\kappa) = (\mathsf{Kg}_c^{dc}(1^\kappa), \mathsf{Kg}_s^{dc}(1^\kappa))$. Define F_{hs} to take as input the pair $((pk, sk, spk, ssk), (r, m, m'))$, compute $\sigma = \mathsf{Sig}_{ssk,r}^{dc}(m)$, and output $((m', \sigma, \mathsf{Vf}_{spk}^{dc}(m', \mathsf{Con}_{sk}^{dc}(\sigma)), pk, spk), sk, \emptyset)$. Let $T_{cs}(1^\kappa)$ simply output $\mathsf{Kg}_c^{dc}(1^\kappa)$. Define F_{cs} to take input $((pk, sk), (m, \sigma, c, spk))$ and output the tuple $((m, \sigma, c, pk, spk), sk, \mathsf{Con}_{sk}^{dc}(\sigma, spk))$.

Theorem 1. [1] *Suppose that $\mathcal{CS}$ is Δ-CCA2-secure and that $\mathcal{SS}$ is CMA-secure and collision-free, and that π_{wf}, π_c, π_e, and π_v are proofs of knowledge for the relations $\mathcal{R}_{wf}$, $\mathcal{R}_c$, $\mathcal{R}_e$, and $\mathcal{R}_v$. Suppose that π_{wf} and π_c are Kg_c^{dc}-zero-knowledge*

---

[1] Camenisch and Michels [2] claim a similar, but weaker, theorem according to their definition, but as explained above their definition can not be satisfied, and only a proof sketch is given.

for the relations $\mathcal{R}_{wf}$ and $\mathcal{R}_c$ respectively. Suppose that π_e is both (T_{hs}, F_{hs})-zero-knowledge and (T_{cs}, F_{cs})-zero-knowledge for the relation $\mathcal{R}_e$. Suppose π_v is Kg_s^{dc}-zero-knowledge for the relation $\mathcal{R}_v$. Then $\mathcal{DCS}$ is secure.

4.1 On the Use of Two Distinct Weak Simulators

Perhaps the most interesting of our techniques is the use of two simulators for the same protocol, of which one requires additional advice. Consider the problem of constructing a black box-reduction of a successful attacker A against non-transferability into a successful attacker A' against the Δ-CCA2-security of the underlying cryptosystem. The Δ-CCA2-attacker A' takes a public key pk as input and must simulate the non-transferability experiment to the adversary without using the secret key sk. At some point A' outputs a random bit string r and two messages (r_0, m_0) and (r_1, m_1) to the Δ-CCA2-experiment, and it is given a ciphertext $\sigma = \mathsf{Enc}_{pk}(spk, \mathsf{Sig}_{ssk, r_b}(m_b))$ for a random $b \in \{0, 1\}$, where $(spk, ssk) = \mathsf{Kg}_r(1^\kappa)$. The ciphertext σ is used in the simulation somehow, and finally A' outputs a bit. The simulation involves converting signatures, but A' may use its decryption oracle to answer such queries, as long as it never asks for a decryption of σ.

We observe that when the protocol π_e is simulated for some DC-signature $\sigma' \neq \sigma$ computed by A, the simulator is free to invoke the decryption oracle on σ', i.e., a (T_{cs}, F_{cs})-zero-knowledge simulator is sufficient. On the other hand, for the particular signature σ we can not proceed in this way, since that would violate the rules of the Δ-CCA2-experiment, but since σ is computed honestly using honestly formed signature keys a (T_{hs}, F_{hs})-zero-knowledge simulator suffices.

5   Concrete Tools

In this section we present the tools we need to instantiate the generic DC-signature scheme with an efficient concrete scheme under standard complexity assumptions.

5.1 A Twin-Moduli Signature Scheme

To prove the existence of the scheme presented below we must assume that an arbitrary bit-string can be embedded into a prime in an efficient way. We assume that there is an efficient algorithm $\mathsf{Emb}_f^{f'}$ that given $n \in [0, 2^\kappa - 1]$ with overwhelming probability finds $s \in [2^{f(\kappa)-1}, 2^{f(\kappa)-1} + 2^{f(\kappa)-f'(\kappa)} - 1]$ such that $e = 2^{f(\kappa)} n + s$ is prime. We call this assumption the (f, f')-Embedding Assumption. In practice this is not a problem for reasonable f and f'. The twin-moduli signature scheme, $\mathcal{SS}^2 = (\mathsf{Kg}^2, \mathsf{Sig}^2, \mathsf{Vf}^2)$, is based on using two sets of RSA-parameters and the embedding algorithm $\mathsf{Emb}_f^{f'}$. Denote by κ_r a security parameter such that $2^{-\kappa_r}$ is negligible in κ. On input 1^κ the key generator Kg^2 chooses $\kappa/2$-bit safe primes p_0, q_0, p_1, and q_1 randomly, defines $N_0 = p_0 q_0$ and $N_1 = p_1 q_1$, chooses $g_0 \in SQ_{N_0}$ and $g_1 \in SQ_{N_1}$ randomly, and outputs $((N_0, g_0, N_1, g_1), (p_0, q_0, g_0, p_1, q_1, g_1))$. Set $\kappa_p = f(2\kappa_r + \kappa_m + 1)$

and $\kappa'_p = f'(2\kappa_r + \kappa_m + 1)$. The signature algorithm Sig^2 takes as input a private key $(p_0, q_0, g_0, p_1, q_1, g_1)$ and a message $m \in [0, 2^{\kappa_m} - 1]$. It chooses $r \in [2^{2\kappa_r + \kappa_m}, 2^{2\kappa_r + \kappa_m} + 2^{\kappa_r + \kappa_m} - 1]$ randomly. Then it computes

$$s_0 = \mathsf{Emb}_f^{f'}(r + m) \qquad e_0 = 2^{\kappa_p}(r + m) + s_0 \qquad z_0 = g_0^{1/e_0} \bmod N_0$$

$$s_1 = \mathsf{Emb}_f^{f'}(r) \qquad e_1 = 2^{\kappa_p}r + s_1 \qquad z_1 = g_1^{1/e_1} \bmod N_1 \ .$$

Finally, it outputs (r, z_0, s_0, z_1, s_1). The verification algorithm Vf^2 takes as input a public key (N_0, g_0, N_1, g_1), a message $m \in \{0, 1\}^{\kappa_m}$, and a candidate signature (r, z_0, s_0, z_1, s_1). It computes $e_0 = 2^{\kappa_p}(r + m) + s_0$ and $e_1 = 2^{\kappa_p}r + s_1$, and verifies that $r \in [1, 2^{2\kappa_r + \kappa_m + 1} - 1]$, $s_0, s_1 \in [0, 2^{\kappa_p} - 1]$, that e_0 and e_1 are odd, and that $z_0^{e_0} = g_0 \bmod N_0$ and $z_1^{e_1} = g_1 \bmod N_1$. The basic idea of the scheme is similar to an idea of Cramer et al. [7], but the proposition below does not follow from their work. Using a collision-free hash function $H : \{0, 1\}^* \to [0, 2^{\kappa_m} - 1]$ it can be used to sign messages of any length.

Proposition 1. *The scheme exists under the (f, f')-Embedding assumption and is CMA-secure and collision-free under the strong RSA-assumption.*

5.2   The Cramer-Shoup-Pailler Cryptosystem

The cryptosystem we use is based on Cramer and Shoup's [8] CCA2-secure version of the Pailler [20] as described in [3], i.e., it is a cryptosystem with labels. This cryptosystem is special in that plaintexts "live in the exponent", which simplifies the construction of Schnorr-like proofs about the plaintext.

If we think of our scheme as the combination of a Paillier ciphertext and a hash proof and write Enc for Paillier encryption we may explain the cryptosystem we use as follows. An encryption of a twin-moduli signature (r, z_0, z_1, s_0, s_1) essentially consists of a tuple

$$(E_a, E_{r_0}, c'_0, E_{r_1}, c'_1) = (\mathsf{Enc}(\mathsf{Pack}(r, s_0, s_1)), \mathsf{Enc}(r_0), g_0^{r_0}z_0, \mathsf{Enc}(r_1), g_1^{r_1}z_1) \ ,$$

where r_0 and r_1 are random and Pack is an invertible function that can be computed using only multiplication by constants and addition. Decryption is done in the obvious way. Note that $g_0^{r_0}z_0$ and $g_1^{r_1}z_1$ leaks information about the public key of the twin-moduli signature scheme, but encryptions of any two signatures of the same signer are indistinguishable. A single hash proof ties the components of a ciphertext together and the result is Δ-CCA2-secure when Δ is defined using the twin-moduli signature scheme.

5.3   Proofs of Knowledge of Equality Relations

There are various protocols in the literature [12,1,9,3] for proving equality of integer exponents over groups of unknown order based on variations of Fujisaki-Okamoto commitments, under the strong RSA-assumption. These protocols are strictly speaking not proofs of knowledge, since extraction may fail with negligible probability over the choice of commitment parameters, but they can be used as proofs of knowledge provided that there are trusted commitment parameters present during the execution of the protocols. Furthermore, they are usually given as honest verifier zero-knowledge protocols, but it is easy to make them zero-knowledge for a malicious verifier. A useful feature of these protocols is that they bound the bit-size of the exponents.

5.4   Verifiable Generation of Hiding Commitment Scheme

The problem with the proofs of equal integer exponents in a two party setting is that it is difficult to generate the Fujisaki-Okamoto commitment parameters efficiently. Recall that the commitment parameters consist of a random RSA-modulus $N = pq$, where p and q are safe primes, a random $g \in SQ_N$, and $h = g^x$ for a random $x \in [0, N2^{\kappa_r}]$. A commitment C of $m \in \mathbb{Z}$ is formed as $C = g^r h^m$ for an $r \in [0, N2^{\kappa_r}]$. The problem is that if (N, g, h) are generated by the prover, then the commitments are not binding. On the other hand, if they are generated by the verifier, then h may not be of the form g^x, and then the commitments are not hiding. As far as we know there is no truly efficient solution to this problem.

We now sketch our solution to this problem. The prover generates a pair (N_r, g_r), where N_r is an RSA-modulus of two safe primes and $g_r \in SQ_{N_r}$. This is done only once and is part of the public key of the prover in our application. The verifier then generates (N, g, h) as above, except that it defines $h = g^x$ for a random $x \in [0, NN_r 2^{\kappa_r}]$. It then computes a "commitment" $h_r = g_r^x$ of x and executes a Schnorr-like zero-knowledge "proof of knowledge" that the same integer x was used for both h_r and h. Extraction of x is possible with high probability provided that (N_r, g_r) are chosen correctly. This ensures that the parameters (N, g, h) can be used safely by a prover. On the other hand, provided $|SQ_{N_r}|$ and $|SQ_N|$ are relatively prime, h_r is essentially independently generated from h. This implies that the parameters (N, g, h) can be used safely by the verifier, since essentially no knowledge of x is leaked. To ensure that $|SQ_{N_r}|$ and $|SQ_N|$ are relatively prime with overwhelming probability we assume that N is generated independently from N_r. In practice this is very reasonable.

In the full paper we show that the parameters (N, g, h) output by the protocol below can be used to execute the proof of equal exponents that assumes trusted commitment parameters. We call the two players in the proof the generator G and the receiver R to distinguish them from their roles in a larger protocol. Denote by $\pi_{pl} = (P_{pl}, V_{pl})$ the zero-knowledge proof of knowledge of a logarithm for prime order groups described by Cramer et al. [6].

Protocol 1 (Secure Generation of Integer Commitment Scheme)
COMMON INPUT: A κ-bit integer N_r and $g_r \in \mathbb{Z}_{N_r}^*$ to both parties.

1. The receiver chooses a safe κ-bit prime $P = 2Q + 1$, and $H \in G_Q$ randomly, where G_Q is the unique subgroup of order Q. Then it chooses $z \in \mathbb{Z}_Q$ randomly, computes $K = H^z \bmod P$, hands (P, H, K) to the generator, and executes π_{pl} as the prover on common input (P, H, K) and private input z. If the verifier rejects, then the generator hands $\perp$ to the receiver and halts.

2. The generator verifies that P is a safe prime and that $H, K \in G_Q$. Then it chooses an RSA-modulus N, $g \in SQ_N$, and $x \in [0, 2^{2\kappa + \kappa_r} - 1]$ randomly, and computes $h = g^x \bmod N$ and $h_r = g_r^x \bmod N_r$. Then it chooses $c_g \in [0, 2^{\kappa_c} - 1]$, $r_g \in \mathbb{Z}_q$, and $r \in [0, 2^{2\kappa + 2\kappa_r + \kappa_c} - 1]$ randomly, defines $w = H^{r_g} K^{c_g}$, $\alpha = g^r \bmod N$ and $\alpha_r = g_r^r \bmod N_r$, and hands $(h, h_r, w, \alpha, \alpha_r)$ to the receiver.
3. The receiver chooses $c_r \in [0, 2^{\kappa_c} - 1]$ randomly and hands c_r to the generator.
4. The generator computes $c = c_g \oplus c_r$ and $d = cx + r \bmod 2^{2\kappa + 2\kappa_r + \kappa_c}$, hands (d, c_g, r_g) to the receiver, and outputs (N, g, h).
5. The receiver outputs (N, g, h) if $H^{r_g} K^{c_g} = w$, $h^c \alpha = g^d \bmod N$ and $h_r^c \alpha_r = g_r^d \bmod N_r$. Otherwise it outputs $\perp$.

We denote the above protocol by $\pi_{tp} = (G, R)$, where G is the generator and R the receiver.

Remark 6. Although the receiver can use the same prime P in every protocol instance, the generator must check that P is of the expected form to be confident that it can run the protocol π_{pl}, which is only sound if G_Q has prime order.

Checking for primality is expensive, i.e., it requires $O(\kappa_r)$ exponentiations. If one assumes that it is infeasible to find a specific safe prime P such that the discrete logarithm problem is feasible in G_Q, then any party can choose a prime P that is used in every protocol instance. Then each party performs the primality test only once. This is a natural assumption in practice, where one can use a prime from a cryptographic standard.

Proposition 2. *For every pair (N_r, g_r) with $N_r \in \mathbb{N}$ and $g_r \in \mathbb{Z}_{N_r}^*$ the probability $\Pr[\langle R, G \rangle (N_r, g_r) \neq \perp]$ is overwhelming.*

Proposition 3. *Let (N, g, h) be randomly distributed Fujisaki-Okamoto parameters. Define $[R^*, G](N_r, g_r)$ to be a pair consisting of the output of R^* and G.*

Then for every receiver $R^* \in$ EPPT there exists a simulator $M \in$ EPPT such that for every pair (N_r, g_r) with $N_r \in \mathbb{N}$ and $g_r \in \mathbb{Z}_{N_r}^*$ the distributions of $[R^*, G](N_r, g_r)$ and $M(N_r, g_r, N, g, h)$ are statistically indistinguishable and $M(N_r, g_r, N, g, h)$ is always on the form $(\cdot, out_G)$ with $out_G \in \{(N, g, h), \perp\}$.

Informally, this simply means that we can simulate the protocol in such a way that a particular set of parameters are used. Since the generator does not have any secret input, it is not meaningful to say that the protocol is zero-knowledge. However, one may view the proposition as saying that the protocol leaks no knowledge to the receiver about the exponent x that is chosen by the generator within the protocol. In this sense the protocol is zero-knowledge.

Denote by T_{srsa} the algorithm that on input 1^κ outputs (N_r, g_r), such that N_r is a product of two random safe $\kappa/2$-bit safe primes and g_r is randomly chosen in SQ_{N_r}.

Proposition 4. *Suppose that $(N_r, g_r) = T_{srsa}(1^\kappa)$. Then the probability that the receiver outputs (N, g, h), where h is not in the subgroup of $\mathbb{Z}_N^*$ generated by g, is negligible under the strong RSA-assumption and the discrete logarithm assumption.*

Remark 7. Even with the modification of Remark 6 the protocol requires a non-constant number of exponentiations, since the generator may have to generate a new RSA-modulus to ensure that its modulus is generated independently of N_r. If the reader find this annoying, please note that if the generator chooses an RSA-modulus N with at least $2\kappa + 2$ bits it can reuse the *same* modulus in any proof, since the orders of g and g_r are coprime. However, the size of the random exponents in the protocol above, and in all protocols that use the modulus must then be doubled, and this gives a far less efficient protocol in practice. Thus, we detail the solution above.

6   An Efficient Instantiation

In the full paper we show that there is an instantiation of the generic DC-signature scheme which is secure under the DCR-assumption, the strong RSA-assumption, and the DDH-assumption. We sketch this solution below.

Given the generic description in Section 4 and Theorem 1 all the algorithms of our instantiation follow from setting the signature scheme equal to our twin-moduli signature scheme, and the Δ-CCA2-secure scheme equal to our variation of the Cramer-Shoup-Paillier scheme, provided that we define a splitting of the key generator and a well-formedness function. The public key of the cryptosystem contains an RSA-modulus n, but the scheme functions properly for any *integer* n with some minor modifications. The key generator outputs a commitment based on the El Gamal cryptosystem of the secret key that is unconditionally committing to ensure that the uniqueness property of well-formedness. Thus, we define the splitting such that it is not necessary to execute an expensive proof that n is correctly formed. The well-formedness function for signatures is based on the fact that for any honestly computed signature (r, z_0, z_1, s_0, s_1) it holds that $r \in [2^{2\kappa_r + \kappa_m}, 2^{2\kappa_r + \kappa_m} + 2^{\kappa_r + \kappa_m} - 1]$, $s_0, s_1 \in [2^{\kappa_p - 1}, 2^{\kappa_p - 1} + 2^{\kappa_p - \kappa_p'} - 1]$, and $z_1^{e_1} = g_1 \bmod N_1$. Recall that the signature verification algorithm only requires that $r \in [1, 2^{2\kappa_r + \kappa_m + 1} - 1]$ and $s_0, s_1 \in [0, 2^{\kappa_p} - 1]$. The slack is exploited to avoid costly proofs of membership in intervals.

The idea of the twin-moduli signature scheme is loosely speaking that all that is needed to verify a signature can be done "in the exponent". Recall that a verification involves multiplication by constants, adding, checking for interval-membership and oddity, and then checking the roots of the signature. Let us write $C(m)$ for a Fujisaki-Okamoto commitment of the form $g^l h^m \bmod N$ for some random l, and simply write $\mathsf{Enc}(m)$ for a Paillier-part of a crypto-text, i.e., we ignore the encoding and the third component that guarantees CCA2-security. Then the proof of validity of a signature can be explained as follows. A DC-signature essentially consists of a tuple $(E_a, E_{r_0}, c_0', E_{r_1}, c_1')$ on the form

$$(\mathsf{Enc}(\mathsf{Pack}(r, s_0, s_1)), \mathsf{Enc}(r_0), g_0^{r_0} z_0, \mathsf{Enc}(r_1), g_1^{r_1} z_1) \ .$$

The prover forms commitments

$$C'_r = C(r - 2^{2\kappa_r + \kappa_m})$$
$$C'_{s_0} = C((s_0 - 2^{\kappa_p - 1} - 1)/2)$$
$$C'_{s_1} = C((s_1 - 2^{\kappa_p - 1} - 1)/2)$$

and proves knowledge of the committed values. The protocol used to do this also implies that $r - 2^{2\kappa_r + \kappa_m} \in [-2^{2\kappa_r + \kappa_m} + 1, 2^{2\kappa_r + \kappa_m} - 1]$ and $(s_0 - 2^{\kappa_p - 1} - 1)/2, (s_1 - 2^{\kappa_p - 1} - 1)/2 \in [-2^{\kappa_p - 2} + 1, 2^{\kappa_p - 2} - 1]$. Then the verifier computes

$$C_r = C'_r C(2^{2\kappa_r + \kappa_m})$$
$$C_{s_0} = (C'_{s_0})^2 C(2^{\kappa_p - 1} + 1)$$
$$C_{s_1} = (C'_{s_1})^2 C(2^{\kappa_p - 1} + 1)$$
$$C_a = C_r^{2^{2\kappa_p}} C_{s_0}^{2^{\kappa_p}} C_{s_1} ,$$

and the prover proves that the value committed to in C_a equal the value encrypted in E_a. Note that C_{s_0} and C_{s_1} are commitments to odd integers s_0 and s_1 in $[0, 2^{\kappa_p} - 1]$ and C_r is a commitment to an integer $r \in [1, 2^{2\kappa_r + \kappa_m + 1} - 1]$. Thus, part of the verification has already been executed.

To complete the verification the verifier computes commitments of the integers e_0 and e_1 induced by the values r, s_0, and s_1 by forming

$$C_{e_0} = (C_r C(m))^{2^{\kappa_p}} C_{s_0} \quad \text{and} \quad C_{e_1} = C_r^{2^{\kappa_p}} C_{s_1} .$$

All that then remains is to prove that if e_0 and e_1 are committed to in C_{e_0} and C_{e_1} and r_0 and r_1 are encrypted in E_{r_0} and E_{r_1}, then

$$(c'_0)^{e_0} = g_0^{e_0 r_0} g_0 \bmod N_0 \quad \text{and} \quad (c'_1)^{e_1} = g_1^{e_1 r_1} g_1 \bmod N_1 .$$

This shows that the encrypted signature is a valid twin-moduli signature.

The proof of invalidity for well-formed signatures is similar, but more complicated in that at some point the prover must show that $z_0^{e_0}/g_0 \neq 1$ without revealing this value. A standard trick to solve this problem is to randomize the result, i.e., revealing $(z_0^{e_0}/g_0)^l$ for a randomly chosen l. However, in general it may happen that $z_0^{e_0}/g_0$ is contained in some particular subgroup of $\mathbb{Z}^*_{N_0}$ and the simulator clearly does not know if this is the case.

When the public signature key is chosen honestly and the malicious verifier does not know the factorization of N_0, it is infeasible to find any element that generates a non-trivial subgroup of SQ_{N_0}. Thus, in this case the above idea works straightforwardly and there is no problem. In other words we have a (T_{hs}, F_{hs})-zero-knowledge simulator.

For maliciously generated N_0, g_0, z_0, and e_0 the above approach does not work at all, and it seems difficult to come up with an efficient approach that does work. Fortunately, we know that it suffices to have a simulator that is given the values z_0 and e_0 as an additional advice string, and given these it is obviously trivial to generate $(z_0^{e_0}/g_0)^l$ with the right distribution. In other words we have a (T_{cs}, F_{cs})-simulator.

The complexity of our scheme for some practical parameters is given below.

Table 1. The estimated average complexity of the algorithms and the protocols in terms of κ-bit exponentiations when $\kappa = 1024$, $\kappa_r = \kappa_c = 50$, and $\kappa_m = 160$

Operation	Alg./Prot.	Signer	Confirmer	Verifier
Signing	Sig^{dc}	140		
Converting	Con^{dc}		66	
Verifying	Vf^{dc}			1
Well-Formedness	π_{wf}		61	59
Correctness of conversion	π_c		327	227
Validity/Invalidity	π_e		189	169
Validity	π_v	166		151

Acknowledgments

I thank Ronald Cramer and Ivan Damgård for answering my questions about their work, I thank Dominik Raub and Stefano Tessaro for discussions on non-transferability, and I thank the anonymous reviewers of TCC 2007 for helpful comments.

References

1. F. Boudot. Efficient proofs that a committed number lies in an interval. In *Advances in Cryptology – Eurocrypt 2000*, volume 1807 of *Lecture Notes in Computer Science*, pages 431–444. Springer Verlag, 2000.
2. J. Camenisch and Markus Michels. Confirmer signature schemes secure against adaptive adversaries. In *Advances in Cryptology – Eurocrypt 2000*, Lecture Notes in Computer Science, pages 243–258. Springer Verlag, 2000.
3. J. Camensisch and V. Shoup. Practical verifiable encryption and decryption of discrete logarithms. In *Advances in Cryptology – Crypto 2003*, volume 2729 of *Lecture Notes in Computer Science*, pages 126–144. Springer Verlag, 2003.
4. D. Chaum. Designated confirmer signatures. In *Advances in Cryptology – Eurocrypt '94*, volume 950 of *Lecture Notes in Computer Science*, pages 86–91. Springer Verlag, 1994.
5. D. Chaum and H. van Antwerpen. Undeniable signatures. In *Advances in Cryptology – Crypto '89*, volume 435 of *Lecture Notes in Computer Science*, pages 212–216. Springer Verlag, 1990.
6. R. Cramer, I. Damgård, and P. D. MacKenzie. Efficient zero-knowledge proofs of knowledge without intractability assumptions. In *Public Key Cryptography – PKC 2000*, volume 1751, pages 354–372. Springer Verlag, 2000.
7. R Cramer, I. Damgård, and T. P. Pedersen. Efficient and provable security amplifications. In *Security Protocols, International Workshop, Cambridge, United Kingdom, April 10-12, 1996, Proceedings*, volume 1189 of *Lecture Notes in Computer Science*, pages 101–109. Springer, 1996.
8. R. Cramer and V. Shoup. Universal hash proofs and a paradigm for adaptive chosen ciphertext secure public-key encryption. http://homepages.cwi.nl/cramer/, June 1999.

9. I. Damgård and E. Fujisaki. A statistically-hiding integer commitment scheme based on groups with hidden order. In *Advances in Cryptology – Asiacrypt 2002*, volume 2501 of *Lecture Notes in Computer Science*, pages 125–142. Springer Verlag, 2002.

10. W. Diffie and M. E. Hellman. New directions in cryptography. *IEEE Transactions on Information Theory*, 22(6):644–654, 1976.

11. D. Dolev, C. Dwork, and M. Naor. Non-malleable cryptography. In *23rd ACM Symposium on the Theory of Computing (STOC)*, pages 542–552. ACM Press, 1991.

12. E. Fujisaki and T. Okamoto. Statistical zero knowledge protocols to prove modular polynomial relations. In *Advances in Cryptology – Crypto '97*, volume 1294 of *Lecture Notes in Computer Science*, pages 16–30. Springer Verlag, 1997.

13. C. Gentry, D. Molnar, and Z. Ramzan. Efficient designated confirmer signatures without random oracles or zero-knowledge proofs (extended abstract). In *Advances in Cryptology – Asiacrypt 2005*, volume 3788 of *Lecture Notes in Computer Science*, pages 662–681. Springer Verlag, 2005.

14. O. Goldreich. A uniform-complexity treatment of encryption and zeroknowledge. *Journal of Cryptology*, 6(1):21–53, 1993.

15. S. Goldwasser, S. Micali, and R. Rivest. A digital signature scheme secure against adaptive chosen-message attacks. *SIAM Journal on Computing*, 17(2):281–308, 1988.

16. S. Goldwasser and E. Waisbard. Transformation of digital signature schemes into designated confirmer signatures. In *1st Theory of Cryptography Conference (TCC)*, volume 2951 of *Lecture Notes in Computer Science*, pages 77–100. Springer Verlag, 2004.

17. M. Jakobsson, K. Sako, and R. Impagliazzo. Designated verifier proofs and their applications. In *Advances in Cryptology – Eurocrypt '96*, volume 1070 of *Lecture Notes in Computer Science*, pages 143–154. Springer Verlag, 1996.

18. M. Michels and M. Stadler. Generic constructions for secure and efficient confirmer signature schemes. In *Advances in Cryptology – Eurocrypt 1998*, volume 1403 of *Lecture Notes in Computer Science*, pages 406–421. Springer Verlag, 1998.

19. T. Okamoto. Designated confirmer signatures and public key encryption are equivalent. In *Advances in Cryptology – Crypto '94*, volume 839 of *Lecture Notes in Computer Science*, pages 61–74. Springer Verlag, 1994.

20. P. Paillier. Public-key cryptosystems based on composite degree residuosity classes. In *Advances in Cryptology – Eurocrypt '99*, volume 1592 of *Lecture Notes in Computer Science*, pages 223–238. Springer Verlag, 1999.

21. R. Pass and A. Rosen. New and improved constructions of non-malleable cryptographic protocols. In *37th ACM Symposium on the Theory of Computing (STOC)*, pages 533–542. ACM Press, 2005.

22. V. Shoup and R. Gennaro. Securing threshold cryptosystems against chosen ciphertext attack. In *Advances in Cryptology – Eurocrypt '98*, volume 1403 of *Lecture Notes in Computer Science*, pages 1–16. Springer Verlag, 1998.

23. Douglas Wikström. Designated confirmer signatures revisited. Cryptology ePrint Archive, Report 2006/123, 2006. http://eprint.iacr.org/.

From Weak to Strong Watermarking

Nicholas Hopper[1], David Molnar[2], and David Wagner[2]

[1] University of Minnesota, Minneapolis MN 55455, USA
hopper@cs.umn.edu
[2] University of California, Berkeley, Berkeley CA 94720, USA
{dmolnar, daw}@eecs.berkeley.edu

Abstract. The informal goal of a watermarking scheme is to "mark" a digital object, such as a picture or video, in such a way that it is difficult for an adversary to remove the mark without destroying the content of the object. Although there has been considerable work proposing and breaking watermarking schemes, there has been little attention given to the formal security goals of such a scheme. In this work, we provide a new complexity-theoretic definition of security for watermarking schemes. We describe some shortcomings of previous attempts at defining watermarking security, and show that security under our definition also implies security under previous definitions. We also propose two weaker security conditions that seem to capture the security goals of practice-oriented work on watermarking and show how schemes satisfying these weaker goals can be strengthened to satisfy our definition.

1 Introduction

Informally, a digital watermarking scheme is a procedure which embeds a "mark" in an object so that it is hard to remove the mark without "damaging" the object. These procedures have a wide variety of applications to digital rights management, including detection of unauthorized copies, limitations on media copying, tracing of information leaks, and resolution of ownership disputes over digital content; for further exposition on various applications see, for example [1, ch. 20]. As a result, watermarking schemes have seen intense research efforts; for example, see [2] and the references therein, or the proceedings [3,4,5,6,7,8,9,10,11,12,13,14,15,16]. Most of this work is focused on the construction of schemes for various digital media and attacks on these schemes, where there is a long history of schemes being broken almost immediately after they are proposed.

Given this history, it is not surprising that in the security community, there is a perception that secure watermarking is "theoretically impossible," as expressed, for instance, in [1,17,18]. While this idea is intuitively appealing, it is difficult to prove something is (im)possible without first formally defining the notion. Consider for instance, the related notions of program obfuscation and steganography, which were both widely believed to be impossible. Program obfuscation was formalized and shown to be impossible in general [19], but subsequently some progress has been made in limited cases [20,21]. Steganography, in contrast, was formalized and shown to be possible, but at limited rates [22,23,24].

S.P. Vadhan (Ed.): TCC 2007, LNCS 4392, pp. 362–382, 2007.

Surprisingly, formal definitions for watermarking security have only recently appeared in the literature. The state of the art focuses on defining schemes secure against specific "protocol attacks," which attack the protocols that use a watermark rather than removing a mark from an object [25]; these very powerful attacks changed researchers' understanding of what it means for a watermark to be "secure." For example, Kutter *et al.* [26] introduced the *copy attack,* in which a watermark is copied from an object O_1 into an object O_2 to form an object O_2' that appears marked even though it was never legitimately watermarked. This makes it impossible to use the attacked watermarking scheme for various applications, such as resolving ownership disputes.

Later Adelsbach, Katzenbeisser, and Veith formalized copy attacks and a different protocol attack known as an ambiguity attack. They then showed protocols intended to be provably secure against these attacks [27]. Several other authors have also produced schemes claimed to be provably resistant to copy attacks or other protocol attacks [28,29].[1] While this line of work has led to interesting results, there are some limitations, which we summarize in [30]. Additionally, this approach leads to an "arms race," in which, as new protocol attacks are discovered, new watermarking schemes must be designed and proven secure.

The primary contribution of this work is to initiate the systematic study of watermarking security definitions. We define a "strong watermarking" security condition with respect to a metric space on objects, which compares a watermark to an *ideal functionality* in which an object is marked if and only if it is similar to some object previously marked by the functionality. We show that this definition implies security against previously known protocol attacks, and explore the question of proving impossibility. We also explore weaker security conditions and show how, under some conditions, schemes satisfying these weaker definitions can be strengthened or amplified to produce strong watermarks.

We stress that in these latter results, we explicitly do not construct "secure" watermarking schemes from scratch. Instead, we show that watermark designers can achieve a strong notion of security from weak constructions that are not secure against protocol attacks. These results have two implications. First, impossibility results for strong watermarking in a metric space will also imply impossibility of these weaker goals. Second, this means that watermark designers need not complicate their schemes by attempting to rule out protocol attacks. Instead, they need only achieve the weaker notion and then apply our results; put another way, it is enough to build schemes that heuristically satisfy these goals and apply our constructions to build (heuristically) strong watermarking schemes, similar to results that say we can build (heuristically) strong secret-key encryption schemes from (heuristically) strong block ciphers.

Overview of our results. In Section 3 we propose a new definition of secure watermarking schemes, that we call *strong watermarking*, in the case that the marking and detecting procedures share a secret key. Our definition allows the

---

[1] We stress that these constructions, similarly to our own, do not attempt to construct a provably secure watermark "from scratch" but rather try to build something "secure against X" from a watermark that is not assumed to be secure in this sense.

adversary to make adaptive queries to oracles for both marking an object and detecting whether an object is marked. The main idea of the definition is that a strong watermarking scheme (in which there is *no communication between the marking and detection procedures*) should simulate an "ideal watermarking functionality," which we define. We show that strong watermarking implies security against all known protocol attacks, and argue that the definition will imply security against future protocol attacks. Furthermore, we show that security in our model depends critically on both the notion of similarity and the distribution on objects to be marked; specifically, we show an example of these settings under which strong watermarking is impossible, and an example where strong watermarking exists, relative to an oracle.

In Section 4 we introduce a "weaker" notion of watermark, which we call a *non-removable embedding*. This is a weak notion because it only requires that the watermark cannot be removed; we explicitly allow copy and ambiguity attacks to succeed against non-removable embeddings. We formalize this notion, prove a separation between the notion and our proposed strong definition, and point out that many watermarking schemes in the literature use a security metric closely related to this notion. We also introduce a notion of "limited" adversaries, who only create new objects based on some limited set of transformations. This notion is interesting since there are some techniques in the watermarking literature which seem to imply provable security against "limited" attacks such as Gaussian noise. Additionally, some applications of watermarking only require watermarks to be "robust" against distortions caused by physical processes; these can be modeled by limited adversaries. We note that all of our results on amplification can be easily extended to the limited adversarial setting. We then show how schemes that are provably secure under the strong watermarking definition can be constructed from non-removable embeddings plus a semi-offline trusted third party, a standard digital signature scheme, and a semantically secure symmetric encryption scheme. This shows that our notion of strong watermarking can be built on the "weak" primitive of non-removable embeddings. While we do require a third party, this party is not required during watermark detection.

In Section 5 we study an alternative method for producing a strong watermarking scheme. Specifically, we consider the question of *security amplification* of watermarking schemes. We formally specify two new notions that correspond to a weaker version of strong watermarking and show how schemes which satisfy these natural conditions can be efficiently composed to produce strong watermarking schemes. Note that this construction can be seen as an heuristic method to create strong watermarking schemes as well as a way to extend impossibility results for a given notion of similarity.

2   Preliminaries

We will work with discrete metric spaces. A *discrete metric space* $\mathcal{M}$ is a finite space equipped with a distance function $d : \mathcal{M} \times \mathcal{M} \to \mathbb{Z}^+ \cup \{0\}$. The distance function is symmetric, obeys the triangle inequality and has the property that if $d(x,y) = 0$ then $x = y$. We will associate with a metric space a similarity relation $\sim$ defined by $x \sim_\delta y \equiv d(x,y) \leq \delta$ for some fixed δ. When the meaning is clear from context, we will drop the δ and simply write $\sim$. For simplicity, we will assume that all parties can efficiently evaluate $\sim$. Finally, we denote by $\mathcal{D}$ a distribution on $\mathcal{M}$. Unless otherwise specified, we assume that all parties can efficiently sample from $\mathcal{D}$ and we denote by $O \leftarrow_R \mathcal{D}$ an object $O \in \mathcal{M}$ sampled according to the distribution $\mathcal{D}$.

We will also make use of a digital signature scheme $\mathcal{S} = \{\mathsf{SGen}, \mathsf{Sig}, \mathsf{Ver}\}$. We say that a signature scheme is (t, q, ϵ)-existentially unforgeable under adaptive chosen message attack [31] if all adversaries running in time at most t making at most q queries to a signature oracle have chance at most ϵ of obtaining a signature on a message not previously queried.

We will use a symmetric encryption scheme $\mathcal{SE} = \{\mathsf{Encrypt}, \mathsf{Decrypt}\}$. We say that a symmetric encryption scheme is (t, q, ϵ)-secure in the left-or-right sense [32] if every time t adversary, given q queries to a "left-or-right" oracle $LOR_K(b, x_0, x_1) = \mathsf{Encrypt}(K, x_b)$ cannot distinguish between the case that $b = 0$ and $b = 1$ with advantage better than ϵ.

Finally, we will need a pseudorandom function ensemble $\left\{F : \{0,1\}^k \times \{0,1\}^{L(k)} \rightarrow \{0,1\}^{\ell(k)}\right\}_{k \in \mathbb{N}}$ [33]. We say that a function F is (t, q, ϵ)-pseudorandom if any adversary running in time at most t and making at most q queries to a function oracle can distinguish an oracle for $F(U_k, \cdot)$ from an oracle for a random function $f : \{0,1\}^{L(k)} \rightarrow \{0,1\}^{\ell(k)}$ with advantage at most ϵ.

3 Strong Watermarking

As previously mentioned, the informal notion of a watermarking scheme requires the ability to somehow "mark" digital objects, such as pictures, sound, video, or text. The scheme should also satisfy several additional requirements:

- The result, O', of marking an object, O, should be "similar" to O.
- An adversary, given O', should not be able to find an object O'' that is similar to O' but unmarked; this prevents removal of the mark except by "damaging" the object.
- Most objects O must not be marked. If this is not the case, then certain desirable uses of watermarks, such as searching for copies of O' and proving ownership of O', are not possible.
- There should be no communication required between the marking procedure and the detecting procedure; or this communication should be minimized. This is necessary for many applications, for example, a media player that may not have a network connection.

We will model the notion of similarity or damage by postulating the existence of a "perceptual metric" that measures the distance between objects of a given type. Thus such a metric would assign a small distance between two pictures that look alike and a large distance between two very different pictures. In practice it is difficult to characterize such a metric space, so researchers typically focus on Euclidean or weighted L_1 distance in some "perceptually significant" space such as the Fourier [34], Wavelet [35], or Fourier Mellin [36] transforms. Once we fix a metric d, the natural notion of similarity is the relation $\sim_\delta$ defined previously, that is, we will say that objects O_1 and O_2 are similar if $d(O_1, O_2) \leq \delta$.

Given this formalization of similarity, we can construct a perfectly secure watermarking scheme that optimally satisfies the above requirements. To mark an object O with key K, the ideal scheme simply adds O to its list of objects marked with K; to test whether an object O' is marked with K, the ideal scheme simply searches the appropriate list of marked objects and returns true if it finds an object similar to O' and false otherwise. This "ideal" scheme does not allow an adversary to succeed in "unmarking" a marked object but leaves the largest possible set of objects unmarked subject to this constraint. The ideal scheme is undesirable in that it requires unbounded, online communication between the marking and detection algorithms; our intent is to compare a real-world watermarking scheme (which does not allow any online communication between the marking and detection procedures) to this ideal.

An informal statement of our definition allows an adversary access to a marking oracle and a detection oracle for a watermarking scheme. The adversary then attempts to attack the scheme by finding an object such that the results of the actual detection algorithm and the ideal detection procedure differ: either the object is marked and should not be, or it is unmarked and should be. Unfortunately, any watermarking scheme that produces objects that are similar to its input and has a static detection scheme would be insecure under this definition. The intuition is that the following attack would succeed with very high probability:

1. The adversary samples an object $O \in \mathcal{M}$. Since it has not been queried to the marking procedure, it is not yet marked under the ideal scheme.
2. Next the adversary queries $\mathsf{Mark}(O)$, to get an object O' similar to O.
3. Finally, the adversary queries $\mathsf{Detect}(O)$. In the watermarking scheme under attack, O should not be marked (since it was not marked in step 1, and there is no communication between marking and detection schemes). But in the ideal scheme, it is close to O', which *is* marked. Thus the adversary has succeeded in finding an object on which the real and ideal schemes differ.

We give a formal proof of this in [30], where we also show that a cryptographically natural alternative definition also rules out secure schemes that distort originals by less than half the similarity radius. Our solution is to introduce a third, *challenge* oracle that selects objects to watermark from some probability distribution; the performance of the watermarking scheme is only compared to that of the ideal scheme on these challenge objects.

3.1   Definition of Strong Watermarking Schemes

A *secret-key watermarking scheme* $\mathcal{W} = \{\mathsf{WMGen}, \mathsf{Mark}, \mathsf{Detect}\}$ consists of three algorithms: $\mathsf{WMGen} : 1^* \rightarrow \mathsf{Keys}$ generates a secret key to be used in marking and detection; $\mathsf{Mark} : \mathsf{Keys} \times \mathcal{M} \rightarrow \mathcal{M}$ takes a key and an object to

Oracle Mark*(O):	Oracle Detect*(O):	Oracle Challenge$_D^*()$
1. $O' \leftarrow$ Mark(K, O)	1. $b \leftarrow$ Detect(K, O)	1. $O \leftarrow_R D$
2. Marked $\leftarrow$ Marked $\cup \{O'\}$	2. $B' \leftarrow$ IdealDetect(O)	2. $O' \leftarrow$ Mark(K, O)
3. return(O')	3. if $b \notin B'$	3. chains $\leftarrow$ chains $\cup \{O'\}$
	4. then bad $\leftarrow$ true	4. Marked $\leftarrow$ Marked $\cup \{O'\}$
	5. return(b)	5. return(O')

Fig. 1. Definition of Mark*, Challenge*, and Detect* oracles for strong watermarking. The global variables K, Marked, chains, and bad are initialized in Figure 2.

Experiment $\mathbf{Exp}_{D,W}^{strong}(A)$:	**Procedure** IdealDetect(O):
1. $K \leftarrow$ WMGen(1^k)	1. if $(\exists O' \in$ chains $: O \sim O')$
2. bad $\leftarrow$ false	2. then return $\{$true$\}$
3. Marked $\leftarrow \emptyset$	3. else if $(\exists O' \in$ Marked $: O \sim O')$
4. chains $\leftarrow \emptyset$	4. then return $\{$true, false$\}$
5. $A^{\text{Mark}^*,\text{Challenge}^*,\text{Detect}^*}()$	5. else
6. return (bad)	6. return $\{$false$\}$

$$\mathbf{Adv}_{D,W}^{strong}(A) = \Pr[\text{bad} = \text{true}]$$

Fig. 2. Definition of security experiment for strong watermarking

mark and returns a new object; and Detect : Keys $\times \mathcal{M} \to \{$true, false$\}$. Notice that *we do not explicitly allow any online communication between the* Detect *and* Mark *procedures*, since in many applications the devices detecting and marking objects may not have any means by which to communicate.

We can now define *strong watermark security*. Our definition formalizes the informal discussion above. An adversary is given access to oracles for Mark and Detect, and a special Challenge* oracle that samples and marks objects from an efficiently sampleable distribution $\mathcal{D}$ over $\mathcal{M}$. The adversary wins if he calls Detect* on an object that is either marked, but not similar to the result of a Mark* or Challenge* query, or unmarked, but similar to the result of some Challenge* query. Notice that unlike in the hypothetical discussion above, we only require the objects near the *result* of Mark (rather than the input) to be marked, since these are (presumably) the ones that the adversary will be able to access. The formal security experiment has four global variables: Marked and chains, sets of objects; bad, a boolean flag; and K, a key. In Figures 1 and 2 we show pseudocode for initializing the security experiment and the ideal detection functionality, as well as for oracles Mark*, Challenge*, and Detect*. We note that some of our reductions require the ability to sample from the distribution $\mathcal{D}$ on $\mathcal{M}$.

We say that a watermark is ρ-*preserving for* $\mathcal{D}$ if $\Pr[K \leftarrow$ WMGen$(1^k); O \leftarrow_R \mathcal{D}; O' \leftarrow$ Mark$(K, O) : d(O, O') > \rho]$ is negligible in k; that is, if the marked version of an object is almost always within distance ρ of the original. This "bounded distortion" requirement is not strictly necessary for security in all applications, but is typically vital to the utility of a watermarking scheme.

The advantage of an adversary A_{Strong} is $\mathbf{Adv}_{D,W}^{strong}(A_{Strong})$ as defined in Figure 2. The scheme is a $(\mathcal{D}, t, q_M, q_D, q_C, \epsilon, \delta)$-*strong watermarking scheme* if for all adversaries A_{Strong} running in time at most t, making at most q_M queries to Mark*, at most q_D queries to Detect*, and at most q_C queries to Challenge* , the advantage of A_{Strong} is at most ϵ with respect to similarity relation $\sim_\delta$.

Philosophically, one may think of the above experiment as a game between, say, a "hacker" and a "studio." The hacker can "give" movies to the studio to see how they look when marked, and he can check, using his personal DVD player, whether any particular object is marked. Meanwhile, the studio will release other videos not created by the hacker; it is the hacker's goal to "unmark" one of these movies, or alternatively, to create a movie that appears to be marked but was never marked by the studio. If the hacker cannot do this, the studio can have good confidence that a movie will appear marked iff it was produced by them.

Dependence on $\sim$ and $\mathcal{D}$. It should be clear that the existence of strong watermarks depends critically on both the similarity relation $\sim$ and the distribution on challenge objects, $\mathcal{D}$. For instance, if an attacker can deduce, given the result of a query to Challenge$^*_{\mathcal{D}}$, the object $O \leftarrow_R \mathcal{D}$ from line 1 of Figure 1, then as pointed out in our earlier discussion, the scheme cannot be secure for $\mathcal{D}$ and $\sim$. Thus $\mathcal{D}$ must have high entropy, and Mark must be "one-way" for most keys. Likewise, if for any given O, enumerating the set $N_\delta(O) = \{O' : O' \sim O\}$ is feasible, then a watermarking scheme cannot be secure. In this work, we do not explore all the necessary conditions on $\sim$ and $\mathcal{D}$; it seems to be a difficult challenge to even identify the correct similarity metric and distribution for many of the applications of watermarking. Here we briefly give two results that show that even when the previous two conditions are satisfied, there cannot be a "generic" argument for the existence or impossibility of strong watermarks.

Proposition 1. Let $\mathcal{D}$ be the uniform distribution on k-bit strings and let $d(x,y)$ be the hamming distance metric on k-bit strings. Then there is no δ-preserving, $(\mathcal{D}, O(k), 1, 1, 1, 1/2^{\delta+1}, \delta)$-strong watermarking scheme.

Notice that for $\delta(k) = O(\log k)$, the neighbor set has size superpolynomial in k, and $\mathcal{D}$ has k bits of entropy, yet no watermarking scheme can have security better than $1/2k$. The proposition can be seen to be true as follows. Suppose we uniformly pick a point $x \in \{0,1\}^k$; consider the point y returned by Mark*(x), and let z and w be uniformly chosen points in $N_\delta(y)$ and $N_\delta(x)$, respectively. Now we know that if a watermarking scheme is to be ε-secure, it must be that $\Pr[\text{Detect}^*(z) = \text{false}] \leq \varepsilon$, since otherwise an adversary can remove a mark with probability greater than ϵ by sampling a random point in the neighborhood of a marked object. It can also be shown that $\Pr[z \in N_\delta(x)] \geq 1/2^\delta$. This gives us that $\Pr[z \in N_\delta(x) \wedge \text{Detect}^*(z) = \text{true}] \geq 1-(\Pr[z \notin N_\delta(x)]+\Pr[\text{Detect}^*(z) = \text{false}]) \geq 2^{-\delta} - \varepsilon$. Note that ε security also requires that $\Pr[\text{Detect}^*(w) = \text{true}] \leq \varepsilon$, since otherwise we can easily find a marked point – by randomly sampling an object in the neighborhood of a random point – breaking the watermark. Thus we also have that $\varepsilon \geq \Pr[\text{Detect}^*(w) = \text{true} \wedge w \in N_\delta(y)]$. But by symmetry, for any fixed

Experiment $\mathbf{Exp}_{\mathcal{D},W}^{cp}(B)$:
1. $K \leftarrow$ WMGen(1^k)
2. $O_1 \leftarrow_R \mathcal{D}$
3. $O_1' \leftarrow$ Mark(K, O_1)
4. $O_2 \leftarrow_R \mathcal{D}$
5. $O_2' \leftarrow B(O_1', O_2)$
6. if Detect(K, O_2')
7.   and $O_2 \sim O_2' \not\sim O_1'$
8.   then $b =$ true
9.   else $b =$ false
10. return(b)

$\mathbf{Adv}_{\mathcal{D},W}^{cp}(B) = \Pr[b = \text{true}]$

Experiment $\mathbf{Exp}_{\mathcal{D},W}^{amb}(B)$:
1. $K \leftarrow$ WMGen
2. repeat
3. $O_1 \leftarrow_R \mathcal{D}$
4. until Detect$(K, O_1) =$ false
5. $O_1' \leftarrow B(O_1)$
6. if Detect(K, O_1') and $O_1 \sim O_1'$
7.   then $b =$ true
8.   else $b =$ false
9. return(b)

$\mathbf{Adv}_{\mathcal{D},W}^{amb}(B) = \Pr[b = \text{true}]$

Adversary $A_{cp}^B()$:
1. $O_1 \leftarrow$ Mark$^*(O \leftarrow_R \mathcal{D})$
2. $O_2 \leftarrow_R \mathcal{D}$
3. $O_2' \leftarrow B(O_1', O_2)$
4. Detect$^*(O_2')$

Adversary A_{amb}^B:
1. $O_1 \leftarrow_R \mathcal{D}$
2. Detect$^*(O_1)$
3. $O_1' \leftarrow B(O_1)$
4. Detect$^*(O_1')$

Fig. 3. Experiments for copy and ambiguity attacks and the corresponding strong watermark adversary

choice of K, x, y, we have $\Pr[\text{Detect}^*(w) = \text{true} \wedge w \in N_\delta(y)] = \Pr[\text{Detect}^*(z) = \text{true} \wedge z \in N_\delta(x)]$. This gives $\varepsilon \geq 2^{-\delta} - \varepsilon$, or $\varepsilon \geq 2^{-\delta-1}$.

Notice that a similar argument applies to any metric space, distribution and marking function such that (i) the neighborhood of an object and its marked version are symmetric, (ii) these neighborhoods have noticeable intersection, and (iii) it is possible to uniformly sample from the neighborhood set of an object. Thus to rule out an impossibility result, we seek to violate these properties.

Proposition 2. *There exists an oracle Π, relative to which there exists a distribution $\mathcal{D}_\Pi$, a metric d_Π, and a 1-preserving watermarking scheme W^Π such that W^Π is $(\mathcal{D}_\Pi, t, t, t, t, t^2/2^k, 1)$-strong.*

Intuitively, we will choose Π, d_Π and $\mathcal{D}_\Pi$ so that for most strings x it will be very hard to even find a string y such that $d_\Pi(x, y) = 1$, but the oracle gives us a way to sample from a set of "special" strings x' that violate this property. Once we mark an object x' it is no longer in this special set, so it is hard for the adversary to remove the mark. Formally, the oracle Π "knows" a uniformly chosen bijection $\pi : \{0,1\}^{2k} \rightarrow \{0,1\}^k \times \{0,1\}^k$ for each k and answers three types of queries: sample, dist, and move. $\Pi(\text{sample}, y)$ returns $\pi^{-1}(y, 0^k)$. $\Pi(\text{dist}, x_0, x_1)$ computes $(y_b, z_b) = \pi(x_b)$, and then returns 0 if $x_0 = x_1$, 1 if $y_0 = y_1$ and some $z_b = 0^k$, 2 if $y_0 = y_1$, and 3 otherwise. $\Pi(\text{move}, x, z')$ computes $(y, z) = \pi(x)$; if $z = 0^k$ then it returns $\pi^{-1}(y, z')$; if $z = z'$ it returns $\pi^{-1}(y, 0^k)$, and otherwise it returns x. The distribution $\mathcal{D}_\Pi$ is defined as $\Pi(\text{sample}, U_k)$ and the metric $d_\Pi(x, y) = \Pi(\text{dist}, x, y)$, so that for most $2k$-bit strings x, there is only one string at distance 1 from x. The marking scheme W^Π uses k-bit keys, and computes $\text{Mark}^\Pi(K, x) = \Pi(\text{move}, x, K)$, while $\text{Detect}^\Pi(K, x)$ returns true iff $\Pi(\text{move}, x, K) \neq x$.

We remark that, obviously, the oracle distribution Π does not prove that strong watermarks exist. It merely shows that there cannot be a "black-box" proof that rules out all possible strong watermarking schemes without considering the details of $\mathcal{D}$ and $\sim$. We believe it is an interesting open question to find any $\mathcal{D}$ and $\sim$, even if they are contrived, that provably admit a strong watermarking scheme without reference to an oracle, or even with small values (q_M, q_C, q_D).

3.2   Strong Watermarks Are Secure Against Protocol Attacks

Adelsbach et al. provided the first formal definition of copy attacks and ambiguity attacks [27]. We adapt their definitions to our setting, in which we consider only the presence of a mark rather than its content. We show that strong watermarks are secure against copy and ambiguity attacks.

First we consider copy attacks. Informally, a copy attack occurs when an adversary can "copy" a watermark from a marked object O_1' to a second object O_2. In our watermarking model, "copy" means that the adversary, given a marked object O_1', can cause an object O_2 to return true for Detect* despite never having been queried to Mark. More formally, we say a watermarking scheme is $(\mathcal{D}, t, \epsilon_{cp}, \delta_{cp})$-secure against copy attacks if all adversaries B running in time at most t have advantage $\mathbf{Adv}_{\mathcal{D}, \mathcal{W}}^{cp}(B) \leq \epsilon_{cp}$ with respect to similarity relation $\sim_{\delta_{cp}}$. Notice that in this definition (and in the original definition of Adelsbach et al. [27]) the copy adversary is not afforded access to a Mark* or Detect* oracle. We can prove that a $\mathcal{D}$-strong watermarking scheme is not vulnerable to copy attacks for any sampleable distribution $\mathcal{D}'$: if there exists an adversary B that successfully carries out a copy attack, then the adversary A_{cp}^B in Figure 3 succeeds at breaking the strong watermark. A formal theorem statement and proof are in [30].

Next, we consider ambiguity attacks. A classical ambiguity attack takes an unmarked object O_1, and produces a new "original" object O_2 such that O_1 appears to be marked with O_2 as the original. In our model, we can recast ambiguity attacks as, given an unmarked object O_1, find an object O_2 such that $O_2 \sim O_1$ and O_2 appears to be marked, without legitimately marking O_2. Strong watermarking implies security against ambiguity attacks: if B succeeds at carrying out an ambiguity attack, then the adversary A_{amb}^B shown in Figure 3 breaks the strong watermark. Details are in [30].

Remark. We note that some works on protocol attacks describe attacks where the adversary is allowed to choose the key to the watermarking scheme. While it is important to eventually address such *chosen-key attacks*, we believe it is an interesting and important first step to concentrate on getting the definitions right for the more basic scenario. Thus in this paper we do not consider attacks that involve manipulating the keys of the marking and detection procedures.

4   Non-removable Embeddings and Strong Watermarks

Many watermarking schemes in the literature actually provide a somewhat different interface from the watermarking primitive described in the previous section. Instead, these schemes focus on embedding a short string within an object so that if the adversary does not distort the object too much, the embedded string can be recovered. Typical schemes do not attempt to prevent "insertion" of strings into an object, which is the reason that many protocol attacks succeed. In this section, we give a formal notion of a primitive, the *non-removable embedding* (NRE), that seems to capture this design goal. We will demonstrate that NREs

Experiment Exp$_D^{NRE}(A)$:
1. $(z, z') \leftarrow$ EMGen(1^k)
2. Embedded $\leftarrow \emptyset$
3. $O_A \leftarrow A^{\mathsf{Embed}(z,\cdot,\cdot),\mathsf{Challenge}^*}(z')$

Oracle Challenge*(m):
1. $O \leftarrow_R \mathcal{D}$
2. $O' \leftarrow$ Embed(z, O, m)
3. Embedded $\leftarrow$ Embedded $\cup \{(O', m)\}$
4. **return** O'

$$\mathbf{Adv}_D^{NRE}(A) = \Pr[\exists (O_i, m_i) \in \mathsf{Embedded} : O_A \sim O_i \wedge \mathsf{Extract}(z', O_A) \neq m_i]$$

Fig. 4. Security experiment and Embed* oracle for non-removable embeddings

are provably weaker objects than strong watermarks: if NREs exist at all, then there are NREs that allow copy attacks. After separating the notions of NREs and strong watermarks, we give a construction which makes limited use of a semitrusted third party to construct a strong watermarking scheme from a NRE.

The notion of an NRE is closely related to a security notion widespread in the watermarking literature. Many schemes presented in the watermarking literature, for example [37,38,39,40,41], take as their evaluation metric the bit error rate for a watermarked message given a specified constraint on the distortion allowed the adversary, or "watermark to noise ratio." Essentially, these schemes attempt to bound the rate of bit errors in the embedded string for a given amount of distortion induced by the adversary. One of the interesting properties of the NRE notion is that we can easily build an NRE from such schemes. Because we deal with probabilistic polynomial time adversaries, we can assume that the bit errors follow a computationally bounded distribution. Therefore, we can use the coding methods of Micali et al. to obtain an NRE from up to a bit error rate of one half: we simply encode the message before embedding and decode on extraction [42].

To begin, an embedding scheme (Embed, Extract, EMGen) is a triple of algorithms with the following signatures: Embed : Aux $\times \mathcal{M} \times \{0,1\}^k \to \mathcal{M}$, Extract : Aux$' \times \mathcal{M} \to \{0,1\}^k \cup \perp$, and EMGen : $1^* \to$ Aux $\times$ Aux$'$ for some fixed k. Here $\mathcal{M}$ is a metric space, and Aux and Aux$'$ are sets of possible auxiliary inputs. For example, Aux might be a set of secret keys, while Aux$'$ might be a set of public keys. k is the length of strings to be embedded in objects.

We further require that embedded messages can be extracted, i.e. for $(z, z') \leftarrow$ EMGen(1^k), we have Extract(z', Embed(z, O, x)) $= x$ with high probability. An embedding scheme is ρ-preserving for $\mathcal{D}$ if for all $m \in \{0,1\}^k$, $d($Embed(O, m), O) $\leq \rho$ with high probability over $O \leftarrow_R \mathcal{D}$. Together, these give a correctness and a bounded distortion requirement for a non-removable embedding.

We define security of embedding scheme NRE by saying it is $(\mathcal{D}, t, q_E, q_C, \epsilon, \delta)$ *non-removable for distribution* $\mathcal{D}$ if for all A running in time at most t, that make at most q_E queries to an Embed oracle and at most q_C queries to the Challenge* oracle, the advantage $\mathbf{Adv}_D^{NRE}(A)$ defined in Figure 4 is at most ϵ.

Remarks. This definition does not rule out the protocol attacks we have discussed: in particular, if there is a ρ-preserving non-removable embedding for the metric space $\mathcal{M}$ with metric d, we can construct a 2ρ-preserving non-removable embedding for the metric space $\mathcal{M} \times \{0,1\}^k$ with metric d', that allows copy attacks to succeed, as follows. We define the metric $d'((O_1, y_1), (O_2, y_2))$ to be $d(O_1, O_2)$ if $y_1 = y_2$ and $d(O_1, O_2) + \rho$ otherwise; define $\mathsf{Embed}'(z, (O, y), x) = (\mathsf{Embed}(z, O, x), x)$, and $\mathsf{Extract}'(z', (O, x)) = \mathsf{Extract}(z', O)$ if $\mathsf{Extract}(z', O) \neq \perp$ and $\mathsf{Extract}'(z', (O, x)) = x$ otherwise. Then it is easy to see that, as long as $\rho < \delta$, given a marked object $O = (O_1, x)$ and an unmarked object $O' = (O_2, y)$ we can "copy" the mark from O onto O' by setting $O'' = (O_2, x)$; yet it is still hard to remove x from O.

Although we do not explicitly require it, we note that typical applications will require that $\rho < \delta$ and in many cases, $\rho \ll \delta$. We also note that it is trivial to construct a ρ-preserving non-removable embedding for the case that $\rho = \sup_{(x,y) \in \mathcal{M} \times \mathcal{M}} d(x, y)$, using an error correcting code with minimum distance 2δ, if one exists for the metric space $\mathcal{M}$.[2] Thus the interesting question, for a given metric space, becomes "for what values of (ρ, δ) is a NRE possible?"

Barak et al. [43] defined watermarking for circuits, showing there are families of circuits for which such watermarking is impossible, and that the notion is incompatible with obfuscation even for watermarks that only succeed on some circuits. They briefly discuss how allowing "approximate implementations" may change their results. Our definition, in contrast, places these decisions in the choice of $\sim$ and the distribution $\mathcal{D}$.

We also note that many "public-key" watermarking schemes in the literature seem to target $(\mathcal{D}, t, q_E, 1, \epsilon, \delta)$ non-removability, expressed in terms of bit error rate for the watermarked message as noted above. A simple hybrid argument implies such schemes also have $(\mathcal{D}, t, q_E, q_C, q_C \epsilon, \delta)$ non-removability [44,41]. Thus while we are not aware of any strong candidate NREs, the existence of such a scheme seems to be a natural assumption if watermarking can be feasible at all.

We note that Moulin and Wang have shown that quantization index modulation (QIM) techniques provide provably good watermarks against an adversarial *memoryless channel.* The restriction to memoryless channels, together with an assumption that the host signal is Gaussian, allows them to analytically derive the "worst possible" channel and evaluate the bit error rate for a watermark signal under a specified bound on the mean squared error introduced by the adversary. Therefore, we can view their result as showing that QIM techniques yield a non-removable embedding for the class of memoryless adversary channels. While this is a severely limited class of adversaries, it shows that our notion is realizable at least under "toy" circumstances.

Finally, the StirMark benchmark [45,46] performs transformations such as resampling, resizing, and "jitter" in images; this benchmark is widely used to evaluate watermarks. We can capture both Moulin and Wang's result and the StirMark benchmark in our framework. If $\mathcal{C}$ is a set of object transformations, we define an attacker from class $\mathcal{C}$ to be an adversary who can only create objects via

---

[2] We let $\mathsf{Embed}(O, x) = \mathtt{encode}(x)$ and $\mathsf{Extract}(O) = \mathtt{decode}(O)$. If the code's minimum distance is 2δ then clearly any distortion by distance δ or less will result in extraction of the "embedded" message, but the worst-case distortion of this procedure is the maximum possible distance between two objects in $\mathcal{M}$.

sampling from $\mathcal{D}$, queries to oracles, and applying transformations from $\mathcal{C}$ to objects he has already created. Then it is a straightforward extension of our results to show that if there is an NRE that is secure against all attackers from class $\mathcal{C}$, there is a strong watermarking scheme that is secure against all attackers from $\mathcal{C}$.

4.1   Building Strong Watermarks from Embeddings

We now show how to build ideal watermarking schemes from non-removable embeddings, digital signature schemes, and a trusted third party (TTP). The main benefit of our scheme is that the TTP need not be present during watermark detection; anyone can check whether an object is marked without needing to contact the TTP in a wide variety of cases. Our scheme requires digital signatures in addition to a TTP because the underlying embeddings are not assumed secure against insertion of watermarks or copy attacks. The nonremovable embedding is necessary to allow offline detection, because otherwise an adversary could remove any metadata that might be attached to an object as a mark.

The TTP has well-known public keys and provides two services over secure channels: $\mathsf{Register}(O,\ K, x)$ picks a unique identifier i, checks that $x = \mathsf{Encrypt}(K, O)$, and returns $(i, \mathsf{Sig}_{TTP}(i, x))$; $\mathsf{Retrieve}(i)$ returns the x associated with i if any exists, or $\bot$ otherwise; we assume that neither call returns until a correctly authenticated response is received. We require that parties who execute Mark can communicate with the TTP as necessary. However, $\mathsf{Retrieve}$ is implemented in a semi-offline manner. Unique identifiers are assigned in ascending order, and the TTP publishes a signed list, $\mathsf{TTPList}$, of all (i, x) pairs each day. Consequently, $\mathsf{Retrieve}(i; \mathsf{TTPList})$ only needs to contact the TTP if $i > \mathsf{TTPList.length}$. Standard measures (such as substituting a zero-knowledge proof of knowledge of (O, K) for (O, K); maintaining an ordered, signed $\mathsf{TTPList}$; checking for consistency of TTP lists between updates; *et cetera*) can be taken to reduce the level of trust required in the TTP; we omit them for clarity of presentation, and because they do not affect the security proof.

Now let $\mathcal{E} = (\mathsf{Embed}, \mathsf{Extract}, \mathsf{EMGen})$ be an embedding; and let $\mathcal{SE} = (\mathsf{Encrypt}, \mathsf{Decrypt})$ be a symmetric encryption scheme. We then define a new watermarking scheme $\mathcal{W_E} = (\mathsf{WMGen}_{\mathcal{E},\mathcal{SE}}, \mathsf{Mark}_{\mathcal{E},\mathcal{SE}}, \mathsf{Detect}_{\mathcal{E},\mathcal{SE}})$ as shown in Figure 5. $\mathsf{Mark}(O)$ encrypts O, registers the ciphertext with the TTP, and embeds the TTP's identifier and signature in O. $\mathsf{Detect}(O; \mathsf{TTPList})$ extracts the TTP identifier and signature, retrieves the associated ciphertext, and checks that O is close to the result of Embed applied to the plaintext.

The main result of this section is that if the underlying embedding is non-removable, then the scheme $\mathcal{W_E}$ satisfies our notion of strong watermarking. Formally, we can state the following theorem, whose proof is in [30].

Theorem 1. *Suppose $\mathcal{E}$ is a $(\mathcal{D}, t_E, q_{EM}, q_{EC}, \epsilon_E, \delta)$-secure non-removable embedding, $S = (\mathsf{SGen}, \mathsf{Sig}, \mathsf{Ver})$ is (t_S, q_S, ϵ_S)-existentially unforgeable under chosen message attack, and $\mathcal{SE} = (\mathsf{Encrypt}, \mathsf{Decrypt})$ is $(t, q_{en}, \epsilon_{en})$ left-or-right secure under chosen plaintext attack. Then $\mathcal{W_E}$ is a $(t', q_M, q_D, q_C, \epsilon', \delta)$-strong watermarking scheme, where $\epsilon' = 2\epsilon_S + \epsilon_{en} + \epsilon_E$, $q_M + q_C \leq \min(q_{en}, q_S)$, $q_M \leq q_{EM}$, and $q_C \leq q_{EC}$.*

Algorithm Mark$_{\mathcal{E}}((z, z', K), O)$
1. $x \leftarrow$ Encrypt(K, O)
2. $(i, \sigma) \leftarrow$ Register(O, K, x)
3. $O' \leftarrow$ Embed$(z, O, (i, \sigma))$
4. return O'

Algorithm WMGen$_{\mathcal{E}}(1^k)$
1. $(z, z') \leftarrow$ EMGen(1^k)
2. $K \leftarrow_R \{0, 1\}^k$
3. return (z, z', K)

Algorithm Detect$_{\mathcal{E}}((z, z', K), O^*;$ TTPList):
1. if (Extract$(z', O^*) = \perp$) then return false
2. $(i^*, \sigma^*) \leftarrow$ Extract(z', O^*)
3. $x^* \leftarrow$ Retrieve$(i^*;$ TTPList)
4. $O \leftarrow$ Decrypt(K, x^*)
5. if $(x^* = \perp$ or $O = \perp$ or Ver$_{TTP}((i^*, x^*), \sigma^*) =$ false)
6.   then return false
7. if Embed$(z, O, (i^*, \sigma^*)) \sim O^*$
8.   then return true
9. else return false

Fig. 5. Pseudocode for WMGen$_{\mathcal{E}}$, Mark$_{\mathcal{E}}$, and Detect$_{\mathcal{E}}$

Remarks. We note that the scheme as written requires the Embed procedure to be deterministic; this is without loss of generality because the shared symmetric key between Mark and Detect can include a seed for a pseudorandom function that is used to generate the random bits used by Embed in a deterministic way from its arguments, without changing the security properties of the scheme.

We also note that if the distribution $\mathcal{D}$ has Shannon entropy less than k – the length of strings embedded by $\mathcal{E}$ – then in principle the TTP can be removed from this scheme. In this case, the marking scheme first losslessly compresses the object O into a short string x of length less than k, and the string x is then encrypted and authenticated using standard cryptographic techniques to get a ciphertext c which is embedded into O. The detection scheme recovers c, checks it for authenticity and if it passes, decrypts c to obtain x, then expands x to the original object O before comparing it to the input object. Thus our TTP can be seen as implementing a compression algorithm for unknown or incompressible distributions $\mathcal{D}$.

5   Strengthening Watermarks by Composition

Suppose we are given a watermarking scheme with known attacks that succeed at insertion or removal of a watermark with high probability, for example 90%, but retains some weak sense of security, in that it is not known how to defeat it with probability 1. In this section, we show that this sense of security is essentially enough for strong watermarking. Given an offline watermarking scheme W that satisfies two weak properties, we can construct an (offline) strong watermarking scheme in the sense of Section 3. The first property is that the scheme is secure in this weak sense – every adversary fails to defeat the scheme with some constant probability. The second property is that marking an object many times preserves some similarity to the original.

As mentioned previously, we believe this results has both positive and negative applications. Many of the heuristic watermarking schemes in the literature are broken, but frequently the known attacks do not succeed with probability 1. Thus applying our amplification scheme could heuristically create schemes which are, in some sense, secure "against known attacks." On the other hand, our results show that in order to rule out even weakly secure watermarking schemes for a given metric and distribution, it is sufficient to concentrate on showing the impossibility of a strong watermarking scheme.

5.1 Weakly Secure Watermarking Schemes

Our scheme will work by applying the Mark function to its own output several times. Because our security notions depend on the probability distribution on the inputs to Mark, we will need some assumption on the distribution of the outputs of Mark. The strongest assumption is that these distributions are identical, but in general this amounts to assuming that Mark is the identity function. Thus, instead, we assume that the (weak) security of a watermark holds even if we make some small distortions to an object before marking it. Formally, we say that a randomized algorithm D is a (t, r)-*perturbation of* $\mathcal{D}$ if D runs in time t and $\Pr[O \leftarrow_R \mathcal{D}; O' \leftarrow D(O) : d_\mathcal{M}(O, O') > r]$ is negligible. We will say that our watermarking schemes are weakly secure for $\mathcal{D}$ if they are weakly secure for any (t, r)-perturbation of $\mathcal{D}$.

(WEAK) SECURITY AGAINST REMOVAL. We define the removal advantage of an adversary against a watermarking scheme to be the probability that an adversary can produce, given a watermarked object drawn from a (t, r)-perturbation of $\mathcal{D}$, a similar object that is not marked. Formally, define

$$\mathbf{Adv}_{W,D}^{rm}(\mathcal{A}) = \Pr[K \leftarrow W.\mathsf{WMGen}(1^k); O \leftarrow_R \mathcal{D}; O' \leftarrow W.\mathsf{Mark}_K(D(O));$$
$$O'' \leftarrow \mathcal{A}(O') : \quad W.\mathsf{Detect}_K(O'') = \mathsf{false} \wedge O'' \sim_\delta O'] .$$

Then, we say that a watermark W is $(t, \epsilon_{rm}, \delta, \mathcal{D}, r)$-*secure against removal* if for every time-t adversary $\mathcal{A}$, and every (t, r)-perturbation D of $\mathcal{D}$, $\mathbf{Adv}_{W,D}^{rm}(\mathcal{A}) \leq \epsilon_{rm}$. Informally, this definition says that every adversary who runs in time at most t fails to remove the watermark of an object drawn from a (t, r)-perturbation of $\mathcal{D}$ with probability at least $1 - \epsilon_{rm}$.

We remark that this experiment captures the intuitive notion of trying to remove a watermark without damaging some challenge object, a common goal of attacks on watermarking schemes found in the literature. We also note that the goal of our scheme is to strengthen a watermark with only *constant* security against removal – meaning that we explicitly allow a watermarking scheme that can be removed, say, 99% of the time.

(WEAK) SECURITY AGAINST INSERTION. We informally define the insertion advantage of an adversary against a watermarking scheme to be the probability that an adversary can produce, given a single watermarked object, another watermarked object. Formally, define

$$\mathbf{Adv}_{W,D}^{ins}(\mathcal{A}) = \Pr[K \leftarrow \mathsf{WMGen}(1^k); O \leftarrow \mathcal{A}(1^k); O' \leftarrow W.\mathsf{Mark}_K(O);$$
$$O'' \leftarrow \mathcal{A}(O') : \quad W.\mathsf{Detect}_K(O'') = \mathsf{true} \wedge O'' \not\sim_\delta O'] .$$

Then, we say that a watermark W is $(t, \epsilon_{ins}, \delta)$-*secure against insertion* if for every time-t adversary $\mathcal{A}$, $\mathbf{Adv}_{W,D}^{ins}(\mathcal{A}) \leq \epsilon_{ins}$. Informally, this definition says that every adversary who runs in time t must fail to produce a (new) watermarked object with probability at least $1 - \epsilon_{ins}$. We remark that security against insertion is essentially an adversarial notion of the "false positive rate" of a watermark [2,27].

We can now state the main result of this section; the proof depends on several additional results proved in the remainder of the section:

Theorem 2. *Suppose there exists a watermarking scheme W such that:*

- *W is ρ-preserving;*
- *W is both $(t, \epsilon_{rm}, \delta, \mathcal{D}, k^{O(1)}\rho)$-secure against removal and $(t, \epsilon_{ins}, \delta)$-secure against insertion; and*
- *$\epsilon_{rm}, \epsilon_{ins}$ are constants such that $4\epsilon_{ins} \lg \frac{1}{\epsilon_{rm}} < 1$; and $t = k^{\omega(1)}$*

Then there exists a $(\mathcal{D}, t', q_M, 1, q_D, \nu, \delta)$-strong watermarking scheme W', where $t' = k^{\omega(1)}$ and $\nu = 1/k^{\omega(1)}$. The scheme W' is $k^{O(1)}\rho$-preserving.

Proof. The new watermark W' is constructed from W using the techniques developed in the remainder of this section: first the "alternating" composition ALT_ℓ with $\ell = O(\lg k)$ levels, from Section 5.3 is applied to W. By repeated application of Theorem 3 the resulting scheme $\mathsf{S}(W)$ is ν-secure against removal and insertion, for negligible ν. Lemma 1 implies that this scheme is also a $(\mathcal{D}, t', q_M, 1, q_D, \nu, \delta)$-strong watermark, for $q_M + q_D = 1$. To achieve arbitrary q_M and q_D, we construct the scheme $\mathsf{S}'(W)$ described in Section 5.4 with $m = q_M + q_D$. By Theorem 4 the resulting scheme is a $(\mathcal{D}, t', q_M, 1, q_D, \nu, \delta)$-strong watermark.

5.2   Single-Property Amplification

Let $\mathbb{K} = (K_1, K_2, \ldots, K_m)$ be a set of independently chosen secret keys. We define $\mathsf{Mark}_{\mathbb{K}}^W(O) := W.\mathsf{Mark}_{K_m}(W.\mathsf{Mark}_{K_{m-1}}(\ldots W.\mathsf{Mark}_{K_1}(O)\ldots))$, i.e. $\mathsf{Mark}_{\mathbb{K}}^W$ is the sequential marking of an object O with each secret key in the vector $\mathbb{K}$. We now have two choices for defining the $\mathsf{Detect}_{\mathbb{K}}^W(O')$ algorithm, each resulting in a different watermarking scheme. Define the schemes as follows:

$$\mathsf{AND}(m, W).\mathsf{Detect}_{\mathbb{K}}(O') = \bigwedge_{1 \leq i \leq m} W.\mathsf{Detect}_{K_i}(O')$$

$$\mathsf{OR}(m, W).\mathsf{Detect}_{\mathbb{K}}(O') = \bigvee_{1 \leq i \leq m} W.\mathsf{Detect}_{K_i}(O')$$

Intuitively, we expect that $\mathsf{AND}(m, W)$ will improve the insertion security of watermark W while impeding the removal security. This is because to insert a watermark one must insert m copies of W, while to delete a watermark one need only delete 1 out of m. Likewise, we intuitively would expect that $\mathsf{OR}(m, W)$ will decrease the insertion security while increasing the removal security. We can write this formally in the following theorem, whose proof is in [30].

Theorem 3. *Let W be ρ-preserving, $(t, \epsilon_{ins}, \delta)$-secure against insertion, and $(t, \epsilon_{rm}, \delta, \mathcal{D}, r)$-secure against removal. Then:*

(a) $\mathsf{OR}(m, W)$ is $(t', m\epsilon_{ins}, \delta - m\rho)$ secure against insertion.
(b) $\mathsf{AND}(m, W)$ is $(t', m\epsilon_{rm}, \delta - m\rho, \mathcal{D}, r - m\rho)$ secure against removal.

Where $t' = t - mT_M - O(1)$ *if* T_M *is the time to mark an object. Furthermore, for any* $q(k) \in k^{O(1)}$,

(c) $\mathsf{AND}(m, W)$ *is* $(t', \epsilon_{ins}^m + 1/q, \delta - m\rho)$ *secure against insertion.*
(d) $\mathsf{OR}(m, W)$ *is* $(t', \epsilon_{rm}^m + 1/q, \delta - m\rho, \mathcal{D}, r - m\rho)$ *secure against removal.*

Where $t' = t/poly(q, m)$.

5.3 Simultaneous Amplification

Let W be a watermarking scheme with key space K and define the scheme $\mathsf{ALT}(W)$ with key space K^4 by $\mathsf{ALT}(W) = \mathsf{AND}(2, \mathsf{OR}(2, W))$. Then by the previous theorem, if W is $(k^{\omega(1)}, c/2, \delta, \mathcal{D}, r)$ secure against removal and $(k^{\omega(1)}, d/4, \delta)$ secure against insertion, then $\mathsf{ALT}(W)$ is $(k^{\omega(1)}, c^2/2, \delta - 4\rho, \mathcal{D}, r - 4\rho)$-secure against removal and $(k^{\omega(1)}, d^2/4, \delta - 4\rho)$-secure against insertion. If we define the scheme $\mathsf{ALT}_\ell(W)$ by $\mathsf{ALT}_1(W) = \mathsf{ALT}(W)$ and $\mathsf{ALT}_\ell(W) = \mathsf{ALT}(\mathsf{ALT}_{\ell-1}(W))$, we see that $\mathsf{ALT}_\ell(W)$ is $(k^{\omega(1)}, d^{2^\ell}/4, \delta - 4^\ell\rho)$-secure against insertion and $(k^{\omega(1)}, c^{2^\ell}/2, \delta - 4^\ell\rho, \mathcal{D}, r - 4^\ell\rho)$-secure against removal, for $\ell = O(\log k)$. By setting $\ell = \lceil \log k \rceil$ and letting $\mathsf{S}(W) = \mathsf{OR}(2, \mathsf{ALT}_\ell(W))$ we obtain a scheme that inserts $poly(k)$ marks such that any $poly(k)$-time adversary has negligible advantage for both removal and insertion, if the original scheme is weakly secure against (for example) subexponential time adversaries.

Intuitively, we can think of this scheme as building a tree of marking schemes over the object O to be marked. By building the tree appropriately, alternating AND and OR at each level, we can reduce both the insertion and deletion probabilities for the resulting detection scheme. Each leaf of the tree corresponds to an independently keyed insertion of a watermark. Suppose we have a depth t tree comprising 2^t independent keys. The top gate, an OR, will recursively compute $\mathsf{AND.Detect}(O, k[1]...k[2^{t-1}])$ and $\mathsf{AND.Detect}(O, k[2^{t-1}]...k[2^t])$ and return true if at least one recursive branch returns true. OR is defined analogously. Alternatively, from the bottom-up view, there is one object in which we may have embedded $n = 2^t$ marks; we check if each mark is present and then compute a formula based on these truth values to decide whether the composed mark is present.

We note that the full alternating binary tree only exponentially reduces the insertion and removal probabilities if we start with $\epsilon_{rm} < 1/2$ and $\epsilon_{ins} < 1/4$. For many watermarking schemes in the literature, however, we might expect that the insertion probability is low, say $\epsilon_{ins} < 1/100$, while the removal probability is high, say $\epsilon_{rm} = 0.9$. In this case, we can make the lowest level of the tree consist of an OR of 20 marks to get $\epsilon_{rm}' = 1/e^2 < 1/2$ and $\epsilon_{ins}' < 1/5$. We can then build a binary tree on top of the resulting watermark.

It remains to show that the scheme $\mathsf{S}(W)$ is *correct*, i.e. that $\mathsf{S.Detect}_{\mathbb{K}}$ $(\mathsf{S.Mark}_{\mathbb{K}}(\mathcal{D})) = \mathsf{true}$ except with negligible probability. Notice, however, that $\mathsf{S.Detect}$ returns true if either its left branch or its right branch return true. But the insertion of the marks in the right branch is just one particular instance of an adversary (against the left branch) that returns an output that is distorted by distance at most $4^\ell \rho$ from its input, so if $\delta > 4^\ell \rho$, the probability that this "adversary" succeeds in removing the mark inserted by the left branch is negligible.

5.4 Strong Watermark Security from Insertion and Removal Security

Notice that the definition of $(t, \epsilon_{ins}, \delta)$ security against insertion implies $(\mathcal{D}, t, 1, 1, 0, \epsilon_{ins}, \delta)$-strong watermark security: any strong watermark adversary $\mathcal{A}$ who makes one Mark* query and one Detect* query can be converted into a weak insertion adversary $\mathcal{B}$: $\mathcal{B}(1^k)$ simply runs $\mathcal{A}$ until $\mathcal{A}$ makes a query to Mark*, say O, and outputs O; $\mathcal{B}(O')$ returns O' to $\mathcal{A}$ and outputs the object O'' that $\mathcal{A}$ queries to Detect*. Since the list chalns is empty, submitting an unmarked O'' will give $b =$ false and $b' =$ false, so $\mathcal{A}$ can only win by "inserting" a watermark. Additionally satisfying $(t, \epsilon_{rm}, \delta, \mathcal{D}, r)$-security against removal implies $(D(\mathcal{D}), t, 0, 1, 1, \epsilon, \delta)$ strong watermark security for any D that perturbs $\mathcal{D}$ by at most r, because an adversary who makes only a single query $O' \leftarrow$ Challenge*$(D(\mathcal{D}))$ can only win by querying Detect*(O'') such that:

- $O'' \sim O'$ and Detect$_K(O'') =$ false; if this happens with probability greater than ϵ_{rm} then the removal security of the scheme is contradicted.
- $d(O'', O') > \delta$ and Detect$_K(O'') =$ true; if this happens with probability greater than ϵ_{ins} then the insertion security is violated: an insertion adversary can always draw his challenge object $O' \leftarrow D(\mathcal{D})$.

This observation leads to the following lemma:

Lemma 1. *If W is $(t, \epsilon_{ins}, \delta)$-secure against insertion and $(t, \epsilon_{rm}, \delta, \mathcal{D}, r)$-secure against removal then W is a $(D(\mathcal{D}), t, q_M, 1, q_C, \epsilon_{ins} + \epsilon_{rm}, \delta)$-strong watermarking scheme, for any distortion function $D \in \mathrm{TIME}(t)$ that perturbs $\mathcal{D}$ by distance at most r, and any $q_C \leq 1 - q_M$.*

Suppose that we extend the definition of a strong watermark to allow Mark to maintain a local state. Then we can generically increase the number of (mark and challenge) queries we are secure against by a factor of n while also increasing the running time of Detect by a factor of n as follows. We require that Mark$'_K$ keeps a count, i, of the number of objects it has marked (say modulo n). When Mark$'_K(O)$ marks a new object, it computes the entire set of keys to use as $\mathbb{K}_i = F_K(i)$, where F is a pseudorandom function of the appropriate output size, and then calls Mark$_{\mathbb{K}_i}(O)$. Then in Detect$'_K(O)$ we try $\mathbb{K} = F_K(1), F_K(2) \dots F_K(n)$ and output true if any of these watermarks is detected. This increases the insertion probability by at most a factor of n. We make this more formal in the following theorem, whose proof is in [30].

Theorem 4. *Let $W =$ (Mark, Detect) be a $(\mathcal{D}, t, q_M, 1, 1 - q_M, \epsilon_{wm}, \delta)$-strong watermarking scheme and let $W' =$ (Mark$'$, Detect$'$) be a watermarking scheme with the stateful Mark$'$ algorithm described above, and let F be a (t, n, ϵ_{prf})-pseudorandom function. Then W' is a $(\mathcal{D}, t, q_M, 1, n - q_M, n\epsilon_{wm} + \epsilon_{prf}, \delta)$-strong watermarking scheme.*

6   Conclusions

In this paper we have initiated the scientific study of complexity-based security of watermarking schemes. We define a notion of watermarking security based on comparison to an ideal scheme, and give evidence that this is the right notion of security for watermarks in two ways. First, we show that security in our sense implies previous definitions of security, while the converse is not true. Second, we have shown how to construct a watermark which is secure in our sense from several weaker primitives, which seem to capture the goals of research in watermarking primitives. Our intent is not to introduce new watermarking protocols, but to suggest that security in the "strong watermark" sense is the "right definition": if secure watermarks (in any sense) are feasible at all, then so are strong watermarking schemes. A key question left open by our work, therefore, is the construction of similarity-preserving strong watermarking schemes that are provably-secure under standard cryptographic assumptions; even a construction for a contrived metric space would be an interesting first step in this direction.

Acknowledgments

We thank Andre Adelsbach, Luis von Ahn, Hayley Iben, Pierre Moulin, and Ahmad-Reza Sadeghi for helpful discussions. We also thank Kannan Ramchandran, Dan Schonberg, and the Berkeley Information Hiding Group for feedback on early versions of this work. Nicholas Hopper was supported by the National Science Foundation, under grant CNS-0546162. David Molnar was supported by a graduate research fellowship from the National Science Foundation and by NSF CCF-0325311. David Wagner was supported by NSF CCF-0325311.

References

1. Anderson, R.J.: Security Engineering: A Guide to Building Dependable Distributed Systems. John Wiley & Sons, Inc., New York, NY, USA (2001)
2. Cox, I., Miller, M.L., Bloom, J.A.: Digital watermarking. Morgan Kaufmann Publishers Inc., San Francisco, CA, USA (2002)
3. Wong, P.W., Delp, E.J., eds.: Security and Watermarking of Multimedia Contents, Proceedings. In Wong, P.W., Delp, E.J., eds.: Security and Watermarking of Multimedia Contents. Volume 3657 of Proceedings of SPIE., SPIE (1999)
4. Wong, P.W., Delp, E.J., eds.: Security and Watermarking of Multimedia Contents II, 2000, Proceedings. In Wong, P.W., Delp, E.J., eds.: Security and Watermarking of Multimedia Contents. Volume 3971 of Proceedings of SPIE., SPIE (2000)
5. Wong, P.W., Delp, E.J., eds.: Security and Watermarking of Multimedia Contents III, 2001, Proceedings. In Wong, P.W., Delp, E.J., eds.: Security and Watermarking of Multimedia Contents. Volume 4314 of Proceedings of SPIE., SPIE (2001)
6. Delp, E.J., Wong, P.W., eds.: Security and Watermarking of Multimedia Contents IV, 2002, Proceedings. In Delp, E.J., Wong, P.W., eds.: Security and Watermarking of Multimedia Contents. Volume 4675 of Proceedings of SPIE., SPIE (2002)

7. Delp, E.J., Wong, P.W., eds.: Security, Steganography, and Watermarking of Multimedia Contents V, 2003, Proceedings. In Delp, E.J., Wong, P.W., eds.: Security and Watermarking of Multimedia Contents. Volume 5020 of Proceedings of SPIE., SPIE (2003)

8. Delp, E.J., Wong, P.W., eds.: Security, Steganography, and Watermarking of Multimedia Contents VI, San Jose, California, USA, January 18-22, 2004, Proceedings. In Delp, E.J., Wong, P.W., eds.: Security, Steganography, and Watermarking of Multimedia Contents. Volume 5306 of Proceedings of SPIE., SPIE (2004)

9. Delp, E.J., Wong, P.W., eds.: Security, Steganography, and Watermarking of Multimedia Contents VII, San Jose, California, USA, January 17-20, 2005, Proceedings. In Delp, E.J., Wong, P.W., eds.: Security, Steganography, and Watermarking of Multimedia Contents. Volume 5681 of Proceedings of SPIE., SPIE (2005)

10. Anderson, R.J., ed.: Information Hiding, First International Workshop, Cambridge, U.K., May 30 - June 1, 1996, Proceedings. In Anderson, R.J., ed.: Information Hiding. Volume 1174 of Lecture Notes in Computer Science., Springer (1996)

11. Aucsmith, D., ed.: Information Hiding, Second International Workshop, Portland, Oregon, USA, April 14-17, 1998, Proceedings. In Aucsmith, D., ed.: Information Hiding. Volume 1525 of Lecture Notes in Computer Science., Springer (1998)

12. Barni, M., Herrera-Joancomartí, J., Katzenbeisser, S., Pérez-González, F., eds.: Information Hiding, 7th International Workshop, IH 2005, Barcelona, Spain, June 6-8, 2005, Revised Selected Papers. In Barni, M., Herrera-Joancomartí, J., Katzenbeisser, S., Pérez-González, F., eds.: Information Hiding. Volume 3727 of Lecture Notes in Computer Science., Springer (2005)

13. Pfitzmann, A., ed.: Information Hiding, Third International Workshop, IH'99, Dresden, Germany, September 29 - October 1, 1999, Proceedings. In Pfitzmann, A., ed.: Information Hiding. Volume 1768 of Lecture Notes in Computer Science., Springer (2000)

14. Moskowitz, I.S., ed.: Information Hiding, 4th International Workshop, IHW 2001, Pittsburgh, PA, USA, April 25-27, 2001, Proceedings. In Moskowitz, I.S., ed.: Information Hiding. Volume 2137 of Lecture Notes in Computer Science., Springer (2001)

15. Petitcolas, F.A.P., ed.: Information Hiding, 5th International Workshop, IH 2002, Noordwijkerhout, The Netherlands, October 7-9, 2002, Revised Papers. In Petitcolas, F.A.P., ed.: Information Hiding. Volume 2578 of Lecture Notes in Computer Science., Springer (2003)

16. Fridrich, J.J., ed.: Information Hiding, 6th International Workshop, IH 2004, Toronto, Canada, May 23-25, 2004, Revised Selected Papers. In Fridrich, J.J., ed.: Information Hiding. Volume 3200 of Lecture Notes in Computer Science., Springer (2004)

17. Gollmann, D.: Computer security. John Wiley & Sons, Inc., New York, NY, USA (1999)

18. Wikipedia: Digital watermarking — wikipedia, the free encyclopedia (2006) [Online; accessed 31-July-2006].

19. Barak, B., Goldreich, O., Impagliazzo, R., Rudich, S., Sahai, A., Vadhan, S.P., Yang, K.: On the (im)possibility of obfuscating programs. In Kilian, J., ed.: CRYPTO. Volume 2139 of Lecture Notes in Computer Science., Springer (2001) 1–18

20. Lynn, B., Prabhakaran, M., Sahai, A.: Positive results and techniques for obfuscation. In Cachin, C., Camenisch, J., eds.: EUROCRYPT. Volume 3027 of Lecture Notes in Computer Science., Springer (2004) 20–39

21. Wee, H.: On obfuscating point functions. In Gabow, H.N., Fagin, R., eds.: STOC, ACM (2005) 523–532
22. Cachin, C.: An information-theoretic model for steganography. In Aucsmith, D., ed.: Information Hiding. Volume 1525 of Lecture Notes in Computer Science., Springer (1998) 306–318
23. Hopper, N.J., Langford, J., von Ahn, L.: Provably secure steganography. In Yung, M., ed.: CRYPTO. Volume 2442 of Lecture Notes in Computer Science., Springer (2002) 77–92
24. Dedic, N., Itkis, G., Reyzin, L., Russell, S.: Upper and lower bounds on black-box steganography. In Kilian, J., ed.: TCC. Volume 3378 of Lecture Notes in Computer Science., Springer (2005) 227–244
25. Craver, S., Memon, N., Yeo, B.L., Yeung, M.M.: Resolving rightful ownerships with invisible watermarking techniques: Limitations, attacks, and implications. IEEE J. SAC 16(4) (May 1998) 573–586 Special issue on copyright & privacy protection.
26. Kutter, M., Voloshynovskiy, S., Herrigel, A.: The watermark copy attack. In: Proceedings of the SPIE vol. 3971, Security and Watermarking of Multimedia Contents II. (2000) 371–380
27. Adelsbach, A., Katzenbeisser, S., Veith, H.: Watermarking schemes provably secure against copy and ambiguity attacks. In: DRM '03: Proceedings of the 2003 ACM workshop on Digital rights management, ACM Press (2003) 111–119
28. Dittmann, J., Katzenbeisser, S., Schallhart, C., Veith, H.: Provably secure authentication of digital media through invertible watermarks. Cryptology ePrint Archive, Report 2004/293 (2004) http://eprint.iacr.org/.
29. Li, Q., Chang, E.C.: On the possibility of non-invertible watermarking schemes. In: Information Hiding Workshop, Springer-Verlag LNCS 3200. (2004)
30. Hopper, N., Molnar, D., Wagner, D.: From weak to strong watermarking. IACR Cryptology ePrint Archive, Report 2006/XXX (2006) http://eprint.iacr.org/2006/XXX.
31. Goldwasser, S., Micali, S., Rivest, R.L.: A digital signature scheme secure against adaptive chosen-message attacks. SIAM J. Comput. 17(2) (1988) 281–308
32. Bellare, M., Desai, A., Jokipii, E., Rogaway, P.: A concrete security treatment of symmetric encryption. In: FOCS '97: Proceedings of the 38th Annual Symposium on Foundations of Computer Science (FOCS '97), Washington, DC, USA, IEEE Computer Society (1997) 394
33. Goldreich, O., Goldwasser, S., Micali, S.: How to construct random functions. J. ACM 33(4) (1986) 792–807
34. Li, L., Pan, Z., Zhang, M., Ye, K.: Watermarking subdivision surfaces based on addition property of fourier transform. In: GRAPHITE '04: Proceedings of the 2nd international conference on Computer graphics and interactive techniques in Australasia and South East Asia, New York, NY, USA, ACM Press (2004) 46–49
35. Meerwald, P., Uhl, A.: A survey of wavelet-domain watermarking algorithms. In Wong, P.W., Delp, E.J., eds.: Proceedings of SPIE, Electronic Imaging, Security and Watermarking of Multimedia Contents III. Volume 4314., San Jose, CA, USA, SPIE (January 2001)
36. Ruanaidh, J.J.O., Pereira, S.: A secure robust digital image watermark. In: International Symposium on Advanced Imaging and Network Technologies - Conference on Electronic Imaging: Processing, Printing, and Publishing in Colour. (1998)
37. Moulin, P., Wang, Y.: Improved QIM strategies for gaussian watermarking. In: International Workshop on Digital Watermarking (IWDW '05). (2005)

38. Bas, P.: A quantization watermarking technique robust to linear and non-linear valumetric distortions using a fractal set of floating quantifiers. In: Information Hiding Workshop (IHW) '05. (2005)
39. Martin, V., Chabert, M., Lacaze, B.: A spread spectrum watermarking scheme based on periodic clock changes for digital images. In: Information Hiding Workshop (IHW) '05. (2005)
40. Doerr, G., Dugelay, J.L.: A quantization watermarking technique robust to linear and non-linear valumetric distortions using a fractal set of floating quantifiers. In: Information Hiding Workshop (IHW) '05. (2005)
41. Hartung, F., Girod, B.: Fast public-key watermarking of compressed video. In: International Conference on Image Processing (ICIP'97). Volume I., Santa Barbara, California, U.S.A. (1997) 528–531
42. Micali, S., Peikert, C., Sudan, M., Wilson, D.: Optimal error correction against computationally bounded noise. In: Theory of Cryptography Conference (TCC) '05. (2005) http://theory.lcs.mit.edu/~cpeikert/pubs/mpsw.ps.
43. Barak, B., Goldreich, O., Impagliazzo, R., Rudich, S., Sahai, A., Vadhan, S., Yang, K.: On the (im)possibility of obfuscating programs. In: CRYPTO. (2001)
44. Wong, P.W., Memon, N.: Secret and public key image watermarking schemes for image authentication and ownership verification. IEEE Trans. Image Processing 10(10) (October 2001) 1593–1601
45. Petitcolas, F.A., Anderson, R.J., Kuhn, M.G.: Attacks on copyright marking systems. In: Information Hiding Workshop (IHW) 1998, Springer-Verlag Lecture Notes in Computer Science 1525. (1998) 219–239
46. Petitcolas, F.A.: Watermarking schemes evaluation. IEEE Signal Processing 17(4) (September 2000) 58–64

Private Approximation of Clustering and Vertex Cover*

Amos Beimel, Renen Hallak, and Kobbi Nissim

Department of Computer Science, Ben-Gurion University of the Negev

Abstract. Private approximation of search problems deals with finding approximate solutions to search problems while disclosing as little information as possible. The focus of this work is on private approximation of the vertex cover problem and two well studied clustering problems – k-center and k-median. Vertex cover was considered in [Beimel, Carmi, Nissim, and Weinreb, *STOC*, 2006] and we improve their infeasibility results. Clustering algorithms are frequently applied to sensitive data, and hence are of interest in the contexts of secure computation and private approximation. We show that these problems do not admit private approximations, or even approximation algorithms that leak significant number of bits. For the vertex cover problem we show a tight infeasibility result: every algorithm that $\rho(n)$-approximates vertex-cover must leak $\Omega(n/\rho(n))$ bits (where n is the number of vertices in the graph). For the clustering problems we prove that even approximation algorithms with a poor approximation ratio must leak $\Omega(n)$ bits (where n is the number of points in the instance). For these results we develop new proof techniques, which are more simple and intuitive than those in Beimel et al., and yet allow stronger infeasibility results. Our proofs rely on the hardness of the promise problem where a unique optimal solution exists [Valiant and Vazirani, *Theoretical Computer Science*, 1986], on the hardness of approximating witnesses for NP-hard problems ([Kumar and Sivakumar, *CCC*, 1999] and [Feige, Langberg, and Nissim, *APPROX*, 2000]), and on a simple random embedding of instances into bigger instances.

1 Introduction

In secure multiparty computation two or more parties wish to perform a computation over their joint data without leaking any other information. By the general feasibility results of [22,8,2], this task is well defined and completely solved for polynomial time computable functions. When what the parties wish to compute is *not a function*, or infeasible to compute (or both) one cannot directly apply the feasibility results, and special care has to be taken in choosing the function that

---

* Research partially supported by the Israel Science Foundation (grant No. 860/06), and by the Frankel Center for Computer Science. Research partly done when the first and third authors were at the Institute for Pure and Applied Mathematics, UCLA.

S.P. Vadhan (Ed.): TCC 2007, LNCS 4392, pp. 383–403, 2007.

is computed securely, as the outcome of the secure computation may leak information. We deal with such problems – vertex-cover and clustering that are NP-complete problems – and check the consequences of choosing to compute *private approximations* for these search problems, i.e., approximation algorithms that do not leak more information than the collection of solutions for the specific instance.

The notion of private approximation was first put forward and researched in the context of approximating *functions* [6,10], and was recently extended to search problems [1]. These works also consider relaxations of private approximations, which allow for a bounded leakage. The research of private approximations yielded mixed results: (i) private approximation algorithms or algorithms that leak very little were presented for well studied problems [6,10,7,15,13,1], but (ii) it was shown that some natural functions do not admit private approximations, unless some (small) leakage is allowed [10]; and some search problems do not even admit approximation algorithms with significant leakage [1]. We continue the later line of research and prove that vertex-cover and two clustering problems – k-center and k-median – do not admit private approximation algorithms, or even approximation algorithms that leak significant number of bits.

1.1   Previous Works

Feigenbaum et al. [6] noted that an approximation to a function may reveal information on the instance that is not revealed by the exact (or optimal) function outcome. Hence, they formulated, , via the simulation paradigm, a notion of private approximations that prevents exactly this leakage. Their definition implies that if applied to instances x, y such that $f(x) = f(y)$, the outcome of an approximation algorithm $\hat{f}(x), \hat{f}(y)$ are indistinguishable. Under their definition of private approximations, Feigenbaum et al. provided a protocol for approximating the Hamming distance of two n-bit strings with communication complexity $\tilde{O}(\sqrt{n})$, and polynomial solutions for approximating the permanent and other natural #P problems. Subsequent work on private approximations improved the communication complexity for the Hamming distance to polylog(n) [13]. Other works on private approximations for specific functions include [15,7].

Attempts to constructs private approximations of the objective functions of certain NP-complete problems were unsuccessful. This phenomenon was explained by Halevi, Krauthgamer, Kushilevitz, and Nissim [10] proving strong inapproximability results for computing the size of a minimum vertex cover even within approximation ratio $n^{1-\epsilon}$. They, therefore, presented a relaxation, allowing the leakage of a deterministic predicate of the input. Fortunately, this slight compromise in privacy allows fairly good approximations for any problem that admits a good deterministic approximation. For example, minimum vertex cover may be approximated within a ratio of 4 leaking just one bit of approximation.

Recently, Beimel, Carmi, Nissim, and Weinreb [1] extended the privacy requirement of [6] from *functions* to *search problems*, giving a (seemingly) lenient definition which only allows leaking whatever is implied by the set of all exact solutions to the problem. A little more formally, if applied to instances x, y that share *exactly* the same set of (optimal) solutions, the outcome of the approximation algorithm $\mathcal{A}(x)$ on x should be indistinguishable from $\mathcal{A}(y)$. They showed that even under this definition it is not feasible to privately approximate the search problems of vertex-cover and 3SAT. Adopting the relaxation of [10] to the context of private search, Beimel et al. showed for max exact 3SAT an approximation algorithm with a near optimal approximation ratio of $7/8 - \epsilon$ that leaks only $O(\log \log n)$ bits. For vertex-cover, the improvement is more modest – there exists an approximation algorithm within ratio $\rho(n)$ that leaks $\ell(n)$ bits where $\rho(n) \cdot \ell(n) = 2n$. On the other hand, they proved that an algorithm for vertex-cover that leaks $O(\log n)$ bits cannot achieve n^ϵ approximation. We close this gap up to constant factors. A different relaxation of private approximation was presented in the context of near neighbor search by Indyk and Woodruff [13], and we refer to a generalization of this relaxation in Section 4.

1.2   Our Contributions

The main part of this work investigates how the notion of private approximations and its variants combine with well studied NP-complete *search* problems – vertex-cover, k-center, and k-median. We give strong infeasibility results for these problems that hold with respect to a more lenient privacy definition than in [1] – that only requires that $\mathcal{A}(x)$ is indistinguishable from $\mathcal{A}(y)$ on instances x, y that have the same *unique* solution. To prove our results, we introduce new strong techniques for proving the infeasibility of private approximations, even with many bits of leakage.

Vertex Cover. As noted above, the feasibility of private approximation of vertex-cover was researched in [1]. Their analysis left an exponential gap between the infeasibility and feasibility results. We close this gap, and show that, unless RP = NP, any approximation algorithm that leaks at most $\ell(n)$ bits of information and is within approximation ratio $\rho(n)$ satisfies $\rho(n) \cdot \ell(n) = \Omega(n)$. This result is tight (up to constant factors) by a result described in [1]: for every constant $\epsilon > 0$, there is an $n^{1-\epsilon}$-approximation algorithm for vertex-cover that leaks $2n^\epsilon$ bits.

Clustering. Clustering is the problem of partitioning n data points into disjoint sets in order to minimize a cost objective related to the distances within each point set. Variants of clustering are the focus of much research in data mining and machine learning as well as pattern recognition, image analysis, and bioinformatics. We consider two variants: (i) k-center, where the cost of a clustering is taken to be the longest distance of a point from its closest center; and (ii) k-median, where the cost is taken to be the average distance of points from their closest centers. Both problems are NP-complete [12,14,18]. Furthermore, we consider two versions of each problem, the one outputting the indices of the centers and the second outputting the coordinates of the solutions. For private algorithms these two versions are not equivalent since different information can be learned from the output.

We prove that, unless RP = NP, every approximation algorithm for the indices version of these problems must leak $\Omega(n)$ bits even if its approximation ratio as poor as $2^{\text{poly}(n)}$. As there is a 2-approximation algorithm that leaks at most n bits (the incidence vector of the set of centers), our result is tight up to a constant factor. Similar results are proved in the full version of the paper for the coordinate version of these problems (using a "perturbable" property of the metric).

Trying to get around the impossibility results, we examine a generalization of a privacy definition by Indyk and Woodruff [13], originally presented in the context of near neighbor search. In the modified definition, the approximation algorithm is allowed to leak the set of η-approximated solutions to an instance for a given η. We consider the coordinate version of k-center, and show that there exists a private 2-approximation under this definition for every $\eta \geq 2$, and there is no approximation algorithm under this definition when $\eta < 2$.

New Techniques. The basic idea of our infeasibility proofs is to assume that there exists an efficient private approximation algorithm $\mathcal{A}$ for some NP-complete problem, and use this algorithm to efficiently find an *optimal* solution of the problem contradicting the NP-hardness of the problem. Specifically, in our proofs we take an instance x of the NP-complete problem, transform it to a new instance x', execute $y' \leftarrow \mathcal{A}(x')$ once getting an *approximate solution* for x', and then efficiently reconstruct from y' an *optimal solution* for x. Thus, we construct a Karp-reduction from the original NP-complete problem to the private approximation version of the problem. This should be compared to the reduction in [1] which used many calls to $\mathcal{A}$, where the inputs to $\mathcal{A}$ are chosen adaptively, according to the previous answers of $\mathcal{A}$.

Our techniques differ significantly from those of [1], and are very intuitive and rather simple. The main difference is that we deal with the promise versions of vertex cover and clustering, where a *unique* optimal solution exists. These problems are also NP-hard under randomized reductions [21]. Analyzing how a private approximation algorithms operate on instances of the promise problem, we clearly identify a source for hardness in an attempt to create such an algorithm – it, essentially, has to output the optimal solution. Furthermore, proving the infeasibility result for instances of the unique problems shows that hardness of private approximation stems from instances we are trying to approximate a "function" – given an instance the function returns its unique optimal solution. Thus, our impossibility results are for inputs with unique solutions where the privacy requirement is even more minimal than the definition of [1].

To get our strongest impossibility results, we use the results of Kumar and Sivakumar [16] and Feige, Langberg, and Nissim [5] that, for many NP-complete problems, it is NP-hard to approximate the witnesses (that is, viewing a witness and an approximation as sets, we require that their symmetric difference is small). These results embed a redundant encoding of the optimal solution, so that seeing a "noisy" version of the optimal solution allows recovering it. In our infeasibility proofs, we assume that there exists an approximation algorithm $\mathcal{A}$ for some unique problem, and use this algorithm to find a solution close to the optimal solution. Thus, the NP-hardness results of [16,5] imply that such efficient algorithm $\mathcal{A}$ cannot exist.

Our last technique is a simple random embedding of an instance into a bigger instance. Let us demonstrate this idea for the unique-vertex-cover problem. In this case, we take a graph, add polynomially many isolated vertices, and then randomly permute the names of the vertices. We assume that there exists a private approximation algorithm $\mathcal{A}$ for vertex-cover and we execute $\mathcal{A}$ on the bigger instance. We show that, with high probability, the only vertices from the original graph that appear in the output of $\mathcal{A}$ are the vertices of the unique vertex cover of the original graph. The intuition behind this phenomenon is that, by the privacy requirement, $\mathcal{A}$ has to give the same answer for many instances generated by different random permutations of the names, hence, if a vertex is in the answer of $\mathcal{A}$, then with high probability it corresponds to an isolated vertex.

Organization. Section 2 contains the main definitions used in this paper and essential background. Section 3 includes our impossibility result for almost private algorithms for the index version of k-center, based on the hardness of unique-k-center. Section 4 discusses an alternative definition of private approximation of the coordinate version of k-center, and contains possibility and impossibility results for this definition. Section 5 describes our impossibility result for almost private algorithms for vertex-cover. Finally, Section 6 discusses some questions arising from our work.

2   Preliminaries

In this section we give definitions and background needed for this paper. We start with the definitions of private search algorithms from [1]. Thereafter, we discuss the problems we focus on: the clustering problems – k-center and k-median – and vertex cover. We then define a simple property of the underlying metrics that will allow us to present our results in a metric independent manner. Finally, we discuss two tools we use to prove infeasibility results: (1) hardness of unique problems and parsimonious reductions, and (2) error correcting reductions.

2.1   Private Approximation of Search Problems

Beimel et al. [1] define the privacy of search algorithms with respect to some underlying privacy structure $\mathcal{R} \subseteq \{0,1\}^* \times \{0,1\}^*$ that is an equivalence relation on instances. The notation $x \equiv_{\mathcal{R}} y$ denotes $\langle x, y \rangle \in \mathcal{R}$. The equivalence relation determines which instances should not be told apart by a private search algorithm $\mathcal{A}$:

Definition 1 (Private Search Algorithm [1]). *Let $\mathcal{R}$ be a privacy structure. A probabilistic polynomial time algorithm $\mathcal{A}$ is private with respect to $\mathcal{R}$ if for every polynomial-time algorithm $\mathcal{D}$ and for every positive polynomial $p(\cdot)$, there exists some $n_0 \in \mathbb{N}$ such that for every $x, y \in \{0,1\}^*$ such that $x \equiv_{\mathcal{R}} y$ and $|x| = |y| \geq n_0$*

$$\left| \Pr[\mathcal{D}(\mathcal{A}(x), x, y) = 1] - \Pr[\mathcal{D}(\mathcal{A}(y), x, y) = 1] \right| \leq \frac{1}{p(|x|)},$$

where the probabilities are taken over the random choices of $\mathcal{A}$ and $\mathcal{D}$.

For every search problem, a related privacy structure is defined in [1], where two inputs are equivalent if they have the same set of optimal solutions. In Section 2.2 we give the specific definitions for the problems we consider.

We will also use the relaxed version of Definition 1 that allows a (bounded) leakage. An equivalence relation $\mathcal{R}'$ is said to ℓ-refine an equivalence relation $\mathcal{R}$ if $\mathcal{R}' \subseteq \mathcal{R}$ and every equivalence class of $\mathcal{R}$ is a union of at most 2^ℓ equivalence classes of $\mathcal{R}'$.

Definition 2 ([1]). *Let $\mathcal{R}$ be a privacy structure. A probabilistic polynomial time algorithm $\mathcal{A}$ leaks at most ℓ bits with respect to $\mathcal{R}$ if there exists a privacy structure $\mathcal{R}'$ such that (i) $\mathcal{R}'$ is a ℓ-refinement of $\mathcal{R}$, and (ii) $\mathcal{A}$ is private with respect to $\mathcal{R}'$.*

2.2 k-Center and k-Median Clustering

The k-center and k-median clustering problems are well researched problems, both known to be NP-complete [12,14,18]. In both problems, the input is a collection P of points in some metric space and a parameter c. The output is a collection of c of the points in P – the cluster centers – specified by their indices or by their coordinates. The partition into clusters follows by assigning each point to its closest center (breaking ties arbitrarily). The difference between k-center and k-median is in the cost function: in k-center the cost is taken to be the maximum distance of a point in P from its nearest center; in k-median it is taken to be the average distance of points from their closest centers. For private algorithms, the choice of outputting indices or coordinates may be significant (different information can be learned from each), and hence we define two versions of each problem.

Definition 3 (k-center – outputting indices (k-center-I)). *Given a set $P = \{p_1, \ldots, p_n\}$ of n points in a metric space and a parameter c, return the indices of c cluster centers $I = \{i_1, \ldots, i_c\}$ that minimize the maximum cluster radius.*

Definition 4 (k-center – outputting coordinates (k-center-C)). *Given a set $P = \{p_1, p_2, \ldots, p_n\}$ of n points in a metric space and a parameter c, return the coordinates of c cluster centers $C = \{p_{i_1}, \ldots, p_{i_c}\}$ that minimize the maximum cluster radius $(C \subseteq P)$.*[1]

The k-median-I and k-median-C problems are defined analogously.

Theorem 1 ([12,14,18]). *In a general metric space, k-center (k-median) is NP-hard. Furthermore, the problem of finding a $(2-\epsilon)$-approximation of k-center in a general metric space is NP-hard for every $\epsilon > 0$.*

Proof (sketch): The reduction is from dominating set. Given a graph $G = (V, E)$, transform each vertex $v \in V$ to a point $p \in P$. For every two points

---

[1] We do not consider versions of the problem where the centers do not need to be points in P.

$p_1, p_2 \in P$ let $\text{dist}(p_1, p_2) = 1$ if $(v_1, v_2) \in E$, otherwise $\text{dist}(p_1, p_2) = 2$. As the distances are 1 and 2, they satisfy the triangle inequality. There is a dominating set of size c in G iff there is a k-center clustering of size c and cost 1 (k-median clustering of cost $\frac{n-c}{n}$) in P. Furthermore, every solution to k-center with cost less than 2 in the constructed instance has cost 1, which implies the hardness of $(2 - \epsilon)$-approximation for k-center.

There is a greedy 2-approximation algorithm for k-center [9,11]: select a first center arbitrarily, and iteratively selects the other $c - 1$ points each time maximizing the distance to the previously selected centers. We will make use of the above reduction, as well as the 2-approximation algorithm for this problem, in the sequel.

We next define the privacy structures related to k-center. Only instances $(P_1, c_1), (P_2, c_2)$ were $|P_1| = |P_2|$ and $c_1 = c_2$ are equivalent, provided they satisfy the following conditions:

Definition 5. *Let P_1, P_2 be sets of n points and $c < n$ a parameter determining the number of cluster centers.*

- *Instances (P_1, c) and (P_2, c) are equivalent under the relation $\mathcal{R}_{k\text{-center-I}}$ if for every set $I = \{i_1, \ldots, i_c\}$ of c point indices, I minimizes the maximum cluster radius for (P_1, c) iff it minimizes the maximum cluster radius for (P_2, c).*
- *Instances (P_1, c) and (P_2, c) are equivalent under the relation $\mathcal{R}_{k\text{-center-C}}$ if (i) for every set $C \subseteq P_1$ of c points, if C minimizes the maximum cluster radius for (P_1, c) then $C \subseteq P_2$ and it minimizes the maximum cluster radius for (P_2, c); and similarly (ii) for every set $C \subseteq P_2$ of c points, if C minimizes the maximum cluster radius for (P_2, c) then $C \subseteq P_1$ and it minimizes the maximum cluster radius for (P_1, c)*

Definition 6 (Private Approximation of k-center). *A randomized algorithm $\mathcal{A}$ is a private $\rho(n)$-approximation algorithm for k-center-I (respectively k-center-C) if: (i) the algorithm $\mathcal{A}$ is a $\rho(n)$-approximation algorithm for k-center, that is, for every instance (P, c) with n points, it returns a solution – a set of c points – such that the expected cluster radius of the solution is at most $\rho(n)$ times the radius of the optimal solution of (P, c). (ii) $\mathcal{A}$ is private with respect to $\mathcal{R}_{k\text{-center-I}}$ (respectively k-center-C).*

The definitions for vertex-cover are analogous and can be found in [1].

2.3   Distance Metric Spaces

In the infeasibility results for clustering problems we use a simple property of the metric spaces, which we state below. This allows us to keep the results general and metric independent. One should be aware that clustering problems may have varying degrees of difficulty depending on the underlying metric used. Our impossibility results will show that unique-k-center and unique-k-median may be exactly solved in *randomized polynomial time* if private algorithms for these problems exist. When using metric spaces for which the problems are NP-hard, this implies RP = NP.

The property states that given a collection of points, it is possible to add to it new points that are "far away":

Definition 7 (Expandable Metric). *Let $\mathcal{M}$ be a family of metric spaces. A family of metric spaces $\mathcal{M}$ is (ρ, m)-expandable if there exists an algorithm* EXPAND *that given a metric $M = \langle P, \text{dist} \rangle \in \mathcal{M}$, where $P = \{p_1, \ldots, p_n\}$, runs in time polynomial in n, m, and the description of M, and outputs a metric $M' = \langle P', \text{dist}' \rangle \in \mathcal{M}$, where $P' = \{p_1, \ldots, p_n, p_{n+1}, \ldots, p_{n+m}\}$, such that*

- $\text{dist}'(p_i, p_j) = \text{dist}(p_i, p_j)$ *for every $i, j \in [n]$, and*
- $\text{dist}'(p_i, p_j) \geq \rho d$ *for all $n < i \leq n + m$ and $1 \leq j < i$, where $d = \max_{i,j \in [n]}(\text{dist}(p_i, p_j))$ is the maximum distance within the original n points.*

General Metric Spaces. Given a connected undirected graph $G = (V, E)$ where every edge $e \in E$ has a positive length $w(e)$, define the metric induced by G whose points are the vertices and $\text{dist}_G(u, v)$ is the length of the shortest path in G between u and v. The family $\mathcal{M}$ of general metric spaces is the family of all metric spaces induced by graphs. This family is expandable: Given a graph G, we construct a new graph G' by adding to G a path of m new vertices connected to an arbitrary vertex, where the length of every new edge is $\rho(n) \cdot d$. The metric induced by G' is the desired expansion of the metric induced by G. The expansion algorithm is polynomial when $\rho(n)$ is bounded by $2^{\text{poly}(n)}$.

Observation 1. *Let $\rho(n) = 2^{\text{poly}(n)}$. The family of general metric spaces is $(\rho(n), m)$-expandable for every m.*

Similarly, the family of metric spaces induced by a finite set of points in the plain with Euclidean distance is expandable.

2.4   Parsimonious Reductions and Unique Problems

Parsimonious reductions are reductions that preserve the number of solutions. It was observed that among the well known NP-complete problems, such reductions can be found [3,19,20]. Indeed, one can easily show that such reductions also exist for our problems:

Lemma 1. *SAT and 3-SAT are parsimoniously reducible to the vertex-cover, k-center, and k-median problems (the general metric version).*

The existence of such parsimonious reductions allows us to base our negative results on a promise version of the problems – where only a unique optimal solution exists. We use the results of Valiant and Vazirani [21] that the promise version unique-SAT is NP-hard under randomized reductions. Therefore, if there exists a parsimonious reduction from SAT to an NP-complete (search) problem $\mathcal{S}$, then its promise version unique-$\mathcal{S}$ is NP-hard under randomized reductions.

Corollary 1. Vertex-cover, unique-k-center, *and* unique-k-median *(general metric version) are NP-hard under randomized reductions.*

2.5 Error Correcting Reductions

An important tool in our proofs are error correcting reductions – reductions that encode, in a redundant manner, the witness for one NP-complete problem inside the witness for another. Such reductions were shown by Kumar and Sivakumar [16] and Feige, Langberg, and Nissim [5] – proving that for certain NP-complete problems it is hard to approximate witnesses (that is, when viewed as sets, the symmetric difference between the approximation and a witness is small). For example, such result is proved in [5] for vertex-cover. We observe that the proof in [5] applies to unique-vertex-cover and we present a similar result for unique-k-center and unique-k-median. We start by describing the result of [5] for unique-vertex-cover.

Definition 8 (Close to a minimum vertex cover). *A set S is δ-close to a minimum vertex cover of G if there exists a minimum vertex cover C of G such that $|S \triangle C| \leq (1 - \delta)n$.*

Theorem 2 ([21,5]). *If* RP $\neq$ NP, *then for every constant $\delta > 1/2$ there is no efficient algorithm that, on input a graph G and an integer t where G has a unique vertex cover of size t, returns a set S that is δ-close to the minimum vertex cover of G.*

We next describe the result for unique-k-center.

Definition 9 (Close to an optimal solution of unique-k-center). *A set S is δ-close an optimal solution of an instance (P, c) of unique-k-center if there exists an optimal solution I of (P, c) such that $|S \triangle I| \leq (1 - \delta)n$.*

Theorem 3. *If* RP $\neq$ NP, *then, for every constant $\delta > 2/3$, there is no efficient algorithm that for every instance (P, c) of unique-k-center finds a set δ-close to the optimal solution of (P, c). The same result holds for instances of unique-k-median.*

The proof technique of Theorem 3 is similar to the proofs in [5]. The proof is described in the full version of this paper.

3 Infeasibility of Almost Private Approximation of Clustering

In this section, we prove that if RP $\neq$ NP, then every approximation algorithm for the clustering problems is not private (and, in fact, must leak $\Omega(n)$ bits). We will give a complete treatment for k-center-I (assuming the underlying metric is expandable according to Definition 7). The modifications needed for

k-median-I are small. The proof for k-center-C and k-median-C are different and use a "perturbable" property of the metric. The proofs for the 3 latter problems appear in the full version of this paper. We will start our proof for k-center-I by describing the infeasibility result for private algorithms, and then we consider deterministic almost private algorithms. The infeasibility result for randomized almost private algorithms appears in the full version of this paper.

3.1 Infeasibility of Private Approximation of Clustering Problems

In this section, we demonstrate that the existence of a private approximation algorithm for k-center-I implies that unique-k-center is in RP. Using the hardness of the promise version unique-k-center, we get our infeasibility result.

We will now show that any private $\rho(n)$-approximation algorithm must essentially return all the points in the unique solution of an instance. We use the fact that the underlying metric is $(2n \cdot \rho(n+1), 1)$-*expandable*. Given an instance $(P, c) = (\{p_1, \ldots, p_n\}, c)$ for k-center-I we use Algorithm EXPAND with parameters $(2n \cdot \rho(n+1), 1)$ to create an instance $(P', c+1)$ by adding the point p^∞ returned by EXPAND, i.e. $p_{n+1} = p^\infty$ and $\mathrm{dist}'(p_i, p^\infty) \geq \rho(n+1) \cdot d$. Any optimal solution I' for $(P', c+1)$ includes the new point p^∞ (if $p^\infty \notin I'$ then this solution's cost is at least $2n \cdot \rho(n+1) \cdot d$ whereas if $p^\infty \in I'$ the cost is at most d). Hence, the unique optimal solution I' consists of the optimal solution I for (P, c) plus the index $n+1$ of the point p^∞.

Lemma 2. *Let $\mathcal{A}$ be a private $\rho(n)$-approximation algorithm for k-center-I, let (P, c) be an instance of k-center-I and construct $(P', c+1)$ as above. Then*

$$\Pr[\mathcal{A}(P', c+1) \text{ returns the indices of all critical points of } (P, c)] \geq 1/3 .$$

The probability is taken over the random coins of algorithm $\mathcal{A}$.

Proof. Let $p_{i_1}, \ldots, p_{i_c}$ be the points of the unique optimal solution of (P, c) (hence $p_{i_1}, \ldots, p_{i_c}, p_{n+1}$ are the points of the unique optimal solution of $(P', c+1)$). Consider an instance $(P'', c+1)$ where P'' is identical to P', except for the points p_{i_1} and p^∞ whose indices (i_i and $n+1$) are swapped.[2] As both p_{i_1} and p^∞ are the optimal solution in P', swapping them does not change the optimal solution, and hence $(P'', c+1) \equiv_{\mathcal{R}_{k\text{-center-I}}} (P', c+1)$.

Let $\tilde{I}'$ and $\tilde{I}''$ denote the random variables $\mathcal{A}(P', c+1)$ and $\mathcal{A}(P'', c+1)$ respectively. Note that the optimal cost of $(P'', c+1)$ is bounded by d. Whereas if $i_1 \notin \tilde{I}''$ we get a clustering cost of $2n \cdot \rho(n+1) \cdot d$. Hence, if $\Pr[i_1 \notin \tilde{I}''] > 1/(2n)$ algorithm $\mathcal{A}$ cannot maintain an approximation ratio of $\rho(n+1)$. This implies that $\Pr[i_1 \notin \tilde{I}'] < 2/(3n)$, otherwise, it is easy to construct a polynomial time procedure that would distinguish $(\tilde{I}', P', P'')$ from $(\tilde{I}'', P', P'')$ with advantage $\Omega(1/n)$. A similar argument holds for indices $i_2, \ldots, i_c$.

To conclude the proof, we use the union bound and get that $\Pr[\{i_1, \ldots, i_m\} \subset \tilde{I}'] \geq 1 - 2c/3n \geq 1/3$. □

---

[2] Note that while P' can be efficiently constructed from P, the construction of P'' is only a thought experiment.

We now get our infeasibility result:

Theorem 4. *Let $\rho(n) \leq 2^{\mathrm{poly}(n)}$. The k-center-I problem does not admit a polynomial time private $\rho(n)$-approximation unless unique-k-center can be solved in probabilistic polynomial time.*

Proof. Let $\mathcal{A}$ be a polynomial time private $\rho(n)$-approximation for k-center-I. Let $(P, c) = (\{p_1, \ldots, p_n\}, c)$ be an instance of unique-k-center and let I be the indices of the centers in its unique solution. Construct the instance $(P', c+1)$ as above by adding the point $p_{n+1} = p^{\infty}$. As $\rho(n) \leq 2^{\mathrm{poly}(n)}$, constructing P' using Algorithm EXPAND is efficient. By Lemma 2, $\mathcal{A}(P')$ includes every index in I with probability at least $1/3$. With high probability, $\mathcal{A}(P', c+1)$ contains exactly c points from P, and the set $\mathcal{A}(P') \setminus \{n+1\}$ is the unique optimal solution for (P, c). □

Combining Theorem 4 with Corollary 1 we get:

Corollary 2. *Let $\rho(n) \leq 2^{\mathrm{poly}(n)}$. The k-center-I problem (general metric version) cannot be privately $\rho(n)$-approximated in polynomial time unless $\mathrm{RP} \neq \mathrm{NP}$.*

3.2 Infeasibility of Deterministic Approximation of Clustering Problems That Leaks Many Bits

In this section we prove that even if $\mathrm{RP} \neq \mathrm{NP}$, then for every $\rho(n) \leq 2^{\mathrm{poly}(n)}$ there is no efficient *deterministic* $\rho(n)$-approximation algorithm of k-center-I that leaks $0.015n$ bits (as in Definition 2).[3] As in the previous section, we assume the underlying distance metric is expandable. To prove the infeasibility of almost private approximation of k-center-I, we assume towards contradiction that there exists an efficient deterministic $\rho(n)$-approximation algorithm $\mathcal{A}$ that leaks $0.015n$ bits. We use this algorithm to find a set close to the solution of a unique-k-center instance.

In the proof of the infeasibility result for private algorithms, described in Section 3.1, we started with an instance P of unique-k-center and generated a new instance P' by adding to P a "far" point. We considered an instance P'' that is equivalent to P' and argued that, since the instances are equivalent, a deterministic private algorithm must return the same output on the two instances. For almost private algorithms, we cannot use the same proof. Although the instances P' and P'' are equivalent, even an algorithm that leaks one bit can give different answers on P' and P''.

The first idea to overcome this problem is to add linearly many new "far" points (using Algorithm EXPAND). Thus, any deterministic approximation algorithm must return all "far" points and a subset of the original points. However, there is no guarantee that this subset is the optimal solution to the original instance. The second idea is using a random renaming of the indices of the instance. We will prove that with high probability (over the random choice of the renaming), the output of the almost private algorithm is close to the optimal solution

---

[3] Throughout this paper, constants are shamelessly not optimized.

of unique-k-center. This contradicts the NP-hardness, described in Section 2.5, of finding a set close to the exact solution for unique-k-center instances.

We next formally define the construction of adding "far" points and permuting the names. Given an instance (P, c) of unique-k-center with distance function dist, we use Algorithm EXPAND with parameters $(2 \cdot \rho(10n), 9n)$ to create an instance $(P', 9n + c)$ with distance function dist$'$ by adding $9n$ "far" points. Let $N \stackrel{\text{def}}{=} 10n$ be the number of points in P' and $c' \stackrel{\text{def}}{=} c + 9n$. We next choose a permutation $\pi : [N] \to [N]$ to create a new instance $(P_\pi, 9n + c)$ with distance function dist$_\pi$, where dist$_\pi(p_{\pi(i)}, p_{\pi(j)}) \stackrel{\text{def}}{=}$ dist$'(p_i, p_j)$.

We start with some notation. Let I be the the set of indices of the points in the unique optimal solution for (P, c) and $S \stackrel{\text{def}}{=} [n] \setminus I$ (that is, S is the set of indices of the points in the original instance P not in the optimal solution). Note that $|I| = c$ and $|S| = n - c$. For any set $A \subseteq [N]$, we denote $\pi(A) \stackrel{\text{def}}{=} \{\pi(i) : i \in A\}$. The construction of P_π and the sets S and I are illustrated in Fig. 1.

It is easy to see that an optimal solution I_π for (P_π, c') includes the $9n$ "far" points, that is, $\{p_{\pi(i)} : n + 1 \le i \le 10n\}$ (if not, then this solution's cost is at least $2 \cdot \rho(N) \cdot d$ whereas if $\{\pi(n + 1), \dots, \pi(10n)\} \subset I_\pi$ the cost is at most d). Thus, I_π contains exactly c points from $\{p_{\pi(i)} : 1 \le i \le n\}$ which must be $\pi(I)$. That is, the unique optimal solution I_π of (P_π, c') consists of the indices in $[N] \setminus \pi(S)$.

Observation 2. *Let π_1, π_2 be two permutations such that $\pi_1(S) = \pi_2(S)$. Then, $(P_{\pi_1}, c') \equiv_{\mathcal{R}_{k\text{-center-I}}} (P_{\pi_2}, c')$.*

In Fig. 2 we describe Algorithm CLOSE TO UNIQUE k-CENTER that finds a set close to the unique minimum solution of an instance of unique-k-center assuming the existence of a deterministic $\rho(N)$-approximation algorithm $\mathcal{A}$ for k-center-I that leaks $0.015N$-bits. Notice that in this algorithm we execute the approximation algorithm $\mathcal{A}$ on (P_π, c') – an instance with $N = 10n$ points – hence the approximation ratio of $\mathcal{A}$ (and its leakage) is a function of N.

We next prove that, with high probability, Algorithm CLOSE TO UNIQUE k-CENTER returns a set that is close to the optimal solution. In the *analysis*, we partition the set of permutations $\pi : [N] \to [N]$ to disjoint subsets. We prove that in every subset, with high probability, Algorithm CLOSE TO UNIQUE k-CENTER returns a set that is close to the optimal solution, provided that it chose a permutation in the subset. Specifically, for every $D \subset [N]$, we consider the subset of the permutations π such that $\pi(S) = D$.

In the rest of the proof we fix an instance (P, c) with a unique optimal solution I and define $S \stackrel{\text{def}}{=} [n] \setminus I$. Furthermore, we fix a set $D \subset [N]$ such that $|D| = |S|$ and consider only permutations such that $\pi(S) = D$. (The algorithm does not need know S and D; these sets are used for the analysis.) We prove in Lemma 4 that with high probability $[N] \setminus \mathcal{A}(P_\pi, c')$ is close to D, and we show in Lemma 3 that in this case Algorithm CLOSE TO UNIQUE k-CENTER succeeds.

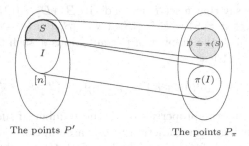

The points P' The points P_π

Fig. 1. The construction of P_π

Algorithm CLOSE TO UNIQUE k-CENTER:
Input: An instance $(P = \{p_1,\ldots,p_n\}, c)$ and an integer t.
Promise: (P,c) has a unique set of c cluster centers with maximum cluster radius at most t.
Output: A set 0.7-close to the unique set of c cluster centers with maximum cluster radius at most t.

1. Use algorithm EXPAND with parameters $(2 \cdot \rho(10n), 9n)$ to create a set of points $P' = \{p_1,\ldots,p_n, p_{n+1},\ldots,p_{10n}\}$.
2. Choose a permutation $\pi : [N] \to [N]$ uniformly at random and construct P_π.
3. Let $B \leftarrow \mathcal{A}(P_\pi, c+9n)$ and $B^{-1} \leftarrow \{i \in [n] : \pi(i) \in B\}$.
4. Return B^{-1}.

Fig. 2. An algorithm that finds a set 0.7-close to the unique minimum solution of an instance of unique-k-center assuming that $\mathcal{A}$ is an almost private approximation algorithm for k-center-I

Lemma 3. *Let B be a set such that $|B \cap D| \leq 0.15n$ and π is a permutation such that $\mathcal{A}(P_\pi, c') = B$. Then, Algorithm* CLOSE TO UNIQUE k-CENTER *returns a set 0.7-close to I when it chooses the permutation π in Step (2).*

Proof. When choosing π, Algorithm CLOSE TO UNIQUE k-CENTER returns the set

$$B^{-1} = \{i \in [n] : \pi(i) \in B\} \; = \; \{i \in I : \pi(i) \in B\} \cup \{i \in S : \pi(i) \in B\}$$
$$= \{i \in I : \pi(i) \in B\} \cup \{i \; : \; \pi(i) \in B \cap D\}.$$

Thus, $|B^{-1} \setminus I| = |B \cap D| \leq 0.15n$. As $|B^{-1}| = |I|$, we get $|I \setminus B^{-1}| = |B^{-1} \setminus I| \leq 0.15n$. Therefore, $|B^{-1} \triangle I| \leq 0.3n$, and B^{-1} is 0.7-close to I. $\qquad\square$

Lemma 4. *Let $pr \overset{\text{def}}{=} \Pr[\,|\mathcal{A}(P_\pi, c') \cap D| \leq 0.15n\,]$, where the probability is taken over the uniform choice of π subject to $\pi(S) = D$. Then, $pr \geq 3/4$.*

Proof. We prove that if $pr < 3/4$, there is a permutation π such that $\mathcal{A}$ does not $\rho(N)$-approximate k-center-I on (P_π, c'), in a contradiction to the definition of $\mathcal{A}$.

In this proof, we say that a set B is "bad" if $|B \cap D| > 0.15n$. The number of permutations such that $\pi(S) = D$ is $(|S|)!(N - |S|)! = (n - c)!(9n + c)!$. As we assumed that pr $< 3/4$, the number of permutations π such that $\pi(S) = D$ and $\mathcal{A}(P_\pi, c')$ is "bad" is at least

$$0.25(n - c)!(9n + c)! \geq (n - c)!\sqrt{n} \left(\tfrac{9n+c}{e}\right)^{9n+c}. \tag{1}$$

We will prove that, by the properties of $\mathcal{A}$, the number of such permutations is much smaller achieving a contradiction to our assumption that pr $< 3/4$.

We first upper bound, for a given "bad" set B, the number of permutations π such that $\pi(S) = D$ and $\mathcal{A}(P_\pi, c') = B$. Notice that the output of the deterministic algorithm $\mathcal{A}(P_\pi, c')$ must contain all points in $\{p_{\pi(i)} : n+1 \leq i \leq 10n\}$ (otherwise the radius of the approximated solution is at least $2 \cdot \rho(N) \cdot d$, compared to at most d when taking all points in $\{p_{\pi(i)} : n + 1 \leq i \leq 10n\}$ and additional c points). Thus, if a permutation π satisfies $\pi(S) = D$ and $\mathcal{A}(P_\pi, c') = B$, then $[N] \setminus B \subset D \cup \pi(I)$, which implies $[N] \setminus (B \cup D) \subset \pi(I)$. Letting $b \stackrel{\text{def}}{=} |B \cap D| \geq 0.15n$,

$$|[N] \setminus (B \cup D)| = N - |B| - |D| + |B \cap D| = 10n - (9n + c) - (n - c) + b = b.$$

Every permutation π satisfying $\pi(S) = D$ and $\mathcal{A}(P_\pi, c') = B$ has a fixed set of size b contained in $\pi(I)$, thus, the number of such permutations is at most

$$(|S|)! \binom{|I|}{b} b!(N - |S| - b)! = (n - c)! \binom{c}{b} b!(9n + c - b)!.$$

Taking $b = 0.15n$ can only increase this expression (as we require that a smaller set is contained in $\pi(I)$). Thus, noting that $c \leq n$, the number of permutations such that $\pi(S) = D$ and $\mathcal{A}(P_\pi, c') = B$ is at most $(n - c)! \binom{n}{0.15n} (0.15n)!(8.85n + c)!$. First, $\binom{n}{0.15n} \leq 2^{H(0.15)n} \leq (16)^{0.15n}$, where $H(0.15) \leq 0.61$ is the Shannon entropy. Thus, using Stirling approximation, the number of such permutations is at most

$$O\left(\sqrt{n}(0.3)^{0.15n}\right) \cdot \left((n - c)!\sqrt{n} \left(\frac{9n + c}{e}\right)^{9n+c}\right). \tag{2}$$

By Observation 2, all instances (P_π, c') for permutations π such that $\pi(S) = D$ are equivalent according to $\mathcal{R}_{k\text{-center-I}}$. Thus, since $\mathcal{A}$ leaks at most $0.015N$ bits, there are at most $2^{0.015N}$ possible answers of $\mathcal{A}$ on these instances, in particular, there are at most $2^{0.015N} = 2^{0.15n}$ "bad" answers. Thus, by (2), the number of permutations such that $\pi(S) = D$ and $\mathcal{A}(P_\pi, c')$ is a "bad" set is at most

$$O\left(2^{0.15n}\sqrt{n}(0.3)^{0.15n}\right) \cdot \left((n - c)!\sqrt{n} \left(\frac{9n + c}{e}\right)^{9n+c}\right) \tag{3}$$

As the number of permutations in (3) is smaller than the number of permutations in (1), we conclude that pr $\geq 3/4$. $\square$

Combining Lemma 3 and Lemma 4, if $\mathcal{A}$ is a $\rho(N)$-approximation algorithm for k-center-I that leaks $0.015N$ bits, then Algorithm CLOSE TO UNIQUE k-CENTER returns a set that is 0.7-close to the optimal solution with probability at least $3/4$, and by Theorem 3, this is impossible unless RP $=$ NP.

In the full version of the paper we show that Algorithm CLOSE TO UNIQUE k-CENTER finds a set close to the optimal solution even when $\mathcal{A}$ is randomize.

Theorem 5. *Let* $\rho(n) \leq 2^{\text{poly}(n)}$. *If* RP $\neq$ NP, *every efficient* $\rho(n)$-*approximation algorithm for* k-center-I *(in the general metric version) must leak* $\Omega(n)$ *bits.*

4 Privacy of Clustering with Respect to the Definition of [13]

Trying to get around the impossibility results, we examine a generalization of a definition by Indyk and Woodruff [13], originally presented in the context of near neighbor search. In the modified definition, the approximation algorithm is allowed to leak the set of approximated solutions to an instance. More formally, we use Definition 1, and set the equivalence relation $\mathcal{R}^\eta$ to include η-approximate solutions as well:

Definition 10. *Let* L *be a minimization problem with cost function* cost. *A solution* w *is an* η-*approximation for* x *if* $\text{cost}_x(w) \leq \eta \cdot \min_{w'}(\text{cost}_x(w'))$. *Let* $\text{appx}(x) \stackrel{\text{def}}{=} \{w : w \text{ is an } \eta\text{-approximation for } x\}$. *Define the equivalence relation* $\mathcal{R}_L^\eta$ *as follows:* $x \equiv_{\mathcal{R}_L^\eta} y$ *iff* $\text{appx}(x) = \text{appx}(y)$.

Note that Definition 10 results in a range of equivalence relations, parameterized by η. When $\eta = 1$ we get the same equivalence relation as before.

We consider the *coordinate* version of k-center. In the full version of this paper we show a threshold at $\eta = 2$ for k-center-C: (1) When $\eta \geq 2$, every approximation algorithm is private with respect to $\mathcal{R}_{k\text{-center-C}}^\eta$. (2) For $\eta < 2$ the problem is as hard as when $\eta = 1$.

5 Infeasibility of Approximation of Vertex Cover That Leaks Information

In [1], it was proven that if RP $\neq$ NP, then for every constant $\epsilon > 0$, every algorithm that $n^{1-\epsilon}$ approximates vertex cover must leak $\Omega(\log n)$ bits. In this paper we strengthen this result showing that if RP $\neq$ NP, then every algorithm that $n^{1-\epsilon}$-approximates vertex cover must leak $\Omega(n^\epsilon)$ bits. We note that this results is nearly tight: In [1], an algorithms that $n^{1-\epsilon}$-approximates vertex cover and leaks $2n^\epsilon$ bits is described. We will describe the infeasibility result in stages. We will start by describing a new proof of the infeasibility of deterministic private approximation of vertex cover, then we will describe the infeasibility of deterministic $n^{1-\epsilon}$-approximation of vertex cover that leaks at most αn^ϵ bits (where $\alpha < 1$ is a specific constant). In the full version of the paper we show the same infeasibility result for randomized algorithms.

5.1   Infeasibility of Deterministic Private Approximation of Vertex Cover

We assume the existence of a deterministic private approximation algorithm for vertex-cover and show that such algorithm implies that RP = NP. The idea of the proof is to start with an instance G of unique-vertex-cover and construct a new graph G_π. First, polynomially many isolated vertices are added to the graph. This means that any approximation algorithm must return a small fraction of the vertices of the graph. Next, the names of the vertices in the graph are randomly permuted. The resulting graph is G_π. Consider two permutations that agree on the mapping of the vertices of the unique-vertex-cover. The two resulting graphs are equivalent and the private algorithm must return the same answer when executed on the two graphs. However, with high probability on the choice of the renaming of the vertices, this answer will contain the (renamed) vertices that consisted the minimum vertex cover in G, some isolated vertices, and no other non-isolated vertices. Thus, given the answer of the private algorithm, we take the non-isolated vertices and these vertices are the unique minimum vertex cover. As unique-vertex-cover is NP-hard [21], we conclude that no deterministic private approximation algorithm for vertex exists (unless RP = NP).

The structure of this proof is similar to the proof of infeasibility of k-center-I, presented in Section 3.2. There are two main differences implied by the characteristics of the problems. First, the size of the set returned by an approximation algorithm for vertex-cover is bigger than the size of the minimum vertex cover as opposed to k-center where the approximation algorithm always returns a set of c centers (whose objective function can be sub-optimal). This results in somewhat different combinatorial arguments used in the proof. Second, it turns out that the roll of the vertices in the unique vertex cover of the graph is similar to the roll of the points *not* in the optimal solution of k-center. For example, we construct a new graph by adding isolated vertices which are not in the minimum vertex cover of the new graph.

We next formally define the construction of adding vertices and permuting the names. Given a graph $G = (V, E)$, where $|V| = n$, an integer $N > n$, and an injection $\pi : V \to [N]$ (that is, $\pi(u) \neq \pi(v)$ for every $u \neq v$), we construct a graph $G_\pi = ([N], E_\pi)$, where $E_\pi = \{(\pi(u), \pi(v)) : (u, v) \in E\}$. That is, the graph G_π is constructed by adding $N - n$ isolated vertices to G and choosing random names for the original n vertices. Throughout this section, the number of vertices in G is denoted by n, and the number of vertices in G_π is denoted by N. We execute the approximation algorithm on G_π, hence its approximation ratio and its leakage are functions of N. Notice that if G has a unique vertex cover C, then G_π has a unique vertex cover $\pi(C) \overset{\text{def}}{=} \{\pi(u) : u \in C\}$. In particular,

Observation 3. *Let G be a graph with a unique minimum vertex cover C, where $k \overset{\text{def}}{=} |C|$, and $\pi_1, \pi_2 : V \to [N]$ be two injections such that $\pi_1(C) = \pi_2(C)$. Then, $(G_{\pi_1}, k) \equiv_{\mathcal{R}_{\mathrm{VC}}} (G_{\pi_2}, k)$.*

In Fig. 3, we describe an algorithm that uses this observation to find the unique minimum vertex cover assuming the existence of a private approximation algorithm for vertex cover. In the next lemma, we prove that Algorithm VERTEX COVER solves the unique-vertex-cover problem.

Algorithm VERTEX COVER:
Input: A Graph $G = (V, E)$ and an integer t.
Promise: G has a unique vertex cover of size t.
Output: The unique vertex cover of G of size t.

1. Let $N \leftarrow (4n)^{2/\epsilon}$.
2. Choose an injection $\pi : V \rightarrow [N]$ uniformly at random and construct the graph G_π.
3. Let $B \leftarrow \mathcal{A}(G_\pi)$ and $B^{-1} \leftarrow \{u \in V : \pi(u) \in B\}$.
4. Return B^{-1}.

Fig. 3. An algorithm that finds the unique minimum vertex cover

Lemma 5. *Let $\epsilon > 0$ be a constant. If $\mathcal{A}$ is a deterministic $N^{1-\epsilon}$-private approximation algorithm for vertex cover and G has a unique vertex cover of size t, then, with probability at least $3/4$, Algorithm VERTEX COVER returns the unique vertex cover of G of size t.*

Proof. First, observe that B^{-1} is a vertex cover of G: For every $(u, v) \in E$ the edge $(\pi(u), \pi(v))$ is in E_π, thus at least one of $\pi(u), \pi(v)$ is in B and at least one of u, v is in B^{-1}. Notice that if $\pi(v) \notin \mathcal{A}(G_\pi)$ for every $v \in V \setminus C$, then Algorithm VERTEX COVER returns the vertex cover C. We will show that the probability of this event is at least $3/4$.

We say that an injection $\pi : V \rightarrow [N]$ *avoids* a set B if $\pi(v) \notin B$ for every $v \in V \setminus C$. See Fig. 4. By Obseration 3, the output B of the deterministic algorithm $\mathcal{A}$ depends only on $\pi(C)$. Thus, it suffices to show that for every possible value of D, the probability that a random injection π such that $\pi(C) = D$ avoids $B = \mathcal{A}(G_\pi)$ is at least $3/4$. As G_π has a cover of size at most n, and $\mathcal{A}$ is an $N^{1-\epsilon}$-approximation algorithm, $|B| \leq nN^{1-\epsilon}$. Thus, since $N = (4n)^{2/\epsilon}$,

$$\Pr[\pi \text{ avoids } B | \pi(C) = D] \geq \prod_{i=1}^{|V|-|C|} \left(1 - \frac{|B|}{N - n}\right) \geq \left(1 - \frac{nN^{1-\epsilon}}{N/2}\right)^n$$

$$= \left(1 - \frac{2n}{N^\epsilon}\right)^n = \left(1 - \frac{1}{8n}\right)^n > \frac{3}{4}.$$

To conclude, the probability that the random π avoids $\mathcal{A}(G_\pi)$ is at least $3/4$. In this case $B^{-1} = C$ (as B^{-1} is a vertex cover of G that does not contain any vertices in $V \setminus C$) and the algorithm succeeds. $\qquad\square$

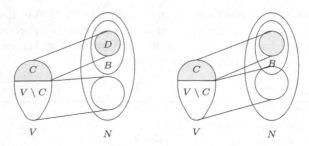

An injection π that avoids B An injection π that does not avoid B

Fig. 4. Injections that avoid and do not avoid the output of $\mathcal{A}$

Infeasibility of leaking $O(\log n)$ bits. Now, assume that Algorithm $\mathcal{A}$ is a deterministic $N^{1-\epsilon}$-approximation algorithm that leaks at most $(\epsilon \log N)/2$ bits. In this case, for every equivalence class of $\equiv_{\mathcal{R}_{\mathrm{VC}}}$, there are at most $2^{(\epsilon \log N)/2} = N^{\epsilon/2}$ possible answers. In particular, for every possible value of D, there are at most $N^{\epsilon/2}$ answers for all graphs G_π such that the injection π satisfies $\pi(C) = D$. If the injection π avoids the union of these answers, then Algorithm VERTEX COVER succeeds for a graph G that has a unique vertex cover of size t. The size of the union of the answers is at most $N^{\epsilon/2} \cdot nN^{1-\epsilon} = nN^{1-\epsilon/2}$, and if we take $N = (4n)^{4/\epsilon}$ in Algorithm VERTEX COVER, then with probability at least $3/4$ the algorithm succeeds for a graph G that has a unique vertex cover of size t. However, we want to go beyond this leakage.

5.2 Infeasibility of Approximation of Vertex Cover That Leaks Many Bits

Our goal is to prove that there exists a constant α such that for every constant $\epsilon > 0$, if RP $\neq$ NP, then there is no efficient algorithm that $N^{1-\epsilon}$-approximates the vertex cover problem while leaking at most $\alpha N^{1-\epsilon}$ bits. This is done by using the results of [16,5] that shows that it is NP-hard to produce a set that is close to a minimal vertex cover as defined in Section 2.5. Using this result, we only need that B^{-1} is close to the minimum vertex cover. We show that, even if $\mathcal{A}$ leaks many bits, for a random injection, the set B^{-1} is close to the minimum vertex cover.

In Fig. 5, we describe Algorithm CLOSE TO UNIQUE k-CENTER that finds a set close to the unique vertex cover of G assuming the existence of a deterministic $N^{1-\epsilon}$-approximation algorithm for vertex cover that leaks αN^ϵ bits. (In the full version of the paper we show how to generalize the analysis to deal with a randomized $N^{1-\epsilon}$-approximation algorithm.) To prove the correctness of the algorithm we need the following definition and lemma.

Definition 11. *Let $C \subset V$ be the unique minimum vertex cover of a graph G, and $\pi : V \to [N]$ be an injection. We say that π δ-avoids a set B if $|\{v \in V \setminus C : \pi(v) \in B\}| \le \delta |V|$.*

Algorithm CLOSE TO VERTEX COVER:
Input: A Graph $G = (V, E)$ and an integer t.
Promise: G has a unique vertex cover of size t.
Output: A set S that is δ-close to the unique vertex cover of G of size t for some constant $\delta > 1/2$.

1. Let $N \leftarrow (100n)^{1/\epsilon}$.
2. Choose a random injection $\pi : V \to [N]$ with uniform distribution and construct the graph G_π.
3. Let $B \leftarrow \mathcal{A}(G_\pi)$ and $B^{-1} \leftarrow \{u \in V : \pi(u) \in B\}$.
4. Return B^{-1}.

Fig. 5. An algorithm that returns a set close to a unique minimum vertex cover

Lemma 6. *Let $\epsilon > 0$ be a constant, and $B \subset [N]$, $D \subset [N]$ be sets, where $|B| \leq nN^{1-\epsilon}$. If $N = (100n)^{1/\epsilon}$ and an injection $\pi : V \to [N]$ is chosen at random with uniform distribution, then $\Pr[\pi$ does not 0.2-avoid $B | \pi(C) = D] \leq e^{-0.2n}$.*

The lemma is proved by using the Chernoff bound noting that the events $\pi(u) \in B$ and $\pi(v) \in B$ are "nearly" independent for $u \neq v$.

Lemma 7. *There exists a constant $\alpha < 1$ such that, for every constant $\epsilon > 0$, if $\mathcal{A}$ is a deterministic $N^{1-\epsilon}$-approximation algorithm for vertex cover that leaks at most αN^ϵ bits, then for every G and t such that G has a unique vertex cover of size t, with probability at least $3/4$, Algorithm CLOSE TO VERTEX COVER returns a set that is 0.6-close to the minimum vertex cover of G.*

Proof (sketch): Let G and t be such that G has a unique vertex cover of size t; denote this vertex cover by C. We fix a set D and consider only injections π such that $\pi(C) = D$. Let $\alpha = 0.002$ and assume that $\mathcal{A}$ leaks at most $\alpha N^\epsilon = 0.2n$ bits (since $N = (100n)^{1/\epsilon}$). By Obseration 3, if we restrict ourself to such injections, then the output of $\mathcal{A}$ has at $2^{0.2n}$ options. Denote these answers by $B_1, \ldots, B_\ell$ for $\ell \leq 2^{0.2n}$. By Lemma 6, for every possible value of B, the probability that a random injection π such that $\pi(C) = D$ does not 0.2-avoid B is at most $e^{-0.2n}$. Thus, by the union bound, the probability that a random injection π such that $\pi(C) = D$ 0.2-avoids $\mathcal{A}(G_\pi)$ is at least $1 - (2/e)^{0.2n} \gg 3/4$. In this case B^{-1} contains at most $0.2n$ vertices not from the minimum vertex cover C. Recall that B^{-1} is a vertex cover of G. Therefore, $|C \setminus B^{-1}| \leq 0.2n$ (as $|B^{-1}| > |C|$ and $|B^{-1} \setminus C| \leq 0.2n$). We conclude that B^{-1} is 0.6-close to a vertex cover of G as claimed. $\qquad\square$

Theorem 6. *There exists a constant $\alpha > 0$ such that, if $\mathrm{RP} \neq \mathrm{NP}$, there is no efficient $N^{1-\epsilon}$-approximation algorithm for vertex cover that leaks αN^ϵ bits.*

6   Discussion

The generic nature of our techniques suggests that, even if the notion of private approximations would be found useful for some NP-complete problems, it would be infeasible for many other problems. Hence, there is a need for alternative formulations of private approximations for search problems.

The definitional framework of [1] allows for such formulations, by choosing the appropriate equivalence relation on input instances. Considering vertex-cover for concreteness, the choice in [1] and the current work was to protect against distinguishing between inputs with the same set of vertex covers. A different choice, that could have been made, is to protect against distinguishing between inputs that have the same lexicographically first maximal matching. (In fact, the latter is feasible and allows a factor 2 approximation).

A different incomparable notion of privacy was pursued in recent work on private data analysis. For example, [4] present a variant on the k-means clustering algorithm that is applied to a database, where each row contains a point corresponding to an individual's information. This algorithm satisfies a privacy definition devised to protect individual information.

Finally, a note about leakage of information as discussed in this work. It is clear that introduction of leakage may be problematic in many applications (to say the least). In particular, leakage is problematic when composing protocols. However, faced by the impossibility results, it is important to understand whether a well defined small amount of leakage can help. For some functionalities allowing a small amount of leakage bypasses an impossibility result – approximating the size of the vertex cover [10], and finding an assignment that satisfies $7/8 - \epsilon$ of the clauses for exact max 3SAT [1]. Unfortunately, this is not the case for the problems discussed in this work.

Acknowledgments. We thank Enav Weinreb and Yuval Ishai for interesting discussions on this subjects and we thank the TCC program committee for their helpful comments.

References

1. A. Beimel, P. Carmi, K. Nissim, and E. Weinreb. Private approximation of search problems. In *Proc. of the 38th STOC*, pages 119–128, 2006.
2. M. Ben-Or, S. Goldwasser, and A. Wigderson. Completeness theorems for non-cryptographic fault-tolerant distributed computations. In *Proc. of the 20th STOC* pages 1–10, 1988.
3. L. Berman and J. Hartmanis. On isomorphisms and density of NP and other complete sets. *SICOMP*, 6:305–322, 1977.
4. A. Blum, C. Dwork, F. McSherry, and K. Nissim. Practical privacy: the SuLQ framework. In *Proc. of the 24th PODS*, pages 128–138, 2005.
5. U. Feige, M. Langberg, and K. Nissim. On the hardness of approximating NP witnesses. In *3rd APPROX*, volume 1913 of *LNCS*, pages 120–131. 2000.
6. J. Feigenbaum, Y. Ishai, T. Malkin, K. Nissim, M. J. Strauss, and R. N. Wright. Secure multiparty computation of approximations. *TALG*, 2(3):435–472, 2006.

7. M. J. Freedman, K. Nissim, and B. Pinkas. Efficient private matching and set intersection. In *EUROCRYPT 2004*, volume 3027 of *LNCS*, pages 1–19. 2004.
8. O. Goldreich, S. Micali, and A. Wigderson. How to play any mental game. In *Proc. of the 19th STOC*, pages 218–229, 1987.
9. T. F. Gonzalez. Clustering to minimize the maximum inter-cluster distance. *TCS*, 38:293–306, 1985.
10. S. Halevi, R. Krauthgamer, E. Kushilevitz, and K. Nissim. Private approximation of NP-hard functions. In *Proc. of the 33th STOC*, pages 550–559, 2001.
11. D. S. Hochbaum and D. B. Shmoys. A unified approach to approximation algorithms for bottleneck problems. *JACM*, 533-550:33, 1986.
12. W. L. Hsu and G. L. Nemhauser. Easy and hard bottleneck location problems. *DAM*, 1:209–216, 1979.
13. P. Indyk and D. Woodruff. Polylogarithmic private approximations and efficient matching. In *TCC 2006*, volume 3876 of *LNCS*, pages 245–264. 2006.
14. O. Kariv and S. L. Hakimi. An algorithmic approach to network location problems, part I: the p-centers. *SIAM J. Appl. Math*, 37:513–538, 1979.
15. E. Kiltz, G. Leander, and J. Malone-Lee. Secure computation of the mean and related statistics. In , *TCC 2005*, volume 3378 of *LNCS*, pages 283–302. 2005.
16. R. Kumar and D. Sivakumar. Proofs, codes, and polynomial-time reducibilities. In *Proc. of the 14th CCC*, pages 46–53, 1999.
17. M. Mitzenmacher and E. Upfal. *Probability and Computing*. Cambridge University Press, 2005.
18. J. Plesnik. On the computational complexity of centers locating in a graph. *Aplikace Matematiky*, 25:445–452, 1980.
19. J. Simon. On the difference between one and many. In *Proc. of the 4th ICALP*, volume 52 of *LNCS*, pages 480–491. 1977.
20. L. G. Valiant. A reduction from satisfiability to Hamiltonian circuits that preserves the number of solutions. Manuscript, Leeds, 1974.
21. L. G. Valiant and V. V. Vazirani. NP is as easy as detecting unique solutions. *TCS*, 47:85–93, 1986.
22. A. C. Yao. Protocols for secure computations. In *Proc. of the 23th FOCS*, pages 160–164, 1982.

Robuster Combiners for Oblivious Transfer

Remo Meier, Bartosz Przydatek, and Jürg Wullschleger

Department of Computer Science, ETH Zurich
8092 Zurich, Switzerland
remmeier@tik.ee.ethz.ch, {przydatek, wjuerg}@inf.ethz.ch

Abstract. A $(k; n)$-*robust combiner* for a primitive $\mathcal{F}$ takes as input n candidate implementations of $\mathcal{F}$ and constructs an implementation of $\mathcal{F}$, which is secure assuming that at least k of the input candidates are secure. Such constructions provide robustness against insecure implementations and wrong assumptions underlying the candidate schemes. In a recent work Harnik *et al.* (Eurocrypt 2005) have proposed a $(2; 3)$-robust combiner for oblivious transfer (OT), and have shown that $(1; 2)$-robust OT-combiners of a certain type are impossible.

In this paper we propose new, generalized notions of combiners for two-party primitives, which capture the fact that in many two-party protocols the security of one of the parties is unconditional, or is based on an assumption independent of the assumption underlying the security of the other party. This fine-grained approach results in OT-combiners *strictly stronger* than the constructions known before. In particular, we propose an OT-combiner which guarantees secure OT even when only one candidate is secure for both parties, and every remaining candidate is flawed for one of the parties. Furthermore, we present an efficient *uniform* OT-combiner, i.e., a single combiner which is secure *simultaneously* for a wide range of candidates' failures. Finally, our definition allows for a very simple impossibility result, which shows that the proposed OT-combiners achieve optimal robustness.

Keywords: robust combiners, oblivious transfer, weak oblivious transfer.

1 Introduction

Many cryptographic schemes are based on unproven assumptions about the difficulty of some computational problems. While there exist assumptions whose validity is supported by decades of research (e.g., factoring or discrete logarithm), many new assumptions offering new possibilities are being proposed in literature, and it is unclear how to decide which assumptions are trustworthy. Therefore, given multiple implementations of some cryptographic primitive, each based on different assumptions, it is often difficult to decide which implementation is the most secure one.

Robust combiners offer a method for coping with such difficulties: they take as input multiple candidate schemes based on various assumptions, and construct a scheme whose security is guaranteed if at least *some* candidates are secure. That

S.P. Vadhan (Ed.): TCC 2007, LNCS 4392, pp. 404–418, 2007.

is, the resulting scheme is secure as long as sufficiently many of the assumptions underlying the input candidates are valid. This provides tolerance against wrong assumptions since even a breakthrough algorithm for breaking one (or some) of the assumptions does not necessarily make the combined scheme insecure.

Actually, the concept of robust combiners is not new, and many constructions of this type have been used in various cryptographic schemes to improve the security guarantees, e.g., cascading of block ciphers. However, a rigorous study of robust combiners was initiated only recently [Her05, HKN+05]. More formally, a $(k; n)$-robust $\mathcal{F}$-combiner is a construction which takes as input n implementations of a primitive $\mathcal{F}$, and yields an implementation of $\mathcal{F}$ which is guaranteed to be secure as long as at least k input implementations are secure. Robust combiners for some primitives, like one-way functions or pseudorandom generators, are rather simple, while for others, e.g., for oblivious transfer (OT), the construction of combiners seems considerably harder. In particular, Harnik *et al.* [HKN+05] show that there exists no "transparent black-box" $(1; 2)$-robust OT-combiner. In the same paper they propose also a very simple and efficient $(2; 3)$-robust OT-combiner.

Contributions. We propose stronger and more general definitions of robust combiners for two-party primitives, which enable a more fine-grained approach to the design of combiners. In particular, the new definitions capture scenarios where in the candidate implementations the security of Alice is based on an assumption different from the assumption underlying Bob's security, or where the security of one party is unconditional. This finer distinction can then be exploited in constructions of combiners.

For this new definition we propose OT-combiners yielding secure OT when the total number of candidates' failures on either side is strictly smaller than the number of candidates. In particular, we propose an OT-combiner which guarantees secure OT even when only one candidate is secure for both parties, and every remaining candidate is insecure for one of the parties. Moreover, we propose also an efficient *uniform* OT-combiner, i.e., a single combiner which is secure *simultaneously* for a wide range of candidates' failures.

We show also that the proposed combiners are optimal in terms of achieved robustness. Specifically, we prove the impossibility of black-box OT-combiners achieving better robustness, and also we show that any (possibly even non-black-box) OT-combiner achieving better robustness would in fact implement OT from scratch. This is in contrast to the impossibility proof from [HKN+05] where only the existence of *transparent* black-box combiners was excluded. However, our impossibility results are not directly comparable with the previous one: on one hand our results are stronger, since they are not limited to the *transparent black-box* combiners, but on the other hand they are weaker, since they exclude a primitive which is stronger then the one considered in [HKN+05] (cf. Section 3).

Finally, since our definition is stronger than the previous definition, all constructions satisfy also the latter, and as a corollary we obtain also tight bounds for the previous definition.

Related work. As mentioned above, there are numerous implicit uses and constructions of combiners in the literature (e.g., [AB81, EG85, MM93, DK05, HL05]), but a more rigorous study of robust combiners was initiated only recently, by Herzberg [Her05] and by Harnik *et al.* [HKN+05], who formalized the notion of combiners, and have shown constructions of combiners for various primitives. Moreover, Harnik *et al.* [HKN+05] have shown also that not all primitives are easy to combine, by proving that there is no *transparent black-box* $(1; 2)$-robust OT-combiner. Boneh and Boyen [BB06] studied the efficiency of combiners for collision resistant hash functions. In [MP06] robust combiners for private information retrieval were proposed, and also *cross-primitive* combiners have been studied. Such combiners can be viewed as generalized reductions between primitives, and their study, in addition to be of practical value, offers insights into relations between cryptographic primitives.

The problem of strengthening imperfect oblivious transfer, which is closely related to OT-combiners (see below), was first considered by [CK88], and has been studied in many subsequent works (e.g.,[BCW03, Cac98, DFSS06]). In particular Damgård, Kilian and Salvail [DKS99] defined the notion of *weak* oblivious transfer (WOT) and provided algorithms for strengthening it. The use of techniques for strengthening WOT in the construction of combiners has been suggested by Harnik *et al.* [HKN+05] as an alternative way of obtaining a $(2; 3)$-robust OT-combiner.

Organization. In the next section we review the primitives used in the rest of the paper, and present generalized definitions of robust combiners for two-party primitives. Then in Section 3 we propose combiners for oblivious transfer tolerating an insecure minority, and prove that they achieve optimal robustness. In Section 4 we exploit the symmetry of oblivious transfer to obtain *uniform* OT-combiners with optimal robustness. Finally, in Section 5 we conclude and discuss some open problems.

2   Preliminaries and Definitions

2.1   Primitives

We review shortly the primitives relevant in this work. For more formal definitions we refer to the literature. The parties participating in the protocols and the adversary are assumed to be probabilistic polynomial time Turing machines, (PPTMs).

Oblivious transfer[1] (OT) is a protocol between a sender holding two bits b_0 and b_1, and a receiver holding a choice-bit c. The protocol allows the receiver to get bit b_c so that the sender does not learn any information about receiver's choice c, and the receiver does not learn any information about bit b_{1-c}.

---

[1] The version of oblivious transfer described here and used in this paper is more precisely denoted as *1-out-of-2 bit-OT* [EGL85]. There are several other versions of OT, e.g., *Rabin's OT*, *1-out-of-n bit-OT*, or *1-out-of-n string-OT*, but all are known to be equivalent [Rab81, Cré87, CK88].

Weak oblivious transfer. ((p,q)-WOT) is an oblivious transfer with relaxed privacy guarantees for the participants [DKS99]: with probability at most p a cheating sender will learn which bit the receiver chose to receive, and with probability q a cheating receiver will learn both of the sender's input bits.

Secret sharing. [Bla79, Sha79] allows a party to distribute a secret among a group of parties, by providing each party with a *share*, such that only authorized subsets of parties can collectively reconstruct the secret from their shares. We say that a sharing among n parties is a ℓ-out-of-n secret sharing, if any ℓ correct shares are sufficient to reconstruct the secret, but any subset of less than ℓ shares gives no information about the secret. A simple method for ℓ-out-of-n secret sharing was proposed by Shamir [Sha79]: a party P having a secret value $s \in \mathbb{F}_q$ where $q > n$, picks a random polynomial $f(x)$ over $\mathbb{F}_q$, such that $f(0) = s$ and the degree of $f(x)$ is (at most) $\ell - 1$. A share for party P_i is then computed as $s_i := f(z_i)$, where $z_1, \ldots, z_n$ are fixed, publicly known, distinct non-zero values from $\mathbb{F}_q$. Since the degree of $f(x)$ is at most $\ell - 1$, any ℓ shares are sufficient to reconstruct $f(x)$ and compute $s = f(0)$ (via Lagrange interpolation). On the other hand, any $\ell - 1$ or fewer shares give no information about s, since they can be consistently completed to yield a sharing of any arbitrary $\overline{s} \in F[q]$, and the number of possible completions is the same for every $\overline{s}$.

Bit Commitment. (BC) is a two-phase protocol between a sender, holding an input bit b, and a receiver, who has no input. In the *commit* phase the sender commits to bit b without revealing it, by sending to the receiver a *commitment to b*, i.e., an "encrypted" representation e of b. Later, in the *decommit* phase, the sender gives to the receiver a decommitment string d, allowing the receiver to "open" e and obtain b. In addition to correctness, a bit commitment scheme must satisfy two properties: *hiding*, i.e., the receiver does not learn the bit b before the decommit phase, and *binding*, i.e., the sender cannot come up with decommitment strings d, d' which lead to opening the commitment as different bits.

2.2   Robust Combiners

In this section we recall definitions of robust combiners, and present some generalizations, which allow for a more fine-grained approach to the design of combiners. These generalizations are motivated by the fact, that in many implementations of cryptographic primitives various security properties are based on different, often independent, computational assumptions, or even hold unconditionally, and cannot be broken. Thus when designing combiners, whose main goal is to protect against wrong assumptions, it can be worthwhile to consider these security guarantees explicitly, as it can potentially lead to more efficient practical constructions (cf. [HKN+05, MP06]). Moreover, the proposed generalizations lead to combiners which are strictly stronger than the constructions known before, and also allow for easier impossibility proofs.

Definition 1 ($(k;n)$-robust $\mathcal{F}$-combiner [HKN⁺05]). *Let $\mathcal{F}$ be a cryptographic primitive. A $(k;n)$-robust $\mathcal{F}$-combiner is a PPTM which gets n candidate schemes implementing $\mathcal{F}$ as inputs and implements $\mathcal{F}$ while satisfying the following two properties:*

1. *If at least k candidates securely implement $\mathcal{F}$, then the combiner securely implements $\mathcal{F}$.*
2. *The running time of the combiner is polynomial in the security parameter κ, in n, and in the lengths of the inputs to $\mathcal{F}$.*[2]

If the primitive for which one wishes to construct a combiner is a two-party primitive between Alice and Bob (like for example OT or bit commitment), we can make a finer characterization of the security required from the candidates. That is, we can distinguish cases when in the candidate implementations the security of Alice is based on an assumption different from the assumption underlying Bob's security, or when the security of one party is unconditional. For such candidates breaking one assumption does not necessarily imply a total loss of security (for both parties) and this property can be exploited for the construction of combiners.

Definition 2 ($(\alpha, \beta; n)$-robust $\mathcal{F}$-combiner). *Let $\mathcal{F}$ be a cryptographic primitive for two parties Alice and Bob. A $(\alpha, \beta; n)$-robust $\mathcal{F}$-combiner is a PPTM which gets n candidate schemes implementing $\mathcal{F}$ as inputs and implements $\mathcal{F}$ while satisfying the following two properties:*

1. *If at least α candidates implement $\mathcal{F}$ securely for Alice, and at least β candidates implement $\mathcal{F}$ securely for Bob, then the combiner securely implements $\mathcal{F}$.*
2. *The running time of the combiner is polynomial in the security parameter κ, in n, and in the lengths of the inputs to $\mathcal{F}$.*

Note that a $(k;n)$-robust combiner is a special case of a $(k, k; n)$-robust combiner, but they are not equivalent. For example, a $(2,2;3)$-robust combiner tolerates input candidates C_1, C_2, C_3, where one C_1 is secure for Alice only, C_2 is secure for Bob only, and C_3 is secure for both parties, while a $(2;3)$-robust combiner can fail for such candidates. In other words, the notion of a $(k, k; n)$-robust combiner (and hence of a $(\alpha, \beta; n)$-robust combiner), is strictly stronger then that of a $(k;n)$-robust combiner, and it provides better security guarantees.

Another difference between $(k;n)$- and $(\alpha, \beta; n)$-robust combiners is that for the new definition it is possible to have "non-uniform" constructions with explicit dependence on α and β. This motivates an even stronger notion of *uniform* combiners. For example, even if there exists a $(\alpha, \beta; n)$-robust combiner for every $\alpha, \beta \geq 0$ satisfying $\alpha + \beta \geq \delta$, where δ is some threshold, it might be the case that the combiner makes explicit use of the values α and β, and thus works differently for every particular pair values (α, β). In such a scenario more desirable would

---

[2] Here an implicit assumption is made, that the candidates themselves run in polynomial time.

be a *uniform* construction, i.e., a single combiner that is secure under the sole assumption that $\alpha + \beta \geq \delta$. In particular, a uniform combiner does not obtain the values of α and β as parameters.

Definition 3 ($\{\delta; n\}$-robust uniform $\mathcal{F}$-combiner). *Let $\mathcal{F}$ be a two-party primitive. We say that an $\mathcal{F}$-combiner is a $\{\delta; n\}$-robust uniform $\mathcal{F}$-combiner if it is a $(\alpha, \beta; n)$-robust $\mathcal{F}$-combiner, simultaneously for all α and β satisfying $\alpha + \beta \geq \delta$.*

Note that the parameter δ is a bound on *the sum* of the number of candidates secure for Alice and the number of candidates secure for Bob, hence given n candidates δ is from the range $0 \ldots 2n$. As an example consider a $\{4; 3\}$-robust *uniform* combiner. Such a combiner is a (regular) $(2; 3)$-robust combiner, but at the same time it is also a $(3, 1; 3)$-robust combiner, i.e., it tolerates input candidates C_1, C_2, C_3, where one C_i is secure for both parties, and the remaining two candidates are secure for Alice only. It is not hard to see that not every $(k; n)$-robust combiner is automatically also a $\{\delta; n\}$-robust *uniform* combiner with $\delta = 2k$. In particular, the $(2; 3)$-robust OT-combiner from [HKN+05] breaks on inputs of the type described above for $(3, 1; 3)$-robust combiner.

For completeness, we recall three more definitions from [HKN+05]. Note that these definitions extend naturally to the generalized combiners from Definitions 2 and 3.

Definition 4 (Black-box combiner [HKN+05]). *A $(1; 2)$-robust combiner is called a* black-box combiner *if the following two conditions hold:*

BLACK-BOX IMPLEMENTATION: *The combiner is an oracle PPTM given access to the candidates via oracle calls to their implementation function.*

BLACK-BOX PROOF: *For every candidate there exists an oracle PPTM R^A (with access to A) such that if adversary A breaks the combiner, then R^A breaks the candidate.*[3]

Definition 5 (Transparent black-box combiner [HKN+05]). *A* transparent black-box combiner *is a black-box combiner for an interactive primitive where every call to a candidate's next message function is followed by this message being sent to the other party.*

Definition 6 (Third-party black-box combiner [HKN+05]). *A* third-party black-box combiner *is a black-box combiner where the input candidates behave like trusted third parties. The candidates give no transcript to the players but rather take their inputs and return outputs.*

Since the primary goal of robust combiners is to protect against wrong assumptions, in our constructions we require that the candidates input to a combiner provide the desired functionality and the underlying assumptions can affect only

---

[3] For $(k; n)$-robust combiners there are at least $n - k + 1$ candidates that can be broken in this manner.

the security properties (e.g. secrecy). This approach is justified by the fact that in cryptographic schemes the security is usually based on some assumptions, while the functionality properties are straightforward and hold unconditionally. Moreover, in some cases a possible way of dealing with unknown implementations of primitives is to test them for the desired functionality, hence, even if the candidate input primitives are given as black-boxes, one can test them before applying a combiner (cf. Section 3.1 in [HKN+05]).

3   OT-Combiner with Secure Majority

The impossibility result for transparent black-box $(1; 2)$-robust OT-combiners [HKN+05] implies directly the impossibility of transparent black-box $(n; 2n)$-robust OT-combiners, as from their existence would follow the existence of transparent black-box $(1; 2)$-robust OT-combiners. Similarly, it implies also the impossibility of transparent black-box $(\alpha, \beta; n)$-robust combiners for $\alpha + \beta \leq n$. However, since $(k, k; n)$-robust combiners are stronger than $(k; n)$-robust combiners, we can show very simple impossibility results, which essentially exclude $(\alpha, \beta; n)$-robust OT-combiners of any type[4], that would work for $\alpha + \beta \leq n$: in Lemma 1 we prove that there are no *black-box* OT-combiners with such robustness, and in Lemma 2 we show that constructing an OT-combiner of *any* type (for $\alpha + \beta \leq n$) is at least as hard as constructing an OT protocol without any assumptions.

As mentioned previously, these results are not directly comparable with the impossibility result from [HKN+05]: on one hand our results are stronger, since they go beyond *transparent black-box* combiners, but on the other hand they are weaker, since they exclude a primitive which is stronger then the one considered in [HKN+05].

Lemma 1. *There does not exist a black-box $(\alpha, \beta; n)$-robust OT-combiner for $\alpha + \beta \leq n$.*

Proof. Assume that such a combiner exists, for some values α, β, and n such that $\alpha + \beta \leq n$. Let OT_1 be the trivial instance of OT where the sender sends both values to the receiver, and let OT_2 be the trivial instance where the receiver sends his choice bit to the sender, who sends the receiver the value of his choice. Observe that OT_1 is information-theoretically secure for the receiver, and OT_2 is information-theoretically secure for the sender.

Consider calling the combiner with input consisting of β instances of OT_1 and $n - \beta \geq \alpha$ instances of OT_2, and let $\overline{\mathsf{OT}}$ denote the resulting OT protocol. By assumption, $\overline{\mathsf{OT}}$ is secure for both parties. Since it is impossible to construct an OT protocol information-theoretically secure for *both* the sender and the receiver, there exists an adversary A (possibly inefficient), which breaks the protocol $\overline{\mathsf{OT}}$. By the definition of a black-box combiner, it follows that given oracle access to A one can break $2n - \alpha - \beta + 1 > n$ "sides" of the candidates. However, since

---

[4] I.e., not only *transparent black-box* combiners.

one side of each candidate is information-theoretically secure, we can break at most n sides. A contradiction.   □

Lemma 2. *Any $(\alpha, \beta; n)$-robust OT-combiner for $\alpha + \beta \leq n$ implies the existence of OT.*

Proof. Assume that such a combiner exists, for some values α, β, and n such that $\alpha + \beta \leq n$. Let OT_1 and OT_2 be trivial instances of OT, as in the proof of Lemma 1. Calling the combiner using β instances of OT_1 and $n - \beta \geq \alpha$ instances of OT_2 as input yields a secure OT protocol without any assumption.   □

We will now show that the bound of Lemmas 1 and 2 is tight, by presenting constructions of $(\alpha, \beta; n)$-robust OT-combiners for any α, β, and n, if $\alpha + \beta > n$. First we describe a combiner, which is very simple but not fully satisfactory, as it is not efficient.[5]

Lemma 3. *For every $\alpha \geq 0$, $\beta \geq 0$ and $\alpha + \beta > n$ there exists an inefficient third-party black-box $(\alpha, \beta; n)$-robust OT-combiner.*

Proof. The combiner is a straightforward generalization of the $(2; 3)$-robust OT-combiner from [HKN+05], which is based on two "special-purpose" combiners, combiner **R** for protecting the receiver, and combiner **S** for protecting the sender.[6]

The $(\alpha, \beta; n)$-robust combiner works in two phases: in the first phase subsets of the candidates of size α are combined using the combiner **R**, resulting in $n' = \binom{n}{\alpha}$ OT schemes. Each of resulting instance is secure for the receiver and at least one is secure for both parties. In the second phase the n' OTs are combined using the combiner **S** to yield a final scheme protecting both the sender and the receiver.   □

The combiner presented in the above proof is perfect in the sense that it does not introduce any additional error. However, it is inefficient in n, since the value of n', i.e., the number of OTs resulting from the first phase, would be superpolynomial in n. Lemma 5 presents a combiner that is not perfect, but efficient in n and other parameters, as required in Definition 1. In the construction we use a third-party combiner for bit commitment, which is an adaptation of a secret-sharing-based BC-combiner, due to Herzberg [Her05], to our generalized definition. As this may be of independent interest, we describe it separately.

Lemma 4. *For every $n \geq 2$ and for every $\alpha, \beta > 0$ satisfying $\alpha + \beta > n$ there exists a third-party black-box $(\alpha, \beta; n)$-robust BC-combiner.*

Proof. We describe a string commitment that lets a sender commit to a value $s \in \{0, 1\}^m$, for an arbitrary m such that $2^m > n$, using n candidate implementations of bit-commitment from which at least α are hiding and at least β are binding.

---

[5] Due to its inefficiency, this is strictly speaking not a robust combiner (cf. Def. 1). In a slight abuse of terminology, we call it an *inefficient* combiner.

[6] For completeness, we recall these special-purpose combiners in the appendix.

The sender computes[7] an ℓ-out-of-n Shamir's secret sharing of s over $\mathbb{F}_{2^m}$, for $\ell := n - \alpha + 1$, resulting in shares $s_1, \ldots, s_n$. Then the sender uses the n instances of bit-commitment to commit to the share s_i bit-by-bit, for each $i \in [n]$. In the opening phase the sender opens the commitments to all the shares, and the receiver reconstructs the secret s.

To see that this commitment is hiding, notice that at least α shares are guaranteed to be hidden from the receiver, since at least α candidate bit-commitments are hiding. Therefore before the opening phase the receiver sees at most $n - \alpha < \ell$ shares, which give no information about the secret. On the other hand, since at least β candidates of the bit-commitments are binding, the sender is indeed committed to at least β shares. Since $\alpha + \beta > n$, i.e., $\beta > n - \alpha$, the sharing polynomial, which has degree at most $n - \alpha$, is uniquely determined by these β shares, and so the commitment to s is also binding.   □

Lemma 5. *For every $n \geq 2$ and for every $\alpha, \beta > 0$ satisfying $\alpha + \beta > n$ there exists a third-party black-box $(\alpha, \beta; n)$-robust OT-combiner.*

Proof. First assume that the sender and the receiver have a common random string r at their disposal. Later we describe how this additional assumption can be dropped.

Using r, the combiner works as follows: it simulates (p, q)-WOT with $p + q \leq 1 - 1/n$ by picking each time an input OT-candidate uniformly at random. This is possible, since we are having n candidates, α of which are secure for the sender and β are secure for the receiver, with $\alpha + \beta \geq n + 1$. By picking one candidate at random we obtain a probability $p \leq (n - \beta)/n$ that the sender learns the receiver's choice, and a probability $q \leq (n - \alpha)/n$ that the receiver learns both bits input by the sender, hence $p + q \leq ((n - \alpha) + (n - \beta))/n = (2n - \alpha - \beta)/n \leq 1 - 1/n$, as required. Given such a (p, q)-WOT, use the (efficient) amplification algorithm of Damgård *et al.* [DKS99] to obtain a secure OT.

To complete the argument, we have to show how the sender and the receiver can generate a common random string r. It is well-known that OT implies bit-commitment [Cré87], and bit-commitment implies coin-toss [Blu82].[8] Therefore, we can convert our n candidate implementation of OT into n candidate implementations of bit-commitment, and then use the bit-commitment-combiner of Lemma 4 to obtain a secure implementation of bit-commitment, provided that $\alpha + \beta > n$. This implementation can then be used to implement coin-toss, i.e., the parties can generate a common random string r using the input candidates only, without additional assumptions. Finally, it is easy to verify that all the described protocols use the candidates in a third-party black-box manner, and that the combiner is efficient.   □

From Lemmas 1 and 5, we get immediately the following theorem about $(\alpha, \beta; n)$-robust OT-combiner.

---

[7] In this computation we view bit-strings from $\{0, 1\}^m$ as elements of $\mathbb{F}_{2^m}$.

[8] For completeness, we describe both protocols in the appendix.

Theorem 1. *There exists a black-box $(\alpha, \beta; n)$-robust OT-combiner if and only if $\alpha + \beta > n$ holds. The construction is third-party black-box and efficient.*

Furthermore, the impossibility result of [HKN+05] together with Lemma 5 yield the following corollary about $(k; n)$-robust OT-combiner.

Corollary 1. *There exists a transparent black-box $(k; n)$-robust OT-combiner if and only if $2k > n$. The construction is third-party black-box and efficient.*

4 OT-Combiners Based on the Symmetry of OT

A closer look at the combiners from the proofs of Lemmas 3 and 5 shows that these are "non-uniform" combiners (cf. Sect. 2.2). Namely, the proofs show that for given $\alpha, \beta > 0$ with $\alpha + \beta > n$ there exists a $(\alpha, \beta; n)$-robust combiner, i.e., the actions of the combiner are different for different values α, β. (For Lemma 3 it is explicit in the construction, and for Lemma 5 it is due to the fact that the amplification algorithm from [DKS99] makes explicit use of parameters p, q.) More desireable would be an *uniform* construction, which would have a switched order of quantifiers, i.e., we would like a single combiner that is secure for every $\alpha, \beta > 0$ with $\alpha + \beta > n$, and would therefore be *strictly stronger* than any of the special combiners. In this section we show how to construct such a combiner by exploiting the *symmetry* of OT, i.e., the fact that given OT with sender Alice and receiver Bob, we can perfectly *logically* reverse it to obtain OT with receiver Alice and sender Bob. That OT can be reversed has first been discovered independently in [CS91, OVY93]. A simpler and more efficient protocol has been proposed in [WW06].

Our construction is based on a simple trick, which is somehow non-standard, yet plausible in most scenarios: we require that the parties can *swap their roles* when executing candidate protocols, i.e., any input candidate OT_i can be executed in such a way that the sender (of the main OT-protocol) plays the role of the receiver in OT_i, and the receiver plays the role of the sender in OT_i. Moreover, we require that we have at our disposal *multiple copies of each candidate implementation* (in particular, our protocols use the candidates both in the original setting as well in the swapped configuration). For example, if the input candidates are given as software packages, these requirements are not a problem, as it means only calling different functions, but if a candidate is given as a pair of physical devices implementing the primitive, the swapping operation can be problematic, as it may require a real physical swap of the corresponding devices. However, it is difficult to come up with a primitive that cannot be swapped or duplicated *in principle*. Such a primitive would need to make use of some kind of a physical phenomenon, only available to one of the parties, but not to the other.

We use the swapping of the roles in OT — which can be viewed as a "physical" reversal — together with a *logical* reversal of OT [CS91, OVY93, WW06] to obtain an OT in the original direction (from the original sender to the original receiver), but *with swapped security properties*. More precisely, let swap be this

SENDER'S INPUT: two bits b_0, b_1
RECEIVERS'S INPUT: choice bit c
INPUT OT PROTOCOLS: OT_1, OT_2

Note: Auxiliary combiners **R** and **S** [CK88,HKN+05] are described in the appendix.

1. Parties apply swap to obtain $OT_1^* = $ swap(OT_1) and $OT_2^* = $ swap(OT_2).
2. Parties define $OT' = \mathbf{R}(OT_1, OT_2)$ and $OT'' = \mathbf{R}(OT_1^*, OT_2^*)$.
3. Parties invoke $\mathbf{S}(OT', OT'')(b_0, b_1; c)$.

Fig. 1. A $\{3; 2\}$-robust *uniform* OT-combiner

two-step process, i.e., physical swap followed by logical reversal, and consider an implementation OT and its swapped-and-reversed version, $OT^* = $ swap(OT). If OT is a correct OT-protocol, then so is OT^*. Moreover, if in OT the security of the sender is based on assumption $\mathcal{A}$, and the security of the receiver is based on assumption $\mathcal{B}$, then in OT^* we have the opposite situation: the security of the sender is based on assumption $\mathcal{B}$, and the security of the receiver is based on assumption $\mathcal{A}$. In particular, if OT is an implementation unconditionally secure for the sender, then OT^* is an implementation unconditionally secure for the receiver.

As a first application of this swapping trick we show a $\{3; 2\}$-robust *uniform* OT-combiner, i.e., a combiner which is simultaneously $(\alpha, \beta; 2)$-robust for any α, β satisfying $\alpha + \beta \geq 3$. Recall that if it is known in advance that the security of one party is guaranteed (e.g. $\alpha = 2$), then the corresponding combiner is very simple [HKN+05]. However, the combiner for the case $\alpha = 2$ is quite different from the combiner for the case $\beta = 2$, hence these simple combiners are not uniform.

The idea behind our uniform combiner is to use *both*, the two candidate OT_1, OT_2, and their swapped counterparts $OT_1^* = $ swap(OT_1) and $OT_2^* = $ swap(OT_2). Since $\alpha + \beta \geq 3$, at least two of OT_1, OT_2, OT_1^*, OT_2^* are secure for both parties, at most one is insecure for the sender, and at most one is insecure for the receiver. This is sufficient to implement a secure OT. The construction makes use of the two "special-purpose" OT-combiners we have used previously in the proof of Lemma 3, i.e., combiner **S** for protecting the sender, and combiner **R** for protecting the receiver (cf. Appendix). Figure 1 presents the entire construction in more detail, and the following theorem summarizes its properties.

Theorem 2. *There exists a third-party black-box $\{3; 2\}$-robust uniform OT-combiner using the* swap*-operation.*

Proof. (sketch) Consider the protocol in Figure 1. Let $\overline{OT}$ denote the resulting OT protocol. $\overline{OT}$ has to satisfy correctness, privacy for the sender, and privacy for the receiver. Correctness is trivially given due to the correctness of the candidates OT_1, OT_2, the symmetric schemes OT_1^*, OT_2^*, and the combiners **R** and

S. Given the symmetry of OT, if the privacy of one party is compromised for one candidate, then the privacy of the other party is compromised for the corresponding swapped candidate. Combining OT_1, OT_2, respectively OT_1^*, OT_2^*, with **R** ensures that the receiver's privacy is protected in both OT' and OT'', and the sender's privacy in at least one of them. Hence $S(OT', OT'')$ protects the sender from a possible security break of one of the input canditates. Finally, is easy to verify that this is a third-party black-box combiner. □

The next lemma gives a general construction to obtain a uniform combiner from a non-uniform one. This construction makes use of the swap-operation, and can be used for combiners of any symmetric two-party primitive.

Lemma 6. *If there exists a $(k, k; 2n)$-robust OT-combiner, then there exists a $\{k; n\}$-robust uniform OT-combiner using the* swap-*operation.*

Proof. (sketch) The (k, n)-robust uniform OT-combiner works as follows: given n candidate instances of OT, satisfying $\alpha + \beta \geq k$, we duplicate all instances, and apply the swap-operation to the duplicates. In this way we obtain $2n$ candidate instances, where at least k of them are secure for the sender, and at least k are secure for the receiver. Now we can apply the $(k, k; 2n)$-robust OT-combiner to these $2n$ instances, and get a secure implementation of OT. □

Lemma 6 together with Theorem 1 give us the following theorem.

Theorem 3. *For any $n \geq 2$ and $\delta > n$, there exists a third-party black-box $\{\delta; n\}$-robust uniform OT-combiner using the* swap-*operation.*

Although the presented uniform OT-combiner works with all OT protocols proposed in the literature, it naturally raises the question whether the role-swapping technique can be dropped. Sommer [Som06] has recently pointed out that for transparent black-box OT-combiners the use of the candidates in the swapped direction is in fact necessary. More precisely, he observed that the impossibility proof of Harnik *et al.* [HKN+05] can be adapted to exclude transparent black-box $\{3; 2\}$-robust uniform OT-combiners using the candidates in the prescribed direction only.

5   Conclusions and Open Problems

We proposed stronger definitions of robust combiners for two-party protocols, which yield robuster, more general combiners for oblivious transfer. The observation that a partially broken candidate implementation can still provide security for one of the parties leads to OT-combiners strictly stronger than the constructions known previously. Furthermore, we have shown that for symmetric two-party primitives even stronger combiners are possible if the parties can swap their roles in the candidate protocols.

As we mentioned above, there is currently a trade-off between the *perfect* security and the efficiency of a combiner: we do not know whether there exists an efficient (in the number of candidates) perfect uniform OT-combiner. Moreover, it would be interesting to find other settings, in which the introduced swapping-trick could be useful.

Acknowledgments. We would like to thank Thomas Holenstein and Christian Sommer for interesting discussions, and anonymous referees for useful comments. Jürg Wullschleger was supported by a grant from the Swiss National Science Foundation (SNF).

References

[AB81]   Charles A. Asmuth and George R. Blakely. An effcient algorithm for constructing a cryptosystem which is harder to break than two other cryptosystems. *Computers and Mathematics with Applications*, 7:447–450, 1981.

[BB06]   Dan Boneh and Xavier Boyen. On the impossibility of efficiently combining collision resistant hash functions. In *Proc. CRYPTO '06*, pages 570–583, August 2006.

[BCW03]   Gilles Brassard, Claude Crépeau, and Stefan Wolf. Oblivious transfers and privacy amplification. *Journal of Cryptology: the journal of the International Association for Cryptologic Research*, 16(4):219–237, 2003.

[Bla79]   George R. Blakley. Safeguarding cryptographic keys. In *Proceedings of the National Computer Conference*, pages 313–317. American Federation of Information Processing Societies, 1979.

[Blu82]   Manuel Blum. Coin flipping by telephone - a protocol for solving impossible problems. In *COMPCON, Proc. Twenty-Fourth IEEE Computer Society International Conference*, pages 133–137, 1982.

[Cac98]   Christian Cachin. On the foundations of oblivious transfer. In *Proc. EUROCRYPT'98*, pages 361–374, May 1998.

[CK88]   Claude Crépeau and Joe Kilian. Achieving oblivious transfer using weakened security assumptions (extended abstract). In *Proc. IEEE FOCS '88*, pages 42–52, 1988.

[Cré87]   Claude Crépeau. Equivalence between two flavours of oblivious transfers. In *Proc. CRYPTO '87*, pages 350–354, 1987.

[CS91]   Claude Crépeau and Miklós Sántha. On the reversibility of oblivious transfer. In *Proc. EUROCRYPT '91*, volume 547, pages 106–113, 1991.

[DFSS06]   Ivan Damgård, Serge Fehr, Louis Salvail, and Christian Schaffner. Oblivious transfer and linear functions. In *Proc. CRYPTO '06*, pages 427–444, August 2006.

[DK05]   Yevgeniy Dodis and Jonathan Katz. Chosen-ciphertext security of multiple encryption. In *Proc. TCC '05*, pages 188–209, 2005.

[DKS99]   Ivan Damgård, Joe Kilian, and Louis Salvail. On the (im)possibility of basing oblivious transfer and bit commitment on weakened security assumptions. In *Proc. EUROCRYPT '99*, pages 56–73, 1999.

[EG85]   Shimon Even and Oded Goldreich. On the power of cascade ciphers. *ACM Trans. Comput. Syst.*, 3(2):108–116, 1985.

[EGL85]   Shimon Even, Oded Goldreich, and Abraham Lempel. A randomized protocol for signing contracts. *Communications of the ACM*, 28(6):637–647, 1985.

[Her05]   Amir Herzberg.   On tolerant cryptographic constructions.   In *CT-RSA*, pages 172–190, 2005. full version on Cryptology ePrint Archive, eprint.iacr.org/2002/135.

[HKN+05] Danny Harnik, Joe Kilian, Moni Naor, Omer Reingold, and Alon Rosen. On robust combiners for oblivious transfer and other primitives.   In *Proc. EUROCRYPT '05*, pages 96–113, 2005.

[HL05]   Susan Hohenberger and Anna Lysyanskaya. How to securely outsource cryptographic computations. In *Proc. TCC '05*, pages 264–282, 2005.

[MM93]   Ueli Maurer and James L. Massey. Cascade ciphers: The importance of being first. *Journal of Cryptology*, 6(1):55–61, 1993. preliminary version in Proc. IEEE Symposium on Information Theory, 1990.

[MP06]   Remo Meier and Bartosz Przydatek.   On robust combiners for private information retrieval and other primitives. In *Proc. CRYPTO '06*, pages 555–569, August 2006.

[OVY93]   Rafail Ostrovsky, Ramarathnam Venkatesan, and Moti Yung. Fair games against an all-powerful adversary. In *Advances in Computational Complexity Theory*, volume 13 of *AMS DIMACS Series in Discrete Mathematics and Theoretical Computer Science*, pages 155–169. AMS, 1993.

[Rab81]   Michael O. Rabin.   How to exchange secrets by oblivious transfer., 1981. Tech. Memo TR-81, Aiken Computation Laboratory, available at eprint.iacr.org/2005/187.

[Sha79]   Adi Shamir. How to share a secret. *Commun. ACM*, 22(11):612–613, 1979.

[Som06]   Christian Sommer. Robust combiners for cryptographic protocols, 2006. Master Thesis, Computer Science Department, ETH Zurich.

[WW06]   Stefan Wolf and Jürg Wullschleger. Oblivious transfer is symmetric. In *Proc. EUROCRYPT '06*, pages 222–232, 2006.

Appendix

For completeness, we recall some constructions used in the proposed combiners. First we describe the "special-purpose" combiners $\mathbf{R}$ and $\mathbf{S}$ from [CK88, HKN+05]. Combiner $\mathbf{R}$ takes n OT candidates, and guarantees security of the receiver if at least one of the candidates is secure for the receiver:

$\mathbf{R}(\mathsf{OT}_1, \ldots, \mathsf{OT}_n)(b_0, b_1; c)$

1. The sender picks random bits $r_1^0, r_2^0, \ldots, r_n^0$, such that $b_0 = r_1^0 \oplus r_2^0 \oplus \cdots \oplus r_n^0$, and sets $r_i^1 := r_i^0 \oplus b_0 \oplus b_1$, for every $i = 1 \ldots n$.

2. The receiver picks random bits $c_1, c_2, \ldots, c_n$ such that $c = c_1 \oplus c_2 \oplus \cdots \oplus c_n$.

3. For every $i = 1 \ldots n$ parties run $\mathsf{OT}_i(r_i^0, r_i^1; c_i)$.
   From i-th execution the receiver obtains output $r_i^{c_i}$.

4. The receiver outputs b_c computed as the XOR of his outputs from all executions, i.e.

$$b_c = r_1^{c_1} \oplus r_2^{c_2} \oplus \cdots \oplus r_n^{c_n} .$$

Combiner $\mathbf{S}$ takes n OT candidtates, and guarantees security of the sender if at least one of the candidates is secure for the sender:

$S(OT_1, \ldots, OT_n)(b_0, b_1; c)$

1. The sender picks random bits $r_1^0, r_2^0, \ldots, r_n^0$, and $r_1^1, r_2^1, \ldots, r_n^1$, such that

$$b_0 = r_1^0 \oplus r_2^0 \oplus \cdots \oplus r_n^0 \quad \text{and} \quad b_1 = r_1^1 \oplus r_2^1 \oplus \cdots \oplus r_n^1 .$$

2. For every $i = 1 \ldots n$ parties run $OT_i(r_i^0, r_i^1; c)$.
   From i-th execution the receiver obtains output r_i^c.
3. The receiver outputs b_c computed as the XOR of his outputs from all executions, i.e.

$$b_c = r_1^c \oplus r_2^c \oplus \cdots \oplus r_n^c .$$

The following protocol generates a random bit-string using bit-commitment. Let $m > 0$.

Coin-toss

1. The sender picks a random $s' \in \{0,1\}^m$ and commits to it.
2. The receiver picks a random $s'' \in \{0,1\}^m$ and sends it to the sender.
3. The sender opens the commitment to s', and both parties output $s = s' \oplus s''$.

The following protocol implements bit-commitment using OT. Let $\kappa > 0$ be a security parameter.

Commit(v)

1. The sender picks random $r \in \{0,1\}^\kappa$, and the receiver a $c \in \{0,1\}^\kappa$.
2. The sender inputs $x_0 = r_i$ and $x_1 = r_i \oplus v$ and the receiver c_i to the i-th instance of OT.
3. The receiver obtains y_i from the i-th instance of OT.

Open

1. The sender sends v and r to the receiver.
2. The receiver verifies whether for all i we have $y_i = r_i \oplus c_i v$.

One-Way Permutations, Interactive Hashing and Statistically Hiding Commitments

Hoeteck Wee*

Computer Science Division,
University of California,
Berkeley
hoeteck@cs.berkeley.edu

Abstract. We present a lower bound on the round complexity of a natural class of black-box constructions of statistically hiding commitments from one-way permutations. This implies a $\Omega(\frac{n}{\log n})$ lower bound on the round complexity of a computational form of interactive hashing, which has been used to construct statistically hiding commitments (and related primitives) from various classes of one-way functions, starting with the work of Naor, Ostrovsky, Venkatesan and Yung (J. Cryptology, 1998). Our lower bound matches the round complexity of the protocol studied by Naor et al.

Keywords: Interactive hashing, statistically hiding commitments, black-box lower bounds.

1 Introduction

A *zero-knowledge proof* is a protocol wherein one party, the prover, convinces another party, the verifier, of the validity of an assertion while revealing no additional knowledge. Introduced by Goldwasser, Micali and Rackoff in the 1980s [GMR89], zero-knowledge proofs have played a central role in the design and study of cryptographic protocols. In these applications, it is important to construct constant-round zero-knowledge protocols for NP under minimal assumptions. In many cases, a computational zero-knowledge argument system suffices, and we know how to construct such protocols for NP under the (essentially) minimal assumption of one-way functions [BJY97, OW93]. On the other hand, there are cases wherein we need stronger guarantees, namely a computational zero-knowledge proof system, or a statistical zero-knowledge argument system.[1] Surprisingly, the main bottleneck to reducing the assumptions for known constructions of both constant-round computational

---

* Work done while visiting Tsinghua University, Beijing; IBM T.J. Watson Research Center; and IPAM, Los Angeles.
[1] It is unlikely that every language in NP has a statistical zero-knowledge proof system [F89, AH91, BHZ87].

S.P. Vadhan (Ed.): TCC 2007, LNCS 4392, pp. 419–433, 2007.

zero-knowledge proof systems and statistical zero-knowledge argument systems [BCY91, GK96a] is statistically hiding commitments.[2]

We know how to construct constant-round statistically-hiding commitments from collision-resistant hash functions [DPP98, NY89] and from claw-free permutations [GK96a]. In 1992, Naor, Ostrovsky, Venkatesan and Yung [NOVY98] showed that one-way permutations are sufficient for statistically hiding commitments wherein the round complexity is linear in the security parameter. This was very recently extended to one-way functions by Haitner and Reingold [HR06b]. Both works use the powerful tool of interactive hashing [OVY93], a 2-party protocol for choosing a small set of strings, with binding and hiding requirements similar to those in commitment schemes. An intriguing open problem (posed in [NOVY98] and reiterated in [DHRS04, KS06, HR06a]) is whether some variant of interactive hashing could yield a constant-round statistically hiding commitment from one-way permutations. In fact, even a $n^{o(1)}$-round commitment would be interesting. The restriction to interactive hashing may seem limiting, but it is the only technique that we presently know of. Moreover, Ding, et al. [DHRS04] exhibited a constant-round interactive hashing protocol satisfying a weaker binding guarantee, which indicates that interactive hashing may not be the bottleneck.

1.1   Our Contributions and Techniques

We study a natural class of black-box constructions of statistically hiding commitments from one-way permutations that include several generalizations of the NOVY construction, and show that any such construction yields a commitment scheme with at least $\Omega(n/\log n)$ rounds. This matches the round complexity of a variant of the main NOVY construction ([KS06, HR06a]). Specifically, our lower bound holds for constructions wherein the sender (in the commitment scheme) evaluates the one-way permutation only at the start of the commit phase, and does so on independent random inputs. The sender then uses the output values, her private input to the commitment scheme, and possibly additional randomness in the rest of the commit phase and does not use the inputs to the one-way permutation until the reveal phase.

We derive as a corollary, a $\Omega(n/\log n)$ lower bound on a computational form of interactive hashing presented in [NOV06, HR06a], based on an abstraction of the way interactive hashing is used in the NOVY construction and the subsequent works of Haitner et al. [HHK+05, NOV06, HR06b]. The same abstraction also applies to the use of interactive hashing in the transformation of honest-verifier zero-knowledge arguments into cheating-verifier zero-knowledge arguments [D93, OVY93]. The lower bound tells us that we need to avoid the standard notion of interactive hashing if we want round-efficient versions of these applications.

---

[2] It is not surprising that we need statistically hiding commitments for statistical zero-knowledge arguments; what is surprising is that the only known approach for constructing constant-round zero-knowledge proof systems [GK96a] requires statistically hiding commitments to guarantee soundness, because the verifier begins by committing to her challenges.

Our lower bound for statistically hiding commitments only holds for fully black-box reductions [RTV04], namely, we require not only that the construction treats the one-way permutation as a black-box, but also that the reduction in the proof of security uses black-box access to a cheating sender that breaks the binding property to invert the permutation with noticeable probability. At a high level, our lower bound follows the paradigm of Gennaro and Trevisan [GT00] for proving lower bounds on efficiency of black-box cryptographic constructions, which is in turn based on the Impagliazzo-Rudich framework [IR89] for separating cryptographic primitives. The proof techniques and ideas are otherwise largely inspired by lower bounds for black-box zero-knowledge from the work of Goldreich and Krawczyk [GK96b].

Roughly speaking, a fully black-box reduction guarantees an efficient procedure that by interacting and rewinding the cheating sender, produces transcripts of the commitment scheme with a certain outcome. Using the repeated sampling technique from [IR89], we can ensure that the probability that a partial transcript has the outcome is exponentially small in the length of the sender's last message. This means that the sender sends $O(\log n)$ bits in each round of protocol. On the other hand, the sender must send a total of $\Omega(n)$ bits in the protocol (so that there is a different transcript for every possible challenge for the one-way permutation), which means the protocol must have $\Omega(n/\log n)$ rounds. This simplified and slightly inaccurate sketch overlooks several technical difficulties.

1.2 Perspective

Notions and limitations of interactive hashing. The last few years has witnessed a lot of work on the use of interactive hashing protocols in cryptography with two main notions of security: computationally binding, and binding for static sets [NOV06]. The latter is used in building and studying oblivious transfer protocols in the bounded storage model and over noisy channels [CCM98, DHRS04, CS06], in constructing variants of statistically binding commitments [NV06], and in transforming honest-verifier zero-knowledge proofs into cheating-verifier zero-knowledge proofs [D93, DGOW95, GSV98]. It was noted in [NOV06, CCM98] that the computational binding implies binding for static sets; our lower bound implies that the converse is not true. Specifically, the constant-round protocol of [DHRS04] does not satisfy the computational formulation (which answers an open problem in [DHRS04] in the negative).

Efficiency of cryptographic reductions. Previous work establishing lower bounds for efficiency of black-box cryptographic reductions has focused on the query complexity and randomness complexity of these reductions [KST99, GGKT05, LTW05, HK05] whereas our work focuses on round complexity. Upon closer inspection, our work is also qualitatively very different (apart from studying a different computational resource) as the works of [GGKT05, LTW05, HK05] rule out weakly black-box reductions (unless P = NP), wherein the proof of security may exploit the code of the adversary (in a non-black-box manner). As mentioned earlier, our main result only rules out fully black-box reductions and uses fairly different techniques. We stress that all known reductions between cryptographic primitives - with the exception of the non-black-box techniques used in zero-knowledge and multi-party protocols, e.g. [B01], but including the non-black-box constructions in [AIK04] - do not exploit the code of the adversary in the proof of security. As such, ruling out fully black-box reductions is almost as meaningful as ruling weakly black-box reductions.

Information-theoretic analogues. Many black-box cryptographic constructions apart from interactive hashing-based commitments have an information-theoretic analogue which is easier to achieve, in that it does not have some kind of "simulateable" requirement, namely, an efficient procedure for simulating random transcripts with a certain outcome. This was articulated in [DGW95], using random selection as a case study. Such connections been exploited in both directions, the most remarkable being the construction of extractors from Nisan-Wigderson pseudo-random generators [T01]. In [LTW05], the connection between hardness amplification and combinatorial hitters was used to derive lower bounds on query and randomness complexity of the former. While the resulting lower bounds on query complexity are tight, those for randomness complexity are far from the best-known constructions. The information-theoretic analogue for computational interactive hashing would be interactive hashing with binding for static sets, for which we cannot expect to prove a super-constant lower bound (again, due to the constant-round protocol in [DHRS04]). Indeed, we exploit the "simulateable" requirement for our main result.

Trade-offs between interaction and assumptions. The NOVY construction demonstrated the feasibility of trading off higher interaction costs in order to build a cryptographic primitive under weaker cryptographic assumptions (specifically, perfectly hiding commitments with a linear number of rounds assuming one-way permutations, versus a constant number of rounds assuming claw-free permutations). Rudich's work [R91] shows that this trade-off is necessary in relation to secret key agreement and trapdoor functions. Our main result shows that the trade-off is also necessary for the NOVY construction. While the trade-off is an additive constant in Rudich's work, our lower bound yields a gap between constant and almost-linear number of rounds.

Implications for protocol design. One could view this work quite broadly as providing a simple informal criterion for reasoning about the round complexity of classes of fully-black-box constructions (of protocols with a "simulatable" requirement) and formal techniques towards establishing a lower bound. The former is especially useful for protocol design in identifying and ruling out inefficient constructions. We stress here that our lower bounds do not apply to the black-box constructions of commitments from various classes of one-way functions in the works of Haitner et al. [HHK+05, NOV06, HR06b], in two different ways. One is the use of one-way functions in [HHK+05] to implement coin-tossing and zero-knowledge proofs to transform commitments that are hiding against honest receivers into commitments that are hiding against arbitrary receivers. We note that our lower bound holds assuming merely hiding against honest receivers. The second is that the inputs to the one-way functions are used again in the commit phase. This is only needed to handle the lack of structure in general one-way functions. In particular, all the constructions are much simpler and requires fewer rounds when optimized for one-way permutations - they "collapse" to the NOVY construction. In short, the ways in which these constructions bypass our lower bounds do not provide much insight into how we may bypass the lower bounds for one-way permutations.

1.3   Additional Related Work

Fischlin [F02] showed that there is no black-box construction of 2-message statistically hiding from one-way permutations (or even trapdoor permutations). The result follows quite readily from Simon's oracle separating collision-resistant hash functions and one-way permutations [S98]. On the other hand, Harnik and Naor [HN06] gave a non-black-box construction of a 2-message statistically hiding commitment from one-way functions under a non-standard assumption on compressibility of NP instances. From what we understand, there is no strong evidence either supporting or refuting the assumption.

2   Definitions and Preliminaries

We use PPT to denote both probabilistic polynomial-time Turing machines and probabilistic polynomial-time interactive Turing machines. The *round complexity* of a 2-party protocol is number of pairs of messages exchanged by both parties (in both directions). Unless otherwise stated, we use 1^n as the security parameter.

2.1   One-Way Permutations

Definition 1. *A function* $f : \{0,1\}^* \rightarrow \{0,1\}^*$ *is a* $s(n)$-secure one-way function *if* f *is computable in polynomial time and for every nonuniform PPT* A,

$$\Pr_{x \in \{0,1\}^n} [A(1^n, f(x)) \in f^{-1}(f(x))] < 1/s(n)$$

A function f *is a* one-way permutation *if for every* n, f *restricted to* $\{0,1\}^n$ *is a permutation, and for all polynomials* $s(n)$ *and all sufficiently large* n, f *is* $s(n)$-secure.

A random permutation π is exponentially one-way even if the adversary is given access to a π^{-1} oracle, as long as it cannot query π^{-1} on the challenge. Here, $\pi^{-1}_{\neq y}$ is an oracle that on input y', returns $\pi^{-1}(y')$ if $y' \neq y$, and $\perp$ otherwise.

Lemma 1 (implicit in [GT00]). *Fix* $s(n) = 2^{n/5}$. *For all sufficiently large* n, *there exists a permutation* π *on* $\{0,1\}^n$ *such that for all circuits* A *of size* $s(n)$,

$$\Pr_{y \in \{0,1\}^n} [A^{\pi, \pi^{-1}_{\neq y}}(y) = \pi^{-1}(y)] < \frac{1}{s(n)}$$

Moreover, the statement relativizes.

2.2 Statistically Hiding Commitments

We present the definition for bit commitment. To commit to multiple bits, we may simply run a bit commitment scheme in parallel.

Definition 2. *A (bit) commitment scheme* (S, R) *is an efficient two-party protocol consisting of two stages. Throughout, both parties receive the security parameter* 1^n *as input.*

> COMMIT. *The sender* S *has a private input* $b \in \{0, 1\}$, *which she wishes to commit to the receiver* R, *and a sequence of coin tosses* σ. *At the end of this stage, both parties receive as common output a commitment* z.

> REVEAL. *Both parties receive as input a commitment* z. S *also receives the private input* b *and coin tosses* σ *for* z. *This stage is non-interactive:* S *sends a single message to* R, *and* R *either outputs a bit and accepts or rejects.*

Definition 3. *A commitment scheme* (S, R) *is* perfectly hiding *if*

> COMPLETENESS. *If both parties are honest, then for any input bit* $b \in \{0, 1\}$ *that* S *gets,* R *outputs* b *and accepts at the end of the decommit stage.*
>
> STATISTICALLY HIDING. *For every unbounded deterministic strategy* R^*, *the distributions of the view of* R^* *in the commit stage while interacting with an honest* S *are identical for* $b = 0$ *and* $b = 1$. *If the distributions are statistically indistinguishable, we obtain a* statistically hiding *commitment.*
>
> COMPUTATIONALLY BINDING. *For every nonuniform PPT* S^*, S^* *succeeds in the following game (breaks the commitment) with negligible probability:*
>
> - S^* *interacts with an honest* R *and outputs a commitment* z.
> - S^* *outputs two messages* τ_0, τ_1 *such that for both* $b = 0$ *and* $b = 1$, R *on input* (z, τ_b) *accepts and outputs* b.

3 Constructing Commitments from One-Way Permutations

In this section, we provide formal definitions of the various classes of constructions of commitments from one-way permutations we consider in this paper.

3.1 Fully Black-Box Constructions

Definition 4. *A fully black-box construction of a statistically hiding commitment scheme from one-way permutations is a triplet of polynomial time computable oracle procedures* (S, R, M) *for which there exists a polynomial* T *and a constant* c *satisfying the following properties:*

EFFICIENCY. *The running times of $\mathcal{S}, \mathcal{R}, M$ are bounded by T.*

FUNCTIONALITY. *For every family of permutations π, $(\mathcal{S}^\pi, \mathcal{R}^\pi)$ is a statistically hiding commitment scheme.*

SECURITY. *For every $\varepsilon = 1/\operatorname{poly}(n)$, for all sufficiently large n, every permutation $\pi : \{0,1\}^n \to \{0,1\}^n$ and every adversary $\mathcal{S}^*$, if $\mathcal{S}^*$ breaks $(\mathcal{S}^\pi, \mathcal{R}^\pi)$ with probability ε, then*

$$\Pr_{y \in \{0,1\}^n}[M^{\mathcal{S}^*,\pi}(y) = \pi^{-1}(y)] \geq \left(\frac{\varepsilon}{T}\right)^c$$

3.2   Interactive Hashing

Interactive hashing is a 2-party protocol between a sender and a receiver, similar to a commitment scheme. The sender begins with a private input $y \in \{0,1\}^q$ and goal is for both parties to select a set of 2^k strings in $\{0,1\}^q$ (specified by a circuit $C : \{0,1\}^k \to \{0,1\}^q$) containing y. The hiding property stipulates that the receiver does not learn which of the 2^k strings equals y, and the binding property stipulates that the sender can "control" at most one of the 2^k strings. The computational formulation (introduced explicitly in [NOV06] along with selecting many instead of merely 2 outputs) guarantees an efficient reduction from breaking the binding property to solving some computational problem on random instances.

Definition 5 ([NOV06]). *A computational interactive hashing scheme (with multiple outputs) is an efficient protocol $(\mathcal{S}_{\mathsf{IH}}, \mathcal{R}_{\mathsf{IH}})$ where both parties receive common inputs $(1^q, 1^k)$, $\mathcal{S}_{\mathsf{IH}}$ receives a private input $y \in \{0,1\}^q$, with the common output being a circuit $C : \{0,1\}^k \to \{0,1\}^q$ and the private output of $\mathcal{S}_{\mathsf{IH}}$ being a string $z \in \{0,1\}^k$. The protocol satisfies the following properties:*

CORRECTNESS. *For all $\mathcal{R}^*$ and all $y \in \{0,1\}^q$, let C, z be the common and private output of $\mathcal{S}_{\mathsf{IH}}$ in the protocol $(\mathcal{S}_{\mathsf{IH}}, R^*)(1^q, 1^k)$. Then, $C(z) = y$.*

PERFECTLY HIDING. *For all $\mathcal{R}^*$, (V, Z) is distributed identically to (V, U_k), where $V = \operatorname{view}_{\mathcal{R}^*}(\mathcal{S}_{\mathsf{IH}}(U_q,), \mathcal{R}^*)$.*

COMPUTATIONALLY BINDING. *There exists an oracle PPT A such that for every $\mathcal{S}^*$ and any relation W, letting $C, ((x_0, z_0), (x_1, z_1))$ be the common and private output of $\mathcal{S}_{\mathsf{IH}}$ in the protocol $(\mathcal{S}_{\mathsf{IH}}, R^*)(1^q, 1^k)$, if it holds that*

$$\Pr[(x_0, C(z_0)) \in W \wedge (x_1, C(z_1)) \in W \wedge z_0 \neq z_1] > \varepsilon,$$

where the above probability is over the coin tosses of $\mathcal{R}_{\mathsf{IH}}$ and $\mathcal{S}^*$, then we have that

$$\Pr_{y \in \{0,1\}^q}[(A^{\mathcal{S}^*}(y, 1^q, 1^k, \varepsilon), y) \in W] > 2^{-k} \cdot (\varepsilon/q)^{O(1)}.$$

Nguyen et al. [NOV06] presented a protocol satisfying the above definition with $q - k$ rounds, obtained by ending the NOVY protocol $k - 1$ rounds earlier. The protocol is very simple: the receiver chooses $q - k$ linearly independent vectors $v_1, \ldots, v_k$ over $\{0, 1\}^q$. In round i, the receiver sends v_i and the sender responds with bit-wise dot product $v_i \cdot y$. We may reduce the round complexity by a factor of $O(\log q)$ by having the receiver send a pairwise independent hash function $h_i :$ $\{0, 1\}^q \to \{0, 1\}^{O(\log q)}$ in round i and the sender responding with $h_i(y)$ [HR06a]. Note that the sender is deterministic, and the protocol is public-coin. Our lower bound shows that using a randomized sender or a private-coin protocol or q-wise independent hash functions will not further improve the round complexity (beyond constant factors).

Returning to the above definition, note that it refers to general relations W that may not be polynomial-time computable, and it does not give A oracle access to the relation W, which strengthens the security guarantee of the [NOV06] protocol. Our lower bound holds even if A has oracle access to the relation W, which is a weaker guarantee and thus a stronger lower bound. We also note that we may use the techniques in [LTW05] to show that this weaker guarantee also implies binding for static sets, thereby strengthening an observation made in [NOV06].

Naor et al. [NOVY98] showed that any computational interactive hashing scheme $(\mathcal{S}_{\mathsf{IH}}, \mathcal{R}_{\mathsf{IH}})$ yields a fully black-box construction of a perfectly hiding commitment scheme $(\mathcal{S}, \mathcal{R})$ from any one-way permutation π with essentially the same round complexity.[3] The construction is as follows:

COMMIT. To commit to a bit b, $\mathcal{S}$ chooses a random $\sigma \in \{0, 1\}^n$, where n is the security parameter. Then, $\mathcal{S}$ and $\mathcal{R}$ run as a sub-protocol $(\mathcal{S}_{\mathsf{IH}}(\pi(\sigma)), \mathcal{R}_{\mathsf{IH}})(1^n, 1^1)$, playing the roles $\mathcal{S}_{\mathsf{IH}}, \mathcal{R}_{\mathsf{IH}}$ respectively. Let C, z be the common and private outputs of $\mathcal{S}$ in the sub-protocol. $\mathcal{S}$ then sends $b' = b \oplus z$.

DECOMMIT. $\mathcal{S}$ sends (b, σ). $\mathcal{R}$ accepts and outputs b if $C(b \oplus b') = \pi(\sigma)$, and rejects otherwise.

We stress that in the construction, $\mathcal{S}$ queries π exactly once, to compute $\pi(\sigma)$, and does not need σ again except for decommitment.

As noted in the introduction, Damgård [D93] showed how any computational interactive hashing scheme can be used to transform constant-round honest-verifier public-coin zero-knowledge arguments into cheating-verifier public-coin zero-knowledge arguments unconditionally. The transformation may also be made more efficient by exploiting interactive hashing with multiple outputs so that a single application of interactive hashing yields a cheating-verifier zero-knowledge argument with soundness to $1/\operatorname{poly}(n)$ (instead of $1/2$).

---

[3] More precisely, Naor et al. showed how to construct a perfectly hiding commitment scheme from any one-way permutation using the interactive hashing protocol in [OVY93]. Implicit in the proof of correctness and security is a proof that the [OVY93] protocol satisfies Definition 5 for $k = 1$.

3.3 π-Oblivious Constructions

We describe the syntactic constraints on the class of fully black-box constructions for which we prove a lower bound. We consider constructions wherein the sender evaluates the one-way permutation only at the start of the commit phase, and does so on independent random inputs. The sender then uses the values (and not the inputs to the permutation), its input bit and possibly additional randomness in the rest of the commit phase. To decommit, the sender sends its input bit and its random tape, including the inputs to the permutation. We allow the receiver to query the permutation at any point in the protocol.

More formally,

Definition 6. *A fully black-box construction* $(\mathcal{S}, \mathcal{R}, M)$ *of a statistically hiding commitments from one-way permutations is* π-*oblivious if there exists some interactive PPT* $\mathcal{S}_{ob}$ *such that for any permutation* π *on* $\{0,1\}^n$, *to commit to a bit* b *with coin tosses* σ, $\mathcal{S}$ *parses* $\sigma = (\mathbf{z}, \tilde{\sigma})$, *where* $\mathbf{z} = (z_1, \ldots, z_t) \in (\{0,1\}^n)^t$, *and proceeds according to* $\mathcal{S}_{ob}(b, \sigma')$, *where* $\sigma' = (\mathbf{z}', \tilde{\sigma})$ *and* $\mathbf{z}' = \pi(\mathbf{z}) = (\pi(z_1), \ldots, \pi(z_t))$. *In particular,* $\mathcal{S}_{ob}$ *never queries* π. *To decommit,* $\mathcal{S}$ *sends a single message* (b, σ).

Clearly, the NOVY construction is a π-oblivious; there, $t = 1$ and $\mathcal{S}_{ob} = \mathcal{S}_{IH}$ gets input $\pi(z_1)$, and $\tilde{\sigma}$ is the empty string since $\mathcal{S}_{IH}$ is deterministic. Other candidates of π-oblivious constructions include variants of the NOVY construction wherein we run n^2 copies of some variant of interactive hashing in parallel either on the same $t = 1$ input $\pi(z_1)$ or on $t = n^2$ independent inputs $\pi(z_1), \ldots, \pi(z_t)$, or a single copy of interactive hashing on the tn-bit string $\pi(z_1), \ldots, \pi(z_t)$.

On the other hand, the construction of statistically hiding commitments from one-way functions in [HR06b] is not π-oblivious. This is because the sender will query π at some point z_1 and send both $h_1(\pi(z_1))$ and $h_2(z_1)$ during the commit phase, for some hash functions h_1, h_2.

4   Main Result: Lower Bound for Commitments

Now, we state and prove our main result:

Theorem 1. *Any* π-*oblivious fully black-box construction of a statistically hiding commitment scheme from one-way permutations yields a commitment scheme with* $\Omega(\frac{n}{\log n})$ *rounds. This holds even if the hiding property for commitment scheme only holds for the honest receiver. More generally, if we assume that permutation is* s-*secure one-way, then we have an* $\Omega(\frac{n}{\log s})$ *lower bound.*

Our lower bound is tight:

Theorem 2 ([NOVY98, KS06, HR06a]). *There is a* π-*oblivious fully black-box construction of a perfectly hiding commitment scheme from* s-*secure one-way permutations with* $O(\frac{n}{\log s})$ *rounds.*

4.1   Proof Intuition

First, we point out at a high level how we exploit the fact that the construction is fully black-box. We use as the one-way permutation the one guaranteed by Lemma 1, which remains one-way even under a "chosen challenge" attack. This means that in order for the reduction M to successfully invert a challenge y, it must get a cheating sender $\mathcal{S}^*$ to invert π on y itself. However, M is only given black-box access to $\mathcal{S}^*$, so it is limited to sending $\mathcal{S}^*$ different inputs and possibly rewinding $\mathcal{S}^*$.

For concreteness, consider the NOVY construction of commitment schemes from one-way permutation using computational interactive hashing as a sub-protocol. When trying to invert a challenge y, the reduction M tries to get the sender to generate a commitment that is consistent with her input to interactive hashing protocol being y (otherwise, the decommitments will not help to invert y). At each round of commit phase, the honest $\mathcal{S}_{\mathsf{IH}}$ reveals some information about her input $\pi(\sigma)$. At the end of the commit phase, she should have revealed $n-1$ bits of information about her input (since we're using interactive hashing to choose 2 strings). We claim, at each round, she can only reveal $O(\log n)$ bits of information about her input, which yields a $\Omega(n/\log n)$ lower bound on the number of rounds. Suppose there is some round where $\mathcal{S}_{\mathsf{IH}}$ reveals $\omega(\log n)$ bits of information. This means that there are $n^{\omega(1)}$ inputs to the interactive hashing protocol that are consistent with the partial transcript. Consider a cheating sender that at each round samples a random input y' that is consistent with the partial transcript and responds as though her input to the interactive hashing protocol is y', then the probability that the reduction observes a transcript that is consistent with y is negligible. It is important that $\mathcal{S}_{\mathsf{IH}}$ does not query π, so that we may sample consistent partial transcripts using a PSPACE oracle. If $\mathcal{S}_{\mathsf{IH}}$ is deterministic, it is straight-forward to quantify "information" about the sender's input and turn this outline into a proof.

For general π-oblivious constructions, we construct the cheating sender in essentially the same way: at each round (for both the commit and reveal phases), the sender samples a random (b, σ') that is consistent with the partial transcript and responds as though her input to $\mathcal{S}_{\mathsf{ob}}$ is (b, σ') (where $\sigma' = (\mathbf{z}', \tilde{\sigma})$). The main technical difficulty in the analysis is in quantifying "information" about the sender's input. Indeed, how much information a message reveals about $\mathbf{z}$ depends on both b and $\tilde{\sigma}$. Also, for a fixed partial transcript, the set (and number) of $\mathbf{z}'$'s that are consistent with the given transcript may vary with different choices of $b, \tilde{\sigma}$.

4.2   Proof of Theorem 1

We may assume that the commitment scheme $(\mathcal{S}, \mathcal{R})$ runs in r rounds, with $\mathcal{R}$ going first. Let T, c be the polynomial and constant guaranteed by the fully black-box reduction. We will show that $r \gtrsim \frac{n - \log t}{8c \log T} = \Omega(\frac{n}{\log n})$. Suppose otherwise, and take π to be the permutation guaranteed by Lemma 1.

Conventions regarding M. Recall that the reduction M has oracle access to a sender $\mathcal{S}^*$ with which it inverts the permutation π. It can query $\mathcal{S}^*$ on sequences of messages of the form $\mathbf{q}_i = (q_1, \ldots, q_i)$ corresponding to the first i messages from $\mathcal{R}$ in the commit phase, or a message of the form $(\mathbf{q}_r, \texttt{decommit})$, requesting for a decommit to a previous commitment. M runs for at most T steps, and therefore makes at most T queries to $\mathcal{S}^*$. In addition, we may adopt WLOG the following simplifying assumptions on M by modifying M appropriately (as is the case with lower bounds for black-box zero-knowledge [GK96b]):

1. It never asks the same query twice.
2. If M queries the oracle with $\mathbf{q}_i$, it has queried the oracle with all proper prefixes of $\mathbf{q}_i$ (namely all sequences of the form $(q_1, \ldots, q_j)$ for $j \leq i$.)

Notations. We introduce some notations:

- $\mathcal{S}_{ob}(b, \sigma', \mathbf{q}_i)$ denotes the $\mathcal{S}_{ob}$'s response with input b, σ' and and the first i messages from $\mathcal{R}$ being $\mathbf{q}_i$.
- Given a partial transcript $(\mathbf{q}_i, \mathbf{a}_i) = (q_1, \ldots, q_i, a_1, \ldots, a_i)$ and $y \in \{0,1\}^n$, $\text{Con}(\mathbf{q}_i, \mathbf{a}_i)$ is the set of inputs (b, σ') to $\mathcal{S}_{ob}$ that would yield the transcript $(\mathbf{q}_i, \mathbf{a}_i)$; formally,

$$\text{Con}(\mathbf{q}_i, \mathbf{a}_i) = \{(b, \sigma') \mid \mathcal{S}_{ob}(b, \sigma', q_1, \ldots, q_j) = a_j, \forall j = 1, 2, \ldots, i\}$$

and

$$\text{Con}_y(\mathbf{q}_i, \mathbf{a}_i) = \{(b, \mathbf{z}', \tilde{\sigma}) \in \text{Con}(\mathbf{q}_i, \mathbf{a}_i) \mid \exists j : z'_j = y\}$$

In particular, $|\text{Con}_y(\epsilon)|/|\text{Con}(\epsilon)| = 1 - (1 - 2^{-n})^t \leq t2^{-n}$, where ϵ is the empty string (transcript).

Sender strategy $\mathcal{S}^*$. Consider the following sender strategy $\mathcal{S}^*$:

- Upon receiving a query of the form $(\mathbf{q}_{i-1}, q_i)$, look up previous replies $\mathbf{a}_{i-1}$. (For $i = 1$, $(\mathbf{q}_{i-1}, \mathbf{a}_{i-1}) = \epsilon$.) Sample uniformly at random[4] (b, σ') from the set $\text{Con}(\mathbf{q}_{i-1}, \mathbf{a}_{i-1})$, and respond with $a_i = \mathcal{S}_{ob}(b, \sigma', \mathbf{q}_i)$.
- Upon receiving a query of the form $(\mathbf{q}_r, \texttt{decommit})$, look up previous replies $\mathbf{a}_r$. Sample uniformly and independently at random $(b_0, \mathbf{z}_0, \tilde{\sigma}_0), (b_1, \mathbf{z}_1, \tilde{\sigma}_1)$ from the set $\text{Con}(\mathbf{q}_r, \mathbf{a}_r)$, and send $(b_0, \pi^{-1}(\mathbf{z}_0), \tilde{\sigma}_0), (b_1, \pi^{-1}(\mathbf{z}_1), \tilde{\sigma}_1)$.

Note that in an interaction with an honest receiver $\mathcal{R}$, $\mathcal{S}^*$ breaks the commitment with probability $1/2 - \text{neg}(n) > 1/4$. This is because the hiding property of the commitment scheme guarantees that a random decommitment is almost equally likely to be a 0 and a 1. Hence,

$$\Pr_{y \in \{0,1\}^n} \left[M^{\mathcal{S}^*, \pi}(y) = \pi^{-1}(y) \right] > \left(\frac{1}{4T} \right)^c$$

---

[4] $\mathcal{S}^*$ can be made stateless by using a rT-wise independent family of hash functions, namely apply a hash function to the queries and use the output as randomness for uniform sampling [GK96b].

Analysis. Note that a PSPACE oracle suffices for simulating S^* in the commit phase, whereas a PSPACE oracle and a π^{-1} oracle suffice in the reveal phase. Fix an input y to M. We want to show that with high probability, we may efficiently simulate the the computation $M^{S^*,\pi}(y)$ given oracle access to PSPACE, $\pi, \pi^{-1}_{\neq y}$.

We say that a partial transcript $(\mathbf{q}_i, \mathbf{a}_i)$ is *heavy* if

$$\frac{|\mathrm{Con}_y(\mathbf{q}_i, \mathbf{a}_i)|}{|\mathrm{Con}(\mathbf{q}_i, \mathbf{a}_i)|} > \gamma^{r+1-i}, \quad \text{where } \gamma = \left(\frac{t}{2^n}\right)^{\frac{1}{r+1}};$$

otherwise, we say that $(\mathbf{q}_i, \mathbf{a}_i)$ is *light*. In particular, ϵ is light, since $\frac{|\mathrm{Con}_y(\epsilon)|}{|\mathrm{Con}(\epsilon)|} \leq \gamma^{r+1}$. Informally, the quantity $\frac{|\mathrm{Con}_y(\cdot)|}{|\mathrm{Con}(\cdot)|}$ applied to a transcript $(\mathbf{q}_i, \mathbf{a}_i)$ is the density of "favorable" outcomes for the reduction M, wherein an outcome is favorable if in the decommitment, S^* inverts π on y. We want to show that with high probability, every transcript generated by S^* (in its interaction with M) is light, that is, the density of favorable outcomes is low.

Consider the queries M makes to S^*:

- A commit phase query of the form $\mathbf{q}_i = (\mathbf{q}_{i-1}, q_i)$. Let $\mathbf{a}_{i-1}$ be S^*'s answers to the prefixes. Observe that

$$\frac{|\mathrm{Con}_y(\mathbf{q}_{i-1}, \mathbf{a}_{i-1})|}{|\mathrm{Con}(\mathbf{q}_{i-1}, \mathbf{a}_{i-1})|} = \sum_{a_i} \frac{|\mathrm{Con}(\mathbf{q}_i, \mathbf{a}_{i-1}, a_i)|}{|\mathrm{Con}(\mathbf{q}_{i-1}, \mathbf{a}_{i-1})|} \cdot \frac{|\mathrm{Con}_y(\mathbf{q}_i, \mathbf{a}_{i-1}, a_i)|}{|\mathrm{Con}(\mathbf{q}_i, \mathbf{a}_{i-1}, a_i)|}$$

$$= \sum_{a_i} \Pr[S^*(\mathbf{q}_i) = a_i] \cdot \frac{|\mathrm{Con}_y(\mathbf{q}_i, \mathbf{a}_{i-1}, a_i|}{|\mathrm{Con}(\mathbf{q}_i, \mathbf{a}_{i-1}, a_i)|}$$

$$> \Pr[S^*(\mathbf{q}_i) \to a_i; (\mathbf{q}_i, \mathbf{a}_{i-1}, a_i) \text{ is heavy}] \cdot \gamma^{r+1-i}$$

This implies

$$\Pr[S^*(\mathbf{q}_i) \to a_i; (\mathbf{q}_i, \mathbf{a}_{i-1}, a_i) \text{ is heavy} \mid (\mathbf{q}_{i-1}, \mathbf{a}_{i-1}) \text{ is light}] < \gamma$$

- A reveal phase query of the form $(\mathbf{q}_r, \mathtt{decommit})$. Let $\mathbf{a}_r$ be S^*'s answers to $\mathbf{q}_r$. If $(\mathbf{q}_r, \mathbf{a}_r)$ is light, that is, $\frac{|\mathrm{Con}_y(\mathbf{q}_r, \mathbf{a}_r)|}{|\mathrm{Con}(\mathbf{q}_r, \mathbf{a}_r)|} \leq \gamma$, then with probability $1 - 2\gamma$, we can generate two independent random decommitments without inverting π on y.

Applying a union bound over that rT commit phase queries that M makes to S^*, we have: with probability at least $1 - rT\gamma$, in every reveal phase query $(\mathbf{q}_r, \mathtt{decommit})$ that M makes to S^*, the transcript $(\mathbf{q}_r, \mathbf{a}_r)$ is light. Taking another union bound, we deduce that with probability $1 - (r+2)T\gamma$, we may efficiently simulate $M^{S^*,\pi}$ on input y with oracle access to PSPACE, $\pi, \pi^{-1}_{\neq y}$. Hence, there is an oracle PPT $\tilde{M}$ running in time $\mathrm{poly}(T, n)$ such that

$$\Pr_{y \in \{0,1\}^n}\left[\tilde{M}^{\mathsf{PSPACE}, \pi, \pi^{-1}_{\neq y}}(y) = \pi^{-1}(y)\right] > \left(\frac{1}{4T}\right)^c - (r+2)T\gamma > \frac{1}{2}\left(\frac{1}{4T}\right)^c$$

a contradiction to π being one-way. $\qquad\square$

4.3   Lower Bounds for Interactive Hashing

Using the connection between commitment schemes and computational interactive hashing described in Section 3.2, we derive a tight lower bound for the latter [NOV06, HR06a]:

Theorem 3. *Any computational interactive hashing scheme on common input* $(1^n, 1^k)$ *has* $\Omega(\frac{n}{\log n})$ *rounds, for* $k = o(1)$.

We believe that our techniques and analysis extend readily to yield lower bounds on efficiency of the security reduction for computational interactive hashing (an open problem posed in [HR06a]) and the round complexity of random selection [DGW95, DGOW95, GSV98]. We will explore these extensions in the full version of this paper.

Acknowledgements

I would like to thank Rosario Gennaro and Luca Trevisan for very helpful technical discussions, as well as Oded Goldreich and Chiu-Yuen Koo for pointers and discussions on previous related work. I would also like to thank Salil Vadhan and the anonymous referees for helpful suggestions and constructive feedback on this work. Finally, I am grateful towards Andy Yao, Luca, the IBM Cryptography Group, and the organizers of the IPAM Workshop on Securing Cyberspace for hosting my delightful stays in Beijing, San Francisco, New York and Los Angeles, during which this work was conceived, continued, and completed.

References

[AH91]   W. Aiello and J. Håstad. Statistical zero-knowledge languages can be recognized in two rounds. *JCSS*, 42(3):327–345, 1991.

[AIK04]   B. Applebaum, Y. Ishai, and E. Kushilevitz. Cryptography in NC⁰. In *Proc. 45th FOCS*, 2004.

[B01]   B. Barak. How to go beyond the black-box simulation barrier. In *Proc. 42nd FOCS*, 2001.

[BCY91]   G. Brassard, C. Crépeau, and M. Yung. Constant-round perfect zero-knowledge computationally convincing protocols. *Theoretical Computer Science*, 84(1):23–52, 1991.

[BHZ87]   R. B. Boppana, J. Håstad, and S. Zachos. Does co-NP have short interactive proofs? *IPL*, 25(2):127–132, 1987.

[BJY97]   M. Bellare, M. Jakobsson, and M. Yung. Round-optimal zero-knowledge arguments based on any one-way function. In *Proc. Eurocrypt '97*, 1997.

[CCM98]   C. Cachin, C. Crépeau, and J. Marcil. Oblivious transfer with a memory-bounded receiver. In *Proc. 39th FOCS*, 1998.

[CS06]   C. Crépeau and G. Savvides. Optimal reductions between oblivious transfers using interactive hashing. In *Proc. Eurocrypt '06*, 2006.

[D93]   I. Damgård. Interactive hashing can simplify zero-knowledge protocol design without computational assumptions. In *Proc. Crypto '93*, 1993.

[DGOW95]   I. Damgård, O. Goldreich, T. Okamoto, and A. Wigderson. Honest verifier vs dishonest verifier in public-coin zero-knowledge proofs. In *Proc. Crypto '95*, 1995.

[DGW95]   I. Damgård, O. Goldreich, and A. Wigderson. Information theory versus complexity theory: Another test case. manuscript, 1995.

[DHRS04]   Y. Z. Ding, D. Harnik, A. Rosen, and R. Shaltiel. Constant-round oblivious transfer in the bounded storage model. In *Proc. 1st TCC*, 2004.

[DPP98]   I. Damgård, T. P. Pedersen, and B. Pfitzmann. Statistical secrecy and multibit commitments. *IEEE Transactions on Information Theory*, 4(3):1143–1151, 1998.

[F89]   L. Fortnow. The complexity of perfect zero-knowledge. *Advances in Computing Research*, 5:429–442, 1989.

[F02]   M. Fischlin. On the impossibility of constructing non-interactive statistically-secret protocols from any trapdoor one-way function. In *Proc. CT-RSA '02*, 2002.

[GGKT05]   R. Gennaro, Y. Gertner, J. Katz, and L. Trevisan. Bounds on the efficiency of generic cryptographic constructions. *SIAM Journal on Computing*, 35(1):217–246, 2005.

[GK96a]   O. Goldreich and A. Kahan. How to construct constant-round zero-knowledge proof systems for NP. *J. Cryptology*, 9(3):167–190, 1996.

[GK96b]   O. Goldreich and H. Krawczyk. On the composition of zero-knowledge proof systems. *SIAM Journal on Computing*, 25(1):169–192, 1996.

[GMR89]   S. Goldwasser, S. Micali, and C. Rackoff. The knowledge complexity of interactive proof systems. *SIAM Journal on Computing*, 18(1):186–208, 1989.

[GSV98]   O. Goldreich, A. Sahai, and S. Vadhan. Honest-verifier statistical zero-knowledge equals general statistical zero-knowledge. In *Proc. 30th STOC*, 1998.

[GT00]   R. Gennaro and L. Trevisan. Lower bounds on efficiency of generic cryptographic constructions. In *Proc. 41st FOCS*, 2000.

[HHK+05]   I. Haitner, O. Horvitz, J. Katz, C.-Y. Koo, R. Morselli, and R. Shaltiel. Reducing complexity assumptions for statistically-hiding commitment. In *Proc. Eurocrypt '05*, 2005.

[HK05]   O. Horvitz and J. Katz. Bounds on the efficiency of "black-box" commitment schemes. In *Proc. 32nd ICALP*, 2005.

[HN06]   D. Harnik and M. Naor. On the compressibility of NP instances and cryptographic applications. In *Proc. 47th FOCS*, 2006.

[HR06a]   I. Haitner and O. Reingold. A new interactive hashing theorem. ECCC TR06-096, 2006.

[HR06b]   I. Haitner and O. Reingold. Statistically-hiding commitment from any one-way function. Cryptology ePrint Archive, Report 2006/436, 2006.

[IR89]   R. Impagliazzo and S. Rudich. Limits on the provable consequences of one-way permutations. In *Proc. 21st STOC*, 1989.

[KS06]   T. Koshiba and Y. Seri. Round-efficient one-way permutation based perfectly concealing bit commitment scheme. ECCC TR06-093, 2006.

[KST99]   J. H. Kim, D. R. Simon, and P. Tetali. Limits on the efficiency of one-way permutation-based hash functions. In *Proc. 40th FOCS*, 1999.

[LTW05]   H. Lin, L. Trevisan, and H. Wee. On hardness amplification of one-way functions. In *Proc. 2nd TCC*, 2005.

[NOV06]   M.-H. Nguyen, S. J. Ong, and S. Vadhan. Statistical zero-knowledge arguments for NP from any one-way function. In *Proc. 47th FOCS*, 2006.

[NOVY98]   M. Naor, R. Ostrovsky, R. Venkatesan, and M. Yung. Perfect zero-knowledge arguments for NP using any one-way permutation.   *J. Cryptology*, 11(2):87–108, 1998.

[NV06]   M.-H. Nguyen and S. Vadhan. Zero knowledge with efficient provers. In *Proc. 38th STOC*, 2006.

[NY89]   M. Naor and M. Yung.   Universal one-way hash functions and their cryptographic applications. In *Proc. 20th STOC*, 1989.

[OVY93]   R. Ostrovsky, R. Venkatesan, and M. Yung. Fair games against an all-powerful adversary. In *AMS DIMACS Series in Discrete Mathematics and Theoretical Computer Science*, 1993.

[OW93]   R. Ostrovsky and A. Wigderson. One-way fuctions are essential for non-trivial zero-knowledge. In *ISTCS*, 1993.

[R91]   S. Rudich. The use of interaction in public cryptosystems. In *Proc. Crypto '91*, 1991.

[RTV04]   O. Reingold, L. Trevisan, and S. Vadhan. Notions of reducibility between cryptographic primitives. In *Proc. 1st TCC*, 2004.

[S98]   D. R. Simon. Finding collisions on a one-way street: Can secure hash functions be based on general assumptions? In *Proc. Eurocrypt '98*, 1998.

[T01]   L. Trevisan. Extractors and pseudorandom generators. *JACM*, 48(4):860–879, 2001.

Towards a Separation of Semantic and CCA Security for Public Key Encryption

Yael Gertner[1,*], Tal Malkin[2,**], and Steven Myers[3,***]

[1] Department of Psychology, University of Illinois at Urbana-Champaign
ygertner@cyrus.psych.uiuc.edu
[2] Department of Computer Science, Columbia University
tal@cs.columbia.edu
[3] School of Informatics, Indiana University
samyers@indiana.edu

Abstract. We address the question of whether or not semantically se-cure public-key encryption primitives imply the existence of chosen ci-phertext attack (CCA) secure primitives. We show a black-box separa-tion, following the methodology introduced by Impagliazzo and Rudich [23], for a large non-trivial class of constructions. In particular, we show that if the proposed CCA construction's decryption algorithm does not query the semantically secure primitive's encryption algorithm, then the proposed construction cannot be CCA secure.

1 Introduction

Public-key encryption primitives (PKEP) are used in numerous cryptographic protocols. Two frequently used definitions of security for PKEP in the crypto-graphic literature are semantic and chosen ciphertext attack security. Semantic security (SS) was introduced by Goldwasser and Micali [21] and guarantees that encrypted messages sent over a network are confidential to *passive adversaries* that are limited to eavesdropping (we provide formal definitions of this and the following notions in the next section). Unfortunately, in practice most adver-saries are not limited to passive eavesdropping, and they can actively control and manipulate network traffic. This is especially true on the modern Internet, where it is particularly easy to manipulate traffic. Therefore, a strengthened se-curity definition was needed. Naor and Yung [29] introduced Chosen Ciphertext Attack (CCA1) security, in which the adversary is allowed temporary access to a decryption oracle prior to the adversary's attempt to decrypt a message of interest. While this definition is substantially stronger than that of semantic se-curity, it is still not strong enough for many network purposes. Therefore, an even stronger definition of CCA security was introduced by Rackoff and Simon [31] that gives the adversary continuous access to a deprecated decryption oracle that

---

* This work was done while at the CS Dept. at U. Penn.
** This work was partially done while at AT&T Labs.
*** This work was partially done while at the CS Dept. of the University of Toronto.

S.P. Vadhan (Ed.): TCC 2007, LNCS 4392, pp. 434–455, 2007.
© International Association for Cryptologic Research 2007

is restricted only in that it will not decrypt ciphertexts of direct interest to the adversary. This security is called CCA2 (or adaptive chosen ciphertext attack) security, and is the security standard that most PKEP need to meet in many of today's cryptographic protocols. The first CCA2 secure PKEP was given by Dolev, Dwork, and Naor [11], followed by a large body of research on developing such protocols and understanding the security notion (c.f. [35,10,27,7,12]).

There are many known constructions of SS PKEPs based on general cryptographic assumptions such as trapdoor predicates[21], trapdoor functions[20,21], and trapdoor permutations[9]. In addition, these constructions are black-box and are relatively efficient. In contrast, *all known* constructions of CCA1 [29] and CCA2 [11,27,35] secure PKEPs from *general cryptographic assumptions* are based on only the existence of enhanced trapdoor permutations and are both non-black-box and inefficient due to their use of ZK or WI proofs.

In this paper we address the question of whether the weaker security requirement (semantic security) for public-key encryption, is in fact equivalent to the stronger requirement (chosen ciphertext attack security). That is, can any SS PKEP be used (without any further assumptions) to construct a CCA PKEP?

This is a natural question which is one of major open problems in cryptography in the last several years. To the best of our knowledge, the first explicit published posing of this as a problem is by Bellare et al. [4], while the most recent one is by Pass, shelat, and Vaikuntanathan [30]. In fact, the latter work addresses a similar problem, and establishes a reduction from any SS PKEP to non-malleable SS PKEP, without any further assumptions (and in a non-black box way). Non-malleable PKEP is a somewhat weaker security requirement than that of CCA2 (in particular, it is equivalent [8] to a definition where the adversary is allowed a single, parallel CCA2 query). As the authors of [30] discuss, their result does *not* generalize to a construction for general CCA security, which remains an interesting open question.

In sum, the current state of knowledge regarding the question we study, is that there is a construction of CCA PKEP from SS PKEP *with additional assumptions*, as well as a (non-black-box) construction of (the weaker) NM PKEP from SS PKEP without any further assumptions. It is not known whether there is an equivalence (whether through a black-box or a non-black-box construction) between SS PKEP and CCA PKEP.

As will be explained below, we show a black-box separation between semantic and CCA1 security for a large interesting class of constructions. This can be interpreted as evidence toward a *negative* answer to our question, or as guidance toward a *positive* answer (a reduction).

1.1   Black-Box Reductions and Separations

The existence of most modern cryptographic primitives implies $\mathcal{P} \neq \mathcal{NP}$, and thus is currently too difficult to prove unconditionally. Instead, cryptographers put a great deal of effort into constructing more complex primitives from simpler ones that are assumed to exist. In such constructions (reductions), if we assume primitives of type P exist and wish to show that a primitive of type Q exists, then we provide a construction C such that $C(M_P)$ is an implementation of Q whenever M_P is an implementation of P. This is proved by showing that any supposed adversary A_Q breaking $C(M_P)$ as an implementation of Q, can be used for an adversary algorithm A_P breaking M_P as an implementation of P.

However, almost all constructions in modern cryptography are *black-box* (for example, the equivalence of one-way functions, weak one-way functions, PRNG's, PRFG's, PRPG's and digital signatures [19,22,26,28,33].) This means, intuitively,[1] that the construction C of Q uses the implementation M_P of P as a black box (or oracle), without using the algorithmic description (actual code) of the construction. Moreover, the proof constructs the adversary A_P which uses the adversary A_Q in a black-box manner (again, using it just as an oracle, without looking at its actual code).

While it is not clear how to prove a *negative* result, namely that there exist no reduction of primitive Q to primitive P, Impagliazzo and Rudich [23] initiated a methodology for proving that no *black-box* reductions exist. Specifically, their methodology involved proving that no *relativizing* reduction exists (note that black box reductions must relativize). This is done by exhibiting an oracle relative to which an implementation of P exists, while an implementation of Q does not.[2] Using this methodology, [23] proved a black-box separation between key agreement and one-way functions. A line of subsequent works used this methodology or new variants to show black-box separations among various other cryptographic primitives (c.f. [34,6,36,24,16,17]), and to show that black-box constructions suffer from inherent efficiency limitations [25,14,15].

Non-Black-Box Constructions. We note that while the vast majority of constructions in cryptography are black-box, there are several results that are non-black-box (importantly, all known constructions of CCA secure PKEP from generic assumption are non-black-box). Many of these constructions are based on using Zero-Knowledge (ZK) or Witness Indistinguishable (WI) proofs (both interactive and non-interactive) in the construction.[3] These proofs are often used

---

[1] There are actually several subtleties and different types of black-box reductions of varying strengths, c.f. [32]. However, this intuitive description suffices for our presentation purposes here.

[2] Even here it's not immediately clear how to make this approach work, since the construction and its proof of security could always ignore the presence of the oracle and independently realize the primitive Q. To address this problem, Impagliazzo and Rudich [23] give a model in which one can prove separations modulo some major results in complexity theory. In their model they begin by assuming that $\mathcal{P} = \mathcal{NP}$, and adding an oracle O relative to which P exists and Q does not, implying that a black-box reduction would yield a proof that $\mathcal{P} \neq \mathcal{NP}$. Subsequent work, starting with Simon [36], used a stronger approach that embeds a PSPACE complete portion into the oracle O before proving that relative to O P exists but Q does not. This yields an unconditional proof that no relativizing (and thus no black-box) reductions exist. Other subsequent work (e.g. [17]) relaxed this approach to obtain a weaker black-box separation methodology.

[3] Perhaps the only exception is the works of Barak [2,3] who has shown the existence of some protocols that are non-black-box, and that do not make use of ZK techniques.

to prove some property about the circuit description of a cryptographic primitive, and thus require the primitive to have a circuit description, and so are not black-box. Examples of such constructions include the development of PKEP that are secure against chosen ciphertext attacks [29,35], assuming (enhanced) trapdoor permutations exist[4]. Unfortunately, the protocols that perform such proofs are invariably far too inefficient for practical deployment of the resulting cryptographic primitive (although, they are still polynomial time, they are of a degree that is too large to be practical), thus further justifying the quest for black-box constructions.

The Meaning of Black-Box Separations in Cryptography and Our Scenario. In general, a black-box separation can be interpreted as evidence that a reduction of Q to P is unlikely using current techniques, or at least that it is unlikely to be efficient (as black-box reductions seem to be much more efficient than non-black-box ones). Such results may also be viewed as guiding which approaches to take when trying to actually prove a reduction exists. We refer the reader to the previous literature on black-box separations, e.g. [23,32], for a more in-depth discussion of the meaning and importance of black-box separations in cryptography.

In the particular scenario of the black-box constructions of CCA secure PKEP from SS secure ones, we can view a separation as pointing to several possibilities:

- The need to develop some form of appropriate ZK or WI proofs based on semantic security (and such a direction is attempted in [30]), but such constructions are still likely to be inefficient.
- The need to develop more non-black-box techniques that are more efficient and applicable to the scenario of public-key encryption.
- In the failure of the latter two points, any construction of a CCA secure primitive derived solely from the hardness of a SS secure PKEP will be inefficient, or need to take into account specifics of the assumption that are not generic. For instance, any CCA cryptosystem based on SS PKEP proposed by Ajtai and Dwork [1], that results from the assumed hardness of a lattice problem, will either be too inefficient to be practically useful due to the need to use inefficient non-black-box techniques, or will require a unique construction whose proof of security relies on specific properties of the lattice assumption. This direction might include finding efficient ZK or WI proofs based on the specific hardness assumption under consideration.

1.2 Our Contributions

We prove the following:

Theorem (informal statement): There exists no black box reduction that from a given SS PKEP (g, e, d) constructs a CCA1 secure scheme

---

[4] Both of these results actually only need the requirement that certain types of non-interactive zero-knowledge proofs exist, and these proofs are known to exist relative to enhanced trapdoor permutations.

$(G^{g,e,d}, E^{g,e,d}, D^{g,d})$ We call such reductions (where the new primitive's decryption algorithm does not query the underlying primitive's encryption algorithm) *shielding reductions*. Our result, then, rules out any shielding black box reductions of CCA1 PKEP to SS PKEP. Consequently, the only possible constructions of a CCA1 (and thus also of CCA2) secure PKEP from a SS PKEP must either be non-black-box, or have its decryption algorithm use the encryption algorithm of the underlying scheme in an essential way.

Our Model and Proof Technique. The proof essentially follows the IR [23] methodology, showing that there is no (shielding) black-box reduction. This is done by introducing an oracle $\mathcal{O}$ relative to which there exists a SS PKEP $\mathbf{O} = (\mathbf{g}, \mathbf{e}, \mathbf{d})$, but no CCA secure PKEP $(\mathbf{G}^{\mathbf{O}}, \mathbf{E}^{\mathbf{O}}, \mathbf{D}^{\mathbf{g,d}})$ exists relative to $\mathcal{O}$.[5] Our oracle $\mathcal{O}$ includes $(\mathbf{g}, \mathbf{e}, \mathbf{d})$, where $\mathbf{g}, \mathbf{e}$ are random functions. If there were no other parts to the oracle, the proof of semantic security would be immediate, but then $\mathbf{O}$ would in fact be CCA secure as well. Thus, we add more "weakening" components to $\mathcal{O}$, which make the proof of semantic security a little harder but still relatively simple, but make $\mathbf{O}$ and any other candidate scheme $(\mathbf{G}^{\mathbf{O}}, \mathbf{E}^{\mathbf{O}}, \mathbf{D}^{\mathbf{g,d}})$ vulnerable to CCA1 attacks. The latter is the technical heart of the proof, which is quite complex. We chose to expand on the intuition and main ideas of the proof. The full proof with all technical detail appears at [18].

For clarity of presentation, we start by thinking of all participants as being computationally unlimited, but restricted to making a polynomial number of polynomial sized oracle queries to the oracle $\mathcal{O}$. This already gives an interesting result, and encompasses all the main issues in the proof. Because the constructed adversary in the proof only uses more than a polynomial amount of time (i.e. its computationally unlimited powers) to search for and randomly choose efficiently verifiable strings, it is therefore possible to remove the requirement of computationally unlimited parties and replace it with the ability of randomly choosing $\mathcal{NP}$ witnesses. The proof can then be extended to support computationally bounded parties, by adding a PSPACE complete component to the oracle (or assuming $\mathcal{P} = \mathcal{NP}$), achieving the standard separation model of [23] and most subsequent work.

It may seem that if a construction of a CCA secure scheme $(\mathbf{G}, \mathbf{E}, \mathbf{D})$ from any SS scheme $(\mathbf{g}, \mathbf{e}, \mathbf{d})$ exists, it would be unnatural for $\mathbf{D}$ to call $\mathbf{e}$. After all, e is intended to be used by parties that do not require knowledge of any secret keys, thus using it in an essential way for a decryption algorithm seems counter intuitive.

However, we show that relative to our oracle $\mathcal{O}$, there *is* in fact a CCA2 secure scheme, where $\mathbf{D}$ uses $\mathbf{e}$ (namely a non-shielding black-box construction).

---

[5] This does not exactly follow the IR methodology, because it does not rule out any relativizing reductions (as the new primitive's $(\mathbf{G}, \mathbf{E}, \mathbf{D})$ algorithms do not have access to the entire oracle $\mathcal{O}$, only to the underlying primitive's algorithms). Nonetheless, it rules out all (shielding) black box reductions.

The basic idea behind this scheme is the following (full details can be found in the full version [18]). To encrypt a message bit b with a random string r, first encrypt b using e with a public-key pk (and the randomness provided by the string r), and then encrypt all the individual bits of r as well using the same public-key, using new random strings derived *deterministically* from r (for example $r + 1, r + 2, \ldots$).

This CCA2 secure (relative to our oracle) primitive implies that the shielding limitation on the decryption algorithm in our theorem is inherent for our oracle (and not just a gap in our proof analysis). On the other hand, note that this scheme is *artificial*, and makes heavy use of the fact that e is a random function, by using new random strings deterministically derived from r (this technique is legitimate when the encryption function is *truly random*, but does not work in general). In fact, based on standard hardness assumptions, it is easy to show that there exist semantically secure PKEP relative to which the above construction does not achieve CCA2 security. Similarly, the construction of Fujisaki and Okamoto [13], which is a black-box construction of CCA PKEP from SS PKEP in the random oracle model, is also non-shielding, but this again heavily relies on the random oracle property.

This leaves open the possibility of using the weaker form of black-box separation of [17] to separate CCA1 security from semantic security without any restrictions on the black box reduction.[6]

We feel that closing this gap and answering whether a black box reduction where the CCA decryption algorithm does invoke the SS encryption algorithm exists, is a very interesting and non-trivial problem for future research. While our work does not completely answer the question of whether CCA secure PKEP can be constructed from SS ones without any further assumptions, we do make significant progress toward that direction.

ORGANIZATION. In Section 2 we formally define the notion of PKEP and the definitions of semantic, CCA1 and CCA2 security. This is followed in Section 3 by a description of our oracle construction, and a proof sketch that relative to such oracles with overwhelming probability there is an SS PKEP. In Section 4 we present our separation theorem and sketch its proof. In Section 5 we provide an example construction on which the various parts of the CCA1 attack are demonstrated, and in Section 6 we briefly discuss why our result transfers to the more tradition model that assumes $\mathcal{P} = \mathcal{NP}$ or that includes a PSPACE oracle.

---

[6] In fact, using this weaker separation model of [17], we can show that there are no black-box reductions of CCA1 to semantic security for another non-trivial class of constructions, which includes the artificial example mentioned above. Specifically, this is the class where **D** does invoke e in a certain way, where for every successful decryption query $\mathbf{d}(sk, c) \in \{0, 1\}$ there is a corresponding invocation of $\mathbf{e}(\mathbf{g}(sk), *, *) = c$ (or very roughly, when D invokes e "in every possible opportunity"). The difficult case for which we do not know how to prove a separation, is the intermediate case where D (roughly) must invoke e in an essential way sometimes, but not other times.

2   Preliminaries and Definitions

2.1   Notation

Given a set S we use the notation $x \in_R S$ to denote the process of choosing x uniformly at random from S. Given a function $f : \mathbb{N} \to \mathbb{R}$, we say it is *negligible* if for all sufficiently large $n \in \mathbb{N}$ and for all $c \in \mathbb{N}$: $f(n) \leq n^{-c}$.

2.2   Definitions of PKEPs

Below we give the formal definitions of PKEPs and the notions of semantic, CCA1 and CCA2 security.

Definition 1 (PKEP). *A public-key encryption primitive is a triple of (G, E, D) of algorithms: G and E are probabilistic while D is deterministic. Let p_1 and p_2 be polynomials specified by the PKEP.*

- *for every n, for every $r \in \{0,1\}^n$ $G(r)$ outputs a pair of keys (sk, pk).*
- *for every $m \in \{0,1\}^{p_1(n)}$, each string $r' \in \{0,1\}^{p_2(n)}$ of coin tosses of E and pair (sk, pk) output by G on some input $r \in \{0,1\}^n$, it holds that $D(sk, E(pk, m, r')) = b$.*

We note that while this definition requires correct decryption our proof can be easily modified to allow for some error in decryption in any purported CCA1 construction.

Next, we give the definitions of semantic, CCA1 and CCA2 security. The definitions are presented concurrently.

Definition 2. *Let $\mathcal{EP} = (G, E, D)$ be a PKEP. Let $A = (A_1, A_2)$ be a probabilistic adversary that is described in two parts, each of which has access to an oracle.*

The PKEP $\mathcal{EP}$ is atk-secure, where $atk \in \{SS, CCA1, CCA2\}$, if there exists a negligible function μ such that for every adversary $A = (A_1, A_2)$ and for all sufficiently large $n \in \mathbb{N}$:

$$\Pr_{\substack{s \in_R \{0,1\}^n,(pk,sk) \leftarrow G(s) \\ (x_0,x_1,\sigma) \leftarrow A_1^{O_1}(pk) \\ b \in_R \{0,1\};r \in_R \{0,1\}^{p_2(n)} c \leftarrow E(pk,x_b,r)}} [A_2^{O_2}(\sigma, c) = b] \leq \frac{1}{2} + \mu(n),$$

*where σ represents state information communicated between the parts of the adversary, c represents a **challenge ciphertext** and :*

- *if atk=SS then O_1 and O_2 are the null oracle: the oracles give the empty response, $\bot$, to all queries;*
- *if atk=CCA1 then $O_1(\cdot) = D(sk, \cdot)$, and O_2 is the null oracle;*
- *if atk=CCA2 then $O_1(\cdot) = D(sk, \cdot)$, and $O_2(\cdot) = D(sk, \cdot)$ but modified on the encryption challenge so that $O_2(c) = \bot$.*

In the case of SS and CCA1 security it is known that there are black-box reductions in both directions between PKEP that encrypt 1-bit messages and PKEP that encrypt n-bit messages (for the direction going from encrypting 1 to n bits, it is easy to see that the concatenation of independent encryptions works as a construction). We make use of this fact in our result, and focus on primitives that encrypt the message space of only one bit. Clearly the above definitions simplify slightly in this case (i.e. $x_0 = 0$ and $x_1 = 1$).

3   The Oracle

We define an experiment that produces an oracle that effectively implements a PKEP that is semantically secure but not CCA1 secure. We think of the oracle as consisting of 5 sub-oracles $(\mathbf{g}, \mathbf{e}, \mathbf{d}, \mathbf{w}, \mathbf{u})$, but this can easily be unified into one oracle by appropriate coding. This security of the oracle if achieved by effectively defining g, e to be appropriate random length increasing functions, and defining d appropriately, so that it can appropriately decrypt these function. This easily gives a secure PKEP, alas it is too secure (CCA2). Therefore, in order to weaken its security a fourth component of the oracle $\mathbf{w}$ is added which given a public-key pk for (g, e, d) will output an encrypted version of the secret-key. This is of no use to the adversary in the SS definition of security, but makes it trivial for a CCA adversary to break the primitive's security. Finally, a fifth sub-oracle $\mathbf{u}$ is added that gives the adversary the ability to determine the legitimacy of public-keys and ciphertexts (i.e., those that could legitimately be output by $\mathbf{g}$ and $\mathbf{e}$); this sub-oracle is not necessary for the result, but substantially simplifies an already technical proof.

Definition 3 (Oracle Distribution). *Let* $\mathcal{O} = (\mathbf{g}, \mathbf{e}, \mathbf{d}, \mathbf{w}, \mathbf{u}) \leftarrow \Upsilon$ *denote an oracle that is chosen randomly according to the distribution described below. For each* $n \in \mathbb{N}$ *let:*

$\mathbf{g}$: $\{0,1\}^n \rightarrow \{0,1\}^{3n}$ *be a random one-to-one function. (*$\mathbf{g}$ *generates public-keys given secret-keys.)*

$\mathbf{e}$: $\{0,1\}^{3n} \times \{0,1\} \times \{0,1\}^n \rightarrow \{0,1\}^{3n}$ *where for every* pk, *the function* $\mathbf{e}(pk, \cdot, \cdot)$ *is a uniformly at random selected one-to-one function. (*$\mathbf{e}$ *takes a public-key, a message bit and a random string, and outputs a ciphertext.)*

$\mathbf{d}$: $\{0,1\}^n \times \{0,1\}^{3n} \rightarrow \{0,1,\perp\}$ *where for every* sk, c *and* b *set* $\mathbf{d}(sk, c) = b$ *if there exists an* r *such that* $\mathbf{e}(\mathbf{g}(sk), b, r) = c$; *otherwise set* $\mathbf{d}(sk, c) = \perp$. *(*$\mathbf{d}$ *takes a secret-key and ciphertext and outputs the corresponding decryption.)*

$\mathbf{w}$: $\{0,1\}^{3n} \times \{0,1\}^n \rightarrow \{0,1\}^{3n \times n}$ *where for each* pk *and* j *set* $\mathbf{w}(pk, j) = \perp$ *if* $\mathbf{g}^{-1}(pk)$ *is undefined; otherwise, if* $\mathbf{g}^{-1}(pk) = sk \stackrel{defn}{=} (sk_1, ..., sk_n)$, *set* $\mathbf{w}(pk, j) = \mathbf{e}(pk, sk_1, r_{pk,1,j}), \ldots, \mathbf{e}(pk, sk_n, r_{pk,n,j})$, *where for* $1 \leq k \leq n$ *let* $r_{pk,k,j} \in_R \{0,1\}^n$. *(*$\mathbf{w}$ *takes a public-key and an index as input, and outputs a bit-by-bit encryption of the public-key's corresponding secret-key.)*

$\mathbf{u}$: $\{0,1\}^{3n} \times \{0,1\}^{3n} \rightarrow \{\top, \perp\}$ *where for each* pk *and* c *set* $\mathbf{u}(pk, c) = \top$ *if there exists an* sk, b *and* r *such that* $\mathbf{g}(sk) = pk$ *and* $\mathbf{e}(pk, b, r) = c$; *otherwise, set* $\mathbf{u}(pk, c) = \perp$. *(*$\mathbf{u}$ *takes a public-key and a string, and determines if the string corresponds to an encryption relative to the public-key.)*

NOTATION. In order to ease discussions of queries to an oracle $\mathcal{O}$, we briefly introduce some notation. Given an oracle $\mathcal{O}$ we often say that $\mathcal{O} = (\mathbf{O}, R)$ where $\mathbf{O} = (\mathbf{g}, \mathbf{e}, \mathbf{d})$ denotes the sub-oracles corresponding to the encryption primitive, and $R = (\mathbf{u}, \mathbf{w})$ corresponds to the security weakening sub-oracle $\mathbf{w}$ and the helper oracle $\mathbf{u}$. We denote by (o, q) the query q to the sub-oracle $o \in \{\mathbf{g}, \mathbf{e}, \mathbf{d}, \mathbf{w}, \mathbf{u}\}$ in $\mathcal{O}$. For example, we denote by $(\mathbf{g}, sk)$ the query $\mathbf{g}(sk)$. Similarly, we denote by the pair $(< o, q >, r)$ the response r to the query q made to the sub-oracle o. We call such a pair a query/response, and say a query/response $(< o, q >, r)$ is consistent with o if $o(q) = r$. In cases where a query $q = (v_1, .., v_i)$ is represented by several semantically different strings $v_1, .., v_i$ we denote by $(< o, v_1, .., v_{j-1}, *, v_{j+1}, .. v_i >, r)$ the fact that there exists a v_j such that the oracle query $\mathbf{o}(v_1, v_2, ..., v_i)$ was made and the response was r. For example $(< \mathbf{e}, (pk, *, r) >, c)$ represents the notion that there exists a bit $b \in \{0, 1\}$ such that $(< \mathbf{e}, (pk, b, r) >, c)$ represents a query/response consistent with the sub-oracle $\mathbf{e}$.

The following theorem states that this oracle provides semantic security for the PKEP $(\mathbf{g}, \mathbf{e}, \mathbf{d})$.

Theorem 1. *For every oracle adversary A limited to a polynomial number of oracle queries, there exists a negligible function μ such that for all sufficiently large n:*

$$\Pr_{\mathcal{O} \leftarrow \Upsilon} \left[\Pr[A^{\mathcal{O}}(pk, c) = b] \leq 1/2 + \mu(n) \right] \geq 1 - 1/2^{n/2}$$

where the interior probability is over the choice of $sk \in_R \{0, 1\}^n, b \in_R \{0, 1\}$, $r \in_R \{0, 1\}^n$ and any coin flips performed by A. Further, $pk = \mathbf{g}(sk)$ and $c = \mathbf{e}(pk, b, r)$.

Proof Sketch: If $\mathcal{O}$ consisted of only the sub-oracles $\mathbf{g}, \mathbf{e}$ and $\mathbf{d}$, then security would follow directly from their probabilistic construction (in a way which is by now standard, c.f. [23,16]). To ensure that $\mathbf{w}$ and $\mathbf{u}$ do not destroy this security, it is shown that the adversary can effectively simulate the responses of these oracles. An adversary can simulate the response to a query $\mathbf{u}(pk, c)$ by outputting b if there has been a previous query/response $(< \mathbf{e}, pk, b, * >, c)$, and otherwise outputting $\perp$. When b is output the simulation is clearly correct, and when outputting $\perp$ the simulation is correct with high probability, as the ability of the adversary to find a value c such that $\mathbf{e}(pk, *, *)^{-1}(c) \neq \emptyset$ is negligible (in n) due to the random selection of $\mathbf{e}$ (again, following a standard argument). Similarly, $\mathbf{w}(pk, i)$ can be simulated if there has previously been a query/response of the form $(< \mathbf{g}, sk >, pk)$ by outputting a random encryption of sk, and otherwise outputting $\perp$.

It is not hard to verify that $(\mathbf{g}, \mathbf{e}, \mathbf{d})$ is not secure against a CCA1 attack: The adversary A_1 takes the input pk, queries $\mathbf{w}(pk, 0)$ to get an encrypted version of sk, and then uses its CCA1 access to the decryption oracle to decrypt sk. sk is then passed to A_2, which uses it to evaluate and output $\mathbf{d}(sk, c)$. In the next section we show that in fact any shielding construction is vulnerable to a (possibly much more complex) CCA1 attack.

4   The Separation

4.1   A Large Class of Constructions: Shielding Reductions

In order to state a proper theorem that provably restricts the class of black-box constructions capable of being CCA1 secure, this class needs to be formally defined. Let $\mathbf{O} = (\mathbf{g}, \mathbf{e}, \mathbf{d})$ be a semantically secure PKEP. We will consider constructions $(\mathbf{G}^{\mathbf{O}}, \mathbf{E}^{\mathbf{O}}, \mathbf{D}^{\mathbf{O}})$ of PKEPs that are purportedly CCA1 secure. We require that there exist constants ρ_0, ρ_1, ρ_2 and ρ_3 such that for all sufficiently large $n \in \mathbb{N}$ we have:

- $\mathbf{G}^{\mathbf{O}} : \{0,1\}^n \to \{0,1\}^{n^{\rho_0}} \times \{0,1\}^{n^{\rho_1}}$. $(\mathbf{G}^{\mathbf{O}}(S) = (SK, PK))$
- $\mathbf{E}^{\mathbf{O}} : \{0,1\}^{n^{\rho_1}} \times \{0,1\} \times \{0,1\}^{n^{\rho_2}} \to \{0,1\}^{n^{\rho_3}}$ $(\mathbf{E}^{\mathbf{O}}(PK, M, R) = C)$
- $\mathbf{D}^{\mathbf{O}} : \{0,1\}^{n^{\rho_0}} \times \{0,1\}^{n^{\rho_3}} \to \{0,1\} \cup \{\bot\}$. $(\mathbf{D}^{\mathbf{O}}(SK, C) = M)$

In the above definition we consider n the security parameter for the PKEP. We make several assumptions without loss of generality: each of the algorithms on inputs corresponding to security parameter n make exactly n^q queries to $\mathbf{O}$ of size at most n^s, that no duplicate queries are made, that $\mathbf{G}$ never queries $\mathbf{d}$ (it can predict the responses itself), and that the triple satisfies the PKEP correctness property so long as $\mathbf{O}$ does (i.e., all ciphertexts decrypt properly, but again this assumption can be weakened so that random encryptions decrypt properly with some probability greater than $1/2$.).

The important assumption we make is that $\mathbf{D}$ does not query $\mathbf{e}$; this is formally what we mean by a shielding construction. This assumption does result in loss of generality and is what is responsible for the restriction in our separation of CCA1 and Semantic Security. This assumption is required in order for latter hybridization experiments to go through. Further, as discussed in the introduction, using the oracle given in this paper, it is possible to construct a non-shielding CCA2 secure PKEP, implying that this assumption is necessary for our oracle.

4.2   Separation Theorem

From this point on, fix an arbitrary PKEP construction $(\mathbf{G}, \mathbf{E}, \mathbf{D})$ that satisfies all of the assumptions of Section 4.1 (in particular, it is shielding).

Theorem 2. *There exists a CCA1 adversary $A = (A_1, A_2)$ for which it's the case that for all sufficiently large n:*

$$\Pr_{\substack{\mathcal{O}=(\mathbf{O},R)\leftarrow \Upsilon \\ S\in_R\{0,1\}^n, M\in_R\{0,1\}, R\in_R\{0,1\}^{n^{\rho_2}(n)} \\ (PK,SK)\leftarrow G^{\mathbf{O}}(S), C\leftarrow E^{\mathbf{O}}(PK,M,R)}} \left[A_1^{\mathbf{D}^{\mathbf{O}}(SK,\cdot),\mathcal{O}}(PK) \to \sigma; A_2^{\mathcal{O}}(\sigma, C) = M \right] \geq 1-1/n.$$

A simple averaging argument then shows that for almost every selection of $\mathcal{O}$, the adversary breaks the $CCA1$ security of the PKEP. Combining this with a simple counting argument shows that there exists a specific oracle relative to which $\mathbf{O} = (\mathbf{g}, \mathbf{e}, \mathbf{d})$ is semantically secure, but where $(\mathbf{G}, \mathbf{E}, \mathbf{D})$ is not.

The main idea behind our oracle separation is as follows. since we want to construct a CCA1 attack, where the adversary only has access to the decryption oracle before it receives the challenge ciphertext, this access cannot be used to learn something specific about the challenge. Therefore, the goal of our adversary will be to use the decryption oracle access to learn enough information on the secret key, that will allow it to later decrypt the challenge ciphertext with good probability. This will be done by reconstructing a secret-key SK', corresponding to its public-key; which will later be used with the decryption algorithm $\mathbf{D}$ to decrypt the challenge ciphertext.

In our black-box model, where parties are computationally unlimited but limited in the number of oracle queries they can make, all security of the constructed primitive $(\mathbf{G}, \mathbf{E}, \mathbf{D})$ must stem from the oracle PKEP $(\mathbf{g}, \mathbf{e}, \mathbf{d})$. Therefore, it seems intuitive that the only secret and usable information that an execution of $G^{\mathbf{O}}(S) \rightarrow (PK, SK)$ embeds in SK are the strings sk for which the corresponding strings $pk = \mathbf{g}(sk)$ have been embedded in PK (It is known by the work of Impagliazzo and Rudich [23] that the construction needs to use the 'trapdoorness' of the oracle if it hopes to be secure, as a random-oracle —such as that provided simply by using only the sub-oracles $\mathbf{g}$ and $\mathbf{e}$— is insufficient to achieve even semantic security). Therefore, our adversary's goal will be to retrieve such sk strings by using the decryption oracle. Clearly, the adversary will additionally have to make use of the sub-oracle $\mathbf{w}$, for without the presence of this oracle, the scheme $(\mathbf{g}, \mathbf{e}, \mathbf{d})$ is CCA1 secure. Once such embedded sk are retrieved, the adversary must learn how to use them to actually construct an appropriate SK' and decrypt the challenge ciphertext. Unfortunately, most of these steps are non-trivial, and the adversary is not able to generate a key SK' that can decrypt every ciphertext. Instead, we focus on the ability of finding an SK' that can be used to decrypt the average ciphertext generated by an execution of $\mathbf{E}^{\mathbf{O}}(PK, M, R)$ for randomly chosen M and R, as this is exactly the distribution from which the adversary's challenge ciphertext will come. Below we give a very high-level description of the steps an adversary must perform to decrypt a challenge ciphertext for the given PKEP.

The large probabilistic experiment (CCA1 attack) that the adversary will perform is broken to the following three parts (given in more detail below).

- In the first part, the adversary uses its input PK to learn the relevant public keys pk and ciphertexts c that are embedded in it.
- In the second part, the adversary's goal is to learn secret keys sk corresponding to the keys pk recovered in the first stage. Note that for each pk, access to $\mathbf{w}$ gives the adversary an encrypted version of sk (encrypted with respect to $\mathbf{e}(pk, *, *)$). The adversary also has access to a decryption oracle $\mathbf{D}(SK, *, *)$. Thus, the goal in this stage will be to 'embed' the encryptions of sk into useful ciphertexts C that can be fed to the decryption oracle, and whose decryptions can be translated back to decryptions of sk. We do not necessarily achieve this goal for all the sk that correspond to pk collected in

the first part. However, we achieve it for enough such sk that make the third part go through.

- In the third part, the adversary uses the keys sk constructed in the second part, in order to construct an SK' such that $\mathbf{D}(SK', C)$ decrypts correctly with high probability for randomly chosen ciphertexts C.

The technical heart of the proof is in the second and third parts, where enough sk should be retrieved to enable a good construction of SK'. Below we provide more technical detail on each of these parts, sketching some of the obstacles encountered and their solutions. The experiment is then described in Section 5 for a specific PKEP construction example that demonstrates several different cases in the proof.

A Caveat. We point out that the description here assumes that certain highly unlikely probabilistic events never occur. Examples of such events are the adversary making queries of the form $\mathbf{d}(sk, c) \neq \bot$ when there has never previously been a query of the form $\mathbf{g}(sk) = pk$ or a query $\mathbf{e}(pk, *, *) = c$; or that estimations of specific values retrieved through sampling deviate substantially from the actual value they estimate. In the full version, these bad events are specified, and their possibility of occurring is taken into account in the analysis. To simplify presentation here, it is simply assumed they do not occur.

The Environment and the First Part of the Experiment: Learning about PK. Define the environment that the adversary is operating in to consist of the oracle $\mathcal{O} = (\mathbf{O}, R)$ that was chosen by Υ in the probabilistic statement of the theorem, as well as the seed S selected to generate the public- and secret-key pair $(PK, SK) = G^{\mathbf{O}}(S)$, where PK is given to the adversary, and access to the decryption oracle $\mathbf{D}^{\mathbf{O}}(SK, \cdot)$ is initially given to the adversary. These are fixed for the remainder of the description of all three parts.

The first part of the experiment learns some basic facts about the semantically secure PKEP $\mathbf{O}$, and it learns which $pk \in \mathbf{g}(\{0,1\}^*)$ are 'embedded' in the public-key PK. The determination of these pk is done by sampling a large number of executions of $\mathbf{E}^{\mathbf{O}}(PK, M, R)$ for randomly chosen M and R and looking for queries of the form $(\mathbf{e}, pk, *, *)$. If such queries are made, then it is reasonable to assume that pk *might* be embedded into PK. Note there are two issues that immediately arise here: first, there might be values of pk retrieved that have been arrived at during the execution of $\mathbf{E}$ by the response to some query $\mathbf{g}(sk)$ (rather than being embedded in PK). However, such values can easily by filtered out by monitoring queries to $\mathbf{g}$. The other issue is that there might very well be pk embedded in PK that are never retrieved by this sampling process, but we can safely ignore them, as the fact that they do not show up in this sampling suggests that they are not used during most encryptions of $\mathbf{E}^{\mathbf{O}}(PK, M, R)$ for randomly chosen M and R. Let KS be the set of public-keys pk retrieved in the first part of the experiment.

The final thing done in this part of the experiment is that a set $\mathcal{E}$ of specific encryptions output by $\mathbf{e}$ during the executions of $\mathbf{E}$ is created. This is done because some specific encryption c output by $\mathbf{e}$ may be consistently embedded into encryptions C produced by $\mathbf{E}$ (i.e, this information is encoded into PK). Later, the decryption algorithm $\mathbf{D}(SK, \cdot)$ may check for the presence of the embedding of c in C, and refuse to decrypt C if c is missing. Knowledge of such $c \in \mathcal{E}$ will be necessary in the second and third parts of the experiment.

To summarize, at the end of the first stage the adversary has a list KS of public keys pk and a list $\mathcal{E}$ of ciphertexts c (with respect to the system $\mathbf{O} = (\mathbf{g}, \mathbf{e}, \mathbf{d})$), that were encountered during a large number of random executions of the encryption protocol $\mathbf{E}^{\mathbf{O}}(PK, *, *)$. Intuitively, KS corresponds to the public keys pk embedded into PK, and $\mathcal{E}$ include ciphertexts embedded into PK.

The Second Part of the Experiment: Retrieving sk Embedded in SK.
In the second part of the experiment the adversary attempts to retrieve a subset of $\mathbf{g}^{-1}(KS)$ to be used to later construct the alternate secret-key SK'. Again, the intuition is that the values in $\mathbf{g}^{-1}(KS)$ that are embedded in SK must be responsible for the purported security of the primitive $(\mathbf{G}, \mathbf{E}, \mathbf{D})$.

Note that for any $pk \in KS$, the adversary can query $\mathbf{w}(pk, 0) = (e_1, ..., e_n)$, where the response $(e_1, \ldots, e_n)$ represents the bit-wise encryption of sk (with respect to pk). Thus, the adversary's goal is to embed these bit encryptions e_i into ciphertexts C whose decryption (obtained from the decryption oracle) helps decrypt e_i to obtain sk. This is done using the following idea (presented in an over simplified form).

Imagine that during the execution of a random encryption of the message M made by $\mathbf{E}^{\mathbf{O}}(PK, M, R)$ there is a query made to $\mathbf{e}(pk, b, r)$ in order to encrypt a bit b for a $pk \in KS$, but which has the property that when one replaces the query's response with a random encryption $\mathbf{e}(pk, 1 - b, *)$ of the bit $1 - b$, the resulting ciphertext C' output by $\mathbf{E}$ will decrypt to something other than M (we say that it decrypts *improperly* since M is not output); but when one replace the query's response with a random encryption $\mathbf{e}(pk, b, *)$ the resulting ciphertext C'' decrypts to M (respectively, we say it decrypts *properly*). Call such a query $\mathbf{e}(pk, b, r)$ *decisive* with respect to pk. If we can find such decisive queries, then the adversary can use the decryption oracle in conjunction with the encryptions $(e_1, \ldots, e_n)$ obtained from $\mathbf{w}$, to retrieve $sk = \mathbf{g}^{-1}(pk)$. This is done by re-executing $\mathbf{E}(PK, M, R)$ n times, where in the ith iteration it replaces the response to the query $\mathbf{e}(pk, b, r)$ with e_i. In the ith case call the output of $\mathbf{E}$ C_i. If C_i decrypts to M (as discovered with the adversary's decryption oracle), then the adversary knows that the ith bit of sk is b and otherwise it is $1 - b$. Therefore, it can retrieve $sk = \mathbf{g}^{-1}(pk)$.

The question is how does the adversary find such decisive queries. There are actually two issues here: how does the adversary know which pk have decisive queries, and assuming it knows that a pk has decisive queries, which query $\mathbf{e}(pk, *, *)$ made during a random encryption $\mathbf{E}^{\mathbf{O}}(PK, M, R)$ is decisive. Assume for the moment that we know that with high probability over the choice of M and R that there is (on average) a decisive query with respect to pk made during an execution of $\mathbf{E}^{\mathbf{O}}(PK, M, R)$. The adversary can perform n^q (the largest number of queries made by $\mathbf{E}$) hybridization experiments, where in the ith experiment a large number of encryptions $\mathbf{E}(PK, M', R')$ are performed (for randomly chosen M' and R') but in each of these the first i responses to queries of the form $\mathbf{e}(pk, b, *)$ are replaced with random encryptions of bits $\mathbf{e}(pk, b', r')$ (b' and r' randomly chosen), and the responses to the remainder of the queries $\mathbf{e}(pk, *, *)$ are left unaltered. Since we have assumed that such a decisive query must exist, then there will be an $i < n^q$ such that there is a significant increase in the fraction of improper decryptions in the ith and the $(i+1)^{\text{th}}$ experiments. In this case, we can think of the ith query of the form $\mathbf{e}(pk, *, *)$ as being decisive in an execution of $\mathbf{E}^{\mathbf{O}}(PK, M, R)$. Of course this is only true on average, so we cannot deduce the value of any bit of $\mathbf{g}^{-1}(pk)$ with a single call to the decryption oracle using the decisive encryption. However, for each bit of sk, we can perform a sampling experiment to retrieve it.

Note that it is the creation of these hybrid ciphertexts that requires us to introduce the shielding restriction. If the construction were not shielding, there is no guarantee that any of the hybrid ciphertexts would decrypt, and therefore the experiment would be useless for retrieving keys. The reason is that, roughly speaking, the reduction being shielding means that in the process of encrypting $\mathbf{E}(PK, M, R)$, replacing one random encryption of $\mathbf{e}(pk, b, r)$ with another random encryption $\mathbf{e}(pk, b, r')$ of the same bit b, should not be noticeable to the decryption algorithm $\mathbf{D}$, and thus still result in a proper decryption (while replacing with an encryption of another bit may result in an improper decryption).

The above explanation assumes that the adversary already knows that a particular $pk \in KS$ will have (on average) a *decisive* query during a random execution of $\mathbf{E}^{\mathbf{O}}(PK, M, R)$. But this presents a serious challenge: how does the adversary even know whether any pk has a decisive query, let alone figure out which pk has one? And what to do about pk that do not have decisive queries? In fact, it is possible that no individual pk has any decisive queries. Consider for example a scheme $\mathbf{E}$ where the message is encrypted three times with respect to three different public keys pk_1, pk_2, pk_3, and the decryption algorithm $\mathbf{D}$ calls $\mathbf{d}$ on each of them, and outputs the majority.[7] Then, it is clear that for any pk_i, even if we replace all encryptions $\mathbf{e}(pk_i, b, *)$ with arbitrary encryptions, the final ciphertext will decrypt correctly, as long as encryptions with respect to the other two public keys are unaltered.

We solve the above problems by adding a layer of an hybridization experiment performed on the set of pk, where at each stage encryptions with respect to all public keys up to the current one are replaced with random encryptions, while other encryptions are executed correctly. Specifically, we consider two sets of keys: a bad key set BKS and a good key set GKS. BKS contains pk that are embedded in PK, but for which $\mathbf{g}^{-1}(pk)$ is unknown. Initially, this is set to be the set KS. GKS contains those pk that were initially in BKS, but for which $sk = \mathbf{g}^{-1}(pk)$ has been previously retrieved by the adversary. Initially, $GKS = \emptyset$. Given BKS we perform the following hybridization experiments over keys in BKS to find a decisive key pk, and then using the methodology described

---

[7] A more detailed example, including this and several other parts demonstrating various aspects of the overall experiment, is presented in Section 5.

earlier to retrieve $sk = \mathbf{g}^{-1}(pk)$. We can then remove pk from BKS and insert it in GKS. The hybridization experiment over BKS is then repeated until enough secret-keys corresponding to decisive pk embedded into PK have been retrieved.

Suppose $BKS = \{pk_1, ..., pk_\ell\}$, then l hybridization experiments are performed where in the ith experiment we sample the percentage of times a modified execution of $\mathbf{E}^O(PK, M, R)$ produces a ciphertext that *decrypts properly*, when all queries of the form $\mathbf{e}(pk_k, b, r)$ for $(k \le i)$ are replaced with queries of random encryptions $\mathbf{e}(pk_k, b', r')$ for randomly chosen b' and r'. Clearly in the zeroth experiment, by the correctness property of the PKEP, all encryptions will properly decrypt, and we expect that as we go through the experiments there will be some experiment i, where the percentage of encryptions that decrypt properly drops substantially. This is because we expect that some bits that $\mathbf{E}$ is using to encode M are encoded in encryption $\mathbf{e}(pk, *, *)$ for $pk \in BKS$. If there is no such substantial drop in the percentage of proper decryptions by the final hybridization experiment, then this intuitively corresponds to the case where enough sk that are embedded in SK have been retrieved that are sufficient to construct an alternate decryption key SK'. Note that this does not mean that all of the embedded sk have been retrieved, only that those that have will suffice to construct an SK' that can be used to decrypt an average ciphertext generated by PK.

To illustrate this, consider again the example mentioned above, where the decryption algorithm outputs the majority of decryptions with respect to three keys pk_1, pk_2, pk_3. Here, the first hybridization experiment will identify pk_2 as having some decisive query j, and find $sk_2 = \mathbf{g}^{-1}(pk_2)$ (by replacing encryptions with respect to pk_1, as well as the first $j-1$ encryptions with respect to pk_2, with random encryptions; maintaining encryptions with respect to pk_3, as well as encryptions $j+1$ and on with respect to pk_2 correctly, and replacing the j'th encryption with respect to pk_2 with the encrypted bit of sk_2). It will then move pk_2 to GKS and perform the hybridization experiment again on $BKS = \{pk_1, pk_3\}$ (where encryptions with respect to pk_2 are now always done correctly). It will identify pk_3 as having a decisive query and find sk_3 in a similar manner, ending up with $BKS = \{pk_1\}$, for which it will not succeed to retrieve a corresponding sk_1 (because once encryptions with respect to pk_2, pk_3 are performed correctly, there is no decisive query for pk_1). Nonetheless, having sk_2, sk_3 is sufficient to decrypt the challenge ciphertexts.

Finally, we note that the hybridization experiments described above must take into account the lists obtained in the first stage. In that stage the adversary constructed a set of encryptions $\mathcal{E}$ that had the property that they might be embedded into encryptions $\mathbf{E}(PK, M', R')$ (for random M' and R'), and the decryption algorithm $\mathbf{D}(SK, \cdot)$ checked for the presence of these embeddings. Because of this, when performing the hybridization experiments that were previously described, it is essential that the response to a query $\mathbf{e}(pk, b, r)$ is replace only if $\mathbf{e}(pk, b, r) \notin \mathcal{E}$.

A More Formal Look. We now give a slightly more formal presentation of the second part of the experiment, to give an idea of technicalities involved. In Figure 1 pseudo-code is given for the second part of the experiment and for a helper function $\widehat{\mathbf{E}}$, which is a modified version of the encryption algorithm $\mathbf{E}$. This encryption algorithm takes as additional arguments a set BKS of bad-keys for which responses to encryption queries can be modified, a set $\mathcal{E}$ of encryption queries that cannot be modified, and a series of $2n^q$ ciphertexts for each $pk \in BKS$, where $c_i^{pk,b}$ is used to answer the ith oracle query of the form $\mathbf{e}(pk, b, *)$ made by $\mathbf{E}$, when $pk \in BKS$ and $(\mathbf{e}, pk, b, r) \notin \mathcal{E}$. By properly modifying the input values to this series, hybridization experiments can easily be performed. Recall that n^q is, by assumption, the largest number of queries that $\mathbf{E}$ makes. To aid future discussion, we say that the input $c_\ell^{pk_j,b}$ corresponds to a *correct encrypted bit* if $c_\ell^{pk_j,b} = \mathbf{e}(pk_j, b, *)$, and otherwise it corresponds an *incorrect encrypted bit*. The algorithm $\widehat{\mathbf{E}}$ is not called directly by the second part of the experiment, rather a function called $ApproxErrorRate(t, PK, ES, BKS)$ is introduced to repeatedly call $\widehat{\mathbf{E}}$ on a distribution of inputs—which will be specified momentarily— and returns an estimate of the probability that ciphertexts from that distribution decrypt properly. This estimate is achieved by using the decryption oracle to determine the fraction of ciphertexts output by $\widehat{\mathbf{E}}$ that decrypt properly. The distribution of inputs to $\widehat{\mathbf{E}}$ is specified by calling $\widehat{\mathbf{E}}(PK, M, R, BKS, \mathcal{E}, c_1', ..., c_t', c_{t+1}, ..., c_{2|BKS|n^q})$ where M and R are chosen randomly, $c_1', ..., c_t'$ are encryptions of random-bits under the appropriate pk's, and $c_{t+1}...c_{2|BKS|n^q}$ are encryptions of *correct encrypted bits*. Therefore, by calling $ApproxErrorRate$ with an increasing value of t we can perform a hybridization experiment in the second part of the experiment.

To summarize, at the end of the second stage the adversary has a list GKS of public keys (which is a subset of the list KS from the first stage), together with a corresponding $sk = \mathbf{g}^{-1}(pk)$ for each pk in GKS. Intuitively, these $\mathbf{g}^{-1}(GKS)$ are the 'essential' secret keys sk (with respect to the system $\mathbf{O} = (g, e, d)$) which are embedded into the secret key corresponding to PK and are used for proper decryption (in the system $(\mathbf{G^O}, \mathbf{E^O}, \mathbf{D^O})$).

The Third Part of the Experiment: Constructing SK'. Next, we specify how to use the secret-keys in $\mathbf{g}^{-1}(GKS)$ in order to construct a secret-key SK'. Given a specific example of a PKEP, this can often be a trivial task, but we require a uniform procedure that is guaranteed to work for all possible constructions that are considered by the statement of the theorem. Further, there is no guarantee that $GKS = KS$, so there may very well be a secret-key sk embedded into SK, for which $\mathbf{g}(sk) \notin GKS$. From the second part of the experiment we know that $\mathbf{g}^{-1}(GKS)$ contains enough secret-keys embedded into SK to decrypt properly, but not necessarily those that are necessary to reconstruct SK. For an example of the difficulty of constructing a uniform protocol for constructing SK', consider two PKEP that completely ignore the oracle $\mathcal{O}$, and therefore fall into the theorem's specification of acceptable constructions: an RSA based and a Quadratic Residuosity based PKEP. In both cases there would be no sk embedded in the secret-keys of either PKEP and so this should in theory be an easy case, but based on the public-keys of each respective PKEP the adversary

$\hat{\mathbf{E}}^{\mathcal{O}}(PK, M, R, BKS, \mathcal{E}, c_1^{pk_1,0}, ..., c_{n^q}^{pk_1,0}, c_1^{pk_1,1}, ..., c_{n^q}^{pk_1,1}, ..., c_1^{pk_{|BKS|},1}, ..., c_{n^q}^{pk_{|BKS|},1})$

$\forall\, pk' \in BKS,\ b' \in \{0,1\}$ set $\delta_{pk',b'} \leftarrow 0$

Simulate Execution $\mathbf{E}^{\mathcal{O}}(PK, M, R)$
    On query $(\mathbf{g}, sk)$ reply with $\mathbf{g}(sk)$.
    On query $(\mathbf{d}, sk, c)$ reply with $\mathbf{d}(sk, c)$.
    On query $(\mathbf{e}, pk, b, r)$
        If $pk \notin KS$ or $(\mathbf{e}, pk, b, r) \in \mathcal{E}$ reply with $\mathbf{e}(pk, b, r)$.
        otherwise
            $\delta_{pk,b} \leftarrow \delta_{pk,b} + 1$
            reply with $c_{\delta_{pk,b}}^{pk,b}$

Output result of simulation

$\mathrm{Exp}_2(PK, KS, \mathcal{E})$
(1)   Let $BKS \leftarrow KS$
(2)   Let $GKS \leftarrow \emptyset$
(3)   Repeat $|KS|$ times
(4) $w \leftarrow 2|BKS|n^q$
(5)   **for** $t = 0$ **to** w
(6)   Let $p_t = ApproxErrorRate(t, PK, \mathcal{E}, BKS)$
(7) $\Delta \leftarrow \{j|\ 1 \le j \le w \text{ and } |p_j - p_{j-1}| > \frac{1}{n^{\alpha_6}}\}$ (computed gaps)
(8)   If $\Delta = \emptyset$ then Output (BKS, GKS) and FINISH EXPERIMENT
(9) $\ell \leftarrow \min \Delta$
(10)   Let $\mathbf{e}(pk_j, b, *)$ correspond to the correct encrypted bit of the ℓth ciphertext
            input to $\hat{\mathbf{E}}$.
(11) $\psi \leftarrow |pk|/3$
(12)   For each k $(1 \le k \le \psi)$ set $\mathcal{D}_k \leftarrow \emptyset$
(13)   **for** $z = 1$ **to** $n^{2\alpha}4$
(14) $(d_1, ..., d_\psi) \leftarrow \mathbf{w}(pk, z)$ (we consider z a binary string)
(15)   For each k $(1 \le k \le \psi)$ set $\mathcal{D}_k \leftarrow \mathcal{D}_k \cup \{d_k\}$
(16)   **for** $k = 1$ **to** ψ
(17) $\pi_k \leftarrow ApproxErrorRate'(\ell, PK, \mathcal{E}, BKS, \mathcal{D}_k)$
(18)   If $|\pi_k - p_{\ell-1}| \ge 1/n^{\alpha_5}$ then $\overline{sk}_k \leftarrow 1 - b$ otherwise $\overline{sk}_k \leftarrow b$
(19)   Let $\overline{sk} = (\overline{sk}_1, ..., \overline{sk}_\psi)$
(20) $GKS \leftarrow GKS \cup \{(pk, \overline{sk})\}$
(21) $BKS \leftarrow BKS \setminus \{pk\}$

Fig. 1.

must generate corresponding secret-keys. To solve this problem, in order to find corresponding secret-keys a massive search is used.

We make use of the unlimited computational power of the adversary and have it enumerate all possible pairs of oracles $\mathcal{O}^*$ generated by Υ and seeds S^* that are consistent with our knowledge of $\mathcal{O}$ and SK, and create a set of *Valid Environments*. Note that this step does not actually require the adversary to query the oracle $\mathcal{O}$, for it is simply enumerating all possible environments and checking to see which are consistent. An oracle $\mathcal{O}^*$ and seed S^* are consistent if $\mathbf{G}^{\mathcal{O}^*}(S^*) = (PK, SK^*)$ for some SK^*, this execution of $\mathbf{G}$ queries $\mathbf{g}^*(sk') = pk$ for each $pk \in KS$, and $sk' = \mathbf{g}^{-1}(pk)$ for each $pk \in GKS$. Further, $\mathcal{O}^*$ is consistent with any queries and responses that have been made to $\mathcal{O}$ by the adversary, and $\mathbf{D}^{\mathcal{O}^*}(SK^*, \cdot)$ is consistent with any queries that have been made to the decryption oracle $\mathbf{D}^{\mathcal{O}}(SK, \cdot)$.

Because of the random process Υ by which $\mathcal{O}$ was selected and the random selection of S, each pair $(\mathcal{O}^*, S^*)$ in the set of *Valid Environment* is equally likely to be the environment $(\mathcal{O}, S)$ that the adversary is actually in. Therefore, the adversary uniformly at random selects one such pair $(\mathcal{O}', S')$, and lets SK' be the reconstructed secret-key where $\mathbf{G}^{\mathcal{O}'}(S') = (PK, SK')$. At this point SK' contains the secret-keys in $\mathbf{g}^{-1}(GKS)$, but while $\mathcal{O}$ and $\mathcal{O}'$ agree on all of the queries that have previously been made by the adversary, they probably agree on little else. Therefore, we consolidate the oracles $\mathcal{O}'$ and $\mathcal{O}$ into a new oracle $\widehat{\mathcal{O}}$. This is done so that $\mathcal{O}$ and $\widehat{\mathcal{O}}$ agree on nearly all queries (and in particular any queries that are likely to be made during calls to $C = \mathbf{E}^{\mathcal{O}}(PK, M, R)$ and $\mathbf{D}^{\mathcal{O}}(SK, C)$), but relative to which it is still the case that $\mathbf{G}^{\widehat{\mathcal{O}}}(S') = (SK', PK)$. This is achieved by taking $\mathbf{O}$ and modifying so that it is consistent with any queries that would have been made during the execution of $\mathbf{G}^{\mathcal{O}'}(S') = (SK', PK)$ and $\mathbf{D}^{\mathcal{O}'}(SK', C)$ for every decryption C made by the adversary so far to the decryption oracle $\mathbf{D}^{\mathcal{O}}(SK, \cdot)$.

Since $\mathcal{O}$ and $\widehat{\mathcal{O}}$ agree on nearly all queries, with high probability $\mathbf{E}^{\mathcal{O}}(PK, M, R) = \mathbf{E}^{\widehat{\mathcal{O}}}(PK, M, R) = C$ and therefore $M = \mathbf{D}^{\widehat{\mathcal{O}}}(SK', C) = \mathbf{D}^{\mathcal{O}}(SK, C)$. Therefore, if the adversary could execute $\mathbf{D}^{\widehat{\mathcal{O}}}(SK', \cdot)$ we'd be done, and the adversary could break the CCA1 security of the PKEP with high probability, by simply decrypting the challenge ciphertext. Unfortunately, the adversary cannot construct the oracle $\widehat{\mathcal{O}}$ with a polynomial number of queries to $\mathcal{O}$. It will instead simulate access to $\widehat{\mathbf{O}}$ using $\mathbf{O}$ and $\mathbf{u}$. The largest problem in simulating $\widehat{\mathbf{O}}$ during an execution of $\mathbf{D}^{\widehat{\mathcal{O}}}(SK', C)$ is in simulating queries $\widehat{\mathbf{d}}(sk, c)$ for $\widehat{\mathbf{g}}(sk) = pk \in BKS$, because $\mathbf{e}(pk, b, r) = \widehat{\mathbf{e}}(pk, b, r) = c$ for most b and r, but most likely $sk \neq \mathbf{g}^{-1}(pk)$, and therefore $\widehat{\mathbf{d}}(sk, c) = b$ but $\mathbf{d}(sk, c) = \bot$. However, it is exactly such queries whose responses were found not to be necessary for the decryption algorithm, because $pk \in BKS$. Therefore, on such queries $\widehat{\mathbf{d}}(sk, c)$ the adversary simply flips a coin and outputs the result as the response to the query. This is where we use $\mathbf{u}$ to make sure that indeed a bit in $\{0, 1\}$ should be output, as opposed to $\bot$. Using this simulation, $\mathbf{D}^{\widehat{\mathcal{O}}}(SK, C)$ is likely to decrypt properly for an encryption $\mathbf{E}^{\mathcal{O}}(PK, M, R)$ for randomly chosen M and R, and thus the adversary can decrypt its challenge ciphertext.

5 An Example

We consider an example of a simple (and artificial) PKEP construction to help ground and clarify the different parts of the experiment. Fix $n \in \mathbb{N}$. Define:

- $\mathbf{G}^{\mathcal{O}}(S)$: let $S = (S_0, ..., ..., S_8)$, where each $S_i \in \{0, 1\}^n$. Query $\mathbf{g}(S_i) = pk_i$ for each i, $0 \leq i \leq 6$. Compute $k_1 = \mathbf{e}(pk_6, 0, S_8)$, and outputs $PK = (pk_0, .., pk_5, pk_6, S_8)$ and $SK = (sk_0 = S_0, ..., sk_5 = S_5, sk_6 = S_6, k_1)$.
- $\mathbf{E}^{\mathcal{O}}(PK, M, R)$: let PK be as noted, $M \in \{0, 1\}$ and $R = (R_0, ..., R_6)$ where each $R_i \in \{0, 1\}^n$. Compute $c_i = \mathbf{e}(pk_i, M, R_i)$ for each $1 \leq i \leq 5$. Compute $k_1 = \mathbf{e}(pk_6, 0, S_8)$. If R_6 is the bit-string of all zeros, then query $\mathbf{e}(pk_0, M, R_0) = c_0$ and output $C = (0, k_1, 0^{3n}, 0^{3n}, 0^{3n}, 0^{3n}, c_0)$; otherwise, output $C = (1, k_1, c_1, c_2, \ldots, c_5)$.
- $\mathbf{D}^{\mathcal{O}}(SK, C)$: Let $C = (b, k_1', c_1, c_2, \ldots, c_5)$ where $b \in \{0, 1\}$, $k_1' \in \{0, 1\}^n$ and each $c_i \in \{0, 1\}^n$. Let SK be as noted. If $k_1' \neq k_1$ output $\bot$. Otherwise, if $\mathbf{d}(sk_6, k_1') \neq 0$ output $\bot$. Otherwise, If $b = 0$, then output $\mathbf{d}(sk_0, c_0)$. Otherwise, let $M_i = \mathbf{d}(sk_i, c_i)$ for each $i \leq 5$, and output $Majority(M_1, ..., M_5)$,

Now consider a (PK, SK) generated by $\mathfrak{G}^O(S)$ as described above, and a CCA adversary attempting break the security of the scheme $(\mathfrak{G}, \mathfrak{E}, \mathfrak{D})$ as prescribed by our experiments.

In the first part of our experiment the adversary will perform a large number of encryptions $\mathfrak{E}^O(PK, M, R)$ for randomly chosen M and R, and will observe queries of the form $\mathbf{e}(pk_i, *, *)$ made during such executions for $1 \leq i \leq 6$, but it is unlikely that queries $\mathbf{e}(pk_0, *, *)$ are observed. Thus it is unlikely the adversary will need pk_0 to decrypt the challenge ciphertext and it can be ignored. The adversary also will observe the query $\mathbf{e}(pk_6, 0, S_8)$ with response k_1, and note that it will have to ensure the key it later constructs is consistent with this query/response.

In the second part of the experiment the adversary will attempt to determine sk_i for $1 \leq i \leq 6$. This will be done by encrypting random messages by executing $\mathfrak{E}^O(PK, M, R)$, but replacing responses of queries of the form $\mathbf{e}(pk_i, b, r)$ with responses to $\mathbf{e}(pk_i, b', r')$ where b' and r' are chosen randomly in a hybridization experiment. In this case the hybridization is over the pk_i. In such an experiment, the resulting ciphertexts C' will either decrypt to the appropriate message M that was originally encrypted or it will not (note the adversary uses the decryption oracle to check this).

In our toy example, randomizing only the responses to all queries $\mathbf{e}_{pk_i}$, $i \in \{1, 2\}$, will result in proper decryptions, as the Majority function in $\mathfrak{D}$ acts as a form of error-correcting code. However, when responses to all queries of the form $\mathbf{e}_{pk_i}$, $i \in \{1, 2, 3\}$, are randomized, the result is occasional improper decryptions. The occasional improper decryption allows the adversary to determine sk_3. This is because the oracle $\mathbf{w}$ will provide a number of random encryptions of sk_3 that can injected into modified executions of $\mathbf{E}^O(PK, M, R)$ as in the hybrid experiment. By determining if the ciphertexts produced by these executions of $\mathbf{E}$ decrypt properly the bits of sk_3 can be retrieved. By the end of the second part of the experiment the adversary will have retrieved sk_i for $3 \leq i \leq 6$. Note that sk_i, $0 \leq i \leq 2$ will not be retrieved because of the error-correcting properties of the Majority function in $\mathfrak{D}$. Still, this is sufficient to decrypt on average and thus all the adversary will ask.

In the third part of the experiment the adversary must reconstruct the secret-key. Since it does not know $sk_0, ..., sk_2$ it cannot reconstruct SK, but it can construct an SK' that is satisfactory to decrypt the challenge ciphertext. From observation it is clear that a secret-key of the form $SK' = (sk'_0, sk'_1, sk'_2, sk_3, ..., sk_6, k_1)$ will decrypt the challenge ciphertext with high probability, only possibly failing in the unlikely event that the first bit of the challenge ciphertext is 0. The issue is automating the above construction. In order to do so the adversary essentially searches through all oracle/seed pairs $(\widehat{O}, \widehat{S})$ in which the oracles are consistent with everything the adversary knows about O (i.e. $\mathbf{g}(sk_i) = pk_i$ for $3 \leq i \leq 6$ and $\mathbf{e}(pk_6, 0, S_8) = k_1$) and that $\mathbf{G}^{\widehat{O}}(\widehat{S}) = (\widehat{SK}, PK)$. Such a $\widehat{SK}$ is then used by the adversary to decrypt its challenge ciphertext.

6   The Complexity Theoretic Statements

A quick review of the experiment the adversary performs shows that the only situation in which the adversary uses more than a polynomial amount of computation is when it must select uniformly at random an oracle and seed pair $(\mathcal{O}', S')$ from the set of *Valid Environments*. It selects such oracles and seeds based on them satisfying a polynomial number of local consistency constraints that are efficiently verifiable. Further, once this is done almost all of the oracle $\mathcal{O}'$ is thrown out when the adversary consolidates $\mathcal{O}$ with $\mathcal{O}'$. Therefore, the process of randomly selecting an oracle and seed could alternately be thought of as selecting an oracle 'stub' with corresponding seeds, where the oracle stub only specifies the oracle's values on those queries that are necessary to satisfy the constraints mentioned. Once such a stub had been selected, the oracle can be randomly extended to a full oracle if needed without changing the distribution. However, choosing such stubs can be thought of as uniformly at random selecting an $\mathcal{NP}$ witness. Bellare, Goldreich and Petrank [5] show that if $\mathcal{P} = \mathcal{NP}$ then one can efficiently and uniformly at random select $\mathcal{NP}$-Witnesses. Therefore, we can consider this result in the more traditional model of Impagliazzo and Rudich[23], and state the theorem in the traditional computational model, based on the assumption that $\mathcal{P} = \mathcal{NP}$. Alternatively, following the lead of Simon [36], we can further embed a *PSPACE* oracle into our final oracle $\mathcal{O}$. Since $\mathcal{P}^{PSPACE} = \mathcal{NP}^{PSPACE}$ we get a restriction on black-box result in the standard computational model.

Acknowledgments. We would like to thank Bill Aiello for posing the problem, and participating in the initial stages of this research. The third author would like to thank Charles W. Rackoff and Toniann Pitassi for many useful discussions.

References

1. Miklós Ajtai and Cynthia Dwork.   A public-key cryptosystem with worst-case/ average-case equivalence. In ACM, editor, *Proceedings of the twenty-ninth annual ACM Symposium on the Theory of Computing: El Paso, Texas, May 4–6, 1997*, pages 284–293, 1997. ACM order no. 508970.
2. B. Barak. How to go beyond the black-box simulation barrier. In *Proceedings of the 42nd IEEE Symposium on Foundations of Computer Science*, pages 106–115, 1109 Spring Street, Suite 300, Silver Spring, MD 20910, USA, 2001. IEEE Computer Society Press.
3. Boaz Barak. Constant-round coin-tossing with a man in the middle or realizing the shared random string model. In *FOCS*, pages 345–355. IEEE Computer Society, 2002.
4. M. Bellare, A. Desai, D. Pointcheval, and P. Rogaway. Relations among notions of security for public-key encryption schemes. In *CRYPTO: Proceedings of Crypto*, 1998.
5. Mihir Bellare, Oded Goldreich, and Erez Petrank.   Uniform generation of np-witnesses using an np-oracle. *Electronic Colloquium on Computational Complexity (ECCC)*, 5(32), 1998.

6. Mihir Bellare, Shai Halevi, Amit Sahai, and Salil Vadhan. Many-to-one trapdoor functions and their relation to public-key cryptosystems. Cryptology ePrint Archive, Report 1998/019, 1998. http://eprint.iacr.org/.
7. Mihir Bellare and Phillip Rogaway. Optimal asymmetric encryption. *Lecture Notes in Computer Science*, 950:92–111, 1995.
8. Mihir Bellare and Amit Sahai. Non-malleable encryption: Equivalence between two notions, and an indistinguishability-based characterization. *Lecture Notes in Computer Science*, 1666:519–536, 1999.
9. Manuel Blum and Shafi Goldwasser. An *efficient* probabilistic public-key encryption scheme which hides all partial information. In G. R. Blakley and David Chaum, editors, *Advances in Cryptology: Proceedings of CRYPTO 84*, volume 196 of *Lecture Notes in Computer Science*, pages 289–299. Springer-Verlag, 1985, 19–22 August 1984.
10. Ronald Cramer and Victor Shoup. A practical public key cryptosystem provably secure against adaptive chosen ciphertext attack. In *CRYPTO '98: Proceedings of the 18th Annual International Cryptology Conference on Advances in Cryptology*, pages 13–25, London, UK, 1998. Springer-Verlag.
11. Danny Dolev, Cynthia Dwork, and Moni Naor. Non-malleable cryptography (extended abstract). In *Proceedings of the Twenty Third Annual ACM Symposium on Theory of Computing*, pages 542–552, New Orleans, Louisiana, 6–8 May 1991.
12. E. Elkind and A. Sahai. A unified methodology for constructing publickey encryption schemes secure against adaptive chosen-ciphertext attack, 2004.
13. Eiichiro Fujisaki and Tatsuaki Okamoto. How to enhance the security of public-key encryption at minimum cost. In Hideki Imai and Yuliang Zheng, editors, *Public Key Cryptography*, volume 1560 of *Lecture Notes in Computer Science*, pages 53–68. Springer, 1999.
14. R. Gennaro and L. Trevisan. Lower bounds on the efficiency of generic cryptographic constructions. In *41st Annual Symposium on Foundations of Computer Science*, pages 305–313. IEEE Computer Society Press, 2000.
15. Rosario Gennaro, Yael Gertner, and Jonathan Katz. Lower bounds on the efficiency of encryption and digital signature schemes. In *Proceedings of the thirty-fifth ACM symposium on Theory of computing*, pages 417–425. ACM Press, 2003.
16. Y. Gertner, S. Kannan, T. Malkin, O. Reingold, and M. Viswanathan. The relationship between public key encryption and oblivious transfer. In IEEE, editor, *41st Annual Symposium on Foundations of Computer Science*, pages 325–335. IEEE Computer Society Press, 2000.
17. Y. Gertner, T. Malkin, and O. Reingold. On the impossibility of basing trapdoor functions on trapdoor predicates. In IEEE, editor, *42nd IEEE Symposium on Foundations of Computer Science*, pages 126–135. IEEE Computer Society Press, 2001.
18. Yael Gertner, Tal Malkin, and Steve Myers. Towards a separation of semantic and cca security for public key encryption. Cryptology ePrint Archive, 2006. http://eprint.iacr.org/.
19. O. Goldreich, S. Goldwasser, and S. Micali. How to construct random functions. *Journal of the ACM*, 33(4):792–807, 1986.
20. Oded Goldreich and Leonid A. Levin. A hard-core predicate for all one-way functions. In *Proceedings of the Twenty First Annual ACM Symposium on Theory of Computing*, pages 25–32, Seattle, Washington, 15–17 May 1989.
21. Shafi Goldwasser and Silvio Micali. Probabilistic encryption. *Journal of Computer and System Sciences*, 28(2):270–299, 1984.

22. J. Hastad, R. Impagliazzo, L.A. Levin, and M. Luby. Construction of pseudo-random generator from any one-way function. *Accepted to the SIAM Journal of Computing*, 28(4):1364–1396, 1998.
23. R. Impagliazzo and S. Rudich. Limits on the provable consequences of one-way permutations. In *Proceedings of the 21st Annual ACM Symposium on Theory of Computing*, pages 44–61. ACM Press, 1989.
24. Jeff Kahn, Michael Saks, and Cliff Smyth. A dual version of reimer's inequality and a proof of rudich's conjecture. In *COCO '00: Proceedings of the 15th Annual IEEE Conference on Computational Complexity*, page 98. IEEE Computer Society, 2000.
25. Jeong Han Kim, D. R. Simon, and P. Tetali. Limits on the efficiency of one-way permutation-based hash functions. In *40th Annual Symposium on Foundations of Computer Science*, pages 535–542. IEEE Computer Society Press, 1999.
26. L. A. Levin. One-way functions and pseudorandom generators. In *ACM Symposium on Theory of Computing (STOC '85)*, pages 363–365, Baltimore, USA, May 1985. ACM Press.
27. Lindell. A simpler construction of CCA2-secure public-key encryption under general assumptions. In *EUROCRYPT: Advances in Cryptology: Proceedings of EU-ROCRYPT*, 2003.
28. M. Luby and C. Rackoff. How to construct pseudorandom permutations from pseudorandom functions. *SIAM Journal on Computing*, 17:373–386, 1988.
29. M. Naor and M. Yung. Public-key cryptosystems provably secure against chosen ciphertext attacks. In Baruch Awerbuch, editor, *Proceedings of the 22nd Annual ACM Symposium on the Theory of Computing*, pages 427–437, Baltimore, MY, May 1990. ACM Press.
30. R. Pass, a. shelat, and V. Vaikuntanathan. Construction of a non-malleable encryption scheme from any semantically secure one. In *CRYPTO '06: Proceedings of the 26th Annual International Cryptography Conference on Advances in Cryptology*, 2006.
31. Charles Rackoff and Daniel R. Simon. Non-interactive zero-knowledge proof of knowledge and chosen ciphertext attack. In *CRYPTO '91: Proceedings of the 11th Annual International Cryptology Conference on Advances in Cryptology*, pages 433–444, London, UK, 1992. Springer-Verlag.
32. Omer Reingold, Luca Trevisan, and Salil P. Vadhan. Notions of reducibility between cryptographic primitives. In Moni Naor, editor, *TCC*, volume 2951 of *Lecture Notes in Computer Science*, pages 1–20. Springer, 2004.
33. J. Rompel. One-way functions are necessary and sufficient for secure signatures. In Baruch Awerbuch, editor, *Proceedings of the 22nd Annual ACM Symposium on the Theory of Computing*, pages 387–394, Baltimore, MY, May 1990. ACM Press.
34. Steven Rudich. The use of interaction in public cryptosystems (extended abstract). In Joan Feigenbaum, editor, *CRYPTO*, volume 576 of *Lecture Notes in Computer Science*, pages 242–251. Springer, 1991.
35. A. Sahai. Non-malleable non-interactive zero knowledge and adaptive chosen-ciphertext security. In *40th Annual Symposium on Foundations of Computer Science*, pages 543–553. IEEE Computer Society Press, 1999.
36. D. R. Simon. Finding collisions on a one-way street: Can secure hash functions be based on general assumptions? In *Advances in Cryptology – EUROCRYPT 98*, pages 334–345, 1998.

Unifying Classical and Quantum Key Distillation

Matthias Christandl[1], Artur Ekert[1,2], Michał Horodecki[3], Paweł Horodecki[4],
Jonathan Oppenheim[1], and Renato Renner[1]

[1] Centre for Quantum Computation, University of Cambridge, United Kingdom
{m.christandl,jono,r.renner}@damtp.cam.ac.uk
[2] Department of Physics, National University of Singapore, Singapore
artur.ekert@qubit.org
[3] Institute of Theoretical Physics and Astrophysics, University of Gdańsk, Poland
fizmh@univ.gda.pl
[4] Faculty of Applied Physics and Mathematics, Gdańsk University of Technology,
Poland
pawel@mif.pg.gda.pl

Abstract. Assume that two distant parties, Alice and Bob, as well as
an adversary, Eve, have access to (quantum) systems prepared jointly
according to a tripartite state ρ_{ABE}. In addition, Alice and Bob can
use local operations and authenticated public classical communication.
Their goal is to establish a key which is unknown to Eve. We initiate
the study of this scenario as a unification of two standard scenarios:
(i) key distillation (agreement) from classical correlations and (ii) key
distillation from pure tripartite quantum states.

Firstly, we obtain generalisations of fundamental results related to
scenarios (i) and (ii), including upper bounds on the key rate, i.e., the
number of key bits that can be extracted per copy of ρ_{ABE}. Moreover,
based on an embedding of classical distributions into quantum states, we
are able to find new connections between protocols and quantities in the
standard scenarios (i) and (ii).

Secondly, we study specific properties of key distillation protocols. In
particular, we show that every protocol that makes use of pre-shared
key can be transformed into an equally efficient protocol which needs
no pre-shared key. This result is of practical significance as it applies
to quantum key distribution (QKD) protocols, but it also implies that
the key rate cannot be locked with information on Eve's side. Finally,
we exhibit an arbitrarily large separation between the key rate in the
standard setting where Eve is equipped with quantum memory and the
key rate in a setting where Eve is only given classical memory. This shows
that assumptions on the nature of Eve's memory are important in order
to determine the correct security threshold in QKD.

1 Introduction

Many cryptographic tasks such as message encryption or authentication rely
on *secret keys*,[1] i.e., random strings only known to a restricted set of parties.

---

[1] In the sequel, we will use the term *key* instead of *secret key*.

S.P. Vadhan (Ed.): TCC 2007, LNCS 4392, pp. 456–478, 2007.

In *information-theoretic cryptography*, where no assumptions on the adversary's resources[2] are made, distributing keys between distant parties is impossible if only public classical communication channels are available [1,2]. However, this situation changes dramatically if the parties have access to additional devices such as noisy channels (where also a wiretapper is subject to noise), a noisy source of randomness, a quantum channel, or a pre-shared quantum state. As shown in [2,3,4,5,6], these devices allow the secure distribution of keys.[3]

This work is concerned with information-theoretic key distillation from pre-distributed noisy data. More precisely, we consider a situation where two distant parties, *Alice* and *Bob*, have access to (not necessarily perfectly) correlated pieces of (classical or quantum) information, which might be partially known to an adversary, *Eve*. The goal of Alice and Bob is to *distill* virtually perfect key bits from these data, using only an authentic (but otherwise insecure) classical communication channel.

Generally speaking, key distillation is possible whenever Alice and Bob's data are sufficiently correlated and, at the same time, Eve's uncertainty on these data is sufficiently large. It is one of the goals of this paper to exhibit the properties pre-shared data must have in order to allow key distillation.

In practical applications, the pre-distributed data might be obtained from realistic physical devices such as noisy (classical or quantum) channels or other sources of randomness. Eve's uncertainty on Alice and Bob's data might then be imposed by inevitable noise in the devices due to thermodynamic or quantum effects.

Quantum key distribution (QKD) can be seen as a special case of key distillation where the pre-shared data is generated using a quantum channel. The laws of quantum physics imply that the random values held by one party, say Alice, cannot at the same time be correlated with Bob and Eve. Hence, whenever Alice and Bob's values are strongly correlated (which can be checked easily) then Eve's uncertainty about them must inevitably (by the laws of quantum mechanics) be large, hence, Alice and Bob can distil key. Because of this close relation between key distillation and QKD, many of the results we give here will have direct implications to QKD.

Furthermore, the theory of key distillation has nice parallels with the theory of *entanglement distillation*, where the goal is to distil maximally entangled states (also called *singlets*) from (a sequence of) bipartite quantum states. In fact, the two scenarios have many properties in common. For example, there is a gap between the *key rate* (i.e., the amount of key that can be distilled from some given noisy data) and the *key cost* (the amount of key that is needed to simulate the noisy data, using only public classical communication) [7]. This gap can be seen as the classical analogue of a gap between *distillable entanglement* (the amount of singlets that can be distilled from a given bipartite quantum state) and *entanglement cost* (the amount of singlets needed to generate the state).

---

[2] In this context, the term *resources* typically refers to computational power and memory space.

[3] In certain scenarios, including the one studied in this paper, an authentic classical channel is needed in addition.

1.1 Related Work

The first and basic instance of an information-theoretic key agreement scenario is Wyner's wiretap channel [8]. Here, Alice can send information via a noisy classical channel to Bob. Eve, the eavesdropper, has access to a degraded version of Bob's information. Wyner has calculated the rate at which key generation is possible if only Alice is allowed to send public classical messages to Bob. Wyner's work has later been generalised by Csiszár and Körner, relaxing the restrictions on the type of information given to Eve [3]. Based on these ideas, Maurer and Ahlswede and Csiszár have proposed an extended scenario where key is distilled from arbitrary correlated classical information (specified by a tripartite probability distribution) [2,4]. In particular, Maurer has shown that two-way communication can lead to a strictly positive key rate even though the key rate in the one-way communication scenario might be zero [2].

In parallel to this development quantum cryptography emerged: in 1984 Bennett and Brassard devised a QKD scheme in which quantum channels could be employed in order to generate a secure key without the need to put a restriction on the eavesdropper [5]. In 1991, Ekert discovered that quantum cryptographic schemes could be based on entanglement, that is, on quantum correlations that are strictly stronger than classical correlations [6]. Clearly, this is key distillation from quantum information.

The first to spot a relation between the classical and the quantum development were Gisin and Wolf; in analogy to *bound entanglement* in quantum information theory, they conjectured the existence of *bound information*, namely classical correlation that can only be created from key but from which no key can be distilled [9]. Their conjecture remains unsolved, but has stimulated the community in search for an answer.

To derive lower bounds on the key rate, we will make repeated use of results by Devetak and Winter, who derived a bound on the key rate if the tripartite quantum information consists of many identical and mutually independent pieces, and by Renner and König, who derived privacy amplification results which also hold if this independence condition is not satisfied [10,11].

1.2 Contributions

We initiate the study of a unified key distillation scenario, which includes key distillation from pre-shared *classical* and *quantum* data (Section 2). We then derive a variety of quantitative statements related to this scenario. These unify and extend results from both the quantum and classical world.

There are numerous upper bounds available in the specific scenarios and it is our aim to provide the bigger picture that will put order into this zoo by employing the concept of a *secrecy monotone*, i.e., a function that decreases under local operations and public communication (Section 3), as introduced in [12]. The upper bounds can then roughly be subdivided into two categories: (i) the ones based on classical key distillation [13] and (ii) the ones based on quantum communication or entanglement measures [14].

The unified scenario that we develop does not stop at an evaluation of the key rate but lets us investigate intricate connections between the two extremes. We challenge the viewpoint of Gisin and Wolf who highlight the relation between key distillation from classical correlation and entanglement distillation from this very correlation *embedded* into quantum states [9]: we prove a theorem that relates key distillation from certain classical correlation and key (and not entanglement) distillation from their embedded versions (Section 4). This ties in with recent work which established that key distillation can be possible even from quantum states from which no entanglement can be distilled [15].

A fruitful concept that permeates this work is the concept of *locking of classical information in quantum states*: let Alice choose an n-bit string $x = x_1 \ldots x_n$ with uniform probability and let her either send the state $|x_1\rangle \ldots |x_n\rangle$ or the state $H^{\otimes n}|x_1\rangle \ldots |x_n\rangle$ to Bob, where H is the Hadamard transformation. Not knowing if the string is sent in the computational basis or in the Hadamard basis, it turns out that the optimal measurement that Bob can do in order to maximise the mutual information between the measurement outcome y and Alice's string x is with respect to a randomly chosen basis, in which case he will obtain $I(X;Y) = \frac{n}{2}$. If, however, he has access to the single bit which determines the basis, he will have $I(X;Y) = n$. A *single bit* can therefore *unlock* an arbitrary amount of information. This effect has been termed *locking of classical information in quantum states* or simply *locking* and was first described in [16]. In this paper, we will discuss various types of locking effects and highlight their significance for the design and security of QKD protocols (Section 5).

Finally, we demonstrate that the amount of key that can be distilled from given pre-shared data strongly depends on whether Eve is assumed to store her information in a classical or in a quantum memory. This, again, has direct consequences for the analysis of protocols in quantum cryptography (Section 6).

For a more detailed explanation of the contributions of this paper, we refer to the introductory paragraphs of Sections 3–6.

2   The Unified Key Distillation Scenario

In classical information-theoretic cryptography one considers the problem of distilling key from correlated data specified by a tripartite probability distribution p_{ijk} ($p_{ijk} \geq 0$, $\sum_{i,j,k} p_{ijk} = 1$). Alice and Bob who wish to distil the key have access to i and j, respectively, whereas the eavesdropper Eve knows the value k (see, e.g., [17]). Typically, it is assumed that many independently generated copies of the triples (i, j, k) are available[4]. The *key rate* or *distillable key* of a distribution p_{ijk} is the rate at which key bits can be obtained per realisation of this distribution, if Alice and Bob are restricted to local operations and public but authentic classical communication.

---

[4] Using de Finetti's representation theorem, this assumption can be weakened to the assumption that the overall distribution of all triples is invariant under permutations (see [18] for more details including a treatment of the quantum case).

Before we continue to introduce the quantum version of the key distillation scenario described above, let us quickly note that it will be convenient to regard probability distributions as *classical states*, that is, given probabilities p_i, we consider $\rho = \sum_{i=1}^{d} p_i |i\rangle\langle i|$, where $|i\rangle$ is an orthonormal basis of a d-dimensional Hilbert space; we will assume that $d < \infty$. In the sequel we will encounter not only classical or quantum states, but also states that are distributed over several systems which might be partly classical and partly quantum-mechanical. To make this explicit, we say that a bipartite state ρ_{AB} is *cq (classical-quantum)* if it is of the form $\rho_{AB} = \sum_i p_i |i\rangle\langle i|_A \otimes \rho_B^i$ for quantum states ρ_B^i and a probability distribution p_i. This definition easily extends to three or more parties, for instance:

- a *ccq (classical-classical-quantum) state* ρ_{ABE} is of the form $\sum_{i,j} p_{ij} |i\rangle\langle i|_A \otimes |j\rangle\langle j|_B \otimes \rho_E^{ij}$, where p_{ij} is a probability distribution and ρ_E^{ij} are arbitrary quantum states.
- the distribution p_{ijk} corresponds to a *ccc (classical-classical-classical) state* $\rho_{ABE} = \sum_{i,j,k} p_{ijk} |ijk\rangle\langle ijk|_{ABE}$, where we use $|ijk\rangle_{ABE}$ as a short form for $|i\rangle_A \otimes |j\rangle_B \otimes |k\rangle_E$ (as above, the states $|i\rangle_A$ for different values of i, and likewise $|j\rangle_B$ and $|k\rangle_k$, are normalised and mutually orthogonal).

We will be concerned with key distillation from arbitrary tripartite quantum states ρ_{ABE} shared by Alice, Bob, and an adversary Eve, assisted by *local quantum operations and public classical communication (LOPC)* [10,19,15]. A local quantum operation on Bob's side is of the form

$$\rho_{ABE} \mapsto (I_{AE} \otimes \Lambda_B)(\rho_{ABE}) .$$

Public classical communication from Alice to Bob can be modelled by copying a local classical register, i.e., any state of the form $\rho_{AA'BE} = \sum_i \rho_{ABE}^i \otimes |i\rangle\langle i|_{A'}$ is transformed into $\rho'_{AA'BB'EE'} = \sum_i \rho_{ABE}^i \otimes |iii\rangle\langle iii|_{A'B'E'}$. Similarly, one can define these operations with the roles of Alice and Bob interchanged.

The goal of a key distillation protocol is to transform copies of tripartite states ρ_{ABE} into a state which is close to

$$\tau_{ABE}^{\ell} = \frac{1}{2^{\ell}} \sum_{i=1}^{2^{\ell}} |ii\rangle\langle ii|_{AB} \otimes \tau_E \tag{1}$$

for some arbitrary τ_E. τ_{ABE}^{ℓ} (also denoted τ^{ℓ} for short) corresponds to a perfect *key of length ℓ*, i.e., uniform randomness on an alphabet of size 2^{ℓ} shared by Alice and Bob and independent of Eve's system. We measure *closeness* of two states ρ and σ in terms of the trace norm $\|\rho - \sigma\| := \frac{1}{2}\mathrm{Tr}|\rho - \sigma|$. The trace norm is the natural quantum analogue of the variational distance to which it reduces if ρ and σ are classical.

We will now give the formal definition of an LOPC protocol and of the key rate.

Definition 1. *An LOPC protocol $\mathcal{P}$ is a family $\{\Lambda_n\}_{n\in\mathbb{N}}$ of completely positive trace preserving (CPTP) maps*

$$\Lambda_n : (\mathcal{H}_A \otimes \mathcal{H}_B \otimes \mathcal{H}_E)^{\otimes n} \to \mathcal{H}_A^n \otimes \mathcal{H}_B^n \otimes \mathcal{H}_E^n$$

which are defined by the concatenation of a finite number of local operation and public communication steps.

Definition 2. *We say that an LOPC protocol $\mathcal{P}$ distills key at rate $\mathcal{R}_\mathcal{P}$ if there exists a sequence $\{\ell_n\}_{n\in\mathbb{N}}$ such that*

$$\limsup_{n\to\infty} \frac{\ell_n}{n} = \mathcal{R}_\mathcal{P}$$

$$\lim_{n\to\infty} \|\Lambda_n(\rho_{ABE}^{\otimes n}) - \tau_{ABE}^{\ell_n}\| = 0$$

where $\tau_{ABE}^{\ell_n}$ are the ccq states defined by (1). The key rate *or* distillable key *of a state ρ_{ABE} is defined as $K_D(\rho_{ABE}) := \sup_\mathcal{P} \mathcal{R}_\mathcal{P}$.*

The quantity K_D obviously depends on the partition of the state given as argument into the three parts controlled by Alice, Bob, and Eve, respectively. We thus indicate the assignment of subsystems by semicolons if needed. For instance, we write $\rho_{AD;B;E}$ if Alice holds an additional system D.

It can be shown that the maximisation in the definition of K_D can be restricted to protocols whose communication complexity grows at most linearly in the number of copies of ρ_{ABE}. Hence, if $d = \dim \mathcal{H}_A \otimes \mathcal{H}_B \otimes \mathcal{H}_E < \infty$ then the dimension of the output of the protocol is bounded by $\log \dim \mathcal{H}_A^n \otimes \mathcal{H}_B^n \otimes \mathcal{H}_E^n \leq cn \log d$, for some constant c. (The proof of this statement will appear in a full version of this paper.)

The above security criterion is (strictly) weaker than the one proposed in [10][5], hence $K_D(\rho_{ABE})$ is lower bounded by a lower bound derived in [10]:

$$K_D(\rho_{ABE}) \geq I(A:B)_\rho - I(A:E)_\rho . \tag{2}$$

This expression can be seen as a quantum analogue of the well-known bound of Csiszár, Körner, and Maurer [3,17]. Here $I(A:B)_\rho$ denotes the mutual information defined by $I(A:B)_\rho := S(A)_\rho + S(B)_\rho - S(AB)_\rho$ where $S(A)_\rho := S(\rho^A)$ is the von Neumann entropy of system A (and similarly for B and E). For later reference we also define the *conditional mutual information* $I(A:B|E)_\rho := S(AE)_\rho + S(BE)_\rho - S(ABE)_\rho - S(E)_\rho$.

Note also that the criterion for the quality of the distilled key used in Definition 2 implies that the key is both uniformly distributed and independent of the adversary's knowledge, just as in [11]. Previous works considered uniformity and security separately. Note that, even though weaker than certain alternative criteria such as the one of [10], the security measure of Definition 2 is universally composable [11].

In [20], the question was posed whether the security condition also holds if the accessible information is used instead of the criterion considered here. Recently, it has been shown that this is not the case [21]. More precisely, an

---

[5] The security criterion of [10] implies that, conditioned on *any* value of the key, Eve's state is almost the same. In contrast, according to the above definition, Eve's state might be arbitrary for a small number of values of the key.

example of a family of states was exhibited such that Eve has exponentially small knowledge in terms of accessible information but constant knowledge in terms of the Holevo information. This implies that in this context, security definitions based on the accessible information are problematic. In particular, a key might be insecure even though the accessible information of an adversary on the key is exponentially small (in the key size).

3   Upper Bounds for the Key Rate

In this section, we first derive sufficient conditions that a function has to satisfy in order to be an upper bound for the key rate (Section 3.1). We focus on functions that are *secrecy monotones* [12], i.e., they are monotonically decreasing under LOPC operations. Our approach therefore parallels the situation in classical and quantum information theory where resource transformations are also bounded by monotonic functions; examples include the proofs of converses to coding theorems and entanglement measures (see, e.g., [14]). As a corollary to our characterisation of secrecy monotones, we show how to turn entanglement monotones into secrecy monotones.

In a second part (Section 3.2), we provide a number of concrete secrecy monotones that satisfy the conditions mentioned above. They can be roughly divided into two parts: (i) functions derived from the intrinsic information and (ii) functions based on entanglement monotones. Finally, we will compare different secrecy monotones (Section 3.3) and study a few particular cases in more detail (Section 3.4).

3.1   Secrecy Monotones

Theorem 1. *Let $M(\rho)$ be a function mapping tripartite quantum states $\rho \equiv \rho_{ABE}$ into the positive numbers such that the following holds:*

1. *Monotonicity: $M(\Lambda(\rho)) \le M(\rho)$ for any LOPC operation Λ.*
2. *Asymptotic continuity: for any states ρ^n, σ^n on $\mathcal{H}_A^n \otimes \mathcal{H}_B^n \otimes \mathcal{H}_E^n$, the condition $\|\rho^n - \sigma^n\| \to 0$ implies $\frac{1}{\log r_n}|M(\rho^n) - M(\sigma^n)| \to 0$ where $r_n = \dim(\mathcal{H}_A^n \otimes \mathcal{H}_B^n \otimes \mathcal{H}_E^n)$.*
3. *Normalisation: $M(\tau^\ell) = \ell$.*

Then the regularisation of the function M given by $M^\infty(\rho) = \limsup_{n\to\infty} \frac{M(\rho^{\otimes n})}{n}$ is an upper bound on K_D, i.e., $M^\infty(\rho_{ABE}) \ge K_D(\rho_{ABE})$ for all ρ_{ABE} with $\dim \mathcal{H}_A \otimes \mathcal{H}_B \otimes \mathcal{H}_E < \infty$. If in addition M satisfies

4. *Subadditivity on tensor products: $M(\rho^{\otimes n}) \le nM(\rho)$,*

then M is an upper bound for K_D.

Proof. Consider a key distillation protocol $\mathcal{P}$ that produces output states σ^n such that $\|\sigma^n - \tau^{\ell_n}\| \to 0$. We will show that $M^\infty(\rho) \ge R_\mathcal{P}$. Let us assume without loss of generality that $R_{\mathcal{P}} > 0$. Indeed, by monotonicity we have $M(\rho^{\otimes n}) \geq M(\sigma^n)$, which is equivalent to

$$\frac{1}{n}M(\rho^{\otimes n}) \geq \frac{\ell_n}{n}\left(\frac{M(\sigma^n) - M(\tau^{\ell_n})}{\ell_n} + 1\right), \tag{3}$$

where we have used the normalisation condition. As remarked in Definition 2 there is a constant $c > 0$ such that $\log r_n \leq cn$ and by definition of $R_{\mathcal{P}}$ there exists a $c' > 0$ and n_0 such that for all $n \geq n_0$, $\log d_n \geq c'n$. Hence $\ell_n \geq c'n \geq \frac{c'}{c}\log r_n$, therefore asymptotic continuity implies

$$\lim_{n \to \infty} \frac{1}{\ell_n}\left|M(\sigma^n) - M(\tau^{\ell_n})\right| = 0 \ .$$

Taking the limsup on both sides of (3) gives $M^{\infty}(\rho) \geq \limsup_n \frac{\ell_n}{n} = R_{\mathcal{P}}$. Thus we have shown that M^{∞} is an upper bound for the rate of an arbitrary protocol, so that it is an also upper bound for K_D. $\square$

If we restrict our attention to the special case of key distillation from bipartite states ρ_{AB}, we can immediately identify a well-known class of secrecy monotones, namely entanglement monotones. A convenient formulation is in this case not given by the distillation of states τ^{ℓ} with help of LOPC operations, but rather by the distillation of states γ^{ℓ} via local operations and classical communication (LOCC), where $\gamma^{\ell} = U|\psi\rangle\langle\psi|_{AB}^{\otimes\ell} \otimes \rho_{A'B'}U^{\dagger}$, for some unitary $U = \sum_{i=1}^{2^{\ell}}|ii\rangle\langle ii|_{AB} \otimes U_{A'B'}^{(i)}$ and $|\psi\rangle = \frac{1}{\sqrt{2}}(|00\rangle + |11\rangle)$ [15,22]. Note that measuring the state γ^{ℓ} with respect to the computational bases on Alice and Bob's subsystems results in ℓ key bits.

Corollary 1. *Let $E(\rho)$ be a function mapping bipartite quantum states $\rho \equiv \rho_{AB}$ into the positive numbers such that the following holds:*

1. *Monotonicity: $E(\Lambda(\rho)) \leq E(\rho)$ for any LOCC operation Λ.*
2. *Asymptotic continuity: for any states ρ^n, σ^n on $\mathcal{H}_A^n \otimes \mathcal{H}_B^n$, the condition $\|\rho^n - \sigma^n\| \to 0$ implies $\frac{1}{\log r_n}|E(\rho^n) - E(\sigma^n)| \to 0$ where $r_n = \dim(\mathcal{H}_A^n \otimes \mathcal{H}_B^n)$.*
3. *Normalisation: $E(\gamma^{\ell}) \geq \ell$.*

Then the regularisation of the function E given by $E^{\infty}(\rho) = \limsup_{n \to \infty} \frac{E(\rho^{\otimes n})}{n}$ is an upper bound on K_D, i.e., $E^{\infty}(\rho_{AB}) \geq K_D(|\psi\rangle\langle\psi|_{ABE})$ where $|\psi\rangle\langle\psi|_{ABE}$ is a purification of ρ_{AB}. If in addition E satisfies

4. *Subadditivity on tensor products: $E(\rho^{\otimes n}) \leq nE(\rho)$,*

then E is an upper bound for K_D.

The analogue of this result in the realm of *entanglement* distillation has long been known: namely, every function E satisfying LOCC monotonicity, asymptotic continuity near maximally entangled states as well as normalisation on maximally entangled states $(E(|\psi\rangle\langle\psi|) = \log d$ for $|\psi\rangle = \frac{1}{\sqrt{d}}\sum_i|ii\rangle)$ can be shown to provide an upper bound on distillable entanglement E_D [23,24], that is, $E^\infty(\rho) \geq E_D(\rho)$. Additionally, if E is subadditive, the same inequality holds with E^∞ replaced by E. Indeed this result can be seen as a corollary to Corollary 1 by restricting from distillation of states τ^ℓ to distillation of $|\psi\rangle\langle\psi|^{\otimes\ell}$ and noting that $|\psi\rangle\langle\psi|^{\otimes\ell}$ is of the form τ^ℓ with trivial $A'B'$.

In the above corollary, we have identified asymptotic continuity on *all* states as well as normalisation on the states γ^ℓ (rather than on singlets) as the crucial ingredients in order for an entanglement measure to bound distillable key from above. Note also that we require those additional conditions as, for instance, the *logarithmic negativity* as defined in [25] satisfies the weaker conditions, therefore being an upper bound on distillable entanglement, but fails to be an upper bound on distillable key.

We will now show how to turn this bound for bipartite states (or tripartite pure states) into one for arbitrary tripartite states. The recipe is simple: for a given state ρ_{ABE}, consider a purification $|\psi\rangle\langle\psi|_{AA'BB'E}$ where the purifying system is denoted by $A'B'$ and is split between Alice and Bob. Clearly, for any splitting, $K_D(|\psi\rangle\langle\psi|_{AA'BB'E}) \geq K_D(\rho_{ABE})$. This inequality combined with the previous corollary applied to $|\psi\rangle\langle\psi|_{AA'BB'E}$ proves the following statement.

Corollary 2. *If E satisfies the conditions of Corollary 1 then*

$$K_D(\rho_{ABE}) \leq E^\infty(\rho_{AA'BB'}) \,,$$

where $\rho_{AA'BB'} = \mathrm{Tr}_E |\psi\rangle\langle\psi|_{AA'BB'E}$ and $\rho_{ABE} = \mathrm{Tr}_{A'B'} |\psi\rangle\langle\psi|_{AA'BB'E}$. If E is subadditive, the same inequality holds with E replacing E^∞.

3.2   Examples of Secrecy Monotones

We will now introduce a number of secrecy monotones. We will only briefly comment on the relations between them. A more detailed analysis of how the different bounds on the key rate compare is given in Section 3.3.

Intrinsic Information. The *intrinsic information* of a probability distribution p_{ijk} is given by

$$I(A : B \downarrow E) := \inf I(A : B|E')_\rho \tag{4}$$

where ρ_{ABE} is the ccc state corresponding to p_{ijk}. The infimum is taken over all channels from E to E' specified by a conditional probability distributions $p_{l|m}$. $\rho_{ABE'}$ is the state obtained by applying the channel to E. This quantity has been defined by Maurer and Wolf and provides an upper bound on the key rate from classical correlations [13]. We can extend it in the following way to arbitrary tripartite quantum states ρ_{ABE}.

Definition 3. *The* intrinsic information *of a tripartite quantum state ρ_{ABE} is given by*

$$I(A : B \downarrow E)_\rho := \inf I(A : B|E')_\rho$$

where the infimum is taken over all CPTP maps $\Lambda_{E\to E}$ from E to E' where $\rho_{ABE'} = (I_{AB} \otimes \Lambda_{E\to E})(\rho_{ABE})$.

This definition is compatible with the original definition since it reduces to (4) if the systems A, B and E are classical.

It is straightforward to show that the intrinsic information satisfies the requirements of Theorem 1. Hence we have proved the following theorem.

Theorem 2. *The intrinsic information is an upper bound on distillable key, i.e.,* $K_D(\rho_{ABE}) \leq I(A : B \downarrow E)_\rho$.

Let us note that this bound differs from the bound proposed in [26,19] where instead of all quantum channels, arbitrary measurements were considered. Our present bound can be tighter, as it can take into account Eve's quantum memory.

In the case where ρ_{ABE} is pure, this bound can be improved by a factor of two because $I(A : B \downarrow E)_\rho = 2E_{sq}(\rho_{AB})$, where E_{sq} is the squashed entanglement defined below and because squashed entanglement is an upper bound for the key rate.

Squashed Entanglement

Definition 4. *Squashed entanglement is defined as*

$$E_{sq}(\rho_{AB}) = \frac{1}{2} \inf_{\substack{\rho_{ABE}: \\ \rho_{AB} = \mathrm{Tr}_E \rho_{ABE}}} I(A : B|E)_\rho$$

Squashed entanglement can be shown to be a LOCC monotone, additive [27], and asymptotically continuous [28]. In [29, Proposition 4.19] it was shown to satisfy the normalisation condition and is therefore an upper bound on distillable key according to Corollary 1.

Theorem 3. *Squashed entanglement is an upper bound on distillable key, i.e.,* $K_D(\rho_{ABE}) \leq E_{sq}(\rho_{AA'BB'})$ *where* $\rho_{AA'BB'} = \mathrm{Tr}_E|\psi\rangle\langle\psi|_{AA'BB'E}$ *and* $\rho_{ABE} = \mathrm{Tr}_{A'B'}|\psi\rangle\langle\psi|_{AA'BB'E}$.

Reduced Intrinsic Information. There is another way in which we can find a bound on the key rate which is tighter than the intrinsic information. In [30] it was shown that the classical intrinsic information is *E-lockable*, i.e., it can increase sharply when a single bit is taken away from Eve. Since (classical) distillable key is not E-lockable, the bound that the intrinsic information provides cannot be tight. This was the motivation for defining the *Reduced Intrinsic Information* by $I(AB \downarrow\downarrow E) = \inf I(AB \downarrow EE') + S(E')$ where the infimum is taken over arbitrary classical values E' [30]. We now define the quantum extension of this function.

Definition 5. *Let* $a = 1, 2$. *The* reduced intrinsic information *(with parameter a) is given by*

$$I(A : B \downarrow\downarrow E)_\rho^{(a)} = \inf\{I(AB \downarrow EE')_\rho + aS(E')_\rho\}$$

where the infimum is taken over all extensions $\rho_{ABEE'}$ *with a classical register E' if $a = 1$ and over arbitrary extensions* $\rho_{ABEE'}$ *if $a = 2$.*

The parameter a reflects the different behaviour of the intrinsic information subject to loss of a single bit (qubit). The Reduced Intrinsic Information is an upper bound on distillable key since

$$K_D(\rho_{ABE}) \leq K_D(\rho_{ABEE'}) + aS(E') \leq I(AB \downarrow EE') + aS(E') .$$

The first inequality corresponds to Corollary 4 below.

Theorem 4. *The reduced intrinsic information is an upper bound on distillable key, i.e., $K_D(\rho_{ABE}) \leq I(A : B \downarrow\downarrow E)_\rho^{(a)}$, for $a = 1, 2$.*

Relative Entropy of Entanglement. The relative entropy of entanglement and its regularised version are well-known entanglement measures that serve as important tools in entanglement theory.

Definition 6. *The relative entropy of entanglement is given by [31,32]*

$$E_R(\rho_{AB}) = \inf_{\sigma_{AB}} S(\rho_{AB} \| \sigma_{AB})$$

where $S(\rho_{AB} \| \sigma_{AB}) = \mathrm{Tr}\rho_{AB}[\log \rho_{AB} - \log \sigma_{AB}]$ and the minimisation is taken over all separable states σ_{AB}, i.e. $\sigma_{AB} = \sum_i p_i \rho_A^i \otimes \rho_B^i$.

The relative entropy of entanglement was the first upper bound that has been provided for $K_D(|\psi\rangle\langle\psi|_{ABE})$ [15,22]. We now extend this result to all tripartite quantum states ρ^{ABE}.

Theorem 5. *The relative entropy of entanglement is an upper bound on distillable key, i.e., $K_D(\rho_{ABE}) \leq E_R^\infty(\rho_{AA'BB'}) \leq E_R(\rho_{AA'BB'})$ where $\rho_{AA'BB'} = \mathrm{Tr}_E |\psi\rangle\langle\psi|_{AA'BB'E}$ and $\rho_{ABE} = \mathrm{Tr}_{A'B'} |\psi\rangle\langle\psi|_{AA'BB'E}$.*

It is a particular advantage of E_R in its function as an upper bound that it is not lockable [33].

3.3   Comparison of Secrecy Monotones

Pure Versus Mixed. For entangled states, bounds derived from entanglement measures are usually tighter than the intrinsic information and its reduced version. Consider for example the state $\rho_{ABE} = |\psi\rangle\langle\psi|_{AB} \otimes \rho_E$ where $|\psi\rangle_{AB} = \frac{1}{\sqrt{2}}(|00\rangle + |11\rangle)$. Here we have

$$E_R(\rho_{ABE}) = E_R^\infty(\rho_{ABE}) = E_{sq}(\rho_{ABE}) = K_D(\rho_{ABE}) = 1 ,$$

while

$$I(A : B \downarrow E)_\rho = I(A : B \downarrow\downarrow E)_\rho^{(a)} = 2 ,$$

for $a = 1, 2$. In general, for tripartite pure states, squashed entanglement is a tighter bound on the key rate than the intrinsic information by at least a factor of two:

$$2E_{sq}(|\psi\rangle\langle\psi|_{ABE}) = I(A : B \downarrow E)_{|\psi\rangle\langle\psi|}.$$

The Locking Effect. We will now give a concrete example which shows that there is a purification $|\psi\rangle_{AA'BB'E}$ of ρ_{ABE} such that

$$K_D(\rho_{ABE}) = E_R(\rho_{AA'BB'}) < I(AA' : BB' \downarrow E)_\rho.$$

Consider the distribution p_{ijkl} defined by the following distribution for p_{ij}

j \\ i	0	1	2	3
0	$\frac{1}{8}$	$\frac{1}{8}$	0	0
1	$\frac{1}{8}$	$\frac{1}{8}$	0	0
2	0	0	$\frac{1}{4}$	0
3	0	0	0	$\frac{1}{4}$

and where k and l are uniquely determined by (i,j),

$$k = i + j \,(\mathrm{mod}\ 2) \quad \text{for} \quad i, j \in \{0, 1\}$$
$$k = i \,(\mathrm{mod}\ 2) \qquad \text{for} \quad i \in \{2, 3\}$$
$$l = \lfloor i/2 \rfloor$$

for all (i,j) with $p_{ij} > 0$. We denote the corresponding cccc state by $\rho_{ABEF} = \sum_{ijkl} p_{ijkl} |ijkl\rangle\langle ijkl|$. Clearly $K_D(\rho_{A;B;EF}) = 0$, as Eve can factorise Alice and Bob, by keeping k when $l = 1$ and forgetting it when $l = 0$. In the former case, when $l = 0$, then Alice and Bob have $(i,j) = (2,2)$, and when $l = 1$, then Alice and Bob have $(i,j) = (3,3)$. In the latter case, both Alice and Bob have at random 0 or 1 and they are not correlated.

On the other hand, when Eve does not have access to l, then the key rate is equal to 1, i.e., $K_D(\rho_{A;B;E}) = 1$. Indeed, it cannot be greater, as key cannot increase more than the entropy of the variable that was taken out from Eve. However one finds that the intrinsic information is equal to $3/2$, i.e., $I(A : B \downarrow E)_\rho = 3/2$ [30].

Let us consider the purification of the above state,

$$|\psi_{A'ABEF}\rangle = \frac{1}{2}\big(|0\rangle_{A'}|22\rangle_{AB}|0\rangle_E|0\rangle_F + |0\rangle_{A'}|33\rangle_{AB}|1\rangle_E|0\rangle_F$$
$$+ |\psi\rangle_{A'AB}|0\rangle_E|1\rangle_F + |\phi\rangle_{A'AB}|1\rangle_E|1\rangle_F\big),$$

where

$$|\psi\rangle = \frac{1}{\sqrt{2}}(|0\rangle_{A'}|00\rangle_{AB} + |1\rangle_{A'}|11\rangle_{AB})$$

and

$$|\phi\rangle = \frac{1}{\sqrt{2}}(|0\rangle_{A'}|01\rangle_{AB} + |1\rangle_{A'}|10\rangle_{AB}).$$

Thus when E and F are with Eve, the state $\rho_{AA';B}$ of Alice and Bob is a mixture of four states: $|0\rangle|22\rangle$, $|0\rangle|33\rangle$, $|\phi\rangle$ and $|\psi\rangle$. This state is separable state, hence $E_R(\rho_{AA';B}) = 0$.

Consider now the state $\rho_{AA'F;B}$ where F is controlled by Alice instead of Eve. Measuring F makes the state separable and in [33] it was shown that measuring a single qubit cannot decrease the relative entropy of entanglement by more than 1, thus we obtain

$$E_R(\rho_{AA'F;B}) \leq 1.$$

By Theorem 5 we then have $K_D(\rho_{ABE}) \leq 1$, but indeed one can distil one bit of key from ρ_{ABE}, therefore

$$K_D(\rho_{ABE}) = E_R(\rho_{AA'F;B}) = 1.$$

In [30] the considered distribution was generalised to make the gap between intrinsic information and distillable key arbitrarily large. It is not difficult to see that E_R is still bounded by one. This shows that the bound based on relative entropy of entanglement, though perhaps more complicated in use, can be significantly stronger than intrinsic information bound. We leave it open, whether or not the intrinsic information bound is weaker in general when compared to the relative entropy bound. This parallels the challenge to discover a relation between the relative entropy of entanglement and squashed entanglement. Here it has also been observed that squashed entanglement can exceed the relative entropy of entanglement by a large amount, due to a *locking effect* [34].

3.4   Upper and Lower Bounds When $\rho_{ABE} = \rho_{AB} \otimes \rho_E$

In this section we focus on states of the form $\rho_{ABE} = \rho_{AB} \otimes \rho_E$. Since distillable key cannot increase under Eve's operations, the form of the state ρ_E is not important and we conclude that $K_D(\rho_{AB} \otimes \rho_E)$ is a function of ρ_{AB} only. If the state ρ_{AB} is classical on system A, then it is known that distillable key is equal to the quantum mutual information, $K_D(\rho_{AB} \otimes \rho_E) = I(A : B)_\rho$ [10]. Indeed, we know from Theorem 2 that the key rate can never exceed $I(A : B)_\rho$. For separable quantum states ρ_{AB} we were able to further improve this bound. The upper bounds are summarised in the following theorem. (Its proof will appear in a full version of this paper.)

Theorem 6. *For all states* $\rho_{AB} \otimes \rho_E$,

$$K_D(\rho_{AB} \otimes \rho_E) \leq I(A : B)_\rho$$

with equality if ρ_{AB} *is classical on system A. If ρ_{AB} is separable, i.e.,* $\rho_{AB} = \sum_i p_i \rho_A^i \otimes \rho_B^i$, *then*

$$K_D(\rho_{AB} \otimes \rho_E) \leq I_{\mathrm{acc}}^{\mathrm{LOPC}}(\mathcal{E}) \leq I_{\mathrm{acc}}(\mathcal{E})$$

where $\mathcal{E} = \{p_i, \rho_A^i \otimes \rho_B^i\}$ *and* $I_{\mathrm{acc}}^{\mathrm{LOPC}}(\mathcal{E})$ *is the maximal mutual information that Alice and Bob can obtain about i using LOPC operations (see e.g. [35,36]), whereas* $I_{\mathrm{acc}}(\mathcal{E})$ *denotes the usual accessible information, i.e. maximal mutual information about i obtained by joint measurements.*

We will now derive a general lower bound on the key rate in terms of the distillable common randomness.

Definition 7. *We say that an LOPC protocol $\mathcal{P}$ distills* common randomness *at rate $\mathcal{R}_\mathcal{P}$ if there exists a sequence $\{\ell_n\}_{n \in \mathbb{N}}$ such that*

$$\limsup_{n \to \infty} \frac{\ell_n - m_n}{n} = \mathcal{R}_\mathcal{P}$$

$$\lim_{n \to \infty} \| \Lambda_n(\rho_{AB}^{\otimes n}) - \tau^{\ell_n} \| = 0$$

where m_n is the number of communicated bits. The distillable common random- *ness of a state ρ^{AB} is defined as $D_R(\rho_{AB}) := \sup_\mathcal{P} \mathcal{R}_\mathcal{P}$.*

For some protocols the rate may be negative. However it is immediate that $D_R(\rho_{AB})$ is nonnegative for all ρ_{AB}. The following statement is a direct consequence of the results in [10,11].

Theorem 7. *For the states $\rho_{ABE} = \rho_{AB} \otimes \rho_E$ the distillable key is an upper bound on the distillable common randomness, i.e., $K_D(\rho_{AB} \otimes \rho_E) \geq D_R(\rho_{AB})$ for all ρ_{AB} and ρ_E.*

4   Embedding Classical into Quantum States

The problem of distilling key from a classical tripartite distribution (i.e., ccc states) is closely related to the problem of distilling entanglement from a bipartite quantum state (where the environment takes the role of the adversary), as noted in [9,30]. It thus seems natural to ask whether, in analogy to *bound entangled* quantum states (which have positive entanglement cost but zero distillable entanglement), there might be classical distributions with *bound information*. These are distributions with zero key rate but positive key cost, i.e., no key can be distilled from them, yet key is needed to generate them. The existence of such distributions, however, is still unproved. (There are, however, some partial positive answers, including an asymptotic result [30] as well as a result for scenarios involving more than three parties [37].)

In [9,30], it has been suggested that the classical distribution obtained by measuring bound entangled quantum states might have bound information. Such hope, however, was put into question by the results of [15], showing that there are quantum states with positive key rate but no distillable entanglement (i.e., they are bound entangled). However, the examples of states put forward in [15] have a rather special structure. It is thus still possible that distributions with bound information might be obtained by measuring appropriately chosen bound entangled states.

In the following, we consider a special *embedding* of classical distributions into quantum states as proposed in [9]. We then show how statements about key distillation starting from the original state and from the embedded state are related to each other. Let

$$\rho_{ccc} := \sum_{ijk} p_{ijk} |ijk\rangle\langle ijk|_{ABE} \tag{5}$$

be a ccc state defined relative to fixed orthonormal bases on the three subsystems (in the following called *computational bases*). We then consider the *qqq embedding* $\rho_{qqq} = |\psi\rangle\langle\psi|$ of ρ_{ccc} given by

$$|\psi\rangle = \sum_i \sqrt{p_{ijk}}|ijk\rangle_{ABE} \ .$$

Note that, if Alice and Bob measure ρ_{qqq} in the computational basis, they end up with a state of the form

$$\rho_{ccq} = \sum_{ij} p_{ij}|ij\rangle\langle ij|_{AB} \otimes |\psi^{ij}\rangle\langle\psi^{ij}|_E \tag{6}$$

for some appropriately chosen $|\psi^{ij}\rangle$. We call this state the *ccq embedding* of ρ_{ccc}.

In a similar way as classical distributions can be translated to quantum states, classical protocols have a quantum analogue. To make this more precise, we consider a classical LOPC protocol $\mathcal{P}$ that Alice and Bob wish to apply to a ccc state ρ_{ccc} as in (5). Obviously, $\mathcal{P}$ can equivalently be applied to the corresponding ccq embedding ρ_{ccq} as defined in (6) (because Alice and Bob's parts are the same in both cases). Because Eve might transform the information she has in the ccq case to the information she has in the ccc case by applying a local measurement, security of the key generated by $\mathcal{P}$ when applied to ρ_{ccq} immediately implies security of the key generated by $\mathcal{P}$ when applied to ρ_{ccc}. Note, however, that the opposite of this statement is generally not true.

In general, a classical protocol $\mathcal{P}$ can be subdivided into a sequence of steps of the following form:

1. generating local randomness
2. forgetting information (discarding local subsystems)
3. applying permutations
4. classical communication.

The *coherent version* of $\mathcal{P}$, denoted $\mathcal{P}_q$, is defined as the protocol acting on a qqq state where the above classical operations are replaced by the following quantum operations:

1. attaching subsystems which are in a superposition of fixed basis vectors
2. transferring subsystems to Eve
3. applying unitary transformations that permute fixed basis vectors
4. adding ancilla systems (with fixed initial state) to both the receiver's and Eve's system, and applying controlled not (CNOT) operations to both ancillas, where the CNOTs are controlled by the communication bits.

Consider now a fixed ccc state ρ_{ccc} of the form (5) and let $\mathcal{P}$ be a classical protocol acting on ρ_{ccc}. It is easy to see that the following operations applied to the qqq embedding ρ_{qqq} of ρ_{ccc} result in the same state: (i) measuring in the computational basis and then applying the classical protocol $\mathcal{P}$; or (ii) applying the coherent protocol $\mathcal{P}_q$ and then measuring the resulting state γ^ℓ in the computational basis. This fact can be expressed by a commutative diagram.

$$
\begin{array}{ccc}
|\psi\rangle\langle\psi|^{\otimes n} & \xrightarrow{\ \mathcal{P}_q\ } & \gamma^\ell \\
\text{measurement}\downarrow & & \downarrow\text{measurement} \\
\rho_{ccq}^{\otimes n} & \xrightarrow{\ \mathcal{P}\ } & \tau^\ell
\end{array}
$$

Hence, if the coherent version $\mathcal{P}_q$ of $\mathcal{P}$ acting on ρ_{qqq} distills secure key bits at rate R then so does the protocol $\mathcal{P}$ applied to the original ccc state ρ_{ccc}.

It is natural to ask whether there are cases for which the converse of this statement holds as well. This would mean that security of a classical protocol also implies security of its coherent version. In the following, we exhibit a class of distributions for which this is always true. The key rate of any such distribution is thus equal to the key rate of the corresponding embedded qqq state.

Roughly speaking, the class of distributions we consider is characterised by the property that the information known to Eve is completely determined by the joint information held by Alice and Bob.

Theorem 8. *Let ρ_{ccc} be a ccc state of the form (5) such that, for any pair of values (i,j) held by Alice and Bob there exists at most one value k of Eve with $p_{ijk} > 0$. If a classical protocol $\mathcal{P}$ applied to ρ_{ccc} produces key at rate R then so does its coherent version $\mathcal{P}_q$ applied to the qqq embedding $|\psi\rangle$ of ρ_{ccc} (and followed by a measurement in the computational basis).*

Proof. The ccq embedding of ρ_{ccc} is given by a state of the form

$$
\rho_{ccq} = \sum_{ij} p_{ij} |ij\rangle\langle ij|_{AB} \otimes |\psi^{ij}\rangle\langle\psi^{ij}|_E .
$$

Since, by assumption, every pair (i,j) determines a unique $k = k(i,j)$, $|\psi^{ij}\rangle\langle\psi^{ij}|_E$ equals $|k(i,j)\rangle\langle k(i,j)|$ and, hence, ρ_{ccq} is identical to the original ccc state ρ_{ccc}. The assertion then follows from the fact that measurements in the computational basis applied to Alice and Bob's subsystems commute with the coherent version $\mathcal{P}_q$ of $\mathcal{P}$.   □

Corollary 3. *Let ρ_{ccc} be a ccc state of the form (5) such that, for any pair of values (i,j) held by Alice and Bob there exists at most one value k of Eve with $p_{ijk} > 0$. Then, the key rate for the qqq embedding ρ_{qqq} of ρ_{ccc} satisfies*

$$
K_D(\rho_{qqq}) = K_D(\rho_{ccc}) .
$$

Note that the above statements do not necessarily hold for general distributions. To see this, consider the state

$$
|\psi\rangle_{ABA'E} = |00\rangle_{AB}|+\rangle_{A'}|+\rangle_E + |11\rangle_{AB}|\psi_+\rangle_{A'E}
$$

where $|+\rangle := \frac{1}{\sqrt{2}}(|0\rangle + |1\rangle)$ and $|\psi_+\rangle := \frac{1}{\sqrt{2}}(|0\rangle|0\rangle + |1\rangle|1\rangle)$. Moreover, let ρ_{ccc} be the ccc state obtained by measuring $|\psi\rangle\langle\psi|_{AA';B;E}$ in the computational basis. Because all its coefficient are positive, it is easy to verify that $|\psi\rangle\langle\psi|_{ABA'E}$ can be seen as the qqq embedding of ρ_{ccc}. Observe that, after discarding subsystem A', ρ_{ccc} corresponds to a perfect key bit. However, the ccq state obtained from $|\psi\rangle\langle\psi|_{ABA'E}$ by discarding A' and measuring in the computational basis is of the form $\frac{1}{2}(|00\rangle\langle00|_{AB} \otimes |+\rangle\langle+|_E + |11\rangle\langle11|_{AB} \otimes I_E/2)$. This state, of course, does not correspond to a key bit as Eve might easily distinguish the states $|+\rangle\langle+|$ and $I_E/2$.

We continue with a statement on the relation between the intrinsic information of a ccc state and the so-called *entanglement of formation*[6] E_F of its qqq embedding. More precisely, we show that, under the same condition as in Theorem 8, the first is a lower bound for the latter (see also [39,40]).

Theorem 9. *Let ρ_{ccc} be a ccc state of the form* (5) *such that, for any pair of values* (i,j) *held by Alice and Bob there exists at most one value* k *of Eve with* $p_{ijk} > 0$, *and let ρ_{qqq} be the qqq embedding of this state. Then*

$$I(A : B \downarrow E)_{\rho_{ccc}} \leq E_F(\mathrm{Tr}_E(\rho_{qqq})) \ .$$

Proof. Note first that any decomposition of $\mathrm{Tr}_E(\rho_{qqq})$ into pure states can be induced by an appropriate measurement on the system E. Hence, we have

$$E_F(\mathrm{Tr}_E(\rho_{qqq})) = \min_{\{|\bar{k}\rangle\}} \sum_{\bar{k}} p_{\bar{k}} S(A)_{|\psi_{\bar{k}}\rangle} \tag{7}$$

where the minimum ranges over all families of (not necessarily normalised) vectors $|\bar{k}\rangle$ such that $\sum_{\bar{k}} |\bar{k}\rangle\langle\bar{k}| = I_E$ (this ensures that they form a measurement), $p_{\bar{k}} := |\langle\bar{k}|_E|\psi\rangle_{ABE}|^2$, and $|\psi_{\bar{k}}\rangle := \langle\bar{k}|_E|\psi\rangle_{ABE}/\sqrt{p_{\bar{k}}}$.

For any pair (i,j) of values held by Alice and Bob (with nonzero probability) we have $\mathrm{Tr}_{AB}[\rho_{qqq}(|ij\rangle\langle ij| \otimes I_E)] = p_{ij}|k\rangle\langle k|$, where $k = k(i,j)$ is the corresponding (unique) value held by Eve. Hence, the probability distribution of the state $\bar{\rho}_{ccc}$ obtained by applying the above measurement on Eve's system satisfies

$$q_{ij\bar{k}} := \mathrm{Tr}(|\psi\rangle\langle\psi|_{ABE}|ij\bar{k}\rangle\langle ij\bar{k}|) = p_{ijk}q_{\bar{k}|k} \ ,$$

where $q_{\bar{k}|k} := \mathrm{Tr}(|\bar{k}\rangle\langle\bar{k}||k\rangle\langle k|)$. The intrinsic information is thus bounded by

$$I(A : B \downarrow E)_{\rho_{ccc}} \leq \min_{\{|\bar{k}\rangle\}} I(A : B|\bar{E})_{\bar{\rho}_{ccc}} \ ,$$

where $\bar{\rho}_{ccc}$ is the state defined above (depending on the choice of the vectors $|\bar{k}\rangle$). Moreover, using Holevo's bound, we find

$$I(A : B|\bar{E})_{\bar{\rho}_{ccc}} \leq \min_{\{|\bar{k}\rangle\}} \sum_{\bar{k}} p_{\bar{k}} S(A)_{|\psi_{\bar{k}}\rangle} \ .$$

The assertion then follows from (7).   □

---

[6] The *entanglement of formation* E_F is an entanglement measure defined for bipartite states by $E_F(\sigma_{AB}) := \min \sum_i p_i S(\mathrm{Tr}_B(\sigma_{AB}^i))$ where the minimum is taken over all ensembles $\{p_i, \sigma_{AB}^i\}$ with $\sum_i p_i \sigma_{AB}^i = \sigma_{AB}$ [38].

Because the intrinsic information is additive (i.e., it is equal to its regularised version), Theorem 9 also holds if the entanglement of formation E_F is replaced by the entanglement cost E_C.

The discussion above suggests that classical key distillation from ccc states can indeed by analysed by considering the corresponding qqq embedding of the state, but the original ccc state has to satisfy certain properties. This relation might be particularly useful for the study of bound information as discussed at the beginning of this section. In fact, there exist bound entangled states which satisfy the property required by Theorem 8 above [41].

5 On Locking and Pre-shared Keys

In [30] it was observed that, by adding one bit of information to Eve, the (classical) intrinsic information can decrease by an arbitrarily large amount. In [16] it was shown that classical correlation measures of quantum states can exhibit a similar behaviour; more precisely, the accessible information can drop by an arbitrarily large amount when a single bit of information is lost. This phenomenon has been named *locking of information* or just *locking*. For tripartite states ρ_{ABE}, locking comes in two flavours: i) locking caused by removing information from Eve, ii) locking caused by removing information from Alice and/or Bob (and possibly giving it to Eve). Let us call those variants *E-locking* and *AB-locking*, respectively.

In [33] it was shown that entanglement cost as well as many other entanglement measures can be AB-locked. Further results show that squashed entanglement and entanglement of purification are also AB-lockable [34,42]. So far the only known non-lockable entanglement measure is relative entropy of entanglement.

It was shown in [30] that distillable key is not E-lockable for classical states. In the sequel we extend this result and prove that the distillable key for quantum states ρ_{ABE} is not E-lockable, either. The proof proceeds along the lines of [30], replacing the bound of Csiszár and Körner by its quantum generalisations due to [10] (see also [11]). Let us emphasise that we leave open the question on whether distillable key is AB-lockable (even for ccc states).

Theorem 10. *Consider a state $\rho_{ABEE'}$ and let $\mathcal{P}$ be a key distillation protocol for ρ_{ABE} with rate $R_\mathcal{P}$. Then there exists another protocol $\mathcal{P}'$ for $\rho_{ABEE'}$ with rate $R_{\mathcal{P}'} \geq R_\mathcal{P} - 2S(\rho_{E'})$. If, in addition, E' is classical then $R_{\mathcal{P}'} \geq R_\mathcal{P} - S(\rho_{E'})$.*

Proof. For any fixed $\epsilon > 0$ there exists $n \in \mathbb{N}$ such that the protocol $\mathcal{P}$ transforms $\rho_{ABE}^{\otimes n}$ into a ccq state σ_{ABE} which satisfies the following inequalities:

$$\|\sigma_{ABE} - \tau^\ell\| \leq \epsilon, \quad \frac{\ell}{n} \geq R_\mathcal{P} - \epsilon. \tag{8}$$

Suppose that Alice and Bob apply this map to the state $\rho_{ABEE'}^{\otimes n}$ (i.e., they try to distil key, as if the system E' was not present). The state $\rho_{ABEE'}^{\otimes n}$ is then transformed into some state $\sigma_{ABEE'}$ which traced out over E' is equal to the ccq state σ_{ABE}. Repeating this protocol m times results in $\sigma_{ABEE'}^{\otimes m}$, from which Alice and Bob can draw at least $m(I(A:B) - I(A:EE')) - o(m)$ bits of key by error correction and privacy amplification [10]. This defines a protocol $\mathcal{P}'$. To evaluate its rate, we use subadditivity of entropy which gives the estimate

$$I(A:EE')_\sigma \leq I(A:E)_\sigma + I(AE:E')_\sigma .$$

From (8) and Fannes' inequality we know that[7]

$$I(A:B)_\sigma \geq \ell - 8\epsilon\ell - H(\epsilon)$$
$$I(A:E)_\sigma \leq 8\epsilon\ell + H(\epsilon) .$$

This together with (2) implies

$$K_D(\sigma_{ABEE'}) \geq I(A:B)_\sigma - I(A:EE')_\sigma \geq (1 - 16\epsilon)\ell - 2H(\epsilon) - I(AE:E')_\sigma .$$

To get the key rate of $\mathcal{P}'$, we divide the above by n and use (8),

$$R_{\mathcal{P}'} \geq \frac{1}{n}K_D(\sigma_{ABEE'}) \geq (1 - 16\epsilon)(R_\mathcal{P} - \epsilon) - \frac{1}{n}2H(\epsilon) - \frac{1}{n}I(AE:E')_\sigma .$$

Because this holds for any $\epsilon > 0$, the assertion follows from $I(AE:E')_\sigma \leq 2S(E')_\sigma = 2nS(E')_\rho$ and, if E' is classical, $I(AE:E')_\sigma \leq S(E')_\sigma = nS(E')_\rho$.   □

Applying the above theorem to an optimal protocol leads to the statement that the key rate K_D is not E-lockable.

Corollary 4. *For any state $\rho_{ABEE'}$, $K_D(\rho_{ABEE'}) \geq K_D(\rho_{ABE}) - 2S(\rho_{E'})$ and, if E' is classical, $K_D(\rho_{ABEE'}) \geq K_D(\rho_{ABE}) - S(\rho'_E)$.*

Consider now a situation where Alice and Bob have some pre-shared key U which is not known to Eve.

A major consequence of Theorem 10 is that a pre-shared key cannot be used as a catalyst to increase the key rate. More precisely, the corollary below implies that, for any protocol $\mathcal{P}$ that uses a pre-shared key held by Alice and Bob, there is another protocol $\mathcal{P}'$ which is as efficient as $\mathcal{P}'$ (with respect to the net key rate), but does not need a pre-shared key.

Corollary 5. *Let $\mathcal{P}$ be a key distillation protocol for $\rho_{ABE} \otimes \tau^\ell$ where τ^ℓ is some additional ℓ-bit key shared by Alice and Bob. Then there exists another protocol $\mathcal{P}'$ for ρ_{ABE} with rate $R_{\mathcal{P}'} \geq R_\mathcal{P} - \ell$.*

Proof. Consider the state $\rho_{A'B'EE'}$ where E' is a system containing the value U of a uniformly distributed ℓ-bit key, $A' := (A, U)$, and $B' := (B, U)$. Note

---

[7] $H(\epsilon)$ denotes the binary entropy, i.e., the Shannon entropy of the distribution $[\epsilon, 1 - \epsilon]$.

that $\rho_{A'B'E}$ is equivalent to $\rho_{ABE} \otimes \tau^\ell$. The assertion then follows from the observation that any protocol which produces a secure key starting from $\rho_{A'B'EE'}$ can easily be transformed into an (equally efficient) protocol which starts from ρ_{ABE}, because Alice and Bob can always generate public shared randomness. $\qquad\square$

The following example shows that the factor 2 in Theorem 10 and Corollary 4 is strictly necessary. Let

$$\rho_{ABEE'} = \sum_{i=1}^{4} |i\rangle\langle i|_A \otimes |i\rangle\langle i|_B \otimes |\psi_i\rangle\langle\psi_i|_{EE'}$$

where $|\psi_i\rangle$ are the four Bell states on the bipartite system EE'. Then, obviously, $K_D(\rho_{ABEE'}) = 0$, but if E' (which is only *one* qubit) is lost, then $K_D(\rho_{ABE}) = 2$, since E is then maximally mixed conditioned on i. One recognises here the effect of superdense coding.

6   Classical and Quantum Adversaries in QKD

Up to now, we have considered an adversary with unbounded resources. Of course, if one limits the adversary's capabilities, certain cryptographic tasks might become easier. In the following, we will examine a situation where the adversary cannot store quantum states and, hence, is forced to apply a measurement, turning them into classical data. We will exhibit an example of a $2d$-dimensional ccq state which only has key rate 1, but if Eve is forced to measure her system, the key rate raises up to roughly $\frac{1}{2}\log d$.

Note that upper bounds on the key rate which are defined in terms of an optimal measurement on Eve's system (see, e.g., [26,19] and Section 3) are also upper bounds on the key rate in a setting where Eve has no quantum memory. Hence, our result implies that these upper bounds are generally only rough estimates for the key rate in the unbounded scenario.

Consider the state

$$\rho_{AA'BB'E} = \frac{1}{2d}\sum_{k=1}^{d} |00\rangle\langle 00|_{AB}(|kk\rangle\langle kk|_{A'B'} \otimes |k\rangle\langle k|_E)$$

$$+ |11\rangle\langle 11|_{AB}(|kk\rangle\langle kk|_{A'B'} \otimes U|k\rangle\langle k|_E|U^\dagger)$$

where U is the quantum Fourier transform on d dimensions. (Such a state has been proposed in [16] to exhibit a locking effect of the accessible information. It also corresponds to the *flower state* of [33].)

It is easy to see that the bit in the system AB is uncorrelated to Eve's information and, hence, completely secret, i.e., $K_D(\rho_{AA'BB'E}) = K_D(\rho_{AB}) \geq 1$. On the other hand, if this bit is known to Eve then she has full knowledge on the state in $A'B'$, i.e., $K_D(\rho_{AA'BB'EE'}) \leq I(AA' : BB' \downarrow EE')_\rho = 0$, where E' is a classical system carrying the value of the bit in AB (see Theorem 2). From this and Corollary 4 (or, alternatively, Theorem 4), we conclude that the key rate (relative to an unbounded adversary) is given by

$$K_D(\rho_{AA'BB'E}) = K_D(\rho_{AB}) = 1 .$$

Let us now assume that Eve applies a measurement on her system E, transforming the state defined above into a ccc state $\sigma_{AA'BB'E}$. Because the values of Alice and Bob are maximally correlated, it is easy to see that the key rate of this state satisfies $K_D(\sigma_{AA'BB'E}) = S(A|E)_\sigma = S(A)_\sigma - I(A:E)_\sigma$. Note that $S(A)_\sigma = 1 + \log d$. Moreover, the mutual information $I(A:E)_\sigma$ for an optimal measurement on E corresponds to the so-called accessible information, which equals $\frac{1}{2} \log d$, as shown in [16]. We thus conclude that

$$K_D(\sigma_{AA'BB'E}) = 1 + \frac{1}{2} \log d .$$

Note that the accessible information is additive, so even if the measurements are applied to blocks of states, the amount of key that can be generated is given by this expression.

The above result gives some insights into the strength of attacks considered in the context of quantum key distribution (QKD). A so-called *individual attack* corresponds to a situation where the adversary transforms his information into classical values. In contrast, a *collective attack* is more general and allows the storage of quantum states.

As shown in [18], for most QKD protocols, security against collective attacks implies security against any attack allowed by the laws of quantum physics. The above result implies that the same is not true for individual attacks, i.e., these might be arbitrarily weaker than collective (and, hence, also general) attacks.

Acknowledgment

We are grateful to Karol Horodecki and Norbert Lütkenhaus for their valuable input and many enlightening discussions. We would also like to thank the anonymous reviewers for their helpful comments and suggestions. This work was supported by the European Commission through the FP6-FET Integrated Projects SCALA CT-015714 and QAP, and through SECOQC. MC acknowledges the support of an EPSRC Postdoctoral Fellowship and a Nevile Research Fellowship, which he holds at Magdalene College Cambridge. RR is supported by HP Labs Bristol.

References

1. Shannon, C.E.: Communication theory of secrecy systems. Bell Systems Technical Journal **28** (1949) 656–715
2. Maurer, U.M.: Secret key agreement by public discussion from common information. IEEE Transactions on Information Theory **39** (1993) 733–742

3. Csiszár, I., Körner, J.: Broadcast channels with confidential messages. IEEE Trans. Inf. Theory **24** (1978) 339–348
4. Ahlswede, R., Csiszár, I.: Common randomness in information theory and cryptography. IEEE Transactions on Information Theory **39** (1993) 1121–1132
5. Bennett, C.H., Brassard, G.: Quantum cryptography: Public key distribution and coin tossing. In: Proceedings of the IEEE International Conference on Computers, Systems and Signal Processing, Bangalore, India, December 1984, IEEE Computer Society Press, New York (1984) 175–179
6. Ekert, A.: Quantum cryptography based on Bell's theorem. Phys. Rev. Lett **67** (1991) 661–663
7. Renner, R., Wolf, S.: New bounds in secret-key agreement: the gap between formation and secrecy extraction. In: Proceedings of EUROCRYPT 2003. Lecture Notes in Computer Science, Springer (2003) 562–577
8. Wyner, A.D.: The wire-tap channel. Bell System Technical Journal **54** (1975) 1355–1387
9. Gisin, N., Wolf, S.: Linking classical and quantum key agreement: is there 'bound information'. In: Advances in Cryptology — CRYPTO 2000. Lecture Notes in Computer Science, Springer (2000) 482–500
10. Devetak, I., Winter, A.: Distillation of secret key and entanglement from quantum states. Proc. Roy. Soc. Lond. Ser. A **461** (2004) 207–235
11. Renner, R., König, R.: Universally composable privacy amplification against quantum adversaries. In: Second Theory of Cryptography Conference, TCC 2005. Volume 3378 of Lecture Notes in Computer Science., Springer (2005) 407–425
12. Cerf, N.J., Massar, S., Schneider, S.: Multipartite classical and quantum secrecy monotones. Phys. Rev. A **66** (2002) 042309
13. Maurer, U., Wolf, S.: The intrinsic conditional mutual information and perfect secrecy. In: Proceedings of the 1997 IEEE Symposium on Information Theory. (1997) 88
14. Horodecki, M.: Entanglement measures. Quantum Inf. Comp. **1** (2001) 3–26
15. Horodecki, K., Horodecki, M., Horodecki, P., Oppenheim, J.: Secure key from bound entanglement. Phys. Rev. Lett **94** (2005) 160502
16. DiVincenzo, D., Horodecki, M., Leung, D., Smolin, J., Terhal, B.: Locking classical correlation in quantum states. Phys. Rev. Lett **92** (2004) 067902
17. Maurer, U., Wolf, S.: Information-theoretic key agreement: From weak to strong secrecy for free. In: Advances in Cryptology — EUROCRYPT 2000. Volume 1807 of Lecture Notes in Computer Science., Springer (2000) 351–368
18. Renner, R.: Security of Quantum Key Distribution. PhD thesis, Swiss Federal Institute of Technology (ETH) Zurich (2005) quant-ph/0512258.
19. Christandl, M., Renner, R.: On intrinsic information. In: Proceedings of the 2004 IEEE International Symposion on Information Theory. (2004) 135
20. Ben-Or, M., Horodecki, M., Leung, D.W., Mayers, D., Oppenheim, J.: The universal composable security of quantum key distribution. In: Second Theory of Cryptography Conference, TCC 2005. Lecture Notes in Computer Science (2005) 386–406
21. König, R., Renner, R., Bariska, A., Maurer, U.: Locking of accessible information and implications for the security of quantum cryptography. (quant-ph/0512021)
22. Horodecki, K., Horodecki, M., Horodecki, P., Oppenheim, J.: General paradigm for distilling classical key from quantum states. quant-ph/0506189 (2005)
23. Horodecki, M., Horodecki, P., Horodecki, R.: Limits for entanglement measures. Phys. Rev. Lett **84** (2000) 2014

24. Donald, M., Horodecki, M., Rudolph, O.: The uniqueness theorem for entanglement measures. J. Math. Phys. **43** (2002) 4252–4272
25. Vidal, G., Werner, R.: A computable measure of entanglement. Phys. Rev. A **65** (2002) 032314
26. Moroder, T., Curty, M., Lütkenhaus, N.: Upper bound on the secret key rate distillable from effective quantum correlations with imperfect detectors. Phys. Rev. A **73** (2006) 012311
27. Christandl, M., Winter, A.: Squashed entanglement — an additive entanglement measure. J. Math. Phys. **45** (2004) 829–840
28. Alicki, R., Fannes, M.: Continuity of conditional quantum mutual information. J. Phys. A **37** (2003)
29. Christandl, M.: The Structure of Bipartite Quantum States: Insights from Group Theory and Cryptography. PhD thesis, University of Cambridge (2006)
30. Renner, R., Wolf, S.: New bounds in secret-key agreement: The gap between formation and secrecy extraction. In: Advances in Cryptology - EUROCRYPT 2003, Lecture Notes in Computer Science, Springer (2003)
31. Vedral, V., Plenio, M.B., Rippin, M.A., Knight, P.L.: Quantifying entanglement. Phys. Rev. Lett **78** (1997) 2275–2279
32. Vedral, V., Plenio, M.B.: Entanglement measures and purification procedures. Phys. Rev. A **57** (1998) 1619–1633
33. Horodecki, K., Horodecki, M., Horodecki, P., Oppenheim, J.: Locking entanglement with a single qubit. Phys. Rev. Lett **94** (2005) 200501
34. Christandl, M., Winter, A.: Uncertainty, monogamy and locking of quantum correlations. IEEE Transactions on Information Theory **51** (2005) 3159–3165 quant-ph/0501090.
35. Bennett, C.H., DiVincenzo, D.P., Fuchs, C.A., Mor, T., Rains, E., Shor, P.W., Smolin, J., Wootters, W.K.: Quantum nonlocality without entanglement. Phys. Rev. A **59** (1999) 1070
36. Badziąg, P., Horodecki, M., Sen(De), A., Sen, U.: Universal Holevo-like bound for locally accesible information. Phys. Rev. Lett **91** (2003) 117901
37. Acin, A., Cirac, I., Massanes, L.: Multipartite bound information exists and can be activated. Phys. Rev. Lett. **92** (2004) 107903
38. Bennett, C.H., DiVincenzo, D.P., Smolin, J., Wootters, W.K.: Mixed-state entanglement and quantum error correction. Phys. Rev. A **54** (1997) 3824–3851
39. Christandl, M.: The quantum analog to intrinsic information. Diploma Thesis, Institute for Theoretical Computer Science, ETH Zurich (2002)
40. Renner, R.: Linking information theoretic secret-key agreement and quantum purification. Diploma Thesis, Institute for Theoretical Computer Science, ETH Zurich (2000)
41. Horodecki, P., Lewenstein, M.: Bound entanglement and continuous variables. Phys. Rev. Lett **85** (2000) 2657
42. Winter, A.: Secret, public and quantum correlation cost of triples of random variables. In: Proceedings of the 2005 IEEE International Symposium on Information Theory. (2005) 2270–2274

Intrusion-Resilient Key Exchange in the Bounded Retrieval Model

David Cash[1], Yan Zong Ding[1], Yevgeniy Dodis[2], Wenke Lee[1],
Richard Lipton[1], and Shabsi Walfish[2]

[1] College of Computing, Georgia Institute of Technology
{cdc,ding,wenke,rjl}@cc.gatech.edu
[2] Department of Computer Science, New York University
{dodis,walfish}@cs.nyu.edu

Abstract. We construct an intrusion-resilient symmetric-key authenticated key exchange (AKE) protocol in the bounded retrieval model. The model employs a long shared private key to cope with an active adversary who can repeatedly compromise the user's machine and perform any efficient computation on the entire shared key. However, we assume that the attacker is communication bounded and unable to retrieve too much information during each successive break-in. In contrast, the users read only a small portion of the shared key, making the model quite realistic in situations where storage is much cheaper than bandwidth.

The problem was first studied by Dziembowski [Dzi06a], who constructed a secure AKE protocol using random oracles. We present a general paradigm for constructing intrusion-resilient AKE protocols in this model, and show how to instantiate it *without* random oracles. The main ingredients of our construction are UC-secure password authenticated key exchange and tools from the bounded storage model.

1 Introduction

Robust systems for network security must guarantee resilience to compromises and intrusions. Company laptops, for example, often fall prey to Trojan horse viruses that users inadvertently "install" when they travel (without the protection of their company's firewalls). These viruses can persist until a sysadmin removes them, and then all credentials stored on the laptop must be replaced. A malicious virus could even steal a user's credentials with a key logger and erase itself, compromising all future use of the credentials until they are replaced.

A natural technique to overcome this problem is to use fresh session keys (or other credentials) *dynamically generated* using some secure key exchange protocol (e.g., Diffie-Hellman) between the parties involved. However, the problem with this latter approach is that such keys are not authenticated. So it seems like one must either sacrifice authentication, or lose one's resilience to compromises. The *bounded retrieval model*, introduced in various related contexts by

S.P. Vadhan (Ed.): TCC 2007, LNCS 4392, pp. 479–498, 2007.

[DLL05, CLW06, Dzi06a, Dzi06b],[1] provides an elegant and often very realistic way to resolve the above conflicting requirements. It assumes that storage is cheap and thus users can afford to store large quantities of data; in our context, this means that users can share very long symmetric keys (which can then be used to provide authentication). On the other hand, the bandwidth and retrieval capacities of both the users and the attacker are limited. For the users, this means that they only need to access very small portions of their long-term keys for authentication (and key exchange). On the other hand, we will be more generous to the attacker. We allow it to repeatedly compromise the user's machine for limited periods of time, and, during each such break-in, to perform any efficient computation on the *entire* shared key. However, we assume that the attacker's bandwidth capacity from the user's machine is bounded, so that the attacker is unable to retrieve too much information about the long symmetric key. Thus, in this model it might be possible to construct intrusion-resilient and, yet, authenticated key exchange protocols.

Indeed, this was recently done by Dziembowski [Dzi06a] who constructed such a protocol in the *random oracle model*. However, it is well known by now that a protocol proven secure in the random oracle model *might not* be secure when the random oracle is replaced by *any* secure cryptographic hash function [CGH04]. Therefore, an efficient construction without a random oracle is desired.

OUR MAIN RESULT. We resolve the above open problem and construct an efficient intrusion-resilient authenticated key exchange (AKE) protocol in the bounded retrieval model. In fact, our contribution consists of two parts. First, we present a general paradigm for constructing such intrusion-resilient protocols, and, second, we then show how to instantiate our paradigm without random oracles. At a high level, our general paradigm first performs what we call a *Weak Key Exchange* (WKE), which only guarantees that the session keys output by the participants Alice and Bob are *individually unpredictable* (and equal in case the attacker is passive). After this stage, Alice and Bob use their "weak" keys as passwords for a *Password Authenticated Key Exchange* (PAK) protocol [BM93, BPR00, BMP00, GL01]. Such protocols are typically used in settings where the shared secrets are not uniformly distributed and have low entropy, and the goal is to construct AKE protocols resistant to offline dictionary attacks. Interestingly, in our setting the secrets will actually have high-entropy, and, with little effort, can even be made *individually* random. However, the utility of PAK protocols for our purposes comes from their implicit authentication guarantee: a party A with a password pw will only arrive at a shared session key only when interacting with a party B holding the *same* password pw. Thus, in our setting PAK is used to guarantee that Alice and Bob will agree on a session key after the WKE phase only if their individually unpredictable weak keys match.

Somewhat surprisingly, the security of our construction does not seem to follow from PAK protocols secure under most previously used definitions of PAK

---

[1] These works used slightly different terminology and formalizations. Here we use the model of [Dzi06a], but borrow the nomenclature of [CLW06], since we feel that it reflects the general model most accurately.

(*e.g.,* [BPR00, GL01, BDK$^+$05]). Instead, we need to employ Universally Composable (UC) PAK, as defined in [CHK$^+$05]. This is a very strong definition which guarantees the security of PAK even when used in arbitrary "environments" (see Section 2). Very informally (see the discussion at the end of Section 4 for more details), the strong guarantees of a UC PAK are required because of the extreme weakness of the security provided by the WKE definition described above. WKE may allow the adversary to *adaptively* correlate the session keys of Alice and Bob in an *arbitrary* manner, provided that they remain individually unpredictable. Indeed, the adversary is even provided with some *a priori* partial information about the secret keys of Alice and Bob, obtained via a previous intrusion. To further complicate matters, we require that the security of all future AKE sessions is preserved even after *multiple intrusions*, yet WKE provides no forward security guarantee at all. It is this latter complication which prevents most simulation based definitions of PAK from sufficing for our purposes. Namely, it is difficult to simulate the PAK protocol in a manner consistent with both past and *future* information obtained by the adversary about the secret key X. On the other hand, UC secure PAK protocols already support such simulation, since the environment in the UC experiment can also provide such partial information about party's secrets (which are, indeed, generated by the UC environment) directly to the adversary.

We can instantiate the above "WKE+PAK" paradigm in several ways to get concrete intrusion-resilient AKE protocols in our model. First, we observe that the original protocol of Dziembowski [Dzi06a] could be viewed as a special case our our method. Specifically, the 2-round WKE implicit in [Dzi06a] is built using purely information-theoretic tools used in the bounded storage model [Mau92] (most crucially, averaging samplers [BR94, Vad04]). In our construction, we will use a similar (and slightly simpler) WKE. As for the (high-entropy) PAK protocol, the protocol implicit in [Dzi06a] first applied the random oracle to the password, which simultaneously solved two main difficulties of the PAK setting: non-uniform passwords became uniform, and correlated passwords became independent (unless equal to begin with). Not surprisingly, after this application of the random oracle, a standard symmetric-key AKE protocol (more or less from [BR93]) was sufficient for the implicit PAK protocol of [Dzi06a]. In contrast, PAK protocols are much more complicated in the standard model, especially in the UC-model. Luckily, Canetti et al. [CHK$^+$05] built such an efficient UC PAK protocol in the common reference string (CRS) model. Since a CRS is trivially implementable in our model, — the parties can generate and store it as part of their long shared secret, — we immediately get the first intrusion-resilient AKE protocol without random oracles.

To summarize, we get our main result by first properly abstracting and modularizing the construction of [Dzi06a] (as consisting of "hidden" WKE and UC PAK), and then proving a general composition theorem allowing us to build intrusion-resilient AKE protocols (by composing WKE and UC PAK).

PROACTIVE SECURITY. One obvious issue in the bounded retrieval model is that even a long key will become useless after too many break-ins, as the adversary will eventually steal too much of the key. Of course, to solve this problem we can let Alice and Bob occasionally refresh their long keys as follows. They first run the AKE protocol to obtain a fresh key short key r, then expand r into a long key R using a pseudorandom generator (this operation will take some time, but is feasible overall), then XOR this long R to their currently shared long key (once again, this will take some time), and, finally, erase the short r. Needless to say, the parties should perform these last few steps "offline" and only when being absolutely sure that the AKE phase succeeded. Also, since most pseudorandom generators operate in the stream cipher mode, these periodic updates can be done "in place". Additionally, the final long key will be secure as long as either a large enough portion of the original key is not yet leaked, or if the AKE phase was not compromised.

CORRECTING ERRORS. We can also extend our model to allow the adversary to adaptively cause errors in some fraction of the symmetric keys. This extension allows us to tolerate either accidental hard-drive failures of some small part of the secret storage, or even malicious attacks when a virus might be able to rewrite or damage a small portion of the disk. In fact, our extension for this case is only nominally less efficient than our main construction. As our main tool, we employ *secure sketches* [DORS06] in this construction. The only caveat here is that we need to assume that the attacker cannot inject errors into the CRS. Thus, the CRS must be stored in a protected read-only area or made public in some other way. Details appear in Appendix A.

1.1 Related Works

The *Bounded Storage Model (BSM)* [Mau92] is closely related to the bounded retrieval model. In the BSM, a large random string (analogous to our long secret key) is broadcast publicly at a rate that overwhelms the adversary's ability to compress and store it. This is similar to the bounded retrieval restriction in our model. However, in the BSM the parties are assumed to share a short additional key which can only be compromised after the attacker lost access to a long string. In contrast, in our model the only secret is the long string, and the attacker can adaptively break-in and learn large parts of this long string. This makes the bounded retrieval model considerably more complicated than the BSM. Indeed, it is easy to see that, unlike the BSM model, it is impossible to have information-theoretically secure AKE protocols in our model. However, the techniques from the BSM model are useful in our model as well: indeed, the WKE protocol we use crucially utilizes tools from the BSM (such as averaging samplers).

The utilization of PAK protocols in our solution was influenced by the study AKE protocols using biometric data [BDK+05]. In particular, this work also introduced the idea of using low entropy intermediate keys as inputs to a PAK protocol, which is fundamental to our construction. The usual PAK model was insufficient in that application as well, and was augmented to allow the adversary to specify some correlation between parties' (unequal) passwords. However, this extension was much weaker and much simpler than the one required in our setting. In particular, UC PAK protocols were not needed for that application.

Another related line of work is that of privacy amplification and *authenticated key agreement* using a shared weak secret key [MW03, Wol98, RW03, DKRS06]. As in our model, the secret key is only guaranteed to have some entropy. However, all these information-theoretic protocols require accessing and performing computation on the *entire* key. In contrast, a key feature of our model is that the keys are huge, and participants can only access a tiny portion of the key (we call this property *locality*). Thus, the above methods are inapplicable to our setting. In fact, we already mentioned that information-theoretic solutions (such as the above) are impossible in our setting.

Another related problem is protection against partial key exposure (so-called *exposure-resilient cryptography*; see [CDH+00]), where an adversary can directly access most of the original bits of the secret key without effecting the security of the system. However, the solutions in this model are again non-local, and the adversary is not allowed to compute arbitrary functions of the key, like in our model.

A different direction for dealing with key exposure is the study of *key-evolving schemes*. Such schemes allow the attacker to obtain the *entire* (short) key, but assume the existence of global time, and update their secret key at each time period (thus, unlike our schemes, these schemes are stateful). In *forward secure* schemes [And97, BM99, CHK03, BY03], an adversary is allowed to compromise the system at some point, but is still unable to break the system for previous time periods all of which have not been compromised. Unfortunately, all "future" security is necessarily gone in this model.

The model of *key-insulated cryptography* [DKXY02, DKXY03] fixed this problem, where all past and future periods remain secure after a fixed number of compromises. However, this is achieved by introducing a non-corruptible server which holds a master key and helps the user update its secret key from period to period. So called *intrusion-resilient* cryptosystems [IR02, DFK+03] extend the above modeling and allow the attacker to corrupt both the user and the server, but not in the same time period. From our perspective, such schemes can be viewed as partitioning a key into two parts, allowing the attacker to obtain either part at each period, and updating both parts in between the periods.

Finally, we already mentioned several recent works that introduced the bounded retrieval model in various contexts. Dagon et al. [DLL05] used it for database protection, Di Crescenzo et al. [CLW06] — for password authentication resisting offline dictionary attacks,[2] and Dziembowski — for public-key encryption [Dzi06b] and intrusion-resilient AKE [Dzi06a] (the latter being the subject of this paper).

1.2   Structure of the Paper

In Section 2 we present some technical lemmas and tools. In Section 3 we formally define the model and security for intrusion resilient AKE. In Section 4 we present

---

[2] Here the model is similar to ours, except the adversary may only steal original bits of the large key and not any function like in our model.

a secure construction of intrusion resilient AKE and discuss the use of Universally Composable PAK. Finally we discuss extending the model to allow for errors in Appendix A.

2   Preliminaries

In this section we briefly review some of the facts and tools used in this paper. Throughout, U_n denotes the uniform distribution on bit strings of length n, and $\|$ denotes string concatenation. We assume that the definitions of negligible function (negl(k)) and computational indistinguishability ($X \overset{c}{\equiv} Y$) are familiar.

Definitions and Facts from Probability. The *statistical distance* between two random variables taking values in S is defined to be

$$\Delta(X, Y) = \max_{T \subset S} |\Pr[X \in T] - \Pr[Y \in T]|.$$

The *min-entropy* $H_\infty(X)$ of a discrete random variable X is defined as

$$H_\infty(X) = \min_{x \in \text{supp}(X)} \{-\log \Pr[X = x]\},$$

where $\text{supp}(X)$ is the support of X. Let X be a random variable taking values in $\{0,1\}^n$, and let $k \leq n$. We say that X is a *k-source* if $H_\infty(X) \geq k$, that is, for every $x \in \text{supp}(X)$, $\Pr[X = x] \leq 2^{-k}$. Therefore, informally speaking, that X has high min-entropy means that the value X takes on is *hard to guess* (for an unbounded adversary). For a random variable X taking values in $\{0,1\}^n$ and $\alpha \in [0,1]$, we say that X has *entropy rate* α if X is an αn-source.

The following well known lemma quantifies the intuition that short compressions of long entropy-rich strings must leave out a lot of information. This allows us to tell how random our long secret keys look after some partial compromises.

Lemma 1 (c.f. [NZ96]). *Let X and Y be any two (correlated) random variables. Suppose that X is an n-source, and Y takes values in $\{0,1\}^r$. Then for every $\varepsilon > 0$, with probability at least $1 - \varepsilon$ over $y \leftarrow Y$, $X|_{Y=y}$ is a $(n - r - \log(1/\varepsilon))$-source.*

Averaging Samplers. We also make use of a combinatorial tool called an *Averaging Sampler*. Averaging samplers, first introduced by [BR94], are procedures that approximate the mean of any function from bit strings to $[0,1]$ by taking a limited number of random samples in the domain of the function. We stress that an averaging sampler must work without any information about the function it is trying to approximate.

Definition 1 (Averaging Sampler). *A function* Samp $: \{0,1\}^d \rightarrow [N]^m$ *is a (μ, θ, γ) averaging sampler if for every function $f : [N] \rightarrow [0,1]$ with average value $\frac{1}{N} \sum_i f(i) \geq \mu$, it holds that*

$$\Pr_{\{i_1, \dots, i_m\} \leftarrow \text{Samp}(U_d)} \left[\frac{1}{m} \sum_{j=1}^{m} f(i_j) < \mu - \theta \right] \leq \gamma$$

UC PAK. Our construction makes use of a Universally Composable Password Authenticated Key exchange (UC PAK) protocol, as defined in [CHK+05]. Informally, the UC security notion of [Can01] requires that a protocol implementing an "ideal functionality" cannot be distinguished from the ideal functionality [3] it implements, even by an "environment" that chooses inputs to (and observes outputs from) the parties running the protocol while it is under attack by the adversary. One result of this very strong notion of security is that protocol designers may use an ideal functionality as a subroutine, yet rest assured that later replacing that ideal functionality with a UC secure protocol will not harm the security analysis of the newly designed protocol. In this spirit, we will be making use of the UC PAK ideal functionality (Figure 1) as a subroutine in our protocol.

The UC PAK functionality captures the following intuitive guarantees: (1) If the parties are both honest *and* the adversary does not attempt to "compromise" their session, then they both receive an identical uniformly distributed random key, (2) if the adversary makes a failed attempt to compromise their session, honest parties each receive *independently generated* uniformly distributed random keys, and (3) the adversary is only able to successfully compromise a session by either correctly guessing the password of one of the parties (in a *single attempt*), or by corrupting one of the parties – in either case the adversary is allowed to choose the key(s) received by the honest parties. Unless the adversary successfully compromises a session, the only information provided by the functionality to the adversary is a notification when parties initiate the PAK protocol. (Of course, the adversary is also notified of the success or failure of an attempt to compromise a session.) Note the functionality *does not* guarantee that honest parties receive a shared key, even if the session was not compromised; they are merely assured that they each either have a good shared key, or a completely random one.

Unfortunately, like most non-trivial two-party UC functionalities, UC PAK protocols cannot be implemented in the plain model. Therefore, we rely on the construction of [CHK+05], which assumes that a short Common Reference String (CRS) is publicly available. Since the parties in our AKE protocol already share large secrets, it is a simple matter for them to share a short (public) CRS value as well (alternatively, a small portion of the shared secret can be "sacrificed" to generate the CRS). Thus, the introduction of a CRS does not require any significant alterations to our model. Note that, in the special case where our security model is augmented with adversarial errors, the adversary should not be allowed to introduce errors into the CRS itself. Since a CRS is usually very short, protecting it within (public) "read-only memory" should not be costly.

3   The Model and Definitions of Security

In this section we describe symmetric-key AKE protocols and their security in the BRM. Throughout this section, k is a security parameter. The length of

---

[3] Technically, the behavior of the protocol under any given attack cannot be distinguished from the behavior of some "simulated" attack on the ideal functionality.

When interacting with an adversary S and a set of parties, with security parameter k, the functionality $\mathcal{F}_{pwKE}$ responds to the following queries:

- *NewSession($sid, P_A, P_B, pw, role$): Upon receiving a query of this form from party P_A, the message (NewSession, $sid, P_A, P_B, role$) is sent to the adversary S. If this is the first NewSession query, or if this is the second NewSession query and there is a record (P_B, P_A, pw'), then record (P_A, P_B, pw) and mark this record* fresh.
- *TestPwd(sid, P_A, pw'): Upon receiving this query from the adversary S, if there is a record of the form (P_A, P_B, pw) which is* fresh, *then: If $pw = pw'$, mark the record* compromised *and notify S of a "correct guess". If $pw \neq pw'$, mark the record as* interrupted *and notify S of an "incorrect guess".*
- *NewKey(sid, P_A, sk): Upon receiving this query from the adversary S (where $|sk| = k$), if there exists a record of the form (P_A, P_B, pw) and this is the first NewKey query for P_A, then one of the following actions occurs, after which the record (P_A, P_B, pw) is marked* completed:
  - *If this record is* compromised, *or either P_A or P_B is a corrupt party, then output (sid, sk) to P_A.*
  - *If this record is* fresh, *and there is a record (P_B, P_A, pw') with $pw' = pw$, and a key sk' was sent to P_B, and the record (P_B, P_A, pw) was marked* fresh *at the time, then output (sid, sk') to P_A.*
  - *In any other case, pick a new random key sk' of length k and send (sid, sk') to P_A.*

Fig. 1. Ideal Functionality $\mathcal{F}_{pwKE}$, from [CHK$^+$05]

the large symmetric-key is denoted by N, where $N = N(k)$ is a fixed sufficiently large polynomial. During a setup phase, a symmetric-key $X \in_R \{0, 1\}^N$ is chosen uniformly and shared between two parties Alice and Bob.

An authenticated key exchange protocol Π is a pair of algorithms (A, B) describing the honest behavior of two parties. As described in the introduction, the adversary's limited access to X is vital to the security of our schemes, and we assume that the honest parties are similarly restricted (in fact, the honest parties will use a relatively small amount of their keys compared to the captured portion). We say that Π is *m-local* if A and B each access at most $m = m(k)$ bits in any execution of Π.

Adversaries will be able to steal a total of βN bits over the course of several executions, where $0 \leq \beta < 1$. We call β the *retrieval rate* of the adversary. Note that the adversary can adaptively decide on the size of the data to capture during a particular session, as long as it does not violate the total accumulated retrieval bound.

For the moment, we assume that a key X will be used for at most T sessions, where $T(k)$ is a polynomial determined by the retrieval rate of the adversary and the size of X. The assumption of an upper bound on the number of uses of X will be relaxed; below we show that security for a fixed $T(k)$ implies security for an arbitrary polynomial number of sessions under the assumption that the parties get some predetermined uncompromised sessions.

As with other models for authenticated key exchange, adversaries in our model (denoted C) completely control the channel between A and B. In particular, C can inject, drop, modify and delay messages. We also give C the power to temporarily intrude into the machines of A and B and retrieve some information about their internal state, including X. In this paper, our protocols are only proven secure for sequential executions of sessions. We comment on this limitation later.

At the beginning of each session, the adversary decides whether or not to intrude. From now on, we call a session during which the adversary intrudes a *compromised session*. After declaring a session (say session i) compromised, the adversary C outputs a circuit V (a virus) that will get to see the private information of A and B. This includes r^A and r^B, the private coins of A and B to be used in this session, and the secret key X. The virus V computes $S_i = V(X, r^A, r^B)$,[4] and sends S_i to the the adversary C.

We stress that V can be any polynomial-size circuit adaptively computed by C at the start of session i (so C may incorporate information gained from previous intrusions into V). The only other restriction on V is that the its output S_i is bounded. In particular, we require that $\sum_{i=1}^{T} |S_i| < \beta N$.

Clearly in a compromised session neither security nor correctness can be achieved. The best we can hope for is to construct a protocol that guarantees security and correctness for each uncompromised session. This is reflected in how success is determined in our definition below.

Our definition follows the style of [BPR00] and [Dzi06a].

The Adversarial Model. The power of an active adversary C is modeled by giving C oracle access to the protocol instances run by Alice and Bob. Denote by A and B the prescribed programs of Alice and Bob respectively. Denote by Π_i^A (resp. Π_i^B) the instance of protocol Π that Alice (resp. Bob) runs in the i-th session. For each $P \in \{A, B\}$, the instances Π_i^P are executed *sequentially*, that is, for each i, instance Π_{i+1}^P starts after instance Π_i^P completes. At the end of the i-th session, Π_i^A (resp. Π_i^B) outputs a bit acc_i^A (resp. acc_i^B) indicating whether A (resp. B) accepts or aborts in Session i. As in previous work, we assume that this bit is always known to the adversary C. Denote by sk_i^A (resp. sk_i^B) the session key output by Π_i^A (resp. Π_i^B) in the i-th session. If $\mathsf{acc}_i^A = 0$ (resp. $\mathsf{acc}_i^B = 0$), then $\mathsf{sk}_i^A = \perp$ (resp. $\mathsf{sk}_i^B = \perp$).

We define the following types of oracles that the adversary C is allowed to invoke, all of which except for the Intrude oracle are as defined in [BPR00]. We note that the adversary's retrieval bounded is expressed below in the definition of the Intrude oracle.

- Execute(i): Upon this call, the complete execution between protocol instances Π_i^A and Π_i^B takes place. The output of this call is the transcript, that is,

---

[4] WLOG we assume that the adversary intrudes Alice and Bob simultaneously in a compromised session.

the sequence of all messages exchanged between Π_i^A and Π_i^B. This oracle models passive eavesdropping in Session i.

- Send(P, i, M): This call sends the message M to the instance Π_i^P (where $P \in \{A, B\}$). The output of this call is the message the instance Π_i^P would send after receiving the message M, given its current state. This oracle models active man-in-the-middle attacks in Session i.
- Intrude(i, V): This oracle models intrusion into the machines of both A and B. The second input V to this call is a circuit (the virus) constructed based on the adversary C's current state. Upon this call, the oracle computes and outputs $S_i = V(X, r_i^A, r_i^B)$, where r_i^A and r_i^B are the private coins of A and B in session i. We require that $\sum_{j=1}^{T} |S_j| < \beta N$.
- Test(P, i): The output of this call is either the session key sk_i^P output by Π_i^P, or an independently chosen random string, each case happening with probability $1/2$. The adversary's goal is to distinguish between the two cases. The call Test(P, i) can be invoked any time after Party P concludes Session i. The adversary may not invoke Test(P, i) when $\mathsf{acc}_i^P = 0$ (i.e. when $\mathsf{sk}_i^P = \bot$), or if the adversary has previously invoked Intrude(i, V).

The adversary C's advantage in Session i is defined as

$$\mathsf{Adv}_i(C) = |2 \cdot \Pr[C \text{ Succeeds in } \mathsf{Test}(P, i)] - 1|.$$

Remark: The query Intrude(i, V) is allowed only before the start of Session i and is not allowed during Session i.

Definition 2. *A session key protocol Π is intrusion-resilient for $T = T(k)$ sessions if for every PPT C with retrieval rate β, the following conditions are satisfied:*

- *(Correctness) With probability $1 - \mathsf{negl}(k)$ the following holds: For each $i \in [T]$, if session i is uncompromised and $\mathsf{acc}_i^A = \mathsf{acc}_i^B = 1$, then $\mathsf{sk}_i^A = \mathsf{sk}_i^B$.*
- *(Privacy) For each $i \in [T]$ s.t. the Test oracle is invoked for session i, $\mathsf{Adv}_i(C) = \mathsf{negl}(n)$.*

We now return to the issue of defining security for a fixed polynomial $T(k)$ number of sessions. We observe that an authenticated key exchange protocol Π can be used to refresh its own long secret keys during uncompromised sessions. The idea is that the parties can run Π to obtain a short key r, and then use that key as the seed for a pseudorandom generator. The output of the generator is then XORed with the previous long key, and the value r is erased. This ensures that the final long key will be "as good as new" if the attacker did not break-in right at the end of Π (i.e., r is uncompromised), and still "as good as before" even if the attacker compromised the value of r. Thus, as long as at least one uncompromised key update happened before the attacker obtained too much information about the long key, the long key remains secure. We defer the details to the full version on the paper.

We conclude this section by presenting a lemma that simplifies the analysis of our constructions. We show below that it suffices to construct a protocol that is intrusion-resilient for *three sessions*.

Lemma 2. *Suppose that a session key protocol Π is intrusion-resilient for* three sessions *against every PPT C that is β-retrieval bounded within the three sessions and attacks Π as follows: The adversary compromises the first session as usual; the second session is uncompromised; in the third session the adversary not only compromises the session but also gets the entire key X of Alice and Bob. Then for every polynomial $T = T(k)$, the protocol Π is intrusion-resilient for T sessions.*

The proof of Lemma 2 is a straightforward simulation of several sessions in only 3 sessions. Again, the proof will appear in the full version of this paper.

4 Authenticated Key Exchange from Weak Key Exchange

In this section we describe an approach for constructing AKE protocols in our model. We define the notion of Weak Key Exchange (WKE) and give an efficient construction, and then we show how to compose any WKE protocol with any UC PAK to realize a secure intrusion resilient AKE protocol.

4.1 Weak Key Exchange: Definition

Briefly, WKE provides only the guarantee that the output keys will have a high min-entropy from the viewpoint of the adversary. In particular, the adversary may possibly arrange for the keys to be unequal and correlated in an arbitrary fashion. Of course, we still require that the keys match when the protocol runs with no active interference from the adversary. WKE also provides no security guarantees on past keys once a subsequent WKE session is initiated (*i.e.* there is no forward security, and indeed, no long term security requirement at all). Our definition for WKE is a modification of the definition for authenticated key exchange, and thus we focus on the differences between the definition of WKE and that of AKE (as described above).

We use the same adversarial model as in AKE, but with a few critical weakenings. In particular, we will only allow for 2 sessions (in a similar spirit to Lemma 2), where the first session is compromised, and the second is not. Furthermore, we modify the Test oracle, and the corresponding experiment defining the adversarial advantage, replacing it with the following:

- Test(P, i, sk): The output of this call is 1 if $i = 2$ and $sk = sk_i^P \neq \perp$.

This oracle may only be called once by the adversary (for one party, using Session 2), in addition to the previous restrictions.

The adversary's advantage in the privacy requirement is redefined as

$$\mathsf{Adv}_2(C) = \Pr[C \text{ causes } \mathsf{Test}(i, P, sk) = 1].$$

There is no forward security guarantee for weak key exchange, as the adversary may not corrupt (or even invoke) any session that occurs after the second session (wherein it must query the Test oracle, attempting to break privacy). Furthermore, the adversary can only succeed by guessing the entire key sk_i^P (which is more difficult than merely distinguishing it from random), and thus the privacy guarantee is considerably weakened.

Definition 3 (Weak Key Exchange(WKE)). *A protocol is a weak key exchange if for every PPT C with retrieval rate β, the following conditions are satisfied:*

- *(Weak Correctness)* With probability $1 - \mathsf{negl}(k)$ the following holds: for each $i \in \{1, 2\}$, if Session i runs honestly, (i.e. the adversary does not tamper with any protocol messages) then $\mathsf{sk}_i^A = \mathsf{sk}_i^B$.
- *(Weak Privacy)* The advantage of the adversary is negligible, i.e. $\mathsf{Adv}_2(C) = \mathsf{negl}(n)$, where $\mathsf{Adv}_2(\mathsf{C})$ is as defined above, and in particular depends on the modified Test oracle.

We note that in the event that the adversary chooses to alter the content of protocol flows, A and B may agree to accept a session where they receive differing keys.[5]

Below we present and analyze a construction meeting this definition, with information theoretic security.

4.2   Weak Key Exchange: Construction

Our protocol makes use of an *averaging sampler*, as described in [BR94], which samples a small number of bits from a much larger source (in this case, the shared secret X), while nearly preserving the min-entropy rate of the larger source. It was shown in [Vad04] how to explicitly construct samplers [6] for a δN-source using only $d = \log(N/m) + O(\log(1/\gamma))$ random bits and $m = O(\log(1/\gamma))$ bit samples from the input source to produce output that is γ-close to a $(2\delta/3)m$-source for any $\gamma > \exp(-N/2^{O(\log^* N)})$. Note that, given the practical importance of efficiency, here we obey the requirement that the number of bits of X which are read during the execution of a WKE is small. (This partly motivates the choice of parameters for the averaging sampler, and in particular, $m = O(\log(1/\gamma))$ is essentially the best one can hope for in terms of efficiency.)

Making use of the existence of such samplers, our protocol proceeds as follows (where we use $X_{\mathsf{Samp}(\cdot)}$ to denote the string formed by concatenating the bits of X located at the indices selected by Samp):

Lemma 3. *The Weak Key Exchange protocol described in Figure 2 is secure for appropriate choices of the sampler parameters.*

---

[5] Indeed, the adversary may attempt to guess either sk_2^A or sk_2^B in the attack scenario for the privacy requirement, and they may not be equal.
[6] See Lemma 8.4 and its usage in Theorem 8.5 in [Vad04] for details.

Setup:

- $X \in_R \{0,1\}^N$: A (large) secret key shared between Alice and Bob.
- Samp : $\{0,1\}^d \to \{1,\ldots,N\}^m$: An averaging sampler [Vad04].

Protocol:

1. Alice and Bob choose random values r_A and r_B, respectively.
2. Alice sends r_A to Bob. Bob receives a value r'_A, and then computes $K_B = X_{\mathsf{Samp}(r'_A)} \parallel X_{\mathsf{Samp}(r_B)}$. Bob outputs K_B as his weak session key.
3. Bob sends r_B to Alice. Alice receives a value r'_B, and then computes $K_A = X_{\mathsf{Samp}(r_A)} \parallel X_{\mathsf{Samp}(r'_B)}$. Alice outputs K_A as her weak session key.

Fig. 2. A WKE Protocol

Proof. The proof of security is direct, and the security guarantee is in fact information theoretic. The weak correctness property is obvious. The view of the adversary at the start of the second session is the output of some circuit $V(X)$, obtained during the compromise of the first session. As we are considering only β-retrieval bounded adversaries, the output of the circuit V must be no larger than βN-bits, and thus the adversary knows at most βN-bits of information about X. In particular, we observe that, from the adversary's point of view, X is nearly [7] a $(1 - \beta)N$-source (since X is uniformly random over N bits, but the adversary's view is conditioned over the βN-bit output of $V(X)$ obtained during the first session).

Thus, by Lemma 6.2 of [Vad04] (and using the same sampler parameters as those of the explicit local extractor construction in Theorem 8.5 therein), the pairs $(r_A, X_{\mathsf{Samp}(r_A)})$ and $(r_B, X_{\mathsf{Samp}(r_B)})$ are (very nearly) individually γ-close to the distribution (U_d, W), where for every $r \in \{0,1\}^d$, the distribution $W|_{U_d=r}$ is at least a $(2(1 - \beta)/3)m$-source. Thus, $K_A = X_{\mathsf{Samp}(r_A)} \parallel X_{\mathsf{Samp}(r'_B)} \approx W \parallel X_{\mathsf{Samp}(r'_B)}$, where W has (nearly) min-entropy $(2(1 - \beta)/3)m$ even conditioned on r_A. Therefore, even conditioned on the adversary's view *after* the WKE protocol completes, K_A has min-entropy nearly $(2(1 - \beta)/3)m$, irrespective of the adversary's choice of r'_B. An analogous argument applies to K_B, and thus, for sufficiently large choices of m, the weak privacy security requirement follows. $\square$

4.3   Intrusion-Resilient **AKE** from **WKE** and **UC PAK**

On a high level, the protocol uses a WKE to select "passwords" with a high min-entropy for use by Alice and Bob in a standard PAK protocol. Since PAK

---

[7] Technically, with probability $(1 - 2^{-\lambda})$ taken over the distribution of $S \leftarrow V(X)$, the random variable $X|_{V(x)=S}$ (which is X conditioned on the adversary's view) has min-entropy $(1 - \beta)N - \lambda$ for any choice of λ. This follows directly from Lemma 1.

protocols securely realizing the ideal functionality $\mathcal{F}_{pwKE}$ do not necessarily verify the output keys (*i.e.* check that the key exchange was successful, satisfying correctness), we use MACs to verify them, completing the protocol.

Setup:

- WKE: A Weak Key Exchange protocol between Alice and Bob.
- PAK: A Universally Composable Password Authenticated Key exchange protocol.
- (MAC,Verify): A MAC secure against chosen message attacks.

Protocol:

1. Alice and Bob run the WKE protocol, obtaining the weak keys K_A and K_B, respectively.
2. Alice and Bob exchange nonces, which are concatenated to form a session ID sid
3. Alice and Bob run the PAK protocol with session ID sid, using the passwords K_A and K_B as their inputs, respectively. Alice and Bob obtain their outputs, secret keys $sk_A = sk_A^1 \| sk_A^2$ and $sk_B = sk_B^1 \| sk_B^2$, respectively.
4. Alice sends Bob $tag_A = \mathsf{MAC}_{sk_A^1}(\text{Bob})$, Bob sends Alice $tag_B = \mathsf{MAC}_{sk_B^1}(\text{Alice})$.
5. Alice receives tag_B' and checks that $\mathsf{Verify}_{sk_A^1}(\text{Bob}, tag_B') = 1$. If not, Alice rejects and aborts. Otherwise, Alice outputs sk_A^2 as her session key.
6. Bob receives tag_A' and checks that $\mathsf{Verify}_{sk_B^1}(\text{Alice}, tag_A') = 1$. If not, Bob rejects and aborts. Otherwise, Bob outputs sk_B^2 as his session key.

Fig. 3. An AKE protocol based on WKE and UC PAK

Remark: PAK can be implemented using the protocol of [CHK$^+$05], which is a 6-round protocol that is efficiently implementable under standard number theoretic assumptions. Our WKE construction is a 2-round protocol, and we add four extra rounds in the above construction (two to exchange nonces, and two to verify the keys at the end), for a total of 12 rounds. However, in practice we can move the exchange of nonces in parallel with the WKE messages, reducing the number of rounds by 2, and we can move one of the verification messages in parallel with the last flow of the PAK protocol to save an additional round, bringing the total down to 9 rounds.

Theorem 1. *The AKE protocol described in Figure 3 is intrusion-resilient for three sessions defined in Lemma 2. Therefore, under the assumption that secure update sessions can be periodically scheduled, there is an intrusion-resilient AKE protocol for an unbounded number of sessions.*

Proof sketch: Since we are using a universally composable PAK protocol, by the UC Theorem of [Can01], we may substitute the execution of the PAK protocol with calls to the ideal functionality $\mathcal{F}_{pwKE}$ described above [8] (using K_A and K_B as the passwords for the calls by P_A and P_B, respectively). Clearly, the session keys output by the ideal functionality are indistinguishable from random to the adversary (by definition), unless the adversary makes a successful TestPwd query (which must be issued prior to the completion of the PAK protocol), allowing him to choose the output keys. However, as can be seen from the definition of the ideal functionality, the adversary is allowed only a single query to TestPwd. Following Definition 3, we observe that the adversary cannot guess the output from the WKE with non-negligible probability (even when conditioned on the entire view of the adversary up to and including the point at which the functionality $\mathcal{F}_{pwKE}$ is invoked, which is identical to the view when attacking just the WKE, plus the addition of a random sid and some other innocuous messages that may be sent from $\mathcal{F}_{pwKE}$ to the adversary), the probability that the adversary succeeds in a TestPwd query is at most negligible. Thus, we are guaranteed that sk_A and sk_B are indistinguishable from random, and either equal (in the event that no TestPwd query was issued) or independently generated (in the event that a failed TestPwd query was issued). Furthermore, it is easy to show that the last two flows of the protocol (Steps 4-6) ensure that correctness holds, provided that sk_A and sk_B are either identical, or independently chosen random values. Finally, we observe that the adversary can never distinguish sk_A or sk_B from random once the session has completed, even after being given all [9] of X, since the keys are chosen at random (independently of X) by the ideal functionality $\mathcal{F}_{pwKE}$ and are never revealed to the adversary.   □

Remark: The UC PAK protocol of [CHK+05] is only secure against "static" adversaries in the UC framework. At first glance, it might seem that we require security against "adaptive" adversaries here, since the random coins used by the parties are revealed during an intrusion. However, since we are only dealing with sequential sessions, and since there is no security guarantee provided for compromised sessions, we need not concern ourselves with the ability to simulate attacks on the compromised sessions (which would have necessitated the use of the adaptive security notion). The random coins used by uncompromised sessions are indeed erased, and thus static adversary UC security is sufficient.

The Need for UC-Secure PAK. We begin by remarking that the use of computationally secure tools is unavoidable in our setting (despite the information theoretic security of our WKE). This is due to a combination of the forward security requirement and efficiency requirements for the scheme: if it is efficient

---

[8] The substitution is legitimate since it is possible for an *environment Z* to internally simulate the rest of the protocol, including the setup phase with the shared secret X. Since UC security holds against any environment, in particular, this environment should be unable to distinguish calls to the ideal functionality $\mathcal{F}_{pwKE}$ from calls to the realized protocol.

[9] Indeed, if the adversary were given all of X prior to the completion of the WKE phase, the security property on the WKE would be broken, allowing the adversary to potentially issue a successful TestPwd query. Thus, it is in fact still critical that the adversary be β-retrieval bounded.

to compute a key-exchange, then the output must be information-theoreticallyfc determined by a small number of bits, all of which can be obtained by the adversary during a subsequent intrusion (without violating the retrieval bound).

Given that reliance on a computationally secure tool is unavoidable, if we are going to use a WKE-based approach, one might initially suggest composing it with a standard PAK protocol relying on a traditional "stand-alone" security definition (particularly since we are restricting parties to sequential AKE sessions). Unfortunately, there are serious obstacles to this seemingly natural approach.

First, we observe that the passwords (the weak keys) output by our WKE phase may be correlated in an *arbitrary* manner. Many traditional definitions of PAK (as well as various other cryptographic tools) do not provide security in the scenario where the parties are using unequal but correlated passwords (or, resp., keys). Indeed, the protocol of [Dzi06a] employs random oracles specifically to transform the parties' correlated random secrets into independently random ones. Since we wish to avoid the use of random oracles, we are left to deal with definitions of PAK that allow the adversary to correlate passwords. For instance, the PAK definition of [BDK+05] allows the adversary to specify (*a priori*) the joint distribution from which honest parties choose their passwords, so that they can be forced to obey an arbitrary correlation function.

Surprisingly, it seems difficult (perhaps even impossible) to prove the security of our construction even when composed with the (very strong) notion of PAK from [BDK+05]. In particular, it is hard to construct a reduction from the AKE protocol to the security of the PAK protocol. To see this, consider the issue of simulating the large AKE secret X. If X is chosen directly by the reduction, then either (1) the passwords used by the parties being attacked by the reduction are not be properly derived from X, or (2) the reduction itself is also able to derive the passwords from X. Of course, in case (2) the reduction is not able to break the security of parties with *unknown* passwords, rendering it meaningless. On the other hand, in case (1), the simulation performed by the reduction does not faithfully reproduce the setting of the AKE adversary. In fact, it seems hopeless to prove that such simulations go undetected by the adversary in our setting, since the adversary can eventually obtain all the relevant portions of X (via an intrusion subsequent to the completion of the PAK protocol), and check for consistency. As an alternative approach, the reduction could specify a correlation function for the PAK security game that chooses the value of X, and then generates correlated passwords accordingly. Unfortunately, this too fails, since the reduction itself will not learn the value of X, and thus will be unable to properly simulate the input to the adversary during an intrusion.

Ultimately, we turn to UC PAK to overcome this difficulty. With a UC PAK, the reduction plays the role of the environment in the UC security definition. Here, the environment *is* allowed to choose the passwords used by the parties, and thus the reduction may choose X and generate passwords accordingly, as in case (1) above. This time, the reduction remains meaningful, since the adversary will *not* be privy to the passwords used by honest parties (due to the separation between the UC distinguishing environment, and the UC adversary). In fact, if we make use of the UC composition theorem, the entire security proof seems to follow the natural intuition for combining WKE and PAK.

To the best of the author's knowledge, this setting represents the first instance of protocol which, even when executed sequentially and in isolation, seems to require the use of a UC secure tool. Our setting thus provides a powerful and naturally occuring example of the benefits of UC security in modular protocol design. Here we are able to consider a protocol which, when designed intuitively using standard tools, (seemingly) cannot be proven secure even in a simple "stand-alone" protocol execution setting. Yet, when the same intuitive design is implemented using a UC secure tool, a proof of security far more readily presents itself.

References

[And97]   Ross Anderson. Two remarks on public key cryptology. Invited Lecture. In *4th ACM Conference on Computer and Communications Security*, 1997.

[BM99]   Mihir Bellare and Sara K. Miner. A forward-secure digital signature scheme. In *CRYPTO*, pages 431–448, 1999.

[BPR00]   Mihir Bellare, David Pointcheval, and Phillip Rogaway. Authenticated key exchange secure against dictionary attacks. In *EUROCRYPT*, pages 139–155, 2000.

[BR93]   Mihir Bellare and Phillip Rogaway. Entity authentication and key distribution. In *CRYPTO*, pages 232–249, 1993.

[BR94]   Mihir Bellare and John Rompel. Randomness-efficient oblivious sampling. In *FOCS*, pages 276–287, 1994.

[BY03]   Mihir Bellare and Bennet S. Yee. Forward-security in private-key cryptography. In *CT-RSA*, pages 1–18, 2003.

[BM93]   Steven M. Bellovin and Michael Merritt. Augmented encrypted key exchange: A password-based protocol secure against dictionary attacks and password file compromise. In *ACM Conference on Computer and Communications Security*, pages 244–250, 1993.

[BDK+05]   Xavier Boyen, Yevgeniy Dodis, Jonathan Katz, Rafail Ostrovsky, and Adam Smith. Secure remote authentication using biometric data. In *EUROCRYPT*, pages 147–163, 2005.

[BMP00]   Victor Boyko, Philip D. MacKenzie, and Sarvar Patel. Provably secure password-authenticated key exchange using diffie-hellman. In *EUROCRYPT*, pages 156–171, 2000.

[Can01]   Ran Canetti. Universally composable security: A new paradigm for cryptographic protocols. In *FOCS*, pages 136–145, 2001.

[CDH+00]   Ran Canetti, Yevgeniy Dodis, Shai Halevi, Eyal Kushilevitz, and Amit Sahai. Exposure-resilient functions and all-or-nothing transforms. In *EUROCRYPT*, pages 453–469, 2000.

[CGH04]   Ran Canetti, Oded Goldreich, and Shai Halevi. On the random-oracle methodology as applied to length-restricted signature schemes. In *TCC*, pages 40–57, 2004.

[CHK03]   Ran Canetti, Shai Halevi, and Jonathan Katz. A forward-secure public-key encryption scheme. In *EUROCRYPT*, pages 255–271, 2003.

[CHK+05] Ran Canetti, Shai Halevi, Jonathan Katz, Yehuda Lindell, and Philip D. MacKenzie. Universally composable password-based key exchange. In *EUROCRYPT*, pages 404–421, 2005.

[CLW06]   Giovanni Di Crescenzo, Richard J. Lipton, and Shabsi Walfish. Perfectly secure password protocols in the bounded retrieval model. In *TCC*, pages 225–244, 2006.

[DLL05]   David Dagon, Wenke Lee, and Richard J. Lipton. Protecting secret data from insider attacks. In *Financial Cryptography*, pages 16–30, 2005.

[DFK+03] Yevgeniy Dodis, Matthew K. Franklin, Jonathan Katz, Atsuko Miyaji, and Moti Yung. Intrusion-resilient public-key encryption. In *CT-RSA*, pages 19–32, 2003.

[DKRS06] Yevgeniy Dodis, Jonathan Katz, Leonid Reyzin, and Adam Smith. Robust fuzzy extractors and authenticated key agreement from close secrets. In *CRYPTO*, pages 232–250, 2006.

[DKXY02] Yevgeniy Dodis, Jonathan Katz, Shouhuai Xu, and Moti Yung. Key-insulated public key cryptosystems. In *EUROCRYPT*, pages 65–82, 2002.

[DKXY03] Yevgeniy Dodis, Jonathan Katz, Shouhuai Xu, and Moti Yung. Strong key-insulated signature schemes. In *Public Key Cryptography*, pages 130–144, 2003.

[DORS06] Yevgeniy Dodis, Rafail Ostrovsky, Leonid Reyzin, and Adam Smith. Fuzzy extractors: How to generate strong keys from biometrics and other noisy data. Cryptology ePrint Archive, Report 2003/235, 2006. http://eprint.iacr.org/.

[Dzi06a] Stefan Dziembowski. Intrusion-resilience via the bounded-storage model. In *TCC*, pages 207–224, 2006.

[Dzi06b] Stefan Dziembowski. On forward-secure storage. In *CRYPTO*, pages 251–270, 2006.

[GL01]   Oded Goldreich and Yehuda Lindell. Session-key generation using human passwords only. In *CRYPTO*, pages 408–432, 2001.

[IR02]   Gene Itkis and Leonid Reyzin. Sibir: Signer-base intrusion-resilient signatures. In Moti Yung, editor, *CRYPTO*, volume 2442 of *Lecture Notes in Computer Science*, pages 499–514. Springer, 2002.

[Mau92]   Ueli Maurer. Conditionally-perfect secrecy and a provably-secure randomized cipher. *Journal of Cryptology*, 5(1):53–66, 1992.

[MW03]   Ueli Maurer and Stefan Wolf. Secret-key agreement over unauthenticated public channels iii: Privacy amplification. *IEEE Transactions on Information Theory*, 49(4):839–851, 2003.

[NZ96]   Noam Nisan and David Zuckerman. Randomness is linear in space. *Journal of Computer and System Sciences*, 52(1):43–52, 1996.

[RW03]   Renato Renner and Stefan Wolf. Unconditional authenticity and privacy from an arbitrarily weak secret. In *CRYPTO*, pages 78–95, 2003.

[Vad04]   Salil P. Vadhan. Constructing locally computable extractors and cryptosystems in the bounded storage model. *Journal of Cryptology*, 17(1):43–77, 2004.

[Wol98]   Stefan Wolf. Strong security against active attacks in information-theoretic secret-key agreement. In *ASIACRYPT*, pages 405–419, 1998.

A   Augmenting the Model with Adversarial Errors

As mentioned earlier, we deal with the case where an adversary is allowed to change the party's keys adaptively so that they disagree at some limited number of indices. We first give a formal definition of the model with errors, and then we give the definition of *secure sketches* and show how they can be used to resist errors.

Definition 4. *Again we modify Definition 2. In this case, during a compromised session, the virus V may have two outputs: the first is the information sent back to the adversary, and second is the* error vector *which describes which bits of X to flip. We say that an adversary is γ-error bounded if it flips at most a γN bits of X between key updates.*

Our construction meeting Definition 4 uses *secure sketches*. Intuitively, a secure sketch is a primitive that lets us generate a *sketch* of a string w that will help a user with a noisy version w' of w correct errors, but will not help an outside observer significantly. Originally, secure sketches were defined for a general metric, but we only need the case for Hamming distance.

Definition 5 (Secure Sketch [DORS06]). *An (k, k', t)-secure sketch is a pair of algorithms $(\mathsf{SS}, \mathsf{Rec})$ such that*

1. *(Security) If W is a k-source, then W can be guessed by an adversary who sees $\mathsf{SS}(W)$ with probability at most $2^{k'}$.*[10]
2. *(Correctness) If w and w' differ at less than t indices, then $\mathsf{Rec}(w', \mathsf{SS}(w)) = w$.*

We now show how to construct a WKE protocol that functions correctly in the presence of errors. We use secure sketches to ensure that the strings $X_{\mathsf{Samp}(r_A)}$ and $X_{\mathsf{Samp}(r_B)}$ are corrected in the passive case. The details of the updated protocol appear below in Figure 4

We choose $\beta_1, \beta_2, \varepsilon$ and γ in Figure 4 so that when we have a β_1-retrieval bounded adversary who can flip a total of γN bits during the interaction, the adversary's chance at guessing a password handed to the PAK is at most $2^{-\beta_2 N}$.

Lemma 4. *For an appropriate setting of parameters, the protocol in Figure 4 satisfies the security definition of WKE in the augmented model where the adversary can inject a γ fraction of errors, and transmit $\beta_1 N$ bits during during the first round. (Recall that we only need to prove security in the case of two rounds for WKE).*

Proof sketch: Correctness in the passive case is obvious from the definitions of the primitives used. Privacy follows from the original analysis of our WKE construction, together with the observation that after seeing s_A (resp. s_B), it is still hard to guess $X^A_{\mathsf{Samp}(r_A)}$ (resp. $X^B_{\mathsf{Samp}(r_B)}$). We defer a detailed analysis, including parameter settings, to an expanded version of this work.   □

---

[10] We would like to briefly say that $\mathrm{H}_\infty(W|\mathsf{SS}(W)) > k'$, but what we actually need is that the *average min-entropy* $\widetilde{\mathrm{H}_\infty}(W|\mathsf{SS}(W))$ is greater than k', where $\widetilde{\mathrm{H}_\infty}(A|B) = -\log(\mathbf{E}_{b \leftarrow B}[\max_a \Pr[A = a|B = b]])$, which corresponds to the prose description given.

Setup:

- X^A and X^B where $X \in_R \{0,1\}^N$ and the Hamming distance between X_A and X is at most $\gamma/2N$ (similarly, X_B and X are within $\gamma/2N$ Hamming distance).
- $\mathsf{Samp} : \{0,1\}^d \to \{1,\ldots,N\}^m$: An averaging sampler
- $(\mathsf{SS}, \mathsf{Rec})$: A $(\beta_1 N - \log(1/\varepsilon), \beta_2 N, \gamma N)$-secure sketch.

Protocol:

1. Alice and Bob choose random values r_A and r_B, respectively.
2. Alice and Bob each compute the sketches s_A and s_B of $X^A_{\mathsf{Samp}(r_A)}$ and $X^B_{\mathsf{Samp}(r_B)}$, respectively.
3. Alice sends (r_A, s_A) to Bob, who receives (r'_A, s'_A), and first recovers $X^{A'}_{\mathsf{Samp}(r'_A)} = Rec(X^B_{\mathsf{Samp}(r'_A)}, s'_A)$, and outputs $K_B = X^{A'}_{\mathsf{Samp}(r'_A)} \parallel X^B_{\mathsf{Samp}(r_B)}$.
4. Bob sends (r_B, s_B) to Alice, who receives (r'_B, s'_B), and first recovers $X^{B'}_{\mathsf{Samp}(r'_B)} = Rec(X^A_{\mathsf{Samp}(r'_B)}, s'_B)$, and outputs $K_A = X^A_{\mathsf{Samp}(r_A)} \parallel X^{B'}_{\mathsf{Samp}(r'_B)}$.

Fig. 4. A WKE Protocol For Noisy Keys

Once we have an error-resistant WKE protocol, composing with a UC-secure PAK protocol results in an error-resistant and intrusion resilient AKE protocol. The reason this is true is because the PAK protocol does not access X on its own, and once the WKE property for its input passwords is established, the UC theorem allows us to replace the PAK protocol with an ideal functionality and the original proof goes through unchanged. Again, we defer the full details of the analysis.

(Password) Authenticated Key Establishment: From 2-Party to Group

Michel Abdalla[1], Jens-Matthias Bohli[2], María Isabel González Vasco[3], and Rainer Steinwandt[4]

[1] Departement d'Informatique, École Normale Supérieure, CNRS,
45 Rue d'Ulm, 75230 Paris Cedex 05, France
Michel.Abdalla@ens.fr
[2] Institut für Algorithmen und Kognitive Systeme, Universität Karlsruhe,
Am Fasanengarten 5, 76128 Karlsruhe, Germany
bohli@ira.uka.de
[3] Departamento de Matemática Aplicada, Universidad Rey Juan Carlos,
c/ Tulipán, s/n, 28933, Móstoles, Madrid, Spain
mariaisabel.vasco@urjc.es
[4] Department of Mathematical Sciences, Florida Atlantic University,
777 Glades Road, Boca Raton, FL 33431, USA
rsteinwa@fau.edu

Abstract. A protocol compiler is described, that transforms any provably secure authenticated 2-party key establishment into a provably secure authenticated group key establishment with 2 more rounds of communication. The compiler introduces neither idealizing assumptions nor high-entropy secrets, e. g., for signing. In particular, applying the compiler to a password-authenticated 2-party key establishment without random oracle assumption, yields a password-authenticated group key establishment without random oracle assumption. Our main technical tools are non-interactive and non-malleable commitment schemes that can be implemented in the common reference string (CRS) model.

Keywords: key establishment, protocol compiler, password-based authentication, common reference string model.

1 Introduction

During the last decades, the design of 2-party key establishments has been explored intensively. Certainly not all relevant issues are covered by the available theoretical models, but the techniques at hand proved to be a valuable foundation for the design of practical protocols. On the other hand, the design of group key establishments with $n > 2$ participants is much less understood, and there is a need for significant theoretical progress. In particular for password-authenticated protocols the situation is not very satisfying. A number of protocols have been designed for such a setting, including [24,1,2,28,13], but it seems to be a non-trivial task to establish strong provable security guarantees without making idealized assumptions.

S.P. Vadhan (Ed.): TCC 2007, LNCS 4392, pp. 499–514, 2007.
© International Association for Cryptologic Research 2007

One valuable tool for breaking down the task of designing a group key establishment protocol into conceptually simpler steps are protocol compilers that build on the security of a given 2-party solution: It seems a plausible design approach to start with a 2-party key establishment and then to apply an efficient compiler which derives the desired n-party solution. Indeed, a number of such generic constructions have been discussed in the literature, including [9,25,17]. Remarkably, all proposed constructions rely, to the best of our knowledge, on the use of high-entropy secrets for achieving security against active adversaries. In particular, for the case of password-based authentication in the standard model, no generic 2-to-n compiler seems to be known. The only result in this direction we are aware of is a construction of Abdalla et al. [1,2] to extend a 2-party solution to the 3-party case.

Our contribution. We describe a compiler that enables the derivation of an authenticated group key establishment protocol from an arbitrary authenticated 2-party key establishment (AKE). In particular, for a password-authenticated 2-party key establishment (PAKE) we obtain a password-authenticated group key establishment. Our compiler does not impose idealizing assumptions or high-entropy secrets for authentication. The suggested construction builds on the use of non-interactive and non-malleable commitments, which in the Common Reference String (CRS) model are known to be implementable through IND-CCA2 secure encryption schemes. For the security proof, we build on a model adapted from [18,20,6] which in turn builds on [4,3]. The structure of our compiler is inspired by the constant-round protocol recently proposed by Bohli et al. [6] which in turn builds on [8,14,15]. If the underlying 2-party protocol requires r rounds of communication, the group key establishment output by the compiler takes $r + 2$ rounds.

Organization of the paper. In the next section we recall the basic components of the security framework. We also address some specifics of password-based authentication, a scenario where the application of our protocol compiler seems particularly attractive. Thereafter, we detail the suggested protocol compiler and present the respective security proof. Section 4 indicates some possible applications of our compiler.

2   Security Model and Security Goals

For our compiler, we assume the availability of a common reference string (CRS) which, similarly as in [14,6], encodes

i) the necessary information for implementing a non-interactive and non-malleable commitment scheme,
ii) a uniformly at random chosen element from a family of universal hash functions and
iii) two values v_0, v_1 that will serve as arguments for a pseudorandom function when computing the session identifier and session key.

The total set of users will be denoted by $\mathcal{P}$ and is assumed to be of polynomial size. By $\mathcal{U} = \{U_1, \ldots, U_n\} \subseteq \mathcal{P}$ we denote the set of protocol participants. We assume that shared (low- or high-entropy) secrets needed for authentication are generated in a trusted initialization phase. During this trusted initialization phase, also possibly needed public keys may be distributed to all potential protocol participants. If authentication is based on shared secrets, we may either assume that each pair of protocol participants $U_i, U_j \in \mathcal{U}$ shares such a secret or that the complete set of protocol participants $\mathcal{U}$ shares one common secret (our compiler is provably secure in either case). We assume that all secrets are chosen independently.

Specifics for password-based authentication. In the case of password-authenticated key establishment, we assume a dictionary $\mathcal{D} \subseteq \{0,1\}^*$ to be publicly available. It is supposed to be efficiently recognizable and of constant or polynomial size. In particular, a polynomially bounded adversary is able to exhaust the complete dictionary $\mathcal{D}$. We assume that all passwords are chosen independently and uniformly at random from $\mathcal{D}$.

2.1 Communication Model and Adversarial Capabilities

As mentioned earlier, our security model is essentially adopted from [6] which in turn builds on [8,14,15,5]. Moreover, as we consider forward secrecy, we also include a Corrupt-oracle. As usual, users are modeled as probabilistic polynomial time (ppt) Turing machines. For our proofs, we may either use uniform or non-uniform Turing machines.

Protocol instances. Each protocol participant $U \in \mathcal{U}$ may execute a polynomial number of protocol *instances* in parallel. A single instance $\Pi_i^{s_i}$ can be interpreted as a process executed by protocol participant U_i. Throughout, the notation $\Pi_i^{s_i}$ ($i \in \mathbb{N}$) will be used to refer to instance s_i of protocol participant $U_i \in \mathcal{U}$. To each instance we assign seven variables:

$used_i^{s_i}$ indicates whether this instance is or has been used for a protocol run. The $used_i^{s_i}$ flag can only be set through a protocol message received by the instance due to a call to the Execute- or to the Send-oracle (see below);
$state_i^{s_i}$ keeps the state information needed during the protocol execution;
$term_i^{s_i}$ shows if the execution has terminated;
$sid_i^{s_i}$ denotes a public session identifier that can serve as identifier for the session key $sk_i^{s_i}$. Note that even though we do not construct session identifiers as session transcripts, the adversary is allowed to learn all session identifiers;
$pid_i^{s_i}$ stores the set of identities of those users that $\Pi_i^{s_i}$ aims at establishing a key with—including U_i himself;
$acc_i^{s_i}$ indicates if the protocol instance was successful, i. e., the user accepted the session key;
$sk_i^{s_i}$ stores the session key once it is accepted by $\Pi_i^{s_i}$. Before acceptance, it stores a distinguished NULL value.

For more details on the usage of the variables we refer to the work of Bellare et al. in [3].

Communication network. We assume arbitrary point-to-point connections among users to be available. The network is non-private and fully asynchronous: The adversary may delay, eavesdrop, insert and delete messages at will.

Adversarial capabilities. We consider ppt adversaries only. Let b be a bit chosen uniformly at random. The capabilities of an adversary $\mathcal{A}$ are made explicit through a number of *oracles* allowing $\mathcal{A}$ to communicate with protocol instances run by the users:

Send(U_i, s_i, M). This sends message M to the instance $\Pi_i^{s_i}$ and returns the reply generated by this instance. If $\mathcal{A}$ queries this oracle with an unused instance $\Pi_i^{s_i}$ and $M \subseteq \mathcal{P}$ a set of identities of principals, the used$_i^{s_i}$-flag is set, pid$_i^{s_i}$ initialized with pid$_i^{s_i} := \{U_i\} \cup M$, and the initial protocol message of $\Pi_i^{s_i}$ is returned.

Execute($\{\Pi_{u_1}^{s_{u_1}}, \ldots, \Pi_{u_\mu}^{s_{u_\mu}}\}$). This executes a complete protocol run among the specified unused instances of the respective users. The adversary obtains a transcript of all messages sent over the network. A query to the Execute oracle is supposed to reflect a passive eavesdropping. In particular, for a password-authenticated setting, no online-guess for the secret password can be implemented with a query to this oracle.

Reveal(U_i, s_i). This yields the value stored in sk$_i^{s_i}$.

Test(U_i, s_i). Provided that the session key is defined (i.e. acc$_i^{s_i}$ = true and sk$_i^{s_i} \neq$ NULL) and instance $\Pi_i^{s_i}$ is fresh (see the definition of freshness below), $\mathcal{A}$ can execute this oracle query at any time when being activated. Then, the session key sk$_i^{s_i}$ is returned if $b = 0$ and a uniformly chosen random session key is returned if $b = 1$. In this model, an arbitrary number of Test queries is allowed for the adversary $\mathcal{A}$, but once the Test oracle returned a value for an instance $\Pi_i^{s_i}$, it will return the same value for all instances partnered with $\Pi_i^{s_i}$ (see the definition of partnering below).

Corrupt(U_i). This returns all long-term secrets of user U_i. In case of password-based authentication, all passwords held by U_i are returned. In the case of U_i having long-term private keys, e.g., for signing, these private keys are returned.

Remark 1. The model described above seems apparently stronger than those normally used elsewhere since it allows for multiple Test queries. Nevertheless, one can easily show the two notions to be equivalent via a standard hybrid argument with a loss of a factor q in the reduction, with q being the total number of protocol instances. A similar model was also considered by Abdalla et al. in [2] to prove the security of their password-authenticated 3-party key establishment. Fortunately, as pointed out in [2], the loss of a factor q in the reduction can be avoided in most cases as several of the existing schemes (e.g., [19,20,15]) already meet this apparently stronger notion of security. This is due to the fact that, in their security proofs, they show that all fresh session keys that can be tested by the adversary are indistinguishable from random.

2.2 Correctness, Integrity and Secrecy

Before we define correctness, integrity and secrecy, we introduce *partnering* to express which instances are associated in a common protocol session.

Partnering. We refer to instances $\Pi_i^{s_i}$, $\Pi_j^{s_j}$ as being *partnered* if $\mathsf{pid}_i^{s_i} = \mathsf{pid}_j^{s_j}$, $\mathsf{sid}_i^{s_i} = \mathsf{sid}_j^{s_j}$, $\mathsf{sk}_i^{s_i} = \mathsf{sk}_j^{s_j}$ and $\mathsf{acc}_i^{s_i} = \mathsf{acc}_j^{s_j} = \mathsf{true}$.

To avoid trivial cases, we assume that an instance $\Pi_i^{s_i}$ always accepts the session key constructed at the end of the corresponding protocol run if no deviation from the protocol specification has occurred. Moreover, we want that all users in the same protocol session come up with the same session key, and we capture this in the subsequent notion of correctness.

Correctness. We call a group key establishment protocol P *correct*, if in the presence of a passive adversary $\mathcal{A}$—i.e., $\mathcal{A}$ must neither use the Send nor the Corrupt oracle—the following holds: for all i, j with both $\mathsf{sid}_i^{s_i} = \mathsf{sid}_j^{s_j}$ and $\mathsf{acc}_i^{s_i} = \mathsf{acc}_j^{s_j} = \mathsf{true}$, we have $\mathsf{sk}_i^{s_i} = \mathsf{sk}_j^{s_j} \neq \text{NULL}$ and $\mathsf{pid}_i^{s_i} = \mathsf{pid}_j^{s_j}$.

Key integrity. By definition, correctness takes only passive attacks into account. In contrast, *key integrity* imposes no restrictions on the adversary's oracle access: We say that a correct group key establishment protocol fulfills *key integrity*, if with overwhelming probability all instances of users that have accepted with the same session identifier $\mathsf{sid}_j^{s_j}$ hold identical session keys $\mathsf{sk}_j^{s_j}$ and identical partner identifiers $\mathsf{pid}_j^{s_j}$.

Next, for detailing the security definition, we will have to specify under which conditions a Test-query may be executed.

Freshness. A Test-query should only be allowed to those instances holding a key that is not for trivial reasons known to the adversary. To this aim, an instance $\Pi_i^{s_i}$ is called *fresh* if none of the following holds:

- For some $U_j \in \mathsf{pid}_i^{s_i}$ a query $\mathsf{Corrupt}(U_j)$ was executed before a query of the form $\mathsf{Send}(U_k, s_k, M)$ has taken place, for some message (or set of identities) M and some $U_k \in \mathsf{pid}_i^{s_i}$.
- The adversary earlier queried $\mathsf{Reveal}(U_j, s_j)$ with $\Pi_i^{s_i}$ and $\Pi_j^{s_j}$ being partnered.

The idea of this definition is that revealing a session key from an instance $\Pi_i^{s_i}$ trivially yields the session key of all instances partnered with $\Pi_i^{s_i}$, and hence this kind of "attack" will be excluded in the security definition.

Security/key secrecy. For a secure group key establishment protocol, we have to impose a corresponding bound on the adversary's *advantage*: The advantage $\mathsf{Adv}_{\mathcal{A}}(\ell)$ of a ppt adversary $\mathcal{A}$ in attacking protocol P is a function in the security parameter ℓ, defined as

$$\mathsf{Adv}_{\mathcal{A}} := |2 \cdot \mathsf{Succ} - 1|.$$

Here Succ is the probability that the adversary queries Test only on fresh instances and guesses correctly the bit b used by the Test oracle (without violating the freshness of those instances queried with Test) :

In the case of password-authenticated key establishment, due to the polynomial size of the dictionary $\mathcal{D}$, we cannot prevent an adversary from correctly guessing shared passwords with non-negligible probability. Thus, for the password-authenticated setting, our goal is to restrict the adversary $\mathcal{A}$ to online-verification of password guesses, namely, to prove that $\mathsf{Adv}_{\mathcal{A}}$ is only negligibly above the probability $\mathcal{A}$ has guessed a shared password online. We introduce a function ε to capture such weaknesses that originate in the employed authentication technique. For the password case, ε should bound $\mathcal{A}$'s probability of guessing a shared password, assuming he is not able to test (online) more than a constant number of passwords per protocol instance.

Remark 2. Following the spirit of [14,6], it would be desirable to restrict the number of passwords that can be guessed online to *one* per protocol instance. As described, our compiler accesses the underlying authenticated 2-party key establishment as a black-box only, and our security proof does not guarantee that only one password can be verified per instance. For specific instances a tighter security reduction may be possible, however.

Definition 1. *We say that an authenticated group key establishment protocol* P *is* ε-*secure, if for every ppt adversary* $\mathcal{A}$ *the following inequality holds for some negligible function* negl:

$$\mathsf{Adv}_{\mathcal{A}}(\ell, q_{\mathrm{send}}) \leq \varepsilon(\ell, q_{\mathrm{send}}) + \mathrm{negl}(\ell), \tag{1}$$

where ℓ *is the security parameter and* q_{send} *is the number of different protocol instances* $\mathcal{A}$ *queries the* Send *oracle with. The function* ε *is expected to be at most linear in its second variable, i.e. the number of* Send *queries.*

Forward Secrecy. We follow the spirit of the definition of forward secrecy from [19], yet our definition is weaker: we consider the "weak corruption model" of [3] in which corrupting a principal means only retrieving his long term secret keys. Forward secrecy is then achieved if such corruption does not give the adversary any information about previously agreed session keys. This same approach has also been taken in [7,16].

Remark 3. Note that our definition of freshness allows for Test queries to instances such that their (or their partners') long term secret keys have been revealed to the adversary by a Corrupt query as long as no Send query has been asked to any of these instances (or their partners) after the Corrupt query. Thus, the above definition of ε-security implies forward secrecy in this sense.

3   From Two to Group: A Compiler

In this section, we describe how an n-party AKE can be derived from any 2-party AKE carrying over its essential security properties. Our compiler assumes the availability of a 2-party key establishment that is ε-secure in the sense of Definition 1, where ε is defined according to the authentication method used. Our construction then yields an $\hat{\varepsilon}$-secure n-party AKE where $\hat{\varepsilon}$ is bounded by $4 \cdot \varepsilon$.

3.1   Tools

For the actual compiler, black-box access to the authenticated 2-party key establishment suffices, and Fig. 1 captures this access with an oracle 2-AKE$(\cdot, \cdot)$ that upon input of two principals $U_i, U_j \in \mathcal{P}$ (or rather their identities), returns the respective output of the 2-party protocol. We assume this output to be either a secret key $\kappa \in \{0,1\}^k$ or a special symbol $\top$ indicating that the key establishment failed (due to adversarial interference). Additionally, the tools involved in our construction are:

- **a non-interactive non-malleable commitment scheme** [12] $\mathcal{C}$, fulfilling the following requirements:
  1. it must be *perfectly binding*, i.e., every commitment c defines at most one value decommit(c);
  2. it must achieve *non-malleability for multiple commitments*—if an adversary receives commitments to a (polynomial sized) set of values ν he must not be able to output commitments to a (polynomial sized) set of values β related to ν in a known way.

  Note that in the CRS model with a common reference string ρ, the above commitment schemes $\mathcal{C} = \mathcal{C}_\rho$ can be constructed from any public key encryption scheme that is non-malleable and secure for multiple encryptions (in particular, from any IND-CCA2 secure public key encryption scheme).

- **a collision-resistant pseudorandom function family** $\mathcal{F} = \{F^\ell\}_{\ell \in \mathbb{N}}$ as used by Katz and Shin [21]. We assume $F^\ell = \{F^\ell_\eta\}_{\eta \in \{0,1\}^L}$ to be indexed by a superpolynomial sized set $\{0,1\}^L$ and denote by $v_0 = v_0(\ell)$ a publicly known value such no ppt adversary can find two different indices $\lambda \neq \lambda' \in \{0,1\}^L$ such that $F_\lambda(v_0) = F_{\lambda'}(v_0)$. For deriving the session key we use another public value v_1 which fulfills the above collision-resistance condition as well and is also encoded in the CRS (see [21] for more details).

- **a family of universal hash functions** $\mathcal{UH}$ that maps the concatenation of bitstrings from $\{0,1\}^{kn}$ and a partner pid$_i^{s_i}$ onto $\{0,1\}^L$. The CRS selects one universal hash function UH from this family. We use UH to select an index within the aforementioned collision-resistant pseudorandom function family.

3.2   Design Rationale

The idea of our compiler is inspired in the classical construction of Burmester and Desmedt [8], where the trick of constructing a group key from pairwise agreed keys among the group principals was first introduced. Further, our construction in some sense generalizes the design of [6], that builds an n-party PAKE on

Gennaro and Lindell's 2-party PAKE. Once the pairwise key establishments have been completed, each principal must commit to the XOR-value of the two keys he shares with his neighbors. This value is disclosed in a subsequent round, allowing all principals to derive each of the 2-party keys, from which both the session identifier and the session key will be derived. Intuitively, if an adversary has not been able to pervert the security of any of the 2-party protocol executions involved, neither will he be able to retrieve any information about the resulting group session key (for XORs of "randomly looking" elements should look as well random to him). Moreover, integrity is also provided by an argument similar to the one in [6].

The compiler does not rely on further authentication techniques than those used in the basic 2-party AKE protocol, neither on any further idealization assumption. Also, our design is symmetric in the sense that all users perform the same steps. Fig. 1 shows the three rounds of our construction, adding 2 rounds to those of the underlying 2-party AKE. For the sake of readability, we do not explicitly refer to instances s_i of users. Also, we omit the $pid_i^{s_i}$-values, assuming that when the protocol is initiated (via a Send or Execute call) each participant involved receives a message informing him of the actual pid of the session, which in addition makes him aware of his position in the "cycle" of involved principals and therefore the 2-AKE step (Round 0) can be performed accordingly.

Remark 4. The compiler can be applied to any polynomial number of participants $n \geq 2$. The case $n = 2$ is not excluded, but to some extent pathological: Here the compiler executes the underlying 2-party AKE *twice*, so that each party obtains two independent keys $\overrightarrow{K}_i$, $\overleftarrow{K}_i$, which are then combined to form the actual session key.

3.3 Security Analysis

Assume that we are given a correct and secure authenticated 2-party key establishment protocol. Assume further that $\mathcal{C}$ is a non-interactive non-malleable commitment scheme and $\mathcal{F}$ a collision-resistant pseudorandom function family. In the following, we show that under these assumptions the compiler in Fig. 1 yields a correct and secure group key establishment. In particular, this is true when the underlying 2-party AKE protocol is based on passwords.

Theorem 1. *Let $\mathcal{F}$ be a family of secure collision-resistant pseudorandom functions, let $\mathcal{C}$ be a non-interactive perfectly binding non-malleable commitment scheme, and let 2-AKE be a correct and ε-secure authenticated 2-party key establishment protocol. Then the protocol in Fig. 1 is a correct and $4 \cdot \varepsilon$-secure authenticated group key establishment protocol, which also provides key integrity.*

Proof. Correctness. In an honest execution of the protocol, it is easy to verify that all participants in the protocol will terminate by accepting and computing the same session identifier and session key.

Integrity. Owing to the collision-resistance of the family $\mathcal{F}$, all oracles that accept with identical session identifiers use with overwhelming probability the same

---

Round 0:

  2-AKE: For $i = 1, \ldots, n$ execute 2-AKE(U_i, U_{i+1}).[a] Thus, each user U_i holds
  two keys $\overrightarrow{K}_i$, $\overleftarrow{K}_i$ shared with U_{i+1} respectively U_{i-1}.

Round 1:

  Computation: Each U_i computes

$$X_i := \overrightarrow{K}_i \oplus \overleftarrow{K}_i$$

  and chooses a random r_i to compute a commitment $C_i = C_\rho(i, X_i; r_i)$.
  Broadcast: Each U_i broadcasts $M_i^1 := (U_i, C_i)$

Round 2:

  Broadcast: Each U_i broadcasts $M_i^2 := (U_i, X_i, r_i)$
  Check: Each U_i checks that $X_1 \oplus X_2 \oplus \cdots \oplus X_n = 0$ and the correctness of
  the commitments. If at least one of these checks fails, set $\mathsf{acc}_i := \mathsf{false}$ and
  terminate the protocol execution.
  Computation: Each U_i sets $K_i := \overleftarrow{K}_i$ and computes the $n - 1$ values

$$K_{i-j} := \overleftarrow{K}_i \oplus X_{i-1} \oplus \cdots \oplus X_{i-j} \quad (j = 1, \ldots, n-1),$$

  defines a master key

$$K := (K_1, \ldots, K_n, \mathsf{pid}_i),$$

  and sets $\mathsf{sk}_i := F_{\mathrm{UH}(K)}(v_1)$, $\mathsf{sid}_i := F_{\mathrm{UH}(K)}(v_0)$ and $\mathsf{acc}_i := \mathsf{true}$.

---

[a] All indices are to be taken in a cycle, i. e., $U_{n+1} = U_1$, etc.

Fig. 1. A protocol compiler

index value UH(K) and therewith also derive the same session key and have
identical partner identifiers.

Key secrecy. The proof of key secrecy will proceed in a sequence of games, starting with the real attack against the key secrecy of the group key exchange protocol and ending in a game in which the adversary's advantage is 0, and for
which we can bound the difference in the adversary's advantage between any
two consecutive games. Following standard notation, we denote by $\mathsf{Adv}(\mathcal{A}, G_i)$
the advantage of the adversary $\mathcal{A}$ in Game i. Furthermore, for clarity, we clas-
sify the Send queries into 3 categories, depending on the stage of the protocol
to which the query is associated, starting with Send-0 and ending with Send-2.
Send-t denotes the Send query associated with round t for $t = 0, 1, 2$.

Game 0. This first game corresponds to a real attack, in which all the parameters, such as the public parameters in the common reference string and the
long-term secrets associated with each user, are chosen as in the actual scheme.
By definition, $\mathsf{Adv}(\mathcal{A}, G_0) = \mathsf{Adv}(\mathcal{A})$.

Game 1. In this game, for $i = 1, \ldots, n$, we modify the simulation of the Send and Execute oracles so that, whenever an instance $\Pi_i^{s_i}$ is still considered fresh at the end of Round 0, the keys $\overleftarrow{K}_i$ and $\overrightarrow{K}_i$ that it shares with instances $\Pi_{i-1}^{s_{i-1}}$ and $\Pi_{i+1}^{s_{i+1}}$ are replaced with random values from the appropriate set. An instance $\Pi_i^{s_i}$ is considered fresh at the end of Round 0 if it has not halted or rejected and if no query $\mathsf{Corrupt}(U_j)$ for some $U_j \in \mathrm{pid}_i^{s_i}$ has been asked by the adversary before a query of the form $\mathsf{Send}(U_k, s_k, M)$ for some $U_k \in \mathrm{pid}_i^{s_i}$ and some message M.

Note that the distance between this game and the previous one is bounded by the probability that the adversary breaks the security of any of the underlying 2-AKE protocols. More precisely, we have

$$\left|\mathsf{Adv}(\mathcal{A}, G_1) - \mathsf{Adv}(\mathcal{A}, G_0)\right| \leq 2 \cdot \mathsf{Adv}_{\text{2-AKE}}(\ell, 2 \cdot q_{\text{send}}),$$

where q_{send} represents the number of *different* protocol instances in Send queries. The factor 2 multiplying q_{send} emerges because one instance in the group key protocol builds on two instances of the 2-AKE protocol for the key establishment with the right and left neighbor, respectively. The other factor 2 is due to the security definition which states that the advantage of an adversary is twice its success probability minus 1.

To prove this, we show how an adversary $\mathcal{A}_{\text{2-AKE}}$ is constructed from a given adversary $\mathcal{A}$ distinguishing Game G_1 from Game G_0.

$\mathcal{A}_{\text{2-AKE}}$ is given access to a simulation of the 2-AKE protocol as outlined in Section 2. To answer its queries, $\mathcal{A}_{\text{2-AKE}}$ will associate each user instance $\Pi_i^{s_i}$ in the group protocol with two independent instances of the same user in the 2-AKE protocol. Now, whenever $\mathcal{A}$ makes a Corrupt query, $\mathcal{A}_{\text{2-AKE}}$ answers it by querying the Corrupt oracle of the 2-AKE protocol and returns the same answer. To answer an Execute query, $\mathcal{A}_{\text{2-AKE}}$ first queries the Execute oracle of the 2-AKE protocol with the corresponding instances to obtain the transcript for Round 0. To simulate the following rounds, $\mathcal{A}_{\text{2-AKE}}$ first queries the Test oracle of the 2-AKE protocol with the corresponding instances and uses the returned values as the keys $\overleftarrow{K}_i$ and $\overrightarrow{K}_i$. To answer a Send-0 query, $\mathcal{A}_{\text{2-AKE}}$ queries the Send oracle of the 2-AKE protocol with the corresponding instance and returns its response. To answer Send queries pertaining rounds 1 and 2, $\mathcal{A}_{\text{2-AKE}}$ first sets the values of the keys $\overleftarrow{K}_i$ and $\overrightarrow{K}_i$ by querying either the Test or Reveal oracle of the 2-AKE protocol with the corresponding instances and proceeds with the simulation as in the previous game. More precisely, if an instance $\Pi_i^{s_i}$ in the group protocol is still considered fresh at the beginning of Round 1, then $\mathcal{A}_{\text{2-AKE}}$ queries the Test oracle of the 2-AKE protocol with the corresponding instances in the 2-AKE protocol. Otherwise, $\mathcal{A}_{\text{2-AKE}}$ queries the Reveal oracle.

Finally, one can easily see that the view of $\mathcal{A}$ corresponds to Game G_0 if Test reveals the actually exchanged key and to Game G_1 if Test returns a random element from the key space. Thus, $\mathcal{A}$ succeeds distinguishing Game G_0 and Game G_1 with a probability of at most $\mathsf{Adv}_{\text{2-AKE}}(\ell, 2 \cdot q_{\text{send}})$.

Game 2. In this game, we change the simulation of the Send oracle so that a *fresh* instance $\Pi_i^{s_i}$ does not accept in Round 2 whenever one commitment C_j for $j \neq i$ it receives in Round 1 was generated by the simulator but not generated by the respective instance $\Pi_j^{s_j}$, $j \neq i$ in the same session. At this, we take two instances $\Pi_{\alpha_0}^{s_{\alpha_0}}$, $\Pi_{\alpha_r}^{s_{\alpha_r}}$ for being in the *same session*, if there is a sequence of instances $(\Pi_{\alpha_\mu}^{s_{\alpha_\mu}})_{0 \leq \mu \leq r}$ such that for each $\mu = 0, \ldots, r-1$ the instances $\Pi_{\alpha_\mu}^{s_{\alpha_\mu}}$ and $\Pi_{\alpha_{\mu+1}}^{s_{\alpha_{\mu+1}}}$ are partnered through an execution of the underlying 2-party key establishment (i. e., they hold a common 2-party session key $\overrightarrow{K}_{\alpha_\mu} = \overleftarrow{K}_{\alpha_{\mu+1}}$ associated with the same session identifier and the same two protocol participants).

The adversary $\mathcal{A}$ can detect the difference to Game G_1 if $\mathcal{A}$ replayed a commitment that should have led to acceptance in Round 2 in that game. Because the committed value X_i is a random value independent of previous messages, the probability for this is negligible.

$$\left| \mathsf{Adv}(\mathcal{A}, G_2) - \mathsf{Adv}(\mathcal{A}, G_1) \right| \leq \mathrm{negl}(\ell)$$

To see why, note that given one session, an instance $\Pi_i^{s_i}$ expects commitments C_j to (j, X_j), such that $X_1 \oplus \cdots \oplus X_n = 0$. $\Pi_i^{s_i}$ will only accept with negligible probability if all commitments where generated by the simulator, however, not being exactly the commitments C_j, $j \neq i$ by the respective oracles $\Pi_j^{s_j}$, $j \neq i$ of the session. This can be seen as follows: The equation

$$X_1 \oplus \cdots \oplus X_n = 0$$

results in

$$\overleftarrow{K}_1 \oplus \overrightarrow{K}_1 \oplus \cdots \oplus \overleftarrow{K}_n \oplus \overrightarrow{K}_n = 0.$$

For $\Pi_i^{s_i}$, $X_i = \overleftarrow{K}_i \oplus \overrightarrow{K}_i$ is given where $\overleftarrow{K}_i$ is shared with U_{i-1} and $\overrightarrow{K}_i$ is shared with U_{i+1}. Because the commitment C_j includes the index of user U_j and is perfectly binding, the adversary $\mathcal{A}$ cannot reveal the commitments if they are permuted within the participants of the session. As by now all keys are random values, the probability for any XOR sum of keys not consisting exactly of the keys in one session (thus canceling each other w.r.t. XOR) to be 0 is only $1/2^k$. The adversary $\mathcal{A}$ is at maximum capable of doing this q_{send} times, giving him a probability $q_{\mathrm{send}}/2^k$ of distinguishing the games.

Game 3. This game reproduces the modification also for *adversary-generated* commitments: The simulation of the Send oracle changes so that a *fresh* instance $\Pi_i^{s_i}$ does not accept in Round 2 whenever one commitment C_j for $j \neq i$ it receives in Round 1 was *adversary-generated*. The adversary's advantage diverges only negligibly from the previous game:

$$\left| \mathsf{Adv}(\mathcal{A}, G_3) - \mathsf{Adv}(\mathcal{A}, G_2) \right| \leq \mathrm{negl}(\ell)$$

To prove this, we construct a malleability attacker $\mathcal{A}_{COM}$ to the commitment scheme from an adversary $\mathcal{A}$ that comes up with a commitment C_j to $\Pi_i^{s_i}$ such that $\Pi_i^{s_i}$ would accept in Game G_2 but not in Game G_3. Our goal is to show that the probability with which $\mathcal{A}_{COM}$ succeeds in outputting a related vector of commitments is related to the probability with which $\mathcal{A}$ can distinguish Games G_3 from G_2.

$\mathcal{A}_{COM}$ is given commitments $C_i = C_\rho(i, X_i; r_i)$ for $i = 1, \ldots, n$ where the X_i values are random bitstrings fulfilling $X_1 \oplus \cdots \oplus X_n = 0$. For bitstrings X'_i, $i = 1, \ldots, n$, the $2n$-ary relation is given by

$$\mathcal{R}(X_1, \ldots, X_n, X'_1, \ldots, X'_n) = 1$$

if and only if

$$X'_1 \oplus \cdots \oplus X'_n = 0 \text{ and } X_i = X'_i \text{ for at least one index } i \in \{1, \ldots, n\}.$$

$\mathcal{A}_{COM}$ starts by guessing the first instance $\Pi_i^{s_i}$ to receive from $\mathcal{A}$ a set of commitments C'_j, for $j \neq i$, with at least one of these commitments being *adversary-generated*. For all sessions other than the one in which $\Pi_i^{s_i}$ is involved, $\mathcal{A}_{COM}$ simulate the oracles exactly as it would in Game G_2. For the session in which $\Pi_i^{s_i}$ is involved, $\mathcal{A}_{COM}$ uses the C_i values that it has received as input to answer Send-1 queries. Then, as soon as $\mathcal{A}$ provides $\Pi_i^{s_i}$ with a set of commitments C'_j for $j \neq i$, then $\mathcal{A}_{COM}$ halts the simulation and outputs this set along with C_i.

One can easily see that $\mathcal{A}_{COM}$ will succeed in outputting a set of related commitments satisfying the relation $\mathcal{R}$ if it guesses correctly the first instance to receive a set of commitments containing at least one *adversary-generated* commitment and passing the verification test. This is true because games G_3 and G_2 are indistinguishable up to that point and the simulation of the oracles by $\mathcal{A}_{COM}$ is perfect.

By definition of non-malleability, the success probability of $\mathcal{A}_{COM}$ is only negligibly greater than that of an adversary who does not see the list of commitments C_i for $i = 1, \ldots, n$. If no commitments are given, an adversary's probability to send valid commitments C_j for $j \neq i$ such that $X'_1 \oplus \cdots \oplus X_i \oplus \cdots \oplus X'_n = 0$ is $q_{\text{send}}/2^k$ as in the previous game. As a result, the non-malleability of the commitment scheme guarantees that the adversary's success probability with access to these commitments is negligibly close to $q_{\text{send}}/2^k$, thus, being negligible in total.

Game 4. Now the simulation of the Execute and Send oracles are modified at the point of computing the session key. The simulator keeps a list of assignments $(K_1, \ldots, K_n, \text{sk}_i^{s_i})$. Once an instance receives the last Send-2 query, the simulator computes $K_1, \ldots, K_n$ and checks if for this sequence a master key was already issued and assigns this key to the instance. If no such entry exists in the list, the simulator chooses a session key $\text{sk}_i^{s_i} \in \{0, 1\}^\ell$ uniformly at random.

The master key $K = (K_1, \ldots, K_n, \text{pid}_i^{s_i})$ has, once the X_i are public, sufficient entropy such that the output of the pseudorandom function $F_{\text{UH}(K)}$ is distinguishable from a random $\text{sk}_i^{s_i}$ with negligible probability only.

$$\left| \text{Adv}(\mathcal{A}, G_4) - \text{Adv}(\mathcal{A}, G_3) \right| \leq \text{negl}(\ell).$$

In Game G_4, all session keys are chosen uniformly at random and the adversary has no advantage.

$$\text{Adv}(\mathcal{A}, G_4) = 0.$$

$\square$

4   Applications and Comments

The above compiler allows for the construction of very efficient authenticated group key exchange protocols adding up to the "base" 2-AKE only two rounds of communication. As we have remarked, our compiler adds neither any authentication tool nor any additional idealization assumptions to the base scheme.

Example 1. Applying our compiler to a password-authenticated 2-party key establishment offering forward secrecy, we immediately obtain a forward secure password-authenticated group key establishment. It should be pointed out here, however, that stronger notions of forward secrecy than ours can be considered [19]. Actually, it is an interesting question to explore whether the KOY 2-AKE from [19] (or variants of it) can be proven secure in our model—therewith yielding through application of our compiler the first forward secure password-authenticated group key establishment.

Of course, our compiler can also be applied in the random oracle model—in practice this means to replace the "full-fledged" commitment scheme and the family of collision resistant pseudorandom functions through the (more efficient) use of a cryptographic hash function (cf. [21]). Going one step further, from an engineering perspective it is tempting to apply the compiler to an efficient authenticated 2-party key establishment, even if no security proof in the above model is available. Of course, in this case our security analysis does not yield a provable security statement on the resulting group key establishment.

Example 2. A natural starting point for applying our compiler would be the (H)MQV family discussed in [27,23,26,22]. The resulting scheme could be rather efficient in practice, but the available formal security analysis builds on a model due to Canetti and Krawczyk [11]. We have not attempted to carry out a security analysis in the model underlying the above discussion and consequently cannot claim provable security guarantees of a derived group key establishment.

5   Conclusions

The compiler we presented allows the construction of authenticated group key establishment schemes based on any provably secure authenticated 2-party key establishment. At this forward secrecy is taken into account, and the suggested compiler does not introduce new idealizing assumptions or tools for authentication, like an existentially unforgeable signature scheme. In terms of efficiency, adding only two additional rounds to a 2-party solution seems acceptable, too, and renders the compiler an interesting tool for practical protocol design.

Both from a theoretical and from a practical point of view, it seems worthwhile to explore the tightness of the above security proof more closely, when applying the compiler to specific protocols. In the described form, the compiler restricts to black-box access to the underlying two-party key establishment, but for a specific use case, there is no need for such a restriction.

Also, we have not explored the behaviour of our compiler within the universal composability framework. In particular, it would be interesting to explore the security level achieved applying our compiler to universally composable password based two party key exchange protocols, along the lines of [10].

Acknowledgements

The first author was supported in part by the European Commission through the IST Program under Contract IST-2002-507932 ECRYPT and by France Telecom R&D as part of the contract CIDRE, between France Telecom R&D and École normale supérieure.

References

1. Michel Abdalla, Pierre-Alain Fouque, and David Pointcheval. Password-Based Authenticated Key Exchange in the Three-Party Setting. In Serge Vaudenay, editor, *Public Key Cryptography – PKC 2005*, volume 3386 of *Lecture Notes in Computer Science*, pages 65–84. Springer, 2005.
2. Michel Abdalla, Pierre-Alain Fouque, and David Pointcheval. Password-Based Authenticated Key Exchange in the Three-Party Setting. *IEE Proceedings – Information Security*, 153(1):27–39, March 2006.
3. Mihir Bellare, David Pointcheval, and Phillip Rogaway. Authenticated Key Exchange Secure Against Dictionary Attacks. In Bart Preneel, editor, *Advances in Cryptology – EUROCRYPT 2000*, volume 1807 of *Lecture Notes in Computer Science*, pages 139–155. Springer, 2000.
4. Mihir Bellare and Phillip Rogaway. Entitiy Authentication and Key Distribution. In Douglas R. Stinson, editor, *Advances in Cryptology – CRYPTO '93*, volume 773 of *Lecture Notes in Computer Science*, pages 232–249. Springer, 1994.
5. Jens-Matthias Bohli, María Isabel González Vasco, and Rainer Steinwandt. Secure Group Key Establishment Revisited. Cryptology ePrint Archive, Report 2005/395, 2005. Available at http://eprint.iacr.org/2005/395/.
6. Jens-Matthias Bohli, María Isabel González Vasco, and Rainer Steinwandt. Password-Authenticated Constant-Round Group Key Establishment with a Common Reference String . Cryptology ePrint Archive: Report 2006/214, 2006. Available at http://eprint.iacr.org/2006/214.
7. Victor Boyko, Philip D. MacKenczie, and Sarvar Patel. Provable-Secure Password-Authenticated Key Exchange Using Diffie-Hellman. In Bart Preneel, editor, *Advances in Cryptology – EUROCRYPT 2000*, volume 1807 of *Lecture Notes in Computer Science*, pages 156–171. Springer, 2000.
8. Mike Burmester and Yvo Desmedt. A Secure and Efficient Conference Key Distribution System. In Alfredo De Santis, editor, *Advances in Cryptology – EUROCRYPT'94*, volume 950 of *Lecture Notes in Computer Science*, pages 275–286. Springer, 1995.
9. Mike Burmester and Yvo G. Desmedt. Efficient and Secure Conference-Key Distribution. In T. Mark A. Lomas, editor, *Proceedings of the International Workshop on Security Protocols*, volume 1189 of *Lecture Notes in Computer Science*, pages 119–129. Springer, 1996.

10. Ran Canetti, Shai Halevi, Jonathan Katz, Yehuda Lindell, and Philip MacKenzie. Universally Composable Password-Based Key Exchange. In Ronald Cramer, editor, *Advances in Cryptology – EUROCRYPT 2005*, volume 3495 of *Lecture Notes in Computer Science*, pages 404–421. Springer, 2005.

11. Ran Canetti and Hugo Krawczyk. Analysis of Key-Exchange Protocols and Their Use for Building Secure Channels. In Birgit Pfitzmann, editor, *Advances in Cryptology – EUROCRYPT 2001*, volume 2045 of *Lecture Notes in Computer Science*, pages 453–474. Springer, 2001.

12. Danny Dolev, Cynthia Dwork, and Moni Naor. Non-Malleable Cryptography. *SIAM Journal of Computing*, 30(2):391–437, 2000.

13. Ratna Dutta and Rana Barua. Password-Based Encrypted Group Key Agreement. *International Journal of Network Security*, 3(1):23–34, July 2006.

14. Rosario Gennaro and Yehuda Lindell. A Framework for Password-Based Authenticated Key Exchange. Cryptology ePrint Archive: Report 2003/032, 2003. Available at http://eprint.iacr.org/2003/032.

15. Rosario Gennaro and Yehuda Lindell. A Framework for Password-Based Authenticated Key Exchange (Extended Abstract). In Eli Biham, editor, *Advances in Cryptology – EUROCRYPT 2003*, volume 2656 of *Lecture Notes in Computer Science*, pages 524–543. Springer, 2003.

16. Oded Goldreich and Yehuda Lindell. Session-key generation using human passwords only. In *Advances in Cryptology – CRYPTO '01*, pages 408–432, London, UK, 2001. Springer-Verlag.

17. Jung Yeon Hwang, Su-Mi Lee, and Dong Hoon Lee. Scalable key exchange transformation: from two-party to group. *Electronic Letters*, 40(12):728–729, 2004.

18. Jonathan Katz, Rafail Ostrovsky, and Moti Yung. Efficient Password-Authenticated Key Exchange Using Human-Memorable Passwords. In Birgit Pfitzmann, editor, *Advances in Cryptology – EUROCRYPT 2001*, volume 2045 of *Lecture Notes in Computer Science*, pages 475–494. Springer, 2001.

19. Jonathan Katz, Rafail Ostrovsky, and Moti Yung. Forward Secrecy in Password-Only Key Exchange Protocols. In Stelvio Cimato, Clemente Galdi, and Giuseppe Persiano, editors, *Security in Communication Networks: Third International Conference, SCN 2002*, volume 2576 of *Lecture Notes in Computer Science*, pages 29–44. Springer, 2003.

20. Jonathan Katz, Rafail Ostrovsky, and Moti Yung. Efficient and Secure Authenticated Key Exchange Using Weak Passwords, 2006. Available at http://www.cs.umd.edu/~jkatz/papers/password.pdf.

21. Jonathan Katz and Ji Sun Shin. Modeling Insider Attacks on Group Key-Exchange Protocols. Cryptology ePrint Archive: Report 2005/163, 2005. Available at http://eprint.iacr.org/2005/163.

22. Hugo Kawczyck. HMQV: A High-Performance Secure Diffie-Hellman Protocol. Cryptology ePrint Archive: Report 2005/176, 2005. Available at http://eprint.iacr.org/2005/176.

23. Hugo Krawczyck. HMQV: A High-Performance Secure Diffie-Hellman Protocol. In Victor Shoup, editor, *Advances in Cryptology – CRYPTO '05*, volume 3621 of *Lecture Notes in Computer Science*, pages 546–566. Springer, 2005.

24. Su Mi Lee, Jung Yeon Hwang, and Dong Hoon Lee. Efficient Password-Based Group Key Exchange. In Sokratis Katsikas, Javier Lopez, and Günther Pernul, editors, *Trust and Privacy in Digital Business: First International Conference, TrustBus 2004*, volume 3184 of *Lecture Notes in Computer Science*, pages 191–199. Springer, 2004.

25. Alain Mayer and Moti Yung. Secure Protocol Transformation via "Expansion": From Two-party to Groups. In *Proceedings of the 6th ACM conference on Computer and Communications Security CCS '99*, pages 83–92. ACM Press, 1999.
26. Alfred Menezes. Another look at HMQV. Cryptology ePrint Archive: Report 2005/205, 2005. Available at `http://eprint.iacr.org/2005/205`.
27. Alfred Menezes, Minghua Qu, and Scott A. Vanstone. Some new key agreement protocols providing mutual implicit authentication. In *Workshop on Selected Areas in Cryptography*, pages 22–32, July 1995.
28. Qiang Tang and Kim-Kwang Raymond Choo. Secure password-based authenticated group key agreement for data-sharing peer-to-peer networks. In *ACNS*, volume 3989 of *Lecture Notes in Computer Science*, pages 162–177. Springer, 2006.

Multi-authority Attribute Based Encryption

Melissa Chase

Computer Science Department
Brown University
Providence, RI 02912
mchase@cs.brown.edu

Abstract. In an identity based encryption scheme, each user is identified by a unique identity string. An attribute based encryption scheme (ABE), in contrast, is a scheme in which each user is identified by a set of attributes, and some function of those attributes is used to determine decryption ability for each ciphertext. Sahai and Waters introduced a single authority attribute encryption scheme and left open the question of whether a scheme could be constructed in which multiple authorities were allowed to distribute attributes [SW05]. We answer this question in the affirmative.

Our scheme allows any polynomial number of independent authorities to monitor attributes and distribute secret keys. An encryptor can choose, for each authority, a number d_k and a set of attributes; he can then encrypt a message such that a user can only decrypt if he has at least d_k of the given attributes from each authority k. Our scheme can tolerate an arbitrary number of corrupt authoritites.

We also show how to apply our techniques to achieve a multiauthority version of the large universe fine grained access control ABE presented by Gopal et al. [GPSW06].

1   Introduction

Identity based encryption(IBE), introduced by Shamir [Sha85], is a variant of encryption which allows users to use any string as their public key (for example, an email address). This means that the sender can send messages knowing only the recipient's identity (or email address), thus eliminating the need for a separate infrastructure to distribute public keys. The first IBE systems were given by Boneh and Franklin [BF01] and Cocks [Coc01], and IBE has received a lot of attention in the literature since then [CHK03, BB04, Wat05].

However, this scenario may not be entirely realistic, since we don't necessarily have a unique string identifier for each person. Instead, we often identify people by their attributes. We might want to send a message to the secretary in accounting in charge of travel reimbursements, or send a question to a nurse in a particular hospital who is knowledgeable about prescriptions, or announce a party to anyone living in town who is either a student or between the ages of 18 and 25. Thus, Sahai and Waters gave a fuzzy IBE scheme which could be used for attribute based encryption. In this model, a recipient is defined not by

S.P. Vadhan (Ed.): TCC 2007, LNCS 4392, pp. 515–534, 2007.
© International Association for Cryptologic Research 2007

a single string, but by a set of attributes [SW05]. Sahai and Waters describe a scheme (from here on referred to as SW) in which a sender can encrypt a message specifying an attribute set and a number d so that only a recipient who has at least d of the given attributes can decrypt the message. For example, a sender could encrypt a message to be decryptable by anyone who has 2 out of 3 of: a Rhode Island driver's license, Rhode Island voter registration, or a student ID from Brown University. Thus, their scheme allows the sender to encrypt a message for more than one recipient, and to specify who should be able to decrypt, using attributes alone.

There is, however, one major limitation to the SW scheme. In their scheme, as in every IBE scheme, the user must go to a trusted party and prove his identity in order to obtain a secret key which will allow him to decrypt messages. In this case, each user must go to the trusted server, prove that he has a certain set of attributes, and then receive secret keys corresponding to each of those attributes. However, this means we must have one trusted server who monitors all attributes – who keeps records of driver's licenses, voter registration, and college enrollment. In reality, we have 3 different entities responsible for maintaining this information (the RI DMV, the RI Board of Elections, and the University office), so we would want to be able to entrust each of these to a different (and perhaps not entirely trusted) server. Thus, Sahai and Waters presented the following challenge: Is it possible to construct an attribute based encryption scheme in which many different authorities operate simultaneously, each handing out secret keys for a different set of attributes?

OUR RESULTS. We resolve this problem in the affirmative. We give an efficient scheme for multiauthority attribute based encryption. We allow the sender to specify for each authority k a set of attributes monitored by that authority and a number d_k so that the message can be decrypted only by a user who has at least d_k of the given attributes from every authority. We allow any number of attribute authorities to be corrupted, and guarantee the security of encryption as long as the required attributes cannot be obtained exclusively from those authorities and the trusted authority remains honest.

We also provide several extensions to our basic multiauthority scheme. We describe techniques to allow the encryptor to determine for each ciphertext how many attributes to require from each authority. We also describe a variant of our scheme in which the encryptor can specify a number D such that a user can decrypt if he has sufficient numbers of the given attributes from at least D authorities. It is this variant that would be used to implement the RI example above. In this example, we have 3 authorities, and the ciphertext will include 1 attribute from each. However, we only want to require that a user must have satisfactory attributes from 2 out of the 3 authorities in order to decrypt.

CHALLENGES AND TECHNIQUES. The most challenging aspect of a single authority ABE scheme is preventing collusion. Recall the above example. Now suppose Alice has a RI driver's license and Bob is a Brown University student. Together they have two out of three of the required attributes, but they should not be able to combine their keys and decrypt the ciphertext. In SW each user's keys are generated using different random sharings of a secret, so keys generated for different users cannot be combined.

It might seem that a multiauthority ABE scheme could be formed simply by letting each authority run its own copy of SW and then combining the results. However, here we once again run into the problem of collusion. The SW techniques will prevent collusion within authorities, so different keys obtained from any one authority cannot be combined. However, suppose we have a ciphertext which requires attributes from authority 1 and authority 2. If Alice has all the appropriate attributes from authority 1 and Bob has all the appropriate attributes from authority 2, they still should not be able to combine their keys and decrypt. Note that the SW techniques cannot be directly applied here: in SW, a single authority sees all the attributes requested by a user and gives a secret key, so it can easily rerandomize the secret sharing appropriately. In the multiauthority case, we would again like to split up a secret in a different way for each user, this time dividing it between multiple authorities. However, now we need to do this *without any communication* between the authorities.

We use two main techniques: The first is to require that every user have some kind of a global identifier (GID). We require only two properties from this: (1) no user can claim another user's identifier, and (2) all authorities can verify a user's identifier. Thus, the GID could be a name or SSN or any other identifying string for which a user has provable credentials, and it seems likely that such information would be present when users' attributes are verified. To see why this is necessary, consider the following two scenarios: In the first Bob requests keys for attribute set $\mathcal{A}_1$ from authority 1 and Alice requests keys for attribute set $\mathcal{A}_2$ from authority 2. In the second Bob requests attribute set $\mathcal{A}_1$ from authority 1 and attribute set $\mathcal{A}_2$ from authority 2. If the authorities do not communicate, and Alice and Bob are identified by nothing beyond their attributes, then in the authorities' view these scenarios must be identical. The global identifier allows the authorities to distinguish these two scenarios in order to prevent collusion.

At the same time we still want a user's ability to decrypt to depend only on his attributes (this is what distinguishes ABE from traditional IBE schemes). Thus, we use our second tool: the central authority. Each user will send his GID to the central authority and receive a corresponding key. Note that the authority will not get any information about the users' attributes; it's purpose is simply to give a setup key for the user's GID. We will also require that this authority be trusted: it will hold the master secret for the system, so it will be able to decrypt any message. Note that the presence of a trusted party is a fairly standard requirement: in an IBE scheme, the single authority must obviously be trusted, and even when this is extended to a hierarchical IBE (HIBE) scheme, in which many of the lower level authorities can be corrupted, one must require that the root authority be honest.

Each authority has a pseudorandom function (PRF) which it will use to randomize the secret keys it gives out. A PRF guarantees that, on the one hand, the secret keys for each user are derived deterministically, but, at the same time, that they will appear completely random. When a user requests a secret key, the authority will compute the PRF on the user's GID and then use the result as the secret in SW key generation. A user with sufficient attributes can then use his secret keys to reconstruct this secret for each authority. However, since the outputs of the PRFs will be different for each user, each user will reconstruct a different set of secrets. Thus, we need the central authority, who will know all of the other authorities' PRFs. For each user, it will compute the extra value which, when combined with the secrets the user has reconstructed, will result in a user-independent system decryption value which allows the user to decrypt.

Essentially this lets us break up a constant secret across multiple authorities based on the user's GID, in such a way that each authority can compute his part independently given only the GID. The use of PRFs mean that each user's secret keys are independent of any other user's keys, and collusion is impossible. Then the central authority gives the added keys necessary to ensure that if we compute each PRF on the same GID, we can always combine the results to obtain a fixed value, and thus it allows us to give ciphertexts that are independent of the GID. For a full description and explanation of this technique, see Section 4.

OTHER ABE SCHEMES. As mentioned above, our scheme is an extension of the basic Fuzzy IBE scheme of Sahai and Waters. Their scheme requires that a user have t out of n of the desired attributes in order to decrypt. More recently, Gopal et al. presented a scheme for fine grained access control in the Key-Policy model [GPSW06]. In this model, when a user requests a private key, the authority determines what combinations of attributes must be present in order for this user to decrypt and gives the user the corresponding private key.

The main difference is that in this system, the private key no longer corresponds to a simple set of attributes that the user possesses. Instead, each private key represents a formula describing which sets of attributes must appear on the ciphertext in order for this user to decrypt. Ciphertexts are encrypted with a simple set of attributes.

Our techniques can also be applied to this more complex scheme to form a system in which, in order to decrypt a ciphertext encrypted with a set of attributes for each authority, a user must have received from each authority a policy which allows decryption for that set of attributes. Gopal et al. also present a large universe access structure scheme (an extension of the large universe scheme in SW). This also can be combined with our techniques to create a multiauthority large universe access structure scheme. For details, see Section 5.

2   Preliminaries

In our ABE scheme, we assume that the universe of attributes can be partitioned into K disjoint sets. Each will be monitored by a different authority. As mentioned above, we also have one trusted central authority who does not monitor any attributes.

Note: In the following we use $\mathcal{A}_u$ to denote the attribute set of user u and $\mathcal{A}_C$ to denote the attribute set of a ciphertext. $\mathcal{A}_u^k$ and $\mathcal{A}_C^k$ are the attributes handled by authority k in the attribute sets of the user and the ciphertext respectively.

A MultiAuthority ABE system is composed of K attribute authorities and one central authority. Each attribute authority is also assigned a value d_k. The system uses the following algorithms:

Setup: A randomized algorithm which must be run by some trusted party (e.g. central authority). Takes as input the security parameter. Outputs a public key, secret key pair for each of the attribute authorities, and also outputs a system public key and master secret key which will be used by the central authority.

Attribute Key Generation: A randomized algorithm run by an attribute authority. Takes as input the authority's secret key, the authority's value d_k, a user's GID, and a set of attributes in the authority's domain $\mathcal{A}_C^k$. (We will assume that the user's claim of these attributes has been verified before this algorithm is run). Output secret key for the user.

Central Key Generation: A randomized algorithm run by the central authority. Takes as input the master secret key and a user's GID and outputs secret key for the user.

Encryption: A randomized algorithm run by a sender. Takes as input a set of attributes for each authority, a message, and the system public key. Outputs the ciphertext.

Decryption: A deterministic algorithm run by a user. Takes as input a ciphertext, which was encrypted under attribute set $\mathcal{A}_C$ and decryption keys for an attribute set $\mathcal{A}_u$. Outputs a message m if $|\mathcal{A}_C^k \cap \mathcal{A}_u^k| > d_k$ for all authorities k.

Note that the number of authorities in the system need not be fixed permanently: it is possible to allow the central authority to add additional attribute authorities to the system at any point. For a discussion of this and other possible extensions to this scheme, see Section 6.

As in [SW05], our scheme is proved secure in the selective ID (sid) model, in which the adversary must provide the identity he wishes to attack(the challenge identity) before receiving the parameters of the system.

Let κ be the security parameter. We require that the number of authorities, K, and the number of attributes monitored by each authority, n_k, be upper bounded by a number n which is polynomial in κ.

Consider the following game:

Setup

- The adversary sends a list of attribute sets $\mathcal{A}_C = \mathcal{A}_C^1 \ldots \mathcal{A}_C^K$, one for each authority. He must also provide a list of corrupted authorities which cannot include the central authority.
- The challenger generates parameters for the system and sends them to the adversary. This means the system public key, public keys for all honest authorities, and secret keys for all corrupt authorities.

Secret Key Queries

- The adversary can make as many secret key queries as he wants to the attribute authorities or to the central authority. The only requirements are (1) that for each GID, there must be at least one honest authority k from which the adversary requests fewer than d_k of the attributes given in $\mathcal{A}_C^k$, i.e. the adversary never requests enough attributes to decrypt the challenge ciphertext, and (2) that the adversary never queries the same authority twice with the same GID (see below for discussion).

Challenge

- The adversary sends two messages M_0 and M_1.
- The challenger chooses a bit b, computes the encryption of M_b for attribute set $\mathcal{A}_C$, and sends this encryption to the adversary.

More Secret Key Queries

- The adversary may make more secret key queries subject to the requirements described above.

Guess

- The adversary outputs a guess b' that message $M_{b'}$ has been encrypted.

The adversary is said to succeed if he can correctly identify the encrypted message, i.e. if $b = b'$.

Definition 1. *A multiauthority attribute scheme is sid-secure if there exists a negligible furnction ν such that, in the above game any adversary will succeed with probability at most $1/2 + \nu(\kappa)$.*

Note that our scheme is designed only for static attributes: each authority will only issue one set of secret keys for each GID. If a user later returns with the same GID but a different set of attributes, the authority will refuse the request. However this can easily be converted into a scheme which allows changes in attributes by allowing each user a range of GID instead of just one. Then when a user needs to change his attribute set, he simply moves on to a new GID and requests secret keys from each authority with the new attribute set and new GID (he must however obtain new secret keys from *all* authorities).

We have found no obvious attack when this requirement is removed; it seems to be an artifact of our proof techniques. Essentially, in our reduction, when we give out secret keys from a certain authority, we need to know whether the adversary will request sufficient attributes from that authority to decrypt the challenge ciphertext. Our reduction responses will depend crucially on that factor. (For more details, see Section 4.)

Definition 2 (Bilinear Diffie-Hellman(BDH) Assumption). *Let G be a group of prime order q and generator g where $|q|$ is proportional to the security parameter κ. There exists a negligible function ν such that for all adversaries A, given G, q, g, g^a, g^b, g^c and bilinear map e for randomly chosen $a, b, c \leftarrow Z_q$, A can distinguish $e(g,g)^{abc}$ from $e(g,g)^R$ for random $R \leftarrow Z_q$ with probability at most $\nu(\kappa)$.*

3   Single Authority ABE

We will begin by demonstrating how the simplest attribute based encryption (ABE) scheme, Sahai and Waters' "Fuzzy IBE" or "Threshold ABE" scheme, can be converted into a multiauthority scheme. Then in Section 5 we will describe a multi authority scheme for more complex ABE.

We will now explain some of the intuition behind our scheme. We will incrementally build up to the full robust multiauthority construction. We begin by examining the single authority case, which was considered by Sahai and Waters [SW05]. In their scheme, there is one authority giving out secret keys for all of the attributes. Each encryptor then specifies a list of attributes such that any user with at least d of those attributes will be able to decrypt. They show that the scheme they present is sid secure.

We review these results and attempt to explain a possible derivation, building up to the description of the SW scheme through a series of incomplete schemes. We will then show how this can be easily converted into a multiauthority scheme. We hope that once the intuition behind SW is completely clear, the changes necessary to convert this scheme into a multiauthority one will also be fully intuitive.

STEP ONE – FELDMAN VSS
First, let's consider a very simplified scheme based on the Feldman Verifiable Secret Sharing scheme [Fel87].

Recall that, given d points $p(1), \ldots, p(d)$ on a $d - 1$ degree polynomial, we can use Lagrange interpolation to compute $p(i)$ for any i. However, given only $d - 1$ points, any other points are information theoretically hidden. According to the Lagrange formula, $p(i)$ can be computed as a linear combination of d known points. Let $\Delta_j(i)$ be the coefficient of $p(j)$ in the computation of $p(i)$. Then $p(i) = \sum_{j \in S} p(j) \Delta_j(i)$ where S is a set of any d known points and $\Delta_j(i) = \prod_{k \in S, j \neq k} (i - k)/(j - k)$. Note that any set of d random numbers defines a valid polynomial, and given these numbers we can find any other point on that polynomial.

Furthermore, if we are instead given $g^{p(1)}, \ldots, g^{p(d)}$, we can similarly compute $g^{p(i)}$ for any i, and the hiding property mentioned above still applies.

This suggests a technique for attribute based encryption: If a user has attribute i, his secret key will include $g^{p(i)}$, for some degree $d - 1$ polynomial p. We can encrypt a message m by giving $g^{p(0)}m$. Then any user with at least d attributes can interpolate to obtain the secret $g^{p(0)}$ and thus discover m. However, to any user without d attributes $g^{p(0)}$ is information theoretically hidden and thus finding m will be impossible.

Note that we can easily extend this to prevent collusion: If we give all our users points from the same polynomial, any group with at least d attributes between them would be able to combine their keys to find $p(0)$. However, if we instead give each user u a different polynomial p_u (but still with the same zero point $p_u(0) = p(0)$), then one user's points will give no information on the polynomial held by the other (as long as neither has more than $d - 1$ points). To see this, note that, given any $d-1$ points on polynomial p_1 and any $d-1$ points on polynomial p_2, with the requirement that these polynomials must intersect at 0, it is still the case that any value for $y = p_1(0) = p_2(0)$ will define a valid pair of polynomials. Thus, y is information theoretically hidden. Then our first scheme runs a follows:

Init First fix $y \leftarrow Z_q$.
SK for user u: Choose a random polynomial p such that $p(0) = y$. SK: $\{D_i = g^{p(i)}\}_{\forall i \in \mathcal{A}_u}$.
Encryption: $E = g^y m$.
Decryption: Use d SK elements D_i to interpolate to obtain $Y = g^{p(0)} = g^y$. Then $m = E/Y$.

STEP TWO – SPECIFYING ATTRIBUTES
If we take this approach, any user with any d attributes will be able to decrypt. But we want each encryptor to be able to give a specific subset of attributes such that at least d are necessary for decryption.

In order to do this, we need an extra tool: bilinear maps. Recall that for bilinear map e, $g \in G_1$, and $a, b \in Z_q$, $e(g^a, g^b) = e(g, g)^{ab}$.

Now, suppose instead of giving each user $g^{p(i)}$ for each attribute i, we choose a random value t_i and give $g^{p(i)/t_i}$. If the user knew g^{t_i} for at least d of these attributes, he could compute $e(g, g)^{p(i)}$ for each i and then interpolate to find the secret $e(g, g)^{p(0)}$. Then if our encryption includes $e(g, g)^{p(0)} m$, the user would be able to find m. Thus, the encryptor can specify which attributes are relevant by providing g^{t_i} for each attribute i in the desired set.

Suppose we only give one secret key to one user u. Now, for $i \in \mathcal{A}_u, i \notin \mathcal{A}_C$ the t_i values appear only once: when we give $g^{p(i)/t_i}$. Thus, since t_i was chosen at random, $p(i)$ is still information theoretically hidden. The only attributes i for which user u has any information on $p(i)$ are those where $i \in \mathcal{A}_u \cap \mathcal{A}_C$. As long as there are less than d of these, $p(0)$ (and thus $e(g, g)^{p(0)}$) must be information theoretically hidden.

If we allow multiple secret key queries, this is no longer the case. However given the BDH Assumption, we can show that $e(g, g)^{p(0)}$ is still hidden as long as no user has more than $d-1$ attributes in common with the ciphertext. This will be a special case of the proof in the next step. The resulting scheme is as follows:

Init First fix $y, t_1, \ldots, t_n \leftarrow Z_q$. Let $Y = e(g, g)^y$.
SK for user u: Choose a random polynomial p such that $p(0) = y$. SK:$\{D_i = g^{p(i)/t_i}\}_{\forall i \in \mathcal{A}_u}$.
Encryption for attribute set $\mathcal{A}_C$: $E = Ym$ and $\{E_i = g^{t_i}\}_{\forall i \in \mathcal{A}_C}$.
Decryption: For d attributes $i \in \mathcal{A}_C \cap \mathcal{A}_u$, compute $e(E_i, D_i) = e(g, g)^{p(i)}$. Interpolate to find $Y = e(g, g)^{p(0)} = e(g, g)^y$. Then $m = E/Y$.

STEP THREE: MULTIPLE ENCRYPTIONS
There are several obvious problems with this scheme. First, we would like to be able to encrypt multiple times without the decryptor needing to get a new

secret key each time. But, once a user has obtained $e(g,g)^{p(0)}$, he can decrypt any subsequent encryptions whether or not he has the appropriate attribute set.

What if instead of giving $e(g,g)^{p(0)}m$ in our encryption, we give $e(g,g)^{p(0)s}m$, where s is a different random number for each encryption? If we also give $\{E_i = g^{t_i s}\}_{\forall i \in \mathcal{A}_C}$ instead of $\{E_i = g^{t_i}\}$, the above process will allow a user with the appropriate attributes to find $e(g,g)^{p(0)s}$, and thus to decrypt m. Note that now our secret $e(g,g)^{p(0)s}$ is different for each ciphertext.

This also solves another of our problems: it gives us a way to compute the ciphertext. Before, computing the ciphertext required knowing the g^{t_i} values and $e(g,g)^{p(0)}$, which was in turn enough to decrypt any message. Now, decrypting a message requires knowing $g^{t_i s}$ for the appropriate attributes i and the appropriate s. Thus, we can now publish $T_i = g^{t_i}$ and $Y = e(g,g)^{p(0)}$ as the public key, and each encryptor can choose his own random s to compute $\{E_i = T_i^s\}_{\forall i \in \mathcal{A}_C}$, and $Y^s m$.

Furthermore, we can show that $e(g,g)^{p(0)s}$ is still hidden, even when the user knows T_i for all i, a set $\{T_i^s\}_{\forall i \in \mathcal{A}_C}$, and adaptively chose secret keys for user u with $|\mathcal{A}_u \cap \mathcal{A}_C|$ at most $d-1$. Thus, we can show this scheme is sid secure based on the BDH Assumption. We have now reconstructed the SW scheme:

Init First fix $y, t_1, \ldots, t_n \leftarrow Z_q$.

PK for system $T_1 = g^{t_1} \ldots T_n = g^{t_n}, Y = e(g,g)^y$. $PK = \{T_i\}_{1 \le i \le n}, Y$

SK for user u: Choose a random polynomial p such that $p(0) = y$. SK: $\{D_i = g^{p(i)/t_i}\}_{\forall i \in \mathcal{A}_u}$.

Encryption for attribute set $\mathcal{A}_C$: $E = Y^s = e(g,g)^{ys}m$ and $\{E_i = g^{t_i s}\}_{\forall i \in \mathcal{A}_C}$.

Decryption: For d attributes $i \in \mathcal{A}_C \cap \mathcal{A}_u$, compute $e(E_i, D_i) = e(g,g)^{p(i)s}$. Interpolate to find $Y^s = e(g,g)^{p(0)s} = e(g,g)^{ys}$. Then $m = E/Y^s$.

4   Multiple Authorities

Now we consider the multiauthority case. Once again, we will build up our construction by first considering a series of incomplete schemes.

As a first thought, we might simply have many copies of SW, one for each authority. We want to require that a user be able to decrypt a ciphertext only if he has at least d of the specified attributes from each of the K authorities. Recall that the SW scheme centers around finding enough polynomial shares $e(g,g)^{p(i)s}$ to reconstruct the secret $e(g,g)^{p(0)s} = e(g,g)^{ys}$ which has been used to blind the message. (Recall that the encryption includes $E = e(g,g)^{ys}m$). Now, if we want each authority to give out its own polynomials, one simple solution might be to do an additive secret sharing to form the SW secrets (i.e. the values y such that every random polynomial p is chosen with $p(0) = y$). Thus, we pick a random value for the master secret y_0 and for each authority $k = 1 \ldots K$, y_k is a share of y_0 so $\sum y_k = y_0$. We can output $e(g,g)^{y_0}$ as the entire system's public key. Then to encrypt message m, a user gives $E = e(g,g)^{y_0 s}m$ and $E_{k,i} = T_{k,i}^s$ for all i,k where they wish to allow a decryptor to use attribute

i from authority k. To decrypt, the user has to perform SW decryption for each authority and find $Y_k^s = e(g,g)^{y_k s}$, then multiply the results together to get $\prod_{k=1}^{K} Y_k^s = \prod_{k=1}^{K} e(g,g)^{y_k s} = e(g,g)^{s \sum_{k=1}^{K} y_k} = e(g,g)^{y_0 s}$ and thus obtain m. However, if a user does not have enough of the required attributes from one authority $\hat{k}$, then the SW secret for that authority: $Y_{\hat{k}} = e(g,g)^{y_{\hat{k}} s}$ will remain indistinguishable from random and thus so will $e(g,g)^{y_0 s}$ and m. Thus our first attempt Multi Authority Scheme is as follows:

System

Init First fix $y_1 \ldots y_k, \{t_{k,i}\}_{i=1\ldots n, k=1\ldots K} \leftarrow Z_q$. Let $y_0 = \sum_{k=1}^{K} y_k$.

System Public Key $Y_0 = e(g,g)^{y_0}$.

Attribute Authority k

Authority Secret Key The SW secret key: $y_k, t_{k,1} \ldots t_{k,n}$.

Authority Public Key $T_{k,i}$ from the SW public key: $T_{k,1} \ldots T_{k,n}$ where $T_{k,i} = g^{t_{k,i}}$.

Secret Key for User u **from authority** k Choose random $d-1$ degree polynomial p with $p(0) = y_k$. Secret Key: $\{D_{k,i} = g^{p(i)/t_{k,i}}\}_{i \in \mathcal{A}_u}$.

Encryption for attribute set $\mathcal{A}_C$ Choose random $s \leftarrow Z_q$. Encryption: $E = Y_0^s m, \{E_{k,i} = T_{k,i}^s\}_{i \in \mathcal{A}_C^k, \forall k}$.

Decryption: For each authority k, for d attributes $i \in \mathcal{A}_C^k \cap \mathcal{A}_u$, compute $e(E_{k,i}, D_{k,i}) = e(g,g)^{p(i)s}$. Interpolate to find $Y_k^s = e(g,g)^{p(0)s} = e(g,g)^{y_k s}$. Combine these values to obtain $\prod_{k=1}^{K} Y_k^s = Y_0^s$. Then $m = E/Y_0^s$.

There is a problem with the scheme as described above: Suppose an encryptor encrypts a message to the attribute set $\mathcal{A}_C$ which includes attributes $\mathcal{A}_C^k$ for each authority k. Now suppose we have a set of K users where each user k has attribute set $\mathcal{A}_u = \mathcal{A}_C^k$ from authority k, but no attributes from any other authority. Recall that we want to allow decryption only if the decryptor has enough of the required attributes from *every one* of the authorities. However, if the scheme is as described above, this set of users will be able to collude: Each user k will use his attribute set to find the SW secret for authority k: $Y_k^s = e(g,g)^{y_k s}$. Then the users combine these values to obtain $\prod_{k=1}^{K} Y_k^s = \prod_{k=1}^{K} e(g,g)^{y_k s} = e(g,g)^{y_0 s} = Y_0^s$ and thus m.

Clearly, if there is no way to identify users beyond their attribute sets, then the above collusion is impossible to prevent: to the authorities, k separate users each with attribute set $\mathcal{A}_C^k$ and one user with attribute set $\mathcal{A}_C$ look identical.

We solve this problem by requiring that each user have a unique global identifier (GID), as described in Section 1. A user must present the same GID to each authority in order to receive a coherent set of keys (and presumably prove to each authority that the GID is valid). However, encryption will, as before, only specify a set of attributes of which d will be required to decrypt. Thus, the ability to decrypt is independent of the GID (except in that all secret keys must have been obtained for the same GID).

Now that we can distinguish different users, we need some way to ensure that different users cannot combine their results from different authorities. Suppose we have each authority k choose a different random $y_{k,u}$ value for each user. Let $y_{u,0}$ be the master secret for user u. If user u finds $e(g,g)^{y_{k,u}s}$ and shares it with user u' in an attempt to collude, it won't give user u' any information on his master secret $e(g,g)^{y_{0,u'}s} = \prod_{k=1}^{K} e(g,g)^{y_{k,u'}s}$ (since $y_{k,u}$ is independent of $y_{k,u'}$), so the above collusion will no longer be possible. Recall that the SW scheme uses different polynomials to split up the secret (y) in a different way for each user, thus preventing collusion. Now we are using a similar technique to prevent collusion across authorities: we choose a new set of $y_{k,u}$ values and divide the secret (y_0) among the authorities in a different way for each user.

Note that we need to include $\prod_{k=1}^{K} e(g,g)^{y_{k,u}} = e(g,g)^{y_{0,u}}$ in the public key so that it can be used to form an encryption $(E = e(g,g)^{y_{0,u}s}m)$. Moreover, recall that these ciphertexts must be independent of the identity of the user – we would like the ability to decrypt to depend only on the attributes. This means, in order for the encryption/decryption to work with this new addition, we would need $\prod_{k=1}^{K} e(g,g)^{y_{k,u}} = e(g,g)^{y_{0,u}}$ to be the same for all users.

But if all authorities choose $y_{k,u}$ independently, how can we ensure that $\sum_{k=1}^{K} y_{k,u} = y_0$ for all u? It would seem that we must need some kind of communication between authorities, and our goal of k autonomous authorities is impossible with this approach.

An alternative might be to allow our authorities to share some state. If one of the authorities knew the other authorities' random choices, he could choose his $y_{k,u}$ values to ensure that $\sum_{k=1}^{K} y_{k,u} = y_0$. However, we don't necessarily want to require that any of our attribute authorities be completely trusted, so we may not want them to share this information.

Thus, we add the additional "central" authority (see Section 1), who handles no attributes, but who must be fully trusted. This authority will be allowed to know some of the state of each of the other authorities. In particular, it will know enough of their secret state to reconstruct $y_{k,u}$ for any user u and for all authorities k. It will use this information to provide a secret key which, when combined with a value g^s to be given in the encryption and with the "secrets" $Y_{k,u}^s$ obtained from each of the other authorities, will give user u the "master secret": $Y_0^s = e(g,g)^{y_0s}$ which can then be used to obtain m. Now we only need to trust one authority and it need not be one of the attribute authorities. [1]

Finally, instead of using truly random values, we have each of our K authorities choose the $y_{k,u}$ values using a pseudorandom function (PRF). Thus, now the central authority has only to store the seeds of all of the PRFs.

Final MultiAuthority Scheme: (changes from previous schemes are underlined.)

---

[1] Note, we could require that one of the attribute authorities be trusted and have it maintain the state information of all the other authorities. However, we chose to separate these functions in order to consider a more general case. The central authority could easily be combined with a trusted attribute authority.

System

    Init Fix prime order groups G, G_1, bilinear map $e : G \rightarrow G_1$, and generator $g \in G$. Choose seeds $s_1, \ldots, s_K$ for for all authorities. Also choose $y_0, \{t_{k,i}\}_{k=1\ldots K, i=1\ldots n} \leftarrow Z_q$.

    System Public Key $Y_0 = e(g, g)^{y_0}$.

Attribute Authority k

    Authority Secret Key $s_k, t_{k,1} \ldots t_{k,n}$.

    Authority Public Key $T_{k,1} \ldots T_{k,n}$ where $T_{k,i} = g^{t_{k,i}}$.

    Secret Key for User u Let $y_{k,u} = F_{s_k}(u)$. Choose random $d - 1$ degree polynomial p with $p(0) = y_{k,u}$. Secret Key: $\{D_{k,i} = g^{p(i)/t_{k,i}}\}_{i \in \mathcal{A}_u}$.

Central Authority

    Central Authority Secret Key s_k for all authorities k, y_0.

    Secret Key for User u Let $y_{k,u} = F_{s_k}(u)$ for all k. Secret Key: $D_{CA} = g^{(y_0 - \sum_{k=0}^{K} y_{k,u})}$.

Encryption for attribute set $\mathcal{A}_C$ Choose random $s \leftarrow Z_q$. $E = Y_0^s m$, $E_{CA} = g^s$, $\{E_{k,i} = T_{k,i}^s\}_{i \in \mathcal{A}_C^k, \forall k}$.

Decryption: For each authority k, for d attributes $i \in \mathcal{A}_C^k \cap \mathcal{A}_u$, compute $e(E_{k,i}, D_{k,i}) = e(g, g)^{p(i)s}$. Interpolate to find $Y_{k,u}^s = e(g, g)^{p(0)s} = e(g, g)^{y_{k,u}s}$ for each authority k. Compute $Y_{CA}^s = e(E_{CA}, D_{CA})$. Combine these values to obtain $Y_{CA}^s * \prod_{k=1}^{K} Y_k^s = Y_0^s$. Then $m = E/Y_0^s$.

Theorem 1. *This scheme is sid-secure according to the definition in Section 2.*

First we give some main points of intuition behind the reduction. Then we follow with a more formal proof.

BASIS OF SW REDUCTION. We will show that we can reduce the BDH problem to the problem of breaking our encryption scheme. That means we are given $A = g^a, B = g^b, C = g^c$ and asked to distinguish $e(g, g)^{abc}$ from $e(g, g)^R$ for a random $R \leftarrow Z_q$. We assume there exists an adversary that can break the security properties of our multiauthority system (as defined in Section 2) and we show that we could use such an adversary to solve this problem.

    We want to show that, even given a challenge encryption and adaptively chosen secret key queries, in our challenge encryption, $M_b = E/e(g, g)^{y_0 s}$ is indistinguishable from a random message (which means the adversary can have no more than negligible probability of correctly identifying b). We will show that $e(g, g)^{y_0 s}$ is indistinguishable from a random element of G_2. Since we are basing our reduction on the BDH assumption, this means we want to implicitly set $y_0 s = abc$. We need to be able to output $e(g, g)^{y_0}$ and g^s as part of the central public key, so we will implicitly set $s = c$ and $y_0 = ab$. (These values cannot be computed, but we will use them to determine the other values in our reduction.)

EXTENSION TO MULTIPLE AUTHORITIES. Note that the adversary is allowed to request secret keys for a given user u and attribute set $\mathcal{A}_u$ as long as there remains one honest authority k such that $\mathcal{A}_C^k \cap \mathcal{A}_u^k < d$, i.e. the user has insufficient attributes from this authority to decrypt. Thus, in the worst case, for all but one authority k, the adversary will be able to compute $Y_{k,u}^s = e(g, g)^{y_{k,u}s}$ for additive share $y_{k,u}$. Every user will also be able to compute $Y_{CA,u}^s = e(g^s, D_{CA}) =$

$e(g^s, g^{y_0 - \sum y_{k,u}})$. We need $\prod e(g,g)^{y_{k,u}s} * e(g^s, D_{CA}) = e(g,g)^{y_0 s}$ to be something which the adversary cannot compute (in particular, $e(g,g)^{abc}$ which is indistinguishable from random). Thus, we must "hide" this incomputable value in the share $y_{k,u}$ for the one authority from which the adversary does not have sufficient attributes. Let $\hat{k}$ be this authority. Then we will implicitly set $y_{\hat{k},u} = ab + z_{k,u}b$ for random $z_{k,u}$. For all other honest authorities, we need $e(g,g)^{y_{k,u}s}$ to be computable, so we will implicitly set $y_{k,u} = z_{k,u}b$ for some random $z_{k,u}$ (this particular choice will be explained below). Note that $\hat{k}(u)$ may be a different authority for each user u. This is where the reduction will make use of the PRFs. Since to the adversary the PRFs for honest authorities are indistinguishable from true random functions, we can implicitly replace these PRFs with the necessary values for each user u and authority $k = \hat{k}(u)$ or $k \neq \hat{k}(u)$, and the result will be indistinguishable. (The values will still be randomly distributed).

ANSWERING SECRET KEY QUERIES. Note that in order for our reduction to succeed, we need to generate public keys for honest authorities k so that:

- When $k = \hat{k}(u)$, we can output secret keys such that this authority's secret $(q(0) = F_{s_k}(u))$ is uncomputable (eg. $ab + z_{k,u}$), given that the user u does not have sufficient attributes from this authority. This situation is identical to that in a single authority security reduction.
- When $k \neq \hat{k}(u)$, using the *same* public keys, we can output secret keys for users who know (potentially) all of the attributes from this authority as long as this authority's secret is generated appropriately. In this case we set the secret $q(0) = F_{s_k}(u) = z_{k,u}b$.

For $i \in \mathcal{A}_C^k$, we must be able to output $T_{k,i} = g^{t_{k,i}}$ in the public key and $E_{k,i} = T_{k,i}^s = g^{t_{k,i}s} = g^{t_{k,i}c}$ in the challenge ciphertext. Thus, we choose $t_{k,i} = \beta_{k,i}$ for known random $\beta_{k,i}$. For $i \notin \mathcal{A}_C$, we set $t_{k,i} = \beta_{k,i}b$ for known random, $\beta_{k,i}$. (Note that for these attributes $E_{k,i} = T_{k,i}^s$ is not computable).

Consider the second case: We need to output $D_{k,i} = g^{p(i)/b\beta_{k,i}}$ and $D_{k,i} = g^{p(i)/\beta_{k,i}}$, where we require $p(0) = bz_{k,u}$. If we simply choose p in terms of b, (eg. $p = b\rho$ for known random $d-1$ degree polynomial ρ), this is trivial.

The first case, as mentioned above, follows the original single authority reduction almost exactly. The only difference is that now we have an extra randomizing term added $(F_{s_k}(u) = ab + z_{k,u}b$ instead of $F_{s_k}(u) = ab)$. For up to $d-1$ points $(i \in \mathcal{A}_C^k \cap \mathcal{A}_u^k)$, we need to give secret keys $D_{k,i} = g^{p(i)/t_{k,i}} = g^{p(i)/\beta_{k,i}}$, where $p(0) = ab + z_{k,u}b$. This might seem difficult since we can't compute g^{ab}. However, recall that, using interpolation, we can pick any d points and use them to define a polynomial. So, for the attributes i in the challenge, we will set $p(i)$ to be a random multiple of b. For these attributes $D_{k,i} = g^{p(i)/\beta_{k,i}}$ will be a computable multiple of B. Since we have also fixed $p(0)$, we have now chosen d points on the polynomial, so now for the remaining attributes, we can interpolate to find $g^{p(i)/t_{k,i}}$ as a weighted product of $g^{p(0)/t_{k,i}}$ and $g^{p(j)/t_{k,i}}$ for each of the fixed attributes j (those in the challenge ciphertext). Recall that for these attributes, we have $t_{k,i} = \beta_{k,i}b$. Thus, for each of the fixed attributes j, $g^{p(j)/t_{k,i}}$ will be a computable multiple of B, and for $p(0)$, $g^{p(0)/t_{k,i}} = g^{(ab+z_{k,u}b)/b\beta_{k,i}} = g^{(a+z_{k,u})/\beta_{k,i}}$

is a computable combination of A and g. That means $D_{k,i} = g^{p(i)/t_{k,i}}$ will be also computable for all $i \in \mathcal{A}_u^k - \mathcal{A}_C^k$.

Proof (of Theorem 1). Suppose there exists and adversary that plays the security game as described in Section 2 and succeeds with nonnegligible probability ϵ. Then we will show that we can use such an adversary to break the BDH assumption.

First, we assume that the adversary will still succeed with the same advantage even when the PRF F_{s_k} is replaced by a truly random function for each honest authority k. Note that if this is not the case, then we can distinguish the PRF from random, contradicting the definition of a PRF.

We need to specify how our reduction responds in each stage of the game (as described in Section 2). Our reduction will behave as follows:

- Given $A = g^a, B = g^b, C = g^c$ and $Z = e(g,g)^{abc}$ or $e(g,g)^R$ for random $R \leftarrow Z_q$
- Receive A_C and list of corrupted authorities $Corr$ from adversary.
- Init:
  - System PK : $Y_0 = e(A, B)$ (implicitly set $y_0 = ab$.)
  - Honest Authority PK's: Choose random $\beta_{k,i}$; PK: $\{T_{k,i} = g^{\beta_{k,i}}\}_{i \in \mathcal{A}_u \cap \mathcal{A}_C^k}$, $\{T_{k,i} = B^{\beta_{k,i}}\}_{i \in \mathcal{A}_u - \mathcal{A}_C^k}$
  - Corrupt Authority k SK's: Choose random $t_{k,i} \leftarrow Z_q$, random PRF key s_k. SK:$s_k, \{t_{k,i}\}$
- SK queries: Let $\hat{k}(u)$ be the first authority k queried such that $|\mathcal{A}_u^k \cap \mathcal{A}_C^k| < d$.
  - SK queries for user u to Honest Attribute Authorities $k \neq \hat{k}(u)$: Recall that for these authorities we will implicitly set $p(0) = F_{s_k}(u) = z_{k,u}b$. Choose a random $z_{k,u}$ and choose a random polynomial ρ such that $\rho(0) = z_{k,u}$. We will implicitly set $p(i) = b\rho(i)$. Now for $i \in \mathcal{A}_C^k$, $t_{k,i} = \beta_{k,i}$, so $D_{k,i} = g^{p(i)/t_{k,i}} = g^{b\rho(i)/\beta_{k,i}} = B^{p(i)/\beta_{k,i}}$. For $i \notin \mathcal{A}_C^k$, $t_{k,i} = b\beta_{k,i}$, so $D_{k,i} = g^{p(i)/t_{k,i}} = g^{b\rho(i)/b\beta_{k,i}} = g^{\rho(i)/\beta_{k,i}}$. SK: $\{D_{k,i} = B^{\rho(i)/\beta_{k,i}}\}_{i \in \mathcal{A}_u^k \cap \mathcal{A}_C^k}, \{D_{k,i} = g^{\rho(i)/\beta_{k,i}}\}_{i \in (\mathcal{A}_u^k - \mathcal{A}_C^k)}$
  - SK queries for user u to Honest Attribute Authorities $k = \hat{k}(u)$: Recall that for authority $\hat{k}$ for user u, we will choose random $r_{k,u}$ and implicitly set $p(0) = F_{s_k}(u) = ab + z_{k,u}b$. Choose $d - 1$ random points v_i. For $i \in \mathcal{A}_C^k$, we will implicitly set $p(i) = v_i b$. For these attributes, $t_{k,i} = \beta_{k,i}$, so that means $D_{k,i} = g^{p(i)/t_{k,i}} = g^{bv_i/\beta_{k,i}} = B^{v_i/\beta_{k,i}}$. Recall that we need $p(0) = F_{s_k}(u) = ab + z_{k,u}b$, and we have now set $p(i) = v_i b$ for $d - 1$ other points. Thus, p is fully determined, and by interpolation, for any other attribute i, we have implicitly defined $\Delta_0(i)(ab + z_{k,u}b) + \sum \Delta_j(i)v_j b$. For these attributes $t_{k,i} = b\beta_{k,i}$, so $D_{k,i} = $
  $$g^{p(i)}/t_{k,i} = g^{\frac{\Delta_0(i)(ab+z_{k,u}b)+\sum \Delta_j(i)v_j b}{b\beta_{k,i}}} = g^{\Delta_0(i)a} * g^{\frac{\Delta_0(i)z_{k,u}+\sum \Delta_j(i)v_j}{\beta_{k,i}}} =$$
  $$A^{\Delta_0(i)} * g^{\frac{\Delta_0(i)z_{k,u}+\sum \Delta_j(i)v_j}{\beta_{k,i}}}$$
  SK: $\{D_{k,i} = B^{v_i/\beta_{k,i}}\}_{i \in \mathcal{A}_u^k \cap \mathcal{A}_C^k}$,
  $$\{D_{k,i} = A^{\Delta_0(i)} * g^{\frac{\Delta_0(i)z_{k,u}+\sum \Delta_j(i)v_j}{\beta_{k,i}}}\}_{i \in (\mathcal{A}_u^k - \mathcal{A}_C^k)}.$$

- SK queries for user u to Central Authority:
  $$D_{CA} = g^{\left(\sum_{k \notin Corr} z_{k,u} - \sum_{k \in Corr} F_{s_k}(u)\right)}$$
– Receive M_0, M_1 and pick a random b.
– Challenge Ciphertext: Zm, $E = g^s = C$, $\{C^{\beta_{k,i}}\}_{i \in A_C}$
– Guess for Z: Receive a guess b' and If $b = b'$ guess "$e(g,g)^{abc}$" otherwise guess "$e(g,g)^R$".

After making all his secret key queries, the adversary will send a guess b' that the message encrypted was $M_{b'}$. Note that, if $Z = e(g,g)^{abc}$, then the encryption given was a valid encryption of M_b, and the adversary should have his usual non-negligible advantage ϵ of correctly identifying M_b. However, if $Z = e(g,g)^R$, then the encryption was a completely random value, so the adversary can have no better than $1/2$ probability of guessing correctly. Thus, if the adversary guesses correctly, we guess that $Z = e(g,g)^{abc}$ and if he is wrong, we guess that $Z = e(g,g)^R$. If we analyze the probability that the reduction successfully distinguishes Z, we find that the reduction has an advantage of $\epsilon/2$. Thus an adversary which breaks this encryption scheme with advantage ϵ implies an algorithm for breaking the BDH Assumption with nonegligible advantage $\epsilon/2$. We can conclude that this encryption scheme is sid-secure.

Remark 1. Note that our reduction relies critically on the fact that, for each user u that the adversary queries about, we choose exactly one authority for which that user has less than d of the attributes in the challenge, and this is the authority for which we set $F_{s_k}(u) = a + z_{k,u}b$. Note that, if the adversary requests at least d attributes for u from this authority, even if none of them appears in the challenge, the value of $F_{s_k}(u)$ is completely determined by the secret keys that the authority returns (although it is not known, since the discrete logs t_i of the public key T_i are not known to the adversary.)

If we allowed the adversary to at some later point request a second set of attributes from this authority for this user, such that it overlapped by at least d with the challenge, we would not be able to give these secret keys in such a way that they would be consistent with the previously determined value of F_{s_k}. (Doing so would involve computing g^{ab}.)

To prevent this, we require that in order to change his attribute set, a user must also change his GID. However, our collusion resistance requires that keys from different authorities for different keys be incompatible. Thus, it is not clear how we could allow the adversary to change the attribute set for a user without obtaining a new key from all authorities using a completely new GID. See Section 2 for a discussion of possible ways to get around this problem.

5 MultiAuthority + Large Universe and Complex Access Structures

In [GPSW06], Goyal et al. showed a Large Universe Access Control Structure ABE scheme. We can also apply our techniques to this scheme to achieve a corresponding multiauthority system.

A large universe construction has the advantage that the size of the public keys is dependent only on the maximum number of attributes which can be required in an encryption. (In contrast, the basic SW scheme has public keys size proportional to the total number of attributes in the system.) Thus, this method allows a much larger universe of attributes. A large universe construction was first presented in [SW05].

Goyal et al. showed how to combine the large universe construction with a construction (GPSW) which allows the authority to give secret keys corresponding to a more general policy than simple t-out-of-n threshold . In particular, they allow any policy that can be described by an access tree. In such a tree, each leaf node corresponds to an attribute, and intermediate nodes are t-out-of-n gates for arbitrary values of t and n. (Thus, OR and AND gates are possible as special cases.) If we consider the input at each leaf node to be 1 if and only if the corresponding attribute is present in a given ciphertext, then evaluating this circuit will determine whether or not the user should be able to decrypt this ciphertext.

Below we give a multiauthority large universe access control structure ABE scheme. In this system, each authority will choose an access structure τ_k for the user and give him a corresponding secret key. A user will be able to decrypt a ciphertext with attribute set $\mathcal{A}_C$ if and only if all of his access structures τ_k output 1 when evaluated on the corresponding subset of the attributes, $\mathcal{A}_C^k$.

We use essentially the same techniques as in the simple multi-authority scheme. Each authority runs a copy of the single authority protocol, with a separate copy of the public key (in this case $t_{k,1}, \ldots t_{k,n+1}$). The master secret for each authority (y_k) is replaced by a PRF on the user's ID, so that the master secret for the entire system (y_0) can be divided among all the authorities in a different way for each user. Finally, a central authority is necessary to ensure that, for each user, the PRFs from all authorities can be combined to obtain the system secret.

MultiAuthority Scheme for Large Universe and Complex Access Structures

System

    Init Fix prime order groups G, G_1, bilinear map $e : G \to G_1$, and generator $g \in G$. Choose PRF keys $s_1 \ldots s_K$ and $y_0 \leftarrow Z_q$, and $g_2 \leftarrow G_1$.

    System Public Key $g_1 = g^{y_0}$.

Attribute Authority k

    Authority Secret Key s_k

    Authority Public Key $t_{k,1} \ldots t_{k,n+1} \leftarrow G_1$.

    Let $h(x)$ be the n degree polynomial defined by $t_{k,1}, \ldots t_{k,n+1}$. Also define $T_k(x) = g_2^{x^n} g^{h(x)} = g_2^{x^n} \prod_{i=1}^{n+1} t_{k,i}^{\Delta_i(x)}$

    Secret Key for User u **for access structure** τ_k Let $y_{k,u} = F_{s_k}(u)$. Now run the Key Generation from GPSW but with the master secret $y = y_{k,u}$ and with the points $t_{k,1}, \ldots t_{k,n+1}$, i.e. KeyGeneration($\tau$, $MK = y_{k,u}$ and $PK = g^{y_{k,u}}, g_2, t_{k,1}, \ldots t_{k,n+1}$): We choose a polynomial q_r for the root node of the tree τ_k such that $q_r(0) = y_{k,u}$. Then we choose random polynomials for all other nodes x such that $q_x(0) = q_{parent(x)}(x)$. Finally, for the leaf node x, we choose random $r_{k,x}$ and compute secret key elements: $D_{k,x} = g_2^{q_x(0)} T(i)^{r_{k,x}}$ where $i = att(x)$ and $R_{k,x} = g^{r_{k,x}}$.

SK: $D_{k,x}, R_{k,x}$ for all leaf nodes x in τ_k

Central Authority

  Central Authority Secret Key s_k for all authorities k, y_0.

  Secret Key for User u Let $y_{k,u} = F_{s_k}(u)$ for all k. Secret Key: $D_{CA} = g_2^{(y_0 - \sum_{i=0}^{K} y_{k,u})}$.

Encryption for attribute set $\mathcal{A}_C$ Choose random $s \leftarrow Z_q$. $E = e(g_1, g_2)^s m$, $E' = g^s$, $\{E_{k,i} = T_k(i)^s\}_{i \in \mathcal{A}_C^k, \forall k}$.

Decryption: For each authority k, run the Decryption algorithm as in GPSW, i.e. run DecryptNode on the root of the tree τ_k:

For leaf node x if we have $D_{k,x}$, this algorithm computes $\frac{e(D_{k,x}, E')}{e(R_{k,x}, E_{k,i})} =$

$\frac{e(g_2^{q_x(0)} T_k(i)^{r_{k,x}}, g^s)}{e(g^{r_{k,x}}, T_k(i)^s)} = \frac{e(g_2^{q_x(0)}, g^s) e(T_k(i)^{r_{k,x}}, g^s)}{e(g^{r_{k,x}}, T_k(i)^s)} = e(g, g_2)^{q_x(0)s}$. Then at each level of the tree, if we have successfully computed DecryptNode for sufficient children, it combines the results to obtain $e(g, g_2)^{q_x(0)s}$ for parent node x. Finally, if we have sufficient attributes for the tree τ_k, DecryptNode computes $e(g, g_2)^{q_r(0)s} = e(g, g_2)^{y_{k,u}s}$.

Now decryption proceeds as in the previous multiauthority scheme: If we have sufficient attributes to decrypt, then for every authority we will have computed $Y_{k,u}^s = e(g, g_2)^{y_{k,u}s}$. Next, use the key obtained from the central authority to compute $Y_{CA}^s = e(E', D_{CA}) = e(g^s, g_2^{y_0 - \sum_{i=1}^{K} y_{k,u}})$. Combine these values to obtain $Y_{CA}^s * \prod_{k=1}^{K} Y_k^s = e(g, g_2)^{y_0 s} = Y_0^s$. Then $m = E/Y_0^s$.

We will present only the key intuition for the proof of security. For a full proof, see the full version.

BASIS OF SINGLE AUTHORITY REDUCTION. We again want to show a reduction from BDH, so we want the quantity which we will use to blind the message, in this case $e(g_1, g_2)$, to be equal to $e(g, g)^{abc}$. Thus, we will set $g_1 = a, g_2 = b$, and implicitly, $s = c$.

EXTENSION TO MULTIPLE AUTHORITIES. Again here, as in the simple multiauthority scheme, for each user u, the adversary is allowed to request keys that are sufficient to decrypt the ciphertext from all but one authority. In this scheme, in the worst case, for all but one authority k, the adversary will be able to compute $Y_{k,u}^s = e(g_1, g_2)^{y_{k,u}s}$. Every user will also be able to compute $Y_{CA,u}^s = e(g^s, D_{CA}) = e(g, g_2)^{(y_0 - \sum y_{k,u})s}$. And, as mentioned above, we need $\prod e(g, g_2)^{y_{k,u}s} * e(g^s, D_{CA}) = e(g, g_2)^{y_0 s}$ to be an uncomputable value (in particular $e(g, g)^{abc}$). Thus, again we must "hide" this incomputable value in the share $y_{k,u}$ for the one authority $\hat{k}$ from which this user does not have sufficient attributes. Thus, we set $y_{\hat{k},u} = a + z_{\hat{k},u}$ and for all other honest authorities, $y_{k,u} = z_{k,u}$ for known random $z_{k,u}$. Once again, we make use of the pseudorandomness of the PRFs to claim that, since all these values are distributed randomly, the result will be indistinguishable from computing the values using a true PRF.

ANSWERING SECRET KEY QUERIES. Once again, we must show that, for all users u, we can form public keys for all honest authorities k such that:

- When $k = \hat{k}(u)$, we can output secret keys such that this authority's secret $(q(0) = F_{s_k}(u))$ is not known (eg. $a + z_{k,u}$), given that the user u does not have sufficient attributes from this authority. This situation is identical to that in a single authority security reduction.
- When $k \neq \hat{k}(u)$, using the *same* public keys, we can output secret keys for users who know (potentially) all of the attributes form this authority as long as this authority's secret is generated appropriately. In this case we set the secret $q(0) = F_{s_k}(u) = z_{k,u}$.

We set up the $t_{k,i}$ values for each authority k as in the GPSW reduction. This is done in such a way that if $i \in \mathcal{A}_C^k$, $T_k(i)^s$ is computable (recall that this must be part of the encryption), but $D_{k,x} = g_2^{q_x(0)} T(i)^{r_{k,x}}, R_{k,x} = g^{r_{k,x}}$ is computable if and only if we know $q_x(0)$. If $i \notin \mathcal{A}_C^k$, $T_k(i)^s$ will not be computable, but we will be able to compute $D_{k,x}, R_{k,x}$ when given only $g^{q_x(0)}$.

Now, in the first situation, we can proceed as in the single authority reduction, and run the PolyUnsat algorithm [GPSW06] to set the polynomials for each level of the tree so that at the root r, $q_r(0) = a + z_{k,u}$. At the leaves, this will make $q_x(0)$ completely known for nodes corresponding to $i \in \mathcal{A}_C$ and will make $g^{q_x(0)}$ known for $i \notin \mathcal{A}_C$. According to the setup above, this lets us form $D_{k,x}, R_{k,x}$ for all required leaf nodes x. In the second case, we simply proceed as in the real protocol.

Finally, the challenge encryption can be formed as $E = ZM_b, E' = C, \{E_{k,i} = T(i)^s\}$ for random bit b. The reduction's output and the analysis proceed as in the previous reduction.

Remark 2. Note that we cannot allow changes in a user's access structure (without corresponding changes in the user's GID) for the same reason that in the threshold multiauthority scheme we cannot allow changes in a user's attribute set (see Section 4).

6   Extensions

We briefly describe several possible extensions to our scheme. For security proofs, see full version.

CHANGING d_k SW noted that one could easily extend their scheme to allow d, the number of attributes in the ciphertext required to decrypt, to vary with each encryption. Essentially, the scheme would be instantiated with $d = dmax$, the maximum overlap one might want to require. We would also extend the attribute set by adding $dmax$ dummy attributes and each user would get a secret key element D_i for each of these new attributes. We refer to the set of dummy attributes as $\mathcal{A}_D$. If the encryptor wanted to require $d' < dmax$ of the attributes in the ciphertext, he could include $E_i = T_i^s$ for $dmax - d'$ dummy attributes $i \in \mathcal{A}_D$.

Note that in our mulitauthority scheme a similar approach would allow one to choose exactly how many of the attributes given in the ciphertext to require from each authority, i.e to vary the values d_k. We need only add $dmax_k$ dummy attributes for each authority k and proceed as described above.

LEAVING OUT AUTHORITIES. In the mulitauthority scheme as stated, each user must go to every authority before he can decrypt any message. Using the technique above, an encryptor can allow decryption by someone who has none of the attributes handled by a specific authority by including all of the dummy attributes for that authority in the encryption set. However, a user would still have to go to that authority to obtain secret keys for these dummy values in order to decrypt. There could be an arbitrarily large number of authorities in the system, so this might involve a lot of work.

We can remove this requirement by adding one "authority attribute k" for each authority. We would add a corresponding $T_{Nk} = g^{t_{Nk}}$ to the public key and the central authority would give every user u a secret key for each authority: $D_{Nk} = g^{y_{k,u}/t_{Nk}}$. To encrypt without requiring any attributes from authority k, the encryptor would include T_{Nk}^s in the encryption. A user could then combine this with his D_{Nk} value and obtain $Y_{k,u}^s = e(g,g)^{y_{k,u}s}$, thus bypassing the need for any attributes from authority k.

MORE COMPLICATED FUNCTIONS OF THE AUTHORITIES. Our basic scheme as described in Section 4 requires that a user have sufficient attributes from all of the authorities in order to decrypt. However, we might want to allow a user to decrypt if he had sufficient attributes from at least D of the authorities.

We have to make the following changes to our basic scheme: Our central authority will now choose a random $D-1$ degree polynomial P with $P(0) = y_0$. For each authority k he will compute $P(k)$, and then compute a value which combined with $Y_{k,u}^s = e(g,g)^{y_{k,u}s} = e(g,g)^{F_{s_k}(u)s}$ will give $e(g,g)^{P(k)s}$. If the user obtains D of these values he can interpolate to find $e(g,g)^{P(0)s} = e(g,g)^{y_0s} = Y_0^s$ and then obtain m. Thus, the secret key from the central authority for user u will be $D_{CA} = \{g^{P(k)-F_{s_k}(u)}\}_{k=1...K}$.

ADDING ATTRIBUTE AUTHORITIES. We can also allow the central authority to add additional attribute authorities to the system at any point in the execution of the scheme.

In the basic scheme, where attributes from all authorities are required for decryption, this will occur as follows: The central authority will choose a new system public key $Y_0' = e(g,g)^{y_0'}$, and all future encryption will be relative to this public key. The central authority will store the PRF seed for the new authority, and all secret keys it gives out will be computed using y_0' and the new enlarged set of PRFs. Note that this means that all users will need to obtain a new key from the central authority in order to decrypt any messages encrypted under the new key. However, they will not need to obtain new keys from any of the old attribute authorities.

In the scheme described in the above section, which uses a threshold over authorities, the central authority has simply to give each user an additional value for the new authority k: $e(g,g)^{P(k)-F_{s_k}(u)}$. Note that in this case, if a user does not intend to use any attributes from this new authority, he need not obtain a new key from the central authority.

SINGLE AUTHORITY CNF ATTRIBUTE ENCRYPTION. In [SW05] it was left open whether decryption ability could be determined by a more complicated function of a user's attributes. Looked at differently, our scheme can be viewed as a single authority scheme in which the attributes necessary for decryption are given by a CNF formula chosen by the encryptor. In this case, each authority corresponds to a clause, and $d_k = 1$ for all authorities.

A user would obtain all of his secret keys ($\{D_{k,i}\}$ and D_{CA}) from the authority and then would be able to decrypt any message for which he had at least one of the specified attributes $i \in \mathcal{A}_C^k$ for each clause k.

The only additional complication is that our multiauthority scheme required that each authority's attribute set be disjoint. Thus, the set of attributes allowed in each clause must be disjoint. To get around this, we create a separate copy of each attribute for each clause in which it could possibly appear. Thus, if a user has attribute i, he will have in his secret key $D_{k,i}$ for every clause k in which attribute i could appear. Then the encryptor includes $T_{1,i}^s$ if attribute i appears in the first clause, and $T_{2,i}^s$ if it occurs in the second clause, etc.

Acknowledgements. Thanks to Anna Lysyanskaya for advice and encouragement, and to Brent Waters for helpful comments and suggestions. The author is supported by NSF grant CNS-0374661 and NSF Graduate Research Fellowship.

References

[BB04]   Dan Boneh and Xavier Boyen. Efficient selective-id secure identity based encryption without random oracles. In *Proc. of EUROCRYPT 2004*, volume 3027, *LNCS*, 54–73. Springer.

[BF01]   Dan Boneh and Matthew Franklin. Identity-based encryption from the Weil pairing. In *Proc. of CRYPTO 2001*, volume 2139, *LNCS*, 213–229. Springer.

[CHK03]   Ran Canetti, Shai Halevi, and Jonathan Katz. A forward-secure public-key encryption scheme. In *Proc. of EUROCRYPT 2003*, volume 2656, *LNCS*, 255–271. Springer.

[Coc01]   C. Cocks. An identity based encryption scheme based on quadratic residues. In *Proc. of Cryptography and Coding, 8th IMA International Conference*, volume 2260, *LNCS*, 360–363. Springer, 2001.

[Fel87]   P. Feldman. A practical scheme for non-interactive verifiable secret sharing. In *Proc. of FOCS 1987*, 427–437.

[GPSW06] Vipul Goyal, Omkant Pandey, Amit Sahai, and Brent Waters. Attribute-based encryption for fine-grained access control of encrypted data. In *Proc. of CCS 2006*, 89–98, New York. ACM Press.

[Sha85]   Adi Shamir. Identity-based cryptosystems and signature schemes. In *Proc. of CRYPTO 1984*, volume 196, *LNCS*, 47–53. Springer.

[SW05]   Amit Sahai and Brent Waters. Fuzzy identity-based encryption. In *Proc. of EUROCRYPT 2005*, volume 3494, *LNCS*, 457–473. Springer.

[Wat05]   Brent Waters. Efficent identity based encryption without random oracles. In *Proc. of EUROCRYPT 2005*, volume 3494, *LNCS*, 114–127. Springer.

Conjunctive, Subset, and Range Queries on Encrypted Data

Dan Boneh[1,*] and Brent Waters[2,**]

[1] Stanford University
dabo@cs.stanford.edu
[2] SRI International
bwaters@csl.sri.com

Abstract. We construct public-key systems that support comparison queries $(x \geq a)$ on encrypted data as well as more general queries such as subset queries $(x \in S)$. Furthermore, these systems support arbitrary conjunctive queries $(P_1 \wedge \cdots \wedge P_\ell)$ without leaking information on individual conjuncts. We present a general framework for constructing and analyzing public-key systems supporting queries on encrypted data.

1 Introduction

Queries on encrypted data are easiest to explain with an example. Consider a credit card payment gateway that observes a stream of encrypted transactions, say encrypted under Visa's public key. The gateway needs to flag all transactions satisfying a certain predicate P. Say, all transactions whose value is over $1000. Storing Visa's secret key on the gateway is a bad idea for both security and privacy concerns. Instead, Visa wishes to give the gateway a token TK_P that enables the gateway to identify transactions satisfying P without learning anything else about these transactions. Of course, generating the token TK_P will require Visa's secret key.

As another example, consider a mail server that receives a stream of email messages encrypted under the recipients public key. If the email message satisfies a certain predicate P the mail server should forward the email to the recipient's pager. If the email satisfies some other predicate P' the server should just discard the email. Otherwise, the server should place the email in the recipient's inbox. The recipient does not want to give the mail server the full private key. Instead, she wants to give the server two tokens TK_P and $TK_{P'}$ enabling the server to test for the predicates P and P' without learning any other information about the email.

Our goal is to build a public-key system that supports a rich set of query predicates. In our payment gateway example one can imagine comparison queries such as (value > 1000) or even conjunctions such as (value > 1000) and (Transaction Time > 5pm). The gateway should learn no information other than the value

---

* Supported by NSF and the Packard Foundation.
** Supported by NSF and U.S. Army Research Office under Research Grant No. W911NF-06-1-0316.

S.P. Vadhan (Ed.): TCC 2007, LNCS 4392, pp. 535–554, 2007.

of the conjunctive predicate. In case a conjunction $P_1 \wedge P_2$ is false, the gateway should not learn which of the two conjuncts P_1 or P_2 is false. In our second example involving a mail server one can imagine testing for subset queries such as (sender $\in S$) where S is a set of email addresses. Conjunctive queries such as (sender $\in S$) and (subject = urgent) also make sense. Perhaps in the distant future, when highly complex queries on encrypted data are possible, one can imagine running an anti-virus/anti-spam predicate on encrypted emails. The mail server learns nothing about incoming encrypted email other than its spam status.

Unfortunately, until now, only simple equality queries on encrypted data were possible. Song et al. [19] developed a mechanism for equality tests on data encrypted with a symmetric key system. Boneh et al. [8] constructed equality tests in the public-key settings.

Our results. We present a general framework for analyzing and constructing searchable public-key systems for various families of predicates. We then construct public-key systems that support comparison queries (such as greater-than) and general subset queries. We also support arbitrary conjunctions. We evaluate our results based on ciphertext size and token size. Let $T = \{1, 2, \ldots, n\}$ and suppose we encrypt a tuple $x = (x_1, \ldots, x_w) \in T^w$. Say x_1 is a transaction value, x_2 is a card expiration date, and so on. The following table summarizes our results at a high level.

Query Type	Source	Ciphertext Size	Token Size
Equality query: $(x_i = a)$ for any $a \in T$	[19, 17, 8, 1]	$O(1)$	$O(1)$
Comparison query: $(x_i \geq a)$ for any $a \in T$	[10, 12][1]	$O(\sqrt{n})$	$O(\sqrt{n})$
Subset query: $(x_i \in A)$ for any $A \subseteq T$	This paper	$O(n)$	$O(n)$
Equality conjunction: $(x_1 = a_1) \wedge \ldots \wedge (x_w = a_w)$	This paper	$O(w)$	$O(w)$
Comparison conjunction: $(x_1 \geq a_1) \wedge \ldots \wedge (x_w \geq a_w)$	This paper	$O(nw)$	$O(w)$
Subset conjunction: $(x_1 \in A_1) \wedge \ldots \wedge (x_w \in A_w)$	This paper	$O(nw)$	$O(nw)$

Here $(a_1, \ldots, a_w)$ is an arbitrary vector that defines a conjunctive equality or a comparison predicate. Similarly, $A_1, \ldots, A_w$ are *arbitrary* subsets of $\{1, \ldots, n\}$ that define a conjunctive subset query predicate. We emphasize that when a conjunction predicate is false, the system does not leak which of the w conjuncts caused it.

Prior to these results the best systems for comparison and subset queries were the trivial brute-force systems that we discuss in Section 3. For comparison queries these systems generate a ciphertext of size $O(n^w)$ and for subset queries they generate a ciphertext of size $O(2^{nw})$. Note that even without conjunction,

---

[1] Both papers [10, 12] focus on traitor tracing, but as we show in the full version of our paper [11], their approach directly gives a comparison searching system without conjunctions.

namely for $w = 1$, our subset query construction generates ciphertexts that are exponentially shorter than the best known previous solution ($O(n)$ vs. $O(2^n)$).

The main tool used in these constructions is a new primitive we call *Hidden Vector Encryption* or HVE for short. This primitive can be viewed as an extreme generalization of Anonymous Identity Based Encryption (AnonIBE) [8, 1, 13]. We show how HVE implies all the results in the table.

A natural question is to look for public key systems that support larger classes of predicates, such as regular expressions. Ultimately, one would like a public-key system that supports searches for any predicate computable by a poly-size circuit. Presently, this appears to be a difficult open problem.

Related work. Equality tests on encrypted data were considered in [19, 8]. Equality searches on an encrypted audit log were proposed in [20]. Equality tests in the symmetric key settings are closely related to oblivious RAM techniques [17, 14]. Equality tests in the public key settings are closely related to Anonymous Identity Based Encryption (AnonIBE) [8, 1, 13]. Conjunctive equality queries were first studied in [15]. Equality searches on streaming data that hide the requested predicate were discussed in [18] and [4]. Efficient equality searches in databases were recently presented in [2]. Bethencourt et al. [3] recently gave a construction for efficient range queries in a weaker security model. That is, when the encrypted index falls in the specified range, the search token reveals the index.

2   Definitions

We begin by defining a general framework for queries on encrypted data. Let Σ be a finite set of binary strings. A predicate P over Σ is a function $P : \Sigma \to \{0, 1\}$. We say that $I \in \Sigma$ satisfies the predicate if $P(I) = 1$.

2.1   Searchable Encryption

Let Φ be a set of predicates over Σ. A Φ-**searchable** public key system comprises of the following algorithms:

Setup(λ). A probabilistic algorithm that takes as input a security parameter and outputs a public key PK and secret key SK.

Encrypt(PK, I, M). Encrypts the plaintext pair (I, M) using the public key PK. We view $I \in \Sigma$ as the searchable field, called an **index**, and $M \in \mathcal{M}$ as the data.

GenToken(SK, $\langle P \rangle$). Takes as input a secret key SK and the description of a predicate $P \in \Phi$. It outputs a token TK_P.

Query(TK, C). Takes a token TK for some predicate $P \in \Phi$ as input and a ciphertext C. It outputs a message $M \in \mathcal{M}$ or $\bot$. Roughly speaking, if C is an encryption of (I, M) then the algorithm outputs M when $P(I) = 1$ and outputs $\bot$ otherwise. The precise requirement is captured in the query correctness property below.

Correctness. The system must satisfy the following **correctness property:**

- **Query correctness:** For all $(I, M) \in \Sigma \times \mathcal{M}$ and all predicates $P \in \Phi$:

    Let $(\text{PK}, \text{SK}) \overset{R}{\leftarrow} Setup(\lambda)$, $C \overset{R}{\leftarrow} Encrypt(\text{PK}, I, M)$, and $\text{TK} \overset{R}{\leftarrow} GenToken(\text{SK}, \langle P \rangle)$.

    If $P(I) = 1$ then $Query(\text{TK}, C) = M$.

    If $P(I) = 0$ then $\Pr[Query(\text{TK}, C) = \bot] > 1 - \epsilon(\lambda)$ where $\epsilon(\lambda)$ is a negligible function.

Suppose that given a ciphertext $C \leftarrow Encrypt(\text{PK}, I, M)$ we are only interested in testing whether a predicate $P(I)$ is satisfied. In this case the message space $\mathcal{M}$ can be set to a singleton, say $\mathcal{M} = \{\text{true}\}$. Algorithm $Query(\text{TK}, C)$ will return true when $P(I) = 1$ and $\bot$ otherwise. A larger message space $\mathcal{M}$ is useful if TK is intended to unlock some $M \in \mathcal{M}$ whenever the predicate $P(I) = 1$. For example, when the transaction value is over \$1000 we may want the payment gateway to obtain more information about the transaction. Otherwise, the gateway should learn nothing.

Notice that a Φ-searchable system does not provide a *Decrypt* algorithm that uses SK to decrypt a ciphertext C and outputs (I, M). One can always add this capability by also encrypting (I, M) under a standard public key system. There is no need for the searchable system to explicitly provide this capability.

An example – comparison queries. Before defining security, we first give a motivating example using comparison queries. Let $\Sigma = \{1, \ldots, n\}$ for some integer n. For $\sigma \in \{1, \ldots, n\}$ let P_σ be the following comparison predicate:

$$P_\sigma(x) = \begin{cases} 1 & \text{if } x \geq \sigma, \\ 0 & \text{otherwise} \end{cases}$$

Let $\Phi_n = \{P_1, \ldots, P_n\}$ be the set of all n comparison predicates. Suppose the adversary has the tokens for predicates $P_{\sigma_1}, P_{\sigma_2}, \ldots, P_{\sigma_w}$ where $\sigma_1 < \sigma_2 < \cdots < \sigma_w$. Lets x, y, z be some integers as in Figure 1. Clearly the adversary can distinguish $Encrypt(\text{PK}, x, m)$ from $Encrypt(\text{PK}, y, m)$ using the token for the predicate P_{σ_2}. However, the adversary should not be able to distinguish $Encrypt(\text{PK}, y, m)$ from $Encrypt(\text{PK}, z, m)$. Indeed, separating an encryption of y from an encryption of z is information that should not be exposed by the tokens at the adversary's disposal. Our definition of security captures this property using the general framework.

2.2   Security

We define security of a Φ-searchable system $\mathcal{E}$ using a **query security game** that captures the intuition that tokens TK reveal no unintended information about the plaintext. The game gives the adversary a number of tokens and

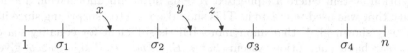

Fig. 1. Tokens for $\sigma_1, \sigma_2, \sigma_3, \sigma_4$ given to the adversary

requires that the adversary cannot use these tokens to deduce unintended information. The game proceeds as follows:

- **Setup.** The challenger runs $Setup(\lambda)$ and gives the adversary PK.
- **Query phase 1.** The adversary adaptively outputs descriptions of predicates $P_1, P_2, \ldots, P_{q_1} \in \Phi$. The challenger responds with the corresponding tokens $TK_j \leftarrow GenToken(SK, \langle P_j \rangle)$. We refer to such queries as **predicate queries**.
- **Challenge.** The adversary outputs two pairs (I_0, M_0) and (I_1, M_1) subject to two restrictions:
  - First, $P_j(I_0) = P_j(I_1)$ for all $j = 1, 2, \ldots, q_1$.
  - Second, if $M_0 \neq M_1$ then $P_j(I_0) = P_j(I_1) = 0$ for all $j = 1, 2, \ldots, q_1$.

  The challenger flips a coin $\beta \in \{0, 1\}$ and gives $C_* \xleftarrow{R} Encrypt(PK, I_\beta, M_\beta)$ to the adversary.

  The two restrictions ensure that the tokens given to the adversary do not trivially break the challenge. The first restriction ensures that tokens given to the adversary do not directly distinguish I_0 from I_1. The second restriction ensures that the tokens do not directly distinguish M_0 from M_1.
- **Query phase 2.** The adversary continues to adaptively request tokens for predicates $P_{q_1+1}, \ldots, P_q \in \Phi$, subject to the two restrictions above. The challenger responds with the corresponding tokens $TK_j \leftarrow GenToken(SK, \langle P_j \rangle)$.
- **Guess.** The adversary returns a guess $\beta' \in \{0, 1\}$ of β.

We define the advantage of adversary $\mathcal{A}$ in attacking $\mathcal{E}$ as the quantity $QU\,Adv_{\mathcal{A}} = |\Pr[\beta' = \beta] - 1/2|$.

Definition 1. *We say that a Φ-searchable system $\mathcal{E}$ is* **secure** *if for all polynomial time adversaries $\mathcal{A}$ attacking $\mathcal{E}$ the function $QU\,Adv_{\mathcal{A}}$ is a negligible function of λ.*

Another example – equality queries. Let Σ be some finite set. For $\sigma \in \Sigma$ let $P_\sigma(x)$ be an equality predicate, namely

$$P_\sigma(x) = \begin{cases} 1 & \text{if } x = \sigma, \\ 0 & \text{otherwise} \end{cases}$$

Let $\Phi_{eq} = \{P_\sigma \text{ for all } \sigma \in \Sigma\}$. Then a Φ_{eq}-searchable encryption supports equality queries on ciphertexts. It is easy to see that a secure Φ_{eq}-searchable encryption is also an anonymous IBE system [8, 1, 13] — an Identity Based

Encryption system where a ciphertext reveals no useful information about the identity that was used to create it. This should not be too surprising since it was previously shown [8, 1] that anonymous IBE is sufficient for equality searches. A Φ_{eq}-searchable encryption system (*Setup*, *Encrypt*, *GenToken*, *Query*) gives an anonymous IBE as follows:

- $Setup_{IBE}(\lambda)$ runs $Setup(\lambda)$ and outputs IBE parameters PK and master key SK.
- $Encrypt_{IBE}(\text{PK}, \mathcal{I}, M)$ where $\mathcal{I} \in \Sigma$ outputs $Encrypt(\text{PK}, \mathcal{I}, M)$.
- $Extract_{IBE}(\text{SK}, \mathcal{I})$ where $\mathcal{I} \in \Sigma$ outputs $\text{TK}_{\mathcal{I}} \leftarrow GenToken(\text{SK}, \langle P_{\mathcal{I}} \rangle)$.
- $Decrypt_{IBE}(\text{TK}_{\mathcal{I}}, C)$ outputs $Query(\text{TK}_{\mathcal{I}}, C)$.

The correctness property ensures that if C is the result of $Encrypt(\text{PK}, \mathcal{I}, M)$ then $Query(\text{TK}_{\mathcal{I}}, C)$ will output M since $P_{\mathcal{I}}(\mathcal{I}) = 1$. It is not difficult to see that the Φ_{eq}-security game ensures semantic security for both the message and the identity. Hence, the resulting system is an anonymous IBE.

By considering larger classes of predicates Φ we obtain more general searching capabilities. The challenge is then to build secure encryption schemes that are Φ-searchable for the most general Φ possible.

Chosen ciphertext security. Definition 1 easily extends to address chosen ciphertext attacks (CCA), but we do not pursue that here.

2.3   Selective Security

We will also need a slightly weaker security definition in which the adversary commits to the search strings I_0, I_1 at the beginning of the game. Everything else remains the same. The game proceeds as follows:

- **Setup.** The adversary outputs two strings $I_0, I_1 \in \Sigma$. The challenger runs $Setup(\lambda)$ and gives the adversary PK.
- **Query phase 1.** The adversary adaptively outputs descriptions of predicates $P_1, P_2, \ldots, P_{q_1} \in \Phi$. The only restriction is that

$$P_j(I_0) = P_j(I_1) \text{ for all } j = 1, 2, \ldots, q_1 \tag{1}$$

  The challenger responds with the corresponding tokens $\text{TK}_j \leftarrow GenToken(\text{SK}, \langle P_j \rangle)$.
- **Challenge.** The adversary outputs two messages $M_0, M_1 \in \mathcal{M}$ subject to the restriction that:

$$\text{if } M_0 \neq M_1 \text{ then } P_j(I_0) = P_j(I_1) = 0 \text{ for all } j = 1, 2, \ldots, q_1 \tag{2}$$

  The challenger flips a coin $\beta \in \{0, 1\}$ and gives $C_* \xleftarrow{R} Encrypt(\text{PK}, I_\beta, M_\beta)$ to the adversary.
- **Query phase 2.** The adversary continues to adaptively request query tokens for predicates $P_{q_1+1}, \ldots, P_q \in \Phi$, subject to the two restrictions (1) and (2). The challenger responds with the corresponding tokens $\text{TK}_j \leftarrow GenToken(\text{SK}, \langle P_j \rangle)$.

– **Guess.** The adversary returns a guess $\beta' \in \{0, 1\}$ of β.

The advantage of adversary $\mathcal{A}$ in attacking $\mathcal{E}$ is the quantity $\mathsf{sQU\,Adv}_{\mathcal{A}} = |\Pr[\beta' = \beta] - 1/2|$.

Definition 2. *We say that a Φ-searchable system $\mathcal{E}$ is* **selectively secure** *if for all polynomial time adversaries $\mathcal{A}$ attacking $\mathcal{E}$ the function $\mathsf{sQU\,Adv}_{\mathcal{A}}$ is a negligible functions of λ.*

3   The Trivial Construction

Let Σ be a finite set of binary strings. We build a Φ-searchable public key system $\mathcal{E}_{\mathrm{TR}}$, for *any* set of (polynomial time computable) predicates Φ. We refer to this system as the brute force Φ-searchable system.

The brute force system. Let $\mathcal{E} = (Setup', Encrypt', Decrypt')$ be a public-key system. Let $\Phi = \{P_1, P_2, \ldots, P_t\}$ The Φ-searchable system $\mathcal{E}_{\mathrm{TR}}$ is defined as follows:

Setup(λ). Run *Setup'*(λ) t times to obtain

$$\mathrm{PK} \leftarrow (\mathrm{PK}_1, \ldots, \mathrm{PK}_t) \quad \text{and} \quad \mathrm{SK} \leftarrow (\mathrm{SK}_1, \ldots, \mathrm{SK}_t)$$

Output PK and SK.

Encrypt(PK, I, M). For $j = 1, \ldots, t$ define:

$$C_j \xleftarrow{R} \begin{cases} Encrypt'(\mathrm{PK}_j, \ M) & \text{if } P_j(I) = 1, \\ Encrypt'(\mathrm{PK}_j, \ \perp) & \text{otherwise.} \end{cases}$$

Output $C \leftarrow (C_1, \ldots, C_t)$. Note that the length of C is linear in n.

GenToken$(\mathrm{SK}, \langle P \rangle)$. Here $\langle P \rangle$ (the description of a predicate P) is the index j of P in Φ. Output $\mathrm{TK} \leftarrow (j, \mathrm{SK}_j)$.

Query(TK, C). Let $C = (C_1, \ldots, C_t)$ and $\mathrm{TK} = (j, \mathrm{SK}_j)$.
Output $Decrypt'(\mathrm{SK}_j, \ C_j)$.

The following lemma proves security of this construction. The proof is a straightforward hybrid argument and is given in Appendix A.

Lemma 1. *The system $\mathcal{E}_{\mathrm{TR}}$ above is a secure Φ-searchable encryption system assuming $\mathcal{E}$ is a semantically secure public key system against chosen plaintext attacks.*

3.1   A Third Example — Conjunctive Comparison Predicates

Suppose $\Sigma = \{1, \ldots, n\}^w$ for some n, w. Let $\Phi_{n,w}$ be the set of n^w predicates

$$P_{a_1 \ldots a_w}(x_1, \ldots, x_w) = \begin{cases} 1 & \text{if } x_j \geq a_j \text{ for all } j = 1, \ldots, w, \\ 0 & \text{otherwise} \end{cases}$$

for all $\bar{a} = (a_1 \ldots a_w) \in \{1, \ldots, n\}^w$. Then $|\Phi_{n,w}| = n^w$.

The trivial system in this case produces ciphertexts of length $O(n^w)$. Essentially, the system uses a unary encoding of the w columns and assigns a private key to each cell in this n by w matrix. We will construct a much better system in Section 6.

4 Background on Pairings and Complexity Assumptions

Our goal is to construct Φ-searchable systems for a large class of predicates Φ that is much better than the trivial construction. To do so we will make use of bilinear maps.

4.1 Bilinear Groups of Composite Order

We review some general notions about bilinear maps and groups, with an emphasis on groups of *composite order*. We follow [9] in which composite order bilinear groups were first introduced.

Let $\mathcal{G}$ be a an algorithm called a *group generator* that takes as input a security parameter $\lambda \in \mathbb{Z}^{>0}$ and outputs a tuple $(p, q, \mathbb{G}, \mathbb{G}_T, e)$ where p, q are two distinct primes, $\mathbb{G}$ and $\mathbb{G}_T$ are two cyclic groups of order $n = pq$, and e is a function $e : \mathbb{G}^2 \to \mathbb{G}_T$ satisfying the following properties:

- (Bilinear) $\forall u, v \in \mathbb{G}, \forall a, b \in \mathbb{Z}, e(u^a, v^b) = e(u, v)^{ab}$.
- (Non-degenerate) $\exists g \in \mathbb{G}$ such that $e(g, g)$ has order n in $\mathbb{G}_T$.

We assume that the group action in $\mathbb{G}$ and $\mathbb{G}_T$ as well as the bilinear map e are all computable in polynomial time in λ. Furthermore, we assume that the description of $\mathbb{G}$ and $\mathbb{G}_T$ includes generators of $\mathbb{G}$ and $\mathbb{G}_T$ respectively.

To summarize, $\mathcal{G}$ outputs the description of a group $\mathbb{G}$ of order $n = pq$ with an efficiently computable bilinear map. We will use the notation $\mathbb{G}_p, \mathbb{G}_q$ to denote the respective subgroups of order p and order q of $\mathbb{G}$ and we will use the notation $\mathbb{G}_{T,p}, \mathbb{G}_{T,q}$ to denote the respective subgroups of order p and order q of $\mathbb{G}_T$.

4.2 The Bilinear Diffie-Hellman Assumption

First we review the standard Bilinear Diffie-Hellman assumption, but in groups of composite order. For a given group generator $\mathcal{G}$ define the following distribution $P(\lambda)$:

$$(p, q, \mathbb{G}, \mathbb{G}_T, e) \xleftarrow{R} \mathcal{G}(\lambda), \quad n \leftarrow pq, \quad g_p \xleftarrow{R} \mathbb{G}_p, \quad g_q \xleftarrow{R} \mathbb{G}_q$$

$$a, b, c \xleftarrow{R} \mathbb{Z}_n$$

$$\bar{Z} \leftarrow \left((n, \mathbb{G}, \mathbb{G}_T, e), \ g_q, \ g_p, \ g_p^a, \ g_p^b, \ g_p^c\right)$$

$$T \leftarrow e(g_p, \ g_p)^{abc}$$

Output $(\bar{Z}, \ T)$

For an algorithm $\mathcal{A}$, define $\mathcal{A}$'s advantage in solving the composite bilinear Diffie-Hellman problem for $\mathcal{G}$ as:

$$\mathsf{cBDH\,Adv}_{\mathcal{G},\mathcal{A}}(\lambda) := \left| \Pr[\mathcal{A}(\bar{Z},T) = 1] - \Pr[\mathcal{A}(\bar{Z},R) = 1] \right|$$

where $(\bar{Z},T) \xleftarrow{R} P(\lambda)$ and $R \xleftarrow{R} \mathbb{G}_{T,p}$.

Definition 3. *We say that $\mathcal{G}$ satisfies the composite bilinear Diffie-Hellman assumption (cBDH) if for any polynomial time algorithm $\mathcal{A}$ we have that the function $\mathsf{cBDH\,Adv}_{\mathcal{G},\mathcal{A}}(\lambda)$ is a negligible function of λ.*

4.3 The Composite 3-Party Diffie-Hellman Assumption

Our construction makes use of an additional assumption in composite bilinear groups. For a given group generator $\mathcal{G}$ define the following distribution $P(\lambda)$:

$$(p,q,\mathbb{G},\mathbb{G}_T,e) \xleftarrow{R} \mathcal{G}(\lambda), \quad n \leftarrow pq, \quad g_p \xleftarrow{R} \mathbb{G}_p, \quad g_q \xleftarrow{R} \mathbb{G}_q$$

$$R_1, R_2, R_3 \xleftarrow{R} \mathbb{G}_q$$

$$a,b,c \xleftarrow{R} \mathbb{Z}_n$$

$$\bar{Z} \leftarrow \left((n,\mathbb{G},\mathbb{G}_T,e),\ g_q,\ g_p,\ g_p^a,\ g_p^b,\ g_p^{ab} \cdot R_1,\ g_p^{abc} \cdot R_2\right)$$

$$T \leftarrow g_p^c \cdot R_3$$

$$\text{Output } (\bar{Z},\ T)$$

For an algorithm $\mathcal{A}$, define $\mathcal{A}$'s advantage in solving the composite 3-party Diffie-Hellman problem for $\mathcal{G}$ as:

$$\mathsf{C3DH\,Adv}_{\mathcal{G},\mathcal{A}}(\lambda) := \left| \Pr[\mathcal{A}(\bar{Z},T) = 1] - \Pr[\mathcal{A}(\bar{Z},R) = 1] \right|$$

where $(\bar{Z},T) \xleftarrow{R} P(\lambda)$ and $R \xleftarrow{R} \mathbb{G}$.

Definition 4. *We say that $\mathcal{G}$ satisfies the composite 3-party Diffie-Hellman assumption (C3DH) if for any polynomial time algorithm $\mathcal{A}$ we have that the function $\mathsf{C3DH\,Adv}_{\mathcal{G},\mathcal{A}}(\lambda)$ is a negligible function of λ.*

The assumption is formed around the intuition that it is hard to test for Diffie-Hellman tuples in the order p subgroup if the elements to be tested have a random order q subgroup component.

5 Hidden Vector Encryption

We construct a Φ-searchable encryption system for a general class of equality predicates. We call such systems Hidden Vector Systems or HVEs for short. We then show in Section 6 that our HVE system leads to comparison and subset queries far more efficient than the trivial system.

5.1   HVE Definition

Let Σ be a finite set and let $*$ be a special symbol not in Σ. Define $\Sigma_* = \Sigma \cup \{*\}$. The star $*$ plays the role of a wildcard or "don't care" value. In our subset and range query applications we typically set $\Sigma = \{0, 1\}$. Note that here we use the symbol Σ differently than how it was used in Section 2.1.

For $\sigma = (\sigma_1, \dots, \sigma_\ell) \in \Sigma_*^\ell$ define a predicate P_σ^{HVE} over Σ^ℓ as follows. For $x = (x_1, \dots, x_\ell) \in \Sigma^\ell$ set:

$$P_\sigma^{\mathrm{HVE}}(x) = \begin{cases} 1 & \text{if for all } i = 1, \dots, \ell : \ (\sigma_i = x_i \text{ or } \sigma_i = *), \\ 0 & \text{otherwise} \end{cases}$$

In other words, the vector x matches σ in all the coordinates where σ is not $*$. Let $\Phi_{\mathrm{HVE}} = \{P_\sigma^{\mathrm{HVE}} \text{ for all } \sigma \in \Sigma_*^\ell\}$. We refer to ℓ as the **width** of the HVE.

Definition 5. *A Hidden Vector System (HVE) over Σ^ℓ is a selectively secure Φ_{HVE}-searchable encryption system.*

The case $\ell = 1$ degenerates to the example discussed in Section 2.2 where we showed equivalence to anonymous IBE [8, 1, 13]. For larger ℓ we obtain a more general concept that is much harder to build. In particular, the wildcard character '$*$' — which is essential for the applications we have in mind — makes it challenging to construct a Φ_{HVE}-searchable system. We construct an HVE with the following parameters:

$$\text{CT-size} = O(\ell) \quad \text{and} \quad \text{TK-size} = O(\text{ weight}(\sigma))$$

where $\text{weight}\big(\sigma = (\sigma_1, \dots, \sigma_\ell)\big)$ is the number of coordinates where $\sigma_i \neq *$.

5.2   Construction

For our particular HVE construction we will let $\Sigma = \mathbb{Z}_m$ for some integer m. We set $\Sigma_* = \mathbb{Z}_m \cup \{*\}$. We describe an HVE where the payload M is in a small subset $\mathcal{M}$ of $\mathbb{G}_T$, namely $|\mathcal{M}| < |\mathbb{G}_T|^{1/4}$. This is not a serious restriction since the payload M is typically a short symmetric message key. Our HVE system works as follows:

Setup(λ). The setup algorithm first chooses random primes $p, q > m$ and creates a bilinear group $\mathbb{G}$ of composite order $n = pq$, as specified in Section 4.1. Next, it picks random elements

$$(u_1, h_1, w_1), \dots, (u_\ell, h_\ell, w_\ell) \in \mathbb{G}_p^3 , \quad g, v \in \mathbb{G}_p , \quad g_q \in \mathbb{G}_q.$$

and an exponent $\alpha \in \mathbb{Z}_p$. It keeps all these as the secret key SK.
It then chooses $3\ell + 1$ random blinding factors in $\mathbb{G}_q$:

$$(R_{u,1}, R_{h,1}, R_{w,1}), \dots, (R_{u,\ell}, R_{h,\ell}, R_{w,\ell}) \in \mathbb{G}_q \text{ and } R_v \in \mathbb{G}_q.$$

For the public key, PK, it publishes the description of the group $\mathbb{G}$ and the values

$$g_q, \quad V = vR_v, \quad A = e(g,v)^\alpha, \quad \begin{pmatrix} U_1 = u_1 R_{u,1}, & H_1 = h_1 R_{h,1}, & W_1 = w_1 R_{w,1} \\ & \vdots & \\ U_\ell = u_\ell R_{u,\ell}, & H_\ell = h_\ell R_{h,\ell}, & W_\ell = w_\ell R_{w,\ell} \end{pmatrix}$$

The message space $\mathcal{M}$ is set to be a subset of $\mathbb{G}_T$ of size less than $n^{1/4}$.

Encrypt(PK, $\mathcal{I} \in \mathbb{Z}_m^\ell$, $M \in \mathcal{M} \subseteq \mathbb{G}_T$). Let $\mathcal{I} = (\mathcal{I}_1, \ldots, \mathcal{I}_\ell) \in \mathbb{Z}_m^\ell$. The encryption algorithm works as follows:

- choose a random $s \in \mathbb{Z}_n$ and random $Z, (Z_{1,1}, Z_{1,2}), \ldots, (Z_{\ell,1}, Z_{\ell,2}) \in \mathbb{G}_q$. (The algorithm picks random elements in $\mathbb{G}_q$ by raising g_q to random exponents from $\mathbb{Z}_n$.)
- Output the ciphertext:

$$C = \left(C' = MA^s, \ C_0 = V^s Z, \quad \begin{pmatrix} C_{1,1} = (U_1^{\mathcal{I}_1} H_1)^s Z_{1,1}, & C_{1,2} = W_1^s Z_{1,2} \\ & \vdots & \\ C_{\ell,1} = (U_\ell^{\mathcal{I}_\ell} H_\ell)^s Z_{\ell,1}, & C_{\ell,2} = W_\ell^s Z_{\ell,2} \end{pmatrix} \right)$$

GenToken(SK, $\mathcal{I}_* \in \Sigma_*^\ell$). The key generation algorithm will take as input the secret key and an ℓ-tuple $\mathcal{I}_* = (\mathcal{I}_1, \ldots, \mathcal{I}_\ell) \in \{\mathbb{Z}_m \cup \{*\}\}^\ell$. Let S be the set of all indexes i such that $\mathcal{I}_i \neq *$. To generate a token for the predicate $P_{\mathcal{I}_*}^{\text{HVE}}$ choose random $(r_{i,1}, r_{i,2}) \in \mathbb{Z}_p^2$ for all $i \in S$ and output:

$$\text{TK} = \left(\ \mathcal{I}_*, \ \ K_0 = g^\alpha \prod_{i \in S} (u_i^{\mathcal{I}_i} h_i)^{r_{i,1}} w_i^{r_{i,2}}, \ \ \forall i \in S: \ \ K_{i,1} = v^{r_{i,1}}, \ \ K_{i,2} = v^{r_{i,2}} \ \right)$$

Query(TK, C). Using the notation in the description of *Encrypt* and *GenToken* do:

- First, compute

$$M \leftarrow C' \ / \ \left(e(C_0, K_0) \ / \ \prod_{i \in S} e(C_{i,1}, K_{i,1}) \, e(C_{i,2}, K_{i,2}) \right) \tag{3}$$

- If $M \notin \mathcal{M}$ output $\bot$. Otherwise, output M.

Correctness. Before proving security we first show that the system satisfies the correctness property defined in Section 2.1. Let $(\mathcal{I}, M)$ be a pair in $\Sigma^\ell \times \mathcal{M}$ and let $B_* \in \Sigma_*^\ell$. This B_* defines a predicate P_{B_*} in Φ_{HVE}.

Let $(\text{PK}, \text{SK}) \xleftarrow{R} Setup(\lambda)$, $C \xleftarrow{R} Encrypt(\text{PK}, \mathcal{I}, M)$, and $\text{TK} \xleftarrow{R} GenToken(\text{SK}, B_*)$.

- If $P_{B_*}(\mathcal{I}) = 1$ then a simple calculation shows that $Query(\text{TK}, C) = M$. This uses in a crucial way the fact that $e(h_p, h_q) = 1$ for all $h_p \in \mathbb{G}_p$ and $h_q \in \mathbb{G}_q$.
- If $P_{B_*}(\mathcal{I}) = 0$ the following lemma shows that when the message space $\mathcal{M}$ satisfies $|\mathcal{M}| < n^{1/4}$ then $\Pr[Query(\text{TK}, C) \neq \bot]$ is negligible. Here the probability is over the random bits used to create the ciphertext.

Lemma 2. *With the notation as above, and assuming* $|\mathcal{M}| < n^{1/4}$, *whenever* $P_{B_*}(\mathcal{I}) = 0$ *the quantity* $\Pr[Query(\mathrm{TK}, C) \neq \bot]$ *is negligible.*
The probability is over the random bits used to create the ciphertext.

Proof. Let $\mathcal{I} = (\mathcal{I}_1, \ldots, \mathcal{I}_\ell) \in \Sigma$ and let $B_* = (B_1, \ldots, B_\ell) \in \Sigma_*^\ell$. Let S be the set of all indexes i such that B_i is not a wildcard $*$ at index i. Since $P_{B_*}(\mathcal{I}) = 0$ we know that there is some $i \in S$ such that $B_i \neq \mathcal{I}_i$. Then the decryption equation (3) contains a factor

$$e(C_0, K_0) \,/\, e(C_{i,1}, K_{i,1})\, e(C_{i,2}, K_{i,2}) = e(v, u_i)^{(B_i - \mathcal{I}_i) \cdot s r_{i,1}}$$

which is a uniformly distributed value in $\mathbb{G}_{T,p}$ and is independent of the rest of the equation. Since the message space is of size $n^{1/4}$ and the size of $\mathbb{G}_{T,p}$ is approximately $n^{1/2}$, the false positive probability is at most $1/n^{1/4}$, which is negligible in the security parameter as required. □

We note that in practice there is no need to use a small message space $\mathcal{M} \subseteq \mathbb{G}_T$ to determine if decryption succeeded. We only use $\mathcal{M}$ to simplify the description of the system. In practice, one could do the following. The encryptor first picks a random $k \in \mathbb{G}_T$ and derives two uniform and independent b-bit symmetric keys (k_0, k_1) from k. It encrypts the payload M using a symmetric encryption system under key k_0 to obtain C_1. Next, it runs our *Encrypt*(PK, $\mathcal{I}$, k) to obtain C. The final ciphertext is the tuple (C, C_1, k_1). Now, our *Query* algorithm works as follows. It first recovers a k' from C using the given token TK. Next, it derives (k_0', k_1') from k' and outputs $\bot$ if $k_1' \neq k_1$. Otherwise, it outputs the decryption of C_1 under k_0' using a symmetric system. Lemma 2 shows that the false error probability is now $1/2^b$. Alternatively, if the symmetric encryption system provides authenticated encryption, then one could decide if *Query* produced the right value based on whether symmetric decryption succeeded.

Extensions. In our description above we limited the index space Σ to be $\mathbb{Z}_m$. We can expand this space to all of $\{0,1\}^*$ by taking a large enough m to contain the range of a collision-resistant hash function. Then *Encrypt*(PK, $\mathcal{I} \in (\{0,1\}^*)^\ell$, $M \in \mathbb{G}_T$) first hashes all the coordinates of $\mathcal{I}$ into $\mathbb{Z}_m$ using the collision resistant hash and then applies the *Encrypt* algorithm described above.

5.3   Proof of Security

We prove our scheme selectively secure (as defined in Section 2.3) under the composite 3-party Diffie-Hellman assumption and the bilinear Diffie-Hellman assumption. We give the high-level arguments of the proof in this section and defer the proofs of some lemmas to the full version of our paper [11].

Suppose the adversary commits to vectors $L_0, L_1 \in \Sigma^\ell$ at the beginning of the game. Let X be the set of indexes i such that $L_{0,i} = L_{1,i}$ and $\overline{X}$ be the set of indexes i such that $L_{0,i} \neq L_{1,i}$.

The proof uses a sequence of $2\ell + 2$ games to argue that the adversary cannot win the original security game of Section 2.3 which we denote by G. We begin by slightly modifying the game G into a game G'. Games G and G' are identical except for how the challenge ciphertext is generated. In G' if $M_0 \neq M_1$ then the adversary multiplies the challenge ciphertext component C' by a random element of $\mathbb{G}_{T,p}$. The rest of the ciphertext is generated as usual. Additionally, if $M_0 = M_1$ then the challenge ciphertext is generated correctly.

Lemma 3. *Assume that the Bilinear Diffie-Hellman assumption holds. Then for any polynomial time adversary $\mathcal{A}$ the difference of advantage of $\mathcal{A}$ in game G and game G' is negligible.*

The proof is in the full version of our paper [11].

Next, we define a game $\tilde{G}$. In this game the adversary will give two challenge messages, M_0, M_1. If $M_0 \neq M_1$ then the challenger outputs a random element of $\mathbb{G}_T$ as the C' component of the challenge ciphertext. The rest of ciphertext is constructed as normal. If $M_0 = M_1$ the challenger outputs the challenge ciphertext as normal.

Lemma 4. *Assume that the Composite 3-party Diffie-Hellman assumption holds. Then for any polynomial time adversary $\mathcal{A}$ the difference of advantage of $\mathcal{A}$ in game G' and game $\tilde{G}$ is negligible.*

The proof is in the full version of our paper [11].

Finally, we define two sequences of hybrid games G_j and G'_j for $j = 1, \ldots, |\overline{X}|$. We define the game G_j as follows. Let $\tilde{X}$ be a set containing the first j indexes in $\overline{X}$. The challenger creates the challenge ciphertext components C_0 and $C_{i,1}, C_{i,2}$ as normal for all $i \notin \tilde{X}$. However, for all $i \in \tilde{X}$ the challenger creates $C_{i,1}, C_{i,2}$ as completely random group elements in $\mathbb{G}$. Additionally, if $M_0 \neq M_1$ then C' is replaced by a completely random element from $\mathbb{G}_T$ (otherwise it is created as normal).

We define a game G'_j as follows. Let $\tilde{X}$ be a set containing the first j indexes in $\overline{X}$ and let δ be the $(j + 1)$-th index in $\overline{X}$. In the challenge ciphertext the challenger creates C_0 and $C_{i,1}, C_{i,2}$ as normal for all $i \notin \tilde{X}$ and $i \neq \delta$. For all $i \in \tilde{X}$ the challenger creates $C_{i,1}, C_{i,2}$ as completely random group elements in $\mathbb{G}$. Finally, the challenger chooses a random s' and creates

$$C_{\delta,1} = (u_p^{\mathcal{I}_\delta} h_p)^{s'} g_q^{z_{\delta,1}}, \quad C_{\delta,2} = g_p^{s'} g_q^{z_{\delta,2}}.$$

Additionally, if $M_0 \neq M_1$ then C' is replaced by a completely random element from $\mathbb{G}_T$ (otherwise it is created as normal).

Observe that for all i in $\tilde{X}$ the challenge ciphertext contains no information about $L_{\beta,i}$. Therefore the adversary's advantage in game $G_{|\overline{X}|}$ is 0. Additionally, game G_0 is equivalent to $\tilde{G}$. We state the following two lemmas whose proofs are given in the full version of our paper [11].

Lemma 5. *Assume the Composite 3-party Diffie-Hellman assumption holds. Then for all j and any polynomial time adversary $\mathcal{A}$ the difference of advantage of $\mathcal{A}$ in game G_j and game G'_j is negligible.*

Lemma 6. *Assume the Composite 3-party Diffie-Hellman assumption holds. Then for all j and any polynomial time adversary $\mathcal{A}$ the difference of advantage of $\mathcal{A}$ in game G'_j and game G_{j+1} is negligible.*

It now follows that if the Composite 3-party Diffie-Hellman and Bilinear Diffie-Hellman assumptions hold then no polynomial-time adversary can break our scheme with non-negligible advantage. This follows from the sequence of hybrid games starting with the original game G:

$$G, \tilde{G}, G'_0, G_1, G_{1'}, G_2, G_{2'}, \ldots, G_{|\overline{X}|}.$$

The adversary's advantage in the game $G_{|\overline{X}|}$ is 0 and the difference in adversary's advantage between any two consecutive hybrid games is negligible by the lemmas above. Hence, no polynomial adversary can win game G with non-negligible advantage.

6   Applications of HVE

We show how HVE leads to efficient systems for subset queries and conjunctive comparison queries. Throughout the section we let $\Sigma_{01} = \{0,1\}$ and $\Sigma_{01*} = \{0,1,*\}$.

Conjunctive comparison queries. In Section 3.1 we defined conjunctive comparison queries and the predicate family $\Phi_{n,w}$. We use HVE to build a $\Phi_{n,w}$-searchable encryption system with ciphertext size $O(nw)$ and token size $O(w)$.

Let $(Setup_{HVE}, Encrypt_{HVE}, GenToken_{HVE}, Query_{HVE})$ be a secure HVE over Σ_{01}^{nw}. Thus, the width of this HVE is $\ell = nw$. We construct a $\Phi_{n,w}$-searchable system as follows:

- $Setup(\lambda)$ is the same as $Setup_{HVE}(\lambda)$.
- $Encrypt(\text{PK}, I, M)$ where $I = (x_1, \ldots, x_w) \in \{1, \ldots, n\}^w$. Build a vector $\sigma(I) = (\sigma_{i,j}) \in \Sigma_{01}^{nw}$ as follows:

$$\sigma_{i,j} = \begin{cases} 1 & \text{if } j \geq x_i, \\ 0 & \text{otherwise} \end{cases} \tag{4}$$

Then output $Encrypt_{HVE}(\text{PK}, \sigma(I), M)$ which gives a ciphertext of size $O(nw)$. For example, for $w = 2$ and $I = (x_1, x_2)$ the vector $\sigma(I)$ looks like:

$$\sigma(S) = \begin{array}{c} \overset{1}{} \quad\quad \overset{x_1}{} \quad\quad \overset{n}{} \overset{1}{} \quad\quad \overset{x_2}{} \quad\quad \overset{n}{} \\ \boxed{0} \cdots \boxed{0} \boxed{1} \boxed{1} \cdots \boxed{1} \boxed{0} \cdots \boxed{0} \boxed{1} \boxed{1} \cdots \boxed{1} \end{array} \in \{0,1\}^{2n}$$

- $GenToken(\text{SK}, \langle P_{\bar{a}} \rangle)$ where $\bar{a} = (a_1, \ldots, a_w) \in \{1, \ldots, n\}^w$. Define $\sigma_*(\bar{a}) = (\sigma_{i,j}) \in \Sigma_{01*}^{nw}$ as follows:

$$\sigma_{i,j} = \begin{cases} 1 & \text{if } x_i = j, \\ * & \text{otherwise} \end{cases} \tag{5}$$

Output $TK_{\bar{a}} \overset{R}{\leftarrow} GenToken_{HVE}(SK, \sigma_*(\bar{a}))$ which gives a token of size $O(w)$. For example, for $w = 2$ and $\bar{a} = (x_1, x_2)$ the vector $\sigma_*(\bar{a})$ looks like:

$$\sigma_*(\bar{a}) = \begin{array}{|c|c|c|c|c|c|c|c|c|c|c|c|c|c|} \hline * & \cdots & * & 1 & * & \cdots & * & * & \cdots & * & 1 & * & \cdots & * \\ \hline \end{array} \quad \in \{0,1,*\}^{2n}$$

with positions labeled 1, x_1, n 1, x_2, n.

- $Query(TK_{\bar{a}}, C)$ output $Query_{HVE}(TK_{\bar{a}}, C)$

To argue correctness and security, observe that for a predicate $P_{\bar{a}} \in \Phi_{n,w}$ and an index $I \in \{1, \ldots, n\}^w$ we have that: $P_{\bar{a}}(I) = 1$ if and only if $P^{HVE}_{\sigma_*(\bar{a})}(\sigma(I)) = 1$. Therefore, correctness and security follow from the properties of the HVE. We thus obtain the following immediate theorem.

Theorem 1. $(Setup, Encrypt, GenToken, Query)$ *is a selectively secure* $\Phi_{n,w}$- *searchable system assuming* $(Setup_{HVE}, Encrypt_{HVE}, GenToken_{HVE}, Query_{HVE})$ *is an HVE over* Σ_{01}^{nw}.

Conjunctive range queries. We note that a system that supports comparison queries can also support range queries. To search for plaintexts where $x \in [a, b]$ the encryptor encrypts the pair (x, x). The predicate then tests $x \geq a \land x \leq b$.

6.1   Subset Queries

Next, we show how to search for general subset predicates. Let T be a set of size n. For a subset $A \subseteq T$ we define a subset predicate as follows:

$$P_A(x) = \begin{cases} 1 & \text{if } x \in A \\ 0 & \text{otherwise} \end{cases}$$

We wish to support searches for any subset predicate. More generally, we wish to support searches for conjunctive subset predicates over T^w. That is, let $\sigma = (A_1, \ldots, A_w)$ be a w-tuple where $A_i \in T$ for all $i = 1, \ldots, w$. Then σ is an elements of $(2^T)^w$. Define the predicate $P_\sigma : T^w \to \{0, 1\}$ as follows:

$$P_\sigma((x_1, \ldots, x_w)) = \begin{cases} 1 & \text{if } x_i \in A_i \text{ for all } i = 1, \ldots, w, \\ 0 & \text{otherwise} \end{cases}$$

Let $\Phi = \{ P_\sigma \text{ for all } \sigma \in (2^T)^w \}$. Note that Φ is huge — its size is 2^{nw}. The Φ-searchable system is as follows:

- $Encrypt(PK, I, M)$ where $I = (x_1, \ldots, x_w) \in T^w$. Build a vector $\sigma(S) = (\sigma_{i,j}) \in \Sigma_{01}^{nw}$ as:

$$\sigma_{i,j} = \begin{cases} 1 & \text{if } x_i = j, \\ 0 & \text{otherwise} \end{cases} \tag{6}$$

Then output $Encrypt_{HVE}(PK, \sigma(I), M)$. The ciphertext size is $O(nw)$ as was the case for comparison queries.

- *GenToken*(SK, $\langle P_\alpha \rangle$) where $\alpha = (A_1, \ldots, A_w)$. Define $\sigma_*(\alpha) = (\sigma_{i,j}) \in \Sigma_{01*}^{nw}$ as follows:

$$\sigma_{i,j} = \begin{cases} 0 & \text{if } j \notin A_i, \\ * & \text{otherwise} \end{cases} \tag{7}$$

Output $\text{TK}_\alpha \xleftarrow{R} GenToken_{HVE}(\text{SK}, \sigma_*(\alpha))$. The token size is $O(nw)$, which is bigger than tokens for comparison queries.

- *Setup* and *Query* are the same algorithms from the HVE system, as for comparison queries.

It is easiest to see how this works in the one dimensional setting, namely $w = 1$. We encrypt a value $x \in T$ using an HVE vector

$$\sigma(x) = \boxed{\begin{array}{c|c|c|c|c|c|c} 0 & \cdots & 0 & 1 & 0 & \cdots & 0 \end{array}} \quad \in \{0,1\}^n$$

Consider a predicate P_A where, for example, $A = \{2, 3, n\} \subseteq T$. We generate a token for P_A by calling $GenToken_{HVE}(\text{SK}, \sigma_*(A))$ using the HVE vector

$$\sigma_*(A) = \boxed{\begin{array}{c|c|c|c|c|c|c} 0 & * & * & 0 & 0 & \cdots & 0 & * \end{array}} \quad \in \{*, 1\}^n$$

The main point is that $x \in A$ if and only if $P_{\sigma_*(A)}^{\text{HVE}}(\sigma(x)) = 1$. Therefore, correctness and security follow from the properties of the HVE. We obtain a secure system for subset queries for *arbitrary subsets*.

Theorem 2. (*Setup, Encrypt, GenToken, Query*) is a selectively secure Φ- searchable system assuming ($Setup_{HVE}, Encrypt_{HVE}, GenToken_{HVE}, Query_{HVE}$) is an HVE over Σ_{01}^{nw}.

Note that the trivial system of Section 3 for subset queries produces ciphertexts of size $O(2^n)$. The construction above generates ciphertexts of size $O(n)$.

Subset queries on large domains using Bloom filters. So far we considered subset queries over a domain of size n. In Section 1 we presented examples where one wishes to test a subset relation over a large domain. For example, we discussed email filtering queries of type (sender $\in S$) where S is a set of email addresses. To use our construction one would first hash email addresses to a set $\{1, \ldots, n\}$ for some n, using a publicly known hash function, and then use the HVE for small domain.

Unfortunately, by hashing into a small domain there is some chance for false positives, namely *Query* may output M even though (sender $\notin S$). False positives result from hash collisions. The false positive probability can be reduced by a standard application of Bloom filters [5]. Instead of using one hash function, we use multiple functions $H_1, \ldots, H_d : \{0,1\}^* \to T$. Again, consider the one-dimensional case, namely $w = 1$. To encrypt a word $W \in \{0,1\}^*$ the encryptor creates a vector $\sigma(W) \in \{0,1\}^n$ that contains a '1' at positions

$H_1(W), \ldots, H_d(W)$ and '0' everywhere else. The encryptor then runs $Encrypt($ PK, $\sigma(W), M)$.

To generate a token for a set $A = \{W_1, \ldots, W_s\}$ the *GenToken* algorithm builds a vector $\sigma_*(A) \in \{0, *\}^n$ that contains $*$ at positions $H_i(W_j)$, for all $i = 1, \ldots, d$ and $j = 1, \ldots, s$, and contains '0' everywhere else. By choosing n and d appropriately, the false positive probability can be made arbitrarily small.

Another subset query application. In our subset query application we identified a ciphertext with an element x and a user's token with a set A. This allowed us to test whether $x \in A$. We observe that we can easily apply HVE to achieve the opposite semantics where a user's key is associated with an element x and the ciphertext with a set A. This could be used by a gateway to test if a particular user was one of the (possibly) many receivers of an email. We expect there to be several other applications that one can build with HVE.

7 Extensions

Privacy for search queries. In some cases one may want the token TK_P not to identify which predicate P is being queried. For example, in the anti-spam example from the introduction, the user may not want to reveal his anti-spam predicate to the server. A similar problem was studied by Ostrovsky and Skeith [18] and is related to Private Information Retrieval [16]. For public-key systems supporting comparison queries this is clearly not possible since, given TK_P the server can identify the threshold in P with a simple binary search. It is an open problem to convert our system to a symmetric-key system where TK_P does not expose P. One approach is to simply keep the public key secret from the server; however, this is not sufficient in our system.

Validating ciphertexts. Throughout the paper we assumed that the encryptor is honestly creating ciphertexts as specified by the encryption system. For some applications discussed in the introduction (e.g. spam filtering) this may not be the case. By creating malformed ciphertexts an attacker may generate false-positive or false-negatives for the server using the tokens.

Fortunately, in some settings including a payment gateway or spam filter, this is easily avoidable. Briefly, one technique is as follows. The recipient who has SK will also publish a regular public-key PK_1 and ask the encryptor to encrypt the plaintext (I, M) with both the searchable system and with PK_1. The resulting ciphertext is the pair $C = \big(Encrypt(\mathrm{PK}, I, M), Encrypt_{PKE}(\mathrm{PK}_1, (I, M))\big)$. When the recipient receives a ciphertext $C = (C_0, C_1)$ it recovers (I, M) from C_1 and uses SK to test that C_0 is a valid encryption of (I, M). If not then the ciphertext is immediately rejected. In doing so, the recipient automatically drops invalid ciphertexts. More precisely, a Φ-searchable system could provide an algorithm $Test(C, I, M, \mathrm{SK})$ that outputs true when C is a valid encryption of (I, M) and false otherwise. Our HVE system supports this type of test.

Alternatively, one could require the encryptor to prove that his ciphertext is well formed, for example to prove that C_0 is consistent with C_1. This can be done using non-interactive proof techniques [6, 7].

8   Conclusion

In public key systems supporting queries on encrypted data a secret key can produce tokens for testing any supported query predicate. The token lets anyone test the predicate on a given ciphertext without learning any other information about the plaintext. We presented a general framework for analyzing security of searching on encrypted data systems. We then constructed systems for comparisons and subset queries as well as conjunctive versions of these predicates.

The underlying tool behind these new constructions is a primitive we call HVE. The one-dimensional version of HVE (namely $\ell = 1$) is essentially an Anonymous IBE system. For large ℓ we obtain a new concept that is extremely useful for a large variety of searching predicates. We note that by setting $\ell = 1$ in our HVE construction we obtain a new simple anonymous IBE system secure without random oracles.

This work posses many challenging open problems. For example, the best non-conjunctive (i.e. $w = 1$) comparison system we currently have requires ciphertexts of size $O(\sqrt{n})$ where n is the domain size. In principal it should be possible to improve this to $O(\log n)$, but this is currently a wide open problem that will require new ideas. Similarly, for non-conjunctive subset queries the best we have requires ciphertexts of size $O(n)$. Again, can this be improved to $O(\log n)$? Our results mostly focus on conjunction. Are there similar results for disjunctive queries? More generally, what other classes of predicates can we search on?

Acknowledgments

We thank Amit Sahai and Alice Silverberg for helpful comments about this work.

References

[1] Michel Abdalla, Mihir Bellare, Dario Catalano, Eike Kiltz, Tadayoshi Kohno, Tanja Lange, John Malone-Lee, Gregory Neven, Pascal Paillier, and Haixia Shi. Searchable encryption revisited: Consistency properties, relation to anonymous ibe, and extensions. In *CRYPTO*, pages 205–222, 2005.

[2] Mihir Bellare, Alexandra Boldyreva, and Adam O'Neill. Efficiently-searchable and deterministic asymmetric encryption. http://eprint.iacr.org/2006/186, 2006.

[3] J. Bethencourt, H. Chan, A. Perrig, E. Shi, and D. Song. Anonymous multi-attribute encryption with range query and conditional decryption. Technical report, C.M.U, 2006. CMU-CS-06-135.

[4] John Bethencourt, Dawn Song, and Brent Waters. New constructions and practical applications for private stream searching. In *Proceeding of 2006 IEEE Symposium on Security and Privacy*, 2006.

[5] Burton H. Bloom. Space/time trade-offs in hash coding with allowable errors. *Communications of the ACM*, 13:422–426, 1970.

[6] Manuel Blum, Paul Feldman, and Silvio Micali. Non-interactive zero-knowledge and its applications (extended abstract). In *STOC*, pages 103–112, 1988.

[7] Manuel Blum, Alfredo De Santis, Silvio Micali, and Giuseppe Persiano. Noninteractive zero-knowledge. *SIAM J. Comput.*, 20(6):1084–1118, 1991.

[8] Dan Boneh, Giovanni Di Crescenzo, Rafial Ostrovsky, and Giuseppe Persiano. Public key encryption with keyword search. In *Proceedings of Eurocrypt '04*, 2004.

[9] Dan Boneh, Eu-Jin Goh, and Kobbi Nissim. Evaluating 2-dnf formulas on ciphertexts. In Joe Kilian, editor, *Proceedings of Theory of Cryptography Conference 2005*, volume 3378 of *LNCS*, pages 325–342. Springer, 2005.

[10] Dan Boneh, Amit Sahai, and Brent Waters. Fully collusion resistant traitor tracing with short ciphertexts and private keys. In *Eurocrypt '06*, 2006.

[11] Dan Boneh and Brent Waters. Conjunctive, subset, and range queries on encrypted data. Cryptology ePrint Archive, Report 2006/287, 2006. http://eprint.iacr.org/.

[12] Dan Boneh and Brent Waters. A fully collusion resistant broadcast trace and revoke system with public traceability. In *ACM Conference on Computer and Communication Security (CCS)*, 2006.

[13] Xavier Boyen and Brent Waters. Anonymous hierarchical identity-based encryption (without random oracles). In *Crypto '06*, 2006.

[14] O. Goldreich and R. Ostrovsky. Software protection and simulation by oblivious rams. *JACM*, 1996.

[15] Philippe Golle, Jessica Staddon, and Brent R. Waters. Secure conjunctive keyword search over encrypted data. In *ACNS*, pages 31–45, 2004.

[16] Eyal Kushilevitz and Rafail Ostrovsky. Replication is not needed: Single database, computationally-private information retrieval. In *FOCS*, pages 364–373, 1997.

[17] Rafail Ostrovsky. *Software protection and simulation on oblivious RAMs*. PhD thesis, M.I.T, 1992. Preliminary version in STOC 1990.

[18] Rafail Ostrovsky and William Skeith. Private searching on streaming data. In *Proceedings of Crypto 2005*, LNCS. Springer, 2005.

[19] Dawn Song, David Wagner, and Adrian Perrig. Practical techniques for searches on encrypted data. In *Proceedings of the 2000 IEEE symposium on Security and Privacy (S&P 2000)*, 2000.

[20] Brent Waters, Dirk Balfanz, Glenn Durfee, and Dianna Smetters. Building an encrypted and searchabe audit log. In *Proceedings of NDSS '04*, 2004.

A   Proof of Lemma 1

We prove that the trivial system presented in Section 3 is secure.

Proof. Showing that $\mathsf{QU\,Adv}_{\mathcal{A}}$ is negligible is a straight forward hybrid argument. Let $\mathcal{A}$ be an adversary playing the query security game. For $i = 1, \ldots, n+1$ we define experiment number i as follows:

– The challenger runs $Setup(\lambda)$ to obtain

$$PK \leftarrow (PK_1, \ldots, PK_n) \quad \text{and} \quad SK \leftarrow (SK_1, \ldots, SK_n)$$

It gives PK to $\mathcal{A}$. Next, $\mathcal{A}$ is given the tokens for any predicates of its choice.

– Then $\mathcal{A}$ outputs two pairs (I_0, M_0) and (I_1, M_1) subject to the restrictions of the query security game challenge phase. For $j = 1, \ldots, n$ the challenger constructs the following ciphertexts:

$$C_j \overset{R}{\leftarrow} \begin{cases} Encrypt'(PK_j, \ M_0) & \text{if } P_j(I_0) = 1 \text{ and } j \geq i, \\ Encrypt'(PK_j, \ M_1) & \text{if } P_j(I_1) = 1 \text{ and } j < i, \\ Encrypt'(PK_j, \ \bot) & \text{otherwise} \end{cases}$$

The challenger gives $C \leftarrow (C_1, \ldots, C_n)$ to $\mathcal{A}$.

– The adversary continues to adaptively request query tokens subject to the restrictions of the query security game. Finally, $\mathcal{A}$ outputs a bit $\beta' \in \{0, 1\}$. We let $\mathsf{EXP}^{(i)}_{\mathrm{QU}}[\mathcal{A}]$ denote the probability that β' equals 1.

This completes the description of experiment i. A standard argument shows that

$$2 \cdot \mathsf{QU\,Adv}_{\mathcal{A}} = \left| \mathsf{EXP}^{(1)}_{\mathrm{QU}}[\mathcal{A}] - \mathsf{EXP}^{(n+1)}_{\mathrm{QU}}[\mathcal{A}] \right| \leq \sum_{i=1}^{n} \left| \mathsf{EXP}^{(i)}_{\mathrm{QU}}[\mathcal{A}] - \mathsf{EXP}^{(i+1)}_{\mathrm{QU}}[\mathcal{A}] \right|$$

But $\left| \mathsf{EXP}^{(i)}_{\mathrm{QU}}[\mathcal{A}] - \mathsf{EXP}^{(i+1)}_{\mathrm{QU}}[\mathcal{A}] \right|$ is clearly negligible assuming $\mathcal{E}$ is semantically secure against chosen plaintext attacks.

How to Shuffle in Public

Ben Adida[1,*] and Douglas Wikström[2]

[1] MIT, Computer Science and Artificial Intelligence Laboratory
ben@mit.edu
[2] ETH Zürich, Department of Computer Science
douglas@inf.ethz.ch

Abstract. We show how to obfuscate a secret shuffle of ciphertexts: shuffling becomes a public operation. Given a trusted party that samples and obfuscates a shuffle *before* any ciphertexts are received, this reduces the problem of constructing a mix-net to verifiable joint decryption.

We construct public-key obfuscations of a decryption shuffle based on the Boneh-Goh-Nissim (BGN) cryptosystem and a re-encryption shuffle based on the Paillier cryptosystem. Both allow *efficient* distributed verifiable decryption.

Finally, we give a distributed protocol for sampling and obfuscating each of the above shuffles and show how it can be used in a trivial way to construct a universally composable mix-net. Our constructions are practical when the number of senders N is small, yet large enough to handle a number of practical cases, e.g. $N = 350$ in the BGN case and $N = 2000$ in the Paillier case.

1 Introduction

Suppose a set of senders $\mathcal{P}_1, \ldots, \mathcal{P}_N$, each with input m_i, want to compute the sorted list $(m_{\pi(1)}, \ldots, m_{\pi(N)})$ of messages while keeping the permutation π secret. A trusted party can provide this service. First, it collects all messages. Then, it sorts the inputs and outputs the result. A protocol, i.e., a list of machines $\mathcal{M}_1, \ldots, \mathcal{M}_k$, that emulates the service of this trusted party is called a *mix-net*, and the parties $\mathcal{M}_1, \ldots, \mathcal{M}_k$ are referred to as *mix servers*. The notion of a mix-net was introduced by Chaum [9] and the main application of mix-nets is to perform electronic elections.

Program obfuscation is the process of "muddling" a program's instructions to prevent reverse-engineering while preserving proper function. Barak et al. [2] first formalized obfuscation as simulatability from black-box access. Goldwasser and Tauman-Kalai [15] extended this definition to consider auxiliary inputs. Some simple programs have been successfully obfuscated [8,26]. However, generalized program obfuscation, though it would be fantastically useful in practice, has been proven impossible in even the weakest of settings for both models (by their respective authors). Ostrovsky and Skeith [21] consider a weaker model, public-key obfuscation, where the obfuscated program's output is encrypted. In this model, they achieve the more complex application of private stream searching.

---

* with support from the Caltech/MIT Voting Technology Project and the Knight Foundation.

1.1 Our Contributions

We show how to obfuscate the shuffle phase of a mix-net: shuffling becomes a public operation, leaving only verifiable decryption to be performed privately. We show how any homomorphic cryptosystem can provide obfuscated mixing, though the resulting mix-net becomes inefficient. We show how special and distinct properties of the Boneh-Goh-Nissim [7] and Paillier [22] cryptosystems enable obfuscated mixing efficient enough to be practical in some settings.

We formalize our constructions in the public-key obfuscation model of Ostrovsky and Skeith, whose indistinguishability property closely matches the security requirements of a mix-net. Of course, in a mix-net setting, one cannot expect a single party to generate a complete and correct obfuscated shuffle: we describe an efficient zero-knowledge proof of the correct obfuscation of a shuffle and a protocol that allows a set of parties to jointly and robustly generate an obfuscated randomly chosen shuffle. Our shuffles require considerably more exponentiations, roughly quadratic in the number of senders instead of linear, than private mix-net techniques, yet they remain reasonably practical for precinct-based elections, where voters are anonymized in smaller batches and all correctness proofs can be carried out in advance.

1.2 Previous Work

Most mix-nets in the literature are based on homomorphic cryptosystems and use the re-encryption-permutation paradigm introduced by Park et al. [23] and made universally verifiable by Sako and Kilian [24]. Each mix server in turn re-encrypts and permutes the ciphertexts. The first efficient zero-knowledge shuffle proofs were given independently by Neff [20] and Furukawa and Sako [14]. Groth [17] generalized Neff's approach and improved its efficiency. A third, different, approach was given recently by Wikström [28]. The first definition of security of a mix-net was given by Abe and Imai [1] and the first proof of security of a mix-net as a whole was given by Wikström [27,28]. Wikström and Groth [30] give the first adaptively secure mix-net.

Multi-candidate election schemes where the set of candidates is predetermined have been proposed using homomorphic encryption schemes, initially by Benaloh [6,5] and subsequently by others to handle multiple races and multiple candidates per race [10,25,11,13,3,17]. Homomorphic tallying is similar to obfuscated shuffles in that, on and after election day, only public computation is required for the anonymization process. However, homomorphic tallying cannot recover the individual input plaintexts, which is required by the election laws in some countries and in the case of write-in votes.

Ostrovsky and Skeith define the notion of public-key obfuscation to describe and analyze their work on streaming-data search using homomorphic encryption [21]. In their definition, an obfuscated program is run on plaintext inputs and provides the outputs of the original program in encrypted form. We use a variation of this definition, where the inputs are encrypted and the unobfuscated program may depend on the public key of the cryptosystem.

1.3   Overview of Techniques

The protocols presented in this work use homomorphic multiplication with a permutation matrix. Roughly, the semantic security of the encryption scheme hides the permutation.

Generic Construction. Consider two semantically secure cryptosystems, $\mathcal{CS} = (\mathcal{G}, \mathcal{E}, \mathcal{D})$ and $\mathcal{CS}' = (\mathcal{G}', \mathcal{E}', \mathcal{D}')$, where $\mathcal{CS}'$ is additively homomorphic and the plaintext space of $\mathcal{CS}'$ can accommodate any ciphertext from $\mathcal{CS}$. Note the interesting properties:

$$\mathcal{E}'_{pk'}(1)^{\mathcal{E}_{pk}(m)} = \mathcal{E}'_{pk'}(\mathcal{E}_{pk}(m)) \ , \quad \mathcal{E}'_{pk'}(0)^{\mathcal{E}_{pk}(m)} = \mathcal{E}'_{pk'}(0) \ , \quad \text{and}$$

$$\mathcal{E}'_{pk'}(0)\mathcal{E}'_{pk'}(\mathcal{E}_{pk}(m)) = \mathcal{E}'_{pk'}(\mathcal{E}_{pk}(m)) \ .$$

Consider the element-wise encryption of a permutation matrix under $\mathcal{E}'$, and consider inputs to the shuffle as ciphertexts under $\mathcal{E}$. Homomorphic matrix multiplication can be performed using the properties above for multiplication and addition. The result is a list of *doubly encrypted* messages, $\mathcal{E}'_{pk'}(\mathcal{E}_{pk}(m_i))$, that must then be decrypted verifiably. Unfortunately, a proof of double decryption is particularly inefficient because revealing any intermediate ciphertext $\mathcal{E}_{pk}(m_i)$ is not an option, as it would immediately leak the permutation.

BGN Construction. The BGN cryptosystem is additively homomorphic and has two encryption algorithms and two decryption algorithms that can be used with the same keys. Both additive and multiplicative homomorphisms are provided in the following sense:

$$\mathcal{E}_{pk}(m_1) \otimes \mathcal{E}_{pk}(m_2) = \mathcal{E}'_{pk}(m_1 m_2) \ , \quad \mathcal{E}_{pk}(m_1)\mathcal{E}_{pk}(m_2) = \mathcal{E}_{pk}(m_1 + m_2) \ ,$$

$$\text{and} \quad \mathcal{E}'_{pk}(m_1)\mathcal{E}'_{pk}(m_2) = \mathcal{E}'_{pk}(m_1 + m_2) \ .$$

Thus, both the matrix and the inputs can be encrypted using the *same* encryption algorithm $\mathcal{E}$ and public key, and the matrix multiplication uses both homomorphisms. The result is a list of singly encrypted ciphertexts under $\mathcal{E}'$, which lends itself to efficient, provable decryption.

Paillier Construction. The Paillier cryptosystem is additively homomorphic and supports layered encryption, where a ciphertext can be encrypted again using the same public key. The homomorphic properties are preserved in the inner layer; in addition to the generic layered homomorphic properties we have the special relation

$$\mathcal{E}'_{pk}(\mathcal{E}_{pk}(0, r))^{\mathcal{E}_{pk}(m,s)} = \mathcal{E}'_{pk}(\mathcal{E}_{pk}(0, r)\mathcal{E}_{pk}(m, s)) = \mathcal{E}'_{pk}(\mathcal{E}_{pk}(m, r + s)) \ .$$

Thus, we can use $\mathcal{E}'$ encryption for the permutation matrix, and $\mathcal{E}$ encryption for the inputs. When representing the permutation matrix under $\mathcal{E}'$, instead of $\mathcal{E}'_{pk}(1)$ to represent a one we use $\mathcal{E}'_{pk}(\mathcal{E}_{pk}(0, r))$ with a random r. During the matrix multiplication, the "inner" $\mathcal{E}_{pk}(0, r)$ performs re-encryption on the inputs, which allows the decryption process to reveal the intermediate ciphertext without leaking the permutation, making the decryption proof much more efficient.

2 Preliminaries

2.1 Notation

We denote by κ the main security parameter and say that a function $\epsilon(\cdot)$ is negligible if for every constant c there exists a constant κ_0 such that $\epsilon(\kappa) < \kappa^{-c}$ for $\kappa > \kappa_0$. We denote by κ_c and κ_r additional security parameters such that $2^{-\kappa_c}$ and $2^{-\kappa_r}$ are negligible, which determines the bit-size of challenges and random paddings in our protocols. We denote by PT, PPT, and PT*, the set of uniform polynomial time, probabilistic uniform polynomial time, and non-uniform polynomial time Turing machines respectively. In interactive protocols we denote by $\mathcal{P}$ the prover and $\mathcal{V}$ the verifier. We understand a proof of knowledge to mean a complete proof of knowledge with overwhelming soundness and negligible knowledge error. We denote by Σ_N the set of permutations of N elements, and we write $\Lambda^\pi = (\lambda_{ij}^\pi)$ for the permutation matrix of $\pi \in \Sigma_N$. We denote by M_{pk}, R_{pk}, and C_{pk}, the plaintext space, the randomizer space, and the ciphertext space induced by the public key pk of some cryptosystem.

2.2 Homomorphic Cryptosystems

In the following definition we mean by abelian group a specific representation of an abelian group for which there exists a polynomial time algorithm for computing the binary operator and inversion.

Definition 1 (Homomorphic). *A cryptosystem $\mathcal{CS} = (\mathcal{G}, \mathcal{E}, \mathcal{D})$ is homomorphic if for every key pair $(pk, sk) \in \mathcal{G}(1^\kappa)$*

1. *The message space M_{pk} is a subset of an abelian group $G(\mathsf{M}_{pk})$ written additively.*
2. *The randomizer space R_{pk} is an abelian group written additively.*
3. *The ciphertext space C_{pk} is a abelian group written multiplicatively.*
4. *For every $m, m' \in \mathsf{M}_{pk}$ and $r, r' \in \mathsf{R}_{pk}$ we have $\mathcal{E}_{pk}(m, r)\mathcal{E}_{pk}(m', r') = \mathcal{E}_{pk}(m + m', r + r')$.*

Furthermore, if $\mathsf{M}_{pk} = G(\mathsf{M}_{pk})$ it is called fully homomorphic, and if $G(\mathsf{M}_{pk}) = \mathbb{Z}_n$ for some integer $n > 0$ it is called additive.

For an additively homomorphic cryptosystem, $\mathcal{RE}_{pk}(c, r) = c\mathcal{E}_{pk}(0, r)$ is called a re-encryption algorithm.

2.3 Functionalities

Definition 2 (Functionality). *A functionality is a family $\mathcal{F} = \{\mathcal{F}_\kappa\}_{\kappa \in \mathbb{N}}$ of sets of circuits such that there exists a polynomial $s(\cdot)$ such that $|F| \leq s(\kappa)$ for every $F \in \mathcal{F}_\kappa$.*

Specifics of the Model. In the original Ostrovsky and Skeith definition [21], plaintext inputs are processed by an obfuscated program into ciphertext outputs. In our setting, inputs are already encrypted. Thus, the original functionality and the obfuscator depend on the *same* public key (and possibly the secret key). The security of the obfuscation—i.e. the indistinguishability property—is then defined separately, following the pattern of Ostrovsky and Skeith.

In addition, as this is a *public-key* obfuscator, the output of the obfuscated program requires a decryption. We call the reader's attention to the *difference between the encryption layers*: though they may use the same public key, the obfuscation-related encryption and the inputs' encryption have distinct purposes.

Definition 3 (Public-Key Obfuscator). *An algorithm $\mathcal{O} \in \mathrm{PPT}$ is a public-key obfuscator for a functionality $\mathcal{F}$ with respect to a cryptosystem $\mathcal{CS} = (\mathcal{G}, \mathcal{E}, \mathcal{D})$ if there exists a decryption algorithm $\mathcal{D}' \in \mathrm{PT}$ and a polynomial $s(\cdot)$ such that for every $\kappa \in \mathbb{N}$, $F \in \mathcal{F}_\kappa$, $(pk, sk) \in \mathcal{G}(1^\kappa)$, and $x \in \{0,1\}^*$,*

1. CORRECTNESS. $\mathcal{D}'_{sk}(\mathcal{O}(1^\kappa, pk, sk, F)(x)) = F(pk, sk, x)$.
2. POLYNOMIAL BLOW-UP. $|\mathcal{O}(1^\kappa, pk, sk, F)| \leq s(|F|)$.

Example 1. Suppose $\mathcal{CS}$ is additively homomorphic, $(pk, sk) \in \mathcal{G}(1^\kappa)$, $a \in \mathsf{M}_{pk}$, and define $F_a(pk, sk, x) = ax$, where $x \in \mathsf{M}_{pk}$. An obfuscated circuit for functionality $\mathcal{F}$ of such circuits can be defined as a circuit with $\mathcal{E}_{pk}(a)$ hardcoded which, on input $x \in \mathsf{M}_{pk}$, outputs $\mathcal{E}_{pk}(a)^x = \mathcal{E}_{pk}(ax)$.

We extend the definition of polynomial indistinguishability (known as IND-CPA security for public-key cryptosystems) to our public-key obfuscator.

Experiment 1 (Indistinguishability, $\mathrm{Exp}^{\mathrm{oind}-b}_{\mathcal{F},\mathcal{CS},\mathcal{O},\mathcal{A}}(\kappa)$)

$$(pk, sk) \leftarrow \mathcal{G}(1^\kappa)$$
$$(F_0, F_1, state) \leftarrow \mathcal{A}(\mathsf{choose}, pk),$$
$$d \leftarrow \mathcal{A}(\mathcal{O}(1^\kappa, pk, sk, F_b), state)$$

If $F_0, F_1 \in \mathcal{F}_\kappa$ return d, otherwise 0.

Definition 4 (Indistinguishability). *A public-key obfuscator $\mathcal{O}$ for a functionality $\mathcal{F}$ with respect to a cryptosystem $\mathcal{CS} = (\mathcal{G}, \mathcal{E}, \mathcal{D})$ is polynomially indistinguishable if $|\Pr[\mathrm{Exp}^{\mathrm{oind}-0}_{\mathcal{F},\mathcal{CS},\mathcal{O},\mathcal{A}}(\kappa) = 1] - \Pr[\mathrm{Exp}^{\mathrm{oind}-1}_{\mathcal{F},\mathcal{CS},\mathcal{O},\mathcal{A}}(\kappa) = 1]|$ is negligible.*

The obfuscator in Example 1 is polynomially indistinguishable if $\mathcal{CS}$ is polynomially indistinguishable (IND-CPA secure.)

2.4   Shuffles

The most basic form of a shuffle is the decryption shuffle. It simply takes a list of ciphertexts, decrypts them and outputs the plaintexts in sorted order. In some sense this is equivalent to a mix-net.

Definition 5 (Decryption Shuffle). *A CS-decryption shuffle, for a cryptosystem $CS = (\mathcal{G}, \mathcal{E}, \mathcal{D})$ is a functionality $\mathcal{DS}_N = \{\mathcal{DS}_{N(\kappa),\kappa}\}_{\kappa \in \mathbb{N}}$, where $N(\kappa)$ is a polynomially bounded and polynomially computable function, such that for every $\kappa \in \mathbb{N}$, $\mathcal{DS}_{N(\kappa),\kappa} = \{DS_\pi\}_{\pi \in \Sigma_{N(\kappa)}}$, and for every $(pk, sk) \in \mathcal{G}(1^\kappa)$, and $c_1, \ldots, c_{N(\kappa)} \in \mathsf{C}_{pk}$ the circuit DS_π is defined by*

$$DS_\pi(pk, sk, (c_1, \ldots, c_{N(\kappa)})) = (\mathcal{D}_{sk}(c_{\pi(1)}), \ldots, \mathcal{D}_{sk}(c_{\pi(N(\kappa))})) \ .$$

Another way to implement a mix-net is to use the re-encryption-permutation paradigm of Park et al. [23]. Using this approach the ciphertexts are first re-encrypted and permuted in a joint way and then decrypted. The re-encryption shuffle below captures the joint re-encryption and permutation phase. Both types of shuffles are illustrated, in their obfuscated form, in Figure 1.

Definition 6 (Re-encryption Shuffle). *A CS-re-encryption shuffle, for a homomorphic cryptosystem CS is a functionality $\mathcal{RS}_N = \{\mathcal{RS}_{N(\kappa),\kappa}\}_{\kappa \in \mathbb{N}}$, where $N(\kappa)$ is a polynomially bounded and polynomially computable function, such that for every $\kappa \in \mathbb{N}$, $\mathcal{RS}_{N(\kappa),\kappa} = \{RS_{\pi,r}\}_{\pi \in \Sigma_{N(\kappa)}, r \in (\{0,1\}^*)^{N(\kappa)}}$, and for every $(pk, sk) \in \mathcal{G}(1^\kappa)$, and $c_1, \ldots, c_{N(\kappa)} \in \mathsf{C}_{pk}$ the circuit $RS_{\pi,r}$ is defined by*

$$RS_\pi(pk, sk, (c_1, \ldots, c_{N(\kappa)})) = (\mathcal{RE}_{pk}(c_{\pi(1)}, r_1), \ldots, \mathcal{RE}_{pk}(c_{\pi(N(\kappa))}, r_{N(\kappa)})) \ .$$

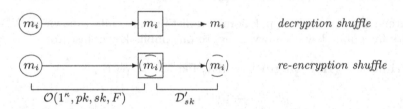

Fig. 1. The obfuscation of two types of shuffles. The circle denotes the encryption scheme under which inputs are encrypted. The square denotes the encryption scheme used by the obfuscator, which may depend on the circle encryption scheme and its keypair. The left-most inputs and right-most outputs do not include the square-layer encryption, which is only used by the public-key obfuscation process. An obfuscated decryption shuffle "swaps" one encryption scheme for the other, while an obfuscated re-encryption shuffle "layers" the two encryption schemes. The dashed circle denotes a re-encryption of the original ciphertext.

3   A Generic Decryption Shuffle

We show that, in principle, all that is needed is an additively homomorphic cryptosystem. Consider two semantically secure cryptosystems, $CS = (\mathcal{G}, \mathcal{E}, \mathcal{D})$ and $CS' = (\mathcal{G}', \mathcal{E}', \mathcal{D}')$, with CS' being additively homomorphic. Suppose that ciphertexts from CS can be encrypted under CS' for all $(pk, sk) \in \mathcal{G}(1^\kappa)$ and $(pk', sk') \in \mathcal{G}'(1^\kappa)$, i.e., $\mathsf{C}_{pk} \subseteq \mathsf{M}'_{pk'}$. The following operations are then possible and, more interestingly, indistinguishable thanks to the semantic security of the first cryptosystem:

$$\mathcal{E}'_{pk'}(1)^{\mathcal{E}_{pk}(m)} = \mathcal{E}'_{pk'}(\mathcal{E}_{pk}(m)) \quad \text{and} \quad \mathcal{E}'_{pk'}(0)^{\mathcal{E}_{pk}(m)} = \mathcal{E}'_{pk'}(0) \ .$$

3.1 The Obfuscator

Consider a permutation matrix $\Lambda^\pi = (\lambda^\pi_{ij})$ corresponding to a permutation π. Consider its element-wise encryption under $\mathcal{CS}'$ with public key pk' and a corresponding matrix of random factors $(r_{ij}) \in \mathsf{R}'^{N^2}_{pk'}$, i.e., $C^\pi = (\mathcal{E}'_{pk'}(\lambda^\pi_{ij}, r_{ij}))$. Then given $d = (d_1, d_2, \ldots, d_N) \in \mathsf{C}^N_{pk}$ it is possible to perform homomorphic matrix multiplication as

$$d \star C^\pi = \left(\prod_{i=1}^N (c^\pi_{ij})^{d_i}\right) \quad \text{giving} \quad \mathcal{D}_{sk}(\mathcal{D}'_{sk'}(d \star C^\pi)) = (m_{\pi(i)})^N_{i=1} \ .$$

Definition 7 (Obfuscator). *The obfuscator $\mathcal{O}$ for the decryption shuffle $\mathcal{DS}_N$ takes input $(1^\kappa, (pk, pk'), (sk, sk'), DS_\pi)$, where $(pk, sk) \in \mathcal{G}(1^\kappa)$, $(pk', sk') \in \mathcal{G}'(1^\kappa)$ and $DS_\pi \in \mathcal{DS}_{N(\kappa),\kappa}$, computes $C^\pi = \mathcal{E}'_{pk'}(\Lambda^\pi)$, and outputs a circuit that hardcodes C^π, and on input $d = (d_1, \ldots, d_{N(\kappa)})$ computes $d' = d \star C^\pi$ as outlined above and outputs d'.*

Technically, this is a decryption shuffle of a new cryptosystem $\mathcal{CS}'' = (\mathcal{G}'', \mathcal{E}, \mathcal{D})$, where $\mathcal{CS}''$ executes the original key generators and outputs $((pk, pk'), (sk, sk'))$ and the original algorithms $\mathcal{E}$ and $\mathcal{D}$ simply ignore (pk', sk'). We give a reduction without any loss in security for the following straight-forward proposition. We also note that $\mathcal{O}$ does *not* use (sk, sk'): obfuscation only requires the public key.

Proposition 1. *If $\mathcal{CS}'$ is polynomially indistinguishable then $\mathcal{O}$ is polynomially indistinguishable.*

The construction can be generalized to the case where the plaintext space of $\mathcal{CS}'$ does not contain the ciphertext space of $\mathcal{CS}$. Each inner ciphertext d_i is split into pieces $(d_{i1}, \ldots, d_{it})$ each fitting in the plaintext space of $\mathcal{CS}'$ and then each list $(d_{1,l}, \ldots, d_{N,l})$ is applied to the encrypted permutation matrix as before. This gives lists $(d'_{1,l}, \ldots, d'_{N,l})$ from which the output $((d'_{1,l})_l, \ldots, (d'_{N,l})_l)$ is constructed.

3.2 Limitations of the Generic Construction

The matrix C^π requires a proof that it is the encryption of a proper permutation matrix. This can be accomplished using more or less general techniques depending on the cryptosystem, but this is prohibitively expensive in general.

Even if we prove that C^π is correctly formed, the post-shuffle verifiable decryption of $\mathcal{E}'_{pk'}(\mathcal{E}_{pk}(m_i))$ to m_i is prohibitively expensive: the inner, intermediate ciphertext $\mathcal{E}_{pk}(m_i)$ is exactly the input ciphertext, which means it cannot be revealed without trivially leaking the permutation. Given this constraint, we know of no efficient way, not even a cut-and-choose approach, to prove correct decryption. Instead, we turn to more efficient constructions.

4   Obfuscating a Boneh-Goh-Nissim Decryption Shuffle

We show how to obfuscate a decryption shuffle for the Boneh-Goh-Nissim (BGN)
cryptosystem [7] by exploiting both its additive homomorphism and its one-time
multiplicative homomorphism.

4.1   The BGN Cryptosystem

We denote the BGN cryptosystem by $\mathcal{CS}^{\mathsf{bgn}} = (\mathcal{G}^{\mathsf{bgn}}, \mathcal{E}^{\mathsf{bgn}}, \mathcal{D}^{\mathsf{bgn}})$. It operates in
two groups $\mathbb{G}_1$ and $\mathbb{G}_2$, both of order $n = q_1 q_2$, where q_1 and q_2 are distinct prime
integers of the same size. We use multiplicative notation in both $\mathbb{G}_1$ and $\mathbb{G}_2$, and
denote by g a generator in $\mathbb{G}_1$. The groups $\mathbb{G}_1$ and $\mathbb{G}_2$ exhibit a polynomial-time
computable bilinear map $e : \mathbb{G}_1 \times \mathbb{G}_1 \to \mathbb{G}_2$ such that $G = e(g, g)$ generates $\mathbb{G}_2$.
Bilinearity implies that $\forall u, v \in \mathbb{G}_1$ and $\forall a, b \in \mathbb{Z}, e(u^a, v^b) = e(u, v)^{ab}$. We refer
the reader to [7] for details on how such groups can be generated and on the
cryptosystem's properties, which we briefly summarize here.

Key generation. On input 1^κ, $\mathcal{G}^{\mathsf{bgn}}$ generates $(q_1, q_2, \mathbb{G}_1, g, \mathbb{G}_2, e(\cdot, \cdot))$ as above
such that $n = q_1 q_2$ is a κ-bit integer. It chooses $u \in \mathbb{G}_1$ randomly, defines
$h = u^{q_2}$, and outputs a public key $pk = (n, \mathbb{G}_1, \mathbb{G}_2, e(\cdot, \cdot), g, h)$ and secret key
$sk = (pk, q_1)$.

Encryption in $\mathbb{G}_1$. On input pk and m, $\mathcal{E}^{\mathsf{bgn}}$ selects $r \in \mathbb{Z}_n$ randomly and
outputs $c = g^m h^r$.

Decryption in $\mathbb{G}_1$. On input $sk = q_1$ and $c \in \mathbb{G}_1$, $\mathcal{D}^{\mathsf{bgn}}$ outputs $\log_{g^{q_1}}(c^{q_1})$.

Since decryption computes a discrete logarithm, the plaintext space must be
restricted considerably. Corresponding algorithms $\mathcal{E}'^{\mathsf{bgn}}$ and $\mathcal{D}'^{\mathsf{bgn}}$ perform encryption and decryption in $\mathbb{G}_2$ using the generators $G = e(g, g)$ and $H = e(g, h)$.
The BGN cryptosystem is semantically secure under the *Subgroup Decision Assumption*, which states that no $\mathcal{A} \in \mathrm{PT}^*$ can distinguish between the uniform
distributions on $\mathbb{G}_1$ and the unique order q_1 subgroup in $\mathbb{G}_1$ respectively.

Homomorphisms. The BGN cryptosystem is additively homomorphic. We need
this property, but we also exploit its one-time multiplicative homomorphism
implemented by the bilinear map:

$$e(\mathcal{E}_{pk}^{\mathsf{bgn}}(m_0, r_0), \mathcal{E}_{pk}^{\mathsf{bgn}}(m_1, r_1)) = \mathcal{E}_{pk}'^{\mathsf{bgn}}(m_0 m_1, m_0 r_1 + m_1 r_0 + (\log_g u) q_2 r_0 r_1)$$

The result is a ciphertext in $\mathbb{G}_2$ which cannot be efficiently converted back to
an equivalent ciphertext in $\mathbb{G}_1$. Thus, the multiplicative homomorphism can be
evaluated only once, after which only homomorphic additions are possible. For
clarity, we write $c_1 \otimes c_2 \stackrel{\text{def}}{=} e(c_1, c_2)$ for ciphertexts in $\mathbb{G}_1$.

4.2   The Obfuscator

Our obfuscator is based on the fact that matrix multiplication only requires
an arithmetic circuit with multiplication depth 1. Thus, the BGN cryptosystem

can be used for homomorphic matrix multiplication. Consider a $N_1 \times N_2$-matrix $C = (c_{ij}) = (\mathcal{E}_{pk}^{\text{bgn}}(a_{ij}))$ and a $N_2 \times N_3$-matrix $C' = (d_{jk}) = (\mathcal{E}_{pk}^{\text{bgn}}(b_{jk}))$, and let $A = (a_{ij})$ and $B = (b_{jk})$. Define homomorphic matrix multiplication by

$$C \star C' = \left(\prod_{j=1}^{N_2} c_{ij} \otimes d_{jk} \right) \quad \text{giving} \quad \mathcal{D}_{sk}^{'\text{bgn}}(C \star C') = \left(\sum_{j=1}^{N_2} a_{ij} b_{jk} \right) = AB .$$

Definition 8 (Obfuscator). *The obfuscator $\mathcal{O}^{\text{bgn}}$ for the decryption shuffle DS_N^{bgn} takes input $(1^\kappa, pk, sk, DS_\pi^{\text{bgn}})$, where $(pk, sk) \in \mathcal{G}^{\text{bgn}}(1^\kappa)$ and $DS_\pi^{\text{bgn}} \in DS_{N(\kappa),\kappa}^{\text{bgn}}$, computes $C^\pi = \mathcal{E}_{pk}^{\text{bgn}}(\Lambda^\pi)$, and outputs a circuit with C^π hard-coded such that, on input $d = (d_1, \ldots, d_{N(\kappa)})$, it outputs $d' = d \star C^\pi$.*

Note that $\mathcal{O}^{\text{bgn}}$ does not use sk.

Proposition 2. *The obfuscator $\mathcal{O}^{\text{bgn}}$ for DS_N^{bgn} is polynomially indistinguishable if the BGN cryptosystem is polynomially indistinguishable.*

Composition. Ciphertexts in $\mathbb{G}_2$ cannot be efficiently converted back into equivalent ciphertexts in $\mathbb{G}_1$. In addition, we do not know how to select groups $\mathbb{G}_1$ and $\mathbb{G}_2$ such that $\mathbb{G}_2$ exhibits a new bilinear map into a third group. Thus, the BGN-based shuffle construction we propose here is not composable: we can only mix once. In Section 7, we explain how to achieve the distributed generation of a BGN-based shuffle.

5   Obfuscating a Paillier Re-encryption Shuffle

We show how to obfuscate a re-encryption shuffle for the Paillier cryptosystem [22] by exploiting its additive homomorphism and its generalization introduced by Damgård et al. [12]. We expose a previously unnoticed homomorphic property of this generalized Paillier construction.

5.1   The Paillier Cryptosystem

We denote the Paillier cryptosystem $\mathcal{CS}^{\text{pai}} = (\mathcal{G}^{\text{pai}}, \mathcal{E}^{\text{pai}}, \mathcal{D}^{\text{pai}})$, defined as:

Key Generation. On input 1^κ, $\mathcal{G}^{\text{pai}}$ chooses safe κ-bit primes $p = 2p' + 1$ and $q = 2q' + 1$ randomly, defines a modulus $n = pq$, defines global parameter $v = n + 1$ and outputs a public key $pk = n$ and a secret key $sk = p$.
Encryption. On input pk and $m \in \mathbb{Z}_n$, $\mathcal{E}^{\text{pai}}$ selects $r \in \mathbb{Z}_n^*$ randomly and outputs $v^m r^n \bmod n^2$.
Decryption. On input sk and c, given e such that $e = 1 \bmod n$ and $e = 0 \bmod \phi(n)$, $\mathcal{D}^{\text{pai}}$ outputs $(c^e - 1)/n$.

The Paillier cryptosystem is polynomially indistinguishable under the *Decision Composite Residuosity Assumption*, which states that no $\mathcal{A} \in \text{PT}^*$ can distinguish the uniform distribution on $\mathbb{Z}_{n^2}^*$ from the uniform distribution on the subgroup of nth residues in $\mathbb{Z}_{n^2}^*$.

Generalized Paillier. Damgård et al. [12] generalize this scheme, replacing computations modulo n^2 with computations modulo n^{s+1} and plaintext space $\mathbb{Z}_n$ with $\mathbb{Z}_{n^s}$. Damgård et al. prove that the security of the generalized scheme follows from the security of the original scheme for $s > 0$ polynomial in the security parameter, though we only exploit the cases $s = 1, 2$. We write $\mathcal{E}^{\mathsf{pai}}_{n^{s+1}}(m) = v^m r^{n^s} \bmod n^{s+1}$ for generalized encryption to make explicit the value of s used in a particular encryption. Similarly we write $\mathcal{D}^{\mathsf{pai}}_{p,s+1}(c)$ for the decryption algorithm (see [12] for details) and we use $\mathsf{M}_{n^{s+1}}$ and $\mathsf{C}_{n^{s+1}}$ to denote the corresponding message and ciphertext spaces.

Alternative Encryption. There are well known alternative encryption algorithms. One can pick the random element $r \in \mathbb{Z}^*_{n^s}$ instead of in $\mathbb{Z}^*_n$. If h_{s+1} is a generator of the group of n^{s+1}th residues, then we may define encryption of a message $m \in \mathbb{Z}_{n^s}$ as $v^m h^r_{s+1} \bmod n^{s+1}$ where r is chosen randomly in $[0, n2^{\kappa_r}]$.

Homomorphisms. The Paillier cryptosystem is additively homomorphic. Furthermore, the recursive structure of the Paillier cryptosystem allows a ciphertext $\mathcal{E}^{\mathsf{pai}}_{n^2}(m) \in \mathsf{C}_{n^2} = \mathbb{Z}^*_{n^2}$ to be viewed as a plaintext in the group $\mathsf{M}_{n^3} = \mathbb{Z}_{n^2}$ that can be encrypted using a generalized version of the cryptosystem, i.e., we can compute $\mathcal{E}^{\mathsf{pai}}_{n^3}(\mathcal{E}^{\mathsf{pai}}_{n^2}(m))$. Interestingly, *the nested cryptosystems preserve the group structures over which they are defined.* In other words we have

$$\mathcal{E}^{\mathsf{pai}}_{n^3}(\mathcal{E}^{\mathsf{pai}}_{n^2}(0,r))^{\mathcal{E}^{\mathsf{pai}}_{n^2}(m,s)} = \mathcal{E}^{\mathsf{pai}}_{n^3}(\mathcal{E}^{\mathsf{pai}}_{n^2}(0,r)\mathcal{E}^{\mathsf{pai}}_{n^2}(m,s)) = \mathcal{E}^{\mathsf{pai}}_{n^3}(\mathcal{E}^{\mathsf{pai}}_{n^2}(m,r+s)) \ .$$

This homomorphic operation is similar to the generic additive operation from Section 3, with the inner "1" replaced by an encryption of 0. As a result, though the output is also a doubly encrypted m_i, a re-encryption has occurred on the inner ciphertext. This technique extends the layered-Paillier homomorphic property first observed by Lipmaa [18].

5.2   The Obfuscator

We use the additive homomorphism and the special homomorphic property exhibited above to define a form of homomorphic matrix multiplication of matrices of ciphertexts. Given an N-permutation matrix $\Lambda^\pi = (\lambda^\pi_{ij})$ and randomness $r, s \in (\mathbb{Z}^*_n)^{N \times N}$, define $C^\pi = (c^\pi_{ij}) = \left(\mathcal{E}^{\mathsf{pai}}_{n^3}(\lambda^\pi_{ij}\mathcal{E}^{\mathsf{pai}}_{n^2}(0,r_{ij}), s_{ij}) \right)$. We define a kind of matrix multiplication of $d = (d_1, \ldots, d_N) \in \mathsf{C}^N_{n^2}$ and C^π:

$$d \star C^\pi = \left(\prod_{i=1}^N (c^\pi_{ij})^{d_i} \right) \text{ giving } \mathcal{D}^{\mathsf{pai}}_{p,2}(\mathcal{D}^{\mathsf{pai}}_{p,3}(d \star C^\pi)) = (m_{\pi(1)}, \ldots, m_{\pi(N)}) \ .$$

In other words, we can do homomorphic matrix multiplication with a permutation matrix using layered Paillier, but we stress that the above matrix multiplication does *not* work for all matrices. We are now ready to define the obfuscator for the Paillier-based shuffle. Again, $\mathcal{O}^{\mathsf{pai}}$ does not use sk.

Definition 9 (Obfuscator). *The obfuscator* $\mathcal{O}^{\mathsf{pai}}$ *for the re-encryption shuffle* $\mathcal{RS}_N^{\mathsf{pai}}$ *takes as input a tuple* $(1^\kappa, n, sk, RS^{\mathsf{pai}})$, *where* $(n, p) \in \mathcal{G}^{\mathsf{pai}}(1^\kappa)$ *and* $RS^{\mathsf{pai}} \in \mathcal{RS}_{N(\kappa), \kappa}^{\mathsf{pai}}$, *computes* $C^\pi = (\mathcal{E}_{n^3}^{\mathsf{pai}}(\lambda_{ij}^\pi \mathcal{E}_{n^2}^{\mathsf{pai}}(0, r_{ij}), s_{ij}))$, *and outputs a circuit with hardcoded* C^π *that, on input* $d = (d_1, \ldots, d_{N(\kappa)})$, *outputs* $d' = d \star C^\pi$.

Proposition 3. *The obfuscator* $\mathcal{O}^{\mathsf{pai}}$ *for* $\mathcal{RS}_N^{\mathsf{pai}}$ *is polynomially indistinguishable if the Paillier cryptosystem is polynomially indistinguishable.*

Composition. It may be possible to compose Paillier re-encryption shuffles using additional layers in the Damgård et. al. Paillier generalization. However, because an extra layer of encryption is added at each step, the re-encryption actions are not truly composed with one another, e.g., the second stage re-encryption acts on the first stage's obfuscation layer, while the innermost ciphertext is re-encrypted only on the first pass. Thus, in Section 7, we explain how to generate and obfuscate a Paillier re-encryption shuffle in a distributed way.

6 Proving Correctness of Obfuscation

We show how to prove the correctness of a BGN or Paillier obfuscation. We assume, for now, that a single party generates the encrypted matrix, though the techniques described here are immediately applicable to the distributed generation and proofs in Section 7. For either cryptosystem, we start with a trivially encrypted "identity matrix", and we let the prover demonstrate that he correctly shuffled the columns of this matrix.

Definition 10. *Denote by* $\mathcal{R}_{mrp}$ *the relation consisting of pairs* $((pk, C, C'), r)$ *such that* $C \in \mathsf{C}_{pk}^{N \times N}$, $C' = (\mathcal{RE}_{pk}(c_{i, \pi(j)}, r_{ij}))$, $r \in \mathsf{R}_{pk}^{N \times N}$, *and* $\pi \in \Sigma_N$.

In the BGN case, the starting identity matrix can be simply $C = \mathcal{E}_{pk}(\Lambda^{\mathsf{id}}, 0^*)$.

Recall that, where the BGN matrix contains encryptions of 1, the Paillier matrix contains outer encryptions of *different* inner encryptions of zero, which need to remain secret. Thus, in the Paillier case, we begin by generating and proving correct a list of N double encryptions of zero. We construct a proof of double-discrete log with $1/2$-soundness that must be repeated a number of times. This repetition remains "efficient enough" because we only need to perform a linear number of sets of repeated proofs. We then use these N doubly encrypted zeros as the diagonal of our identity matrix, completing it with trivial outer encryptions of zero.

In both cases, we then take this identity matrix, shuffle and re-encrypt its columns, and provide a zero-knowledge proof of knowledge of the permutation and re-encryption factors. A verifier is then certain that the resulting matrix is a permutation matrix.

6.1 Proving a Shuffle of the Columns of a Ciphertext Matrix

Consider the simpler and extensively studied problem of proving that ciphertexts have been correctly re-encrypted and permuted, a so-called "proof of shuffle."

Definition 11. *Denote by $\mathcal{R}_{rp}$ the relation consisting of pairs $((pk, d, d'), r)$ such that $d = (d_j) \in \mathsf{C}_{pk}^N$ and $d' = \mathcal{RE}_{pk}((d_{\pi(j)}), r)$ for some $r \in \mathsf{R}_{pk}^N$ and $\pi \in \Sigma_N$.*

There are several known efficient methods [20,14,17,28] for constructing a protocol for this relation. Although these protocols differ slightly in their properties, they all essentially give "honest-verifier zero-knowledge proofs of knowledge." As our protocol can be adapted to the concrete details of these techniques, we assume, for clarity, that there exists an honest-verifier zero-knowledge proof of knowledge π_{rp} for the above relation. These protocols can be extended to prove a shuffle of lists of ciphertexts (which is what we need), but a detailed proof of this fact has not appeared. We present a simple batch proof (see [4]) of a shuffle to allow us to argue more concretely about the complexity of our scheme.

Protocol 1 (Matrix Re-encryption-Permutation)
COMMON INPUT. *A public key pk and $C, C' \in \mathsf{C}_{pk}^{N \times N}$*
PRIVATE INPUT. *$\pi \in \Sigma_N$ and $r \in \mathsf{R}_{pk}^{N \times N}$ such that $C' = \mathcal{RE}_{pk}((c_{i,\pi(j)}), r)$.*

1. *$\mathcal{V}$ chooses $u \in [0, 2^{\kappa_c} - 1]^N$ randomly and hands it to $\mathcal{P}$.*
2. *They both compute $d = (\prod_{i=1}^N c_{ij}^{u_i})$ and $d' = (\prod_{i=1}^N (c'_{ij})^{u_i})$.*
3. *They run the proof of a shuffle π_{rp} on common input (pk, d, d') and private input $\pi, r' = (\sum_{i=1}^N r_{ij} u_i)$.*

Proposition 4. *Protocol 1 is public-coin and honest-verifier zero-knowledge. For inputs with $C = \mathcal{E}_{pk}(\Lambda^\pi)$ for $\pi \in \Sigma_N$ the error probability is negligible and there exists a knowledge extractor.*

Remark 1. When the plaintexts are known, and this is the case when C is an encryption of the identity matrix, slightly more efficient techniques can be used. This is sometimes called a "shuffle of known plaintexts" (see [20,17,28]).

6.2   Proving Double Re-encryption

The following relation captures the problem of proving correctness of a double re-encryption.

Definition 12. *Denote by $\mathcal{R}_{dr}^{\mathsf{pai}}$ the relation consisting of pairs $((n, c, c'), (r, s))$, such that $c' = c^{h_1^r} \bmod n^2 h_2^s \bmod n^3$ with $r, s \in [0, N2^{\kappa_r}]$.*

Protocol 2 (Double Re-encryption)
COMMON INPUT. *A modulus n and $c, c' \in \mathsf{C}_{n^3}$*
PRIVATE INPUT. *$r, s \in [0, n2^{\kappa_r}]$ such that $c' = c^{h_2^r \bmod n^2} h_3^s \bmod n^3$.*

1. *$\mathcal{P}$ chooses $r' \in [0, n2^{2\kappa_r}]$ and $s' \in [0, n^3 2^{2\kappa_r}]$ randomly, computes $\alpha = c^{h_2^{r'} \bmod n^2} h_3^{s'} \bmod n^3$, and hands α to $\mathcal{V}$.*
2. *$\mathcal{V}$ chooses $b \in \{0, 1\}$ randomly and hands b to $\mathcal{P}$.*
3. *$\mathcal{P}$ defines $(e, f) = (r' - br, s' - b(h_2^e \bmod n^2)s)$. Then it hands (e, f) to $\mathcal{V}$.*
4. *$\mathcal{V}$ checks that $\alpha = ((c')^b c^{1-b})^{h_2^e \bmod n^2} h_3^f \bmod n^3$.*

The protocol is iterated in parallel κ_c times to make the error probability negligible. For proving a lists of ciphertexts, we use independent copies of the protocol for each element, but reuse the challenges.

Proposition 5. *Protocol 2 is a public-coin honest verifier zero-knowledge proof of knowledge for $\mathcal{R}_{dr}^{\mathsf{pai}}$.*

7 Distributed Generation and Obfuscation of a Shuffle

Our two constructions can be efficiently generated in a distributed fashion. Roughly, we begin with the trivial encryption of the identity matrix. (In the Paillier case, a sub-protocol is required to generate the inner-layer encryptions of the 0-diagonal using successive re-encryptions by the parties.) We then let each party in turn shuffle and re-encrypt the rows of this matrix. In the end, the resulting permutation matrix captures the composition of the shuffles from each party: it is as if the actions of a mix-net were captured ahead of time into an encrypted matrix, then unleashed onto the ciphertext inputs at shuffle time. The details of this process, including the proofs of correct shuffling and the UC proof of security, are provided in the full version of this paper.

8 Complexity Estimates

Our constructions clearly require $O(N^2)$ exponentiations, but we give estimates that show that the constant hidden in the ordo-notation is reasonably small in some practical settings. For simplicity we assume that the cost of squaring a group element equals the cost of multiplying two group elements and that computing an exponentiation using a κ_e-bit integer modulo a κ-bit integer corresponds to κ_e/κ full exponentiations modulo a κ-bit integer. We optimize using fixed-base exponentiation and simultaneous exponentiation (see [19]). We assume that evaluating the bilinear map corresponds to computing 6 exponentiations in the group $\mathbb{G}_1$ and we assume that such one such exponentiation corresponds to 8 modular exponentiations. This seems reasonable, although we are not aware of any experimental evidence. In the Paillier case we assume that multiplication modulo n^s is s^2 times as costly as multiplication modulo n. We assume that the proof of a shuffle requires $8N$ exponentiations (this is conservative).

Most exponentiations when sampling and obfuscating a shuffle are fixed-base exponentiations. The only exception is a single exponentiation each time an element is doubly re-encrypted, but there are only N such elements. In the proof of correct obfuscation the bit-size κ_c of the elements in the random vector u used in Protocol 1 is much smaller than the security parameter, and simultaneous exponentiation is applicable. In the Paillier case, simultaneous exponentiation is applicable during evaluation, and precomputation lowers the on-line complexity. Unfortunately, this does not work in the BGN case due to the bilinear map. We refer the reader to the Scheme program in the full paper for details on our estimates. For practical parameters we get the estimates in Fig. 2.

Given a single computer, the BGN construction is only practical when $N \approx 350$ and the maximal number of bits in any submitted ciphertext is small. On the other hand, the Paillier construction is practical for normal sized voting precincts in the USA: $N \approx 2000$ full length messages can be accommodated, and, given one week of pre-computing, the obfuscated shuffle can be evaluated overnight. Furthermore, all constructions are easily parallelized.

Construction	Sample & Obfuscate	Prove	Precompute	Evaluate
BGN with $N = 350$	14 (0.5h)	3 (0.1h)	NA	588 (19.6h)
Paillier with $N = 2000$	556 (18.5h)	290 (9.7h)	3800 (127h)	533 (17.8h)

Fig. 2. The table gives the complexity of the operations in terms of 10^4 modular κ-bit exponentiations and in parenthesis the estimated running time in hours assuming that $\kappa = 1024$, $\kappa_c = \kappa_r = 50$, and that one exponentiation takes 12 msec to compute (a 1024-bit exponentiation using GMP [16] takes 12 msec on our 3 GHz PC)

9   Conclusion

It is surprising that a functionality as powerful as a shuffle can be public-key obfuscated in any useful way. It is even more surprising that this can be achieved using the Paillier cryptosystem which, in contrast to the BGN cryptosystem, was not specifically designed to have the kind of "homomorphic" properties we exploit. One intriguing question is whether other useful "homomorphic" properties have been overlooked in existing cryptosystems.

From a practical point of view we stress that, although the performance of our mix-net is much worse than that of known constructions, it exhibits a property which no previous construction has: a relatively small group of mix servers can prepare obfuscated shuffles for voting precincts. The precincts can compute the shuffling without any private key and produce ciphertexts ready for decryption.

Acknowledgments

We thank Ronald L. Rivest, Susan Hohenberger, Rafael Pass, Shafi Goldwasser, and Guy Rothblum for productive and insightful discussions.

References

1. M. Abe and H. Imai. Flaws in some robust optimistic mix-nets. In *Australasian Conference on Information Security and Privacy – ACISP 2003*, volume 2727 of *Lecture Notes in Computer Science*, pages 39–50. Springer Verlag, 2003.
2. B. Barak, O. Goldreich, R. Impagliazzo, S. Rudich, A. Sahai, S. P. Vadhan, and K. Yang. On the (im)possibility of obfuscating programs. In *Advances in Cryptology – Crypto 2001*, volume 2139 of *Lecture Notes in Computer Science*, pages 1–18. Springer Verlag, 2001.

3. O. Baudron, P.-A. Fouque, D. Pointcheval, J. Stern, and G. Poupard. Practical multi-candidate election system. In *20th ACM Symposium on Principles of Distributed Computing – PODC*, pages 274–283. ACM Press, 2001.

4. M. Bellare, J. A. Garay, and T. Rabin. Fast batch verification for modular exponentiation and digital signatures. In *Advances in Cryptology – Eurocrypt '98*, pages 236–250. Springer Verlag, 1998.

5. J. Benaloh and M. Yung. Distributing the power of a government to enhance the privacy of voters. In *5th ACM Symposium on Principles of Distributed Computing – PODC*, pages 52–62. ACM Press, 1986.

6. J. Cohen (Benaloh) and M. Fischer. A robust and verifiable cryptographically secure election scheme. In *28th IEEE Symposium on Foundations of Computer Science (FOCS)*, pages 372–382. IEEE Computer Society Press, 1985.

7. D. Boneh, E.-J. Goh, and K. Nissim. Evaluating 2-DNF formulas on ciphertexts. In *2nd Theory of Cryptography Conference (TCC)*, volume 3378 of *Lecture Notes in Computer Science*, pages 325–342. Springer Verlag, 2005.

8. R. Canetti. Towards realizing random oracles: Hash functions that hide all partial information. In *Advances in Cryptology – Crypto 1997*, volume 1294 of *Lecture Notes in Computer Science*, pages 455–469. Springer Verlag, 1997.

9. D. Chaum. Untraceable electronic mail, return addresses and digital pseudo-nyms. *Communications of the ACM*, 24(2):84–88, 1981.

10. R. Cramer, M. Franklin, L. A.M. Schoenmakers, and M. Yung. Multi-authority secret-ballot elections with linear work. Technical report, CWI (Centre for Mathematics and Computer Science), Amsterdam, The Netherlands, 1995.

11. R. Cramer, R. Gennaro, and B. Schoenmakers. A secure and optimally efficient multi-authority election scheme. In *Advances in Cryptology – Eurocrypt '97*, volume 1233 of *Lecture Notes in Computer Science*, pages 103–118. Springer Verlag, 1997.

12. I. Damgård and M. Jurik. A generalisation, a simplification and some applications of paillier's probabilistic public-key system. In *Public Key Cryptography – PKC 2001*, volume 1992 of *Lecture Notes in Computer Science*, pages 119–136. Springer Verlag, 2001.

13. P.-A. Fouque, G. Poupard, and J. Stern. Sharing decryption in the context of voting or lotteries. In *Financial Cryptography 2000*, volume 2339 of *Lecture Notes in Computer Science*, pages 90–104, London, UK, 2001. Springer-Verlag.

14. J. Furukawa and K. Sako. An efficient scheme for proving a shuffle. In *Advances in Cryptology – Crypto 2001*, volume 2139 of *Lecture Notes in Computer Science*, pages 368–387. Springer Verlag, 2001.

15. S. Goldwasser and Y. Tauman Kalai. On the impossibility of obfuscation with auxiliary input. In *46th IEEE Symposium on Foundations of Computer Science (FOCS)*, pages 553–562. IEEE Computer Society Press, 2005.

16. T. Granlund. Gnu multiple precision arithmetic library (GMP). Software available at http://swox.com/gmp, March 2005.

17. J. Groth. A verifiable secret shuffle of homomorphic encryptions. In *Public Key Cryptography – PKC 2003*, volume 2567 of *Lecture Notes in Computer Science*, pages 145–160. Springer Verlag, 2003.

18. Helger Lipmaa. An oblivious transfer protocol with log-squared communication. In *Information Security – ISC 2005*, volume 3650 of *Lecture Notes in Computer Science*, pages 314–328. Springer Verlag, 2005.

19. A. Menezes, P. Oorschot, and S. Vanstone. *Handbook of Applied Cryptography*. CRC Press, 1997.

20. A. Neff. A verifiable secret shuffle and its application to e-voting. In *8th ACM Conference on Computer and Communications Security (CCS)*, pages 116–125. ACM Press, 2001.
21. R. Ostrovsky and W. E. Skeith III. Private searching on streaming data. Cryptology ePrint Archive, Report 2005/242, 2005. http://eprint.iacr.org/.
22. P. Paillier. Public-key cryptosystems based on composite degree residuosity classes. In *Advances in Cryptology – Eurocrypt '99*, volume 1592 of *Lecture Notes in Computer Science*, pages 223–238. Springer Verlag, 1999.
23. C. Park, K. Itoh, and K. Kurosawa. Efficient anonymous channel and all/nothing election scheme. In *Advances in Cryptology – Eurocrypt '93*, volume 765 of *Lecture Notes in Computer Science*, pages 248–259. Springer Verlag, 1994.
24. K. Sako and J. Kilian. Reciept-free mix-type voting scheme. In *Advances in Cryptology – Eurocrypt '95*, volume 921 of *Lecture Notes in Computer Science*, pages 393–403. Springer Verlag, 1995.
25. B. Schoenmakers. A simple publicly verifiable secret sharing scheme and its application to electronic. In *Advances in Cryptology – Crypto '99*, volume 3027 of *Lecture Notes in Computer Science*, pages 148–164. Springer Verlag, 1999.
26. H. Wee. On obfuscating point functions. In *37th ACM Symposium on the Theory of Computing (STOC)*, pages 523–532. ACM Press, 2005.
27. D. Wikström. A universally composable mix-net. In *1st Theory of Cryptography Conference (TCC)*, volume 2951 of *Lecture Notes in Computer Science*, pages 315–335. Springer Verlag, 2004.
28. D. Wikström. A sender verifiable mix-net and a new proof of a shuffle. In *Advances in Cryptology – Asiacrypt 2005*, volume 3788 of *Lecture Notes in Computer Science*, pages 273–292. Springer Verlag, 2005. (Full version [29]).
29. D. Wikström. A sender verifiable mix-net and a new proof of a shuffle. Cryptology ePrint Archive, Report 2004/137, 2005. http://eprint.iacr.org/.
30. D. Wikström and J. Groth. An adaptively secure mix-net without erasures. In *33rd International Colloquium on Automata, Languages and Programming (ICALP)*, volume 4052 of *Lecture Notes in Computer Science*, pages 276–287. Springer Verlag, 2006.

A   Proofs

Proof (Proposition 1 and Proposition 2). We only detail the first proof, since the second follows by a trivial modification. Denote by $\mathcal{A}$ an arbitrary adversary in the polynomial indistinguishability experiment run with the obfuscator $\mathcal{O}$. Denote by $\mathcal{A}'$ an adversary to the polynomial indistinguishability experiment $\mathsf{Exp}^{\mathsf{ind}-b}_{\mathcal{CS}',\mathcal{A}'}(\kappa)$ with the cryptosystem $\mathcal{CS}'$ defined as follows. It accepts a public key pk as input and forwards it to $\mathcal{A}$. When $\mathcal{A}$ returns (DS_{π_0}, DS_{π_1}), $\mathcal{A}'$ outputs the two messages 0 and 1. Then it is given an encryption $c^{(b)} = \mathcal{E}'_{pk}(b)$. Denote by Λ^{π_0} and Λ^{π_1} the two permutation matrices corresponding to DS_{π_0} and DS_{π_1} respectively. The adversary $\mathcal{A}'$ defines a matrix $C'^{\pi_b} = (c^{\pi}_{ij})$, by setting $c^{\pi}_{ij} = \mathcal{E}'_{pk}(\lambda^{\pi_0}_{ij})$ if $\lambda^{\pi_0}_{ij} = \lambda^{\pi_1}_{ij}$ and c^{π}_{ij} to a reencryption of $c^{(b)}$ if $\lambda^{\pi_0}_{ij} = 0$, or to a reencryption of $c^{(1-b)}$ if $\lambda^{\pi_0}_{ij} = 1$. Note that $c^{(1-b)}$ can be computed homomorphically from $c^{(b)}$. Then $\mathcal{A}'$ continues the simulation using C'^{π_b} to compute the obfuscated circuit, and when $\mathcal{A}$ outputs a bit it gives it as its output. By construction $\Pr[\mathsf{Exp}^{\mathsf{ind}-b}_{\mathcal{CS}',\mathcal{A}}(\kappa) = 1] = \Pr[\mathsf{Exp}^{\mathsf{oind}-b}_{\mathcal{DS}_N,\mathcal{CS}',\mathcal{O},\mathcal{A}}(\kappa) = 1]$.

Proof (Proposition 3). Let $\mathcal{A}$ be any adversary in the polynomial indistinguishability experiment run with the obfuscator $\mathcal{O}^{\mathsf{pai}}$. Denote by $\mathcal{A}_{ind}$ the polynomial indistinguishability adversary that takes a public key pk as input and then simulates this protocol to $\mathcal{A}$. When $\mathcal{A}$ outputs two challenge circuits $(RS_0^{\mathsf{pai}}, RS_1^{\mathsf{pai}})$ with corresponding matrices (M_0, M_1), i.e., the matrices are permutation matrices with the ones replaced by re-encryption factors, $\mathcal{A}_{ind}$ outputs (M_0, M_1). When the experiment returns $\mathcal{E}_{pk}^{\mathsf{pai}}(M_b)$ it forms the obfuscated circuit and hands it to $\mathcal{A}$. Then $\mathcal{A}_{ind}$ outputs the output of $\mathcal{A}$. It follows that the advantage of $\mathcal{A}_{ind}$ in the polynomial indistinguishability experiment with the Paillier cryptosystem and using polynomial length list of ciphertexts is identical to the advantage of $\mathcal{A}$ in the polynomial indistinguishability experiment with $\mathcal{O}^{\mathsf{pai}}$. It now follows from a standard hybrid argument that the polynomial indistinguishability of the Paillier cryptosystem is broken if $\mathcal{O}^{\mathsf{pai}}$ is not polynomially indistinguishable.

Proof (Proposition 4). Completeness and the fact that the protocol is public-coin follow by inspection. We now concentrate on the more interesting properties.

Zero-Knowledge. The honest-verifier zero-knowledge simulator simply picks u randomly as in the protocol and then invokes the honest-verifier zero-knowledge simulator of the subprotocol π_{rp}. It follows that the simulated view is indistinguishable from the real view of the verifier.

Negligible Error Probability. Consider the following intuitively appealing lemma.

Lemma 1. *Let η be a product of $\kappa/2$-bit primes and let N be polynomially bounded in κ. Let $\Lambda = (\lambda_{ij})$ be an $N \times N$-matrix over $\mathbb{Z}_\eta$ and let $u \in [0, 2^{\kappa_c} - 1]^N$ be randomly chosen. Then if Λ is not a permutation matrix $\Pr_u[\exists \pi \in \Sigma_N : u\Lambda^\pi = u\Lambda]$ is negligible.*

Proof. Follows by elementary linear algebra (see [29]).

By assumption $C = \mathcal{E}_{pk}(\Lambda^\pi)$ for some $\pi \in \Sigma_N$. Write $\Lambda = \mathcal{D}_{pk}(C')$. Then the lemma and the soundness of the proof of a shuffle π_{rp} implies the soundness of the protocol.

Knowledge Extraction. For knowledge extraction we may now assume that C' can be formed from C by permuting and re-encrypting its columns. Before we start we state a useful lemma.

Lemma 2. *Let η be a product of $\kappa/2$-bit primes, let N be polynomially bounded in κ, and let $u_1, \ldots, u_{l-1} \in \mathbb{Z}^N$ such that $u_{jj} = 1 \bmod \eta$ and $u_{ji} = 0 \bmod \eta$ for $1 \le i, j \le l - 1 < N$ and $i \ne j$. Let $u_l \in [0, 2^{\kappa_c} - 1]^N$ be randomly chosen, where $2^{-\kappa_c}$ is negligible. Then the probability that there exists $a_1, \ldots, a_l \in \mathbb{Z}$ such that if we define $u'_l = \sum_{j=1}^l a_j u_j \bmod \eta$, then $u'_{l,l} = 1 \bmod \eta$, and $u'_{l,i} = 0 \bmod \eta$ for $i < l$ is overwhelming in κ.*

Proof. Note that $b = u_{l,l} - \sum_{j=1}^{l-1} u_{l,j} u_{j,l}$ is invertible with overwhelming probability, and when it is we view its inverse b^{-1} as an integer and define $a_j = -b^{-1} u_{l,j}$ for $j < l$ and $a_l = b^{-1}$. For $i < l$ this gives $u_{l,i} = \sum_{j=1}^{l} a_j u_{j,i} = b^{-1}(1 - a_i u_{ii}) = 0 \bmod \eta$ and for $i = l$ this gives $u_{l,l} = \sum_{j=1}^{l} a_j u_{j,l} = b^{-1}(u_{l,l} - \sum_{j=1}^{l-1} u_{l,j} u_{j,l}) = 1 \bmod \eta$.

It remains to exhibit a knowledge extractor. By assumption there exists a polynomial $t(\kappa)$ and negligible knowledge error $\epsilon(\kappa)$ such that the extractor of the subprotocol π_{rp} executes in time $T_{\gamma'}(\kappa) = t(\kappa)/(\gamma' - \epsilon(\kappa))$ for every common input (d, d'), induced by a random vector u, to the subprotocol such that the success probability of the subprotocol is γ'. We invoke the extractor, but we must stop it if γ' turns out to be too low and find a new random u that induces a common input to the subprotocol with a larger value of γ'. We assume that the same negligible function $\epsilon(\kappa)$ bounds the failure probability in Lemma 2.

Consider a fixed common input (pk, C, C') and prover $\mathcal{P}$. Denote by γ the probability that $\mathcal{P}$ convinces $\mathcal{V}$. We assume that $\epsilon(\kappa) < \gamma/4$, i.e., the knowledge error will increase somewhat compared to the knowledge error of π_{rp}.

We denote by B the distribution over $\{0, 1\}$ given by $p_B(1) = \gamma/(8t(\kappa))$. Note that this distribution can be sampled for any common input even without knowledge of γ, since we can simply perform a simulation of the protocol, pick element from the space $\{1, \ldots, 8t(\kappa)\}$ randomly, and define the sample to be one if the prover succeeds and the picked element equal one. We are going to use the random variable to implicitly be able to say if an induced common input to the subprotocol gives a too low success probability γ'. We now make this idea precise. The extractor proceeds as follows, where in the BGN case η denotes the modulus n and in the Paillier case η denotes the order of the plaintext space of the outer layer Paillier, i.e., n^2 where n is the modulus.

1. For $l = 1, \ldots, N$ do:
   (a) Start the simulation of an execution between $\mathcal{V}$ and $\mathcal{P}$ and denote by u_l the random vector chosen by the simulator. Denote by (pk, d_l, d'_l) the common input to the subprotocol π_{rp} induced by u_l.
   (b) If $u_{l,j} = u_{l,j'}$ for some $j \neq j'$ or if there does not exists $a_{k,l} \in \mathbb{Z}$ such that $\sum_{l'=1}^{l} a_{k,l'} u_{l',j}$ equals one modulo η if $j = l$ and it equals zero modulo η for $j < l$, then go to Step 1a.
   (c) Invoke the knowledge extractor of the protocol π_{rp} on the common input (pk, d_l, d'_l). However, in between each step executed by the extractor, the distribution B is sampled. If a sample equals one before the extractor halts, then go to Step 1a. Otherwise, denote by π_l and s_l the permutation and extracted randomness such that $((pk, d_l, d'_l), (\pi_l, s_l)) \in \mathcal{R}_{rp}$.

2. Compute $a_{k,l} \in \mathbb{Z}$ such that $\sum_{l=1}^{N} a_{k,l} u_{l,j}$ equals one or zero modulo η depending on if $k = j$ or not. Define $(b_{kj}) = (a_{kl})(u_{lj}) - I$, where I is the identity $N \times N$-matrix and the matrix operations are taken over the integers.

In BGN Case. Compute $r = (r_{k,j}) = (a_{k,l})(s_{l,j})$, and output (π, r).

In Paillier Case. Compute $r = (r_{k,j}) = (\prod_{i=1}^{N}(c_{i,\pi(j)}/c'_{ij})^{b_{ki}/\eta} \prod_{l=1}^{N} s_{l,j}^{a_{k,l}})$, where the division b_{ki}/η is taken over the integers, and output (π, r).

We do the easy part of the analysis first. Consider the correctness of the output given that the extractor halts. Since $u_{l,j} = u_{l,j'}$ for all $j \neq j'$ and both $\mathcal{D}_{pk}(C)$ and $\mathcal{D}_{pk}(C')$ are permutation matrices by assumption, we conclude that $\pi_1 = \ldots = \pi_N = \pi$ for some permutation $\pi \in \Sigma_N$. We have

$$\prod_{i=1}^{N}(c'_{ij})^{u_{li}} = d'_{lj} = d_{l,\pi(j)}\mathcal{E}_{pk}(0, s_{l,j}) = \mathcal{E}_{pk}(0, s_{l,j})\prod_{i=1}^{N} c_{i,\pi(j)}^{u_{li}} . \tag{1}$$

Apply the $a_{k,l}$ as exponents on the left of Equation (1) and take the product over all l. This gives

$$\prod_{l=1}^{N}\left(\prod_{i=1}^{N}(c'_{ij})^{u_{li}}\right)^{a_{k,l}} = \prod_{i=1}^{N}\left(\prod_{l=1}^{N}(c'_{ij})^{a_{k,l}u_{li}}\right) = c'_{kj}\prod_{i=1}^{N}(c'_{ij})^{b_{ki}} .$$

Then apply the exponents $a_{k,l}$ on the right side of Equation (1) and take the product over all l. This gives

$$\prod_{l=1}^{N}\left(\mathcal{E}_{pk}(0, s_{l,j})\prod_{i=1}^{N} c_{i,\pi(j)}^{u_{li}}\right)^{a_{k,l}} = \left(\prod_{l=1}^{N}\mathcal{E}_{pk}(0, s_{l,j})^{a_{kl}}\right)\left(\prod_{i=1}^{N} c_{i,\pi(j)}^{b_{ki}}\right)c_{k,\pi(j)} .$$

To summarize: $c'_{kj} = \left(\prod_{i=1}^{N}(c_{i,\pi(j)}/c'_{ij})^{b_{ki}}\right)\left(\prod_{l=1}^{N}\mathcal{E}_{pk}(0, s_{l,j})^{a_{kl}}\right)c_{k,\pi(j)}$. The argument is concluded differently depending on the cryptosystem used.

In BGN Case. Note that $b_{ki} = 0 \bmod \eta$ for all k and i, and the order of any ciphertext divides η. Thus, the first product equals one in the ciphertext group. Furthermore, the randomizer space is $\mathbb{Z}_\eta$ so we have

$$c'_{kj} = \mathcal{E}_{pk}\left(0, \sum_{l=1}^{N} a_{kl}s_{l,j}\right)c_{k,\pi(j)} .$$

In Paillier Case. Again $b_{ki} = 0 \bmod \eta$ for all k and i, but the order of a ciphertext may be larger than η. However, we may define $b'_{ki} = b_{ki}/\eta$, where division is over the integers, define $s'_j = \prod_{i=1}^{N}(c_{i,\pi(j)}/c'_{ij})^{b'_{ki}}$, and write

$$c'_{kj} = \mathcal{E}_{pk}\left(0, s'_j \prod_{l=1}^{N} s_{l,j}^{a_{kl}}\right)c_{k,\pi(j)} .$$

We remark that s'_j is an element in $\mathbb{Z}_{n^3}^*$ and not in $\mathbb{Z}_n^*$ as expected. However, it is a witness of re-encryption using alternative Paillier encryption.

It remains to prove that the extractor is efficient in terms of the inverse success probability of the prover. Fix an l. Denote by E the event that the prover succeeds to convince the adversary, i.e., $\Pr[E] = \gamma$. Denote by S the set of vectors u such that $\Pr[E \mid u \in S] \geq \gamma/2$. An averaging argument implies that $\Pr[u \in S] \geq \gamma/2$. Denote by E_{u_l} the event that the go to statement in Step 1b is executed. We show that if $u \in S$, then a witness is extracted efficiently with constant probability, and if $u \notin S$, then the extraction algorithm will be stopped relatively quickly.

If $u \notin S$, then we focus only on the distribution B. The expected number of samples from B needed before a sample is equal to one is clearly $1/p_B(1) = 8t(\kappa)/\gamma$. Thus, if we ignore the issue of finding the witness the simulation in Step 1c is efficient in terms of $1/\gamma$.

If $u \in S$, then the expected number of steps needed by the extractor of the subprotocol π_{rp} is bounded by $T_{\gamma/2}(\kappa)$. By Markov's inequality the probability that more than $2T_{\gamma/2}(\kappa)$ steps are needed is bounded by $1/2$. The probability that one of the first $\omega = 2T_{\gamma/2}(\kappa)$ samples of B is one is bounded by $1 - (1 - p_B(1))^{\omega} \leq 1 - e^{\omega(-p_B(1)+p_B(1)^2)} \leq 1 - e^{-1/2}$, since $\epsilon(\kappa) < \gamma/4$.

Thus, Step 1c executes in expected time $8t(\kappa)/\gamma$, and from independence follows that it halts due to the extractor finding a witness with probability at least $1 - \frac{1}{2}(1 - e^{-1/2})$. In other words the expected number of restarts of the lth iteration of Step 1 is constant.

From Lemma 2 and independence of the $u_{l,j}$ follow that the probability that the go to statement of Step 1b is executed is negligible. This means that the extractor runs in expected time $cNt(\kappa)/(\gamma - 4\epsilon(\kappa))$ for some constant c. This concludes the proof, since $cNt(\kappa)$ is polynomial and $4\epsilon(\kappa)$ is negligible.

Proof (Proposition 5). Completeness and the public-coin property follow by inspection. The honest-verifier zero-knowledge simulator simply picks $e \in [0, n2^{\kappa_r}]$ and $f \in [0, n^3 2^{\kappa_r}]$ and $b \in \{0,1\}$ randomly and defines $\alpha = ((c')^b c^{1-b})^{h_2^e} h_3^f \bmod n^3$. The resulting view is statistically close to a real view, since $2^{-\kappa_r}$ is negligible.

For soundness, note that if we have $c^{h_2^e} h_3^f = \alpha = (c')^{h_2^{e'}} h_3^{f'} \bmod n^3$ with $e, f, e', f' \in \mathbb{Z}$, then we can divide by h_3^f and take the h_2^eth root on both sides. This gives $c = (c')^{h_2^{e'-e}} h_3^{(f'-f)/h_2^e} \bmod n^3$, which implies that the basic protocol is special $1/2$-sound. The protocol is then iterated in parallel κ_c times which gives negligible error probability $2^{-\kappa_c}$. The proof of knowledge property follows immediately from special soundness.

Evaluating Branching Programs on Encrypted Data*

Yuval Ishai and Anat Paskin

Computer Science Department, Technion
yuvali@cs.technion.ac.il, anps83@gmail.com

Abstract. We present a public-key encryption scheme with the following properties. Given a branching program P and an encryption c of an input x, it is possible to efficiently compute a *succinct* ciphertext c' from which $P(x)$ can be efficiently decoded using the secret key. The size of c' depends polynomially on the size of x and the *length* of P, but does not further depend on the size of P. As interesting special cases, one can efficiently evaluate finite automata, decision trees, and OBDDs on encrypted data, where the size of the resulting ciphertext c' does not depend on the size of the object being evaluated. These are the first general representation models for which such a feasibility result is shown. Our main construction generalizes the approach of Kushilevitz and Ostrovsky (FOCS 1997) for constructing single-server Private Information Retrieval protocols.

We also show how to strengthen the above so that c' does not contain additional information about P (other than $P(x)$ for some x) even if the public key and the ciphertext c are maliciously formed. This yields a two-message secure protocol for evaluating a length-bounded branching program P held by a server on an input x held by a client. A distinctive feature of this protocol is that it hides the size of the server's input P from the client. In particular, the client's work is independent of the size of P.

1 Introduction

Computing on encrypted data is arguably one of the most intriguing open problems in cryptography. The variant of this problem we are interested in may be illustrated by the following motivating scenario. Suppose that a client, holding a sensitive local input x, wishes to run a remote program P on this input. For instance, x can be the medical history of an individual and P a complex propriety algorithm determining whether to offer insurance coverage to this individual. To the end of evaluating $P(x)$, the client wishes to publish an *encrypted* version of x, denoted by c, while still allowing a server owning P to effectively run its program on the ciphertext c. That is, based on P and c the server should compute in polynomial time a message c', from which the client can recover $P(x)$ using its secret key.

As described so far, the problem can be solved by simply letting c' include a complete description of P. However, this trivial solution has two significant weaknesses. First, it completely reveals P to the client, whereas ideally the client should only be able to learn $P(x)$. Second, when the description size of P is larger than its input and output,

---

* Supported by grants 36/03 and 1310/06 from the Israel Science Foundation and grant 2004361 from the U.S.-Israel Binational Science Foundation.

S.P. Vadhan (Ed.): TCC 2007, LNCS 4392, pp. 575–594, 2007.

this solution is wasteful in terms of communication. Ideally, the communication should be *a-priori* bounded by some polynomial in the size of the input x, the output $P(x)$ and the security parameter, independently of the description size of P. The same holds for the amount of local computation and storage used by the client. To summarize, it is desirable to obtain solutions which satisfy the following two goals:

1. Hide P from the client (to the extent possible).
2. Make the client's work independent of the size of P. In particular, c' should be *succinct* in the sense that its size depends only on the size of the input and output and not on that of P.

Jumping ahead, the main open problem in the area is that of realizing the second goal. This problem is the focus of our work.

Before addressing known methods for realizing the above two goals, it is instructive to further clarify what we mean when referring to a "program" P. A program is a string that represents a function, mapping an input x to an output y. To simplify the exposition, we restrict the attention to finite boolean functions $f : \{0,1\}^n \rightarrow \{0,1\}$. The correspondence between a program P and the function it represents is determined by an underlying *representation model*. Common representation models for finite functions include circuits, formulas, branching programs, OBDDs, finite automata, decision trees, and truth tables. Once the representation model is fixed, every string P has a unique interpretation as a program computing some specific function f. In this work we will be interested in *universal* representation models, in which every function f can be computed by some program P in the model. Note that all of the models in the above list are universal. However, the *complexity* of representing a function can greatly vary between the models. Circuits are the most powerful model in the list, in the sense that a program in any of the other models has an equivalent circuit of essentially the same size. On the other extreme, truth tables are the least powerful of these models, requiring a program of size 2^n for any function f. This makes truth tables useless for all but very small input lengths n.

We return to the question of realizing the above two goals. Goal 1 can be addressed by using techniques from the area of secure computation. Most notably, Yao's garbled circuit technique [36,7,25] can handle any *circuit* P, allowing to computationally hide all information about P other than $P(x)$ and the size of P. A similar result can be obtained for less powerful representation models, such as formulas or various kinds of branching programs, with the additional feature of keeping P information-theoretically private [35,4,18,22]. However, all these techniques inherently fail with respect to Goal 2, as they require the size of c' to be comparable to the size of P. This gives rise to the following question:

> For which natural representation models can we realize Goal 2, namely evaluate an arbitrary program P on an encrypted input so that the client's work does not depend on the size of P?

A positive answer for the case of circuits (hence also for all other models) would easily follow from the existence of a completely malleable encryption scheme — one that allows to freely perform both additions and multiplications on ciphertexts. However, there is yet no candidate for an encryption scheme with this strong property.

The first protocols in which the client's work can go below the size of P were given in the context of Private Information Retrieval (PIR) [10,23]. A single-server PIR protocol can be viewed as a protocol for evaluating a *truth table* P of size $N = 2^n$ on an encrypted input x of size n. There are such protocols in which the client's work is polynomial in n [6,26], thus affirmatively answering the above question for the case of a truth table representation. Extensions to a *set* representation (where P lists the set of inputs on which f evaluates to 1) were given in the context of private keyword search [23,9,13,30]. Recently, an efficient protocol for evaluating 2-DNF formulas and degree-2 polynomials on encrypted data was given by Boneh et al. [5].[1] The question of realizing Goal 2 for more powerful and useful representation models remained open.

1.1   Our Contribution

We obtain an affirmative answer to our main question for the case of *length-bounded branching programs*. To explain the meaning of this result, we give some background on branching programs and their complexity. A (deterministic) branching program P is defined by a directed acyclic graph in which the nodes are labeled by input variables and every nonterminal node has two outgoing edges, labeled by 0 and 1. An input $x \in \{0,1\}^n$ naturally induces a computation path from a distinguished initial node to a terminal node, whose label determines the output $P(x)$. The *size* of P is defined as the number of nodes in the graph and its *length* is the length of the longest path from the initial node to a terminal node. Branching programs are a relatively powerful representation model. In particular, any logarithmic space or NC^1 computation can be carried out by a family of polynomial-size branching programs.

We consider classes of branching programs whose length is bounded by some public parameter ℓ, where $\ell = \ell(n)$ is polynomial in n. Representation by $\ell(n)$-bounded branching programs is universal whenever $\ell(n) \geq n$. Indeed, any function f can be computed by a complete decision tree of length n and size $O(2^n)$. Branching programs of length $\ell(n) = n$ are of special interest, as they can simulate several representation models that are often used in practice. For instance, if f can be computed by a deterministic finite automaton with s states, then it can be computed by a branching program of length n and size $sn + 1$. Other useful models such as decision trees and OBDDs are also special cases of length-n branching programs.

Our main result is a public-key encryption scheme with the following properties. Given a branching program P and an encryption c of an input x, it is possible to efficiently compute a *succinct* randomized ciphertext c' from which $P(x)$ can be efficiently decoded using the secret key. The size of c' and the work required for decrypting it depend polynomially on the size of x and the *length* of P, but do not further depend on the size of P. Thus, whenever the length $\ell(n)$ is some fixed (polynomial) function of n, we realize Goal 2 above. As interesting special cases, one can evaluate finite automata, decision trees, and OBDDs on encrypted data, where the size of the resulting ciphertext c' does not depend on the size of the object being evaluated. These are the first general

---

[1] In fact, the scheme of [5] realizes a stronger form of computing on encrypted data in which the length of the ciphertext c' depends only on the security parameter and not on the length of the input.

representation models for which such a feasibility result is shown. We also strengthen the above protocol to realize Goal 1 in a very strong sense, guaranteeing that c' does not contain additional information about P (other than $P(x)$ for some x) even if the public key and the ciphertext c are maliciously formed.

Size hiding. Our protocols have the following *size hiding* feature: the ciphertext c' does not reveal any information whatsoever about the *size* of P, no matter how large P is.[2] This should be contrasted with previous methods of computing on encrypted data, in which the communication complexity and the client's work directly reflect (an upper bound on) the size of P. Thus, we achieve a stronger version of Goal 1 than in all previous solutions. A similar notion of size hiding was previously considered by Micali et al. in the context of *zero-knowledge sets* [27].

Applications to secure two-party computation. Our technique for computing on encrypted data immediately gives rise to a one-round (two-message) secure protocol for evaluating a length-bounded branching program P held by a server on an input x held by a client. (This also implies a protocol for the setting in which P is public but its inputs are partitioned between the two parties.) In the case of malicious parties, the protocol satisfies the same relaxed security definition used in previous works on one-round secure computation [29,1,13,20,24]. A distinctive feature of our protocol is that the client's work is independent of the size of P and moreover the protocol hides the size of P from the client.[3] The latter size hiding feature demonstrates that while hiding the sizes of *both* inputs is impossible for interesting functions, there are useful special cases where one can hide the size of *one* of the inputs while maintaining security.

As a concrete application, one can obtain a secure one-round protocol for *keyword search* which totally hides from the client the size of the data set held by the server. That is, a client holding a secret keyword x can query a database D held by a server without revealing x and while assuring the server that it cannot learn anything about D (including its size) other than whether $x \in D$. Previous solutions to the secure keyword search problem [9,13,30] fall short of achieving the size hiding goal. A size hiding protocol as above is obtained by representing D as a *trie* data structure, which can be viewed as an instance of a length-n branching program.

We finally note that the one-round protocol obtained using our technique yields a simpler alternative to similar protocols from the literature that provide *unconditional* security to the server [35,4,18,22]. Its complexity improves over previous protocols even in the case of branching programs of unbounded length. For evaluating a branching program of size s over n inputs, the communication complexity of our protocol is $O(kns)$ (where k is a security parameter), improving over the $O(ks^2)$ complexity of the best previous solutions in this setting [18].

---

[2] We note that perfect size hiding cannot be achieved in the physical reality, as the *time* it takes the server to respond reveals an upper bound on the size of P. However, increasing this upper bound on the size of P does not involve additional work. This should be contrasted with the partial size hiding that can be achieved using previous protocols by simply padding the inputs.

[3] A secure two-party protocol in which the client's work is almost independent of the size of P can be obtained using the technique of Naor and Nissim [28]. However, this protocol requires multiple rounds of interaction and does not achieve size hiding.

Techniques. The basic version of our protocol uses a simple generalization of the technique of Kushilevitz and Ostrovsky [23] for constructing single-server PIR protocols. In fact, the protocol of [23] (as well as its variants from [34,26]) can be viewed as an instance of our protocol in which the branching program is a complete (but possibly non-binary) decision tree whose i-th level depends only on the i-th input variable.

Our protocol proceeds roughly as follows. The ciphertext c is obtained by separately encrypting each bit of x using a homomorphic public-key encryption scheme. (For efficiency reasons we rely on the Damgård-Jurik scheme [11]; this scheme was previously used in the context of PIR by Lipmaa [26].) To evaluate P on x we proceed in a bottom up manner. Starting from the terminal nodes, in the i-th iteration we handle all nodes whose distance from the terminal nodes is i. For each such node, we compute a ciphertext containing an (iterated) encryption of its value. Using the homomorphic property, the encryption assigned to every node can be computed from the encryptions assigned to its children (which were computed in previous iterations) and the encryption of the input bit labeling this node. The ciphertext c' is the (iterated) encryption assigned to the initial node. The client can recover $P(x)$ by applying iterated decryptions to c'.

The stronger variant of our protocol which remains secure in the case of malicious clients is more involved, and relies on variants of previous techniques of Aiello et al. [1], Naor and Pinkas [29], Laur and Lipmaa [24], and (especially) Kalai [20].

Organization. In Section 2 we define our general notion of representation models as well as the specific branching program model for which our results apply. In Section 3 we define the problem of computing on encrypted data as well as a variant of Oblivious Transfer on which our solution relies. Our main protocol is presented in Section 4. This protocol guarantees the privacy of the client as well as the privacy of the server against a semi-honest client. The case of malicious clients is discussed in Section 5. For lack of space, some details are deferred to the full version.

2 Preliminaries

We denote by $y \leftarrow A(x)$ the process of invoking the (possibly randomized) algorithm A on input x and assigning the result to y. We say that a function $\epsilon(k)$ is negligible if for every constant $c > 1$ we have $\epsilon(k) < 1/k^c$ for all sufficiently large k. We use the following standard notion of statistical distance:

Definition 1 (Statistical distance). *Let X, Y be random variables over the finite set U. Denote the distance between X and Y by*

$$\mathsf{SD}(X, Y) = max_{U' \subseteq U} \left| \Pr_{x \leftarrow X}[x \in U'] - \Pr_{y \leftarrow Y}[y \in U'] \right|$$

2.1 Representation Models

Loosely speaking, a representation model is a way of interpreting strings as "programs" for evaluating (families of) functions over some finite domain. We are only interested in representation models which are *universal* in the sense that every function has a program evaluating it in that model. For simplicity we restrict the attention to functions defined over a binary input alphabet. An extension to the general case is straightforward.

Definition 2 (Representation model). *A representation model is a polynomial-time computable function* $U : \{0,1\}^* \times \{0,1\}^* \to \{0,1\}^*$, *where* $U(P, x)$ *is referred to as the value returned by a "program"* P *on the input* x. *When* U *is understood from the context, we use* $P(x)$ *to denote* $U(P, x)$. *We say that a function* $f : \{0,1\}^* \to \{0,1\}^*$ *can be implemented in a representation model* U *if there exists an infinite sequence* $(P_0, P_1, \ldots)$, *referred to as an implementation of* f *in* U, *such that* $f(x) = U(P_{|x|}, x)$ *for every* $x \in \{0,1\}^*$.

We now define the branching programs model. This is the representation model for which our main result applies.

Definition 3 (Branching program (BP)). *A (deterministic) branching program over the variables* $x = (x_1, \ldots, x_n)$ *with input domain* I *and output domain* O *is defined by a tuple* $(G = (V, E), v_0, T, \psi_V, \psi_E)$ *where:*

- *G is a directed acyclic graph. Denote by* $\Gamma(v)$ *the children set of a node* v.
- v_0 *is an initial node of indegree 0. We assume without loss of generality that every* $u \in V - \{v_0\}$ *is reachable from* v_0.
- $T \subseteq V$ *is a set of terminal nodes of outdegree 0.*
- $\psi_V : V \to [n] \cup O$ *is a node labeling function assigning an output value to each terminal node in* T, *and a variable index from* $[n]$ *to each nonterminal node in* $V - T$.
- $\psi_E : E \to 2^I$ *is an edge labeling function, such that every edge is mapped to a non-empty set, and for every node* v *the sets labeling the edges to nodes in* $\Gamma(v)$ *form a partition of* I.

BP evaluation. The output $P(x)$ of a branching program P on an input assignment $x \in I^n$ is naturally defined by following the path induced by x from v_0 to a terminal node v_ℓ, where the successor of node v is the unique node v' such that $x_{\psi_V(v)} \in \psi_E(v, v')$. The output is the value $\psi_V(v_\ell)$ labeling the terminal node reached by the path.

BP complexity measures. Let $P = (G(V, E), v_0, T, \psi_V, \psi_E)$ be a BP. The *size* of P is $|E|$. (Note that in the case where $|I|$ is constant we have $|E| = O(|V|)$.) The *height* of a node $v \in V$, denoted height(v), is the length (in edges) of the longest path from v to a node in T. The *length* of P is the height of v_0. We say that an implementation $(P_0, P_1, \ldots)$ of a function f in the branching program model is length-bounded by $\ell(\cdot)$ if the length of each P_n is at most $\ell(n)$.

Remark 1. In the following we will sometimes assume that branching programs have binary inputs and outputs, namely that $I = O = \{0,1\}$. We stress, however, that the generalization to non-binary domains is useful for some of the applications we have in mind. For instance, non-binary input alphabets are useful for casting the PIR protocol from [23] as a special case of our main construction, and large output alphabets are useful for applications such as private retrieval by keywords [9,13].

Our protocols take the simplest form when the branching program being evaluated is *layered* in the following sense.

Definition 4 (Layered BP). *We say that P is a* layered *branching program of length ℓ if the node set V can be partitioned into $\ell + 1$ disjoint levels $V = \bigcup_{i=0}^{\ell} V_i$, such that $V_0 = \{v_0\}$, $V_\ell = T$, and for every $e = (u, v)$ we have $u \in V_i, v \in V_{i+1}$ for some i. We refer to V_i as the i-th level of P.*

Every branching program of size s can be efficiently transformed into a layered branching program of size at most s^2 and same length (cf. [32]). For convenience, we assume in our protocol that the server's BP is layered, which may square the server's work but has no effect on the communication complexity or the client's work. The quadratic overhead in the server's work can be avoided in most useful special cases (e.g., evaluating decision trees or finite automata) and can be avoided in the general case if only client privacy is required.

3 Cryptographic Primitives

In this section we define both our goal of computing on encrypted data and the main cryptographic tool on which we rely.

3.1 Computing on Encrypted Data

We consider a scenario where a client, holding an input x, publishes a public key pk and an encryption c of x under pk. This encryption is used by a server to efficiently evaluate a program P (in some given representation model) on c, obtaining a ciphertext c'. The client then uses its secret key to recover $P(x)$ from c'. This is formalized as follows.

Definition 5 (Computing on encrypted data). *Let $U : \{0, 1\}^* \times \{0, 1\}^* \to \{0, 1\}^*$ be a representation model. A protocol for evaluating programs from U on encrypted data is defined by a tuple of algorithms* (Gen, Enc, Eval, Dec) *and proceeds as follows.*

- SETUP: *Given a security parameter k, the client computes $(pk, sk) \leftarrow$ Gen(1^k) and saves sk for a later use.*
- ENCRYPTION: *The client computes $c \leftarrow$ Enc(pk, x), where x is the input on which a program P should be evaluated.*
- EVALUATION: *Given the public key pk, the ciphertext c, and a program P, the server computes an encrypted output $c' \leftarrow$ Eval$(1^k, pk, c, P)$.*
- DECRYPTION: *Given the encrypted output c', the client outputs $y \leftarrow$ Dec(sk, c').*

We require that if both parties act according to the above protocol, then for every input x, program P, and security parameter $k \in \mathbb{N}$, the output y of the final decryption phase is equal to $U(P, x)$ except, perhaps, with negligible probability in k.

An essential security requirement for computing on encrypted data is *client privacy*, requiring that the pair (pk, c) produced in the above process keep the client's input x semantically secure [17,16].

Definition 6 (Client privacy). *Let $\Pi = $ (Gen, Enc, Eval, Dec) be a protocol for computing on encrypted data. We say that Π satisfies the* client privacy *requirement if the advantage of any PPT adversary* Adv *in the following game is negligible in the security parameter k:*

- Adv *is given 1^k and generates a pair $x_0, x_1 \in \{0, 1\}^*$ such that $|x_0| = |x_1|$.*
- *Let $b \xleftarrow{R} \{0, 1\}$, $(pk, sk) \leftarrow $ Gen(1^k), and $c \leftarrow $ Enc(pk, x_b).*
- Adv *is given the challenge (pk, c) and outputs a guess b'.*

The advantage of Adv *is defined as* $\mathbf{Pr}[b = b'] - 1/2$.

Client privacy alone can be realized by simply letting Eval output P. However, it becomes nontrivial to satisfy when $|P| \gg |x|$ and the communication complexity is required to be sublinear in $|P|$. The latter requirement is in the center of this work.

While client privacy suffices for some applications, we will also be interested in protecting the privacy of the server by hiding the program P to the extent possible. For simplicity we consider here the case of a semi-honest client, who generates a valid public key pk and ciphertext c. The case of malicious clients will be addressed in Section 5.

Definition 7 (Server privacy: semi-honest model). *Let $\Pi = $ (Gen, Enc, Eval, Dec) be a protocol for evaluating programs from a representation model U on encrypted data. We say that Π has* statistical server privacy *in the semi-honest model if there exists a PPT algorithm* Sim *and a negligible function $\epsilon(\cdot)$ such that the following holds. For every security parameter k, input $x \in \{0, 1\}^*$, pair (pk, c) that can be generated by* Gen, Eval *on inputs k, x, and program $P \in \{0, 1\}^*$, we have*

$$\text{SD}(\text{Eval}(1^k, pk, c, P), \text{Sim}(1^k, 1^{|x|}, pk, U(P, x), 1^{|P|})) \leq \epsilon(k).$$

The case of perfect *server privacy is defined similarly, except that $\epsilon(k) = 0$ and* Sim *is allowed to run in expected polynomial time.*

In the case of computational *server privacy,* Sim *should satisfy the following requirement. For every polynomial-size circuit family D there is a negligible function $\epsilon(\cdot)$ such that for any k, x, pk, c, P as above we have*

$$\Pr[D(\text{Eval}(1^k, pk, c, P)) = 1] - \Pr[D(\text{Sim}(1^k, 1^{|x|}, pk, U(P, x), 1^{|P|})) = 1] \leq \epsilon(k).$$

Our main protocol will have perfect server privacy. In fact, it will additionally hide the size of the server's input P from the client. We refer to this property as *size hiding*. This implies, in particular, that the length of c' must be independent of the length of P.

Definition 8 (Size hiding server privacy: semi-honest model). *We say that Π has (perfect, statistical, or computational)* size hiding *server privacy in the semi-honest model if it satisfies the requirements of Definition 7 with the following difference:* Sim *does not get the length of P as an input.*

Remark 2. Protocols Π which satisfy our definitions of client privacy (Definition 6) and standard server privacy (Definition 7) can be easily derived from previous protocols for one-round secure computation. In particular, Yao's protocol [36] yields a protocol for evaluating circuits on encrypted data with computational server privacy, and protocols from [35,21,14,18,4,22] yield protocols for evaluating formulas, branching programs, and even non-deterministic branching programs on encrypted data with perfect or statistical server privacy. However, in all these protocols the length of c' is generally bigger than the length of P. In particular, none of these protocols satisfies the additional size hiding property of Definition 8.

3.2 Oblivious Transfer

It will be convenient to present our main protocol in a modular way, using a variant of one-round Oblivious Transfer (OT) [33,12] as a subprotocol. To this end it will be necessary to rely on a stronger server privacy property than the one implied by standard definitions of OT. As before, we focus here on the case of a semi-honest client and postpone the treatment of malicious clients to Section 5.

A standard one-round OT protocol involves a server, holding a list of t secrets $(s_1, s_2, \ldots, s_t)$, and a client, holding a selection index i. The client sends a query q to the server, who responds with an answer a. Using a and its random input, the client should be able to recover s_i. The standard security requirements include *client privacy*, requiring that q keep i hidden from the server, and *server privacy*, requiring that a keep all secrets other than s_i hidden from the client. Note that the latter server privacy requirement does not rule out the possibility that a reveals information about the query q which is not implied by the output s_i alone. (In fact, a can include the entire query q without violating server privacy.) This might compromise the security of our main protocol, in which the client issues multiple OT queries and each query is used by the server to compute multiple answers. It will be crucial for the security of the protocol that the client be unable to correlate answers with queries, beyond correlations which follow from the outputs. Such correlations will reveal to the client information about the structure of the server's branching program.

Roughly speaking, our notion of strong OT strengthens the above server privacy requirement by requiring the distribution of the answer a conditioned on the output s_i to be independent of the query q. In other words, the distribution of the answer depends on the output alone. It turns out that a natural implementation of one-round OT based on homomorphic encryption [23,34] satisfies the required properties (see Section 4.1). We now formally define strong OT.

Definition 9 (Strong OT). *A strong OT protocol is defined by a tuple of PPT algorithms* $(\mathsf{G_{OT}}, \mathsf{Q_{OT}}, \mathsf{A_{OT}}, \mathsf{D_{OT}})$. *The protocol involves two parties, a client and a server, where the server's input is a t-tuple of strings $(s_1, \ldots, s_t)$ of length τ each, and the client's input is an index $i \in [t]$. The parameters t, τ are given as inputs to both parties. The protocol proceeds as follows:*

- *The client generates $(pk, sk) \leftarrow \mathsf{G_{OT}}(1^k)$, computes a query $q \leftarrow \mathsf{Q_{OT}}(pk, 1^t, 1^\tau, i)$, and sends (pk, q) to the server.*
- *The server computes $a \leftarrow \mathsf{A_{OT}}(pk, q, s_1, \ldots, s_t)$ and sends a to the client.*
- *The client computes and outputs $\mathsf{D_{OT}}(sk, a)$.*

We require that if both parties follow the protocol, the client always outputs s_i. We denote the length of the query q by $\alpha(k, t, \tau)$ and the length of the answer a by $\beta(k, t, \tau)$.

Our main protocol will require $\beta(k, t, \tau) = \tau + \text{poly}(k, t)$ to efficiently accommodate BPs of arbitrary length. (In fact, it suffices that the above holds for $t = 2$.) This will be our default efficiency requirement. However, this requirement can be relaxed if one settles for weaker forms of our main result that apply to shallow BPs, such as constant-length BPs over a polynomial-size input alphabet.

We now define the client privacy and (strong) server privacy requirements.

Definition 10 (Strong OT: client privacy). *We require that the client's query q keep i semantically secure. That is, the advantage of any PPT adversary Adv in the following game is negligible in the security parameter k:*

- *Adv is given 1^k and generates $1^t, 1^\tau$ and i_0, i_1 such that $i_0, i_1 \in [t]$.*
- *Let $b \xleftarrow{R} \{0, 1\}$, $(pk, sk) \leftarrow \mathsf{G_{OT}}(1^k)$, and $q \leftarrow \mathsf{Q_{OT}}(pk, 1^t, 1^\tau, i_b)$.*
- *Adv is given the challenge (pk, q) and outputs a guess b' for b.*

The advantage of Adv is defined as $\mathbf{Pr}[b = b'] - 1/2$.

Our strong variant of perfect server privacy is defined similarly to Definition 7.

Definition 11 (Strong OT: server privacy). *There exists an expected polynomial time simulator $\mathsf{Sim_{OT}}$ such that the following holds. For every $k, t, \tau, i \in [t]$, pair (pk, q) that can be generated by $\mathsf{G_{OT}}, \mathsf{Q_{OT}}$ on inputs k, t, τ, i, and strings $s_0, \ldots, s_{t-1} \in \{0, 1\}^\tau$, the distributions $\mathsf{A_{OT}}(pk, q, s_1, \ldots, s_t)$ and $\mathsf{Sim_{OT}}(pk, 1^t, s_i)$ are identical.*

In the following it will sometimes be convenient to index the server's inputs s_i by $0, 1, \ldots, (t-1)$ instead of $1, 2, \ldots, t$.

4 Main Protocol

In this section we will describe our main protocol for evaluating branching programs on encrypted data. The protocol will provide client privacy, along with size hiding server privacy in the semi-honest model. Extensions that achieve server privacy in the malicious model will be presented in Section 5.

We fix a polynomially bounded length function $\ell(\cdot)$, and assume that if the client's input x is of length n, then the server's BP P is of length $\ell(n)$. (To conform to our general definition of representation models, one may define $P(x) = 0$ for P and x that do not match.) We also view the input domain I and output domain O as being implicitly determined by n. However, in the following it will be convenient to view ℓ, $|I|$, and $|O|$ as separate parameters which are given to both parties, and analyze the complexity of the protocol as a function of these parameters. We will also assume that P is layered (see Definition 4). As discussed in Section 2.1, every BP can be efficiently transformed into an equivalent layered BP without increasing its length.

Our protocol is based on a strong OT protocol as defined in Section 3.2 and proceeds roughly as follows. (For simplicity, assume that the input domain I of P is binary and that every nonterminal node in the graph has outdegree 2.) The client generates, for every input variable x_i and level j, an OT query q_i^j corresponding to a selection of the x_i-th string out of a pair of strings of an appropriate length. (This length will depend on j and will be later understood from the context.) The ℓn queries q_i^j jointly form the encryption c of x.

To evaluate P on c, the server makes a bottom-up pass on P, starting with the terminal nodes T and ending with the initial node v_0. This pass labels each node v in the graph by an OT answer which encrypts the output value to which x leads from this node. The pass consists of $\ell + 1$ iterations, where in iteration j ($0 \leq j \leq \ell$) all nodes of height j are handled. In iteration 0 every terminal node v is labeled by the corresponding output value $\psi_V(v)$. At the onset of the j-th iteration, $j \geq 1$, all nodes of height $j - 1$ have already been labeled. For each node v of height j, we want the labeling of v to encrypt the label of the child of v to which x leads. Such a label is computed by using the OT answering algorithm as follows. Suppose that the children of v are v_0 and v_1, where P branches from v to v_b if $x_i = b$. The label of v then computed by applying the OT answering algorithm to the query q_i^j on the pair of strings $(\mathsf{label}(v_0), \mathsf{label}(v_1))$. Note that since P is layered, the two labels have the same length. Moreover, by the strong server privacy property of the OT protocol, the label of v can be viewed as an encryption of the label of the selected child v_{x_i}. In particular, this label does not contain any information about the identity of the variable x_i that was used to determine the selection. (If a standard one-round OT is used, this is not necessarily be the case.)

Finally, at the end of iteration ℓ, the initial node v_0 is labeled by an OT answer which can be viewed as an (iterated) encryption of the output value $P(x)$. The client decrypts $P(x)$ by applying the OT decryption algorithm ℓ times to the label of v_0.

The above protocol is formally described in Figure 1. Its correctness is implied by the following lemma, which can be easily proved by induction on the height h.

Lemma 1. *For any node v, let $P_v(x)$ denote the output of P on the input x if v is used as the initial node. Then, for every $0 \leq h \leq \ell$ and every node v of height h we have $\mathsf{D_{OT}}^{(h)}(sk, \mathsf{label}(v)) = P_v(x)$, where $\mathsf{D_{OT}}^{(h)}(sk, \cdot)$ denotes the h-th iterate of $\mathsf{D_{OT}}(sk, \cdot)$.*

In particular, $\mathsf{D_{OT}}^{(\ell)}(sk, \mathsf{label}(v_0)) = P(x)$, from which correctness follows. We turn to analyze the protocol's efficiency.

Efficiency. Recall that we denote the length of an OT query by $\alpha(k, t, \tau)$ and the length of an OT answer by $\beta(k, t, \tau)$. Let β_j be as defined in Step 2, namely the result of applying the j-th iterate of $\beta(k, t, \cdot)$ on $\log |O|$. The length of the encryption c computed by the client is then bounded by $\ell n \cdot \alpha(k, t, \beta_\ell)$ and the length of the ciphertext c' computed by the server is $\beta_{\ell+1}$. By default, we assume the strong OT implementation to be such that $\beta(k, t, \tau) = \tau + \mathrm{poly}(k, t)$. (See Section 4.1 for a concrete implementation using the Damgård-Jurik cryptosystem.) In such a case, the overall communication is $\mathrm{poly}(k, n, \ell)$, which is in particular independent of $|P|$ as required. We will later present an optimized instantiation of the main protocol with a total communication of $O(kn\ell)$ (for the case of binary inputs and outputs). Finally, the computation performed by each party is polynomial in the length of its input.

Remark 3. When $\ell(n) \ll n$, the requirement that $\beta(k, t, \tau) = \tau + \mathrm{poly}(k, t)$ can be relaxed. In particular, if $\ell(n) = O(\log n)$ it suffices that $\beta(k, t, \tau) = O(\tau) + \mathrm{poly}(k, t)$. A strong OT protocol with the latter efficiency requirement can be based on homomorphic cryptosystems which expand the ciphertext length by a constant factor, such

---

Main Protocol

- Common inputs: security parameter 1^k, a branching program length parameter 1^ℓ, input domain $I = \{0, 1, \ldots, t-1\}$, output domain $O = \{0, 1\}^\gamma$.
- Client input: an assignment $x = (x_1, \ldots, x_n) \in I^n$.
- Server input: a layered BP $P = (G(V, E), v_0, T, \psi_V, \psi_E)$ of length ℓ.
- Sub-protocol: a strong OT protocol $(\mathsf{G_{OT}}, \mathsf{Q_{OT}}, \mathsf{A_{OT}}, \mathsf{D_{OT}})$ with answer length $\beta(k, t, \tau)$.

1. **Setup** $\mathsf{Gen}(1^k)$:
   - Let $(pk, sk) \leftarrow \mathsf{G_{OT}}(1^k)$.
   - Return (pk, sk).
2. **Encryption** $\mathsf{Enc}(pk, x)$:
   - For $1 \le i \le n$, generate a vector $q_i = (q_i^1, \ldots, q_i^\ell)$, where q_i^j is obtained by:

   $$q_i^j \leftarrow \mathsf{Q_{OT}}(pk, 1^t, 1^{\beta_j}, x_i),$$

   and where the lengths β_j are defined by $\beta_1 = \log |O|$ and $\beta_{j+1} = \beta(k, t, \beta_j)$.
   - Return $c = (q_1, \ldots, q_n)$.
3. **Evaluation** $\mathsf{Eval}(1^k, pk, c = (q_i^j), P)$:
   - Initialization: for each $v \in T$ set $\mathsf{label}(v) \leftarrow \psi_V(v)$.
   - While v_0 isn't labeled:
     - Pick an unlabeled node $v \in V - T$ such that all its children are labeled.
     - Let $i \leftarrow \psi_V(v)$ and $h \leftarrow \mathsf{height}(v)$.
     - Let $\mathsf{label}(v) \leftarrow \mathsf{A_{OT}}(pk, q_i^h, \mathsf{label}(u_0), \ldots, \mathsf{label}(u_{t-1}))$, where u_m is the (unique) node such that $m \in \psi_E(v, u_m)$.
       Note that the nodes u_m are not necessarily distinct.
   - Return $c' = \mathsf{label}(v_0)$.
4. **Decryption** $\mathsf{Dec}(sk, c')$:
   - Let $d_\ell \leftarrow c'$.
   - For $j = \ell$ down to 1, let $d_{j-1} \leftarrow \mathsf{D_{OT}}(sk, d_j)$.
   - Return d_0.

Fig. 1. Evaluating a branching program on encrypted data

as El-Gamal (see Section 4.1). If $\ell(n) = O(1)$, we can rely on an arbitrary strong OT, which in turn can be based on an arbitrary homomorphic encryption scheme (including, for instance, the Goldwasser-Micali cryptosystem [17]).

Remark 4. The PIR protocol of [23] can be viewed as an instance of our construction in which ℓ is set to some constant d, the input domain I is of size $t = N^{1/d}$ (where N is the database size), and the database is represented as a complete decision tree of depth d and degree $N^{1/d}$. Its variant suggested in [34] (resp., [26]) corresponds to a decision tree of depth $\sqrt{\log N}$ and degree $t = 2^{\sqrt{\log N}}$ (resp., depth $\log N$ and degree $t = 2$). These three depth parameters correspond to the different BP length regimes discussed in Remark 3.

We turn to prove the security properties of the main protocol. In the following we assume that the given strong OT subprotocol is secure and that its answer complexity is $\beta(k, t, \tau) = \tau + \text{poly}(k, t)$. In Section 4.1 we will show that this assumption is implied by the DCRA assumption.

Theorem 1. *The protocol described in Figure 1 provides client privacy according to Definition 6 as well as perfect size hiding server privacy in the semi-honest model according to Definition 8.*

Proof sketch: Client privacy readily follows from the client privacy requirement in the underlying OT protocol. The security of sending polynomially many strong OT queries under the same key follows from the security of encrypting multiple messages under the same key in public-key encryption schemes (see [16], Theorem 5.2.11).

To prove size hiding server privacy, we describe a perfect simulator Sim. The idea is to recreate the labels of the computation path from v_0 to a terminal node labeled with $P(x)$ without knowing the nodes traversed by the path. Sim will use the OT simulator Sim_{OT} as a subroutine. On inputs $(1^k, 1^{|x|}, pk, P(x))$ (and given $|I| = t$ as an additional public input), Sim proceeds as follows:

- Let $\ell \leftarrow \ell(|x|)$, $\lambda_0 \leftarrow P(x)$.
- For $j = 1$ to ℓ, let $\lambda_j \leftarrow \text{Sim}_{\text{OT}}(pk, 1^t, \lambda_{j-1})$.
- Return λ_ℓ.

Consider the computation path $v_0, v_1, \ldots, v_\ell$ induced by x. It follows by induction on j that the distribution of λ_j produced by Sim is identical to the distribution of $\text{label}(v_{\ell-j})$ produced by $\text{Eval}(1^k, pk, c, P)$, for every k, x, P and pair (pk, c) which can be generated by Gen, Enc on k, x. In particular, the simulator's output λ_ℓ is distributed identically to $c' = \text{label}(v_0)$. Note that the strong OT requirement allows Sim_{OT} to produce the correct distributions independently of the OT queries included in c.   □

4.1   Implementing Strong OT

Our concrete implementation of strong OT is based on the Damgård-Jurik (DJ) homomorphic public-key cryptosystem [11], which generalizes Paillier's cryptosystem [31]. It is suitable for our needs because it allows us to encrypt a group element of length τ into a ciphertext of length $\tau + O(k)$, where k is a security parameter. This efficiency feature is unique among all known homomorphic encryption schemes and is needed for our main protocol to be efficient for arbitrary length bounds $\ell(n)$. The semantic security of the DJ cryptosystem follows from the Decisional Composite Residuosity Assumption (DCRA) [11].

We now describe the main properties of the DJ cryptosystem that are useful for our purposes (see [11] for further details).

- KEY GENERATION: Given a security parameter k, $Gen(1^k)$ outputs a secret key (p_1, p_2), where p_1, p_2 are random k-bit primes (i.e., $2^{k-1} \geq p_1, p_2 < 2^k$), and a public key $N = p_1 p_2$. The above choice of p_1, p_2 guarantees that $gcd(N, \phi(N)) = 1$. This property will be useful in what follows. We refer to N which can be generated by $Gen(1^k)$ as a *valid DJ key*.

- ENCRYPTION: The DJ cryptosystem is length-flexible in the sense that every fixed key N allows to encrypt plaintexts of an arbitrary (polynomial) length, where the encryption only *adds* $O(k)$ bits to the length of the plaintext. Given a plaintext length parameter e, where $1 \le e < \min(p_1, p_2)$, we define a plaintext group $M_{N,e} = \mathbb{Z}_{N^e}$ and a ciphertext group $C_{N,e} = \mathbb{Z}_{N^{e+1}}^*$. The restriction on e is required for correct decryption, and since we will only use $e \le \mathrm{poly}(k)$ it will always hold. Now fix some valid pair (N, e). To abbreviate notation we denote the ciphertext group $C_{N,e} = \mathbb{Z}_{N^{e+1}}^*$ by C. Let $C_0 = C^{N^e} = \{c^{N^e} | c \in C\}$. Clearly, C_0 is a subgroup of C. Let $g = N + 1 \in C$. The output distribution of the encryption is specified via an injective homomorphism $H : M_{N,e} \to C/C_0$ defined by $H(m) = g^m \cdot C_0$, where $g^m \cdot C_0$ denotes the coset represented by g^m in C/C_0. To encrypt $m \in M_{N,e}$, the encryption function $E_{N,e}(m)$ returns a random element in the coset $H(m)$. This can be done by sampling $r \stackrel{R}{\leftarrow} \mathbb{Z}_N^*$ and outputting $c = g^m \cdot r^{N^e}$ (where all multiplications are in C). In particular, an encryption of 0 is a random element of C_0. Note that the difference between the size of the ciphertext ($\lceil \log N^{e+1} \rceil$) and the size of the plaintext ($\lceil \log N^e \rceil$) is indeed only $O(k)$.
- DECRYPTION: Given $c = g^m \cdot r^{N^e}$ and the factorization (p_1, p_2) of N, it is possible to efficiently decrypt m. We denote the decryption algorithm by $D_{(p_1, p_2), e}(c)$.
- HOMOMORPHISM: Given two ciphertexts $c \in E_{N,e}(m)$ and $c' \in E_{N,e}(m')$, their product $c \cdot c'$ (in the ciphertext group) is a valid encryption of the sum $m + m'$ (in the plaintext group). It follows that c^ρ is an encryption of $\rho \cdot m$. Moreover, multiplying c by a random encryption of 0 *rerandomizes* c into a random encryption of m.

Strong OT from the DJ cryptosystem. The following strong OT protocol is similar to the PIR protocol of [23] and its generalizations from [34,26]. The choice of DJ as the underlying cryptosystem is motivated by the goal of handling branching programs of an arbitrary length. If the length function $\ell(n)$ is small, other homomorphic cryptosystems can be used (see Remark 3).

Construction 2 (Strong OT). Let $(Gen, E_{N,e}, D_{(p_1, p_2), e})$ be the DJ cryptosystem. The OT protocol $(\mathsf{G_{OT}}, \mathsf{A_{OT}}, \mathsf{Q_{OT}}, \mathsf{D_{OT}})$ proceeds as follows.

1. $\mathsf{G_{OT}}(1^k)$:
   - Let $(N, (p_1, p_2)) \leftarrow Gen(1^k)$.
   - Return $(N, (p_1, p_2))$.
2. $\mathsf{Q_{OT}}(N, 1^k, 1^t, 1^\tau, i)$:
   - Let e be the minimal integer such that $N^e > 2^\tau$. We naturally identify strings in $\{0,1\}^\tau$ with integers in $M_{N,e} = \mathbb{Z}_{N^e}$, and assume that elements in the groups $M_{N,e}$ and $C_{N,e}$ are padded so that their representation reveals e.
   - Let $q_i \leftarrow E_{N,e}(1)$ and $q_j \leftarrow E_{N,e}(0)$ for all $j \in [t] \setminus i$.
   - Return $q = (q_1, \ldots, q_{t-1})$.
3. $\mathsf{A_{OT}}(N, q, s_1, \ldots, s_t)$:
   - Infer e from q.
   - Let $q_t \leftarrow E_{N,e}(1) \cdot (\prod_{i=1}^{t-1} q_i^{s_i})^{-1}$ (where all operations are in $C_{N,e}$).

- Let $a \leftarrow \prod_{i=1}^{t} q_i^{s_i} \cdot E_{N,e}(0)$.
- Return a.
4. $\mathsf{D_{OT}}((p_1, p_2), a)$:
   - Infer e from a.
   - Return $D_{(p_1,p_2),e}(a)$.

Analysis. Correctness follows by observing that $(q_1, \ldots, q_t)$ encrypt the i-th unit vector of length t and a encrypts the inner product of $(s_1, \ldots, s_t)$ with this vector, which yields s_i. Client privacy follows from the semantic security of the DJ cryptosystem, which can be based on the DCRA assumption [11]. Server privacy follows from the fact that (due to rerandomization) the server's answer on any valid q is a *random* encryption of s_i, which can be easily generated by $\mathsf{Sim_{OT}}$. The protocol's query length is $\alpha(k, t, \tau) = t \cdot (\tau + O(k))$ and its answer length is $\beta(k, t, \tau) = \tau + O(k)$.

4.2 Optimizations

Optimizing the server's work. Our main protocol requires the branching program P to be layered. Converting an arbitrary BP to an equivalent layered BP of the same length may generally result in a quadratic blowup to its size, which in turn results in a quadratic computational overhead on the server's part. (We note, however, that most "natural" BPs, including ones that arise from other computation models such as finite automata, are either already layered or can be turned into equivalent layered BPs with only a linear overhead.) The quadratic overhead can be easily avoided in general if only client privacy is required. The main protocol can be modified in this case to operate on a non-layered BP by padding the labels that serve as OT inputs to match the size of the longest label.

Optimizing the encryption length. In the main protocol, the length of the encryption c produced by Enc must be bigger than $\sum_{j=1}^{\ell} \beta_j > \ell^2$. It turns out that the quadratic dependence on ℓ can be avoided by exploiting the specific structure of the DJ cryptosystem. The improvement is based on the following observation:

Observation 3. *For every valid DJ key pair* (N, e), $e' < e$, $m \in M_{N,e}$ *and* $c \in E_{N,e}(m)$ *(i.e., c is some valid encryption of m) it holds that*

$$c \bmod N^{e'+1} \in E_{N,e'}(m \bmod N^{e'}).$$

It follows from Observation 3 that the ciphertext c may consist of n encryptions q_i in the largest group (rather than n encryptions q_i^j for every level j of the BP), since the server can convert encryptions from the largest group into encryptions from smaller groups. (Note that since we only encrypt 0's and 1's, the conversion does not modify the encrypted value.) The improved implementation achieves communication complexity of $O(kn\ell)$ bits from the client to the server (instead of $O(kn\ell^2)$ in the original implementation) and $O(k\ell)$ bits from the server to the client (as in the original implementation). Clearly, the optimization doesn't compromise client or server privacy. Thus, we have:

Theorem 4. *Assuming DCRA [11], there is a protocol for evaluating a binary branching program of length ℓ and of arbitrary size on an encrypted input of length n, with a total communication of $O(kn\ell)$ bits (where k is a security parameter). The protocol provides client privacy as well as size hiding server privacy in the semi-honest model.*

5 Handling Malicious Clients

In this section we sketch the required modifications for achieving security against malicious clients. For lack of space we only describe the high level ideas and refer the reader to the full version for further details. For simplicity, we restrict the attention throughout this section to the case of branching programs over binary inputs.

We start by observing that a malicious client can easily break the server privacy of the main protocol even if it honestly generates the public key pk.

Example 1. Consider a client who sends an encryption of 2 (instead of 0 or 1) as an OT query. In this OT invocation, the client can recover both s_0 and s_1. This potentially reveals additional information about the structure of the branching program P. For instance, in the degenerate case where P consists of an initial node and two terminal nodes, the client will learn the values of both terminal nodes.

The above mild form of cheating is relatively easy to handle using previous techniques [15,1,24] and will be addressed in Section 5.1. A more challenging goal is to handle clients that are also free to choose invalid public keys pk. This scenario will be addressed in Section 5.2.

Before describing our solutions, we formalize our notions of server privacy in the malicious model. The following definitions modify Definition 8 in that they allow an unbounded simulator to extract an effective input x^* from a corrupted ciphertext c^* and a (possibly) corrupted public key pk^*. The use of unbounded simulation seems necessary in the "vanilla" one-round malicious model (i.e., without setup assumptions) and was previously made in similar contexts [29,1,13,20,24].

We start by defining the *trusted setup model*, where the client is forced to use a valid public key pk but can cheat by creating invalid ciphertexts c^*. This model is motivated by the fact that the same public key may be reused to encrypt many different inputs. Thus, one can afford an expensive certification procedure (e.g., using interactive zero-knowledge proofs or a trusted party) that is used once and for all.

Definition 12 (Size hiding server privacy: trusted setup model). *Let Π = (Gen, Enc, Eval, Dec) be a protocol for evaluating programs from a representation model U on encrypted data. We say that Π has statistical server privacy in the trusted setup model if there exists a computationally unbounded, randomized algorithm Sim and a negligible function $\epsilon(\cdot)$ such that the following holds. For every security parameter k, valid public key pk that can be generated by $\mathsf{Gen}(1^k)$, and arbitrary ciphertext c^* there exists an "effective" input x^* such that for every program $P \in \{0,1\}^*$, we have*

$$\mathsf{SD}(\mathsf{Eval}(1^k, pk, c, P), \mathsf{Sim}(1^k, pk, c^*, U(P, x^*))) \leq \epsilon(k).$$

The case of computational server privacy is defined in an analogous way (see Definition 7), where statistical indistinguishability is replaced by computational one.

We turn to the fully malicious model.

Definition 13 (Size hiding server privacy: fully malicious model). *We say that Π has (statistical or computational) server privacy in the fully malicious model if it satisfies Definition 12 with the following modification: instead of quantifying over all valid public keys pk, now the quantification is over arbitrary public keys pk^*.*

Protocols for computing on encrypted data in the above model give rise to one-round (two-message) protocols for secure two-party computation of $U(\cdot, \cdot)$ under the relaxed security definitions of [29,1,13].

A natural approach for handling malicious clients would be to leave the main protocol as it is and only upgrade the original strong OT primitive into one that achieves security against malicious clients. Unfortunately, we cannot use this modular approach for several reasons. First, the basic variant of the protocol requires the client to use each input x_i in multiple OT invocations (corresponding to the different levels where x_i appears) and so the client could cheat by simply using inconsistent inputs in these OT invocations. More importantly, we do not know how to construct a strong OT protocol which simultaneously satisfies both our security and efficiency requirements in the malicious model. It is interesting to note that a one-round OT protocol of Kalai [20], which is based on Paillier's cryptosystem and can be generalized to work with the DJ cryptosystem, fails with respect to both security (in that it is not a *strong* OT) and efficiency (in that its answer significantly blows up the length of the selected string). Still, ideas from [20] will be instrumental in our solution for the fully malicious model.

5.1 Trusted Setup Model

We now describe a solution in the trusted setup model. Our starting point is the optimized instantiation of the protocol for the semi-honest model (Section 4.2), where in the case of binary inputs ($t = 2$) the client sends a single encryption for each input. Our goal is to prevent the type of attack described in Example 1, namely to ensure that each encryption sent by the client is indeed an encryption of 0 or 1. To this end one could employ general-purpose zero-knowledge proofs, forcing the client to prove that its queries are well formed. However, this approach requires multiple rounds of interaction which we would like to avoid, and also involves a considerable efficiency overhead.

Instead, we apply the conditional disclosure of secrets (CDS) methodology of [15,1]. The idea is that instead of making the client prove that its queries are well formed, it suffices for the server to disclose its answer c' to the client only under the condition that the queries are well formed. Using the homomorphic property of the encryptions, the latter conditional disclosure can be done without the server even knowing whether the condition is satisfied.

The original CDS solutions from [1] relies on homomorphic encryption over groups of a prime order. An efficient extension to groups of a composite order was suggested in [24], assuming that the order of the group is sufficiently "rough". We employ a similar extension which avoids the roughness assumption and is geared towards the solution in the fully malicious model.

We start by describing the approach of [1]. The simplest setting involves a server holding a (valid) public key pk of a homomorphic cryptosystem, a ciphertext $c \in E_{pk}(m)$ (presumably generated by a client), and a secret s. The client holds the secret key sk corresponding to pk. The goal is for the server to send a single (randomized) ciphertext $\tilde{c}$ such that: (1) if $m = 0$ then s can be recovered from $\tilde{c}$ using the secret key; and (2) if $m \neq 0$ then $\tilde{c}$ reveals (almost) no information about s. The above is referred to as a CDS of the secret s under the condition $m = 0$. A solution to this simple CDS problem can be easily extended to CDS under more general conditions, involving multiple inputs m_i and general predicates over atomic conditions of the form $m_i = b_i$. In particular, $2n$ invocations of the above primitive are sufficient to disclose a secret under the suitable condition here, namely that n ciphertexts c_i *all* encrypt 0/1 values.

The solution of [1] is to let $\tilde{c}$ be a random encryption of $s + \rho m$, where ρ is a random integer between 1 and the order of the plaintext group. Note that $\tilde{c}$ can be efficiently computed using the homomorphic properties of the encryption. Requirement (1) holds because if $m = 0$ then $\tilde{c}$ encrypts $s + \rho \cdot 0 = s$. Requirement (2) holds in the case where the plaintext group is of a prime order; indeed, in that case if $m \neq 0$ then ρm is uniformly distributed over the plaintext group and therefore can be used to hide s.

The next observation is that in the case that the plaintext group has a composite order, not all is lost. In this case, if $m \neq 0$ then ρm is uniformly distributed over a nontrivial subgroup of the plaintext group. If s is chosen uniformly at random from the plaintext group, then s will still have at least one bit of remaining entropy even when conditioned on $\tilde{c}$. This residual randomness can be extracted using standard privacy amplification techniques. Specifically, to disclose an l-bit secret we first repeat the above $l + k$ times with independent secrets s_i, increasing the conditional entropy to $l + k$, and then apply an arbitrary strong randomness extractor (e.g., a pairwise independent hash function) to extract l (almost) perfectly secret bits from the partially leaked secrets.

The above approach (or the similar approach from [24]) solves our problem in the trusted setup model. In this case, every possible string c^* can be interpreted as a valid ciphertext encrypting some message m in the plaintext group Z_{N^e}. Thus, we can use the above to disclose the server's answer under the condition that the n encryptions produced by the client are well formed. This yields a protocol for the trusted setup model whose communication complexity is comparable to that of the optimized version of the original protocol.[4]

5.2   Fully Malicious Model

The previous solution relied on the fact that for a valid key N, every c^* can be interpreted as a valid encryption of some message m. This does not hold in general. In fact, there is an explicit cheating strategy which uses N such that $\gcd(N, \phi(N)) > 1$ (e.g., $N = p_1 p_2$, where p_1, p_2 are odd primes and $p_2 = 2p_1 + 1$) in order to break the previous protocol. The main difficulty arises from the fact that the set of harmful keys N cannot be efficiently recognized. Our high level approach for getting around this problem is to project ciphertexts sent by the client onto a "harmless" subgroup of C by having the server raise them to the power of N^T, where $T = \lceil \log N \rceil$. To maintain correctness, plaintexts are chosen from a subgroup of Z_{N^e} of size N^{e-T}, which requires to moderately increase the values of e used in our protocol. We refer the reader to the full version for further details.

Acknowledgements. We thank Mike Freedman, Benny Pinkas, and Omer Reingold for discussions about secure keyword search from which this work originated. We also thank the anonymous TCC referees for many helpful comments and suggestions.

---

[4] This holds for the case of *computational* server privacy, where we can afford to disclose a short secret s and then encrypt the (long) answer using this key. The statistically private variant involves an additional multiplicative overhead of $O(\ell)$.

References

1. W. Aiello, Y. Ishai and O. Reingold. Priced oblivious transfer: How to sell digital goods. In *Proc. of EUROCRYPT 2001*, pages 119-135.
2. M. Blum, A. De Santis, S. Micali, and G. Persiano. Non-interactive Zero Knowledge. *SIAM Journal of Computing*, 20(6), pages 1084-1118, 1991.
3. B. Barak and O. Goldreich. Universal Arguments and their Applications. In *Proc. CCC 2002*, pages 194-203.
4. D. Beaver. Minimal-Latency Secure Function Evaluation. In *Proc. of EUROCRYPT 2000*, pages 335-350.
5. D. Boneh, E.J. Goh, and K. Nissim. Evaluating 2-DNF formulas on ciphertexts. In *Proc. 2nd TCC*, pages 325–341, 2005.
6. C. Cachin, S. Micali, and M. Stadler. Computationally private information retrieval with polylogarithmic communication. *Proc. of EUROCRYPT '99*, pages 402–414.
7. C. Cachin, J. Camenisch, J. Kilian, and J. Muller. One-round secure computation and secure autonomous mobile agents. In *Proceedings of ICALP '00*.
8. R. Canneti. Security and composition of multiparty cryptographic protocols. *Journal of Cryptology*, 13(1), pages 143-202.
9. B. Chor, N. Gilboa, and M. Naor. Private information retrieval by keywords. Technical Report TR-CS0917, Department of Computer Science, Technion, 1997.
10. B. Chor, O. Goldreich, E. Kushilevitz, and M. Sudan. Private information retrieval. *J. of the ACM*, 45:965–981, 1998. Earlier version in FOCS '95.
11. I. Damgård and M. Jurik. A Generalisation, a Simplification and some Applications of Paillier's Probabilistic Public-Key System. In *Proc. of CT-RSA '02*, pages 79-95.
12. S. Even, O. Goldreich and A. Lempel. A Randomized Protocol for Signing Contracts. In *Communications of the ACM*, 28(6):637–647, 1985.
13. M.J. Freedman, Y. Ishai, B. Pinkas and O. Reingold. Keyword search and oblivious pseudo-random fuctions. In *Proc. of TCC 2005* vol. 3378, pages 303-324.
14. U. Feige, J. Kilian, and M. Naor. A minimal model for secure computation. In *Proc. of 26th STOC*, pages 554-563, 1994.
15. Y. Gertner, Y. Ishai, E. Kushilevitz, and T. Malkin. Protecting Data Privacy in Private Information Retrieval Schemes. In *Proc. of 30th STOC*, pages 151-160, 1998.
16. O. Goldreich. *Foundations of Cryptography: Basic Applications*. Cambridge University Press, 2004.
17. S. Goldwasser and S. Micali. Probabilistic encryption. *JCSS*, 28(2):270–299, 1984. Preliminary version in Proc. STOC '82.
18. Y. Ishai and E. Kushilevitz. Randomizing polynomials: A new representation with applications to round-efficient secure computation. In *Proc. 41st FOCS*, pp. 294–304, 2000.
19. Y. Ishai and E. Kushilevitz. Perfect Constant-Round Secure Computation via Perfect Randomizing Polynomials. In *Proc. of the 29th ICALP*, pages 244-256, 2002.
20. Y. T. Kalai. Smooth Projective Hashing, and two message Oblivious Transfer. In *Proc. of EUROCRYPT 2005*, pages 78-95.
21. J. Kilian. Founding cryptography on oblivious transfer. In *Proc. of the 20th ACM*, pages 20-31, 1998.
22. V. Kolesnikov. Gate Evaluation Secret Sharing and Secure One-Round Two-Party Computation. In *Proc. of ASIACRYPT 2005*, pages 136-155.
23. E. Kushilevitz and R. Ostrovsky. Replication is not needed: single database, computationally-private information retrieval In *Proc. 38th FOCS*, pages 364-273, 1997.
24. S. Laur and H. Lipmaa. Additively homomorphic Conditional Disclosure of Secrets and applications. Eprint report 2005/378.

25. Y. Lindell and B. Pinkas. A Proof of Yao's Protocol for Secure Two-Party Computation. Cryptology ePrint Archive, Report 2004/175, 2004.
26. H. Lipmaa. An Oblivious Transfer Protocol with Log-Squared Communication. In *Proc. 8th ICS*, pages 314-328, 2005. Full version on eprint.
27. S. Micali, M. Rabin and J. Kilian. Zero knowledge sets. In *Proc. 44th FOCS*, pages 80-91, 2003.
28. M. Naor and K. Nissim. Communication Preserving Protocols for Secure Function Evaluation. In *Proc. 33rd STOC*, pages 590–599, 2001.
29. M. Naor and B. Pinkas. Efficient oblivious transfer protocols. In Proc. SODA 2001.
30. R. Ostrovsky and W. E. Skeith III. Private Searching on Streaming Data. In *Proc. Crypto 2005*, pages 223-240.
31. P. Paillier. Public-Key Cryptosystems Based on Composite Degree Residuosity Classes. In *Proc. of EUROCRYPT 1999*, pages 223-238.
32. N. Pippenger. On simultaneous resource bounds. In *Proc. of the 20th FOCS*, pages 307-311, 1979.
33. M. Rabin. How to Exchange Secrets by Oblivious Transfer. Tech. Memo TR-81, Aiken Computation Laboratory, Harvard U., 1981.
34. J.P. Stern. A new and efficient all or nothing Disclosure of Secrets protocol. In *Proc. AsiaCrypt 98*, vol. 1514, pages 357-371.
35. T. Sander, A. Young and M. Yung. Non-interactive cryptocomputing for NC_1. In *Proc. 20th FOCS*, pages 554-566, 1999.
36. A.C. Yao. How to generate and exchange secrets. In *Proc. 18th STOC*, pages 162-167, 1986.

Author Index

Abdalla, Michel   499
Abe, Masayuki   118
Adida, Ben   555
Aumann, Yonatan   137

Backes, Michael   157
Beimel, Amos   253, 383
Bohli, Jens-Matthias   499
Boneh, Dan   535
Bosley, Carl   1

Canetti, Ran   61
Cash, David   479
Chase, Melissa   515
Christandl, Matthias   456

Ding, Yan Zong   479
Dodis, Yevgeniy   1, 61, 479

Ekert, Artur   456

Fehr, Serge   118
Fitzi, Matthias   311
Franklin, Matthew   253, 291, 311

Garay, Juan   311
Gertner, Yael   434
Goldreich, Oded   174
Goldwasser, Shafi   194

Hallak, Renen   383
Hazay, Carmit   323
Hofheinz, Dennis   214
Hohenberger, Susan   233
Hopper, Nicholas   362
Horodecki, Michał   456
Horodecki, Paweł   456

Ishai, Yuval   575

Katz, Jonathan   323
Kiltz, Eike   291
Koo, Chiu-Yuen   323

Lee, Wenke   479
Lindell, Yehuda   137, 323
Lipton, Richard   479

Malkin, Tal   434
Malone-Lee, John   214
Martí-Farré, Jaume   273
Meier, Remo   404
Mohassel, Payman   291
Molnar, David   362
Müller-Quade, Jörn   41, 157
Myers, Steven   434

Nissim, Kobbi   383

Oppenheim, Jonathan   456

Padró, Carles   273
Panjwani, Saurabh   21
Paskin, Anat   575
Pass, Rafael   61
Pietrzak, Krzysztof   86
Przydatek, Bartosz   404

Renner, Renato   456
Rothblum, Guy N.   194, 233

shelat, abhi   233
Stam, Martijn   214
Steinwandt, Rainer   499

Unruh, Dominique   41, 157

Vaikuntanathan, Vinod   233
Vardhan, S. Harsha   311
Vasco, María Isabel González   499

Wagner, David   362
Walfish, Shabsi   61, 479
Waters, Brent   535
Wee, Hoeteck   103, 419
Weinreb, Enav   291
Wikström, Douglas   86, 342, 555
Wullschleger, Jürg   404

Lecture Notes in Computer Science

For information about Vols. 1–4289

please contact your bookseller or Springer

Vol. 4405: L. Padgham, F. Zambonelli (Eds.), Agent-Oriented Software Engineering VII. XII, 225 pages. 2007.

Vol. 4392: S.P. Vadhan (Ed.), Theory of Cryptography. XI, 595 pages. 2007.

Vol. 4390: S.O. Kuznetsov, S. Schmidt (Eds.), Formal Concept Analysis. X, 329 pages. 2007. (Sublibrary LNAI).

Vol. 4385: K. Coninx, K. Luyten, K.A. Schneider (Eds.), Task Models and Diagrams for Users Interface Design. XI, 355 pages. 2007.

Vol. 4384: T. Washio, K. Satoh, H. Takeda, A. Inokuchi (Eds.), New Frontiers in Artificial Intelligence. IX, 401 pages. 2007. (Sublibrary LNAI).

Vol. 4381: J. Akiyama, W.Y.C. Chen, M. Kano, X. Li, Q. Yu (Eds.), Discrete Geometry, Combinatorics and Graph Theory. XI, 289 pages. 2007.

Vol. 4380: S. Spaccapietra, P. Atzeni, F. Fages, M.-S. Hacid, M. Kifer, J. Mylopoulos, B. Pernici, P. Shvaiko, J. Trujillo, I. Zaihrayeu (Eds.), Journal on Data Semantics VIII. XV, 219 pages. 2007.

Vol. 4378: I. Virbitskaite, A. Voronkov (Eds.), Perspectives of Systems Informatics. XIV, 496 pages. 2007.

Vol. 4377: M. Abe (Ed.), Topics in Cryptology – CT-RSA 2007. XI, 403 pages. 2006.

Vol. 4373: K. Langendoen, T. Voigt (Eds.), Wireless Sensor Networks. XIII, 358 pages. 2007.

Vol. 4372: M. Kaufmann, D. Wagner (Eds.), Graph Drawing. XIV, 454 pages. 2007.

Vol. 4371: K. Inoue, K. Satoh, F. Toni (Eds.), Computational Logic in Multi-Agent Systems. X, 315 pages. 2007. (Sublibrary LNAI).

Vol. 4369: M. Umeda, A. Wolf, O. Bartenstein, U. Geske, D. Seipel, O. Takata (Eds.), Declarative Programming for Knowledge Management. X, 229 pages. 2006. (Sublibrary LNAI).

Vol. 4368: T. Erlebach, C. Kaklamanis (Eds.), Approximation and Online Algorithms. X, 345 pages. 2007.

Vol. 4367: K. De Bosschere, D. Kaeli, P. Stenström, D. Whalley, T. Ungerer (Eds.), High Performance Embedded Architectures and Compilers. XI, 307 pages. 2007.

Vol. 4364: T. Kühne (Ed.), Models in Software Engineering. XI, 332 pages. 2007.

Vol. 4362: J. van Leeuwen, G.F. Italiano, W. van der Hoek, C. Meinel, H. Sack, F. Plášil (Eds.), SOFSEM 2007: Theory and Practice of Computer Science. XXI, 937 pages. 2007.

Vol. 4361: H.J. Hoogeboom, G. Păun, G. Rozenberg, A. Salomaa (Eds.), Membrane Computing. IX, 555 pages. 2006.

Vol. 4360: W. Dubitzky, A. Schuster, P.M.A. Sloot, M. Schroeder, M. Romberg (Eds.), Distributed, High-Performance and Grid Computing in Computational Biology. X, 192 pages. 2007. (Sublibrary LNBI).

Vol. 4358: R. Vidal, A. Heyden, Y. Ma (Eds.), Dynamical Vision. IX, 329 pages. 2007.

Vol. 4357: L. Buttyán, V. Gligor, D. Westhoff (Eds.), Security and Privacy in Ad-Hoc and Sensor Networks. X, 193 pages. 2006.

Vol. 4355: J. Julliand, O. Kouchnarenko (Eds.), B 2007: Formal Specification and Development in B. XIII, 293 pages. 2006.

Vol. 4354: M. Hanus (Ed.), Practical Aspects of Declarative Languages. X, 335 pages. 2006.

Vol. 4353: T. Schwentick, D. Suciu (Eds.), Database Theory – ICDT 2007. XI, 419 pages. 2006.

Vol. 4352: T.-J. Cham, J. Cai, C. Dorai, D. Rajan, T.-S. Chua, L.-T. Chia (Eds.), Advances in Multimedia Modeling, Part II. XVIII, 743 pages. 2006.

Vol. 4351: T.-J. Cham, J. Cai, C. Dorai, D. Rajan, T.-S. Chua, L.-T. Chia (Eds.), Advances in Multimedia Modeling, Part I. XIX, 797 pages. 2006.

Vol. 4349: B. Cook, A. Podelski (Eds.), Verification, Model Checking, and Abstract Interpretation. XI, 395 pages. 2007.

Vol. 4348: S.T. Taft, R.A. Duff, R.L. Brukardt, E. Ploedereder, P. Leroy (Eds.), Ada 2005 Reference Manual. XXII, 765 pages. 2006.

Vol. 4347: J. Lopez (Ed.), Critical Information Infrastructures Security. X, 286 pages. 2006.

Vol. 4345: N. Maglaveras, I. Chouvarda, V. Koutkias, R. Brause (Eds.), Biological and Medical Data Analysis. XIII, 496 pages. 2006. (Sublibrary LNBI).

Vol. 4344: V. Gruhn, F. Oquendo (Eds.), Software Architecture. X, 245 pages. 2006.

Vol. 4342: H. de Swart, E. Orłowska, G. Schmidt, M. Roubens (Eds.), Theory and Applications of Relational Structures as Knowledge Instruments II. X, 373 pages. 2006. (Sublibrary LNAI).

Vol. 4341: P.Q. Nguyen (Ed.), Progress in Cryptology - VIETCRYPT 2006. XI, 385 pages. 2006.

Vol. 4340: R. Prodan, T. Fahringer, Grid Computing. XXIII, 317 pages. 2007.

Vol. 4339: E. Ayguadé, G. Baumgartner, J. Ramanujam, P. Sadayappan (Eds.), Languages and Compilers for Parallel Computing. XI, 476 pages. 2006.

Vol. 4338: P. Kalra, S. Peleg (Eds.), Computer Vision, Graphics and Image Processing. XV, 965 pages. 2006.

Vol. 4337: S. Arun-Kumar, N. Garg (Eds.), FSTTCS 2006: Foundations of Software Technology and Theoretical Computer Science. XIII, 430 pages. 2006.

Vol. 4335: S.A. Brueckner, S. Hassas, M. Jelasity, D. Yamins (Eds.), Engineering Self-Organising Systems. XII, 212 pages. 2007. (Sublibrary LNAI).

Vol. 4334: B. Beckert, R. Hähnle, P.H. Schmitt (Eds.), Verification of Object-Oriented Software. XXIX, 658 pages. 2007. (Sublibrary LNAI).

Vol. 4333: U. Reimer, D. Karagiannis (Eds.), Practical Aspects of Knowledge Management. XII, 338 pages. 2006. (Sublibrary LNAI).

Vol. 4332: A. Bagchi, V. Atluri (Eds.), Information Systems Security. XV, 382 pages. 2006.

Vol. 4331: G. Min, B. Di Martino, L.T. Yang, M. Guo, G. Ruenger (Eds.), Frontiers of High Performance Computing and Networking – ISPA 2006 Workshops. XXXVII, 1141 pages. 2006.

Vol. 4330: M. Guo, L.T. Yang, B. Di Martino, H.P. Zima, J. Dongarra, F. Tang (Eds.), Parallel and Distributed Processing and Applications. XVIII, 953 pages. 2006.

Vol. 4329: R. Barua, T. Lange (Eds.), Progress in Cryptology - INDOCRYPT 2006. X, 454 pages. 2006.

Vol. 4328: D. Penkler, M. Reitenspiess, F. Tam (Eds.), Service Availability. X, 289 pages. 2006.

Vol. 4327: M. Baldoni, U. Endriss (Eds.), Declarative Agent Languages and Technologies IV. VIII, 257 pages. 2006. (Sublibrary LNAI).

Vol. 4326: S. Göbel, R. Malkewitz, I. Iurgel (Eds.), Technologies for Interactive Digital Storytelling and Entertainment. X, 384 pages. 2006.

Vol. 4325: J. Cao, I. Stojmenovic, X. Jia, S.K. Das (Eds.), Mobile Ad-hoc and Sensor Networks. XIX, 887 pages. 2006.

Vol. 4323: G. Doherty, A. Blandford (Eds.), Interactive Systems. XI, 269 pages. 2007.

Vol. 4320: R. Gotzhein, R. Reed (Eds.), System Analysis and Modeling: Language Profiles. X, 229 pages. 2006.

Vol. 4319: L.-W. Chang, W.-N. Lie (Eds.), Advances in Image and Video Technology. XXVI, 1347 pages. 2006.

Vol. 4318: H. Lipmaa, M. Yung, D. Lin (Eds.), Information Security and Cryptology. XI, 305 pages. 2006.

Vol. 4317: S.K. Madria, K.T. Claypool, R. Kannan, P. Uppuluri, M.M. Gore (Eds.), Distributed Computing and Internet Technology. XIX, 466 pages. 2006.

Vol. 4316: M.M. Dalkilic, S. Kim, J. Yang (Eds.), Data Mining and Bioinformatics. VIII, 197 pages. 2006. (Sublibrary LNBI).

Vol. 4314: C. Freksa, M. Kohlhase, K. Schill (Eds.), KI 2006: Advances in Artificial Intelligence. XII, 458 pages. 2007. (Sublibrary LNAI).

Vol. 4313: T. Margaria, B. Steffen (Eds.), Leveraging Applications of Formal Methods. IX, 197 pages. 2006.

Vol. 4312: S. Sugimoto, J. Hunter, A. Rauber, A. Morishima (Eds.), Digital Libraries: Achievements, Challenges and Opportunities. XVIII, 571 pages. 2006.

Vol. 4311: K. Cho, P. Jacquet (Eds.), Technologies for Advanced Heterogeneous Networks II. XI, 253 pages. 2006.

Vol. 4309: P. Inverardi, M. Jazayeri (Eds.), Software Engineering Education in the Modern Age. VIII, 207 pages. 2006.

Vol. 4308: S. Chaudhuri, S.R. Das, H.S. Paul, S. Tirthapura (Eds.), Distributed Computing and Networking. XIX, 608 pages. 2006.

Vol. 4307: P. Ning, S. Qing, N. Li (Eds.), Information and Communications Security. XIV, 558 pages. 2006.

Vol. 4306: Y. Avrithis, Y. Kompatsiaris, S. Staab, N.E. O'Connor (Eds.), Semantic Multimedia. XII, 241 pages. 2006.

Vol. 4305: A.A. Shvartsman (Ed.), Principles of Distributed Systems. XIII, 441 pages. 2006.

Vol. 4304: A. Sattar, B.-H. Kang (Eds.), AI 2006: Advances in Artificial Intelligence. XXVII, 1303 pages. 2006. (Sublibrary LNAI).

Vol. 4303: A. Hoffmann, B.-H. Kang, D. Richards, S. Tsumoto (Eds.), Advances in Knowledge Acquisition and Management. XI, 259 pages. 2006. (Sublibrary LNAI).

Vol. 4302: J. Domingo-Ferrer, L. Franconi (Eds.), Privacy in Statistical Databases. XI, 383 pages. 2006.

Vol. 4301: D. Pointcheval, Y. Mu, K. Chen (Eds.), Cryptology and Network Security. XIII, 381 pages. 2006.

Vol. 4300: Y.Q. Shi (Ed.), Transactions on Data Hiding and Multimedia Security I. IX, 139 pages. 2006.

Vol. 4299: S. Renals, S. Bengio, J.G. Fiscus (Eds.), Machine Learning for Multimodal Interaction. XII, 470 pages. 2006.

Vol. 4297: Y. Robert, M. Parashar, R. Badrinath, V.K. Prasanna (Eds.), High Performance Computing - HiPC 2006. XXIV, 642 pages. 2006.

Vol. 4296: M.S. Rhee, B. Lee (Eds.), Information Security and Cryptology – ICISC 2006. XIII, 358 pages. 2006.

Vol. 4295: J.D. Carswell, T. Tezuka (Eds.), Web and Wireless Geographical Information Systems. XI, 269 pages. 2006.

Vol. 4294: A. Dan, W. Lamersdorf (Eds.), Service-Oriented Computing – ICSOC 2006. XIX, 653 pages. 2006.

Vol. 4293: A. Gelbukh, C.A. Reyes-Garcia (Eds.), MICAI 2006: Advances in Artificial Intelligence. XXVIII, 1232 pages. 2006. (Sublibrary LNAI).

Vol. 4292: G. Bebis, R. Boyle, B. Parvin, D. Koracin, P. Remagnino, A. Nefian, G. Meenakshisundaram, V. Pascucci, J. Zara, J. Molineros, H. Theisel, T. Malzbender (Eds.), Advances in Visual Computing, Part II. XXXII, 906 pages. 2006.

Vol. 4291: G. Bebis, R. Boyle, B. Parvin, D. Koracin, P. Remagnino, A. Nefian, G. Meenakshisundaram, V. Pascucci, J. Zara, J. Molineros, H. Theisel, T. Malzbender (Eds.), Advances in Visual Computing, Part I. XXXI, 916 pages. 2006.

Vol. 4290: M. van Steen, M. Henning (Eds.), Middleware 2006. XIII, 425 pages. 2006.